Seismic Design of Industrial Facilities

Sven Klinkel · Christoph Butenweg ·
Gao Lin · Britta Holtschoppen
Editors

Seismic Design of Industrial Facilities

Proceedings of the International Conference on Seismic Design of Industrial Facilities (SeDIF-Conference)

Springer Vieweg

CHAIR OF
STRUCTURAL STATICS AND DYNAMICS
RWTH AACHEN

DALIAN UNIVERSITY OF TECHNOLOGY
1949

Editors

Sven Klinkel
Lehrstuhl für Baustatik u. Baudynamik
RWTH Aachen
Aachen, Germany

Christoph Butenweg
Lehrstuhl für Baustatik u. Baudynamik
RWTH Aachen
Aachen, Germany

Gao Lin
Faculty of Infrastructure Engineering
Dalian University of Technology No.2
Dalian City, China

Britta Holtschoppen
Lehrstuhl für Baustatik u. Baudynamik
RWTH Aachen
Aachen, Germany

ISBN 978-3-658-02809-1 ISBN 978-3-658-02810-7 (eBook)
DOI 10.1007/978-3-658-02810-7

Library of Congress Control Number: 2013947577

Picture Credits for this book cover (from left to right): Lanxess AG, AkzoNobel, BASF SE, LBB (RWTH Aachen Univ.), BASF SE

Springer Vieweg

Printed on acid-free paper.

Springer is part of Springer Science+Business Media
www.springer.com

Preface

Devastating earthquakes in China (2008 and 2010), New Zealand (2011), Japan (2011) and Italy (2012) have tightened the social and the political focus on the seismic risk emanating from industrial facilities. Seismic Design of Industrial Facilities, however, demands a deep knowledge on the seismic behaviour of the individual structural and non-structural components of the facility, possible interactions and last but not least the individual hazard potential of primary and secondary damages.

From 26.–27. September 2013 the International Conference on Seismic Design of Industrial Facilities firstly addresses this broad field of work and research in one specialized conference. It brings together academics, researchers and professional engineers in order to discuss the challenges of seismic design for new and existing industrial facilities and to compile innovative current research.

This volume contains more than 50 contributions to the SeDIF-Conference covering the state of the art of international building codes and guidelines on the seismic design of industrial facilities, seismic design of structural and non-structural components, seismic design of liquid-filled tanks and other self-contained structures, seismic safety evaluation of existing structures, uncertainties and reliability analysis, latest retrofitting measures and innovative seismic protection systems as well as theoretical and practical approaches in the investigation of soil-structure-interaction effects.

We thank all authors for their varied and highly interesting contributions showing the broad field of work and auspicious new research activities regarding the seismic design of industrial facilities.

Aachen, Germany
September 2013

Prof. Sven Klinkel
Prof. Gao Lin

Dr. Christoph Butenweg
Dr. Britta Holtschoppen

Acknowledgements

The SeDIF-Conference is hosted by the Chair of Structural Statics and Dynamics of RWTH Aachen University, Germany, in cooperation with the Institute for Earthquake Engineering of the Dalian University of Technology, China.

Organising Committee of the SeDIF-Conference and Editors of the Proceedings:

Prof. Sven Klinkel	RWTH Aachen University
Prof. Gao Lin	Dalian University of Technology
Dr. Christoph Butenweg	RWTH Aachen University
Dr. Britta Holtschoppen	RWTH Aachen University

The SeDIF-Conference is supported by the DFG (Deutsche Forschungsgemeinschaft). This support is greatly acknowledged by the organising committee.

The oragnising committee would also like to express their gratitude to the scientific committee of the SeDIF-Conference:

Prof. Elena Dumova	(Univ. SS. Cyril and Methodius, Skopje, MK)
Prof. Mario Duran Lillo	(Univ. La Serena, Chile)
Prof. Mihail Garevski	(IZIIS-Skopje, Macedonia)
Prof. Linhai Han	(Tsinghua Univ., Bejing, China)
Prof. Sven Klinkel	(RWTH Aachen University, Germany)
Prof. Carsten Könke	(Bauhaus-Univ. Weimar, Germany)
Prof. Jie Li	(Tongji Univ., China)
Prof. Gao Lin	(Dalian University of Technology, China)
Prof. em. Konstantin Meskouris	(RWTH Aachen University, Germany)
Dr. Philippe Renault	(swissnuclear, Switzerland)
Prof. Stavros Savidis	(TU Berlin, Germany)
Prof. Chongmin Song	(Univ. of New South Wales, Australia)
Prof. Demosthenes Talaslidis	(Aristotle University Thessaloniki, Greece)

Content

Part X Soil-Structure Interaction: Scientific Approaches

Part I

Vulnerability of Industrial Facilities

Earthquake Damage and Fragilities of Industrial Facilities

Mustafa Erdik[1], **Eren Uckan**[1]

[1] Kandilli Observatory and Earthquake Research Institute / Bogazici University, Istanbul, erdik@boun.edu.tr

ABSTRACT:

An industrial facility consists of many integrated components and processes. As such, operation of a facility depends upon the performance of its critical components. The greatest risk from an earthquake is that to life safety. However, in large earthquakes, industrial buildings and related machinery and equipment damaged may be costly to repair and there may be additional damage from fire and chemical spills. As such, the design (or seismic retrofit) of industrial facilities should preferably be based on performance-based methodologies with the objective of controlling structural and non-structural damage and the triggered technological disasters. In this paper industrial damages and losses that took place in past important earthquakes, especially in the 1999 Kocaeli earthquake, will be summarized. A general description of industrial-sector and component based earthquake performance and vulnerabilities will be provided.

Keywords: industry, seismic risk, fragility, damage.

1 Introduction

Earthquakes world over, such as 1994 Northridge-USA, 1995 Kobe-Japan, 1999 Kocaeli-Turkey, 2008 Wenchuan-China, 2010 Chile, 2011 Tohoku-Japan and 2011 Van-Turkey earthquakes, have resulted in significant loss of life and property as well as extensive losses to industry. In all these earthquakes older, heavy industrial facilities, especially those with taller structures that partially to totally collapsed, were more affected by the earthquake than newer facilities. It was observed that any type and quality of anchorage improved the performance of machines and equipment, except very sensitive equipment such as assembly line sensors in the automotive industry and rotary kilns in cement plants. Losses associated with business interruption were more severe for these types of facilities. For light industrial facilities, building damage turned out to be the primary reason for direct and indirect losses. For refineries and other chemical processing facilities, non-building structures turned out to be the most vulnerable, with tanks being the most susceptible to earthquake and fire damage. Large storage tanks, pipelines,

S. Klinkel et al. (Eds.), *Seismic Design of Industrial Facilities*,
DOI 10.1007/978-3-658-02810-7_1, © Springer Fachmedien Wiesbaden 2014

transmission lines and precision machinery were generally susceptible to damage. Port and harbour facilities are particularly susceptible to sub-marine landslides or ground settlement due to liquefaction that may occur during earthquakes. In addition, all processes that involve a substantial risk of explosion such as those in the petrochemical industry and processes involving molten metal.

Fragility functions of an element at risk represent the probability that its response to earthquake excitation exceeds its various performance (damage) limit states based on physical considerations. Fragility assessments are usually based on past earthquake damages (observed damage and, to a lesser degree, on analytical investigations.

The 1999 Kocaeli earthquake (Mw7.4) is considered the largest event to have damaged an industrialized area since the 1906 San Francisco and 1923 Tokyo earthquakes (Unless referenced otherwise, the information regarding the 1999 Kocaeli earthquake is adopted from Erdik and Durukal [6]).

The epicenter of the 1999 Kocaeli earthquake was the main site of Turkey's heavy industry. Major industries exposed included: automobile manufacturing; petrochemicals; motor and railway vehicle manufacture and repair; basic metal works; production and weaving of synthetic fibers and yarns; paint and lacquer production; tire manufacturing; paper mills; steel pipe production; pharmaceuticals; sugar processing; cement production and power plants. It was observed that any type and quality of anchorage improved the performance of machine and equipment except very sensitive equipment, such as assembly line sensors in case of automotive industry and rotary kilns in cement plants. For the case of light industrial facilities in the earthquake area, the building damage turned out to be primary reason for direct and indirect losses. In the case of refineries and other chemical processing facilities, non-building structures turned out to be vulnerable with tanks being the most susceptible ones to earthquake and fire damage. The extend of the damage was attributed to the duration and long period motion of the earthquake MCEER(14).

2 Sector Based Description of Earthquake Performance and Damage

2.1 Petrochemical Industry

In 1999 Kocaeli earthquake an extensive concentration of petrochemical complexes are located within 5 km of the causative fault. The earthquake caused significant structural damages to the Tupras refinery itself and associated tank farm with crude oil and product jetties. The consequent fire in the refinery and tank farm caused extensive damage. There was damage to cooling towers and the port area. Collapse of a 150m high heater stack on the boiler and crude oil processing unit caused significant damage and started a second fire Figure 1. The total damage is estimated to be around US$350 million. Fault rupture and soil failure caused extensive damage to pump station and pipelines at about 20 locations. The failure of the water supply caused problems in controlling the fire. There were at least 15 gas firms with spherical LPG storage tanks in the area. No major structural damage was observable at these plants (EERI [4]). Being unanchored some tanks slided horizontally on their supports.

Figure 1: Damaged tanks at tank farm (left) and collapsed stack at TÜPRAŞ Refinery

2.2 Automotive Industry

The Hyundai car factory experienced significant non-structural damage to its air handling systems, cable trays and shearing of bolted connections in the steel structure EERI [4], Figure 2.

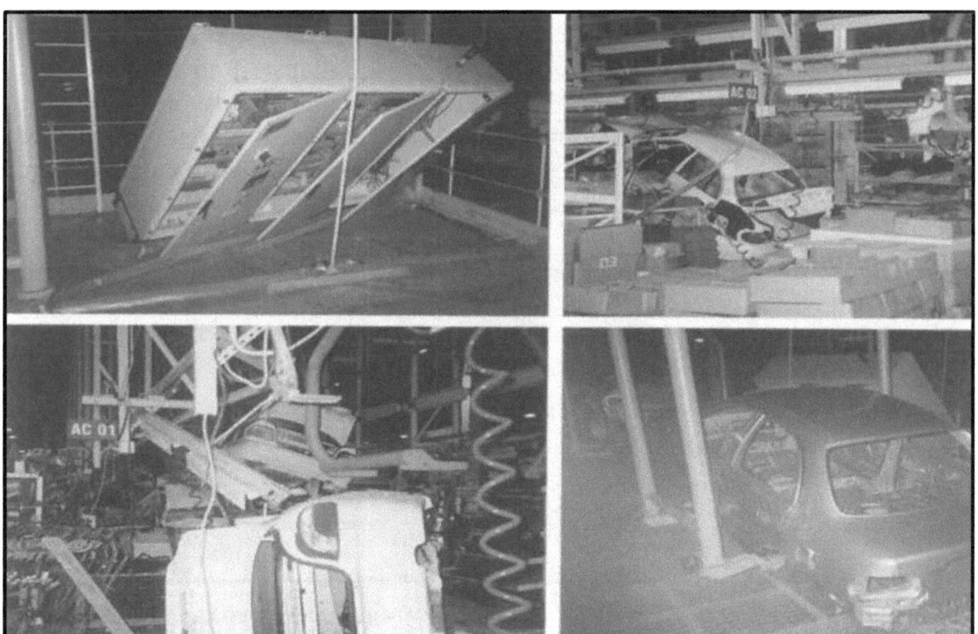

Figure 2: Equipment damage at Hyundai-Assan car factory (after Milli-Re)

In Toyota car factory there was little structural damage to the steel framed buildings, two buildings experienced damage to their columns. Non-structural damage included collapsed storage racks, transformers, cars on the assembly line, sliding of the cooling tower associated with pipe damage. Ford Otosan car factory, under construction during the earthquake, experienced significant terrain subsidence and some structural damage Figure 3.

Figure 3: Damaged prefabricated buildings at Ford Otosan Plant

2.3 Other Industry

In 1999 Kocaeli earthquake the TUVASAS railway wagon, Adapazari sugar and steel production factories have received extensive structural damage. Examples of specific damage included collapse of cranes, roof collapse, transformer damage, silo collapse, toxic releases from mixing chemicals, and collapse of liquid oxygen tank support structures. Some tanks in Aksa chemical installation in Yalova experienced damage, which was associated with leakage of chemicals. Numerous industrial facilities experienced losses of stored items Figure 4.

Figure 4: Damaged steel structure at Adapazari rail car factory (left) and losses of open stored materials

3 Component Based Description of Earthquake Performance and Damage

3.1 Buildings

Most of the buildings in the 1999 Kocaeli earthquake affected region qualify for the fragility class of Cof EMS [5]. The damage to reinforced concrete buildings was attributed to one or more of the following: Failure to meet the design requirements of the code use of poor and inappropriate construction materials; Soft stories at the first-floor level; Strong beams and weak columns; Lack of column confinement and poor detailing practice (Erdik and Aydinoglu [8]).

In 2008 Wenchuan earthquake, building damage and sometimes collapse were omnipresent at industrial facilities Krausmann et al. [13]. This included roof and wall damage, as well as top-storey collapse and pancaking of floors with associated life losses. This concerned mostly concrete structures with insufficient confinement or poor reinforcement that could not withstand the earthquake loads.

3.2 Prefabricated/Precast Reinforced-Concrete Structures

The performance of this building type in the 1999 Kocaeli and 2011Van earthquakes were very poor, with many collapses or partial collapses in areas of intensity VIII-IX. The main reason of damage was the failure of weak joints between the roof beams and columns, lack of bracing or roof diaphragm. Heavy precast-concrete frames with precast roof beams suffered from movements at beam-column connections and lack of steel bonding.

3.3 Steel Frame Structures

In 1999 Kocaeli earthquake steel buildings performed much better than the RC frames. Typical causes for collapses include failure of anchor bolts at column bases and roof trusses and structural instability under overturning forces. For low rise (<5 stories) steel braced frame structures with moderate-code seismic design level the equivalent-PGA structural fragility relationships reported by HAZUS [11] indicate moderate damage starting at 0.26g.

3.4 Electric Power

In 1999 Kocaeli earthquake the heat recovery steam generation facility of the LNG plant was damaged. Nine transmission substations suffered damage or disruption to transformers, switching equipment, and buildings. The transformers mounted on wheels moved in the switchyard some bus bars and high-voltage bushings were broken Figure 5. In the M=7.2 Van Earthquake, 2011, 10% of the total transformer inventory and 600km of interconnecting cables was damaged, Uckan [16].

Figure 5: Damaged transformers at Izmit substation (left) and in Van

3.5 Tanks, Silos, Cooling Towers and Stacks

In 1999 Kocaeli earthquake the majority of damage at the Tupras Refinery was at the storage tank farm area. The sloshing of fluid damaged the perimeter seal producing overtopping and gross damage in near the tops of walls. The shell buckling at tank bases also caused oil leakage. The vertical movement of the floating roof created sparks causing fire. No significant damage to the spherical LPG tanks were has been reported. At the SEKA Paper Factory three reinforced concrete silos containing wastewater completely collapsed (Figure 6). In TÜPRAŞ Facility the upper two thirds of a 110-m-tall reinforced concrete stack collapsed.

Figure 6: Collapsed silo at SEKA Factory in Izmit (left) and cement silo in Van

In 2008 Wenchuan earthquake liquid sloshing may have exacerbated the earthquake impact (Krausmann et al. [13]). Several of the tanks were not anchored to their foundations or otherwise restrained. This made them vulnerable to sliding or uplifting.

In 2011 Van earthquakes, the elevated wheat and cement silos in small size industrial plants collapsed due to weak welding and insufficient seating widths of the supporting concrete. Foundations with a continuous ring beam at the bottom performed better, Uckan [16].

3.6 Pipelines and Piping Systems

In 1999 Kocaeli earthquake the damage to the segmented water and sewage systems included broken distribution pipes, especially in areas of severe permanent ground movement, particularly, along the southern coast of the Izmit Bay Uckan et al. [17]. There was some damage to major welded steel water transmission lines at fault crossings.

In 2008 Wenchuan earthquake much of the loss at the chemical facilities resulted from damage to pipes and equipment,Krausmann et al. [13]. This was caused by direct loading by the earthquake forces or indirectly by falling debris from collapsing buildings.

In the 2011 Van earthquake only the segmented pipes were damaged. No damage was seen at continuous gas pipes. The observed repair rates Uckan [16] were consistent with the estimates from ALA (2005) [1] and O'Rourke and Deyoe [15].

3.7 Ports and Jetties

In 1999 Kocaeli earthquake most of the ports and jetties sustained damage. Damage included failure of piers, mechanical equipment, piping and the collapse of cranes (Figure 7). Derince and Golcuk ports suffered heavy damage to docks, cranes and warehouses, including cracks and severe subsidence.

Figure 7: Damage at navy port in Gölcük (left) and failed column at SEKA port

3.8 Fire Following Earthquake And Hazardous Material Release

Fire following earthquakes is common occurrence, and can cause significant additional damage in industrial facilities. Losses become significant if the fires spread in an uncontrolled manner, Coburn and Spence [2]. The 1999 Kocaeli earthquake caused one of the most important and dangerous fire events of Turkey.Damaged tanks at TUPRAS tank farm and insulated tanks at HABAŞ Facility are shown in Figure 8.

Figure 8: Damaged tanks at TUPRAS and insulated tanks at HABAŞ Facility

The release of hazardous materials may cause physical damages, environmental contamination or temporary health problems in humans, it can also lead to fires. The risk regarding hazardous material release is particularly important in industrialized regions. In the 1999 Kocaeli earthquake damage occurring in several facilities caused toxic releases, Erdik [7].

3.9 Fragility of Non-Structural Components

Critical non-structural equipment in industrial facilites include fire detection, alarm and suppression systems, communication systems, emergency and uninterrupted power supply systems, safe-shut down systems, system control centers and hazardous material suppression systems.During the 1994 Northridge earthquake significant damages and service disruption took place in critical facilities due to primarily non-structural or equipment failures, Gates and McGavin [10]. HAZUS [11] provides fragility relationships for nonstructural components.

4 Intensity Based Vulnerability of Industrial Facilities in Turkey

Table 1 provides the mean damage ratio for the equipment-machinery and stock inventory of different industrial sectors in Turkey, Durukal et al. [3].

Table 1: Mean Loss Ratios for MMI IX

Sector Description	Equipment Loss	Stock Loss
Mining, Const, Ceram, Glass Min	10%	10%
Commercial Facilities, Food and Beverage	10%	10%
Textile, Leather	10%	30%
Wood products and furniture, Agriculture	10%	10%
Chemical and Petroleum Products	30%	35%
Iron- steel and the other metals	2%	2%
Machinery and automotive	2%	2%
Transportation and telecommunication	10%	2%

5 Earthquake Resistant Design Codes for Industrial Facilities

The current seismic design provisions were written predominantly to address commercial and institutional buildings. Industrial buildings have geometries, framing systems, mass characteristics, load types and magnitudes, and stiffness properties that may vary significantly from those of typical commercial or institutional buildings and may require facility (or component) specific earthquake resistant design codes. ASCE (American Society of Civil Engineering) have published Guidelines for Seismic Evaluation and Design of Petrochemical Facilities, Nuclear Facilities and Electric Power Systems.For silos and bins: ACI (American Concrete Institute) have published Guidelines for the Design and Construction of Concrete Silos and Stacking Tubes.One of the few codes that specifically addresses to a broad spectrum of structures, including the non-building structures are the IBC-2009 and ASCE 7-10 (ASCE Standard ASCE/SEI 7-10, 2010, ISBN 978-0-7844-1115-5)Codes.

Earthquake resistant design codes and recommendations for liquid storage tanks that have found widespread international use are the API Standard 650(API Std 650 Welded Tanks for Oil Storage, 11th Edition, Includes Addendum 1 (2008) and Addendum 2 (2009) Edition: 11th, American Petroleum Institute) and FEMA 450-2003 [9].

The international codes used for the earthquake resistant design of liquid hydrocarbon transmission pipelines are: ASME (2012) B31.8 "Gas Transmission and Distribution Piping Systems", API (1999) Recommended Practice (RP) 1111 "Design, Construction, Operation, and Maintenance of Offshore Hydrocarbon Pipelines", PRCI (2004) Seismic Design Guidelines.Among these, 2006 IBC, Eurocode 8, and NZSEE are the national codes, and ACI 350.3, ACI 371, AWWA D-100, AWWA D-110, AWWA D-115, and API 650 are standards from American industries, namely, American Concrete Institute, AmericanWater Works Association, and American Petroleum Institute, Jaiswala [12].

American Lifelines Alliance has prepared Guidelines for the: Design of Buried Steel Pipe, Seismic Design and Retrofit of Piping Systems and Guideline for Assessing the Performance of Oil and Natural Gas Pipeline Systems in Natural Hazard and Human Threat Events, all of which address seismic risk to pipelines. The ALA (2005) [1] guidelines entitled "Seismic Guidelines for Water Pipelines" recommends design earthquakes associated respectively with return periods of 975 and 2475 years for the seismic design of "Critical" and "Essential" pipelines.

6 Final Remarks

In this paper a summary of earthquakefragilities and damages sustained by the industrial facilities during major earthquakes, especially during the 1999 Kocaeli earthquake, arepresented. One general observation is that the earthquake damage observed in Turkey in the 1999 Kocaeli earthquake is not really different from industrial damage observed in worldwide earthquake, particularly for heavy industrial facilities. Small and medium size facilities have their own particularities depending on the socio-economic conditions of a country. Building code requirements in most counties, are set with the intent of protecting the life of the occupants. The building is allowed to experience damage but without any collapse thereby allowing for the safe evacuation of occupants with minimum risk of casualties. However, in large earthquakes, the damage to the industrial buildings and other structures may cause costly to repairs to the machinery and equipment they house and may also lead to consequential damages such as fire and chemical spills. Since most of the revenue generated by industrial facilities is related to the products and services they provide, rather than the physical assets of the company, any significant interruption to the production of these goods and services because of this damage will also have an adverse affect on the business. The risk of business interruption is an important economic reason for controlling the damage from and following earthquakes. As such, the design (or seismic retrofit) of industrial facilities should preferably be based on performance-based methodologies with the intent on controlling the structural and non-structural damage. This requires development and enforcement of structural and non-structural codes and regulations, as well as a thorough understanding of the vulnerabilities associated with the production processes.

REFERENCES

[1] ALA (2005), Guidelines for the Design of Buried Steel Pipe American Lifeline Alliance, ASCE.

[2] Coburn, A., and Spence, R., 1992, Earthquake Protection; John Wiley and Sons Ltd., Chichester, 355 p.

[3] Durukal, E., M. Erdik, E. Uçkan (2008). Earthquake Risk to Industry in Istanbul and its Management, Natural Hazards, 44:199-212.

[4] EERI (1999), The Izmit (Kocaeli), Turkey Earthquake of August 17, 1999, EERI Special Earthquake Report-Learning From Earthquakes-October 1999, Also available online: www.eeri.org/Reconn/Turkey0899/Turkey0899.html

[5] EMS-98 (1998), European Macroseismic Scale 1998, G. Grünthal Editor, European Seismological Commission, Luxembourg.

[6] Erdik, M. and E.Durukal (2002), Earthquake Damage and Vulnerabılıty of Industry in The 1999 Kocaeli, Turkey Earthquake, The World Bank Group Disaster Management Facility, Washington, DC.

[7] Erdik. M. (2000), Report on 1999 Kocaeli and Duzce (Turkey) Earthquakes, Proc. Of the 3rd Intl. Workshop on Structural Control, Paris – France, 6-8 July 2000, p, 149-186.

[8] Erdik. M. and N. Aydinoglu (2002), Earthquake Performance and Vulnerability of Buildings in Turkey, Report Prepared for ProVention Consortium – World Bank, Bogazici University, Istanbul, Turkey.

[9] FEMA-450 (2003) NEHRP Recommendations for New Buildings and Other Structures.

[10] Gates, W.E. and McGavin, G. 1998, "Lessons Learned from the 1994 Northridge Earthquake on the Vulnerability of Nonstructural Systems," Proceedings of the Seminar on Seismic Design, Retrofit, and Performance of Nonstructural Components, ATC 29-1, San Francisco, CA, 93-106.

[11] HAZUS 2012, Earthquake Loss Estimation Methodology, FEMA, USA

[12] Jaiswala O.R., Durgesh C. R. and Sudhir K. J., Review of Seismic Codes on Liquid-Containing Tanks Earthquake Spectra, Volume 23, No. 1, pages 239–260, February 2007.

[13] Krausmann, E., A.M.Cruzand B.Affeltranger (2010), The impact of the 12 May 2008 Wenchuan earthquake on industrial facilities, Journal of Loss Prevention in the Process Industries 23 (2010).

[14] MCEER (2000), The Marmara, Turkey Earthquake of August 17, 1999: Reconnaissance Report, Ed by C. Scawthorn, The Multidisciplinary Center for Earthquake Engineering Research, Universty of Buffalo, NY, USA.

[15] O'Rourke, M., and Deyoe, E., (2004), "Seismic Damage to Segmented Buried Pipe," Earthquake Spectra, Vol. 20, No. 4, pp. 1167-1183.

[16] Uckan, E. "Lifeline Damage Caused In the 23 October (Mw=7.2) 2011 and 9 November (M=5.6) 2011, Van Earthquakes in Eastern Turkey" Sixth China Japan US trilateral Symposium on Lifeline Engineering,May 28th-June 13th, 2013, Chengdu Sichuan,China.

[17] Uckan,E., Durukal, E., Demircioğlu, M., Siyahi B., and Erdik, M. "Observed damage at buried pipelines during the 1999 Kocaeli (Izmit) –Turkey, Earthquake" Proc. European Geosciences Union (EGU) General Assembly, EGU 05-A-10583,Vienna, April 2005.

Part II

Seismic Risk of Industrial Facilities

Seismic Risk Analysis of an Oil-Gas Storage Plant

Roberto W. Romeo[1]

[1] Department of Earth Sciences, Life and Environment
University of Urbino
Scientific Campus, 61029 Urbino, Italy
rwromeo@uniurb.it

ABSTRACT

The wide range of induced effects of earthquakes, from direct damage due to seismic shaking to indirect damage caused by secondary effects (e.g. liquefaction, soil densification and landslides) makes the seismic risk one of the most common cause of structural failures among natural hazards. The degree of vulnerability and the level of exposure of the threatened elements may further amplify such effects. In this sense, the seismic risk induced by an oil-gas storage plant located close to an important commercial harbour in Southern Italy is analyzed. The plant is situated in one of the areas with the highest levels of seismic hazard in Italy, hit in the past by earthquakes as large as 7 in magnitude. Moreover, the plant lies near to the shoreline and the facing seafloor is characterized by the presence of a deep submarine canyon filled by loose, unconsolidated soils coming from the excavation of the harbour channel. Given these conditions the following phenomena have been investigated: local site amplification, liquefaction, submarine landslides and sea-waves run-up. The stability analyses considered both the plant's structure itself and the site. A vulnerability analysis provided the response to the ground motions of the steel tanks forming the structure, while dynamic analyses gave the response of the soils to the wide range of possible ground failures. Joining all the possible effects that could destabilize the plant, an overall probability that the safety of the plant may be affected was computed. The total risk was then assessed considering the effects, in terms of human life losses, produced by the failure of the plant. This risk was then compared with those deriving from other human activities to provide a reasonable basis for risk the acceptability assessment.

Keywords: hazard, fragility, risk, seismic ground motion, secondary effects

1 Introduction

Industrial facilities provide for the needs of developed countries in several activities such as power production, transportation, and so on. Nevertheless, the risk related to their failure under the seismic activity has been under-rated for a

S. Klinkel et al. (Eds.), *Seismic Design of Industrial Facilities*,
DOI 10.1007/978-3-658-02810-7_2, © Springer Fachmedien Wiesbaden 2014

long time, basically due to lack of sufficient knowledge about seismic hazard and/or seismic vulnerability.

In Italy, the recent (2003) seismic classification of the country, highlighted that about one-third of relevant risk plants (317 out of 1024) are located in medium to high seismic areas, where ground accelerations are expected to exceed 0.15g with a probability of 10% in 50 years.

Risk analysis of critical facilities consists in evaluation of potential losses related to relevant accidents. Amongst others, the consequences of a failure of a critical facility due to earthquakes, are given by the complete destruction of the near field, environmental pollution and long-term health effects. Moreover, the collapse of a system can extend the accident to nearby structures triggering an uncontrolled mechanism known as Domino Effect.

The target of a risk analysis is the probabilistic assessment that a given system may not survive all the possible occurrences of the considered source of damage; in other words, it is one minus the probability that the considered system completes its mission successfully (also termed as system reliability). Due to the stochastic nature of risk, it requires to be related to a given timeframe, usually consisting of the lifetime of the structure.

As a case study for the application of QRA, a petro-chemical facility located in a highly seismic area in southern Italy and potentially threatened by strong ground motions and earthquake-induced ground failures is shown (Figure 1).

Figure 1: Aerial view of the critical facility

2 Risk assessment

Since risk is based on the quantification of a failure probability, which is basically a non-dimensional quantity, it can include several failure sources (even airplane or meteorite accidents, or terrorism attacks). Events algebra allows keeping separate procedures for each considered mechanism and then combining the results.

This is why seismic risk, which includes several causes of damage (from ground motion to ground failures) is a failure probability, too. In the simplest way it can be considered as the convolution of the seismic hazard [at the site] with the structural vulnerability [of the system].

Traditional structural reliability methods define hazard and vulnerability in terms of demand and capacity, respectively. In the events algebra approach, risk is the failure probability – which includes vulnerability – given a certain event occurs:

$$\text{Risk} = \mathrm{P}[failure|Hazard] \cdot \mathrm{P}[Hazard|time] \tag{1}$$

and reliability or survival, in turn, is the complementary of risk. Therefore, it is possible to explore the relationships between hazard and vulnerability using a single non-structural parameter, commonly termed as [seismic] intensity measure (IM) or ground motion (GM).

2.1 Hazard and Vulnerability

The goal of probabilistic seismic hazard analysis (PSHA) is to assess the probability of exceeding various ground-motion (GM) levels at a site given all possible earthquakes. A GM parameter commonly adopted in PSHA is the peak ground acceleration (PGA), which is used to define lateral forces and shear stresses in the equivalent-static-force procedures of structural design, as well as in the liquefaction and landslide analyses:

$$Hazard = \mathrm{P}[PGA \geq a|t] = 1 - e^{-\lambda(a)\cdot t} \tag{2}$$

where a is the PGA-value expected to be exceeded in time t, that is the structure's lifetime, and $\lambda(a)$ is the annual frequency of exceedance of a, namely the hazard function. Seismic hazard assessment is commonly performed in a two-stage analysis: on a regional scale, it is carried out through seismological studies (PSHA sensu strictu); at local scale it is based on geophysical and geotechnical investigations (local seismic response analysis, LSRA).

In Figure 2 the site-specific seismic hazard curve for PGA is shown.

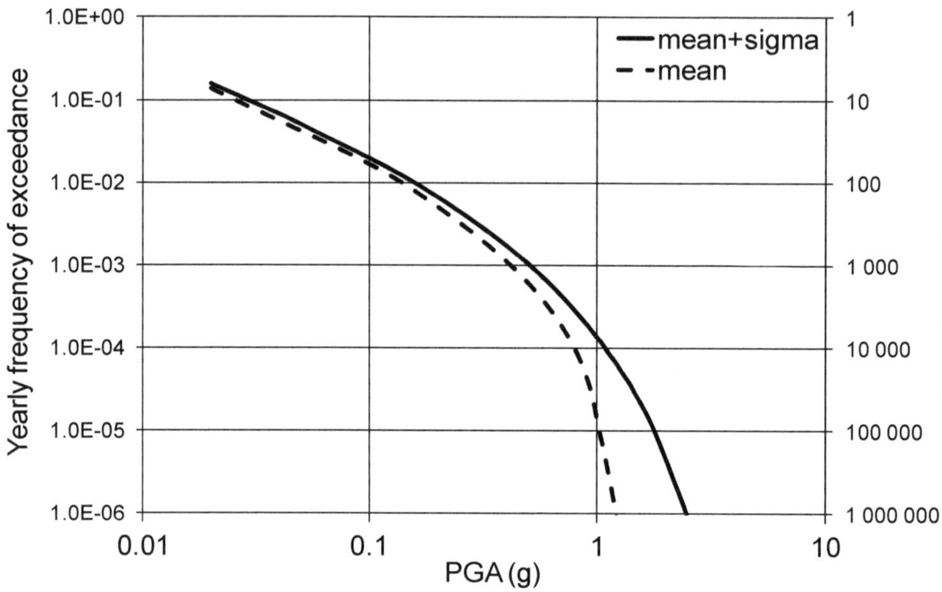

Figure 2: Seismic hazard curves for PGA

Vulnerability can be expressed by failure probability as a function of the same IM as hazard. In other words, probability of an event (=failure) given that an earthquake-related ground motion parameter has just occurred: in such a form vulnerability is called fragility function:

$$Fragility = \mathrm{P}[Failure|PGA] = \Phi\left[\frac{1}{\sigma}\, ln\left(\frac{PGA}{\mu}\right)\right] \quad (3)$$

where Φ is the cumulative normal standard distribution, μ and σ are, respectively, mean and dispersion values of a limit state to be reached or exceeded. Two limit states were analyzed, corresponding to a moderate content loss (Serviceability Limit State, SLS) and an extensive content loss (Ultimate Limit State, ULS), whose fragility functions were derived according to O'Rourke and So [1] and shown in Figure 3.

In structural analysis, hazard and fragility are related to two random variables called load (or demand, S, figure 2) and resistance (or capacity, R, figure 3). Due to their randomness, S and R are completely described by their probability density functions, $f_{S,R}(s,r)$. The probability that the system remains in the safe domain during its lifetime, is the probability that S never exceed R, or, invoking the performance function G=R–S, that G>0, therefore:

$$Risk = \mathrm{P}[G < 0] = P[S > R] = \int [\int \mathrm{f_R(r)\,dr}]\mathrm{f_S(s)ds} \quad (4)$$

where the limits of integration are: S[0÷∞] and R[0÷s].

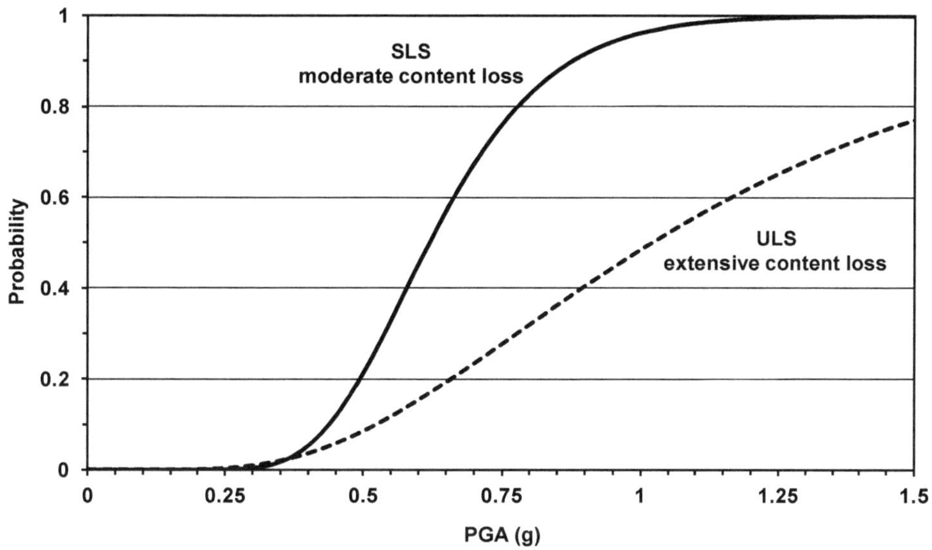

Figure 3: Fragility curves of the steel tanks for serviceability (SLS) and ultimate (ULS) limit states design

In the equation above the integral in *ds* is the hazard function and the integral in *dr* is the fragility function or, respectively, the demand and capacity (McGuire [2]).

2.2 Consequence analysis

The potential consequences strictly depend on the context within which the system is placed. This context defines the exposure of the socio-economical environment. For instance, referring to the potential for a life loss (L) the exposure is given by:

$$\text{Life-Loss Exposure, } \mathrm{E}[L] = \mathrm{P}[C(L)|Risk] \cdot \mathrm{P}[space, time|C] \qquad (5)$$

Life-exposure is given by the probability of a person to lose his/her life due to a consequence (C) of the failure risk times his/her spatial and temporal presence at the moment of the event. The overall assessment of risk is schematically shown in Figure 4.

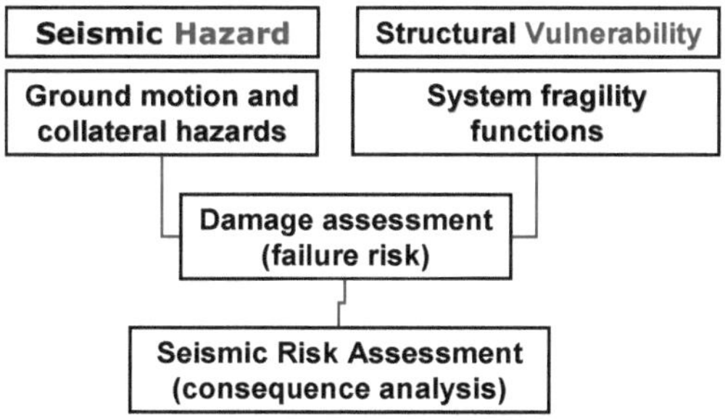

Figure 4: Flow-chart for seismic risk analysis.

3 Collateral hazards (secondary effects)

When dealing with seismic risk analysis, a relevant cause of damage is given by secondary effects induced by the seismic shaking. Many geotechnical hazards can be triggered by earthquakes, such as liquefaction, landslides and ground settlements, among others. Nonetheless, some of them may trigger others, such as flow-failures due to liquefaction, dam-breaks due to lateral spread of embankments, or sea-wave run-up due to submarine landslides. Thus, in addition to the risk of failure given by seismic ground motion, there is also a risk of failure given by seismic geotechnical hazards. Apart from ground settlements that can influence the assessment of manifold limit states, most of the geotechnical hazards can only affect the stability of the structure as a whole, that means they are relevant only for the assessment of the ultimate limit state (e.g., liquefaction and run-up). The approach is not different from that shown for the assessment of the risk of failure due to the ground shaking, provided that in this case the binomial distribution is more consistent than the Poisson distribution to characterize the hazard. For the case-study the stability of the plant can be threatened by liquefaction and induced flow-failures and by the sliding of the adjacent submarine scarp that may trigger, in turn, a sea-wave run-up striking the plant area. These effects are well documented to have occurred in the studied area: during the earthquakes that hit Southern Italy in 1783 several liquefaction were observed throughout the coastline; on July 12, 1977 more than 5 million cubic metres of material slid down the submarine canyon facing the harbour, causing a sea-wave up to 5 metres high that damaged many cranes and other harbour facilities. To investigate these phenomena an extensive survey was carried out, consisting in several onshore and offshore investigations. Equivalent statistic and dynamic analyses (Hungr et al. [3]) were performed to determine the failure probability due to liquefaction (Figure 5) and the initiation of a sea-wave run-up due to a submarine landslide (Figure 6).

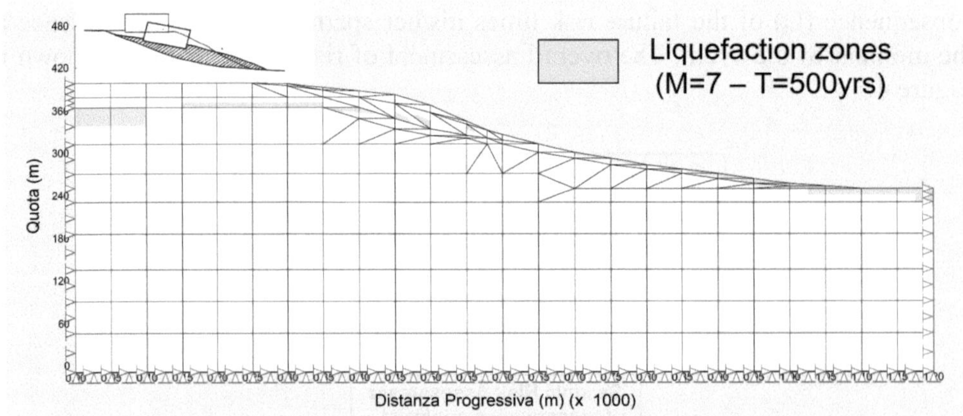

Figure 5: Dynamic analysis for liquefaction and flow-failure

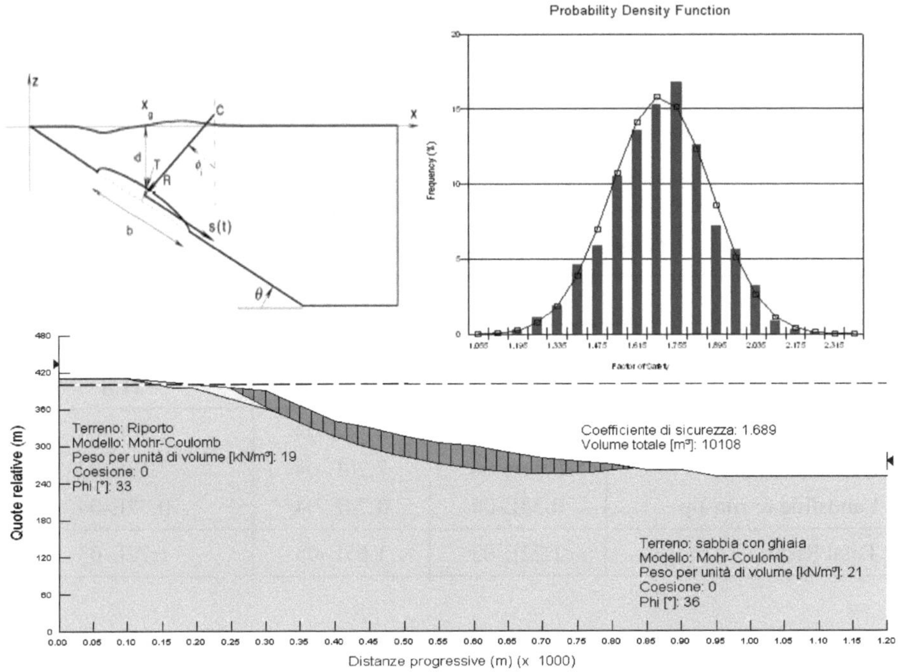

Figure 6: Stability analyses of the submarine scarp carried out for computing the probability distribution of the safety factors (Picarelli et al., 2005) and for modelling the sea wave run-up due to a rapid flow slides

4 Results

Catastrophic failure of the steel tanks may give rise to potential accidents listed in Table 1. Thus, the consequence of an accident is conditional to the spatial presence of a person within the distances shown in Table 1.

Table 1: Spatial extent of potential accidents due to a failure

Accident	Begins of death (m)	High mortality (m)
Pool fire	80	60
Flash fire	220	160
UVCE/BLEVE*	250	190

*vapour cloud explosion

The life-loss vulnerability (P[C(L)] in equation 5) is assumed to be equal to 1 for high mortality and greater than 50% for serious life-threatening injury. Spatial probabilities refer to three work locations, tanks, offices and the whole plant area, depending on the working tasks of employed people. Temporal probabilities are inferred from the employees' working time plan.

Consequence analysis leads to the computation of the probability of an individual to loss his/her life due to an accident is triggered by the occurrence of a failure event (Table 2).

Table 2: Annual probabilities of a life loss

	Offices	Tanks	Whole plant-area
Workers	3	5	2
Exposure (%)	15.4	20.8	21.8
Ground motion	6.30E-04	8.54E-04	8.93E-04
Liquefaction	5.31E-04	7.20E-04	7.53E-04
Landslide & run-up	0.54E-04	0.74E-04	0.77E-04
Total Risk	1.22E-03	1.65E-03	1.72E-03

The table shows, for each place within the plant area, the probability that a worker may loss his/her life due to an accident triggered by a failure of the plant triggered by either ground motion, or liquefaction, or a landslide and induced sea-wave run-up. Despite ground motion is the triggering of liquefaction and landslides, too, each event can take place independently from the others, thus the overall risk of an individual to loss his/her life is given by the total probability theorem:

$$\text{Total Risk} = 1 - \prod_i (1 - P_i) \qquad (6)$$

where P_i is the annual probability of a life loss due to the accident triggered by the i-th event.

May a risk (the negative consequence of an event or activity) be acceptable or not is a social and political choice. Nevertheless, a comparison with other industrial risks may facilitate this choice. In Figure 7 the societal risk of several industrial activities are shown (modified from Whitman [4]), along with the risk computed for the studied facility. Societal risk is defined as the probability that a group of N or more people would get killed due to an accident triggered by a system failure. This is commonly expressed by a frequency – number (FN) curve, representing the annual frequency of exceeding N or more casualties given a failure.

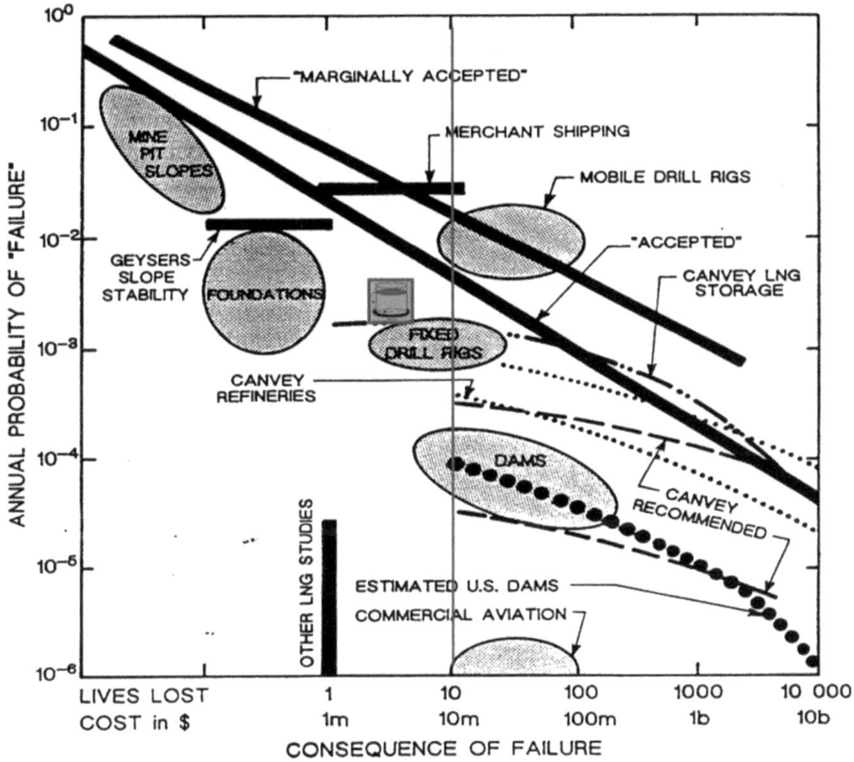

Figure 7: F-N curve for various industrial risk activities. The societal risk of the studied plant is shown in the middle of the figure with the symbol of a cylindrical tank. The vertical red solid line marks the limit of people that could in theory be involved simultaneously in the plant's activities

5 Conclusion

The innovative concepts of Consequence Based Engineering (Abrams et al. [5]) and Performance Based Earthquake Engineering (Porter [6]) are founded on the availability of reliable tools to forecast losses (human, social, economical, etc.) due to the collapse under seismic actions of civil engineering structures.

In the above contexts, deterministic analyses don't represent the best answer, since they aren't able to take into account all the uncertainties regarding the resistance demand and system's capacity. Conversely, a probabilistic approach allows for a rational choice and a consistent risk mitigation management.

In this paper, the main aspects related to the development of a risk assessment procedure taking into account site features (hazard) and structural performance (vulnerability) have been reported. The procedure shown is well suitable for both the retrofitting of existing facilities and the design of new ones. The case-study shown in this paper is a worthwhile example of a multi-hazard based seismic risk analysis of an oil-gas storage plant threatened by seismic ground motion and collateral hazards (earthquake-induced ground failures). The main implications of the study regard the possibility to establish acceptability or not of an industrial activity in relation to the possible negative consequences of a failure, the decision about the feasible countermeasures to be adopted to mitigate the risk, and the establishment of consistent insurance fees to cover the losses eventually resulting from a system's failure.

6 Acknowledgements

The Author thanks the E&G Engineering Consulting for the geotechnical investigations.

REFERENCES

[1] O'Rourke, M.J. and So, P. 2000. Seismic Fragility Curves for On-Grade Steel Tanks. Earthquake Spectra, Vol. 16 (4), pp. 801-815.

[2] McGuire, R.K. 2004. Seismic hazard and risk analysis. Earthquake Engineering Research Institute, pp. 240.

[3] Hungr, O., Evans, S.G., Bovis, M., and Hutchinson, J.N. 2001. Review of the classification of landslides of the flow type. Environmental and Engineering Geoscience, VII, pp. 221-238.

[4] Whitman, R.V. 1984. Evaluating calculated risk in geotechnical engineering. Journal of Geotechnical Engineering, ASCE, Vol. 110(2), pp. 145-188.

[5] Abrams, D.P., Elnashai, A.S. and Beavers, J.E. 2002. A new engineering paradigm: Consequence-Based Engineering. Available on the web.

[6] Porter, K.A. 2003. An Overview of PEER's Performance-Based Earthquake Engineering methodology. 9th International Conference on Applications of Statistics and Probability in Civil Engineering (ICASP), July 6-9, 2003, San Francisco,

Site-Specific Seismic Hazard Assessment

Timo Schmitt [1]

[1] SDA-engineering GmbH
Kaiserstrasse 100, D-52134 Herzogenrath, Germany
schmitt@sda-engineering.de

ABSTRACT:

Seismic design loads for standard buildings are given in seismic building codes. Code response spectra are obtained from generalised spectra for different soil classes and reference hazard parameters, like peak ground acceleration, in order to scale the spectrum according to the hazard at the site (e.g. using earthquake zones). For sites of special facilities and constructions that are designed for longer return periods than standard buildings, a site-specific hazard assessment leads to more realistic seismic loads than code spectra scaled by importance factors. The article presents general methodologies, procedures and approaches for a site-specific seismic hazard assessment, taking into account local soil properties.

Keywords: seismicity, hazard, earthquakes, site-effects

1 Introduction

Earthquakes belong to the most destructive natural disasters in the world, producing significant accelerations at frequencies where buildings are vulnerable. The first step before seismic design is the evaluation of the seismic hazard according to the required safety level. For standard civil engineering structures the seismic loads are specified in national building codes, by generalised response spectra. Usually, the seismic hazard is given for a probability of 10% in 50 years, i.e. a return period of 475 years (e.g. EN 1998-1 [1]). Special facilities with higher risk potential - like industrial facilities or dams - are out of the scope of standard building codes. Regulations for these facilities recommend longer return periods and sometimes a seismic hazard assessment is required.

EN 1998-1 [1] provides an important factor γ_I to transform the reference peak ground acceleration to higher or lower return periods, according to the building importance class. In a note, a formula for scaling the reference peak ground acceleration respectively the response spectrum to other return periods is given. This formula contains an exponent “k” that may be interpreted as a parameter representing the relationship between the occurrence of small and big earthquakes. The value “k” is regionally dependent and therefore a national determined

DOI 10.1007/978-3-658-02810-7_3, © Springer Fachmedien Wiesbaden 2014

parameter. However, a scaling factor can only be a rough estimation to transfer seismic hazard to other return periods, because every seismic source region has its own characteristic seismicity.

In contrast to the generalised code response spectra which are scaled by the reference peak ground acceleration, a site-specific seismic hazard assessment calculates the spectral accelerations at the site for each frequency, according to the surrounding seismicity. So, the magnitude and distance distribution, controlling the hazard, affects the shape of the site response spectra. Furthermore, local site-effects due to the soil profile and soil properties can be considered. Site-specific seismic hazard analyses in combination with soil dynamic studies lead to much more precise seismic load assumptions than building codes can provide. Also, industrial facilities can take advantage of the hazard results for different return periods regarding the required design levels.

2 History

The basic data for every seismic hazard assessment are historical (pre-instrumental) and recent (instrumentally registered) seismicity, compiled in earthquake catalogues and knowledge about geology and tectonics, as source regions and active faults.

The first seismic hazard assessments and hazard maps were based on deterministic procedures. The deterministic seismic hazard considers case scenarios and evaluates ground motion based on the distribution and the strength of historical and recent earthquakes, taking into account tectonic structures.

The probabilistic seismic hazard assessment (PSHA) was presented by the American civil engineer Carl Allin Cornell and the Mexican civil engineer Luis Esteva. In the year 1968, Cornell published a major theoretical work for a probabilistic seismic hazard assessment [2], which estimates the seismic hazard for different probabilities of exceedance. The main part of this work is a total probability theorem, where the probability that the expected earthquake parameter (e.g. maximum ground acceleration) at the site will be reached or exceeded is dependent on earthquake strength, distance and the cumulative distribution functions of these two parameters. Based on this theory, computer programmes were developed in the 70's. It took some more years for probabilistic methods to become popular and used for site-specific hazard assessments. Nowadays, the PSHA is the standard procedure for seismic hazard assessment and seismic hazard maps. Since its first application, PSHA methodologies and the evaluation of parameters have been improved and the assessment and integration of uncertainties in the calculations became more important. The development of PSHA is often driven by the importance to assess the seismic hazard for nuclear facilities.

Basically deterministic and probabilistic methods are the same, except that the PSHA evaluates the earthquakes statistically and provides design accelerations for different probabilities of exceedance.

3 Probabilistic seismic hazard assessment (PSHA)

3.1 Principle of PSHA

In Figure 1 the basic principle of a probabilistic seismic hazard calculation for a site is shown: It is assumed that earthquakes are Poisson distributed and they are statistically independent events. Therefore, it is important to exclude pre- and aftershocks before calculating the regression parameters of the magnitude frequency distribution for each source region. The area around the site is cut in small zones. For each zone the frequency distribution and the activity rate of earthquakes is known according to its source region. Now, for a given ground motion prediction equation the hazard at the site can be calculated. The summation of all contributions from all source regions results in a hazard curve for the site. The hazard curve gives the earthquake impact in terms of peak ground acceleration (PGA) or spectral acceleration (Sa) according to the annual probability of exceedance (P).

The following points of PSHA are presented in this chapter:

- Seismic source regions and faults
- Earthquake statistics (frequency distribution and activity rate)
- Upper bound magnitude
- Ground motion prediction equations
- Local site effects (usually evaluated after PSHA calculation)
- Treatment of uncertainties
- Uniform hazard spectrum and deaggregation

3.2 Seismic source regions and faults

Seismicity is not distributed homogenously. In areas where capable faults are known, the faults can be modelled directly. However, in most of the areas faults are not or rarely known and covered under sediments. In these cases, seismic source regions are defined according to the distribution of seismicity and the tectonic environment. In a seismic source region is assumed a similar seismicity and a homogenous distribution of earthquakes. For each seismic source region the earthquakes are compiled from the earthquake catalogue and the frequency distribution is calculated.

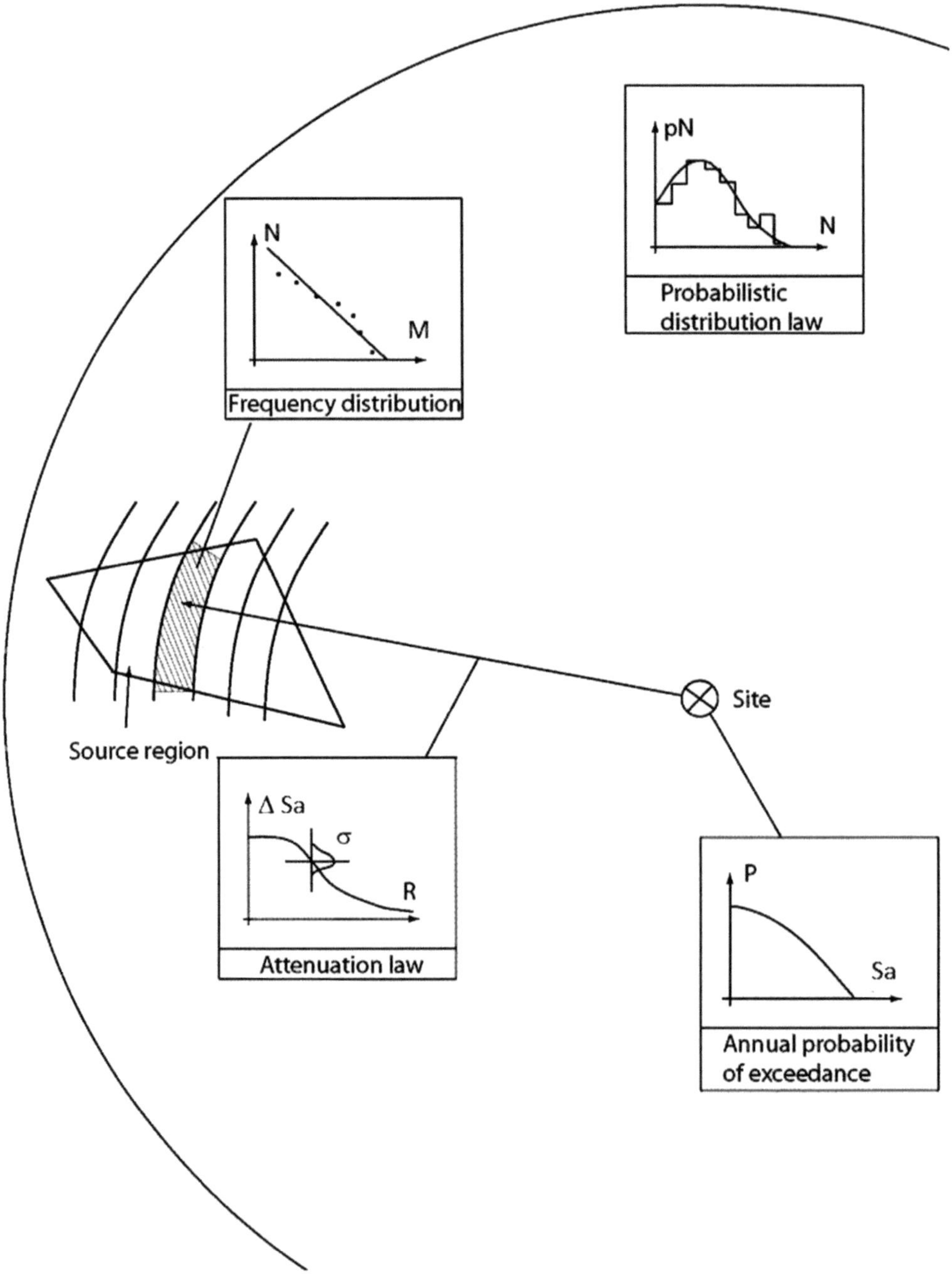

Figure 1: Principle of a probabilistic seismic hazard calculation (N = No. of earthquakes, P = probability, M = magnitude, Sa = spectral acceleration, R = distance)

3.3 Earthquake statistics

Hazard assessment usually considers magnitude relations, but in the case of historical earthquakes only macroseismic intensities are known. Therefore, a good estimation of magnitude values for historical earthquakes is an important task. Most of the relations in seismic hazard assessment refer to moment magnitude (M_W). However, for many earthquakes only local magnitude M_L is determined. For the sake of data homogenisation other magnitude values are often transferred to moment magnitude by empirical relations. Especially for M_L is important to apply an appropriate relation, because M_L may differ significantly among different evaluations from different institutions.

Before the earthquakes are evaluated from the earthquake catalogue, pre- and aftershocks have to be eliminated to fulfil the criteria of independent events. Furthermore, the completeness of the earthquake catalogue should be tested. For former times, catalogues are less complete. The completeness depends on the starting year and is given for magnitudes greater than a minimum magnitude.

The frequency distribution of earthquakes can be calculated according to Gutenberg & Richter (1958) by

$$\log_{10} N = a - b\,M \qquad (1)$$

with N number of earthquakes, magnitude M and the regression parameters a and b.

The annual frequency distribution of a certain magnitude M is calculated by

$$\nu\,(M) = \frac{10^{a-b\,M}}{\text{observation time}}. \qquad (2)$$

3.4 Upper bound magnitude

For each seismic source region or fault, a maximum magnitude M_{max} is estimated as an upper bound value in the hazard model.

In the case that a major fault is known and considered to be capable to produce a strong earthquake, M_{max} can be estimated by empirical relations. Wells & Coppersmith (1994) provide such empirical relations. An estimation of M_{max} from the fault segment length, for instance, is given by Lindenfeld & Leydecker (2004). If paleoseismological studies are available, the results can be used to define M_{max}.

Areal source regions estimation of M_{max} is very difficult and uncertainties are high. Often M_{max} is selected by adding a margin ΔM to the maximum observed earthquake magnitude in the source region. Typically ΔM is determined between 0.5 and 1.0 magnitude units.

The upper bound magnitude becomes more important for PSHA results for low probabilities of exceedance.

3.5 Ground motion prediction equations

Ground motion prediction equations (GMPE) are needed to calculate the vibration attenuation from the earthquake source to the site. These equations are based on empirical evaluations of strong-motion registrations. Many GMPE can be found in literature and an overview is given by Douglas (2011) [6].

The main parameters of the attenuation function are magnitude and distance and a term regarding soil conditions. Some GMPE include further parameters e.g. earthquake source mechanism and fault orientation. A generalised form of the equation is

$$f(\mathrm{Y}) = \mathrm{a} + f_1(\mathrm{M}) + f_2(\mathrm{R}) + f_3(\mathrm{S}) + \varepsilon \qquad (3)$$

where Y is PGA or spectral acceleration, f_1 (M) a function of magnitude, f_2 (R) a function of distance, f_1 (S) a function of soil and ε the dispersion.

Recent GMPE refer to moment magnitude, where the distance measure are used with different definitions: e.g. epicentral distance, hypocentral distance, closest distance to fault rupture or distance from Joyner & Boore (1981) [7]. The soil is considered as soil classes or with the parameter vs_{30}, representing the mean shear wave velocity of the upper 30 m below surface.

The dispersion among different GMPE results is significantly high, especially for short distances to the site. The selection of appropriate GMPE for the target region is an important task, due to the impact on the final PSHA results. Besides the obvious selection criteria that the GMPE should be based on a sufficient dataset, the magnitude and distance distribution cover the range of interest, it is also recognised that the equation should include a non-linear scaling of ground-motion amplitudes with magnitude and magnitude-dependent distance dependence. To define GMPE selection criteria is difficult and so far, no standard procedure exists. An overview of the discussion is given in Bommer et al. (2010) [8] and Graizer (2011) [9]. GMPE are also influenced by the source region of the dataset. A proposal for the adjustment of GMPE from source to target region is provided by Campbell (2003). However, in practice this task is often challenging due to the lack of information about propagation paths.

In the hazard calculation different GMPE are combined in a logic tree. Because the distribution curve of the ground motion attenuation is not limited and in order to avoid unrealistic high accelerations for very low probabilities of exceedance, the distribution curve is truncated, usually at two or three standard deviations.

3.6 Local site effects

Local site-effects can have a strong influence on spectral accelerations at the site. Due to the resonance and the damping of the sediments, accelerations are amplified or deceased. Most severe resonance effects are caused if a strong impedance contrast between two soil layers exists, for instance, in sediment layers on rock. If soil layers are horizontally located, the first resonance frequency can be estimated by the formula

$$f_1 = vs / 4\ h \qquad (4)$$

where vs is the mean shear wave velocity of the upper sediment layer and h is the sediment layer thickness. For example, a site with 50 m thick sediment layer over rock where the shear wave velocity in the sediments is 500 m/s will have a resonance at 2.5 Hz. As a result, around this frequency the free-field response spectrum will have amplified ordinates relative to the base-rock spectrum.

In most of the cases (horizontal soil layers, no basin effects) site amplification can be assessed using a simple 1D soil profile model. The input motion at the model basis (half-space) is derived from the PSHA site response spectrum. Then, the motion at the depth of interest (e.g. free-field or foundation level) is calculated. The most common engineering approach is using linear equivalent calculations in the frequency domain. Input parameters for each soil layer are shear wave velocity and density. Appropriate shear modulus reductions and damping curves have to be selected. Other approaches are nonlinear calculations or random vibration theory.

For sites of industrial facilities the knowledge of local soil layers and its properties derived from the soil expertise report should be used to evaluate potential site-effects and to specify the response spectrum at free-field or foundation level.

3.7 Treatment of uncertainties

In PSHA uncertainties are divided in two groups, depending on the kind of treatment in the hazard calculation: Aleatory variability and epistemic uncertainty. Aleatory variability is an uncertainty due to data dispersion. An example is the variability of a GMPE. The aleatory variability is included in the hazard model by the distribution curve and its deviation. The epistemic uncertainty can be considered as a model uncertainty due to a lack of knowledge or data. Examples for epistemic uncertainties are delineation of seismic source zones, M_{max} or selection of GMPE. Usually, these uncertainties are incorporated in the hazard calculation by a logic tree.

3.8 Uniform hazard spectrum and deaggregation

A result of the PSHA is the uniform hazard spectrum (UHS), derived from the spectral accelerations and for the regarded probability of exceedance. The UHS contains all contributions from all the seismic sources around the site and can be used for seismic design. However, for nonlinear analyses or probabilistic risk assessment (PRA) single earthquake scenarios may be considered. These can be obtained from a deaggregation analysis. A deaggregation evaluates the hazard according to magnitude and distance bins and gives the percentage of its contributions. Hazard controlling scenarios can be identified. Instead of UHS, response spectra for controlling earthquake scenarios can be used, too.

4 Conclusion

The principles and the general procedures of a site-specific probabilistic seismic hazard analysis have been presented. For industrial facilities and constructions designed for longer return periods than standard buildings (e.g. 475 years), a site-specific hazard assessment has many advantages: The shape of the response spectra is more realistic, because the hazard is calculated for the site coordinates and for various spectral accelerations. Building codes just scale generalised response spectra to the hazard level of a single parameter (e.g. PGA), (some code use two points of support). A precision improvement of the site response spectra is recommended performing soil dynamic calculations, taking advantage of the knowledge about local soil properties. Regarding seismic design, PSHA provides seismic loads for all return periods of interest. In combination with soil dynamic calculations, response spectra can be obtained at the free-field or at any other depth level (e.g. foundation). Last but not least, an update of hazard maps in building codes does not affect the validity of a site-specific hazard assessment.

References

[1] EN 1998-1: Eurocode 8: Design of structures for earthquake resistance - Part 1: General rules, seismic actions and rules for buildings.

[2] Cornell, C.A. (1968): Engineering seismic risk analysis. - Bull. Seism. Soc. Am. 58(5), pp. 1583-1606.

[3] Richter, C.F. (1958): Elementary Seismology. – W. H. Freeman & Company, New York, 768 pp.

[4] Wells, D. L. & K. J. Coppersmith (1994): New empirical relationships among magnitude, rupture length, rupture width, rupture area, and surface displacement. Bull. Seismol. Soc. Am., 84: 974-1002.

[5] Lindenfeld, M. & G. Leydecker (2004): Bestimmung des Verhältnisses Bruchlänge zu Störungslänge sowie Ergebnisse der Gleittendenzanalyse entlang neotektonischer Störungen in Norddeutschland. - 100 S., 69 Abb., 7 Tab., 1 Anhang; BGR Hannover, Tagebuch Nr. 10639/04, 5. April 2004.

[6] Douglas, J. (2011): Ground-motion prediction equations 1964 – 2010. – Final Report BRGM/RP-59356-FR, Feb. 2011.

[7] Joyner, W.B. & D.M. Boore (1981): Peak horizontal acceleration and velocity from strong-motion records including records from the 1979 Imperial Valley, California, earthquake. - Bull. Seism. Soc. Am. 71, pp. 2011-2038.

[8] Bommer, J.J., Douglas, J., Scherbaum, F., Cotton F., Bungum, H. & D. Fäh (2010): On the Selection of Ground-Motion Prediction Equations for Seismic Hazard Analysis. - Seismological Research Letters Volume 81, Number 5, September /October 2010.

[9] Graizer (2011): Comment on "On the Selection of Ground-Motion Prediction Equations for Seismic Hazard Analysis" by Julian J. Bommer, John Douglas, Frank Scherbaum, Fabrice Cotton, Hilmar Bungum, and Donat Fäh. - Seismological Research Letters Volume 82, Number 2, March/April 2011.

[10] Campbell, K.W. (2003): Prediction of Strong Ground Motion Using the Hybrid Empirical Method and Its Use in the Development of Ground-Motion (Attenuation) Relations in Eastern North America. - Bulletin of the Seismological Society of America, Vol. 93, No. 3, pp. 1012–1033, June 2003.

Critical Industrial Facilities: Simply Applying Current Importance Factors γ_I is not Enough!

Martin G. Koller[1] **and Ehrfried Kölz**[2]

[1] Résonance Ingénieurs-Conseils SA
21 rue Jacques Grosselin, 1227 Carouge, Switzerland
martin.koller@resonance.ch

[2] Risk & Safety AG
5073 Gipf-Oberfrick, Switzerland
koelz@risksafety.ch

ABSTRACT:

A generic seismic risk study for critical industrial facilities (CIFs) is presented and discussed in detail. The study is focussed on the residual seismic risk of critical facilities supposed to be correctly designed according to Eurocode (EC) 8. The initial objective was to define a design importance factor γ_I in order to achieve sufficiently low probabilities of a major accident. The residual seismic risk is dominated by earthquakes for which the probability of occurrence is typically one or two orders of magnitude lower than for the design earthquake. According to Swiss practice, the annual probability of a major accident with more than 100 fatalities outside the industrial facility must not exceed 10^{-7}. In order to achieve this goal, it turned out that a design earthquake with a return period of the order of 100'000 years should be considered, with an associated importance factor around 8! Such a design, however, would be technically and economically unfeasible. Therefore, it is necessary to adopt a risk based view and first explore all possibilities of reducing the largest possible number of fatalities – by other means than just a strong seismic design. At present, it is not yet clear what will be done by the Swiss authorities once all reasonably practicable measures of reducing the size of the largest possible accidents have been put into action and the residual seismic risk is still too high. In any case, however, it is strongly recommended to also look at what could happen if ground motions (GMs) much above design GM occur, instead of simply design for a given GM level.

Keywords: critical industrial facilities, residual seismic risk, design return period, importance factor, design ground motion

DOI 10.1007/978-3-658-02810-7_4, © Springer Fachmedien Wiesbaden 2014

1 Introduction and objectives

In general, operators of critical industrial facilities (CIFs) – classified Seveso or similarly – have to make sure that the societal risk associated with the operation of their facility complies with some risk acceptance criteria. These criteria vary from country to country. However, for some kinds of risk, depending on the country, the explicit risk analysis is replaced by a deterministic prescriptive regulation. At least in Europe, this seems to be the current practice for dealing with seismic risk.

Since 2003, the Swiss building code, SIA 261, has stated that the importance factor to be applied for seismic design or control of CIFs has to be fixed on the basis of a risk analysis. However, in practice, nobody has ever followed this code prescription; instead, an importance factor of $\gamma_I = 1.4$ has simply been applied, and this has so far been – tacitly – accepted by the safety authorities. It's only recently that a generic risk analysis was undertaken in order to check whether the residual seismic risk of this practice complies with the risk acceptance criteria in use in Switzerland. This study was carried out by the authors of the present article on behalf of the Swiss Federal Office for the Environment. The main results of this study will be presented here.

The described Swiss practice is believed to be a direct consequence of the lack of communication between the earthquake engineering and environmental risk assessment communities. On the one hand, earthquake engineers are used to apply deterministic and conservative design procedures, even if design is done for a hazard level that has previously been fixed – by seismologists – on the basis of a probabilistic seismic hazard assessment (PSHA). Only very few earthquake engineers are familiar with probabilistic risk assessment. On the other hand, most risk analysts have virtually no knowledge in earthquake engineering, and sometimes use rather questionable seismic vulnerability data for their risk analyses. There is an urgent need for improved communication and cooperation. Both worlds have to learn a lot from each other.

The objective of the aforementioned study was to determine the residual seismic risk of CIFs, supposing a correct seismic design for a usual importance factor γ_I and a faultless construction. This residual risk was then compared with the risk acceptance criteria for CIFs in use in Switzerland. From this comparison, it was concluded that simply designing for a fixed importance factor γ_I was not sufficient.

In a simplified manner, two types of residual seismic risks can be distinguished. One is linked with the structural reliability of a code compliant design (what is the residual risk due to an earthquake whose ground motion (GM) at the CIF is at most as strong as the design GM?), and one is linked with earthquakes that produce (much) stronger GMs than the design GM (what is the risk that a stronger than design GM causes a major accident in the CIF in spite of a 'correct' seismic design?). The present article mentions only briefly the first kind of the residual seismic risk and focuses on the second kind.

In order to evaluate the residual seismic risk of the second kind, henceforth simply referred to as 'residual seismic risk', the probability of occurrence of stronger GMs than the design GM must be known first. This information is given by the so-called seismic hazard curve: a (decreasing) probability of exceedance versus an (increase-ing) level of GM at a given site. Second, fragility curves for relevant mechanical failures, leading to the loss of a safety barrier (for instance the tank wall of a storage tank containing toxic gases) must be known, i.e. the conditional probabilities of failure as a function of GM (stronger than the design GM). In the present context, a rough estimation of the fragility curves will turn out to be sufficient. The combination of these probabilities will lead to the (absolute) probability of loss of the corresponding safety barrier. Finally, the physical consequences of the loss of the safety barrier, for instance the propagation of a toxic cloud towards a populated area, must be simulated in order to determine the damage, usually expressed in fatalities, outside the site of the CIF.

All these elements will be discussed in the following. However, first, the risk assessment criteria in use in Switzerland will be presented, together with some indications for analogue criteria in the Netherlands, Germany and France.

2 Societal risk acceptance criteria

It would be far beyond the scope of this article to present and discuss risk acceptance criteria in an exhaustive way. Only a few aspects of societal risk criteria, as far as relevant for the judgment of residual seismic risk, will be discussed here.

Societal risk acceptance criteria are usually expressed in terms of so-called F-N curves (annual frequency F of event versus N or more fatalities). These criteria may or may not incorporate risk aversion. Risk aversion means that one single accident with 100 fatalities is perceived more severely and therefore less tolerated than 100 accidents at different places with one fatality each. Indeed, modern societies react with a strong risk aversion, as is – unfortunately – confirmed every day. A plane crash in Europe with 100 victims will be reported on many newspaper front pages, whereas every day, more than 100 people are killed in road traffic accidents throughout Europe, with very little reaction from society. However, risk aversion is not only a matter of subjective perception, but is also justified by the fact that the society is much better prepared to handle many small accidents than one major event with the same total number of casualties. This becomes evident when looking at injured people: 100 injured persons from car accidents throughout Europe, on the same day, do not saturate hospitals, whereas 100 injured persons at the same place at once will immediately saturate all hospitals in an astonishingly wide area around the accident location so that appropriate care is much more difficult to be given to these people.

The Swiss risk acceptance criteria (FOEN, 1996 [1]) are shown in Figure 1. The upper tolerable probability of exceedance, for more than 10 fatalities, is given by

$10^{-3}/N^2$ per year, and the negligible level by $10^{-5}/N^2$ per year. These limits are straight lines in the F-N (loglog) space with a slope of -2 and therefore incorporate a significant risk aversion. A priori, there is no bonus for existing CIFs. If the risk is between the upper tolerable level and the negligible level, the ALARP (As Low As Reasonably Practicable) principle is essentially applied, i.e. risk is further reduced as far as technically and economically feasible. The risk values correspond to one facility (industrial site), and the cumulative risks from several facilities are not taken into account.

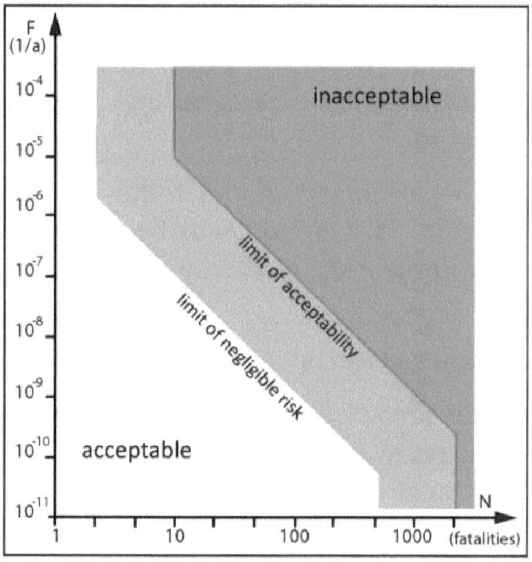

Figure 1: Societal risk acceptance criteria (F-N curve) used in Switzerland

Historically, the Swiss criteria were deduced from Dutch studies, and indeed, are essentially identical with those presently applied in the Netherlands (Trbojevic, 2005 [2], Web-1).

In Germany, the situation is completely different. 'No' risk is allowed outside the boundaries of CIFs [2], and this is assumed to be the case if all DIN codes are satisfied. With respect to earthquake risk, and as a complement to the DIN codes, a VCI guideline (RWTH Aachen, 2012 [3]) specifies the importance factors γ_I that should be used, the highest value being $\gamma_I = 1.6$, applicable to the worst cases where very toxic gases can affect large areas outside the industrial site.

In France, the risk acceptance criteria are formulated with respect to the (conditional) probability to die (lethality > 5% or > 1%) and with respect to irreversible health problems outside the industrial site, due to a major accident. For new CIFs, a non-zero probability of having $\geq$10 people with a lethality of 5%, $\geq$100 people with a lethality of 1% or $\geq$1'000 with irreversible health problems is simply not tolerated, whereas a maximum annual probability of 10^{-5} is accepted for these cases for existing CIFs. Non-zero or higher probabilities are tolerated if fewer people are affected. These criteria are extremely severe for relatively 'small' accidents. However, since there is no further differentiation for more than 10 persons with lethality of 5 %, the same acceptance criteria apply whether the accident causes 10, 100 or 1000 fatalities: such scenarios are not accepted for new CIFs, whatever their probability of occurrence, but would be tolerated for existing CIFs as long as their annual probability of occurrence remains $\leq 10^{-5}$. Therefore, the French acceptance criterion, for existing CIFs, becomes much less severe than the Dutch or Swiss criteria if significantly more than 10 fatalities are possible.

Seismic risk, however, is treated in a deterministic way. The French 'arrêté' of 24 January 2011 (MEDDTL, 2011 [4]) fixes the importance factors that have to be applied to facilities that represent a particular risk beyond the boundaries of the industrial site. Values of $\gamma_I = 2.2$ and $\gamma_I = 1.85$ have to be used for new and existing CIFs, respectively. Applying $\gamma_I = 2.2$ is intended to lead to a design GM with a return period of 5'000 years.

The key question is whether the deterministic seismic design in Switzerland, Germany and France for CIFs achieves the goal that it is meant to achieve, i.e. whether the residual seismic risk associated with correctly designed and constructed (or upgraded) CIFs satisfies the risk acceptance criteria.

3 Possibility of Exceptionally Strong GM

Most structural engineers, and even many specialists in earthquake engineering, are not aware of how much stronger very rare GM can be with respect to GM with a 'standard' return period of 500 or 1'000 years. If confronted with modern PSHA results, showing very strong GM for very low probabilities, they suspect these results of being unrealistic, caused by mathematical artefacts or flaws in the PSHA methodology. That is why this aspect must be thoroughly discussed here.

First of all, let us look at what probabilities we are interested in: If a scenario with a potential of 1000 fatalities cannot be excluded, an annual probability of at most 10^{-9} can be tolerated according to the Swiss risk criteria. Imagine an earthquake that is sufficiently strong so that the conditional probability of causing 1'000 fatalities is 1/100 – in spite of a correct seismic design for, say, $\gamma_I = 1.4$, corresponding to a return period of about 1'000 years in Switzerland. Now, if such a strong earthquake is really possible, it would have to have an annual probability of less than 10^{-7}. However, 10^{-7} is an extremely low probability, corresponding to a return period of 10 million (!) years; nearly no hazard studies exist that cover such low probabilities. In Europe, so far, the probably single exception is the PEGASOS project (NAGRA, 2004 [5]), as well as its follower, the PEGASOS Refinement Project, finishing in 2013.

Figure 2 shows a typical hazard curve resulting from PEGASOS, for a site with relatively low seismicity. Since modern PSHAs use logic trees, whose branches represent epistemic uncertainty (alternative models: different tectonic models, different GM prediction equations, etc.), several fractiles of the hazard curve can be calculated. Typically, these fractiles spread out more and more for decreasing probabilities (look at the increasing range of probabilities for a *given* level of GM). However, the mean hazard is what is usually considered as relevant for engineering purposes (thick line in Fig. 2). Because of the spread of the fractiles, the mean hazard curve 'climbs' across higher and higher fractiles for decreasing probabilities (the fractiles with higher probabilities dominate; be aware of the log-scale in Fig. 2!). Musson, 2005, [6] explains this as follows: going to lower and lower probabilities, one should not be surprised to come closer and closer to the 'worst' case.

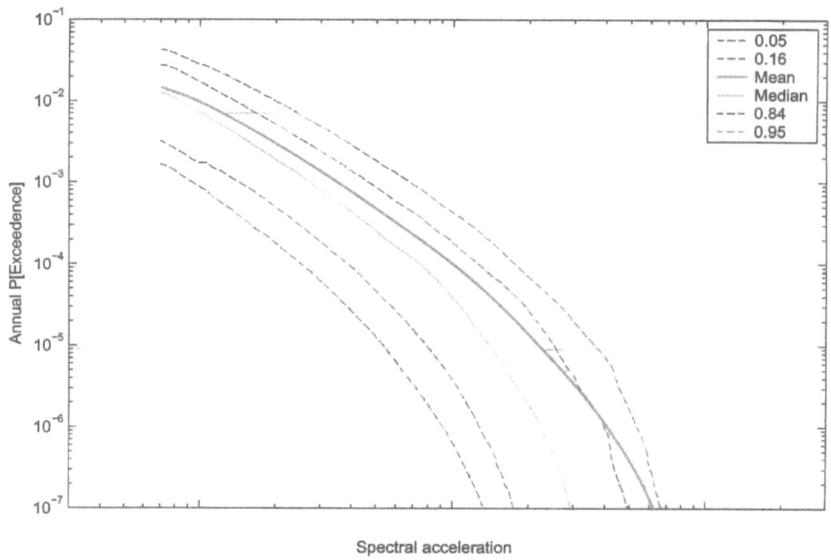

Figure 2: Hazard curve for the spectral acceleration at 2.5 Hz, in loglog scale, for one of the Swiss nuclear power plant sites (from [5]); no acceleration unities are given since in the present context, only ratios between spectral accelerations for different probabilities of exceedance are of interest

Looking at the mean hazard in Figure 2, it appears that the GM (spectral acceleration at 2.5 Hz) is 10 (!) times larger for an annual probability of 10^{-6} instead of 10^{-3} – although a mechanism had been introduced for limiting maximum GM, mainly due to limited soil strength. How is this possible?

The large GMs for low probabilities are primarily due to the high variability of GM for a given earthquake scenario, and much less due to exceptionally large magnitudes. Indeed, GM at a given site for a given magnitude, distance and source depth is lognormally distributed, up to at least 3 standard deviations (Abrahamson, 2006 [7]), one standard deviation corresponding to roughly a factor of 2. Now, GMs that are three standard deviations (i.e. a factor of 8!) above the median are extremely rare, but not sufficiently rare that they would not 'appear' if we are looking at sufficiently low probabilities.

A few decades ago, many people thought that the variability of observed GM was a problem of inhomogeneous data acquisition, instrumental errors, etc., and only partly physical. However, the many more reliable instruments and recordings available since then show that this is not true. High GM variability is physical! This is illustrated by Figure 3 for the 2004 Parkfield earthquake: there are many observations far outside the range of plus minus one standard deviation of a widely used GM prediction equation, even for this single earthquake in a densely instrumented area with high quality instruments.

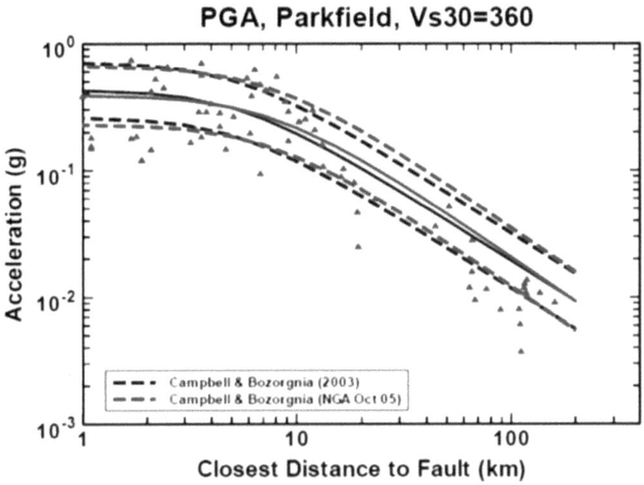

Figure 3: Parkfield 2004 earthquake ($M_w = 6.0$): comparison of measured peak ground acceleration (PGA) with two GM prediction equations; dashed lines correspond to plus minus one standard deviation (from Campbel&Bozorgnia, 2007 [8])

Why do so many engineers not 'assimilate' the high GM variability and their consequences, suspecting flaws in the PSHA methodology instead? One (partial) explanation might be that GM variability was often simply ignored in earlier PSHA studies, which – logically and mathematically - is simply wrong [7]. The hazard integral should integrate over the full GM variability. The consequence of neglectting GM variability was a significant underestimation of hazard, particularly for low probabilities, as illustrated by Figure 4.

Figure 4 shows median hazard curves for the Swiss nuclear power plants (HSK, 2007 [9]): The curves from earlier studies, essentially neglecting GM variability, are much 'steeper' than the PEGASOS curves that correctly account for GM variability, 'steeper' meaning a faster decrease in probability of exceedance for increasing GM. The differences between the mean hazard curves would be even more pronounced, because PEGASOS also took into account more realistic epistemic (model) uncertainties. It seems that most earthquake engineers are still accustomed to the steeper hazard curves and are therefore very sceptical when confronted with modern hazard curves like those from PEGASOS. For a more detailed discussion of the reasons why modern PSHA often lead to increased hazard estimates, the reader is referred to Bommer and Abrahamson, 2006 [10]. The conclusion is that the mean hazard curve of Figure 2, which will be used in the following, is not flawed, but corresponds to the present state-of-the-art in PSHA; older studies, however, were often flawed due to an incorrect treatment of GM variability.

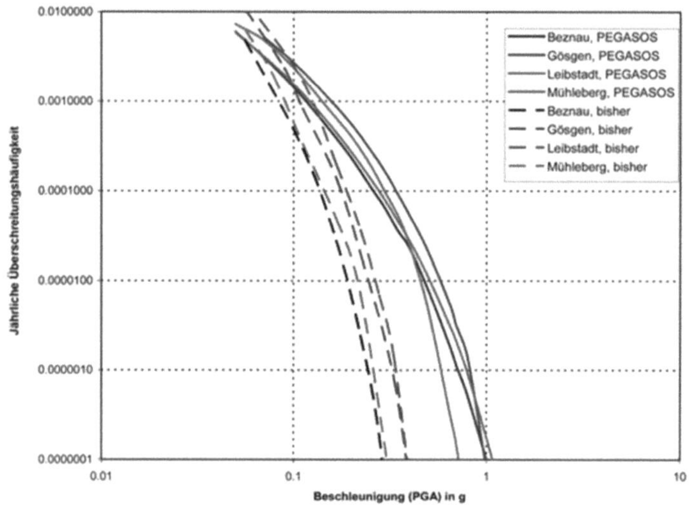

Figure 4: Comparison of earlier median hazard curves (dashed lines) with the PEGASOS median hazard curves (solid lines) for the Swiss nuclear power plants (from HSK, 2007 [9])

Nevertheless, the already mentioned PEGASOS Refinement Project might result in slightly steeper hazard curves. The reason is that great efforts were undertaken to reduce uncertainties. Furthermore, it is also worth noting that the hazard curves would be somewhat steeper for areas of high seismicity. However, qualitatively, the problems discussed in the following would remain the same.

4 Fragility

In order to evaluate typical residual risks, generic fragility curves were estimated, solely based on expert judgment. These curves are expressed as a function of how many times the design GM is exceeded. To get a rough idea of the sensitivity of the resulting risk with respect to the fragility curves, two different curves were used: a 'best-estimate' and a so-called 'optimistic' curve.

These fragility curves might represent the probability of a significant leak occurring in the wall of a pressurised liquid storage tank correctly designed for a given seismic design GM. The best estimate curve assumes a failure probability of 5 % for a GM that reaches twice the design GM, and a failure probability of roughly 50 % for four times the design GM. On the optimistic curve, the corresponding failure probabilities are only 3 % and 25 %, respectively. As it will turn out, the resulting residual risks will only very weakly depend on these fragility curves as long as they remain in a more or less 'reasonable' range. So, if the reader does not like the assumed fragility curves, he is encouraged to introduce his own estimation of fragility in the risk evaluation presented in the next chapter…

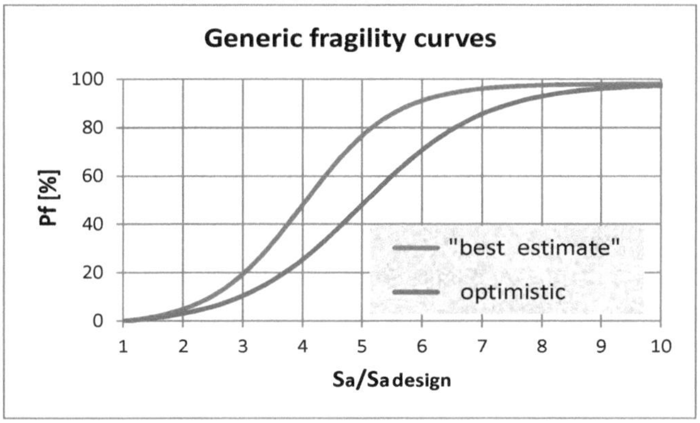

Figure 5: Generic fragility curves used for the assessment of residual risk: conditional failure probability as a function of how many times the design GM (spectral acceleration Sa design) is exceeded

5 Residual seismic risk due to exceptionally strong GM

A generic residual seismic risk will be evaluated with the aid of the mean hazard curve shown in Figure 2 and the generic fragility curves presented in Figure 5. This can be done in a very simple, pragmatic way with sufficient accuracy.

First, it has to be recalled that hazard curves, as shown in Figure 2, give the annual probability of *exceedance* (not occurrence) versus a GM intensity measure. If the GM with a return period of 1'000 years is used for design, the probability that this GM is exceeded is 10^{-3} per year. One might now look at the GM for an annual probability of exceedance of, say, 4 x 10^{-4}. For the hazard curve of Figure 2, reproduced in Figure 6, a GM roughly 1.5 times larger than the design GM corresponds to this probability of exceedance. Therefore, there is an annual probability of 10^{-3} minus 4 10^{-4} = 6 10^{-4} that a GM between the design GM and 1.5 times the design GM occurs. This range of GM can be considered as a class of GM with a probability of *occurrence* of 6 10^{-4} per year. Let's say that the average GM in this class is around 1.2 times the design GM (somewhat closer to the design GM than to 1.5 times the design GM since lower values are slightly more probable). With the formulas defining the fragility curves of Figure 5 – tanh() functions were assumed –, conditional failure probabilities of 0.6 % and 0.4 % can be found for 1.2 times the design GM. Finally, multiplying the annual probability of occurrence of this GM class (6 10^{-4}) with the conditional failure probabilities (0.6 % or 0.4 %) gives the (absolute) probability of failure due to this GM class. This simple calculation is shown on the first line of Table 1.

Now, a second GM class can be considered in an analogous way for annual probabilities of exceedance between, say, 4 10^{-4} and 2 10^{-4}, and so on, as illustrated in Figure 6. The corresponding simple calculations can be found in Table 1. It turns out that the risk contribution of GM with an annual probability of exceedance lower than 10^{-7} remains negligible.

For the generic examples presented here, the total annual probability of (mechanical) failure (assumed identical with the loss of a relevant safety barrier) is 6.6 10^{-5} or 4.2 10^{-5} for the best estimate or the optimistic fragility curves, respectively. From Table 1, it can be seen that the largest contribution to this residual risk stems from the GM classes 3 to 5, i.e. from GMs with return periods between 5'000 and 50'000 years (see Figure 6). The GM classes 1 and 2, with shorter return periods, contribute relatively little to the residual risk because of low conditional failure probabilities. And the GM classes 6 to 11 contribute little as well, in spite of high associated conditional failure probabilities, because the probabilities of occurrence of these GMs are too low.

Let us imagine that the mechanical failure, say the leakage of a tank, can cause 100 or 1'000 fatalities due to the release of a highly toxic gas. Let us further assume a conditional probability of 1/3 that the wind is directed towards the populated area, causing the 100 or 1'000 fatalities, and a probability of 2/3 that the wind is blowing the toxic cloud away from the population. In this case, the scenario with 100 or 1'000 fatalities finally has probabilities of occurrence of 2.2 10^{-5} or 1.4 10^{-5} (for best estimate or optimistic fragility, respectively).

Comparing these values with the Swiss risk acceptance criteria (Fig. 1), it becomes evident that these probabilities of occurrence would only be acceptable for scenarios with less than 10 fatalities. However, if 100 or even 1'000 fatalities can be the consequence of the mechanical failure, the residual risk is far above the upper limit of risk tolerance! Furthermore, this desolate situation hardly changes whether the best estimate or the optimistic fragility curve is used. This is bad news, since it essentially means that a moderate reinforcement, improving the fragility of the tank from the best estimate to the optimistic curve, would have no significant impact on the residual risk.

Comparing these values with the Swiss risk acceptance criteria (Fig. 1), it becomes evident that these probabilities of occurrence would only be acceptable for scenarios with less than 10 fatalities. However, if 100 or even 1'000 fatalities can

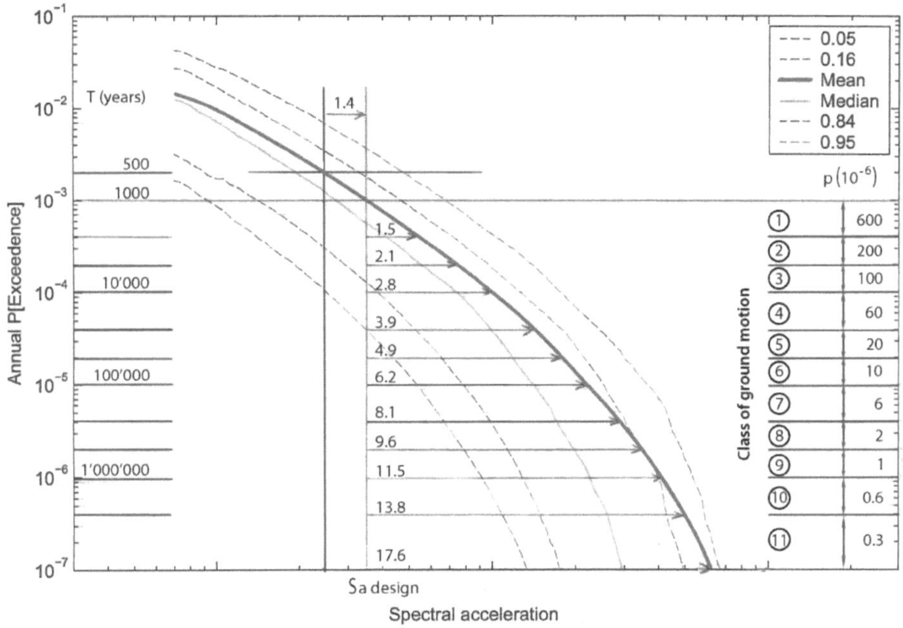

Figure 6: Determination of average GM (how many times the design spectral acceleration Sa design) for different classes of frequency of occurrence, based on the mean hazard curve of Figure 2

be the consequence of the mechanical failure, the residual risk is far above the upper limit of risk tolerance! Furthermore, this desolate situation hardly changes whether the best estimate or the optimistic fragility curve is used. This is bad news, since it essentially means that a moderate reinforcement, improving the fragility of the tank from the best estimate to the optimistic curve, would have no significant impact on the residual risk.

Since designing with an importance factor of $\gamma_I = 1.4$ has turned out to be insufficient if the potential for more than 10 fatalities exists, one might think of applying $\gamma_I = 2.2$, following the French arrêté [3]. According to the hazard curve of Figure 2, representative for large parts of Switzerland with low seismicity, $\gamma_I = 2.2$ corresponds to a return period of 2'500 years, whereas in France, it is supposed to correspond to 5'000 years. This difference is probably due to different assumptions with respect to epistemic uncertainty within the corresponding PSHA studies and much less due to different characteristics of seismicity between the two countries.

Table 1: Approximate evaluation of residual risk for a design with a return period of 1000 years

Class	P $[10^{-6}]$	Sa/Sa design	$P_f\ [10^{-2}]$ best est.	$P\ P_f\ [10^{-8}]$ best est.	$P_f\ [10^{-2}]$ optimistic	$P\ P_f\ [10^{-8}]$ optimistic
1	600	1.0 ÷ 1.5	0.6	360	0.4	240
2	200	1.5 ÷ 2.1	3	600	2	400
3	100	2.1 ÷ 2.8	10	1000	5	500
4	60	2.8 ÷ 3.9	25	1500	13	780
5	20	3.9 ÷ 4.9	65	1300	32	640
6	10	4.9 ÷ 6.2	85	850	78	780
7	6	6.2 ÷ 8.1	95	570	85	510
8	2	8.1 ÷ 9.6	97.5	195	95	190
9	1	9.6 ÷ 11.5	98	98	97.5	97.5
10	0.6	11.5 ÷ 13.8	98	59	98	59
11	0.3	13.8 ÷ 17.6	98	29	98	29
Σ				~ 6600		~ 4200

Performing the same exercise as before, but now assuming a design return period of 2'500 years, leads to a probability of occurrence of 0.7 10^{-5} when using the best estimate fragility curve. This is valid for the two scenarios assumed, as before, to cause 100 or 1'000 fatalities, respectively. This probability indeed meets the French criteria for existing CIFs. However, with respect to the Swiss criteria, the risk reduction by a factor of approximately 3 with respect to the design for $\gamma_I = 1.4$ is nearly negligible with respect to what would in fact be needed: 2 or even 4 orders of magnitude of reduction! Figure 7 illustrates this.

Since nuclear power plants were designed in Switzerland for a return period of 10'000 years (according to older PSHAs!), the same exercise has been repeated here for a design return period of 10'000 years. According to the mean hazard curve in Figure 2, this corresponds to an importance factor as high as $\gamma_I = 4.1$. The resulting probability of occurrence of the considered scenarios is then 10^{-6}, again using the best estimate fragility curve and the conditional probability of 1/3 that the mechanical failure leads to either 100 or 1'000 fatalities. Again, the resulting probability is still orders of magnitudes too high (Fig. 7)!

In order to comply with the Swiss risk criteria, a design return period of nearly 100'000 years would have to be used if 100 fatalities are possible, which means an importance factor of the order of 8. If 1'000 fatalities are possible, it's even worse: a design return period of 1 million years would be necessary, with an importance factor of the order of 16! Such a design, obviously, would not be reasonable neither from a technical nor from an economical point of view.

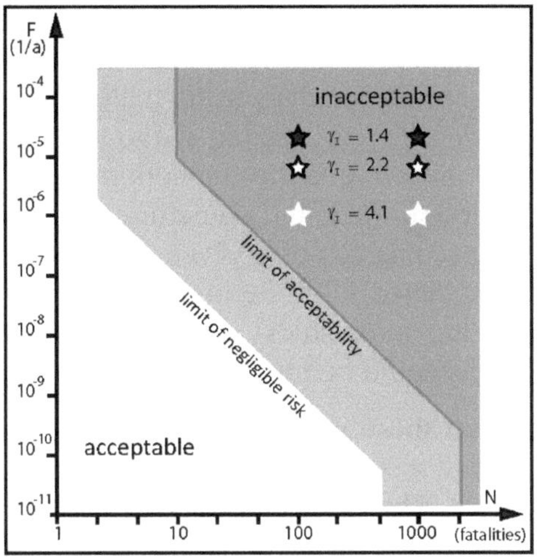

Figure 7: Residual seismic risk due to GM stronger than the design GM, for design return periods of 1'000 ($\gamma_I = 1.4$), 2'500 ($\gamma_I = 2.2$) and 10'000 ($\gamma_I = 4.1$) years

6 What can be done?

First, it might be asked whether the Swiss risk criteria are realistic, even with respect to general structural reliability objectives given by the Eurocodes (EN 1900). From Trbojevic, 2009 [11], it can be concluded that structures correctly designed according to the Eurocodes are expected to have annual probabilities of failure that are a couple of orders of magnitude higher than 10^{-9} – even without earthquakes.

In spite of the results presented here, showing that only unreasonably high importance factors would allow attaining sufficiently small residual risks, the Swiss safety authorities are not ready to relax the risk acceptance criteria for seismically induced major accidents, at least not in a general way. At the same time, they admit that importance factors of 5, 10 or even 15 would not be feasible. So what can be done?

If it is impossible, with reasonable efforts, to further reduce the probability of a given mechanical failure, the only way to lower the associated risk is to reduce the consequences of this failure, i.e. to limit the size of the largest possible accident. Theoretically, there are several measures possible. Probably the most efficient one would be to reduce the volumes of stored dangerous materials, either by changing the processes of production so that less of this material is needed, or by producing it on site – as and when needed – from less dangerous materials. Another possibility, only for new CIFs though, would be to site the CIFs at a larger distance from populated areas. All these measures would have the very appealing advantage of reducing not only the seismic risk, but also other risks due to false manipulation, terrorist attacks, etc. However, are these ideas realistic for industrial practice?

In order to get at least a tentative answer to this question, two pilot studies with two CIFs were undertaken in Switzerland. Needless to say that the first reaction of the industrials was that the quantities of dangerous materials were already minimised and that no further reductions were possible without jeopardising the survival of the industrial facility. And indeed, quantity reductions of dangerous materials were impossible for most production lines, but sometimes, technical measures could be found for reducing the consequences of a given mechanical failure. Nevertheless, for the production line associated with the most dangerous storage tank of one of the CIFs participating in the pilot study, a way of reducing a highly toxic gas storage from 5 tons to 2 tons could be found.

A further lesson learnt from these pilot studies was that the risk analysts of CIFs have the tendency of making a series of conservative assumptions in assessing the consequences of an accident. These assumptions are usually taken in order to avoid more elaborate studies that would be needed otherwise. In the aforementioned case of a tank with highly toxic gas, a worst case scenario with as many as 800 fatalities outside the industrial site was originally considered possible. However, the significant reduction of the stored quantity, together with a more realistic assessment of the consequences of a tank leak, due to an earthquake or whatever, allowed showing that finally less than 10 fatalities would have to be expected in the worst case.

Of course, such a 'success story' is not always possible, and so far, it is not yet clear what will be done by the Swiss authorities once all reasonably practicable measures of reducing the size of the largest possible accidents have been put into action and the residual seismic risk is still too high.

In the case of an extremely rare, exceptionally strong GM, most if not all ordinary buildings with no modern seismic design would collapse, and even among those correctly designed, many would collapse, too. Therefore, it could be argued that so many people would die in the collapsing buildings that 100 or even 1000 more fatalities due to a toxic cloud would not really matter anymore. Hence, it would not make sense to take into account the residual seismic risk associated with GM that is so strong that the collapse of ordinary buildings around the CIF would cause many more fatalities than the seismically induced industrial accident.

This argument, however, is not convincing for two reasons. Firstly, in industrialised countries, only about 10 % of the occupants of collapsing buildings die (Spence et al., 2011 [12]), but many more people would be prisoners among the debris before being rescued, without any possibility to protect themselves from a toxic cloud. Thus, many more people would die, and their number might be even larger than the number of those killed by the collapsing buildings themselves. Secondly, owing to constructive and destructive wave interferences, GM intensity can vary very strongly over short distances. Therefore, GM could be exceptionally strong within a CIF, but rather 'ordinary' within a neighbouring urban area. From Figure 2, it becomes clear that exceptionally strong and weak GM can coexist for a single earthquake (green triangles outside the range of ± one standard deviation).

Probably, a pragmatic solution will have to be adopted. From the point of view of earthquake engineering, the French importance factor of $\gamma_I = 2.2$ seems to be an upper practicable limit which remains economically feasible, at least for new CIFs.

7 Conclusions

The present conclusions are valid for CIFs with the potential of causing significantly more than 10 fatalities outside their site in case of an exceptionally strong earthquake. Simply designing such CIFs with an importance factor, whether with $\gamma_I = 1.4$, 1.6 or 2.2, is by far not sufficient to guarantee an acceptable low residual seismic risk, at least not with respect to risk acceptance criteria similar to those in use in Switzerland, the Netherlands, the Czech Republic, etc. [4]. Much higher, technically and economically unfeasible importance factors would have to be used.

Therefore, it is important to adopt a risk based view and first explore all possibilities of reducing the largest possible number of fatalities – by other means than just a strong seismic design. This is nothing else than the ALARP (as low as reasonably practicable) principle, current practice for risk managers, but much less so for earthquake engineers.

Furthermore, it is strongly recommended to look at what could happen if GMs much above design GM occur, instead of simply design for a given GM level. After all, this is simply what we should have learned from Fukushima. In fact, measures to improve the behaviour of dangerous equipment for GMs above design GM might be more cost-effective in reducing the residual seismic risk than a conventional reinforcement with respect to the design GM level. An example: The aforementioned storage tank, a horizontal cylindrical reservoir, has a pipe connected to its 'bottom'; if the pipe or its connection fails, a safety valve inside the tank immediately shuts, without electricity, simply by gravity. However, for GM much above the design level, the tank might fall from its bases and possibly overturn, the pipe being torn off: the valve would not shut though, since gravity would now act in the wrong direction! Therefore, the operator in charge has replaced the gravity valve by a valve that automatically shuts thanks to a pre-stressed spring. This simple measure probably reduces the residual seismic risk more than a conventional reinforcement of the tank supports, since a much stronger GM than the design GM can *never* be excluded, whether the design was done for $\gamma_I = 1.4$ or for $\gamma_I = 2.2$!

In the aftermath of Fukushima, it is time for us, earthquake engineers, to get rid of our blinkers and to adopt a broader, risk based view of seismic safety instead of only blindly following traditional codes. It is time for a change!

8 Acknowledgements

The authors express their gratitude to Bernard Gay and Blaise Duvernay, both of the Federal Office for the Environment, not only for financing the underlying study, but also for their substantial contributions. The participation of the two CIFs to the pilot study is also greatly acknowledged. Furthermore, the aid of Cécile Dubien (Bureau Veritas Lyon) and Britta Holtschoppen (RWTH Aachen) in obtaining information from France and Germany, respectively, is acknowledged, too.

References

[1] FOEN (Federal Office for the Environment): Assessment Criteria for the 'Ordinance on Protection against Major Accidents, 1991', Berne, 1996

[2] Trbojevic, V.M.: Risk criteria in EU, Proceedings of the ESREL'05; 2005, Poland

[3] RWTH Aachen: Der Lastfall Erdbeben im Anlagenbau (Leitfaden), Verband der chemischen Industrie (VCI), Frankfurt, Oktober 2012

[4] Ministère de l'Ecologie, du Développement Durable, des Transports et du Logement: Arrêté du 24 janvier 2011 fixant les règles parasismiques appliquables à certaines installations classées; Journal Officiel de la République Française, Paris, 2011

[5] NAGRA: Probabilistic Seismic Hazard Analysis for Swiss Nuclear Power Plant Sites (PEGASOS Project); Final Report, Vol. 1, 2004, http://www. swissnuclear.ch/upload/cms/user/PEGASOSProjectReportVolume1.pdf

[6] Musson, R.M.W.: Against Fractiles; Earthquake Spectra; vol. 21, n° 3 (2005), Pages 887-891

[7] Abrahamson, N.A.: Seismic Hazard Assessment: Problems with Current Practice and Future Developments; in: Proceedings of the First European Conference on Earthquake Engineering and Seismology; 2006, keynote lecture

[8] Campbell, K.W.; Bozorgnia, Y.: Campbell-Bozorgnia NGA Gound Motion Relations for the Geometric Mean Horizontal Component of Peak and Spectral Ground Motion Parameters, PEER 2007/02, Pacific Earthquake Engineering Research Center, Berkeley, May 2007

[9] HSK (Hauptabteilung für die Sicherheit der Kernanlagen, today: ENSI): Neubestimmung der Erdbebengefährdung an den Kernkraftwerkstandorten in der Schweiz (Projekt PEGASOS), HSK-AN-6252, Villigen, Juni 2007

[10] Bommer, J.J.; Abrahamson, N.A.: Why Do Modern Probabilistic Seismic-Hazard Analyses Often Lead to Increased Hazard Estimates?; Bull. Seism. Soc.Am., vol. 96, n° 6 (2006), Pages 1967-1977

[11] Trbojevic, V.M.: Another Look at Risk and Structural Reliability Criteria; Sructural Safety, vol. 31, n° 3 (2009), Pages 245-250

[12] Spence, R.; So, E.; Scawthorn, C. (eds.): Human Casualties in Earthquakes, Progress in Modelling and Mitigation, Springer, Dordrecht Heidelberg London New York, 2011, ISBN 978-90-481-9454-4

[Web-1] i_m_-_risk_based_external_safety_regulations_in_the_netherlands-1, R Plarina, R.: Risk based external safety regulations in the Netherlands, Presentation at the Seveso Conference, Stockholm, 2011

Part III

International Building Codes and Guidelines

Overview of Seismic Regulations for French Industrial Facilities

Pecker Alain[1,2]

[1] Géodynamique et Structure
157 rue des Blains, 92220 Bagneux, France
alain.pecker@geodynamique.com

[2] French Association for Earthquake Engineering
15 rue de la Fontaine au Roi, 75011 Paris, France

ABSTRACT:

This paper presents the French regulations for seismic protection of critical industrial facilities. After an overview of the seismic regulations newly enforced to implement the Eurocodes in France, emphasis is put on the scope of the recently published bylaws governing the seismic protection of such installations: scope, definition of the seismic hazard, schedule of implementation. To conclude the guidelines under preparation, defining the technical rules for each type of equipment, are introduced.

Keywords: Regulations, Seismic hazard, technical guidelines,

1 Introduction

Seismic protection in France is enforced by law; therefore, the various documents related to its aspect must be endorsed by the Administration and published via decrees and bylaws. Anticipating the publication, started in 2005, of the various parts of Eurocode 8 dealing with seismic design of constructions, a new seismic zonation map of France had to be established based on the probabilistic framework retained by Eurocode 8 for seismic hazard. Based on this zonation map, several bylaws have been published for buildings, bridges, critical industrial facilities and several others are still under preparation.

The purpose of this paper is to provide an overview of the existing regulations applicable in France and to describe in more details those related to industrial facilities, which are at the heart of this conference. In addition, some information is provided on the various technical guidelines, under preparation, that will accompany the regulatory documents.

S. Klinkel et al. (Eds.), *Seismic Design of Industrial Facilities*,
DOI 10.1007/978-3-658-02810-7_5, © Springer Fachmedien Wiesbaden 2014

2 Background on Eurocode 8 and its implementation in France

For the purpose of EN 1998, national territories shall be subdivided by the National Authorities into seismic zones, depending on the local hazard. By definition, the hazard within each zone is assumed to be constant. For most of the applications of EN 1998, the hazard is described in terms of a single parameter, i.e. the value of the reference peak ground acceleration on rock (type A ground), a_{gR}.

The reference peak ground acceleration, chosen by the National Authorities for each seismic zone, corresponds to the reference return period T_{NCR} of the seismic action for the no-collapse requirement (or equivalently the reference probability of exceedance in 50 years, P_{NCR}) chosen by the National Authorities. An importance factor γ_I equal to 1.0 is assigned to this reference return period. For return periods other than the reference, the design ground acceleration on type A ground a_g is equal to a_{gR} times the importance factor γ_I ($a_g = \gamma_I . a_{gR}$). The reference peak ground acceleration on type A ground, a_{gR} is defined by the National Authorities with a recommended value of 475 years (10% probability of exceedance in 50 years) for T_{NCR}.

Given that until 2010 the zonation map in France was not based on a probabilistic approach, and according to the previous statements, a probabilistic seismic hazard analysis has been entrusted to GEOTER, a private consulting engineering company, under the control of the French Association for Earthquake Engineering (AFPS) and of the Institute for Radioprotection and Nuclear Safety (IRSN). The results have been eventually translated into regulatory documents and published in the form of two decrees in October 2010 (2010-1254 and 2010-1255). The national territory is divided into 5 seismic zones (zone 5 corresponds to the Caribbean islands); each town is allocated to one of these zones.

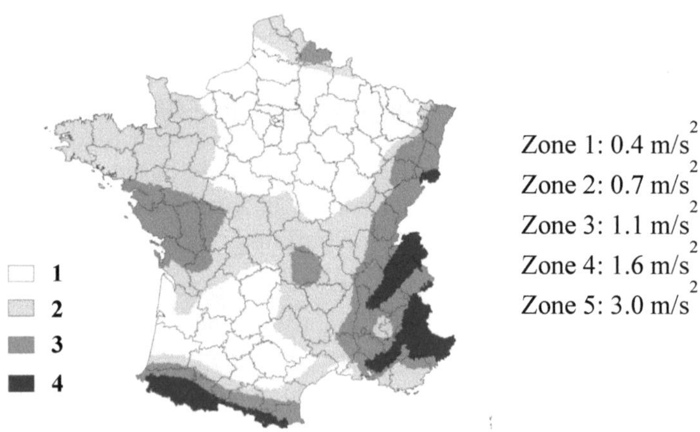

Figure 1 : Seismic zonation map of metropolitan France

3 Overview of existing regulations

Based on the seismic hazard map depicted in figure 1, several regulatory documents were published by the National Authorities in the form of bylaws during the period 2010-2011. These documents define the reference ground acceleration a_{gR} associated to each zone and the importance category of each type of construction. The peak ground accelerations listed in figure 1 are deemed to represent a seismic action with a probability of exceedance of 10% in 50 years (earthquake return period of 475 years). However, this value is not explicitly mentioned in the official documents because several discussions took place in the scientific community and some institutions, like AFPS, considered that the values are overconservative and represent a seismic hazard corresponding to longer return periods, as evidenced by discrepancies with neighbouring countries. Nevertheless, the discussion may seem irrelevant since the choice of the level of seismic protection belongs to the National Authorities, whatever the return period is. However, as it will be discussed later in the paper, this underlying assumption of 475 years has consequences on the hazard level for industrial facilities.

Today, three official documents (bylaws) have been published during the period 2010-2011:

- The first one (October 2010) concerns ordinary buildings for which the reference ground accelerations indicated in figure 1 have been retained. Two different spectral shapes, which depend on the soil classification, are assigned to seismic zones 1 to 4 and to seismic zone 5. For the latter the spectral shape recommended in EN 1998-1 for type I earthquake is chosen. For zones 1 to 4 a modified version of the recommended shape for type II earthquake is provided. Importance coefficients range from 0.8 to 1.4 depending on the importance category of the building (I to IV).
- The second one (October 2011) concerns bridges. The reference ground accelerations in each zone and the spectral shapes are identical to those of the ordinary buildings. Importance coefficients range from 1.0 to 1.4 depending on the importance category of the bridge (II to IV).
- The third one (January 2011) concerns critical industrial facilities. This document will be presented in more details in the following paragraph.

In addition to the previous documents two additional ones are under preparation. They are related to ordinary pipelines, silos, reservoirs and slender structures for the first one and to dams for the second one. It is worth noting that in all documents Eurocode 8 is referenced as the relevant technical document, even if it is supplemented by additional nationally established guidelines, like for industrial facilities or dams.

The spectral shapes defined in the bylaws are presented in figure 2 for ordinary buildings, bridges and ordinary pipelines, silos, reservoirs and slender structures.

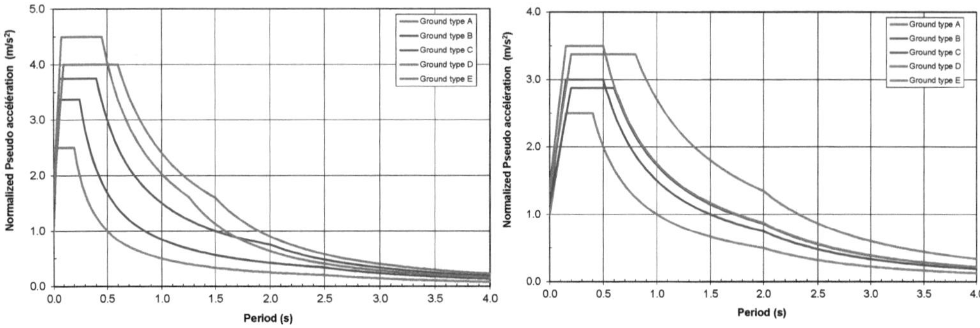

Figure 2 : Spectral shapes for ordinary buildings and bridges seismic zones 1 to 4 (*left*) – seismic zone 5 (*right*)

4 Regulations for Industrial facilities

Until January 2011, industrial facilities were covered by a bylaw dated May 10th 1993. This document has been removed and the new regulations are now provided in the January 24th, 2011 bylaw, which is in fact an amendment to the bylaw of October 4th 2010 that represents the regulations for critical facilities covering, until that date, all aspects but seismic design.

The new regulation defines the scope, the seismic action with reference to the seismic zonation map of figure 1, and the schedule for implementation. Unlike the other regulatory documents for ordinary buildings and bridges, the document also requires that existing structures be assessed.

4.1 Scope of the document

The equipment inside a given facility which are covered by the text are those for which "*seismic failure is susceptible to induce dangerous phenomena for human lives outside the perimeter of the facility, except if there is no human occupation*". In other words areas concerned by the previous sentence are the areas located outside the facility; if those areas are not populated in the sense defined below, the equipment under consideration does not have to comply with the document. Zones of non-permanent human occupation are defined as areas without any public building, inhabitants, permanent workshops, roads with a traffic flow not exceeding 5 000 vehicles per day, and in which new constructions are prohibited. The regulations are applicable both to new facilities and existing ones. New facilities are defined as those for which the administrative authorization has been granted after the 1st of January 2013; all others facilities are considered as existing ones.

4.2 Seismic action

The zonation map of figure 1 is applicable. However, given the consequences of failure of the concerned equipment, the earthquake return period for which the equipment is designed is increased in order to obtain a probability of exceedance of 1% in 50 years for new facilities. Although, the exact value of the target return period is not explicitly mentioned in the regulatory document, it is intended to be 5 000 years. To define the accelerations associated with each seismic zone of the map, it has been assumed that the reference accelerations for ordinary buildings and bridges are for a return period of 475 years. Then, according to Eurocode 8, the value of the importance factor γ_I multiplying the reference seismic action to achieve the same probability of exceedance in T_L years as in the T_{LR} years for which the reference seismic action is defined, may be computed as

$$\gamma_I \sim (T_{LR}/T_L)^{-1/k} \tag{1}$$

The exponent k depends on the seismicity of the area; Eurocode 8 recommends a value of 3. This value is confirmed as a representative value for the French metropolitan territory in the study by Marin et al [1]. Accordingly for a return period T_L = 5 000 years, the importance factor should be equal to 2.2. This is the value that has been chosen to define the reference acceleration for new facilities. For existing facilities an importance factor of 1.85 has been retained corresponding to a return period of approximately 3 000 years. The applicable values for each seismic zone and type of facility are provided in table 1.

Table 1: Reference horizontal acceleration (m/s^2) for critical facilities

Seismic zone	New facility	Existing facility
1	0.88	0.74
2	1.54	1.30
3	2.42	2.04
4	3.52	2.96
5	6.60	5.55

With regards to the spectral shapes associated with each seismic zone, in view of the acceleration levels specified in table 1, which should be linked to higher magnitudes than those linked to the reference seismic action with the return period of 475 years, the so-called type II spectrum, adapted for France, has been retained for zones 1 to 3 (left diagram in figure 2) and the so-called type I spectrum for zones 4 and 5 (right diagram in figure 2).

4.3 Schedule for implementation

The new facilities shall comply with the requirements of the regulatory document at the time of application for the authorization to operate and the required seismic protective measures shall be implemented before starting operating the facility.

For existing facilities, the owner shall produce no later than December 31th, 2015 studies assessing the seismic reliability of the facility and defining the necessary retrofitting to comply with the regulatory document. The schedule for implementation of the needed retrofits will be defined by the Administration before July 31th, 2016 and will not extend beyond January 1st, 2021.

4.4 Future evolutions

It is indicated in the regulatory document that if the seismic zonation happens to be modified, increasing the seismic levels, the owner of the facility shall undertake a new study within 5 years following the modification.

Furthermore, the regulations will be revisited after comments emanating from a relevant committee (CSPRT: Conseil supérieur de la prévention des risques technologiques) upon presentation before July 1st, 2016 of a report, presenting the conclusions of the seismic studies, by the Minister in charge of the facilities.

5 Technical guidelines for seismic design, assessment and retrofit of critical facilities

As evidenced by the presentation made in paragraph 4, the regulatory document for critical facilities only covers, from a technical point of view, the definition of the seismic hazard which the facility must be designed for. Reference to Eurocode is not made explicitly in the document except again for the definition of the spectral shapes. Therefore, in order to help the owners of facilities, who are not necessarily seismic experts, and to provide the owners and the Administration with common reference technical documents for design of new facilities and assessment and retrofit of existing ones, task groups have been set up to write guidelines. When the task is completed, the guidelines will have the status of jointly agreed standards. These tasks groups are composed of experts from AFPS and representative of the concerned Industries. The program is jointly sponsored by the Ministry of Ecology-Sustainable Development and Energy (MEDDE) and the Industries and is placed under the responsibility of GICPER, a professional organization gathering the Industries. The technical aspects of the program are entrusted to AFPS and SNCT (trade union for boilers and industrial piping) and have to be endorsed by these organizations. A representative of the Ministry participates in each of the task group to ensure that the guidelines will be acceptable to the Ministry. A general flowchart of the operational organization is presented in figure 3.

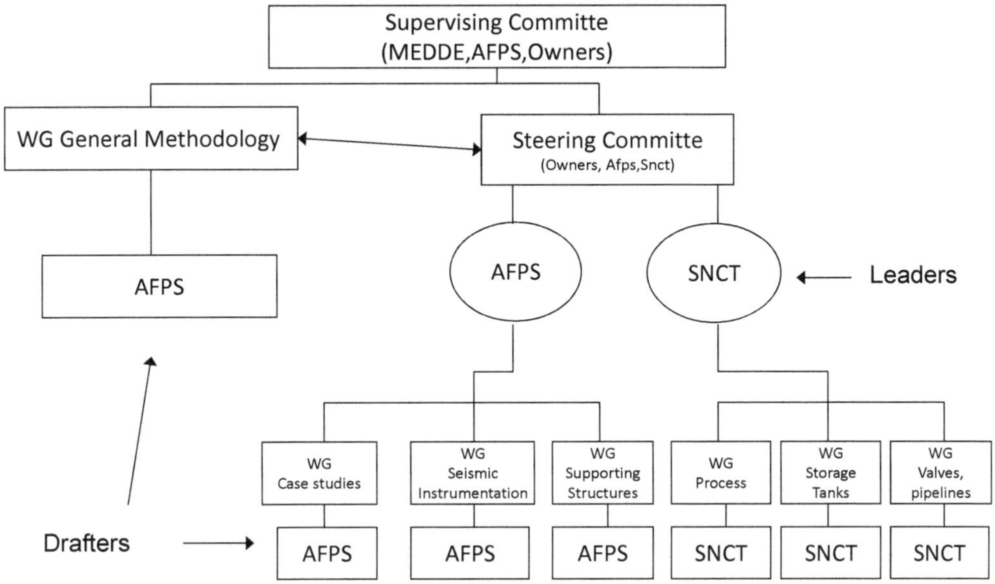

Figure 3: Organization of the task group for guidelines

The guidelines will be composed of several documents:

- A document on the general methodology describing the procedure to be followed and covering transverse topics to all other guides, like load combinations, analyses methods, criteria for verification, etc…
- Several specialized technical guides dealing with specific equipment and allowing the owner to undertake a seismic study aiming at conferring his facility an acceptable seismic behaviour and demonstrating compliance with the regulatory document.

The general document describes the methodology to be followed by the owner, notably for the definition of the regulatory framework, identification of the concerned equipment, the methodology to classify the equipment, the requirements and the justification tools for new and existing facilities. The tools can be based on post earthquake observations, calculations, and/or experiments.

Each of the following topic will be covered by a specific guide:

- 1. Atmospheric storage tanks (july 2013)
- 2. Safe shutdown of an installation based on seismic instrumentation (may 2013)
- 3. Supporting structures (December 2013)
- 4. Pipelines and valves (December 2013)

- 5. Process
- 6. Case studies

The last guide listed above is intended to provide concrete examples of application of the guidelines on a test facility. Another paper in this conference is presenting the content of guideline number 2.

The schedule of publication of the various guidelines is indicated in parenthesis; the general methodology will be ready by the end of June and published next fall. The last two guides will start during summer 2013.

6 Conclusion

This paper has presented a general overview of the implementation of the new seismic regulations in France to accompany the publication of the seismic Eurocode EN 1998. Emphasis has been put on critical facilities describing in details the content of the regulatory (bylaw) document and outlining the additional work under progress for the development of technical guidelines which will have the status of jointly agreed standards between the Administration and the owners and are deemed to comply with the regulatory documents.

References

[1] Marin S., Avouac J.P., Nicolas M., Schlupp A.: A Probabilistic Approach to Seismic Hazard in Metropolitan France; Bulletin of Seismological Society of America; 94 (6), (2004), Pages 2137-2163.

Seismic Design of Industrial Facilities in Germany

Christoph Butenweg[1], **Britta Holtschoppen**[1]

[1] Chair of Structural Statics and Dynamics, RWTH Aachen University
Mies-van-der-Rohe-Str. 1, 52074 Aachen, Germany
Butenweg@LBB.RWTH-Aachen.de, Holtschoppen@LBB.RWTH-Aachen.de

ABSTRACT:

Industrial facilities are typically complex systems consisting of a primary load-carrying structure with multiple technical installations like tanks, vessels and pipes, which are generally designated as secondary structures. Due to high cost of the process engineering components and due to the risk of business interruption and the release of harmful substances into air, water and ground if damages occur, industrial facilities must be designed to safely withstand seismic loading. The design must consider both the primary structure and the secondary structures as well as the dynamic interaction effects between structural and non-structural components. However, in Germany a basis for the seismic design of such facilities is still missing, since the current earthquake code DIN 4149 and the forthcoming code DIN EN 1998-1 are limited to conventional buildings. For this reason a technical guideline for the seismic design of industrial facilities was developed in collaboration with the German Chemical Industry Association (VCI) to close the gap of the design standards. The present paper introduces the guideline with special emphasis on plant specific aspects.

Keywords: Industrial Facilities, Seismic Safety, Eurocode 8, DIN 4149, VCI-Guideline

1 Introduction

Industrial facilities depict a complex composition of diverse components and structures which are linked on a structural or an operational scale (Figure 1). Depending on the type of industry such facilities consist of load-bearing frames and supporting structures for process relevant secondary structures like vessels, pumps and equipment as well as infrastructural components like piping systems, and / or self-contained components like silos for bulk solids, liquid-filled tanks, distillation columns or chimneys.

S. Klinkel et al. (Eds.), *Seismic Design of Industrial Facilities*,
DOI 10.1007/978-3-658-02810-7_6, © Springer Fachmedien Wiesbaden 2014

Figure 1: Typical Industrial Facility

Devastating earthquakes of recent years have sensitized the population and politicians worldwide for the possible hazard emanating from industrial facilities. This has lead to distinct activities regarding the assessment and improvement of seismic safety of systemic structures and infrastructures, also in countries of low seismicity like Germany. Here, the VCI has initiated in 2009 the development of a guideline on the seismic design of industrial facilities, since the legal regulations regarding the seismic design of buildings (DIN 4149:2005) explicitly excluded facilities with particular hazard potential from the scope of application.

As part of the harmonisation of technical regulations in Europe, DIN 4149:2005 [1] will be replaced by DIN EN 1998-1 [2] and will be complemented by further parts of DIN EN 1998. However, industrial facilities with high hazard potential do still not fall into the scope of application, and so, the VCI-Guideline was adapted to DIN EN 1998 in 2012 and updated to the current state of the art [4].

Both editions of the VCI-Guideline have been substantially developed under the leadership of the Chair for Structural Statics and Dynamics (LBB) of RWTH Aachen University and will be presented briefly in the following sections.

2 Structure of the VCI-Guideline

The VCI-Guideline offers design rules and recommendations for the seismic design of new industrial facilities. Beyond that, it covers the evaluation and the possible retrofitting of existing facilities and the use of seismic protection systems. The VCI-Guideline does, at the current date, not represent an official state-wide legal standard. Yet, it is accepted by several state authorities who are responsible for the approval of building measures in Germany and it is widely employed in engineering practice.

For the purpose of good clarity and applicability the VCI-Guideline is split into two documents: The actual guideline [4] comprises all relevant regulations for the seismic design of industrial facilities. As it constantly refers to the corresponding parts of DIN-EN 1998 it is to be used in connection with this legal standard and only names the relevant changes and extensions to account for the special situation of industrial facilities. The second, considerably larger commentary document [5] offers comprehensive information on the scientific background of the regulations and gives numerous recommendations regarding the practical realization.

The structure of both the VCI-Guideline and the commentary document mainly follows the established sections of DIN 4149:2005 [1], but additionally it considers novel sections of DIN EN 1998 and completely new sub-topics: After stating fundamental rules for the constructive design of primary structures, self-contained components and non-structural components both documents cover the determination of the site-specific seismic loading, the actual design regulations for typical types of components of the facility and specific rules for certain construction materials and types. Concluding sections on seismic protection systems and on the evaluation of existing structures complement the VCI-Guideline and its commentary document.

3 Basic principles of conceptual design

Seismically induced damages of structural and non-structural components can be reduced if certain basic principles of conceptual design are considered. Additional to the basic principles that are valid for the design of buildings which are given in all international standard provisions for seismic design (e.g. structural simplicity, regularity in plan and elevation, redundancy, adequate foundation, et al.) certain design principles should be considered when planning and installing an industrial facility:

Due to requirements of process technology a regular distribution of masses in plan and elevation demanded by standard provisions is oftentimes not feasible. Instead, the structural members of the primary structure must account for irregular vertical and horizontal loads in case of an earthquake. As a consequence torsional effects may play a major role in the design of the primary structure. Special attention also needs to be paid to expansion joints especially when (historic) facilities are expanded by additional building parts.

The typical steel frame structures are beneficial in many respects: they are installed quickly, are very versatile and allow for easy modifications of the location of equipment. One has to bear in mind, however, that such constructions in connection with high masses of vessels, agitators or other equipment show rather low eigenfrequencies. This again leads to an increased influence of higher modes of vibration including torsional vibration modes which has to be considered in the design of both primary and secondary structures.

4 Ground conditions and seismic action

4.1 Seismic hazard maps

In seismic design the relevant seismic loading generally depends on the building's importance for the civil infrastructure and population or the potential hazard emanating from it to its surroundings. By nature it must be determined on a statistical basis. The required safety level is typically reflected by the statistical return period of the seismic loading that is taken as basis for the design of the building. Since, due to geological coherencies, modified statistical return periods of seismic events influence the geographic range of impact about the epicentre the VCI-Guideline recommends the use of probabilistic seismic hazard maps considering the adequate return period for the design of an industrial facility (Figure 2).

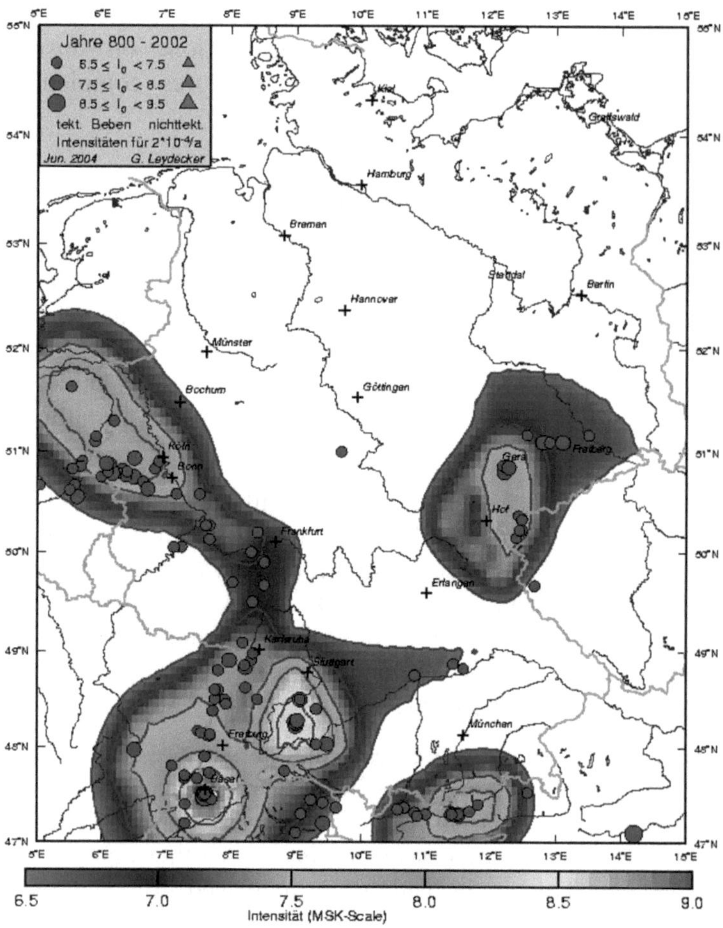

Figure 2: Seismic hazard map for Germany, return period of 2000 years [16]

4.2 Importance factor

As long, however, as such maps do not exist for a sufficiently large number of possible relevant return periods the VCI-Guideline tolerates the use of the established importance factor γ_I which is multiplied to the seismic reference load with a statistical return period of 475 years which is defined in the National Annex DIN EN 1998-1/NA [3]. In order to account for the individual hazard situation of each industrial facility the importance factor γ_I according to the VCI-Guideline is determined in dependence on the handled and processed goods, its damage potential, the possible range of impact and its possible effect on people and the environment. All these factors are weaved into three tables (Table 1 to Table 3). In design of a certain industrial facility or a component the highest of these three importance factors adequate to the considered facility is decisive. It can range from 1.0 to 1.6. A factor of 1.6 would approximately represent a seismic event of a return period of 1950 years or a probability of exceedance in 50 years of 2.5% respectively. It should be noted that the importance factor with respect to protection of human lives (Table 1) is allocated according to hazard categories of substances ("H-Sätze") given in the regulation no. 1272/2008 of the European Parliament and the European Council [6]. The respective hazard categories corresponding to the damage potential of (Table 1) are listed in the VCI-guideline.

Table 1: Importance factor γ_I with respect to protection of human lives

		Consequences				
		Inside plants	Sourrounding area of the plant (block inside of the industrial area)**	Inside the plant / industrial area (with fence)	Outside a plant / industrial area	Large-scale consequences outside a plant / industrial area
Damage Potential*	Non-volatile toxic substances Flammable and oxidizing substances	1.0	1.0	1.0	1.0	1.1
	Non-volatile highly toxic substances Easily and highly flammable substances Oxidizing gas	1.0	1.1	1.2	1.2	1.2
	Volatile toxic substances Volatile highly toxic substances Explosive substances Highly flammable liquefied gas	1.1	1.2	1.3	1.4	1.4
	Medium volatile and highly toxic substances	1.2	1.3	1.4	1.5	1.6

* Flammable, easily flammable and highly flammable and oxidizing substances include only gases and liquids.
** A block inside a plant corresponds to an operational area according to the "Hazardous Incident Ordinance".

Table 2: Importance factor γ_I with respect to protection of the environment

	Consequences		
	No consequences for the environment outside the plant	Minor consequences for the environment outside the plant	Large-scale consequences for the environ. outside the plant
Influence on the environment	1.0	1.2	1.4

Table 3: Importance factor γ_I with respect to protection of lifeline installations

	Requirements		
	Standard requirements regarding the availability	High requirements regarding the availability	Very high requirements regarding the availability
Restraint systems, traffic infrastructure, emergency routes	**1.2**	**1.2**	**1.2**
Lifeline buildings (fire stations, fire-extinguishing systems, rescue-service stations, energy supply, pipe bridges)	**1.3**	**1.4**	**1.4**
Emergency power supply*, safety systems*	**1.4**	**1.5**	**1.6**
*Special systems necessary for shutdown of processes into safe condition			

4.3 Seismic load combinations

When combining the seismic load with other loads like dead loads, variable loads, wind, snow and loads due to temperature differences or ground settlements, again, the special situation of industrial facilities is considered in design. The VCI-Guideline introduces additional load categories and values of ψ_2-factors for the combination of variable actions (Table 4) in accordance with the combination rule of DIN EN 1990. The reduction factor φ of DIN-EN 1998-1 [2] ($\psi_{Ei} = \varphi \cdot \psi_{2i}$) is replaced by the requirement to consider in the design all unfavourable load constellations possible during the production process.

Table 4: Recommended values of ψ_2-factors for industrial facilities (Factors for quasi-permanent values of variable actions)

Action	Combination coefficient ψ_2
Live loads	
Storage areas	0.8
Operations areas	0.15
Office areas	0.3
Vertical crane and trailing loads	0.8
Variable machine loads, vehicle loads	0.5
Brake loads, starting loads (caused by vehicles or cranes etc.)	0
Loads due to assemblage or other short time loads	0
Operational loads	
Variable operational loads	0.6*
Operating pressure	1.0
Operating temperature	1.0
Wind loads	0
External temperature impact (temporary)	0
Snow loads	0.5
Likely differential settlement of the foundation soil	1.0
* Constant operational loads are to be considered as constant load G_k.	

5 Primary Structures

5.1 Modelling

All relevant characteristic features of the dynamic behaviour of industrial facilities (section 3) have to be considered in the computational model. This implies that all possible unfavourable mass constellations resulting from the production process must be represented in adequate design models.

In most cases it will be sufficient to consider secondary structures as point masses in the model of the primary structure. In cases when the vibration behaviour of the secondary structure strongly influences the dynamic behaviour of the primary structure, however, such secondary structure must be modelled in detail (e.g. large masses on soft supports, stiff multi-level components which are horizontally constrained on several floors or strong interaction potential due to other conditions). In analogy to the regulations of DIN EN 1998-1/NA [3] vertical seismic action only needs to be considered in the design of load bearing components that carry high masses, of long horizontal load bearing components (beams), and of pre-stressed components.

5.2 Methods of analysis

The typical seismic design of buildings is based on the response spectrum analysis. This is also the standard method of analysis in the VCI-Guideline. In certain cases, however, nonlinear time history calculations are permitted – the relevant requirements and regulations are stated in the VCI-Guideline and in the commentary document.

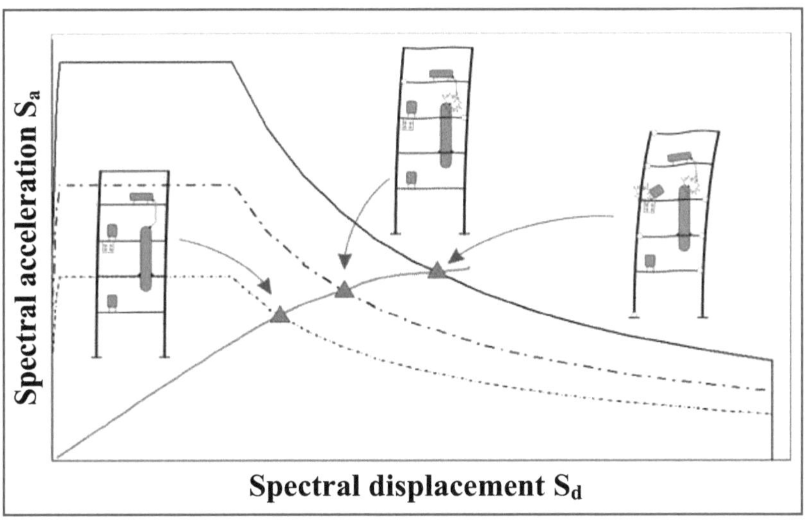

Figure 3: Performance-based design of industrial facilities

The application of nonlinear static analyses provides the opportunity for a global performance based design. This way several different damage states – which may include economical limit states regarding the operational reliability – can be investigated simultaneously (Figure 3). These procedures are assumed to gain influence in the future especially in the investigation of seismic safety of process chains and in the proof of highly loaded primary structures. They are provided in the VCI-Guideline as alternative to the modal response spectrum analysis especially if global reserves are to be bailed.

6 Secondary Structures

Recent earthquakes in highly industrialised countries have shown that the damage to secondary structures and the resulting losses due to operational failures financially exceed the primary damages many times over. Therefore, the proper conceptual design and proof of secondary structures is of high importance.

As already stated in section 3 steel frame structures of facilities of the chemical and other process industry typically show rather low eigenfrequencies. Furthermore, their second and even higher natural modes of vibration often have a notable influence on the overall vibration behaviour of the primary structure (e.g. [9], [7]). This implies that "linear" design rules for secondary structures in buildings (which approximate the first eigenmode of the primary structure) may considerably underestimate the seismically induced force of inertia on secondary structures located in the lower third of the primary structure [7]. Thus, the VCI-Guideline recommends considering the actual vibration behaviour of the primary structure when determining the seismic load on secondary structures. A corresponding design formula, which was developed at the LBB in reference to the North American guideline FEMA 450 [10] is suggested and explained in the commentary document.

For estimate calculations an upper limit value of design force F_a is stated in the VCI-Guideline (eq. 2) considering the plateau value of the elastic acceleration response spectrum $S_{e,max}$, the mass of the equipment m_a, the importance of the equipment γ_a and a dynamic factor of 1.6:

$$F_a = 1{,}6 \cdot S_{e,max} \cdot \gamma_a \cdot m_a \;\; [kN] \qquad (1)$$

This upper limit value is widely used in pre-design when the final location and configuration of the equipment is not yet clear. It is comparable to limit values of international standard provisions like IBC 2006 [11] and DIN EN 1998-1/NA [3].

7 Silos, Tanks and Pipelines

Large liquid filled tanks play an important role in the infrastructure of many industrial facilities assuring the supply with raw material needed for the production process or serving as storage for intermediate products. Due to their oftentimes large dimensions in diameter and height the stored fluid develops high seismic loads to the tank shell induced by the vibration of the liquid (sloshing), the movement of the tank structure (impulsive rigid load component) and the interactive vibration of shell and liquid (impulsive flexible load component). Figure 4 shows the different pressure components of liquid filled tanks subjected to horizontal seismic loading.

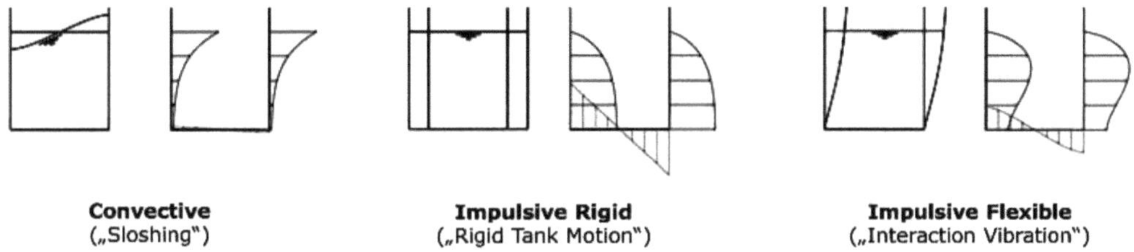

Figure 4: Modes of vibration of liquid filled tanks induced by horizontal seismic excitation

CFD models (computational fluid dynamics) can analyse the tank's response to seismic loading by modelling the shell and the fluid and reproduce all (interaction) effects simultaneously. As such computational analyses, however, are extremely time expensive and require highly sophisticated software tools they are hardly employed in everyday engineering practice. The well established estimate calculation methods according to Housner [12] on the other hand neglects the impulsive flexible load component (interaction of fluid and shell) and, thus, may lead to highly underestimated seismic loads for thin and slender tanks.

Instead, the VCI-Guideline recommends determining the seismic loads to the tank shell using a calculation method based on the velocity potential of the fluid in conjunction with the added mass concept. This method was investigated intensively by Fischer, Rammerstorfer, Scharf, Seeber, Habenberger and others (e.g. [13], [14]) and has been introduced to the informative Annex D of DIN-EN 1998-4. It is based on individual formulae to determine and to superpose the single load components. These formulae yield the seismically induced load on the tank shell in dependence of the cylindrical coordinate (ξ, ζ, θ), but they require the determination of modified Bessel-functions of first order and their derivation and coshyp-terms respectively. In the case of the impulsive flexible load component, an iterative procedure is necessary to calculate the load $p(\xi, \zeta, \theta)$.

In order to simplify the application of the above mentioned method the prefixed coefficients of coshyp- or Bessel-terms can be tabulated in dependence of the geometric parameters of the tank and the loading of the tank shell can be determined easily. Comprehensive details and theoretical background information on this method as well as the mentioned tables are published by Meskouris et al. [8].

Having determined the seismically induced load on the tank shell and its foundation special attention must be paid to the design and deformation compatibility of pipeline connections and fairleads. DIN-EN 1998-4 [4] provides principles and application rules for the seismic design of the structural aspects of above-ground pipeline systems and buried pipeline systems. The VCI-Guideline complements the pipeline design with some simple rules for frequently used pipeline diameters.

8 Seismic Protection Devices

In industrial facilities the process relevant equipment and secondary structures depict the actual value of the facility. Therefore, it might be sensible to reduce the seismic load by individual seismic protection systems. Since the design and installation of such protection systems are extremely individual the VCI-guideline only states general recommendations and refers to respective legal standard provisions. The commentary document explains the basic principles of typical seismic protection systems and shows exemplary constructive details.

9 Seismic safety of existing facilities

Industrial facilities which are subjected to German immission laws and which are "part of an operational area" according to the German law for the protection from immissions (BImSchG [15]) need to be checked on a regular basis regarding the structural safety considering all possible hazards including seismic hazards.

In order to assess the seismic safety of an existing industrial facility the VCI-Guideline recommends a three-stage procedure: In the framework of an intensive inspection possible weak points can be detected by following an exemplary checklist provided in the commentary document. Critical details are noted down and graded according to a given evaluation scheme with respect to its structural deficiency and its hazard potential in case of actual damage. This visual inspection serves as an acquisition of the current status of the facility. On the basis of its result further measures are initiated: For critical details computational analyses or simulations verify or refute the seismic safety mathematically. In these computational analyses and in the determination of the design seismic load the estimated remaining runtime of the facility can be allowed for in coordination with the responsible authorities. Finally constructive improvements or structural retrofitting measures are realized corresponding to the computational verifications.

10 Conclusion

The presented VCI-Guideline provides rules and recommendations for the seismic design of planned industrial facilities, the seismic hazard assessment of existing facilities as well as the application of strengthening measures and seismic protection systems. The importance factors tailored to the specifics of industrial facilities and the recommendations concerning the seismic design of primary structural systems, process relevant secondary structures and infrastructural components lead to a sufficient level of safety of industrial facilities. The alternative use of nonlinear static analyses enables the application of performance based design using serviceability limit states defined in collaboration with the plant operator. The VCI-Guideline is used continuously since the introduction in 2009 by the member companies of the VCI.

REFERENCES

[1] DIN 4149:2005 "Bauten in deutschen Erdbebengebieten, Teil 1: Grundlagen, Erdbebeneinwirkungen und Regeln für Hochbauten", Normenausschuss Bauwesen im DIN e.V., April 2005

[2] DIN EN 1998-1 „Eurocode 8: Auslegung von Bauwerken gegen Erdbeben, Teil1: Grundlagen, Erdbebeneinwirkungen und Regeln für Hochbauten", Normenausschuss Bauwesen (NABau) im DIN, Dezember 2010

[3] DIN EN 1998-1/NA „Nationaler Anhang zu Eurocode 8: Auslegung von Bauwerken gegen Erdbeben, Teil 1: Grundlagen, Erdbebeneinwirkungen und Regeln für Hochbauten", Normenausschuss Bauwesen (NABau) im DIN, 2010

[4] Butenweg, C., Dargel, H.-J., Höchst, T., Holtschppen, B., Schwarz, R., Sippel, M.: "VCI-Leitfaden 'Der Lastfall Erdbeben im Anlagenbau'", 2. Auflage, Frankfurt: Verband der Chemischen Industrie, 20 Seiten, 2012

[5] Butenweg, C., Dargel, H.-J., Höchst, T., Holtschppen, B., Schwarz, R., Sippel, M.: "Erläuterungen zum VCI-Leitfaden 'Der Lastfall Erdbeben im Anlagenbau'", 2. Auflage, Frankfurt: Verband der Chemischen Industrie, 71 Seiten, 2012

[6] Regulation (EC) No. 1272/2008 of the European Parliament and of the Council on classification, labelling and packaging of substances and mixtures, Dec. 2008

[7] Holtschoppen, B: "Beitrag zur Auslegung von Industrianlagen auf seismische Belastungen", Dissertation, RWTH Aachen University, 2009

[8] K. Meskouris, K.-G Hinzen,. C. Butenweg, M. Mistler: „Bauwerke und Erdbeben – Grundlagen – Anwendung – Beispiele", 3. Auflage, Vieweg+Teubner Verlag, 2011

[9] Singh, M.P., Moreschi, L. M., (1998), "Simplified methods for calculating seismic forces for non-structural components", Proceedings of Seminar on Seismic Design, Retrofit, and Performance of Non-structural Components (ATC-29-1), Applied Technology Council

[10] Building Seismic Safety Council, (2003), "NEHRP Recommended Provisions for Seismic Regulations for New Buildings and Other Structures (FEMA 450)", Part 1: Provisions

[11] International Code Council (ICC), (2006), "International Building Code", ISBN: 1-58001-251-5, Whittier, CA.

[12] Housner, G.W., „The dynamic behaviour of water tanks", Bulletin of the Seismological Society of America, Vol. 53, p. 381-387, 1963

[13] Fischer, F.D., Rammerstorfer, F.G., Scharf, K., "Earthquake Resistant Design of Anchored and Unanchored Liquid Storage Tanks under Three-Dimensional Earthquake Excitation", in G.I. Schuëller (Hrsg.) Structural Dynamics, Springer Verlag, S. 317-371, 1991

[14] Habenberger, J., „Beitrag zur Berechnung von nachgiebig gelagerten Behältertragwerken unter seismischen Einwirkungen", Dissertation, Weimar, 2001

[15] Gesetz zum Schutz vor schädlichen Umwelteinwirkungen durch Luftverunreinigungen, Geräusche, Erschütterungen und ähnliche Vorgänge (Bundes-Immissionsschutzgesetz – BImSchG) in der Fassung der Bekanntmachung vom 26. September 2002 (BGBI. I S. 3830), zuletzt geändert durch Artikel 8 des Gesetzes vom 08. November 2011 (BGBL. I S. 2178)

[16] BGR 2005 BGR: Seismische Gefährdungskarten die Bundesrepublik Deutschland - Entwurf (unveröffentlicht). Leydecker, G., Bundesanstalt für Geowissenschaften und Rohstoffe (BGR), Hannover, 2005. – Interne Mitteilung

Precast Industrial Buildings in Italy Current Building Code and New Provisions Since the 2012 Earthquake

Marco Mezzi[1], **Fabrizio Comodini**[2], **Leonardo Rossi**[3]

[1] Dept. of Civil and Environmental Engineering, University of Perugia
via Duranti 93, 06125 Perugia, Italy
marco.mezzi@unipg.it

[2] University eCampus
via Isimbardi 10, 22060 Novedrate (CO), Italy
fabrizio.comodini@uniecampus.it

[3] PhD Student, University of Perugia
via Duranti 93, 06125 Perugia, Italy
leonardo.rossi1@studenti.unipg.it

ABSTRACT

First of all the paper describes the Italian regulatory framework for precast buildings. Then the work focuses on the structural weaknesses most frequently found in existing buildings. It also discusses the changes made to building standards and to the technical specifications following the earthquake that struck the regions Emilia-Romagna, Veneto and Lombardy in May 2012. Finally, it presents the guidelines developed by the Working Group on the Seismic Conformity of Industrial Buildings for the rapid restoration of accessibility and seismic improvement of existing precast buildings.

Keywords: precast building, Italian building code, Emilia-Romagna Earthquake

1 The May 2012 Earthquakes

In May 2012, a large area of north-central Italy, including the regions Emilia-Romagna, Veneto and Lombardy, was struck by a series of earthquakes of medium to high intensity, culminating in the seismic shocks of 20 and 29 May (with respective Richter Magnitudes (M_L) of 5.9 and 5.8). The series of earthquakes, which is commonly referred to as the Emilia-Romagna Earthquake, mainly affected the provinces of Bologna, Modena, Ferrara, Mantua, Reggio-Emilia and Rovigo. As a result, 27 people lost their lives and much damage was done to historical and artistic heritage, buildings in general and to manufacturing activities. Figure 1 shows the INGV ShakeMaps [Web-1] for the earthquakes mentioned.

S. Klinkel et al. (Eds.), *Seismic Design of Industrial Facilities*,
DOI 10.1007/978-3-658-02810-7_7, © Springer Fachmedien Wiesbaden 2014

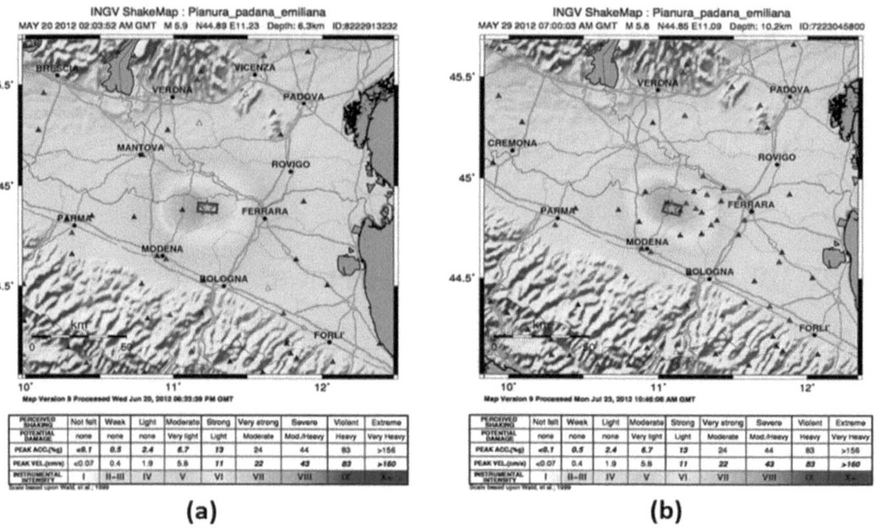

Figure 1: INGV ShakeMaps for of 20 and 29 May 2012 earthquakes

The Emilia-Romagna Earthquake highlighted the high seismic risk associated with precast structures, particularly if built with no reference to seismic design criteria or using outdated construction models.

The paradox that emerged from the events of May 2012 is that technologically advanced productive activities, such as those in the bio-medical sector, were housed in buildings that were structurally very simple, basically designed only for vertical loads. In particular, there were frequent cases of single-storey frames composed of precast elements, with slender isostatic pillars and simply supported beams. Structures of this type are often used for the storage of finished and semi-finished products or include the permanent presence of staff and equipment. The critical aspects noted were the same as those that had emerged after other earthquakes. In 1978, in an article entitled "Considerations on the design of earthquake-resistant precast buildings" [1], Prof. Parducci A. emphasised the "bad design habit" of creating simply supported beam-to-pillar and roof-to-beam connections. The document states that friction grip connections were regularly used in Italy, even on very slender pillars with high lateral deformability. The considerations contained in the article were developed after the Friuli Earthquake in 1976, which caused the collapse of numerous precast industrial buildings.

2 Italian Regulations

To understand the reasons for the numerous structural and non-structural collapses that occurred due to the earthquake in May 2012, it may be helpful to present the basic steps in the development of Italian guidelines in recent decades.

Technical standards and specifications for repairs, reconstruction and new buildings in seismic areas have been in existence since the first decade of the twentieth century. However, the industrial buildings in use today come under the regulations of the following documents, drawn up since the 1970s:

Legge 5 novembre 1971, n.1086 [2]. This document formed the basis for all subsequent technical standards for buildings, including those currently in force in 2013.

Legge 2 febbraio 1974, n°64 [3]. The document specifically refers to horizontal seismic forces, which can be represented as two perpendicular force systems not acting at the same time.

Decreto Ministeriale 3 marzo 1975 [4]. This contains an explicit reference to the evaluation of displacements caused by earthquakes, emphasising that the retention of connections should not be compromised and that hammering should not occur between adjacent independent structures.

Decreto Ministeriale del 24 gennaio 1986 [5]. This document permits the use of beam-to-pillar and beam-to-roof friction grip connections in precast buildings, provided that specific checks are made, "*to be studied on a case-by-case basis in order to ensure that possible sliding does not produce harmful effects*".

Decreto Ministeriale 3 Dicembre 1987 [6]. This fundamental decree provides criteria and calculation methods for safety checks. It provides information about purely technical matters, specifying, for example, that: "*the minimum depth of total support for beams must not be less than 8 cm + L/300, with L being the clear span of the beam*". According to this formula, a support 13 cm in length would be sufficient for a beam with a 15 m span. Finally, it states that: "*the use of supports in which the transmission of horizontal forces depends on friction alone is not permitted in seismic zones. Supports of this type are permitted where the capacity of transmitting horizontal actions is not a relevant factor; the support must allow displacements in accordance with the requirements of seismic regulations*". Connections between elements are also required to have "*sufficiently ductile behaviour*".

Decreto Ministeriale 16 gennaio 1996 [7]. This document specifies that calculated displacements and rotations must not compromise the integrity of hinges and sliding bearings. With sliding bearings, special devices are required to be used to contain the extent of displacement in the event of an earthquake.

Decreto del Presidente della Repubblica del 6 giugno 2001, n. 380 [8]. This contains the fundamental principles, general guidelines and regulations for construction works.

Ordinanza del Presidente del Consiglio dei Ministri del 20 marzo 2003 (OPCM 3274) [9], and subsequent updates. These documents represent a fundamental step forward in terms of the updating of the criteria and methods for the design, evaluation and adaptation of buildings in seismic zones.

Decreto Ministeriale del 14 settembre 2005 [10]. This presents a complete reorganisation of building legislation, although the text was not widely applied, due to extensions of previous regulatory documents.

Decreto Ministeriale del 14 gennaio 2008 [11]. This contains a large section on precast structures (see paragraph 3).

The series of documents mentioned above have produced a regulatory framework that is constantly evolving and improving. Seismic action on structures has been introduced gradually and defined with increasing detail, in response to the natural catastrophic events that have occurred over the decades. There has been a gradual increase in the specifications provided to designers regarding construction details; the concept of structural ductility has also been introduced, in line with the design approach that has become established at international level over the same period. It is important to point out that alongside the changes in the regulatory guidelines, there has been constant modification to the seismic hazard map of Italy. For example, the regions affected by the series of earthquakes in May 2012 were still classified as areas "not prone to seismic hazard" in the 1984 hazard map. Figure 2 shows the 1984 seismic zoning map [Web-2] and the 2004 seismic hazard map [Web-3]. The maximum seismic hazards are represented by "category 1" and "zone 1" respectively. As shown in Figure 2b, major updates to the seismic hazard map of Italy were made in 2004. It can be seen that:

- The Emilia-Romagna region changed from being almost entirely unclassified to become a "zone 3" area (low seismicity),
- The provinces of Mantua and Rovigo, located in the regions of Lombardy and Veneto respectively, changed from being unclassified to become "zone 4" areas (fairly limited seismicity). In "zone 4", the individual regions became responsible for introducing seismic zone design requirements.

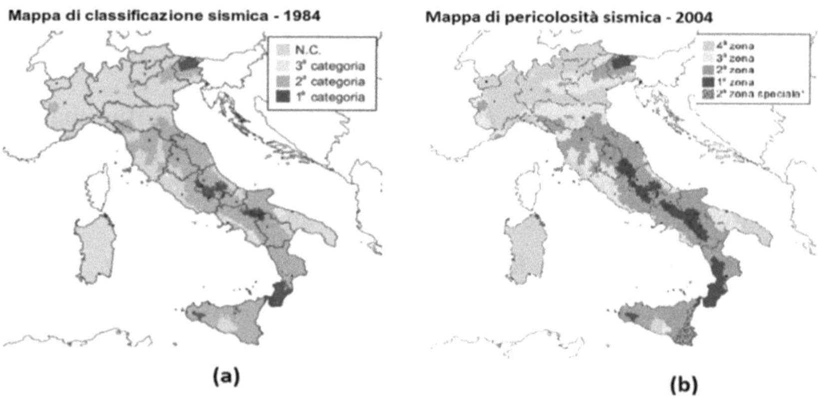

Figure 2: Italian seismic zoning/hazard maps - (a) 1984, (b) 2004

The seismic hazard maps were further updated in 2006. The current status is briefly described in section 3 of this article.

From what has been seen, it may be immediately concluded that the range of precast structures currently in use in Italy features a considerable variety in terms of performance levels for projected seismic action. This is due to the accumulation of various technical provisions for zones considered to be of seismic risk and to the updating of seismic hazard assessments for the various areas.

3 Technical standards for buildings in force in Italy in May 2012

In May 2012 the New Technical Standards for Construction [11] were into force in Italy. These standards still regulate the design, construction and inspection of buildings. Application of the provisions contained in the document, which was published in the Gazette of the Italian Republic No. 29 of 4 February 2008, is mandatory throughout Italy. The New Technical Standards for Construction [11] (often referred to in practice by the acronym NTC2008) came into force on 1 July 2009 and owe their name to the fact that they replaced the previous standards [10].

The main points that distinguish the new provisions from the older documents are:

- the development of performance criteria,
- alignment with EC legislative guidelines,
- greater detail for aspects related to seismic action,
- more detailed guidelines regarding geotechnical aspects.

There has been a significant change in terms of seismic hazard assessment. There are no longer only 4 seismic zones. A grid with 10,751 nodes has been defined, with the bedrock acceleration values a_g for each node determined for 9 different return period values. The parameters necessary to define the response spectrum for analyses in any site in Italy can be determined by taking the weighted average of the values assigned to the 4 nearest grid points. It is important to note that seismic hazard was not reassessed following the Emilia-Romagna Earthquake.

Paragraph 7.4.5 of the NTC2008 provides accurate informations about buildings with precast structures in seismic areas. It covers several categories, including perhaps the most common, described as "isostatic pillar structures". This expression indicates a single-storey structure, with roofing elements supported by fixed bearings resting on isostatic pillars. The use of simply supported pre-cast beams is permitted, provided they are structurally connected to the supporting pillars or walls; the connections must ensure the transmission of lateral forces during an earthquake, without relying on friction. The New Technical Standards for Construction [11] include the use of the structure factor “q” for the reduction of actions obtained through the elastic response spectra. The use of the structure factor is subject to compliance with many specifications for connections and types of structural elements. The minimum expected value for the structure factor "q" is set at 1.5. In regard to construction details, precast structures in seismic zones are subject to the same limits as cast-in-place reinforced concrete structures. There are therefore specific geometric and reinforcement limitations for beam, pillar, wall, coupling beam and node elements.

The text of the NTC2008 is accompanied by a Circular [12], although its application is not mandatory. It contains additional information, clarifications and application instructions for a broader understanding of NTC2008.

These documents are, to date, the main reference for structural designers working in Italy. Although regional as well as municipal building regulations are required in Italy, the main principles of structural design and the basic regulatory requirements are set out in the New Technical Standards for Construction [11] and further elaborated in the Circular [12].

4 Documents issued and adopted due to the May 2012 earthquakes

The severity of the damage found in the territory of Emilia-Romagna, Veneto and Lombardy led to the rapid establishment and adoption of important technical and regulatory documents.

AeDES Forms [Web-4]. Although these already existed before the 2012 earthquake, they were adopted in the immediate aftermath of the emergency as an instrument for the detection of damage to structural and non-structural elements. They are also used to indicate the emergency measures carried out and to provide an overall assessment of the accessibility of structures. There are six different possible summary assessments: A = building accessible; B = building temporarily unusable, but accessible with emergency measures; C = building partially unusable; D = building temporarily unusable; E = building unusable; F = building unusable due to external risk.

Ordinanza del Capo del Dipartimento della Protezione Civile, del 02 giugno 2012 [13]. During the post-earthquake emergency, the Italian Council of Ministers authorised the Head of the Department of Civil Protection to issue decrees overriding other current provisions (although always in compliance with the general principles of law). On the basis of this act, a decree was issued on 2 June 2012 specifying that owners of productive activities, being responsible for safety in the workplace under Italian law, are obliged to obtain seismic conformity certification in order to resume activities. The seismic conformity must be issued by a qualified professional, in accordance with local regulations. The importance of this act is clear in the light of the Decreto Legge n°74 del 6 giugno 2012 [14].

Decreto Legge n°74 del 6 giugno 2012 [14]. The document established a state of emergency until 31 May 2013 and provided for the allocation of reconstruction funds. It also redefined the concept of seismic conformity, no longer referring only to a structure's capacity to effectively resist new shocks. According to the new definition, seismic conformity is the absence in the structure of the serious shortcomings listed in Article 3 of the Decree:

a. Lack of cross-ties between vertical and horizontal structural elements and between separate horizontal structural elements;

b. Precast infill elements not properly anchored to the main structures;

c. Unbraced shelving bearing heavy materials, which, in the event of its collapse, could affect the main structure, causing damage and collapse.

Irrespective of the state of damage, if even one of the shortcomings identified in the decree is found, the production activity is automatically stopped. The owner of the business is not allowed to use the structure, since, according to Italian law, he/she is responsible for safety in the workplace.

This interpretation of the concept of conformity of production facilities may set a precedent in case of future seismic events. For this reason, the seismic hazard assessment of an area becomes a direct risk index for production activities in the case of facilities deemed inadequate under Article 3 of the decree. This risk may be connected with an appropriate assessment of the economic cost over the medium to long term, but that is beyond the scope of this document.

Another very important development of the Decree is the requirement of two intervention phases for damaged structures. A 6-month period is specified for the first intervention (PHASE 1) and a further period of 18 months is established for the second (PHASE 2). The objective of PHASE 1 is the rapid securing of the premises. The goal of PHASE 2 is the attainment of a performance capacity of 60% of that required by the standards for new structures. The two intervention phases are also required to be well integrated with each other. For this reason, it should be possible to directly incorporate the work carried out during the emergency period into the subsequent series of interventions.

Linee di indirizzo per interventi locali e globali su edifici industriali monopiano non progettati con criteri antisismici [Web-5] (hereafter referred to as the "Guidelines"). This is a document without binding force, drawn up by the Working Group on the Seismic Conformity of Industrial Buildings. The document is of considerable importance to engineers and was drawn up in order to:

- Provide an overview of the damage and collapses affecting single-storey precast structures discovered in the aftermath of the events of May 2012
- Clarify the meaning of the two intervention phases specified into Decreto Legge n°74 del 6 giugno 2012 [14]
- Propose general criteria for intervention

- Describe simplified intervention methods
- Provide procedures and intervention plans directly replicable in practice

The structural model considered is very simple. Single-storey buildings are examined. The pillars, normally between 3 and 7 m in height, are fixed at the base and free at the top. The beams, normally with a span varying between 10 and 25 m, have constraints at the ends consisting of simple friction supports. The considerations made obviously also apply to structures with a higher level of structural complexity.

The analysis of cases of damage that occurred in May 2012 allows the identification of distinct categories of structural and non-structural damage in event of an earthquake:

A. Damage to beam-to-pillar and roof-to-beam connections
B. Damage to infill elements
C. Damage to pillars

In view of the materials contained within the structures, the damaging and overturning of shelving is also of considerable importance.

Regarding point A, the most common types of damage are:

- Loss of support due to relevant sliding in friction systems (with undamaged structural elements)
- Loss of support due to damage to one or both of the structural elements involved
- Loss of support due to the collapse of the beam-pillar metal tie element
- Failure of the reinforced concrete fork at the head of the pillars.
- Rigid rotation of the beam on its axis

Regarding point B, the most common types of damage are:

- Collapse of panels due to hammering by horizontal elements, pillars or even perpendicular panels
- Collapse due to differential displacements of the pillars supporting the panel
- Collapse due to failure or opening of the metal tie element between the infill and the pillar
- Failure of the panel in its plane due to actions not envisaged during the design phase
- Tilting out of plane of masonry infills
- Cracking of masonry infills due to in-plane mechanisms

Regarding point C, the most common types of damage are:

- Rigid rotation at the foot of the pillar due to rotation of the entire foundation element
- Rigid rotation at the foot of the pillar due to damage to the sleeve footing components
- Incipient plastic hinge formation at the base of the pillar
- Incipient plastic hinge formation on the pillar, at a height
- Damage to the pillars due to impact by other elements that have collapsed due to loss of support
- Shear brittle failure in stocky elements

In regard to shelving systems inside buildings, the types of damage are essentially due to their collapsing or overturning.

The primary objective set by the Guidelines is to overcome the serious lack of beam-pillar connections. The Guidelines also emphasise the importance of preserving the original static layout and the reallocation of horizontal stiffness between the elements. Once the problem of beam-to-pillar and beam-to-roof connections has been addressed, the importance of the following aspects is emphasised:

- Installation of deformable connections for infills
- Installation of restraint systems for infills
- Increasing the resistance of structural elements (particularly at the base of the pillars)
- Increasing the ductility of structural elements (at the base of the pillars)
- Increasing the load-bearing capacity of foundation plinths
- Installation of anti-tipping systems for beams
- Connections between pillars to contain relevant displacements
- Installation of steel bracing to reduce the deformability of the overall frame system of which the building is composed

In regard to the simplified interventions methods, Guidelines suggest the use of a simple single degree of freedom scheme. Methods for calculate flexural stiffness, seismic mass, design displacement and design force are provided. The concluding section of the Guidelines contains procedures and diagrams for intervention; it also lists the advantages and disadvantages of the use of various solutions and provides a series of specifications to be taken into account when sizing.

5 Conclusions

The series of earthquakes that struck north-central Italy in May 2012 brought to light the significant vulnerability of existing precast structures, designed without consideration for appropriate seismic criteria. Specific regulatory indications have been established for over two decades, even for precast structures. However, much of the building stock predates these provisions or is in areas in which the seismic hazard map has changed significantly in recent years.

The New Technical Standards for Construction [11], issued in 2008, are currently in force in Italy. The seismic events of May 2012 did not lead to the publication of new building codes. Similarly, in regard to the seismic hazard assessment, the reference maps in force in May 2012 are still valid. However, a decree was issued by the Head of the Civil Protection Department [13], together with a specific decree law [14], for the management of emergency and to guide reconstruction. These redefine the concept of seismic conformity and establish times and procedures for intervention in areas affected by the earthquake. Two distinct intervention phases are established, the first to deal with the actual emergency and the second to achieve a higher level of safety in the medium to long term. In order to facilitate the work of the technicians, the Working Group on the Seismic Conformity of Industrial Buildings has also developed guidelines [Web-5] that can be of use to structural designers for the retrofitting of structures.

REFERENCES

[1] Parducci A., "Considerazioni sulla progettazione di edifici pre-fabbricati antisismici", L'INDUSTRIA ITALIANA DEL CEMENTO, XLVIII, February 1978, Pages 103-113.

[2] Legge 5 novembre 1971, n.1086 - "Norme per la disciplina delle opere di conglomerato cementizio armato, normale e precompresso ed a struttura metallica", Gazzetta Ufficiale n.321, 21 December 1971.

[3] Legge 2 febbraio 1974, n. 64 - "Provvedimenti per le costruzioni con particolari prescrizioni per le zone sismiche", Gazzetta Ufficiale n. 76, 21 March 1974.

[4] Decreto Ministeriale 3 marzo 1975 - "Approvazione delle norme tecniche per le costruzioni in zone sismiche", Gazzetta Ufficiale n.93, 8 April 1975.

[5] Decreto Ministeriale del 24 gennaio 1986 - "Norme tecniche relative alle costruzioni antisismiche", Gazzetta Ufficiale n. 108, 12 May 1986.

[6] Decreto Ministeriale 3 Dicembre 1987 - "Norme tecniche per la progettazione, esecuzione e collaudo delle costruzioni prefabbricate", Gazzetta Ufficiale n. 106, 7 May 1988.

[7] Decreto Ministeriale 16 gennaio 1996 – "Norme Tecniche per le Costruzioni in Zone Sismiche", Gazzetta Ufficiale n. 29, 5 February 1996.

[8] Decreto del Presidente della Repubblica del 6 giugno 2001, n. 380 - "Testo unico delle disposizioni legislative e regolamentari in materia edilizia", Gazzetta Ufficiale n. 245, 20 October 2001.

[9] Ordinanza del Presidente del Consiglio dei Ministri del 20 marzo 2003, OPCM 3274 - "Primi elementi in materia di criteri generali per la classificazione sismica del territorio nazionale e di normative tecniche per le costruzioni in zona sismica.", Gazzetta Ufficiale n. 105, 8 May 2003.

[10] Decreto Ministeriale del 14 settembre 2005 - "Norme Tecniche per le Costruzioni", Gazzetta Ufficiale n. 222, 23 September 2005.

[11] Decreto Ministeriale del 14 gennaio 2008 - "Nuove Norme Tecniche per le Costruzioni", Gazzetta Ufficiale n. 29, 4 February 2008.

[12] Circolare 2 febbraio 2009, n. 617 C. S. LL. PP. -"Istruzioni per l'applicazione delle «Nuove norme tecniche per le costruzioni» di cui al decreto ministeriale 14 gennaio 2008". Gazzetta Ufficiale n. 47, 26 February 2009.

[13] Ordinanza del Capo del Dipartimento della Protezione Civile, del 2 giugno 2012, Gazzetta Ufficiale n. 130, 6 June 2012.

[14] Decreto Legge del 6 giugno 2012, n. 74 - "Interventi urgenti in favore delle popolazioni colpite dagli eventi sismici che hanno interessato il territorio delle province di Bologna, Modena, Ferrara, Mantova, Reggio Emilia e Rovigo, il 20 e il 29 maggio 2012", Gazzetta Ufficiale n. 131, 7 June 2012.

[Web-1] http://shakemap.rm.ingv.it/shake/archive/ , publisher : Istituto Nazionale di Geofisica e Vulcanologia, INGV. Data gathered on 5.May.2013

[Web-2] http://mi.ingv.it/class1984.html , publisher : Istituto Nazionale di Geofisica e Vulcanologia, INGV. Data gathered on 5.May.2013

[Web-3] http://mi.ingv.it/class2004.html , publisher : Istituto Nazionale di Geofisica e Vulcanologia, INGV. Data gathered on 5.May.2013

[Web-4] www.protezionecivile.gov.it/resources/cms/documents/Scheda_AEDES.pdf, publisher : Protezione Civile. Data gathered on 5.May.2013

[Web-5] www.reluis.it/images/stories/Linee_di_indirizzo_GDL_Capannoni.pdf, publisher : ReLUIS. Data gathered on 5.May.2013

Part IV

Seismic Safety Evaluation and Retrofitting

Earthquake Assessment of Existing Chemical Production Facilities

T. Drommer[1], **C. Gellert**[2]

[1] BASF SE
Carl-Bosch-Straße 38, 67056 Ludwigshafen, Germany
thomas.drommer@basf.com

[2] SDA-engineering GmbH
Kaiserstr. 100, TPH III-B, 52134 Herzogenrath, Germany
gellert@sda-engineering.de

ABSTRACT:

Due to the introduction of the revised German Earthquake design standard DIN 4149 in 2005 [1] including a re-evaluation of earthquake loadings and the forthcoming introduction of the European Earthquake standard DIN EN 1998 (Eurocode 8) [2] the demands on the overall Earthquake structural design increases.

As plant operators of Chemical production units governed by the Major Accidents Ordinance (Störfallverordnung) are obligated to operate their facilities in accordance to the latest state of the art safety standards the existing production facilities will need to be evaluated in regards to earthquake resistance. The Evaluation is based on the VCI-guideline [3] which provides in addition to DIN 4149 and DIN EN 1998 a basis of an Earthquake assessment and design principles for chemical production facilities due to Earthquake loading.

This paper introduces the Earthquake assessment program of existing chemical production facilities that BASF-SE has undertaken in the past years on their production site in Ludwigshafen, Germany. Based on specific examples the assessment procedure for the initial evaluation of existing Chemical production facilities is presented. Furthermore experiences and results of already finalized assessments of more than 28 production units are summarized and recommendations are derived for further assessments.

Keywords: Earthquake assessment program, existing production facility, Major Accidents Ordinance (Störfallverordnung), BASF

S. Klinkel et al. (Eds.), *Seismic Design of Industrial Facilities*,
DOI 10.1007/978-3-658-02810-7_8, © Springer Fachmedien Wiesbaden 2014

1 Introduction

In general Earthquakes are natural events whose impacts on chemical production facilities may cause hazardous incidents. Due to damage and collapsed structures of production facilities that contain hazardous substances incidents may occur whose consequences go far beyond the expected material impacts.

The BASF-SE Ludwigshafen chemical production site as one of the largest "Verbund" sites in the world comprises more than 160 production facilities and 300 storage facilities (tank farms, warehousing etc.). Most of these production facilities handle large quantities of various chemical substances and compounds.

In 2005 the revised German Earthquake standard DIN 4149 has been introduced and consequently Earthquake loadings have changed for German Earthquake zones. The revised national Earthquake design standard DIN 4149 and the European Earthquake standard DIN EN 1998 (Eurocode 8) suggest higher reference peak ground accelerations within Earthquake zone 1. BASF-SE Ludwigshafen production site is located in zone 1 which now needs to account for a 60% higher reference peak ground acceleration. (from $0.25 m/s^2$ to $0.4 m/s^2$).

Production facilities and storage areas that are subjected to the 12th Federal Immission Protection Law have to be retained to the latest state-of the art standards. This also requires adopting the newest development in applicable codes and standards for existing facilities that have been designed and constructed over the past decades or even centuries.

Consequently BASF-SE has developed a specific Earthquake assessment program in particular for existing facilities. The development of the assessment program has been conducted in close cooperation to responsible authorities.

The earthquake assessment program and its practical application based on one typical production plant are presented within this paper.

2 BASF Earthquake Assessment Program

DIN 4149-2005 excludes the design of facilities (e.g. Chemical production facilities) where an additional hazard for human life, health and the environment is present.

Due to this limitation a group of experts of the German Chemical Industries Association (VCI) has developed a guideline [3] that provides analysis and design principles of how to adopt the basis of DIN 4149 and DIN EN 1998 Earthquake standard to the chemical industry. This guideline is widely recognized with authorities and is regarded as good practice.

Based on the VCI guideline the BASF-SE Earthquake assessment program for existing Chemical Production Facilities was developed for the Ludwigshafen production site.

One of the major aspects of the Earthquake assessment program was the method to establish an appropriate facility selection along with an investigation order. The Earthquake assessment program has been carried out in several stages based on the hazardousness and importance of a chemical facility. The hazardousness and importance of a facility also defines a priority of possible measures of necessary improvements that may to be taken.

2.1 VCI-Guideline

The German Chemical Industries Association VCI has brought together a group of experts consisting of Civil / Structural engineers, Geologists and Safety experts to develop a design guideline and established principle design requirements for chemical facilities that are covered by the major accident regulations.

The VCI-guideline was officially introduced in 2009 and shall be applied in conjunction with German Earthquake code and the European Earthquake code.

Initially the VCI-guideline was intended to be applied in the structural design and construction of new chemical facilities. However it not only covers approved basic design principles but also discusses the assessment of existing chemical facilities and their safety standards. An evaluation form is provided in the commentary to assist in detecting critical areas within chemical facilities. As a result the evaluation form should identify major risks and the need for improvement in terms of the structural integrity of the global structure, their components and installations.

2.2 Assessment Approach

In general the Earthquake assessment program has been conducted in multiple cycles based on the magnitude of the assigned importance factor. The principle selection process for existing chemical production facilities to be evaluated is described in detail in section 2.3.

In summary facilities with a high importance factor were evaluated at high priority the ones with the lowest importance factor at last. The selection of the importance factor for a chemical facility is dependent on the hazardousness of the chemical substances and compounds that are handled, produced, used, filled or stored in a production unit or in a storage area and is given in the VCI-Guideline.

Importance factors can be derived from tables that are provided in the VCI-guideline. Generally importance factors can be chosen for three different impact categories:

- Personnel safety
- Environmental protection
- Effects on lifeline entities.

The maximum value of the three different categories shall apply for the assessment. For the petrochemical industry the values of the Importance factor may vary from 1.0 to 1.6 whereas the later presents the most severe instance.

The implementation of the first round of the BASF-SE Earthquake assessment program was separated into two areas of action. For a range of facilities that have been assigned to the highest Importance factor according to the VCI guideline (γ_I=1.6) a comprehensive site inspection as well as a detailed structural analysis of typical structures have been carried out by experts. Results were documented and final reports were provided to the responsible authorities for review and discussion.

Within the second round of the assessment program facilities that have been assigned to an Importance factor between 1.4 and 1.6 ($1.4 \leq \gamma_I < 1.6$) were evaluated. Therefore site inspections have been carried out by experts. With regards to a reference peak ground acceleration of a_{gR}=0.4 m/s² the detailed analysis results performed in the first round indicate that the combination of all typical loadcases such as self-weight, live - and wind loads as well as stabilizing forces often still produce larger design loads than respective loadcases that include earthquake loads. Otherwise higher design loads that comprise earthquake loads are generally covered by common load bearing reserves of the global structure.

Based upon detailed assessment results of the first round and in agreement with the responsible authorities the second round of assessment did not perform further detailed structural calculations on chemical production facilities of the Ludwigshafen site!

Derived from the experiences and outcome of finalized Earthquake assessment rounds a guidebook for BASF operating personnel has been developed [4]. This guidebook should support a self-evaluation practice on installations and components within production facilities that can be undertaken by competent operating personnel. This procedure was introduced for facilities with low Importance factors $\gamma_I \leq 1.2$.

2.3 Selection process of production facilities and storage areas

A proven method to establish an appropriate facility selection for an Earthquake assessment is based on the classification of hazard characteristics of chemical substances and their quantities. BASF safety experts have surveyed the variety of existing production facilities and storage areas on the Ludwigshafen site in order to classify their hazardousness. The facility selection concept considers a dependency on used, handled or stored quantity limits of chemical substances and their hazard characteristics.

In a second step potential impacts that may arise from a seismic event have been defined in accordance to the damage extend that may occur. Therefore production facilities and storage areas that contain large quantities of highly volatile, very toxic as well as highly and easy flammable substances were assigned the highest priority. Their failure may cause a large impact that could reach far beyond the Ludwigshafen production site. The principle selection process is visualized in Figure 1.

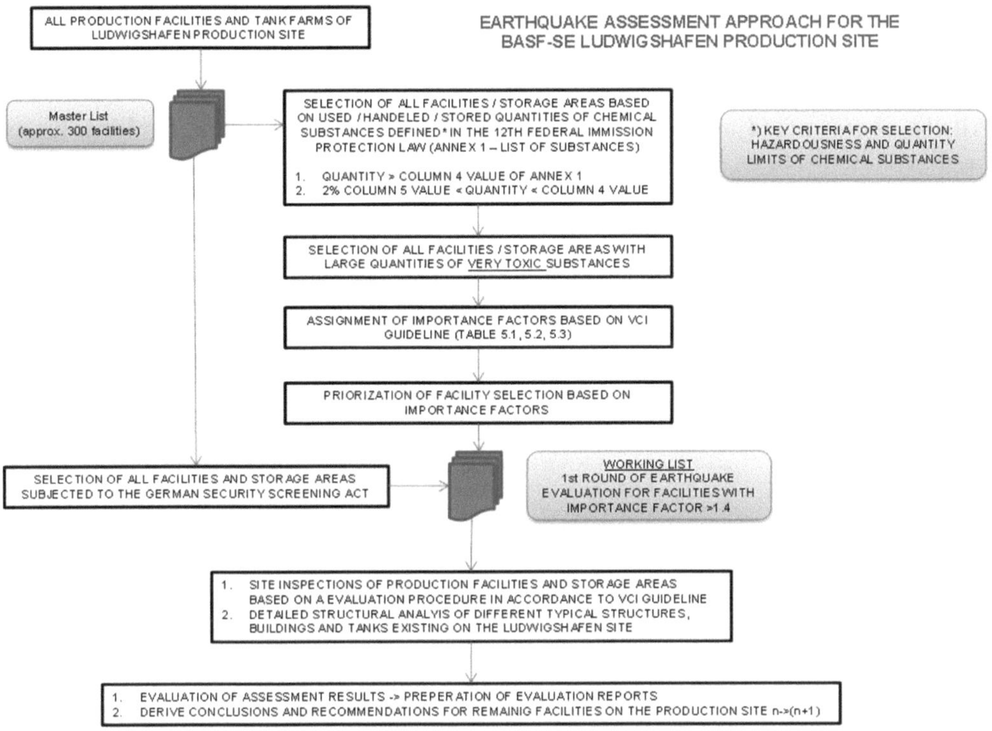

Figure 1: BASF earthquake assessment program

Critical quantity limits of chemical substances and compounds may be derived from column 4 of annex 1 of the 12th federal immission protection law. On the BASF-SE Ludwigshafen site approximately 300 production units or storage areas that contain quantities larger than given in column 4 of annex 1 of the 12th federal immission protection law have initially been determined.

The extent of an impact to what an incident may be assigned to can be as follows. An incident caused by an Earthquake event may have an impact (also refer to table 5.1 of the VCI guideline)

- inside a production or storage facility,
- within a close proximity to the production or storage facility,
- inside the production site,

- outside of the production site,
- a distant beyond the production site.

As a last step of the selection process typical building structures and tanks that may be found throughout the Ludwigshafen production site have been classified. Facilities with similar structural characteristics have been grouped and examined in an exemplary manner in terms of their structural integrity due to higher Earthquake loadings.

All the above noted aspects of the selection process originated a working list of facilities for which an Earthquake assessment should be carried out.

3 Evaluation Example

In the following section an Earthquake assessment procedure is demonstrated based on a typical production facility (see Figure 2). In an exemplary manner some results of a site inspection which was based on the evaluation method provided in the VCI guideline are shown. In addition results of a detailed structural analysis are presented.

Quelle: www.bing.com

Figure 2: One representative production facility

The BASF-SE chemical production site Ludwigshafen is located in a zone of low seismicity with a reference peak ground acceleration of $a_{gR}=0.4$ m/s². Based on the designated classification and selection process discussed in section 2.3 the Importance factor for the production facility under consideration has been assigned to $\gamma_I = 1.6$.

3.1 On-site Earthquake Assessment

In a first phase of the evaluation process a comprehensive site inspection has been conducted in accordance to the VCI guideline. All evident weak points concerning structural issues on installations and components have been documented and evaluated by experts. As a result the production facility has been evaluated with an overall defect-index of 3.5. According to VCI guideline the index can assume values ranging from 0 (optimum) to 25 (worst case). The result of the site inspection and the small defect-index indicates no immediate need for action. The list of details illustrated in the evaluation report should be improved within the course of usual periodic inspections.

To improve the assessment procedure the site inspections may be reasonable supplemented by an approximate calculation of the existing bearing capacity for e.g. anchor bolts of vessels and other installations. Therefore the VCI-guideline provides a rough analysis approach applicable for non-structural components[3].

If a rough estimate of the horizontal seismic force is considered to be sufficient Eqs. (1) provides a conservative value for the equivalent static force:

$$F_a = 1.6 \cdot S_{a,max} \cdot \gamma_a \cdot m_a \text{ [kN]} \qquad (1)$$

The factor $S_{a,max}$ corresponds to $2.5 \cdot S \cdot \eta \cdot a_{gR}$, a_{gR} being the reference peak ground acceleration, η the damping factor (η=1.0 representing 5% viscous damping) and S the soil factor. γ_a is the importance factor and m_a the mass of the investigated component.

Applied to the small vessel shown in Figure 3 the overall mass m_a is calculated to 10t taking into account the weight of the containment and a filling of about 7300l. The damping factor is calculated considering a viscous damping of 2%, the soil factor is 0.75 and the importance factor is 1.4. This leads to a resulting horizontal design force of:

$$F_a = 1.6 \cdot S_{a,max} \cdot \gamma_a \cdot m_a = 1.6 \cdot (2.5 \cdot 0.75 \cdot 1.2 \cdot 0.4) \cdot 1.4 \cdot 10 = 20.16 \text{ kN} \qquad (2)$$

The horizontal force is applied in the center of mass of the equipment. The overall height of the tank is about 3.5m including a height of the feet of about 1.0m. Hence the estimated elevation of the center of mass is about 2.25m. Considering a foot spacing of about 1.0m the resulting overturning moment leads to uplift forces of about:

$$Z = (20.16 \cdot 2.25) \cdot \sqrt{1^2+1^2} = 32.1 \text{ kN} \qquad (3)$$

Taking into account a vertical force of D = 24.5 kN the bolted connection on each vessel foot has to be designed for a uplift force of $F_{t,Ed}$ = 7.6 kN. Furthermore a shear force of $F_{v,Ed}$=20.16/4 = 5.0 kN is considered.

Each foot is anchored to the substructure using one M20 class 4.6 according to DIN EN 1993 [5]. The anchor bolt connection has to be checked according to EN 1993 for a combination of shear and tension forces:

$$\frac{F_{v,Ed}}{F_{v,Rd}} + \frac{F_{t,Ed}}{1.4 \cdot F_{t,Rd}} = \frac{5.0}{47.0} + \frac{7.6}{1.4 \cdot 49.4} = 0.22 \leq 1.0$$

The approximate estimation indicates a sufficient bearing capacity of the anchor bolt connection.

Lfd. Nr.	Bild (Foto, Skizze, Planausschnitt, o.ä.)	Detailbeschreibung (Standort, Problembeschreibung, gehandhabte Stoffe, etc.)	Bed.-Beiwert (Leitf. 5.3)	Index A (Mangel)	Index B (Gefahr)	Index C (Priorität C=A*B)
16	 **Bild PB200078**: Behälter B1448 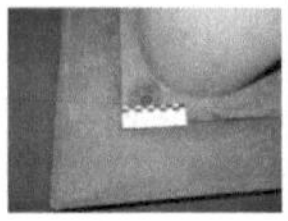 **Bild PB200080**: Mangelhafter Fußpunkt	**Behälter B1448** Volumen: max. 7260 l Inhalt: Trimethylamin Der Behälter B1448 ist vollständig gedämmt und steht auf 4 Rundfüßen mit einer Höhe von ca. 1,00m auf. Jeder Fuß ist augenscheinlich mit 1 Schraube M20 im Sockel verankert und lagegesichert. Die Dimensionierung der Füße und der Lagesicherung erscheint für den Erdbebenfall ausreichend. Eine überschlägige Berechnung der Fußpunktlagerung hat ergeben, dass die Lage- und Kippsicherung sicher gestellt ist (s. Anlage A). Dies setzt allerdings voraus, dass die Verankerung wirksam ist. An den Fußpunkten sind teilweise die Kontermuttern entfernt oder nicht ausreichend verschraubt. ***Die Fußpunkte sind hinsichtlich der konstruktiven Verschraubung zu überprüfen und zu ertüchtigen. Die Wirksamkeit der Lager ist wieder herzustellen.***	1,4	2	4	8

Figure 3: Tank detail - Extract from earthquake assessment report

3.2 Structural Analysis

Besides the comprehensive site inspection a detailed structural analysis has been performed on the global structure of the production facility. Under investigation have been the lateral load-bearing elements of the structure. The production facility may be separated into three independent building parts that all have been examined separately. In this paper the result of building part 2 are summarized.

The global support structure of building part 2 is a structural steel framework. Due to the asymmetric mass distribution and the irregular distribution of the bracing system a three dimensional FE calculation using the modal response spectrum analysis has been carried out. The building part under consideration comprises of 5 column lines with vertical K- and Y bracing (3 N-S, 2 E-W).

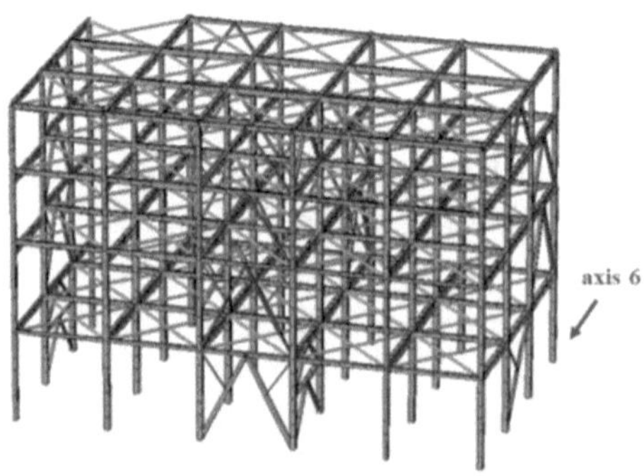

Figure 4: 3-dimensional calculation model: Building Part 2

The mass distribution in plan on each main level has a significant influence on the analysis results. The loads of all equipment and other installations / components and their positions in the structure have been considered accurate each as single point mass.

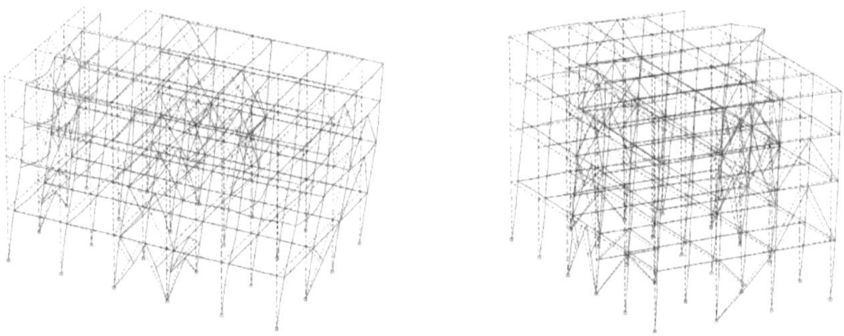

Figure 5: Eigenmodes (left hand T_{1x} = 1.49 s and right hand T_{1y} = 1.05 s)

The first Eigenmode with the corresponding Eigenperiod in each direction is shown in Figure 5. Figure 6 displays the corresponding design spectrum applied to the global support structure of the production site in Ludwigshafen. Assuming a low ductile behaviour of the structure the ductility factor is chosen to be q=1.5.

The design spectrum with the marked Eigenfrequencies indicates that the structure is characterized by relative long eigenperiods with a corresponding low spectral acceleration.

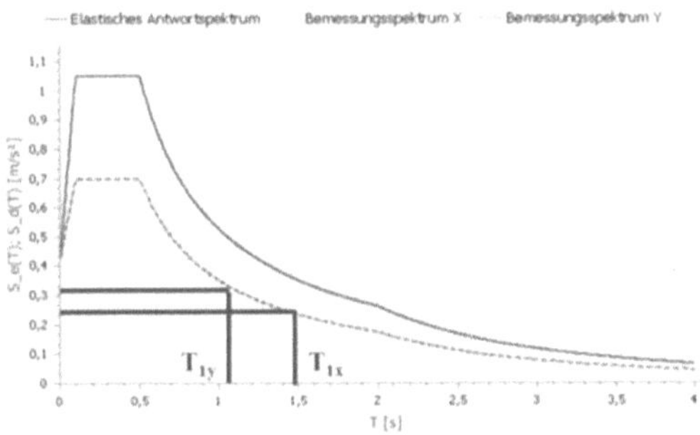

Figure 6: Design spectrum: Production site Ludwigshafen

Based on the results of the modal response spectrum analysis the safety verification is executed for the main bracing elements. In detail stress analysis, stability checks, safety verification for connections and foundations are provided according to applicable EN codes. Figure 7 shows a typical steel frame in column line 6 with the results of the stress analysis. The maximum utilization of 40% suggests a sufficient load capacity for the structural members being considered.

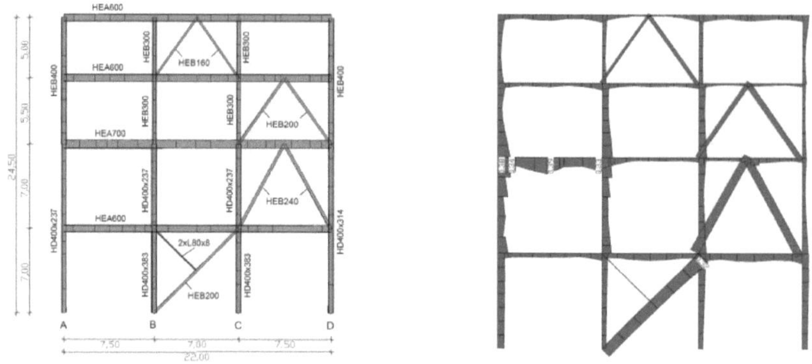

Figure 7: Steel frame in axis 6 with corresponding stress utilization

In summary the design verification of the investigated production facility comprising three building parts in total has been successful. A sufficient earthquake resistance could be demonstrated.

4 Experience

4.1 On-site Earthquake Assessment

Since 2009 in overall 28 production facilities and storage areas with an Importance factor $\gamma_I \geq 1.4$ on-site Earthquake assessments has been undertaken. Thus various equipment and other installations with more than 600 construction details have been evaluated and documented. Over 94% of all detected defects and weak points are categorized as low or moderate and 6% only as severe. In case a defect was categorized as severe further detailed calculations have been carried out to demonstrate a sufficient earthquake resistance or improvement measures have been undertaken at short notice. Very often low and moderate defects are very simple to resolve. In Table 1 below typical low and moderate defects are illustrated in an exemplary manner.

4.2 Structural analysis

The chemical production site Ludwigshafen is located in a low earthquake prone zone in Germany. Taking into account a reference peak ground acceleration a_{gR}=0.4 m/s² detailed structural analysis results show for most cases that the combination of all typical loadcases such as self-weight, live - and wind loads as well as stabilizing forces often produce still larger design actions than respective loadcases that include earthquake loads.

Table 1: Frequently detected defects - low effort in troubleshooting

Image	Description
 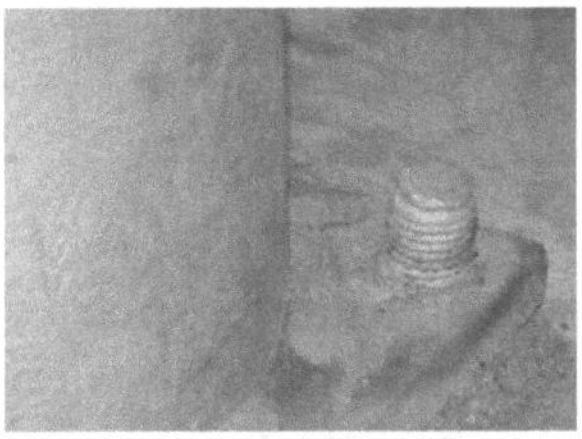	Structural defects e.g.: missing nuts, broken joints, corroded connections.
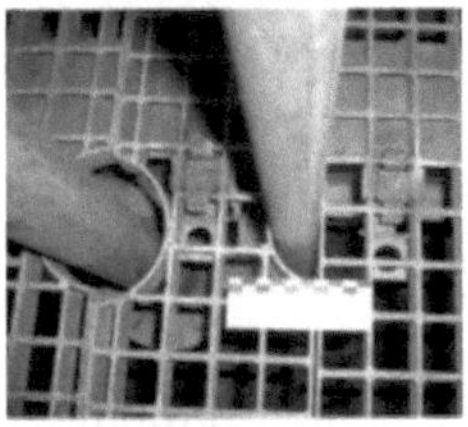 	Undesirable bearing (support) e.g.: pipe penetrations, small or insufficient gaps between structural elements and installations / components
 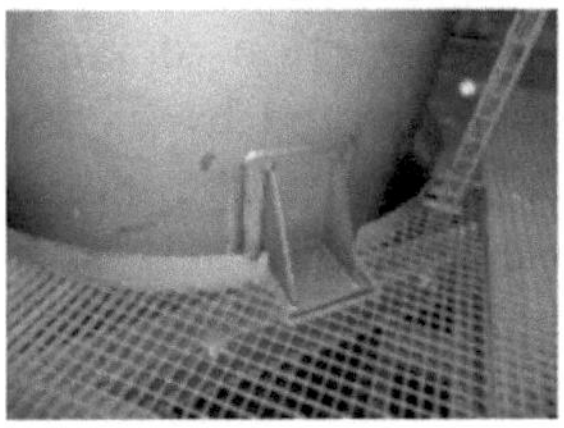	Missing bolts e.g.: Vessel on bearing lugs without sufficient anchoring (missing anchor bolts)
 	Missing lateral bracing e.g.: Vessel on steel support without cross bracing
 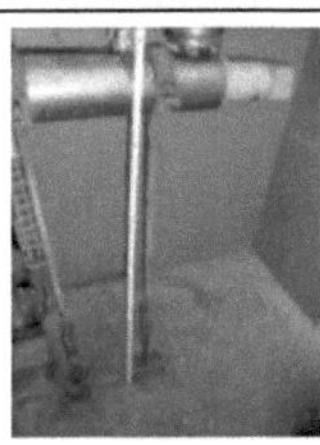	Design defects e.g.: elevated pipes, valves or other installations without lateral bracing or sufficient anchoring
 	Insufficient stability against overturning and sliding e.g.: vessel base without anchoring to foundation

However if structural design actions that include Earthquake loads exceed the design load level computed from typical loadcases without seismic loads then typical load capacity reserves of the global structure that have not been considered in the analysis approach may provide the required additional safety.

5 Conclusion

Chemical production facilities and storage areas have to be kept to the latest state-of the art standards. This also includes adopting the newest development in applicable codes and standards such as the latest Seismic design code. Hence existing facilities will need to be evaluated in regards to earthquake resistance due to increased seismic loads. The key objective for the assessment program was to obtain a clear view about the conditions of existing production facilities and its installations and components due to higher Earthquake loadings.

In the past years for 28 production facilities and storage areas on-site Earthquake assessments has been carried out on the BASF production site in Ludwigshafen, Germany. Thereby more than 600 construction details, vessels and other installations have been evaluated and documented by experts. The Evaluation is based on the VCI-guideline [3] which provides in addition to DIN 4149 and DIN EN 1998 a basis of an appropriate Earthquake assessment. Derived from the experiences and outcome of already finalized assessments a guidebook for BASF operating personnel has been developed [4]. This guidebook should support a self-evaluation practice on installations and components within production facilities.

References

[1] Deutsches Institut für Normung (DIN), DIN 4149:2005-04 Bauten in deutschen Erdbebengebieten, Beuth Verlag, 2005.

[2] DIN EN 1998-1: Eurocode 8: Auslegung von Bauwerken gegen Erdbeben – Teil 1: Grundlagen, Erdbebeneinwirkungen und Regeln für Hochbauten; Deutsche Fassung EN 1998-1:2004 + AC:2009, Dezember 2010.

[3] Earthquake resistance in the design of structures of Chemical production units, layout, design and execution of structures and components for the Chemical Industry based on DIN EN 1998-1, guideline version 3.0 and commentary dated 05.07.2012.

[4] Erdbeben in Chemieanlagen, Sicherung von Apparaten und Rohrleitungen – Handbuch für Betriebspersonal – BASF SE – Werk Ludwigshafen, SDA-engineering GmbH, Stand 24.11.2011.

[5] DIN EN 1993-1-8: Eurocode 3: Design of steel structures, Part 1-8: Design of joints; German version EN 1993-1-8:2005 + AC:2009, December 2010

Probabilistic Seismic Analysis of Existing Industrial Facilities

Hamid Sadegh-Azar[1], Truong-Diep Hasenbank-Kriegbaum[1]

[1] HOCHTIEF Solutions AG, Consult IKS Energy
Lyoner Straße 25, Frankfurt am Main, Germany
Hamid.Sadegh-Azar@hochtief.de
Truong-Diep.Hasenbank-Kriegbaum@hochtief.de

ABSTRACT

This paper presents the application of probabilistic methods for the seismic analysis of existing industrial facilities. First, the main advantages and the rationale for probabilistic (versus deterministic) approaches are discussed for existing structures/facilities. A short overview of existing probabilistic and deterministic seismic analysis approaches follows. Afterwards a simple and efficient probabilistic approach is presented with an example as application on existing industrial facilities. The method involves state-of-the-art probabilistic seismic hazard analysis (PSHA). It covers the whole industrial facility including structures, components, mechanical installations, piping, tanks, etc. In comparison to Monte Carlo Simulation, this method is cost-effective and practical and can be used for risk-informed/performance-based rehabilitation or strengthening.

Keywords: Probabilistic Seismic Analysis, Facility Analysis, Latin Hypercube

1 Why Probabilistic?

Structural engineers are used to apply deterministic design and analysis approaches. The reason for this is mainly because deterministic design and analysis approaches are more convenient to apply, more simple and straightforward and most design codes prescribe deterministic approaches. Also the design result or outcome is very clear and for everybody conceivable and understandable: Design load "is" or "is not" less than design capacity. Black or White.

S. Klinkel et al. (Eds.), *Seismic Design of Industrial Facilities*,
DOI 10.1007/978-3-658-02810-7_9, © Springer Fachmedien Wiesbaden 2014

For seismic design and analysis the deterministic approach is not appropriate because:

- Seismic loads are not deterministic. There is a relationship between the seismic load magnitude and probability of exceedance or return period (s. Figure 1).
- The seismic source and thereby the direction and spatial distribution of seismic loads are not deterministic. There are numerous probable seismic sources and fault mechanisms. For the majority of sites or locations the major earthquake direction and the spatial distribution of the load components are not known.
- The seismic (load bearing) capacity of a building is also not deterministic. The capacity depends on many parameters (e.g. nonlinear behaviour and ductility of materials, actual strength and overstrength of materials, influence of non-structural elements, etc.). It is very difficult to characterize all these parameters in a deterministic manner for a deterministic design.

These reasons together with the fact that each deterministic approach has a probabilistic basis lead to the conclusion that a probabilistic approach is more reasonable and appropriate for seismic design and analysis.

2 Introduction

The purpose of a seismic probabilistic risk assessment (SPRA) is to determine the probability distribution of the frequency of occurrence of exceeding various damage states or performance limits due to the potential effects of earthquakes. In contrast to a deterministic analysis that considers single-parameter values for seismic-induced forces and capacities. SPRA considers the total variability in seismic input, structure response, and material capacity variables. In simple terms, SPRA is the formal process in which the randomness and uncertainty in the various physical variables are propagated through an engineering model leading to a probability distribution of frequency of occurrence of failure or other damage states. Seismic risk analysis which is one of the facets of a SPRA can be performed for many different reasons. It can be used to compute the frequency of occurrence of failure due to seismic effects in order to compare these to similar results for other hazards. It is a useful tool to identify weak links in a system or facility. In this context, it can guide the efficient allocation of funds to strengthen or modify an existing industrial facility. It also can be used as part of the design process to size members to comply with a performance standard [1]. A SPRA consists of the following main parts: Seismic Hazard Analysis and Seismic Fragility Evaluation.

2.1 Seismic Hazard Analysis

The seismic hazard gives the relationship between seismic intensity (SI) and the corresponding probability of exceedance. There are plenty of parameters to quantify the seismic intensity. A summary of some of these parameters is given in the following:

- Damage-based Intensity Values: It is based on a qualitative description of the local effects of the earthquake at a site, for example using the Modified Mercalli Intensity.
- Seismological Intensity Values: The earthquake magnitude and the closest distance to the rupture zone can also be employed to express the SI.
- Engineer-seismological parameters: Time-domain or frequency-domain parameters and characteristic values of accelerograms, like peak ground acceleration (PGA), effective peak acceleration, spectral acceleration value (S_a) and etc. Typical seismic hazard curves based on PGA are shown in Figure 1.

The seismic hazard produces a connection between the intensity of an earthquake quantified by these parameters and the probability of its appearance. The curve, which gives the relation between the intensities of earthquakes at a location and the belonging exceeding probability, is called Site-Specific Hazard Curve, for example see Figure 2. This curve has to be determined for each location for different structural eigenfrequencies.

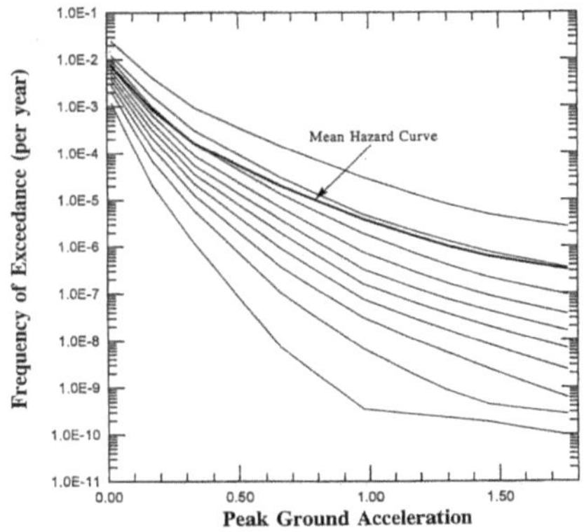

Figure 1: Seismic Hazard Curves [1]

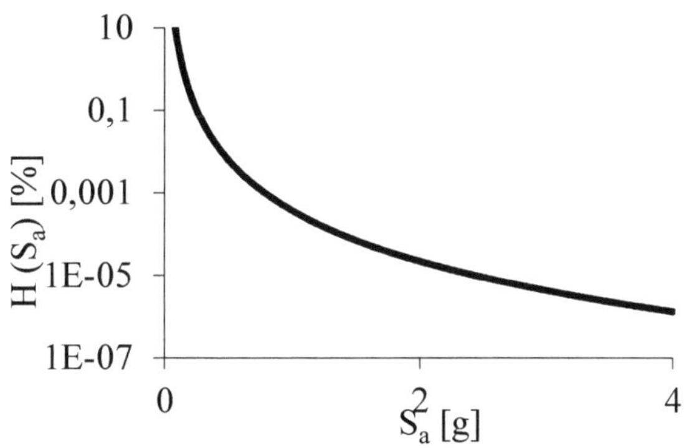

Figure 2: Site Specific Hazard Curve of Istanbul, Turkey (T =2.2 s)

2.2 Seismic Fragility Evaluation

The seismic fragility of a structure or equipment is defined as the conditional probability of its failure (or exceeding a given damage state) at a given seismic intensity (i.e., PGA or S_a at different frequencies).

Typical seismic fragility curves are given in Figure 3. These are developed using plant design information and realistic response analysis. The databases used for fragility analysis include simulations, earthquake experience data, generic equipment ruggedness spectra and fragility test results.

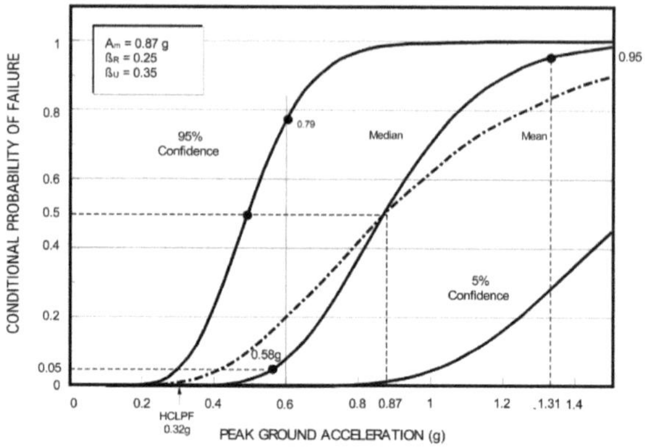

Figure 3: Typical seismic fragility curves [3]

3 Conservative Deterministic Failure Margin Approach

The Conservative Deterministic Failure Margin (CDFM) Method was first proposed in [1] as a deterministic method for estimating seismic capacity and was aimed at achieving a seismic capacity corresponding to about the 1% non-exceedance probability (NEP) for a specified target response spectrum [4]:

Table 1: Summary of Conservative Deterministic Failure Margin Approach

Load Combination	Normal + Seismic Margin Earthquake (SME)
Ground Response Spectrum	Anchor CDFM Capacity to defined response spectrum shape without consideration of spectral shape variability
Seismic Demand	Perform seismic demand analysis according to ASCE 4 ([7])
Damping	Conservative estimate of median damping
Structural Model	Best Estimate (Median) + uncertainty variation in frequency
Soil-Structure-Interaction (SSI)	Best Estimate (Median) + Parameter Variation
Material Strength	Code specified minimum strength or 95% exceedance actual strength if test data are available.
Static Strength Equations	Code ultimate strength (ACI), maximum strength (AISC), Service Level D (ASME), or functional limits or using 84% exceedance of test data for strength equation.
Inelastic Energy Absorption	For non-brittle failure modes and linear analysis, use appropriate inelastic energy absorption factor from ASCE 43-05 ([6]), or perform nonlinear analysis and go to 95% exceedance ductility levels.
In-Structure (Floor) Spectra Generation	Use frequency shifting rather than peak broadening to account for uncertainty plus use conservative estimate of median damping

4 Simplified Probabilistic Approach

The ground acceleration capacity is a random variable that can be described completely by its probability distribution. However, there is uncertainty in the estimation of the parameters of this distribution, the exact shape of this distribution, and in the appropriate failure model for the structural or mechanical component. For any postulated failure mode and set of parameter values describing the ground acceleration capacity and shape of the probability distribution, a fragility curve depicting the conditional probability of failure as a function of PGA can be obtained (s. Figure 3).

At any acceleration value, the component fragility (i.e., conditional probability of failure) varies from 0 to 1; this variation is represented by a subjective probability distribution. On this distribution we can find a fragility value (say, 0.05) that corresponds to the cumulative subjective probability of 95%. We have 95% cumulative subjective probability (confidence) that the fragility (failure or exceeding probability) is less than 0.05. On the high confidence curve, we can locate the fragility value of 5%; the acceleration corresponding to this fragility on the high confidence curve is the so-called "high-confidence-of-low-probability-of-failure" (HCLPF) capacity of the component. Development of the family of fragility curves using different failure models and parameters for a large number of components in a SPRA is impractical if it is done as described above. Hence, a simple model for the fragility was proposed. In the following section this fragility model is described.

The entire family of fragility curves for an element corresponding to a particular failure mode can be expressed in terms of the best estimate of the median ground acceleration capacity, A_m, and two random variables. Thus, the ground acceleration capacity, A, is given by:

$$A = A_m \cdot e_R \cdot e_U \tag{1}$$

in which e_R and e_U are random variables with median values of 1.0, representing, respectively, the inherent randomness about the median and the uncertainty in the median value. In this model, we assume that both e_R and e_U are lognormally distributed with logarithmic standard deviations, β_R and β_U, respectively. The formulation for fragility given by Eq. (1) and the assumption of a lognormal distribution allow easy development of the family of fragility curves that appropriately represent fragility uncertainty.

With perfect knowledge of the failure mode and parameters describing the ground acceleration capacity (i.e., only accounting for the random variability, β_R), the conditional probability of failure, f_0, for a given PGA level, a, is given by:

$$f_0 = \Phi\left[\ln\left(\frac{a}{A_m}\right) \cdot \frac{1}{\beta_R}\right] \tag{2}$$

where $\Phi[.]$ is the standard Gaussian cumulative distribution of the term in brackets. The relationship between f_0 and a is the median fragility curve plotted in Figure 3 for a component with a median ground acceleration capacity $A_m = 0.87g$ and $\beta_R = 0.25$. For the median conditional probability of failure range of 5% to 95%, the ground acceleration capacity would range from $A_m \cdot \exp(-1.65\ \beta_R)$ to $A_m \cdot \exp(1.65\ \beta_R)$, i.e., 0.58g to 1.31g as shown in Figure 3.

When the modelling uncertainty β_U is included, the fragility becomes a random variable (uncertain). At each acceleration value, the fragility f can be represented by a subjective probability density function. The subjective probability, Q (also known as "confidence") of not exceeding a fragility f' is related to f' by:

$$f' = \Phi\left[\left\{\ln\left(\frac{a}{A_m}\right) + \beta_U \Phi^{-1}(Q)\right\} \cdot \frac{1}{\beta_R}\right] \tag{3}$$

where: $Q = P[f < f'|a]$; i.e., the subjective probability that the conditional probability of failure, f, is less than f' for a PGA a, and $\Phi^{-1}[\ ]$ = the inverse of the standard Gaussian cumulative distribution of the term in brackets.

In estimating fragility parameters, it is convenient to work in terms of an intermediate random variable called the factor of safety. The factor of safety, F, on ground acceleration capacity, A, above a reference level earthquake specified for design; e.g., the safe shutdown earthquake (SSE) level specified for design, A_{SSE}, is defined as follows:

$$A = F \cdot A_{SSE} \rightarrow F = \frac{\text{Actual seismic capacity of element}}{\text{Actual response due to SSE}} \tag{4}$$

This relationship is typically expanded to identify the conservatism or factor of safety in both the strength and the response.

$$F = \frac{\text{Actual capacity}}{\text{Design response due to SSE}} \cdot \frac{\text{Design response due to SSE}}{\text{Actual response due to RE}}$$
$$F = F_C \cdot F_{SR} \tag{5}$$

where F_C is the capacity factor, F_{SR} is the structural response factor and RE is the reference earthquake spectrum derived from the probabilistic hazard study, anchored to the same PGA as the SSE.

The median factor of safety, F_m, can be directly related to the median ground acceleration capacity, A_m, as:

$$F_m = \frac{A_m}{A_{SSE}} \tag{6}$$

The logarithmic standard deviations of F, representing inherent randomness and uncertainty, are identical to those for the ground acceleration capacity A.

In seismic margin studies, an index of seismic margin is the HCLPF capacity of the component. This quantity considers both the uncertainty and randomness variabilities and is the acceleration value for which the analyst has 95% confidence that the failure probability is less than 5%. For example, Figure 3 shows a HCLPF of 0.32g for a fragility description of $A_m = 0.87g$, $\beta_R = 0.25$, $\beta_U = 0.35$. That is, it is an acceleration value for the component for which we are highly confident there is only a small chance of failure given this ground acceleration level:

$$\text{HCLPF Capacity} = A_m \exp\{-1.65(\beta_R + \beta_U)\} \tag{7}$$

The HCLPF capacity is approximately defined as: a 1% conditional probability of failure (-2.33 log standard deviation below the mean), where β_C is the composite variability [3].

$$\text{HCLPF Capacity} = A_m \exp(-2.33\beta_C) \tag{8}$$

5 Plant Level Fragility

It is sometimes useful to develop the plant level fragility curves. They depict the conditional probability of failure / collaps (or other damage indicators) for different levels of ground motion input. The plant level fragility curves can be generated by quantifying the accident sequences consisting of component and structural successes and failures. By entering the plant level fragility curves corresponding to 95% confidence at 5% conditional probability of failure, the plant HCLPF capacity can be obtained. In this case the plant HCLPF capacity is determined from the detailed modelling of the plant systems and structures responses for an earthquake [3].

6 Monte Carlo Method and Latin Hypercube Procedure

Monte Carlo (MC) sampling refers to the traditional technique for using random or pseudo-random numbers to sample from a probability distribution. MC sampling techniques are entirely random – that is, any given sample may fall anywhere within the range of the input distribution. Samples, of course, are more likely to be drawn in areas of the distribution which have higher probabilities of occurrence. To include the effects of the low probability outcomes, a large number of MC iterations have to be performed. Otherwise the impact of the values in the outer ranges of the distribution is not included in the simulation output [10].

The Latin Hypercube (LH) procedure ensures that the full ranges of uncertainties of important variables are utilized but requires considerably fewer simulations than the classic MC simulation procedure, which usually requires thousands of simulations. The techniques being used during LH sampling is "sampling without replacement". The number of stratifications of the cumulative distributions is equal to the number of iterations performed. However, once a sample is taken from a stratification, this stratification is not sampled from again – its value is already represented in the sampled set [4].

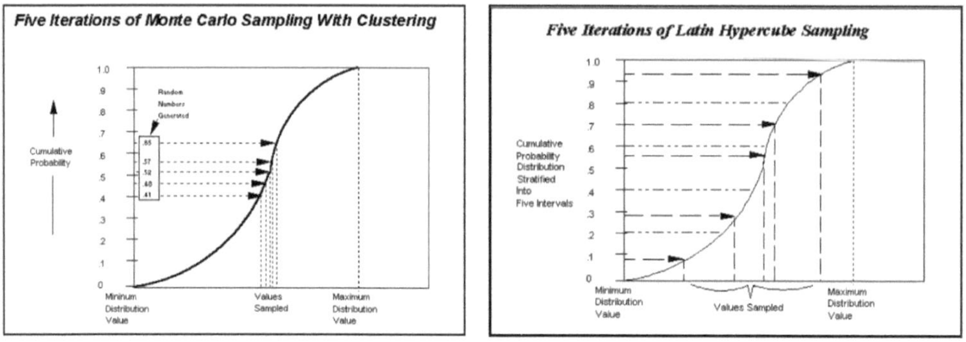

Figure 4: Monte Carlo Sampling / Latin Hypercube Sampling [10]

Probabilistic response is required for fragility analysis. When probabilistic response analysis is conducted for the development of response spectra or structural loads, all the important variables that affect the structural response are included. This probabilistic response analysis is based on a LH stratified sampling simulation process that requires significantly fewer simulations (about 30) than a MC process. In this approach, the variables that affect response are assumed to be lognormally distributed and the probability distribution of the variable is broken up into equal parts, equal to the number of simulations. Combinations of each variable are randomly selected for inclusion in an analysis. Once value of a variable is selected, it is not used again. In this manner it is assured that the 30 or so simulations include the total distribution defined for each variable. Statistics are then applied to the results (e. g. response spectra) in order to define median and 84th percentile response spectra [3]. The variation of results (also in spectral shape) is simulated by utilizing 30 scaled natural and synthetic time histories with median and 84th percentile response spectra ordinates that match the median and 84th percentile ground motion spectra - Uniform Hazard Spectra (UHS). In developing the time histories, the UHS may be first modified to incorporate ground motion incoherence (GMI) effects and high frequency spectral reduction to account for limited ductility of components. Other variables included in the probabilistic analysis are structural stiffness, structural damping, soil stiffness and soil damping.

7 Application of the simplified probabilistic approach on a frame structure

The application of the forementioned simplified probabilistic approach on a framed structure is decribed in the following.

The structure is a five storey building, which is mainly a reinforced concrete frame in cross direction. In the longitudinal direction the building is stiffened by two reinforced concrete shear walls. Therefore, the lateral earthquake and wind loads in the longitudinal direction can be resisted by these shear walls. The main structural system in cross direction is shown in Figure 5. This frame will be used for the nonlinear dynamic analysis.

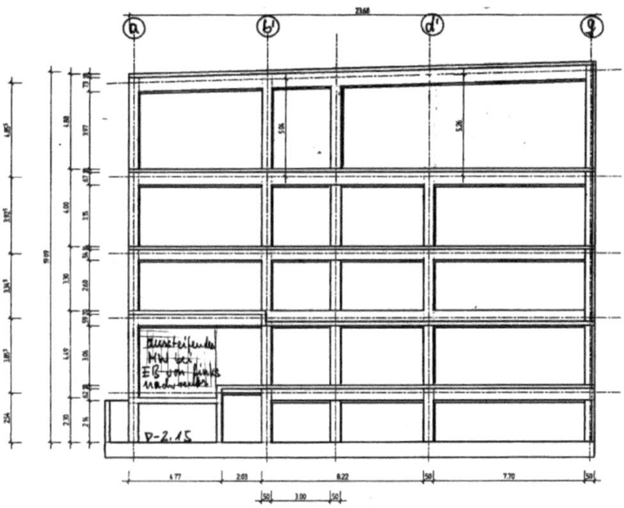

Figure 5: Framed structure in the axis 150

For the nonlinear time history analysis, 30 sets of time histories (TH) are used to represent the reference earthquake, which has a median PGA equal to 0.337g (horizontal) and 0.204g (vertical) with a probability of exceedance of 10^{-4}/a and a damping of 5%.

7.1 Latin Hypercube Variations

As described in the section above, the LH Sampling technique is more advanced and efficient than Direct MC Simulation (DMCS) methods. Using the methodology of LH Variation for the probabilistic analysis, five parameters are taken into account:

- Seismic excitation with 30 combinations of horizontal and vertical time histories covering the whole spectrum of seismic load variations including seismic load magnitude, seismic source, directional and spatial distribution of seismic loads for an given probability of exceedance, herein 10^{-4}/a.
- The seismic capacity of a building: The capacity depends on many parameters (e.g. nonlinear behaviour and ductility of materials, actual strength and overstrength of materials, influence of non-structural elements, etc.). Herein four parameters are taken into account. These are concrete strength, f_c, Young's modulus of concrete, E_c, steel strength of reinforcement, f_s, and damping of the structure, D. The median values and the corresponding variations of these parameters have been determined from available tests.

- Table 2 shows the resulting scaling factors which are determined by the LH Method. These scaling factors are applied on the median model.

Table 2: Latin Hypercube Sampling

	f_c	E_c	f_s	D
Contribution	lognormal	lognormal	lognormal	lognormal
median	1.00	1.00	1.00	1.00
COV	0.14	0.5	0.06	0.35
1	0.92775	0.51740	0.92715	0.93833
2	0.93711	1.13528	0.93968	0.36489
3	0.94301	0.82401	0.88220	0.94837
...	...	...	...	...
28	0.86455	1.27017	0.95632	0.75999
29	0.72856	0.61368	1.03930	1.59551
30	0.81336	0.67313	1.05451	0.99911

7.2 Capacity factor of safety

The capacity factor of safety is estimated as the product of the strength factor times the inelastic energy absorption factor. Based on all results of the nonlinear time history analysis, the capacity factor of safety, F_C, was estimated to have a median value of 3.6 with a logarithmic standard deviation of 0.23.

7.3 Building response factor

The structure response factor, F_{SR}, is modelled as a product of factors influencing the response variability:

$$F_{SR} = F_{SA} \cdot F_{\delta} \cdot F_M \cdot F_{MC} \cdot F_{EC} \cdot F_{GMI} \cdot F_{SSI} \qquad (9)$$

where F_{SA} accounts for the difference between the safe shutdown earthquake and the reference earthquake spectrum from probabilistic hazard study, F_{δ} accounts for the effects of actual damping versus design damping, F_M accounts for the effects of dynamic modelling uncertainty, F_{MC} represents response effects introduced by combination of modes, F_{EC} represents the effects of earthquake component combination, F_{GMI} accounts for the fact that a travelling seismic wave does not excite a large foundation uniformly, and F_{SSI} represents the effects of SSI.

7.3.1 Structural effects

For the analysis, 30 sets of TH representing the site-specific response spectra are used. Therefore, median value for spectral shape F_{SA} is 1.0. The time histories in horizontal and vertical direction are applied simultaneously. Therefore, the median value for the earthquake component combination F_{EC} is 1.0. The corresponding variabilities, β_R and β_U, equal to 0. The distribution of damping and material parameters (concrete strength, Young's modulus of concrete, and steel strength) are covered by the 30 analysed models using nonlinear direct-integration time-history analysis. Thus, the median values for damping factor F_δ, and mode combination factor F_{MC} are equal to 1.0 and their variablities are equal to 0. The median value of the modelling factor F_M is 1.0 and the corresponding variability β_U is 0.

7.3.2 Soil-structure interaction effect

The interaction between the structure and the supporting foundation includes consideration of ground motion incoherence, vertical spatial variation of ground motion, and soil-structure-interaction analysis. In general all of these have an influence on the response of structures at soil sites, while only ground motion incoherence has a significant effect at stiff rock sites. Here, it is assumed that the structure lays on very stiff soil, therefore a fixed-base analysis of the structure is used. The effect of ground motion incoherence in reducing the seismic excitation of the foundation has been characterized by a function of foundation size and the frequency. According to EPRI [2] the reduction factor is conservatively 1.0 for structures with a fundamental frequency below 5.0 Hz. The uncertainty β_U is 0.0.

7.4 HCLPF - Results of frame structure

For structures, the factor of safety consists of a capacity factor, F_C, and a structure response factor, F_{RS}, see Eq. (5). The median factor of safety F_m for the analysed frame structure subjected to design basis earthquake (0.337g) is found to be 3.6, the corresponding composite variability β_C is 0.23. Hence, the median ground acceleration capacity, A_m is 1.21g. As shown by eq. (5), the HCLPF value can be estimated to be:

$$\text{HCLPF Capacity} = A_m \exp(-2.33\beta_C) = 1.21\text{g} \cdot \exp(-2.33 \cdot 0.23) = 0.71\text{g}$$

8 Conclusion

A simple and efficient probabilistic approach to estimate the failure probabilities and safety margins of existing and new industrial facilities is presented and discussed in this paper. It covers the whole industrial facility including structures, components, mechanical installations, piping, tanks, etc.

In comparison to sophisticated Monte Carlo simulations, the presented method is cost-effective and practical and can be used for risk-informed/performance-based rehabilitation or strengthening. The required effort is relatively low compared to other probabilistic approaches and the results can be explained and compared easily.

It is also a useful tool to identify weak links in a system or whole facility. In this context, it can guide the efficient allocation of funds to strengthen or modify an existing industrial facility. It also can be used as part of the design process to size members to comply with a given performance standard.

REFERENCES

[1] EPRI: A Methodology for Assessment of Nuclear Power Plant Seismic Margin, EPRI NP-6041SL, Revision 1, Electric Power Research Institute (EPRI), Palo Alto, California, June 1991

[2] EPRI: Methodology for Developing Seismic Fragilities, EPRI TR 103959, Electric Power Research Institute (EPRI), Palo Alto, California, June 1994

[3] EPRI: Seismic Fragility Application Guide, EPRI 1002988, Electric Power Research Institute (EPRI), Palo Alto, California, December 2002

[4] EPRI: Seismic Fragility Application Guide Update, EPRI 1019200, Electric

[5] Kennedy, R. P. et al.: Probabilistic Seismic Safety Study of an Existing Nuclear Power Plant, Nuclear Engineering and Design, 59 (1980), Pages 315-338, North-Holland Publishing Company

[6] ASCE: Seismic Design Criteria for Structures, Systems, and Components in Nuclear Facilities, ASCE/SEI 43-05, American Society of Civil Engineers, 2005

[7] ASCE: Seismic Analysis of Safety-Related Nuclear Structures and Commentary, ASCE 4-98, American Society of Civil Engineers, 2000

[8] FEMA 356, Prestandard and commentary for the seismic rehabilitation of buildings, November 2000

[9] SAP 2000: Three Dimensional Static and Dynamic Finite Element Analysis and Design of Structures, Version 15.1.0, Computers and Structures Inc., Berkeley, CA , 2011

[10] @RISK: Risk Analysis Add-in for Microsoft Excel, Version 6.0.1: Industrial Edition, Palisade Corporation, West Drayton, 2012

Uncertainty Propagation in Engineering Systems: Probability Density Evolution Theory and Its Recent Developments

Jie Li[1]

[1] School of Civil Engineering, Tongji University
1239 Siping Rd, Shanghai, P. R. China
lijie@tongji.edu.cn

ABSTRACT:

Developments of modern science and technology have greatly enhanced the ability of engineering community in understanding the phenomena, mechanism and performance of engineering structures and systems. Meanwhile, the defect and inadequacy of deterministic methodologies in modelling and analysis of engineering systems also expose the importance of uncertainty analysis. As a matter of fact, it is recognized more and more clearly that the randomness propagation in a physical system plays an important role in understanding and controlling many phenomena and behaviours of engineering structures and systems, particularly those emerging in nonlinear mechanics and systems.

On the basis of the principle of probability preservation and its random event description, a new kind of general probability density evolution equation (GPDEE) is introduced which could capture the randomness propagation in a dynamical system. Then this kind of equation is extended to general physical systems and therefore reveals the essence of randomness propagation in a physical system. Some recent developments using GPDEE are summarized, including: (1) the physical random models for dynamic excitations, especially taking seismic ground motion as an example; (2) the multi-scale stochastic damage model for concrete materials and structures; (3) a physical approach to the global reliability of structures, respectively. Besides, some typical engineering applications are illustrated as well.

Keywords: physical system, randomness propagation, probability preservation, general probability density evolution equation, engineering application

1 Introduction

Stochastic dynamical systems have been studied in mathematics, physics, chemises and many engineering disciplines for over a century and their developments have greatly enhanced the ability of humans in understanding the phenomena, mechanism and performance of engineering systems. Meanwhile, in the process of developing approaches, people recognized more and more clearly that the randomness propagation in a physical system plays an important role in understanding and controlling many phenomena and behaviors of engineering systems, particularly those emerging in nonlinear mechanics and systems.

It is generally believed that the stochastic dynamics is originally from Albert Einstein's investigation on the Brownian motion. In 1905, Einstein induced the irregular collisions between the molecules and the Brownian particles and deduced the evolution equation of the density of the particles and found that this equation belongs to the diffusion equation [1]. This thought was then boosted by Fokker in 1914 and by Planck in 1917, leading to the probability density equation well known as the Fokker-Planck equation in the physicist circle [2][3]. In 1931, the Soviet mathematician Kolmogorov derived the same equation independently; simultaneously he gave a backward equation. This investigation set a rigorous mathematical foot for the equation [4]. Thus, the Fokker-Planck equation is also referred to as the Fokker-Planck-Kolmogorov (FPK) equation. It is noticeable that although Einstein started with the physical mechanism of irregular collisions of the molecules, the crux of his tactics is to view the evolution of the particle group in a phenomenological way. Due to the Kolmogorov's work, the analysis of stochastic dynamical systems can be transformed to the problem of a deterministic partial differential equation. Afterwards, far more emphasis was put on the mathematical aspects than on the physical aspects. On this background, the methodology originated from Einstein, along the path of Einstein-Fokker-Planck-Kolmogorov might be referred to as the phenomenological tradition in the studies of stochastic dynamical systems.

On the other hand, to study the Brownian motion, Langevin applied the Newton's law to a single Brownian particle almost simultaneously [5]. In Langevin investigation, the resultant force induced by the collisions of the around molecules in the fluids becomes an irregular (random) force acted on the Brownian particle. It is interesting that although along a way completely different from Einstein's, with some simple assumptions on the nature of the irregular forces, Langevin obtained the dissipation-diffusion relation identical to Einstein's in a much more concise and straight forward manner. This result is so impressing that scientists believed that Langevin's method is an effective and independent new method although the assumptions on the irregular force were somewhat strange. In the early 1920s, Wiener studied the features of Brownian motion deeply [6], which built the foot for correctly understanding the meanings of Langevin's assumption. In the early and middle 1940s, Itô made systematic studies on stochastic processes and stochastic integral, resulted in rigorous definition of the Itô calculus [7][8]. This clarified the meanings of Langevin's random forces and the related operations, demonstrating that the Langevin's forces can be modeled by the mathematical white noise. In the early 1960s, Stratonovich came up with the physical interpretation of the white noise [9]. In the methodology originated from Langevin, along the path of Langevin-Itô-Stratonovich, the stochastic differential equations arising from physical laws are the central entities. Thus, it is reasonable to refer to this methodology as the physical tradition in the studies of stochastic dynamical systems [10].

For the studies of multi-dimensional nonlinear stochastic dynamical systems, the comprehensive understanding of the above two methodologies are in need, based on which a new path should be developed. In the investigations, we found that it is the evolution of the physical state induced by the physical laws results in the probability density evolution of a stochastic system [11]. This understanding or new finding therefore established a direct relationship between general physical system and stochastic systems. It is based on the clarification of the principle of preservation of probability, deeper understanding of the traditional probability density equations is achieved and then a family of generalized density evolution equation is reached. This widens the way of studying the probability density evolution analysis of nonlinear stochastic dynamical system [12][13][14].

2 Generalized probability density evolution equation

2.1 The random event description of the principle of preservation of probability

For convenience, consider an n-dimensional stochastic dynamical system

$$\dot{\boldsymbol{Y}} = \boldsymbol{A}(\boldsymbol{Y}, t), \boldsymbol{Y}(t_0) = \boldsymbol{Y}_0 \tag{1}$$

where $\boldsymbol{Y} = (Y_1, Y_2, \dots, Y_n)^T$ is the n-dimensional state vector, $\boldsymbol{Y}_0 = (Y_{0,1}, Y_{0,2}, \dots, Y_{0,n})^T$ is the corresponding initial vector, $\boldsymbol{A}(\cdot)$ is a deterministic operator vector. Evidently, in the case $\boldsymbol{Y}_0$ is a random vector, $\boldsymbol{Y}(t)$ will be a stochastic process vector.

The state equation (1) essentially establishes a mapping from $\boldsymbol{Y}_0$ to $\boldsymbol{Y}(t)$, which can be expressed as

$$\boldsymbol{Y}(t) = \mathcal{g}(\boldsymbol{Y}_0, t) = G_t(\boldsymbol{Y}_0) \tag{2}$$

Note that $\boldsymbol{Y}_0$ is a random vector, thus $\{\boldsymbol{Y}_0 \in \Omega_0\}$ is a random event. Here Ω_0 is an arbitrary domain in the distribution domain of $\boldsymbol{Y}_0$. According to the stochastic state equation (1), $\boldsymbol{Y}_0$ will be changed to $\boldsymbol{Y}(t)$ at time t. The domain Ω_0 to which $\boldsymbol{Y}_0$ belongs at time t_0 is accordingly changed to Ω_t to which $\boldsymbol{Y}(t)$ belongs at time t (Figure 1), i.e.

$$\Omega_t = \mathcal{g}(\Omega_0, t) = G_t(\Omega_0) \tag{3}$$

Hence, the random event $\{\boldsymbol{Y}_0 \in \Omega_0\}$ is represented as $\{\boldsymbol{Y}(t) \in \Omega_t\}$ at time t. In other words, because in the evolution process there are no new random sources,

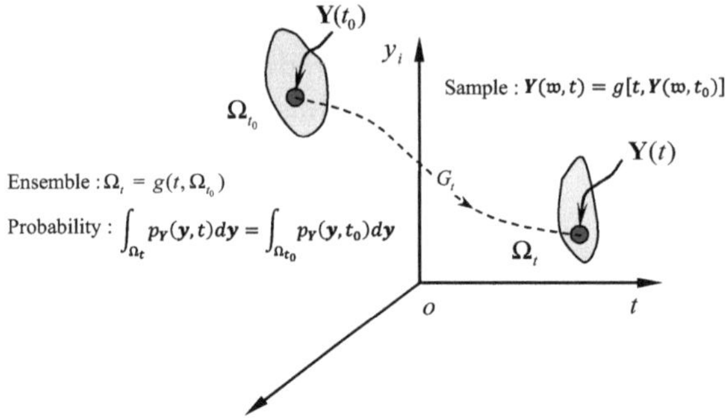

Figure 1: Dynamical system, mapping and probability evolution

$\{\boldsymbol{Y}_0 \in \Omega_0\}$ and $\{\boldsymbol{Y}(t) \in \Omega_t\}$ are essentially the same random event at different time, consequently, the probability of the random event must be identical, i.e.

$$\Pr\{\boldsymbol{Y}_0 \in \Omega_0\} = \Pr\{\boldsymbol{Y}(t) \in \Omega_t\} \tag{4}$$

where $\Pr\{\cdot\}$ denotes the probability of a random event.

Denote the joint probability density function (PDF) of $\boldsymbol{Y}_0$ by $p_{\boldsymbol{Y}_0}(\boldsymbol{y}_0)$, of $\boldsymbol{Y}(t)$ by $p_{\boldsymbol{Y}}(\boldsymbol{y}, t)$, in which $\boldsymbol{y}_0 = \left(y_{0,1}, y_{0,2}, \ldots, y_{0,n}\right)^{\mathrm{T}}$, $\boldsymbol{y} = (y_1, y_2, \ldots, y_n)^{\mathrm{T}}$, then Equation (4) means that

$$\int_{\Omega_0} p_{\boldsymbol{Y}_0}(\boldsymbol{y}_0)\mathrm{d}\boldsymbol{y}_0 = \int_{\Omega_t} p_{\boldsymbol{Y}}(\boldsymbol{y}, t)\mathrm{d}\boldsymbol{y} \tag{5}$$

To be clearer, denoting Ω_0 by Ω_{t_0} and noting that $p_{\boldsymbol{Y}}(\boldsymbol{y}, t_0) = p_{\boldsymbol{Y}_0}(\boldsymbol{y})$, Equation (5) becomes

$$\int_{\Omega_{t_0}} p_{\boldsymbol{Y}}(\boldsymbol{y}, t_0)\mathrm{d}\boldsymbol{y} = \int_{\Omega_t} p_{\boldsymbol{Y}}(\boldsymbol{y}, t)\mathrm{d}\boldsymbol{y} \tag{6}$$

Evidently, the above equation also holds at the time $t + \Delta t$, which will then result in

$$\frac{\mathrm{D}}{\mathrm{D}t}\int_{\Omega_t} p_{\boldsymbol{Y}}(\boldsymbol{y}, t)\mathrm{d}\boldsymbol{y} = 0 \tag{7}$$

where $\mathrm{D}(\cdot)/\mathrm{D}t$ denotes the total derivative.

It should be stressed here that in Equation (7) both the integrand $p_{\boldsymbol{Y}}(\boldsymbol{y}, t)$ and the integral domain Ω_t are time variant. This can be seen clearly from Equation (4), while the underlying reason is that the evolution of $\boldsymbol{Y}(t)$ is governed by the state equation (1). Therefore, the exact meaning of the total derivative $\mathrm{D}(\cdot)/\mathrm{D}t$ is that

$$\frac{\mathrm{D}}{\mathrm{D}t}\int_{\Omega_t} p_{\boldsymbol{Y}}(\boldsymbol{y}, t)\mathrm{d}\boldsymbol{y}$$
$$= \lim_{\Delta t \to \infty} \frac{1}{\Delta t}\left(\int_{\Omega_{t+\Delta t}} p_{\boldsymbol{Y}}(\boldsymbol{y}, t + \Delta t)\mathrm{d}\boldsymbol{y} - \int_{\Omega_t} p_{\boldsymbol{Y}}(\boldsymbol{y}, t)\mathrm{d}\boldsymbol{y}\right) \tag{8a}$$

or equivalently

$$\frac{\mathrm{D}}{\mathrm{D}t}\int_{\Omega_t} p_{\boldsymbol{Y}}(\boldsymbol{y},t)\mathrm{d}\boldsymbol{y}$$

$$= \lim_{t'\to t}\frac{1}{t'\to t}\left(\int_{\Omega_{t'}} p_{\boldsymbol{Y}}(\boldsymbol{y}',t')\mathrm{d}\boldsymbol{y}' - \int_{\Omega_t} p_{\boldsymbol{Y}}(\boldsymbol{y},t)\mathrm{d}\boldsymbol{y}\right) \tag{8b}$$

Equation (7) is clearly the embodiment of the principle of preservation of probability in the stochastic dynamical systems. Since it is the result from the perspective that **the probability of a random event is invariant**, we refer to it as the random event description of the principle of preservation of probability [14].

Since a random event could be a compound event consisting of elementary events, i.e. the random events satisfy the σ - algebra, hence there is a possibility of decomposing a random event. It is this possibility that makes the way to view a physical problem in an uncoupled manner whereas the probability is still preserved.

2.2 From the equations of motion to uncoupled physical equations

Consider the equation of motion of a MDOF system

$$\boldsymbol{M}(\boldsymbol{\eta})\ddot{\boldsymbol{X}} + \boldsymbol{C}(\boldsymbol{\eta})\dot{\boldsymbol{X}} + \boldsymbol{f}(\boldsymbol{\eta},\boldsymbol{X}) = \boldsymbol{\Gamma}\boldsymbol{\xi}(t) \tag{9}$$

where $\boldsymbol{M}$ and $\boldsymbol{C}$ are the $n_{\mathrm{d}} \times n_{\mathrm{d}}$ mass and damping matrices, respectively, $\boldsymbol{f}$ is the n_{d} -dimensional restoring force vector, $\boldsymbol{\eta} = \eta_1, \eta_2, \dots, \eta_{S_1}$ are the random parameters characterizing the randomness involved in the physical properties of the system, $\boldsymbol{X}$, $\dot{\boldsymbol{X}}$ and $\ddot{\boldsymbol{X}}$ are the n_{d}-dimensional displacement, velocity and acceleration vectors, respectively, $\xi(t)$ is the r-dimensional excitation vector, $\boldsymbol{\Gamma}$ is the $n_{\mathrm{d}} \times r$ load influence matrix; for instance, if $\xi(t)$ is a one-dimensional ground motion acceleration $\ddot{x}_g(t)$, then $\boldsymbol{\Gamma} = -\boldsymbol{MI}$ where $\boldsymbol{I} = \{1,1,\dots,1\}^T$ is an n_{d}-dimensional column vector. Clearly, in the case $\boldsymbol{f}(\boldsymbol{X}) = \boldsymbol{KX}$ where $\boldsymbol{K}$ is a $n_{\mathrm{d}} \times n_{\mathrm{d}}$ stiffness matrix Equation (9) is a linear system. For simplicity, we consider the case involving only one-dimensional random excitation. Now Equation (9) becomes

$$\boldsymbol{M}(\boldsymbol{\eta})\ddot{\boldsymbol{X}} + \boldsymbol{C}(\boldsymbol{\eta})\dot{\boldsymbol{X}} + \boldsymbol{f}(\boldsymbol{\eta},\boldsymbol{X}) = \boldsymbol{\Gamma}\xi(t) \tag{10}$$

In the modeling of the stochastic dynamic excitations such as earthquakes, strong wind and sea waves, the concept of physical stochastic process can be employed and from which a rational stochastic physical model could be derived [15][16].

For general stochastic processes or random fields, the Karhunen-Loeve decomposition can be adopted to represent them as combinations of random functions [17]. Investigations show that most stochastic processes can be reasonably represented with only several terms by employing an approach by combining the orthogonal expansion using a family of Hartley functions with the decomposition of the covariance matrix [18].

$$\xi(\boldsymbol{\zeta},t) = \sum_{j=1}^{S_2} \zeta_j\sqrt{\lambda_j}\, f_j(t) \tag{11}$$

where $\boldsymbol{\zeta} = (\zeta_1, \zeta_2, \dots, \zeta_{s_2})$ are uncorrelated random variables, i.e. $E[\zeta_i \zeta_j] = \delta_{ij}$, δ_{ij} is the Kronecker delta, $\sqrt{\lambda_j}\, f_j(t)$ are deterministic functions.

Recently, a new stochastic harmonic function was presented to approach general stochastic processes or random fields [19].

For the consistency of the symbols, denote

$$\boldsymbol{\Theta} = (\boldsymbol{\eta}, \boldsymbol{\zeta}) = (\eta_1, \eta_2, \dots, \eta_{s1}, \zeta_1, \zeta_2, \dots, \zeta_{s_2}) = (\Theta_1, \Theta_2, \Theta_s) \tag{12}$$

in which $s = s_1 + s_2$ is the total number of basic random variables involved in the system.

Then equation (10) can be rewritten as

$$\boldsymbol{M}(\boldsymbol{\Theta})\ddot{\boldsymbol{X}} + \boldsymbol{C}(\boldsymbol{\Theta})\dot{\boldsymbol{X}} + \boldsymbol{f}(\boldsymbol{\Theta}, \boldsymbol{X}) = \mathbf{F}(\boldsymbol{\Theta}, t) \tag{13}$$

It should be noticed that, although the randomness in the initial condition is not considered here, in the case the initial conditions are random, the random variables can be introduced into $\boldsymbol{\Theta}$. Hence, in the following it is subsumed that all the randomness involved in the initial conditions, system properties and excitations have been taken into account in Equation (13). In other words, for the dynamical system (13), the randomness is treated in a unified way. This is in contrast to what have been done in the Liouville system or the Itô system, where the randomness is separately treated according to the phenomenological different sources.

Generally, most physical systems in engineering are well-posed. For such systems, the solutions exist, and are unique and dependent continuously on the system parameters and initial conditions. In this case, for the system (13), the solution $\boldsymbol{X}(t)$ must depend on and be a function of $\boldsymbol{\Theta}$, and can thus be denoted by[1]

$$\boldsymbol{X}(t) = \boldsymbol{G}(\boldsymbol{\Theta}, t) \tag{14a}$$

of which the scalar form can be written by

$$\boldsymbol{X}_l(t) = \boldsymbol{G}_l(\boldsymbol{\Theta}, t), l = 1, 2, \dots, n_{\mathrm{d}} \tag{14b}$$

Likewise, the velocity is also a function of $\boldsymbol{\Theta}$,

$$\dot{\boldsymbol{X}}(t) = \boldsymbol{H}(\boldsymbol{\Theta}, t) \tag{15}$$

Evidently, there exists $\boldsymbol{H}(\boldsymbol{\Theta}, t) = \partial \boldsymbol{G}(\boldsymbol{\Theta}, t)/\partial t$.

In engineering practice, usually not only the displacements, velocities and accelerations of the structure are of interest, but also are some other important physical quantities such as the stress and strain at critical points, the internal forces and deformations at critical sections, etc. Generally speaking, these physical quantities are determined once the states of the structure (displacements and

[1] If the initial conditions are deterministic, they need not explicitly occur for simplicity of writing. In the case the initial conditions are random, the randomness can be involved in $\boldsymbol{\Theta}$.

velocities) are known [20]. For instance, the strain at some point can be obtained as the partial derivative of the displacement. Denote the physical quantity of interest by $\boldsymbol{Z} = (Z_1, Z_2, \dots, Z_m)^T$, then

$$\dot{\boldsymbol{Z}}(t) = \psi[\boldsymbol{X}(t), \dot{\boldsymbol{X}}(t)] \tag{16}$$

Where $\psi(\cdot)$ is the transform operator from the state vector to the target physical quantities. It is linear for linear structures, whereas for nonlinear structures, it might either be a linear or nonlinear operator. For example, if $\boldsymbol{Z}$ is the strain at some point, then if small deformation is considered, even if material nonlinearity is involved $\psi(\cdot)$ is a linear operator, whereas if geometrical nonlinearity is involved, then even if the material is linear, $\psi(\cdot)$ is a nonlinear operator. Specifically, if $\boldsymbol{Z}$ is the displacements at some nodes, then $\psi(\cdot)$ is the selection operator, a matrix in which only a few elements are 1 whereas all the others are zeros.

Inserting Equations (14) and (15) into (16) yields

$$\dot{\boldsymbol{Z}}(t) = \psi[\boldsymbol{G}(\boldsymbol{\Theta}, t), \boldsymbol{H}(\boldsymbol{\Theta}, t)] = \mathbf{h}(\boldsymbol{\Theta}, t) \tag{17a}$$

Due to the randomness of $\boldsymbol{\Theta}$, this is a stochastic state equation, of which the components are

$$\dot{Z}_1(t) = h_1(\boldsymbol{\Theta}, t), l = 1,2, \dots, m \tag{17b}$$

It should be stressed that the state equation (17a, b) is an uncoupled equation, i.e. to delineate the physical quantities of interest separately other than to view them together with the coupling state vector.

2.3 Generalized Probability density evolution equation

As shown above, for the stochastic dynamical system (9), what is really of concern is the physical quantity $\boldsymbol{Z}(t)$, while $\boldsymbol{Z}(t)$ itself satisfies the stochastic state equation (17). Hence, to capture the probabilistic information of $\boldsymbol{Z}(t)$, we will start with Equation (17) directly instead of (9).

Consider a random event $\{(\boldsymbol{Z}(t), \boldsymbol{\Theta}) \in \Omega_t \times \Omega_\theta\}$, where Ω_Θ is an arbitrary domain in the distribution domain of $\boldsymbol{\Theta}$, Ω_t is a domain at time t in the distribution domain of $\boldsymbol{Z}$. After a short time $\mathrm{d}t$, at the time instant $t + \mathrm{d}t$, this random event becomes $\{(\boldsymbol{Z}(t + \mathrm{d}t), \boldsymbol{\Theta}) \in \Omega_{t+\mathrm{d}t} \times \Omega_\theta\}$. Clearly,

$$\begin{aligned} &Pr\{(\boldsymbol{Z}(t), \boldsymbol{\Theta}) \in \Omega_t \times \Omega_\theta\} \\ &= Pr\{(\boldsymbol{Z}(t + \mathrm{d}t), \boldsymbol{\Theta}) \in \Omega_{t+\mathrm{d}t} \times \Omega_\theta\} \end{aligned} \tag{18}$$

i.e.

$$\int_{\Omega_t \times \Omega_\theta} p_{\boldsymbol{Z\Theta}}(\boldsymbol{z}, \boldsymbol{\theta}, t) \mathrm{d}\boldsymbol{z} \mathrm{d}\boldsymbol{\theta} = \int_{\Omega_{t+\mathrm{d}t} \times \Omega_\theta} p_{\boldsymbol{Z\Theta}}(\boldsymbol{z}, \boldsymbol{\theta}, t + \mathrm{d}t) \mathrm{d}\boldsymbol{z} \mathrm{d}\boldsymbol{\theta} \tag{19}$$

Simultaneously, the domain $\Omega_{t+\mathrm{d}t}$ is the superposition of the domain Ω_t and the motion of the boundary, i.e.

$$\Omega_{t+\mathrm{d}t} = \Omega_t + \int_{\partial\Omega_t} (\boldsymbol{v}\mathrm{d}t) \cdot \boldsymbol{n}\mathrm{d}S$$

$$= \Omega_t + \int_{\partial\Omega_t} (\boldsymbol{h}(\boldsymbol{\theta}, t)\mathrm{d}t) \cdot \boldsymbol{n}\mathrm{d}S \tag{20}$$

Note that use has been made of the velocity determined by the physical equation (17a), again demonstrating that the evolution of the probability density is the result of the evolution of the physical system.

It is also seen that no matter whether Ω_t is dependent on Ω_θ or not, Ω_{t+dt} is dependent on Ω_θ. Thus, generally , for $t \neq t_0$, Ω_t should be dependent on Ω_θ. Hence, rigorously, Ω_t should be written as $\Omega_t(\Omega_\theta)$. This is also why to make sure the probability is preserved the augmented system $(\boldsymbol{Z}(t), \boldsymbol{\Theta})$ should be examined instead of the evolution of the original system $\boldsymbol{Z}(t)$.

Inserting Equations (20) into (19) and examining the right hand side yield

$$\int_{\Omega_{t+\mathrm{d}t}\times\Omega_\theta} p_{\boldsymbol{Z\Theta}}(\boldsymbol{z}, \boldsymbol{\theta}, t + \mathrm{d}t)\mathrm{d}\boldsymbol{z}\mathrm{d}\boldsymbol{\theta}$$

$$= \int_{\Omega_t\times\Omega_\theta} \left(p_{\boldsymbol{Z\Theta}}(\boldsymbol{z}, \boldsymbol{\theta}, t) + \frac{\partial p_{\boldsymbol{Z\Theta}}(\boldsymbol{z},\boldsymbol{\theta},t)}{\partial t}\mathrm{d}t\right)\mathrm{d}\boldsymbol{z}\mathrm{d}\boldsymbol{\theta}$$

$$+ \int_{\partial\Omega_t\times\Omega_\theta} \left(p_{\boldsymbol{Z\Theta}}(\boldsymbol{z}, \boldsymbol{\theta}, t) + \frac{\partial p_{\boldsymbol{Z\Theta}}(\boldsymbol{z},\boldsymbol{\theta},t)}{\partial t}\mathrm{d}t\right)(\boldsymbol{h}(\boldsymbol{\theta}, t)\mathrm{d}t) \cdot \boldsymbol{n}\mathrm{d}S\mathrm{d}\boldsymbol{\theta} \tag{21}$$

where $p_{\boldsymbol{Z\Theta}}(\boldsymbol{z}, \boldsymbol{\theta}, t + \mathrm{d}t) = p_{\boldsymbol{Z\Theta}}(\boldsymbol{z}, \boldsymbol{\theta}, t) + (\partial p_{\boldsymbol{Z\Theta}}(\boldsymbol{z}, \boldsymbol{\theta}, t)/\partial t)\mathrm{d}t$ has been used.

Substituting Equation (21) into the right hand side of (19) and canceling the identical terms give

$$\int_{\Omega_t\times\Omega_\theta} \left(\frac{\partial p_{\boldsymbol{Z\Theta}}(\boldsymbol{z},\boldsymbol{\theta},t)}{\partial t}\mathrm{d}t\right)\mathrm{d}\boldsymbol{z}\mathrm{d}\boldsymbol{\theta}$$

$$= -\int_{\Omega_t\times\Omega_\theta} \left(p_{\boldsymbol{Z\Theta}}(\boldsymbol{z}, \boldsymbol{\theta}, t) + \frac{\partial p_{\boldsymbol{Z\Theta}}(\boldsymbol{z},\boldsymbol{\theta},t)}{\partial t}\mathrm{d}t\right)(\boldsymbol{h}(\boldsymbol{\theta}, t)\mathrm{d}t) \cdot \boldsymbol{n}\mathrm{d}S\mathrm{d}\boldsymbol{\theta} \tag{22}$$

Clearly, the first line of the equality is the increment of the probability during dt, while the second line is the probability entering the domain through the boundary. Therefore, this is just the fact that the preservation of probability when it is observed from the state space description during $[t, t + \mathrm{d}t]$. Thus, here we have changed from the random event description to the state space description. We can thus see the equivalence between the two descriptions of the principle of preservation of probability.

Applying the divergence theorem to the boundary integral in the right hand side of Equation (22) and neglecting the quantity of higher order of dt yield

$$\int_{\Omega_t\times\Omega_\theta}\left(\frac{\partial p_{\mathbf{Z}\boldsymbol{\Theta}}(\mathbf{z},\boldsymbol{\theta},t)}{\partial t}\mathrm{d}t\right)\mathrm{d}\mathbf{z}\mathrm{d}\boldsymbol{\theta}$$

$$= -\int_{\partial\Omega_t\times\Omega_\theta}\sum_{j=1}^{m}\frac{\left[\partial p_{\mathbf{Z}\boldsymbol{\Theta}}(\mathbf{z},\boldsymbol{\theta},t)h_j(\boldsymbol{\theta},t)\mathrm{d}t\right]}{\partial z_j}\mathrm{d}\mathbf{z}\mathrm{d}\boldsymbol{\theta} \quad (23)$$

Noting the arbitrariness of $\Omega_t\times\Omega_\theta$ and canceling $\mathrm{d}t$ on both sides give rise to

$$\frac{\partial p_{\mathbf{Z}\boldsymbol{\Theta}}(\mathbf{z},\boldsymbol{\theta},t)}{\partial t}+\sum_{j=1}^{m}h_j(\boldsymbol{\theta},t)\mathrm{d}t\frac{[\partial p_{\mathbf{Z}\boldsymbol{\Theta}}(\mathbf{z},\boldsymbol{\theta},t)]}{\partial z_j}=0 \quad (24a)$$

In view of Equation (17b), this equation can be equivalently rewritten as

$$\frac{\partial p_{\mathbf{Z}\boldsymbol{\Theta}}(\mathbf{z},\boldsymbol{\theta},t)}{\partial t}+\sum_{j=1}^{m}\dot{Z}_j(\boldsymbol{\theta},t)\frac{[\partial p_{\mathbf{Z}\boldsymbol{\Theta}}(\mathbf{z},\boldsymbol{\theta},t)]}{\partial z_j}=0 \quad (24b)$$

This is the generalized density evolution equation (GPDEE) [10][12][13][14]. Specifically, as $m=1$ the GPDEE becomes

$$\frac{\partial p_{\mathbf{Z}\boldsymbol{\Theta}}(\mathbf{z},\boldsymbol{\theta},t)}{\partial t}+\dot{Z}(\boldsymbol{\theta},t)\frac{[\partial p_{\mathbf{Z}\boldsymbol{\Theta}}(\mathbf{z},\boldsymbol{\theta},t)]}{\partial \mathrm{z}}=0 \quad (25)$$

which is a one-dimensional partial differential equation.

Generally, the boundary condition for Equation (25) can be

$$p_{\mathbf{Z}\boldsymbol{\Theta}}(\mathbf{z},\boldsymbol{\theta},t)|_{z_j\to\pm\infty}=0, j=1,2,\dots,m \quad (26)$$

while the initial condition is usually

$$p_{\mathbf{Z}\boldsymbol{\Theta}}(\mathbf{z},\boldsymbol{\theta},t)|_{t=t_0}=\delta(\mathbf{z}-\mathbf{z}_0)p_{\boldsymbol{\Theta}}(\boldsymbol{\theta}) \quad (27)$$

where $\mathbf{z}_0$ is the deterministic initial values.

Solving the generalized density evolution equation, finally, the PDF of $\boldsymbol{Z}(t)$ can be obtained through

$$p_{\mathbf{Z}}(\mathbf{z},t)=\int p_{\mathbf{Z}\boldsymbol{\Theta}}(\mathbf{z},\boldsymbol{\theta},t)\mathrm{d}\boldsymbol{\theta} \quad (28)$$

In history, the GPDEE was firstly obtained as the uncoupled version of parametric Liouville equation for linear systems [21][22]. Then for nonlinear systems we derive the GPDEE when the formal solution is employed [12][13]. Obviously, the GPDEE is the natural result from the possibility of observing individual physical quantities separately and by the random event description of the principle of preservation of probability.

2.4 The extension of GPDEE to general physical systems

The above GPDEE can be extended to general physical system. Without loss of generality, consider a generic physical system

$$\boldsymbol{L}(\boldsymbol{Y}, \partial^{(j)}\boldsymbol{Y}, \boldsymbol{\Theta}, \tau, x, t) = 0 \tag{29}$$

where $\boldsymbol{L}(\cdot)$ is a general operator and $\boldsymbol{\Theta}$ is a random vector.

Regarding τ as an "evolution parameter", then the joint PDF of $(Y, \boldsymbol{\Theta})$ will be governed by the following probability density evolution equation

$$\frac{\partial p_{Y_l\boldsymbol{\Theta}}(y_l,\boldsymbol{\theta},\tau)}{\partial \tau} + \sum_{j=1}^{m} \dot{Y}_l(\boldsymbol{\theta}, \tau)\frac{\partial p_{Y_l\boldsymbol{\Theta}}(y_l,\boldsymbol{\theta},\tau)}{\partial y_l} = 0 \tag{30}$$

For one-dimensional case, there exist

$$\frac{\partial p_{Y_l\boldsymbol{\Theta}}(y_l,\boldsymbol{\theta},\tau)}{\partial \tau} + \dot{Y}_l(\boldsymbol{\theta}, \tau)\frac{\partial p_{Y_l\boldsymbol{\Theta}}(y_l,\boldsymbol{\theta},\tau)}{\partial y_l} = 0 \tag{31}$$

Obviously, this provides a basic framework for generalizing GPDEE to generic physical systems.

More important is that, such a progress gives us a new understanding on the relationship between the physical world and random world. Actually, if rewriting GDEE as follows

$$\frac{\partial p_{Y_l\boldsymbol{\Theta}}(y_l,\boldsymbol{\theta},\tau)}{\partial \tau} = -\dot{Y}_l(\boldsymbol{\theta}, \tau)\frac{\partial p_{Y_l\boldsymbol{\Theta}}(y_l,\boldsymbol{\theta},\tau)}{\partial y_l} \tag{32}$$

We will find such an important fact immediately: the transition of probability structures is determined by the change of physical state of the system! This demonstrates strongly that the evolution of probability density is not irregular, it obeys restrict physical law. Actually, this fact tells us the relationship between a deterministic system and the counterpart stochastic system, and why the statistical rules exist. Obviously, such an understanding gives us a new world perspective.

3 Applications of GPDEE to general physical systems

On the basis of above background, a series of developments using GPDEE to research physical system have been carried out in recent years. Some of them will be summarized as following which including: (1) the physical random models for dynamic excitations, especially taking seismic ground motion as an example; (2) the multi-scale stochastic damage model for concrete materials and structures; (3) a physical approach to the global reliability of structures, respectively.

3.1 Physical random models for seismic ground motions

According to the viewpoint of stochastic physical system, the reasonable model of seismic ground motions should be derived from their embedded physical mechanisms. Generally speaking, an acceleration time history of seismic ground motion could be expressed in a combination form of Fourier amplitude and phase spectrums, which is written as follows:

$$a_R(t) = \frac{1}{2\pi}\int_{-\infty}^{+\infty} A_R(\omega)\cdot cos[\omega t + \Phi_R(\omega)]\mathrm{d}\omega \tag{33}$$

where $a_R(t)$ is the acceleration time history of the seismic ground motion with epicentral distance R, $A_R(\omega)$ and $\Phi_R(\omega)$ are the Fourier amplitude and phase spectrums.

From the point of view of physics, the seismic ground motions are formed in such a way: source-path-site mechanism. On this background, using the dislocation source model proposed by Brune [23], and considering the damping and frequency dispersion effects in path and fitting effect in local site, we could express $A_R(\omega)$ and $\Phi_R(\omega)$ as following respectively [15][24],

$$A_R(\omega) = \frac{A_0\omega\cdot e^{-K\omega R}}{\sqrt{\omega^2+\left(\frac{1}{\tau}\right)^2}}\cdot\sqrt{\frac{1+4\xi_g^2(\omega/\omega_g)^2}{\left[1-(\omega/\omega_g)^2\right]^2+4\xi_g^2(\omega/\omega_g)^2}} \tag{34}$$

$$\Phi_R(\omega) = \arctan\left(\frac{1}{\tau\omega}\right) - R\cdot\mathrm{d}\cdot\ln\left[(a+0.5)\omega + b + \frac{1}{4c}\sin(2c\omega)\right] \tag{35}$$

where A_0 is the amplitude parameter, τ is introduced by Brune to describe the rupture process of earthquake fault, K is the attenuation parameter of the seismic wave propagation media, ξ_g is the equivalent damping ratio and ω_g is the predominate circular frequency. An empirical frequency-wavenumber formulas is applied to reflect the frequency dispersion effect and a, b, c, d are parameters in formula (35).

If only the randomness of seismic source and local site is considered, then A_0, τ, ξ_g, ω_g will be random variables. 4438 seismic acceleration records were adopted to identify the sample values of the above random variables. Figure 2 shows the probability distribution functions of A_0, τ, ξ_g, and ω_g on site class C of ASCE7-2010.

Obviously, for acceleration of ground motion, there exit following GPDEE,

$$\frac{\partial p_{A\Theta}(a,\boldsymbol{\theta},t)}{\partial t} + \dot{A}(\boldsymbol{\theta},\tau)\frac{\partial p_{A\Theta}(a,\boldsymbol{\theta},\tau)}{\partial a} = 0 \tag{36}$$

According to this equation, it is easier to get the probability density evolution process of acceleration of ground motions. Figure 3 shows the comparison of the probability density distribution of theoretical results with the statistical results of realistic ground motion record set.

Most recently, this model was extended to a ground motion field model that captures spatial variation [25].

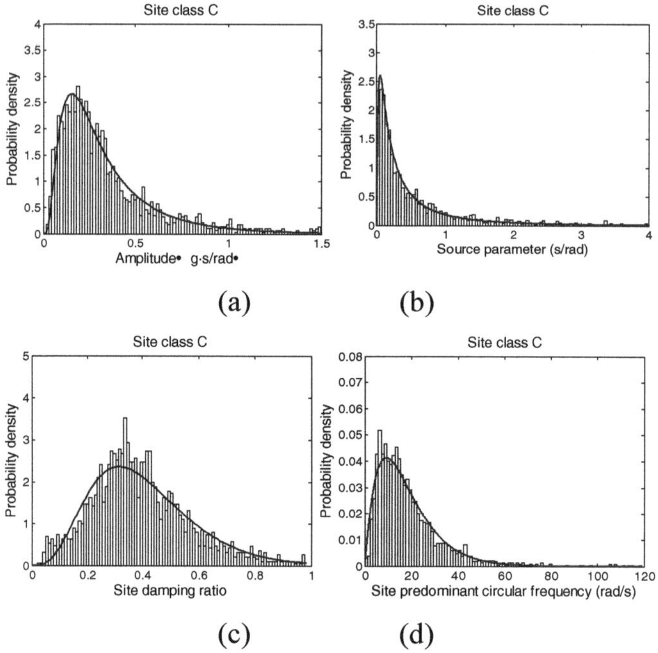

Figure 2: Probability distribution functions of the physical random variables on site class C of ASCE7-2010

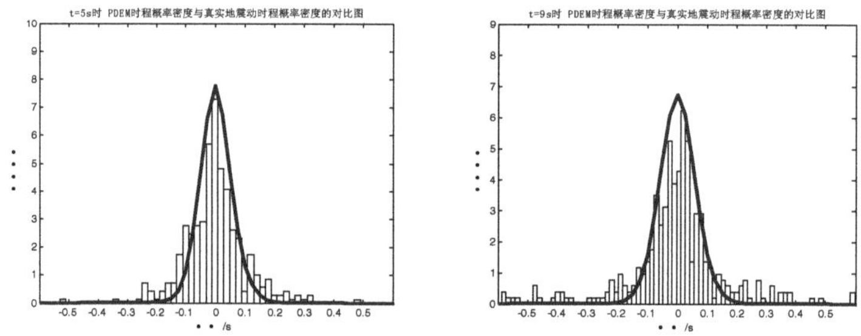

Figure 3: Comparison of probability density distribution at typical time instants

3.2 Multi-scale stochastic damage model for concrete

The complicated behaviors of concrete under external loading are induced by the initiation and propagation of cracks in mesoscale. When subjected to external loading, the cracks may initiate by the stress concentration induced by the initial defects, and thereafter the stress-strain curve of concrete diverges from the linear elastic trend. On the other hand, concrete material possesses evident randomness. Actually, the micro strengths of ingredient of concrete or the micro infects in concrete all possess uncontrollable characteristics. Therefore, the stochastic damage process is essential for the nonlinear mechanical evolution of concrete structures. Actually, the coupling between the randomness and nonlinearity plays an important role for the performance of concrete structures, such as its constitutive relationship in material level as well as the resistance in structural level [26].

Therefore, a multi-scale viewpoint is introduced to investigate the random damage behavior of concrete. In the mesoscale, the cracks and defects initiate and propagate in stochastic ways due to the material heterogeneity. In the macro scale (structural level), the structure degrades and fails in a continuous way with random responses. The analytical method adopted in each level is quite different from each other. In the meso-level, the random heterogeneity should be considered. In the macroscale, the detailed cracks and defects are too complicated to be simulated in an explicit way, thus the continuum damage model is adopted for the structural simulation.

In meso-level, the damage evolution process could be reflected by two basic damage mechanisms: tensile damage and shear damage [27][28]. Each kind of micro-damage possesses random fracture strain following specified probability distribution. Meanwhile, the plastic deformation should be considered in these micro models. On this basic idea, two kinds of micro stochastic rupture-sliding models were suggested based on the classic parallel bundle model (Figure 4(a)). Here the tensile element represents the tensile damage by the direct tensile rupture of a micro element, while the shear element experiences shear fracture of a micro element under compressive loading (Figure 4(b)). The sliding part is introduced for both of the elements to describe the remnant deformation of concrete induced by the plastic deformation of cement matrix.

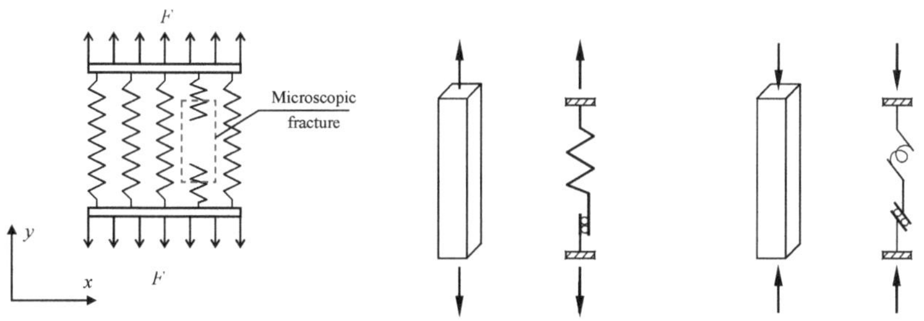

(a) Parallel bundle model (b) Microscopic tensile and shear elements

Figure 4: Micro stochastic rupture-sliding models

The derived stochastic damage evolution function is expressed as

$$D(\varepsilon) = \int_0^1 H[\varepsilon - \Delta(x)]\mathrm{d}x \tag{37}$$

where $\Delta(x)$ is a 1-D randomness field defined on coordinate x.

Then by introducing the Helmholtz free energy and using the law of thermodynamics, a damage model in multi-dimension could be derived. The stress-strain relation is

$$\boldsymbol{\sigma} = (\boldsymbol{I} - \boldsymbol{D}) : \boldsymbol{C}_0 : (\boldsymbol{\varepsilon} - \boldsymbol{\varepsilon}^p) \tag{38}$$

The evolution of plastic strain $\boldsymbol{\varepsilon}^p$ could be defined by the effective space plasticity and the fourth order damage tensor is

$$\boldsymbol{D} = d^+ \boldsymbol{P}^+ + d^- \boldsymbol{P}^- \tag{39}$$

where $\boldsymbol{P}^+$ and $\boldsymbol{P}^-$ are the projection tensors; d^+ and d^- are the tensile and compressive damage variables. The evolution of damage variables could be derived through Equation (39) by replacing the tensile strain and compressive strain by the energy equivalent strain ε_{eq}^+ and ε_{eq}^- calculated by [29][30]

$$\varepsilon_{eq}^+ = \sqrt{\frac{2Y^+}{E_0}}, \quad \varepsilon_{eq}^- = \frac{1}{(\alpha-1)}\sqrt{\frac{Y^-}{b_0}} \tag{40}$$

Corresponding to the stress-strain relation (38), there exist following GPDEE

$$\frac{\partial p_{\sigma_l \boldsymbol{\Theta}}(\sigma_l, \boldsymbol{\theta}, \tau)}{\partial \tau} + \dot{\sigma}_l(\boldsymbol{\theta}, \tau) \frac{\partial p_{\sigma_l \boldsymbol{\Theta}}(\sigma_l, \boldsymbol{\theta}, \tau)}{\partial \sigma_l} \tag{41}$$

Using this equation, it is easier to get the probability density evolution process of stress with loading. Figure 5 shows the calculated mean value and standard deviation of the stress-strain curve. The agreements between the simulated results and the experimental results suggest the validation of the proposed model.

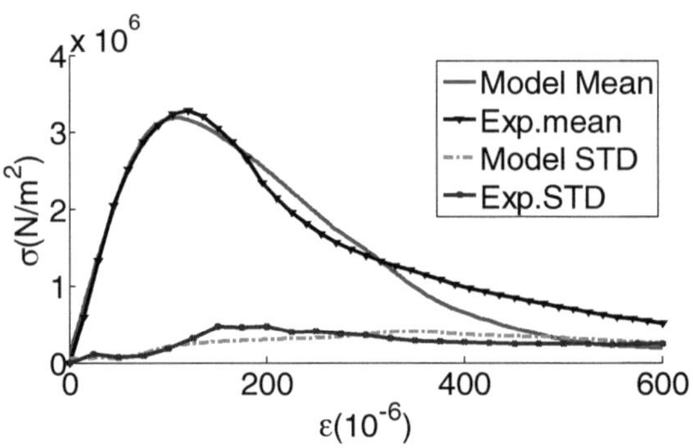

Figure 5: Mean and STD of stress-strain curves.

3.3 Global reliability of structures

The global reliability analysis for complex structures is another important example to use GPDEE in engineering system. In order to describe the basic idea, let us take a series system as an example. As is well known, by introducing weakest link assumption, the reliability of the a series system will be

$$P_f = \max_{1 \le j \le n} P(E_j) \tag{42}$$

where P_f denotes the failure probability of the system; $P(\cdot)$ denotes the probability of failure elements.

It is seen that the main concern of the above approach is the failure probability of system element. However, the problem may be approached from a physical point of view: find the element which has minimum strength. This leads to the concept of equivalent extreme-value event [31]. In fact, for the above series system, if we try to find the element which possess the minimum strength of the system, there will exist an equivalent extreme-value event as following

$$Z_{\min} = \min_{1 \le j \le m} \mathcal{g}_j(\boldsymbol{\Theta}) \tag{43}$$

where $\mathcal{g}_j(\boldsymbol{\Theta})$ is the limit state function (performance function) of the ith element of the system.

Then the reliability of the series system could be given by

$$R = \Pr\left\{\cap_{j=1}^{m} \mathcal{g}_j(\boldsymbol{\Theta}) > 0\right\} = \Pr\{Z_{\min} > 0\} \tag{44}$$

For a general series-parallel system, one could get a similar result

$$R = \Pr\left\{\cup_{\mathrm{i}=1}^{\mathrm{n}}\left[\cap_{j=1}^{m} \mathcal{g}_{ij}(\boldsymbol{\Theta}) > 0\right]\right\} = \Pr\{Z_{\mathrm{ext}} > 0\} \tag{45}$$

It is indicated that the system reliability analysis involving multiple performance functions can be recast in terms of an equivalent extreme function involving the maximum (or minimum) of all performance functions. In other words, the inherent correlation information in the original performance functions is retained in the equivalent extreme-value event.

Then the global structural reliability could be determined in the analysis process of structural performances according to the specific demand [32]. Actually, denoting $X_i(\boldsymbol{\Theta}, \tau)$ as the response of structures of interest, an equivalent extreme-value process could be constructed as follows:

$$Z(\Theta, t) = \max_{0 \le \tau \le t}\{X_1(\boldsymbol{\Theta}, \tau), X_2(\boldsymbol{\Theta}, \tau), \dots\} \tag{46}$$

For this process, a generalized probability density evolution equation could be derived as following

$$\frac{\partial p_{Z\boldsymbol{\Theta}}(z,\boldsymbol{\theta},\tau)}{\partial \tau} + \dot{Z}(\boldsymbol{\theta}, \tau)\frac{\partial p_{Z\boldsymbol{\Theta}}(z,\boldsymbol{\theta},\tau)}{\partial z} = 0 \tag{47}$$

Furthermore, by introducing an absorbing boundary condition

$$p_{Z\boldsymbol{\Theta}}(\boldsymbol{z}, \boldsymbol{\theta}, \tau) = 0, \quad \text{for } z \in \Omega_f \tag{48}$$

The global reliability of structures could be obtained as

$$R(\tau) = \int_{\Omega} \breve{p}_{Z\boldsymbol{\Theta}}(\boldsymbol{z}, \boldsymbol{\theta}, \tau)\, d\theta \tag{49}$$

As an example, consider a two-bay two-story RC frame, of which the dimensional details are show in Figure 6. The ratio of Q to F_0 is constant and equal to 3.0. The load Q is a random variable with normal distribution (μ=105kN, δ=20%). The strength of the reinforcement is a deterministic value f_y= 350MPa, and the strength of the concrete is a random variable with normal distribution (μ_{fc}=25MPa, δ_{fc}=10%).

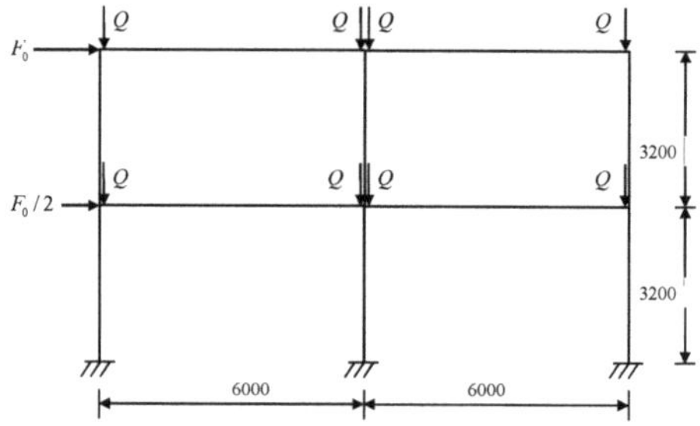

Figure 6: Two-bay two-story RC frame

Employing the proposed method described as above, it is found that P_f= 8.7%. For the purpose of verification, Monte-Carlo simulations are carried out and it shown a randomly convergent process. Actually, the failure probability of the structure varies from 8.1% to 9.2% in case that the number of simulation varies from 2500 to 20000.

4 Conclusion

A family of generalized density evolution equation (GPDEE) is derived based on the principle of preservation of probability incorporated with the uncoupled physical equations, which takes advantages over traditional probability density evolution equations. These progresses provide an important tool in understanding many phenomena and behaviors of engineering structures and systems, especially the randomness propagation in nonlinear dynamical systems even for general physical systems.

The recent progress of using GPDEE show that it not only provides an efficient scheme for stochastic dynamical response analysis, and first-passage reliability or optimum control of stochastic systems, but also can be used for the modeling dynamic excitations of structures, the stochastic damage of concrete materials and structures, the global reliability evaluation of complex structural systems, the time-dependent reliability of a life-cycle engineering system, involving deterioration of materials, degradation of components and rehabilitation or maintenance process. All these progress indicates a common principle: the general probability density evolution equation reveals the essence of randomness propagation in physical systems.

REFERENCES

[1] Einstein, A.: über die von der molecular-kinetischen Theorie der Wärme geforderte Bewegung von in rhuenden Flüssigkeiten suspendierten Teilchen; Ann. Phys. (Leipzig); 17. (1905), Pages 549-560

[2] Fokker, A.D.: Die mittlere Energie rotierender elektrischer Dipole im Strahlungsfeld; Annalen der Physik (Leipzig); 43. (1914), Pages 810-820

[3] Planck, M.: Uber einen Satz der statistichen Dynamik und eine Erweiterung in der Quantumtheorie; Sitzungberichte der Preussischen Akadademie der Wissenschaften, (1917), Pages 324-341

[4] Kolmogorov, A.: Über die analytischen Methoden in der Wahrscheinlichkeitsrechnung; Mathematische Annalen, 104. (1931), Pages 415-458

[5] Langevin, P.: Sur la theorie du mouvement brownien; C. R. Acad. Sci. Paris, (1908), Pages 530-532

[6] Wiener, N.: Differential space; J. Math. Phys., 58. (1923), Pages 131-174

[7] Itô, K.: Differential equations determining a Markoff process; Zenkoku Sizyo Sugaku Danwakasi, (1942), Pages 1077

[8] Itô, K.: Stochastic integral; Proc. Imp. Acad. Tokyo, 20. (1944), Pages 519-524

[9] Stratonovich, R.L.: Introduction to the Theory of Random Noise, Gordon and Breack, New York, 1963

[10] Li, J.; Chen, J.B.: Stochastic Dynamics of Structures, John Wiley & Sons, Singapore, 2009.

[11] Li, J: A physical approach to stochastic dynamical systems; Science Paper Online, 2. (2006), Pages 95-104 [in Chinese]

[12] Li, J.; Chen, J.B.: The probability density evolution method for dynamic response analysis of non-linear stochastic structures; International Journal for Numerical Methods in Engineering, 65. (2006a), Pages 882-903

[13] Li, J.; Chen, J.B.: Generalized density evolution equation in stochastic systems; Progress in Natural Science, 6. (2006b), Pages 712-719 [in Chinese]

[14] Li, J.; Chen, J.B.: The principle of preservation of probability and the generalized density evolution equation; Structural Safety, 30. (2008), Pages 65-77

[15] Li, J.; Ai, X.Q.: Study on random model of earthquake ground motion based on physical process; Earthquake Engineering and Engineering Vibration, 5. (2006), Pages 21-26 [in Chinese]

[16] Li, J.; Zhang, L.L.: Random Fourier spectrum of the measured wind field; Journal of Vibration Engineering, 1. (2007), Pages 66-72 [in Chinese]

[17] Loeve, M.: Probability Theory, Springer-Verlag, Berlin, 1977

[18] Li, J.; Liu, Z.J.: Decomposition of stochastic processes based on standard orthogonal basis; Journal of Tongji University, 10. (2006), Pages 1279-1283 [in Chinese]

[19] Chen, J.B.; Li, J.: Stochastic harmonic function and spectral representations; Acta Mechanica Sinica; 5. (2012), Pages 516-522 [in Chinese]

[20] Fung, Y.C.: A First Course in Continuum Mechanics, Prentice-Hall, Inc., New Jersey,1994

[21] Li, J.; Chen, J.B.: The probability density evolution method for analysis of dynamic nonlinear response of stochastic structures; Acta Mechanica Sinica, 6. (2003), Pages 716-722 [in Chinese]

[22] Li, J.; Chen, J.B.: Probability density evolution method for dynamic response analysis of structures with uncertain parameters; Computational Mechanics, 34. (2004), Pages 400-409

[23] Brune, J.N.: Tectonic stress and the spectra of seismic shear waves from earthquakes; Journal of Geophysical Research; 26. (1970), Pages 4997-5009

[24] Wang, D.; Li, J.: Physical random function model of ground motions for engineering purpose; Science China: Technological Sciences, 54. (2011), Pages 175-183

[25] Wang, D.; Li, J.: A random physical model of seismic ground motion field on local engineering site; Science China: Technological Sciences, 55. (2013), Pages 2057-2065

[26] Li, J: Research on the stochastic damage mechanics for concrete materials and structures; Journal of Tongji University, 10. (2004), Pages 1270-1277 [in Chinese]

[27] Li, J.; Zhang, Q.Y.: Study of stochastic damage constitutive relationship for concrete material; Journal of Tongji University, 10. (2001), Pages 1135-1141 [in Chinese]

[28] Li, J.; Wu, J.Y.: Elasto-plastic damage constitutive model for concrete based on damage energy release rates, Part I: basic formulations; China Civil Engineering Journal, 9. (2005), Pages 14-20 [in Chinese]

[29] Wu, J.Y.; Li, J.; Faria, R.: An energy release rate-based plastic-damage model for concrete; International Journal of Solids and Structures, 3-4. (2006), Pages 583-612

[30] Li, J.; Ren, X.D.: Stochastic damage model for concrete based on energy equivalent strain; International Journal of Solids and Structures, 11-12. (2009), Pages 2407-2419

[31] Li, J.; Chen, J.B.; Fan, W.L.: The equivalent extreme-value event and evaluation of the structural system reliability; Structural Safety, 29. (2007), Pages 112-131

[32] Chen, J.B.; Li, J.: Development-process-of-nonlinearity-based reliability evaluation of structures; Probabilistic Engineering Mechanics; 3. (2007), Pages 267-275

S e D I F **Conference**

International Conference on
Seismic Design of Industrial Facilities
2013, RWTH Aachen University

Elliptical Response Envelopes for the Design of Reinforced Concrete Structures: New Developments and Application to Nuclear Power Plant Buildings

Quang Sang Nguyen[1], Silvano Erlicher[1], François Martin[1]

[1] EGIS Industries
4 rue Dolorès Ibarruri TSA 50012
93188 Montreuil cedex – France
quang-sang.nguyen@egis.fr; silvano.erlicher@egis.fr; francois.martin@egis.fr

ABSTRACT:

Seismic analysis is one of the main steps in the structural design of Nuclear Power Plants (NPPs). Design is usually made by assuming linear structural behavior and using the so-called response spectrum analysis. This method is based on the calculation of the response peaks for each earthquake component (X, Y or Z) of several single-degree-of-freedom oscillators representing the modes of the analyzed structure. Then, the modal peaks of each response parameter for each earthquake direction are combined using, for instance, the so-called Complete Quadratic Combination-CQC (Der Kiureghian [1]). The superposed responses are, by definition, positive quantities. Hence, their sign must be defined according to a fundamental mode shape or another reference structural configuration. Actually, signature of CQC of modal responses is not required for the approach based on the notion of "peak response hyper-ellipsoid envelope" (e.g. [2]). This is one of the interesting advantages of this method. For this reason, in this paper we discuss two developments based on the notion of "response envelopes". The first one is an "equivalent static method" (ASCE [3], Nguyen et al. [4]-[5]) based on the theory of the "response envelopes". The second development is an improved procedure for the definition of the signs of the CQC of modal peaks. Some of these proposed methods are applied to a NPP building and results are then compared with those coming from a standard response spectrum analysis.

Keywords: Response spectrum method for seismic analysis, CQC, Hyper-ellipsoid response envelope, equivalent static loads

1 Hyper-ellipsoid response envelopes

Let us consider an N-degree-of-freedom linear and classically damped structure, for which N real eigenmodes can be calculated. For seismic applications, only

DOI 10.1007/978-3-658-02810-7_11, © Springer Fachmedien Wiesbaden 2014

$n \leq N$ modes are usually retained, by either guarantying that the sum of effective masses of the n modes is high enough or introducing a pseudo-mode. The seismic effects are estimated by considering three earthquakes, one per direction ($k = x, y, z$). For an earthquake in direction k, the displacement vector $\underline{u}_k(t)$ can be written as a linear combination of the modal peak displacement vectors $\underline{U}_{i,k}$:

$$\underline{u}_k(t) = \sum_i \alpha_{i,k}(t)\underline{U}_{i,k} \text{ with } -1 \leq \alpha_{i,k}(t) = r_{i,k}(t)/R_{i,k} \leq 1 \quad (1)$$

where $R_{i,k} = \max_t |r_{i,k}(t)| = S_{d,k}(\omega_i, \xi_i)$ is the maximum displacement amplitude; the time-function $r_{i,k}(t)$ is the solution of dynamics equation of the single-degree-of-freedom oscillator representing the mode i undergoing the ground acceleration $\ddot{s}_{g,k}(t)$ associated with $S_{d,k}(\omega_i, \xi_i)$. The total displacement due to the earthquake in the three directions reads:

$$\underline{u}(t) = \sum_k \underline{u}_k(t) = \sum_k \sum_i \alpha_{i,k}(t)\underline{U}_{i,k} \quad (2)$$

1.1 Hyper-ellipsoid of linear combination coefficients (α-ellipsoid): case of a single seismic response

Let us consider a seismic response $f(t)$, e.g. a displacement, an axial force, a moment, etc. in a node, section or element of the structure. By virtue of linearity, one can always find a vector $\underline{d}$ such that:

$$f(t) = \underline{d}^T \underline{u}(t) = \sum_k \sum_i \alpha_{i,k}(t)\underline{d}^T \underline{U}_{i,k} = \sum_k \sum_i \alpha_{i,k}(t) F_{i,k} = \sum_k f_k(t) \quad (3)$$

where $F_{i,k} = \underline{d}^T \underline{U}_{i,k}$ is the value of the seismic response corresponding to the peak displacement vector $\underline{U}_{i,k}$ for the mode i and direction k.

In the sense of probability, the maximum value of $f_k(t)$ can be estimated using the Complete Quadratic Combination-CQC of modal peaks (Der Kiureghian [1]) $F_k^{CQC} = \sqrt{\sum_{ij} \rho_{ij} F_{i,k} F_{j,k}}$, where ρ_{ij} is the modal cross-correlation coefficient between modes i and j. Thus, in order for a linear combination $f_k(t)$ of modal responses to be probable, the following condition must hold:

$$f_k(t) = \sum_i \alpha_{i,k}(t) F_{i,k} \leq F_k^{CQC} \text{ or } \underline{\alpha}_k^T(t) \underline{F}_k \leq \sqrt{\underline{F}_k^T \underline{\underline{H}} \, \underline{F}_k} \quad (4a \,\&\, 4b)$$

where $\underline{\alpha}_k(t) = [\alpha_{1,k}(t), \alpha_{2,k}(t), \ldots, \alpha_{n,k}(t)]^T$ is the vector of the combination coefficients for all modes and direction k, $\underline{F}_k = [F_{1,k}, F_{2,k}, \ldots, F_{n,k}]^T$ is the vector of the modal force peaks and $\underline{\underline{H}} = [\rho_{ij}]$ is the $n \times n$ matrix of the modal correlation coefficients. The inequalities in Eqs. (4a)-(4b) can be extended to the case of three earthquake directions using a quadratic combination of F_k^{CQC} (e.g. Menun and Der Kiureghian [2]):

$$f(t) = \sum_k f_k(t) = \sum_k \sum_i \alpha_{i,k}(t) F_{i,k} \leq F^{CQC} = \sqrt{\sum_k \left(F_k^{CQC} \right)^2} = \sqrt{\sum_k \sum_{ij} \rho_{ij} F_{i,k} F_{j,k}} \quad (5a)$$

or $$\underline{\alpha}(t)^T \underline{F} \leq \sqrt{\underline{F}^T \underline{\underline{\widetilde{H}}}\, \underline{F}}, \quad (5b)$$

with $\underline{\alpha} = \left[\underline{\alpha}_x^T, \underline{\alpha}_y^T, \underline{\alpha}_z^T\right]^T$, $\underline{F} = \left[\underline{F}_x^T, \underline{F}_y^T, \underline{F}_z^T\right]^T$ and $\underline{\underline{\widetilde{H}}} = \mathrm{diag}\left[\underline{\underline{H}}, \underline{\underline{H}}, \underline{\underline{H}}\right]$.

From Eqs. (4b) – (5b), supposing that the matrix $\underline{\underline{H}}$ is invertible, one can prove that (Martin [8]):

$$\underline{\alpha}_k^T \underline{\underline{H}}^{-1} \underline{\alpha}_k \leq 1 \text{ and } \underline{\alpha}^T \underline{\underline{\widetilde{H}}}^{-1} \underline{\alpha} \leq 1 \quad (6a \,\&\, 6b)$$

Eqs. (6a) – (6b) mean that the coefficients $\underline{\alpha}_k$ and $\underline{\alpha}$ define probable combinations of peak modal responses when they belong to an n and 3n-dimension hyper-ellipsoid (named α_k-ellipsoid and α-ellipsoid), respectively.

1.2 α-ellipsoid and f-ellipsoid: case of n_r different seismic responses

Let $\underline{x}_k(t) = \left[f_{1,k}(t), f_{2,k}(t), \ldots, f_{n_r,k}(t)\right]^T$ be a vector of n_r simultaneous seismic responses due to an earthquake in direction k, and let $\underline{\underline{R}}_k = \left[\underline{F}_{1,k}, \underline{F}_{2,k}, \ldots, \underline{F}_{n_r,k}\right]$ be a $n \times n_r$ matrix whose columns $\underline{F}_{r,k} = \left[F_{r,1,k}, F_{r,2,k}, \ldots, F_{r,n,k}\right]^T$ are the vectors of peak modal values of the responses $f_{r,k}(t)$. By virtue of structural linearity, one has $\underline{x}_k = \underline{\underline{R}}_k^T \underline{\alpha}_k$. Moreover, supposing that the matrix $\underline{\underline{H}}$ is invertible, one can prove that:

$$\underline{x}_k^T \underline{\underline{X}}_k^{-1} \underline{x}_k = \underline{\alpha}_k^T \underline{\underline{H}}^{-1} \underline{\alpha}_k \leq 1 \quad (7a)$$

where $\underline{\underline{X}}_k = \underline{\underline{R}}_k^T \underline{\underline{H}} \underline{\underline{R}}_k$ ($n_r \times n_r$ matrix). The inequality follows from Eq. (6a). In the case of three earthquake directions, one can also prove that:

$$\underline{x}^T \underline{\underline{X}}^{-1} \underline{x} = \underline{\alpha}^T \underline{\underline{\widetilde{H}}}^{-1} \underline{\alpha} \leq 1 \quad (7b)$$

where $\underline{\underline{X}} = \sum_k \underline{\underline{X}}_k$ and $\underline{x} = \sum_k \underline{x}_k$. Eqs. (7a) – (7b), considered as identities, define two hyper-ellipsoids of dimension n_r associated with the matrix $\underline{\underline{X}}$ and $\underline{\underline{X}}_k$, that we name f_k-ellipsoid and f-ellipsoid, respectively. These matrices are those of the classical definition of the hyper-ellipsoids [2]. Each point inside or on the boundary of f-ellipsoid corresponds to a probable combination of the n_r simultaneous seismic responses $f_1(t), f_2(t), \ldots, f_{n_r}(t)$. Eq. (7b) implies that a point $\underline{x}$ of the f-ellipsoid corresponds to one and only one point $\underline{\alpha}$ of the α-ellipsoid.

1.3 Discretization of hyper-ellipsoid envelopes

For practical application purposes, a finite number of probable combinations of simultaneous seismic responses must be considered for the seismic analysis. They correspond to a finite number of points on the *f-ellipsoid* surface. Several discretization methods exist (e.g. ASCE [3]). A procedure discretizing the hyper-ellipsoid by a polyhedron envelope was proposed by Leblond [6] and Vézin et al. [7], improved and optimized by Nguyen et al. [4] – [5]. According to Nguyen et al. [5], the hyper-ellipsoid can be discretized using $n_r \times 2^{n_r}$ points (approach A) or $(n_r - 1) \times n_r \times 2^{n_r}$ points (approach B), where n_r is the number of simultaneous seismic responses.

2 Definition of equivalent static loads

For some applications, it may be useful to represent the seismic action on a structure by one (or several) equivalent static load field(s), usually defined at each node of the structural model as the product between the nodal mass and suitable nodal accelerations. However, the definition of the nodal accelerations is often based on approximate procedures. It is nonetheless possible to avoid these approximations by defining the acceleration field using the notion of "peak response envelopes", as it has been explained by the authors in reference [4].

Some details of this procedure are recalled hereinafter. From Eqs. (1) – (2), one can define the displacement $u_x^l(t)$, the pseudo-acceleration $a_x^l(t)$ and the force $q_x^l(t)$ in the direction x for the node l of the structure:

$$u_x^l(t) = \sum_k \sum_i \alpha_{i,k}(t) U_{i,k,x}^l \tag{8a}$$

$$a_x^l(t) = \sum_k \sum_i \alpha_{i,k}(t) \omega_i^2 U_{i,k,x}^l = \sum_k \sum_i \alpha_{i,k}(t) A_{i,k,x}^l \tag{8b}$$

$$q_x^l(t) = \sum_k \sum_i \alpha_{i,k}(t) m^l \omega_i^2 U_{i,k,x}^l = \sum_k \sum_i \alpha_{i,k}(t) Q_{i,k,x}^l \tag{8c}$$

where $U_{i,k,x}^l$, $A_{i,k,x}^l$ and $Q_{i,k,x}^l$ are respectively the peak displacement of node l and the corresponding pseudo-acceleration and force for mode i, in the direction x and due to the earthquake direction k; m^l is the mass of the node l. Analogous expressions can be written for directions y and z, leading to the following nodal force field at the generic time t:

$$\underline{q}(t) = \left[q_x^1(t), \ldots, q_x^N(t), q_y^1(t), \ldots, q_y^N(t), q_z^1(t), \ldots, q_z^N(t)\right]^T = \sum_k \sum_i \alpha_{i,k}(t) \underline{Q}_{i,k} \tag{9}$$

where $\underline{Q}_{i,k} = \left[Q_{i,k,x}^1, Q_{i,k,x}^2, \ldots, Q_{i,k,x}^N, Q_{i,k,y}^1, \ldots, Q_{i,k,y}^N, Q_{i,k,z}^1, Q_{i,k,z}^2, \ldots, Q_{i,k,z}^N\right]^T$ is the vector of the modal peak forces defined in Eq. (8c).

In general, the combination coefficients $\alpha_{i,k}$ are not known. However, if a dominant mode exist for each direction k (we can indicate these three modes with the indices (1,x), (1,y) and (1,z)), one has $\underline{q}_k(t) \approx \underline{Q}_{1,k}$. The dominant mode is considered representative for the earthquake in direction k ($\alpha_{1,k} \approx 1$, $\alpha_{i \neq 1,k} \approx 0$).

An alternative procedure is based on the use of the Complete Quadratic Combination: for each earthquake direction k, the force field $\underline{q}_k = [q_k^{1,CQC}, q_k^{2,CQC}, \dots, q_k^{N,CQC}]^T$ is defined, with $q_k^{l,CQC} = \sqrt{\sum_{ij} \rho_{ij} Q_{i,k,k}^{l} Q_{j,k,k}^{l}} = m^l \sqrt{\sum_{ij} \rho_{ij} A_{i,k,k}^{l} A_{j,k,k}^{l}}$. In these two approaches, the Newmark's rule is used to combine the force fields $\underline{q}_k$ associated with the different earthquake directions. In these cases, the linear combination in Eq. (9) is not used. These approaches are commonly used in engineering applications, but they are characterized by some approximations, since the computation of the combination coefficients $\alpha_{i,k}$ is not performed.

A rigorous definition of these coefficients has been proposed by Nguyen et al. [4]. This definition is based on the use of the α-ellipsoid and a particular case of f-ellipsoid (see the definitions in Section 1.2). This procedure can be summarised as follows:

(a) At a given time t, the vector of nodal forces $\underline{q}(t)$ in Eq. (9) depends on the 3n-component vector (point) $\underline{\alpha}(t)$ and corresponds to one equivalent static load case.

(b) If the instant t changes, the point $\underline{\alpha}(t)$ changes too, but the locus of the probable positions of the points $\underline{\alpha}(t)$ (i.e. the probable values of the combination coefficients) is known: it is the α-ellipsoid defined by Eq. (6b).

(c) For n modes, the α-ellipsoid has dimension 3n. Its polyhedral envelope would have either $3n \times 2^{3n}$ points (approach A) or $(3n-1) \times 3n \times 2^{3n}$ (approach B) (see Nguyen et al. [5]). This number of points is too high for practical calculations

(d) Instead of finding all the points $\underline{\alpha}$ approximating the α-ellipsoid (in order to define all probable force fields $\underline{q}(t)$), a preliminary selection of the most important ones (according to some engineering criteria) is performed.

(e) The vector $\underline{\alpha}$ corresponding to the chosen engineering criterion is computed by using a suitable analytical procedure [4].

Item (d) deserves a further discussion: from the seismic design point of view, the 6 points (load cases) $\underline{\alpha}$ belonging to the α-ellipsoid *and maximizing the total shear seismic forces* $F_x(t), F_y(t), F_z(t)$ *and moments* $M_{xx}(t), M_{yy}(t), M_{zz}(t)$ *at the basis of the building* are very important. However, these six load cases constitute a rather poor description of the set of all probable combinations of forces and moments at the building basis. Actually, a complete description of probable seismic forces at the base of the building is provided by the 6D hyper-ellipsoid (named here T-ellipsoid): each point $\underline{T} = [F_x, F_y, F_z, M_{xx}, M_{yy}, M_{zz}]^T$ of this T-ellipsoid represents one probable combination of the total forces and moments at the base (Figure 1). The T-ellipsoid is a particular case of f-ellipsoid (section 1.2). Hence, it is proposed to look for the points $\underline{\alpha}$ fulfilling Eq. (6b) and such that the corresponding vector of total forces and moments at the base belongs to the T-ellipsoid. In practice, the T-ellipsoid can be approximated by a 6D-polyhedron with 6x26 = 384 vertices and the number of points $\underline{\alpha}$ is 384 (or 5x6x26 = 1920 points for approach B).

For the details about the analytical procedure to compute $\underline{\alpha}$ for a given point $\underline{T}$ (item (e)) the reader is referred to paper [4].

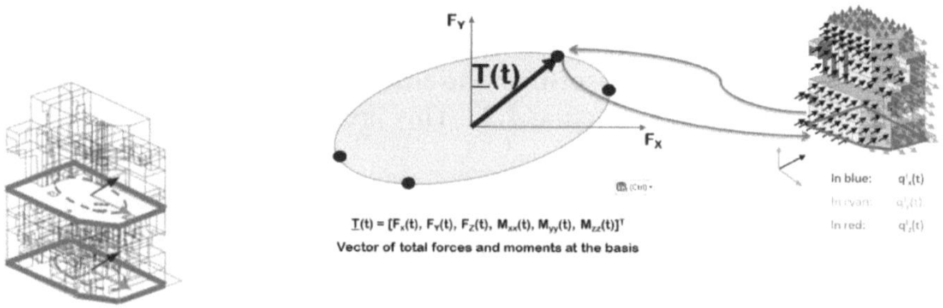

Figure 1: Selection of the most important points $\underline{\alpha}$ according to the total forces and moments at the building basis

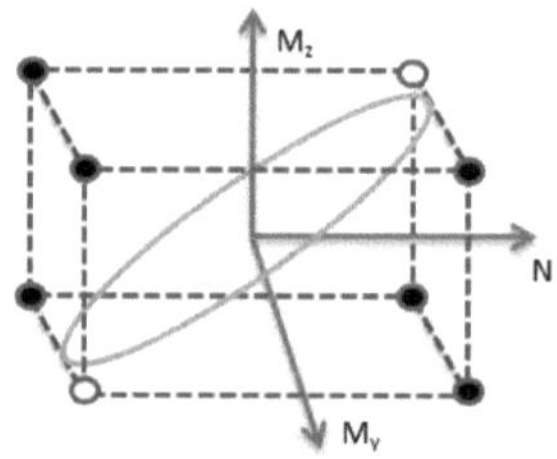

Figure 2: CQCs + Quadratic Combination (case n_r=3: normal effort and two moments in a beam section)

3 Definition of the signs for the Complete Quadratic Combinations (CQC)

In a classical response spectrum procedure, the modal peak responses are often superposed in each earthquake direction using the so-called Complete Quadratic Combination – CQC (Der Kiureghian [1]). Then, the values of the CQCs in the three directions ("directional CQCs") need to be combined to obtain the "global" response. A possible method to combine the three directional CQCs responses is the Quadratic Combination. The result of this combination (named here "global CQC") is a positive quantity, and all sign permutations between the n_r different global CQCs (for instance, the normal effort N and the moments M_y and M_z in a beam section) have to be considered to cover all possible seismic load combinations. There are 2^{n_r} possible sign permutations. When this kind of procedure is applied, no sign combination is omitted (i.e. all black and white points in Figure 2 are considered). However, this method may retain many non-realistic combinations (e.g. black points in Figure 2). This may lead to a large overestimation of the reinforcement area in reinforced concrete structural elements.

Another way of combining directional responses was proposed by Newmark [9]: global responses due to the three earthquake directions are defined by 24 linear combinations of directional CQCs:

$$\pm S_{x,CQC} \pm \beta S_{y,CQC} \pm \beta S_{z,CQC};\ \pm\beta S_{x,CQC} \pm S_{y,CQC} \pm \beta S_{z,CQC};\ \pm\beta S_{x,CQC} \pm \beta S_{y,CQC} \pm S_{z,CQC}$$

where β is a coefficient less than 1 (e.g. $\beta = 0.4$ in ASCE [3]) accounting for the contribution of the two secondary earthquake directions; $S_{x,CQC}, S_{y,CQC}, S_{z,CQC}$ are three directional CQCs of the seismic response S (e.g. the moments or membrane forces in a shell element). The indices x,y,z indicate the earthquake direction.

When $n_r>1$, the relative sign of the directional CQCs responses must be defined before computing the Newmark's combinations. For instance, for each earthquake direction it is necessary to know if a positive normal effort $N_{x,CQC}$ (traction) occurs simultaneously with a positive or a rather a negative bending moment $M_{x,CQC}$, because this affect the area and the position of steel bars in a beam section. A classical procedure for the definition of the signs of directional CQCs is based on the assumption that the sign of each directional CQC of the response S is equal to the sign of the same response when the deformed structural shape coincides with the fundamental mode for the given earthquake direction. It is implicitly assumed that this mode has a high effective mass (e.g. 70%) for the given earthquake direction. Nevertheless, complex structures like NPP buildings are usually multi-modal and represented by several important eigenmodes, which makes it difficult to determine without ambiguity a unique dominant mode to define the sign of directional CQCs.

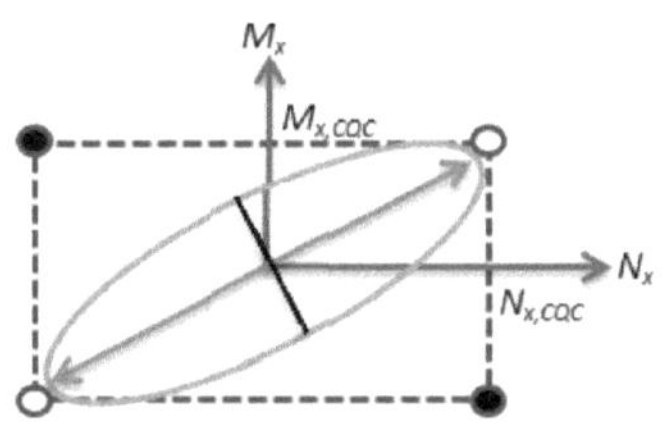

Figure 3: Schematic representation of the couple of directional CQC responses M-N in a beam due to an earthquake in direction x

An alternative procedure can be proposed. Let us consider the example of Figure 3, with a couple of beam efforts (n_r=2) due to the earthquake in direction x: the axial force N_x and the bending moment M_x (where the index x indicates the earthquake direction). Observe that the four vertices of the rectangle correspond to the four possible sign permutations: $\left(\pm N_{x,CQC};\pm M_{x,CQC}\right)$. In this example, attributing the signs before doing Newmark's combinations means that two points must be retained, and the other two are discarded. Looking at Figure 3, one can say that the two white points have the appropriate signs, because of their position with respect to the hyper-ellipsoid envelope which can be considered as the reference solution.

More in general, one can propose to define the signs of directional CQCs according to the direction of the major axis of the hyper-ellipsoid envelope. In other words, the signs of CQC responses for a ground motion in direction k are assumed equal to the signs of the components of the eigenvector associated with the largest eigenvalue of the matrix $\underline{\underline{X}}_k$ in Eq. (7a). The "signed (directional) CQC" points of Figure 4 have been defined by using this assumption. We name this procedure "CQC ellipse - Newmark's Combinations". The hyper-ellipsoid is discretized by a polyhedron whose vertices are the "rhomb-shaped" points depicted in Figure 4. In this example, each "point" is defined by six coordinates (n_r=6), which are the shell efforts $N_{xx}, N_{yy}, N_{xy}, M_{xx}, M_{yy}, M_{xy}$. Figure 4 shows the projections of these efforts in the planes $N_{xx}-N_{yy}$ and $N_{yy}-M_{xx}$.

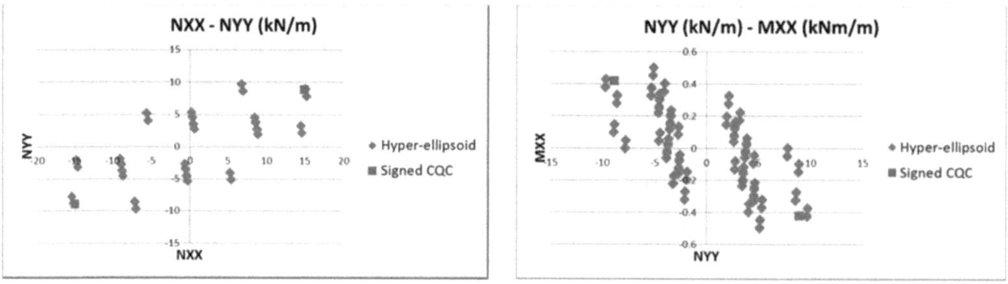

Figure 4: Definition of the signs of directional CQC efforts in a shell element using elliptical envelopes

4 Application to a NPP reinforced concrete building

In this Section, several seismic analysis approaches based on the response-spectrum method are applied to a NPP reinforced concrete building. In particular, two seismic analyses are performed using the equivalent static load methodology based on the response envelopes presented in Section 2. The following analyses are performed:

1. Complete Quadratic Combination of the modal shell efforts for each direction, signs based on the fundamental mode, Newmark's Combinations of three directions (1st procedure: CQC-24 Newmark's Combinations);
2. Complete Quadratic Combination of the modal shell efforts for each direction and Quadratic Combination of three directions (2nd procedure: CQC-Quadratic Combination), 2^6=64 sign permutations;
3. Hyper-ellipsoid envelope of simultaneous shell efforts in each element of the model, i.e. $\underline{x} = [N_{xx}, N_{yy}, N_{xy}, M_{xx}, M_{yy}, M_{xy}]^T$ according to the notation of Section 1.2, where (N_{xx}, N_{yy}, N_{xy}) are membrane efforts and (M_{xx}, M_{yy}, M_{xy}) are bending moments. The approximation of the hyper-ellipsoids is carried out using two procedures: polyhedron 384 vertices (3rd procedure: Ellipsoid 384 points), polyhedron 1920 vertices (4th procedure: Ellipsoid 1920 points);
4. Static load cases using modal linear combinations and considering 384 (5th procedure: Equivalent static 384 points, forces at the basis) and 1920 (6th procedure: Equivalent static 1920 points, forces at the basis) probable combinations of three total forces and three total moments at the base of the structure.

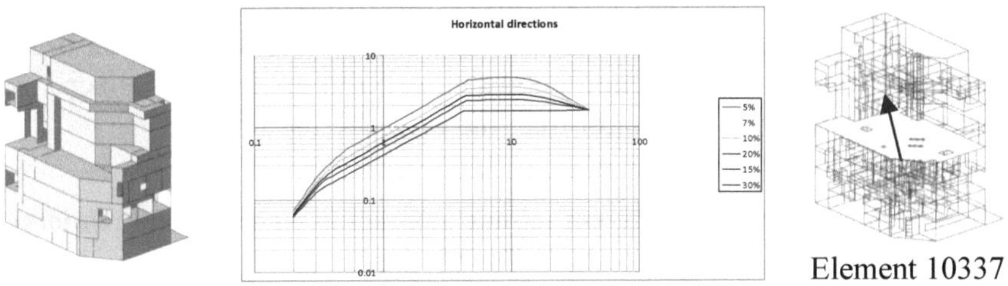

(a) (b) (c)

Figure 5: (a) Finite element model. (b) Pseudo-acceleration spectrum in horizontal directions (acceleration (m/s2) vs. frequency (Hz)). (c) Finite element considered in Figure 6

4.1 Structure description and modal analysis

The reinforced concrete building analysed here has the following dimensions: width 16.5m, length 27.5m, height 31.94m (Figure 5a). The finite element software used for the structural analysis is HERCULE. The number of nodes and elements is 14400 and 16900, respectively. The soil under the foundation raft is modeled by a set of vertical and horizontal linear elastic springs. After the modal analysis, 35 modes plus the pseudo-mode are retained (n=36). A spectrum analysis is then carried out using the pseudo-acceleration spectrum of Figure 5b. For the earthquake in vertical direction, the spectrum ordinate is reduced by a factor equal to 2/3. The load cases used in this example include the permanent load (G) and the seismic load due to earthquakes in directions x, y and z.

4.2 Comparison of several seismic analysis methods in terms of total reinforcement demand

Once the efforts are known for each element of the model, the reinforcement can be determined using the method proposed by Capra and Maury [10], which provides the required reinforcement area for both directions and for both upper and bottom layers of each shell element. The total reinforcement volume is estimated by summing the required reinforcement volumes of all shell elements. Table 1 gives the ratios of the total reinforcement volumes found by the six procedures, considering the "CQC-Newmark's Combinations" as reference method. One observes that the result obtained using the polyhedron enveloping the peak modal response hyper-ellipsoid (procedures 3 and 4) is very close to the reference one. As expected, the "CQC-Quadratic Combination" (procedure 2) gives the maximum reinforcement demand.

The reinforcement volume obtained by the static load cases (5th procedure) is more important than the reference one. However, the result is less conservative with the finer approximation of the 6th procedure. The difference between this approach and the hyper-ellipsoid envelope can be explained by the fact that the 6-dimension T-ellipsoid of the forces and moments at the building basis is discretized by a polyhedral envelope which is larger than the original T-ellipsoid.

Table 1: Comparison of the different seismic analyses in terms of total reinforcement demand

Procedure		Total reinforcement ratio
CQC-Newmark's Combinations (1)		1.00
CQC-Quadratic Combination (2)		1.58
Hyper-ellipsoid response envelope	Ellipsoid 384 points (3)	0.99
	Ellipsoid 1920 points (4)	0.97
Equivalent static loads	384 points, forces at the basis (5)	1.14
	1920 points, forces at the basis (6)	1.05

4.3 Efforts in a single shell element

The difference between the seismic analysis approaches presented in the previous paragraph can also be illustrated by plotting the points representing the combinations of the six efforts Nxx, Nyy, Nxy, Mxx, Myy, Mxy, in a single shell element of the structure. For the finite element indicated in Figure 5c, the projection of these efforts in the plane Nxx – Nyy is shown in Figure 6.

Figure 6 shows that (i) the points obtained by the "equivalent static load" approach, envelope almost all the points of the hyper-ellipsoid envelope of shell efforts and the CQC-Newmark's points. This explains why the reinforcement demand found by the "equivalent static" approach is more important than the ones found by approaches 1 and 3; (ii) the reinforcement quantity obtained by the "CQC-Quadratic Combinations" is the most important one. The efforts are strongly overestimated especially when an important correlation between shell efforts exists; (iii) for both the "hyper-ellipsoid response envelope" and "equivalent static loads" approaches, the reinforcement quantity is reduced when a finer polyhedral approximation is used (discretization with 1920 points).

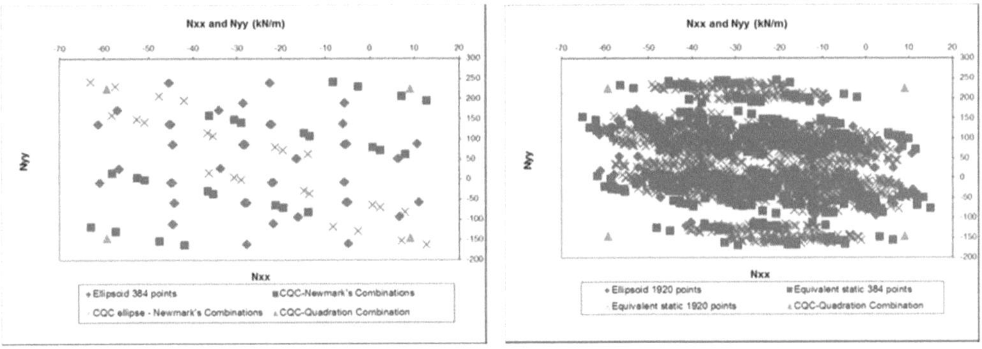

Figure 6: $N_{xx} - N_{yy}$ **: shell efforts obtained by seven different seismic analysis methods**

Figure 6 also shows the points coming from the sign definition discussed at the end of Section 3 ("CQC ellipse – Newmark's combinations"). One observes in the figure on the left that this approach gives results somehow opposed to the ones of the standard approach "CQC-Newmark's combinations" (red points): the points representing the efforts in the two cases are not grouped around the same diagonal. Moreover, the method "CQC ellipse – Newmark's combinations" seems to be closer to the reference solution ("Ellipsoid 384 points" or "Ellipsoid 1920 points") than the standard method. This simple example shows that the definition of the sign of directional CQC responses may affect the final Newmark's combinations. Hence, in order to avoid ambiguity and/or underestimation of efforts and steel reinforcement, the authors' opinion is that the ellipsoid method should be preferred for design calculations.

5 Conclusions

In the first part of the paper, the definition of the peak hyper-ellipsoid response envelope has been recalled. The definition of equivalent static load cases based on the hyper-ellipsoid envelopes has been given in Section 2. The use of response envelopes to define the signs of CQC combinations has been presented in Section 3. In the last Section, several seismic analysis procedures based on the response spectrum method have been applied to a NPP building. The total volume of steel reinforcement has been computed in each case showing that the seismic analysis method has a very important effect on the computed reinforcement, even though all the methods are based on the response-spectrum approach. A brief discussion about the definition of the sign of the directional CQC responses has also been presented.

References

[1] Der Kiureghian, A.: On reponse of structures to stationary excitation; in: Earthquake Engineering Research Center. University of California, Berkeley; 1979, Report No. UCB/EERC-79/32

[2] Menun, C.; Der Kiureghian, A.: Envelopes for seismic response vectors. I: Theory. II: Application; Journal of structural engineering; 2001, Pages 467-473

[3] ASCE: Seismic analysis of safety-related nuclear structures; in: Working *Group* on Revision of ASCE Standard 4; 2009, Draft

[4] Nguyen, Q.S. et al.: Comparison of several variants of the response spectrum method and definition of equivalent static loads from the peak response envelopes; in: 15th World Conference of Earthquake Engineering; 2012, paper 4269

[5] Nguyen, Q.S. et al.: Definition of static load cases equivalent to the seismic action using peak response hyper-ellipsoid envelopes: theoretical formulation, algorithmic aspects and application to a NPP building; in: SMiRT-22; 2013, in press

[6] Leblond, L.: Calcul sismique par la méthode modale. Utilisation des réponses pour le dimensionnement; Théories et méthodes de calcul; N°380. (1980), Pages 119-127 (in French)

[7] Vézin, J.M. et al.: Optimisation du dimensionnement sismique par analyse spectrale en utilisant le domaine de concomitance des sollicitations; in: Colloque AFPS; 2007 (in French)

[8] Martin, F.: Problèmes de concomitance en analyse modale spectrale. Combinations linéaires de modes – Domaine de concomitance; Internal report – IOSIS Industries; (2014), unpublished (in French)

[9] Newmark, N.M.: Seismic design criteria for structures and facilities Trans-Alaska pipeline system; Proceedings of the U.S. National Conference on Earthquake Engineering; (1975), Pages 94-103

[10] Capra, A.; Maury, J.F.: Calcul automatique du ferraillage optimal des plaques ou coques en béton armé; Annales de l'insitut technique du bâtiment et des travaux publics; N° 367 (1978)

Improvement of Seismic Response of an Industrial Structure

Milan Sokol, Rudolf Ároch

Slovak University of Technology
Faculty of Civil Engineering
Radlinského 11, 813 68 Bratislava, Slovakia
milan.sokol@stuba.sk
rudolf.aroch@stuba.sk

ABSTRACT:

An industrial heavy structure subjected to seismic action and its response after a few design improvements is presented. The difficulties of modelling of this industrial structure compared with ordinary structures is discussed, especially the effect of free hanged 250 tons of steel pipes used for medium cooling. The effect of long hanger (about 15m) swaying and its possible bouncing on steel casing should be minimalised. A detailed FEM model was prepared. Seismic effects were calculated via time history analyses. Five different alternatives of design improvements were taken into account. They differ by constructing difficulties and costs needed for achieving the desired effects. An introduction of seismic stoppers and dampers is considered too. Gap closing effects and contact forces calculation between different parts of the relatively moving structure are introduced too. The advantages of the best solution are discussed. The ratio of reduction seismic effects with and without appropriate measures is compared.

Keywords: non-linear time-history analysis, gap-closing effect, damping device

1 Introduction

Earthquake damage to the most common structural framing systems of industrial-facility constructions is summarized in [1].

This article shows some proposed measures that are expected to improve the seismic response of a waste-heat boiler structure. All constructive alternatives were modelled in great detail.

The main aim was to reduce the vibrations of the horizontal pipe bundles, especially in the transverse direction. The possible bouncing of the pipe bundles with the casing can cause serious damage. Each alternative was analysed by means of a non-linear time-history analysis.

S. Klinkel et al. (Eds.), *Seismic Design of Industrial Facilities*,
DOI 10.1007/978-3-658-02810-7_12, © Springer Fachmedien Wiesbaden 2014

2 Investigated structure

The investigated structure is an industrial steel structure of a waste-heat boiler shown on the following figure 1.

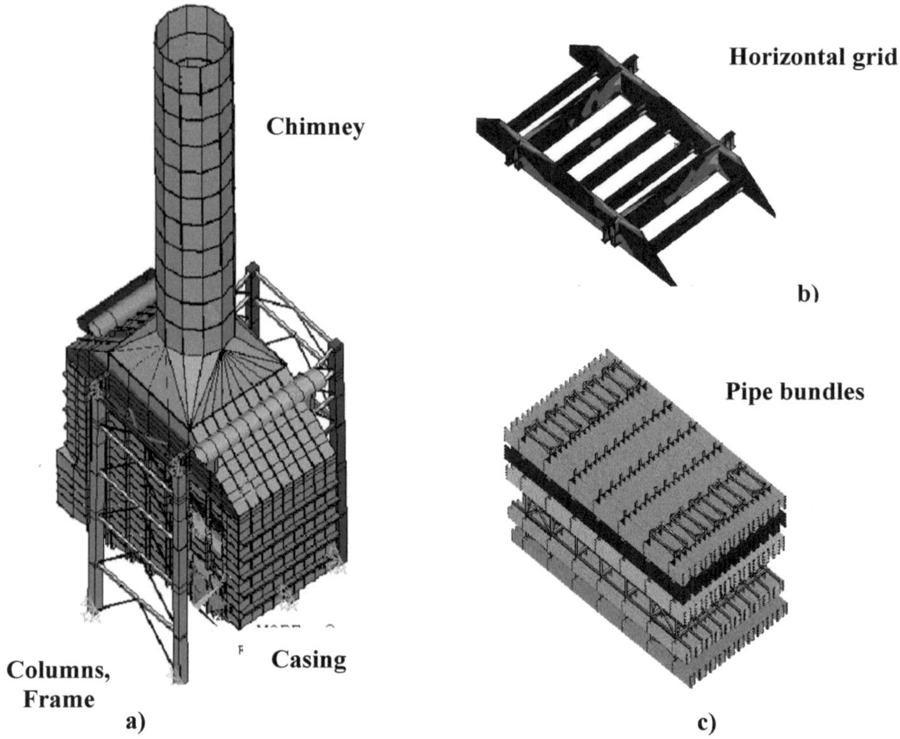

Figure 1: View of the structure (a), detail of horizontal grid (b) and pipe bundles inside casing hanged on the horizontal grid (c)

We can distinguish the following structural parts: **columns and frames, horizontal grid, chimney, casing, pipe bundles,** platforms, outer envelope, foundations, staircase. The parts listed in bold were part of the FEM model.

3 Structural model

Ansys [2] FEM programme was used to model the structure. The following elements were used:

Columns and frames - beam elements. All changes in the profiles were considered.

Horizontal grid (upper part of the boiler) - shell, beam and pipe elements. All important parts of the transverse and longitudinal beams were modelled with shell elements.

Chimney - shell elements. Modelled with no great detail, only to judge the global stiffness and mass.

The Casing – Chimney connection was, on the other hand, exactly modelled. All stiffeners were also considered.

Casing wall - shell elements and the stiffeners – eccentric beam elements. All important details were accounted for.

System of pipes in the boiler. Because the original system consists of 2300 pipes we modelled the system with a much lower number of pipes (105) but having the same global bending rigidity EI_y. The modelled pipe bundles are shown on figure 2 together with the gaps, which were modelled with gap elements.

A Rayleigh damping was assumed with the value of 1% of the critical damping [3].

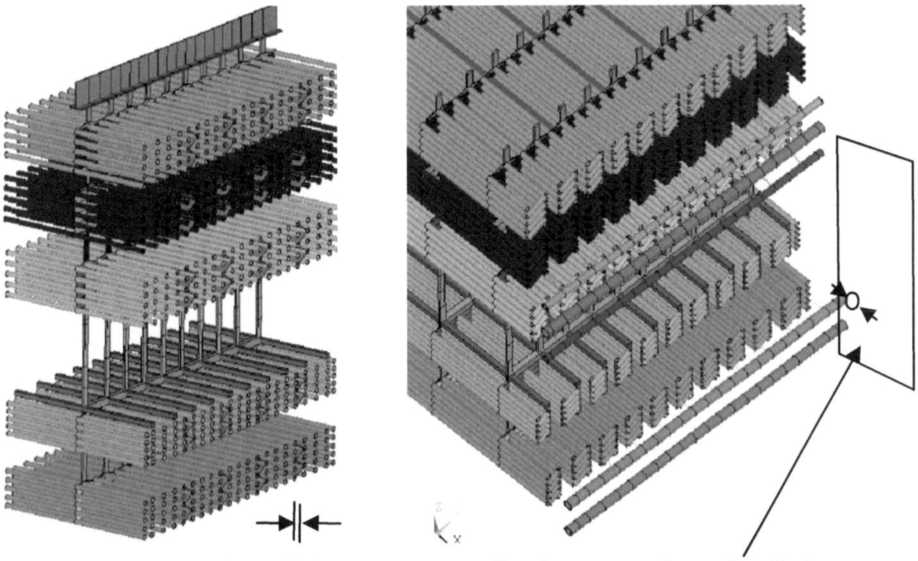

Gap between perforated plates in Y- direction ~ 26 mm

Gap between casing and collector in X- direction~50 mm

Figure 2: Pipe bundles with gaps

4 Seismic actions

We considered three ground types A to C according to EN 1998 [4], [5]. For each ground two peak accelerations $a_g = 1.5 ms^{-2}$ and $3.3 ms^{-2}$ were used. In the vertical direction an acceleration of 2/3 a_g was used.

Artificially generated ground acceleration time histories (compatible with the EN 1998 response spectra) were used (figure 3). A linear material constitutive law was assumed. The non-linearities are caused by opening and closing of the gaps, and using the discrete dampers with non-linear characteristic of the damping force.

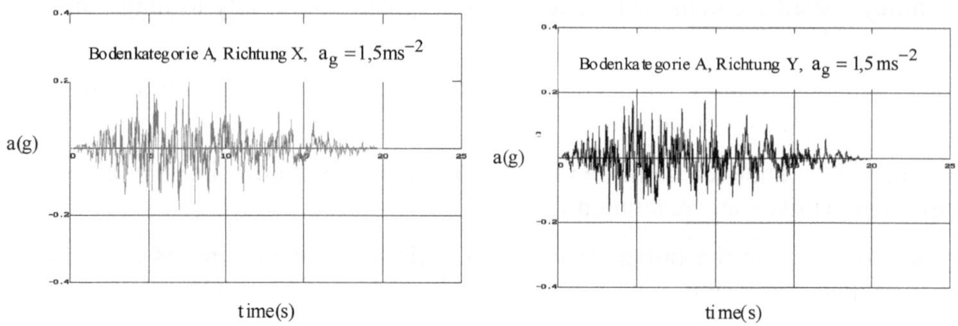

Figure 3: Ground acceleration time-histories (Ground type A)

Five different combinations of accelerations in the x-, y-and z-directions were used. These individual results were then also averaged (Figure 4).

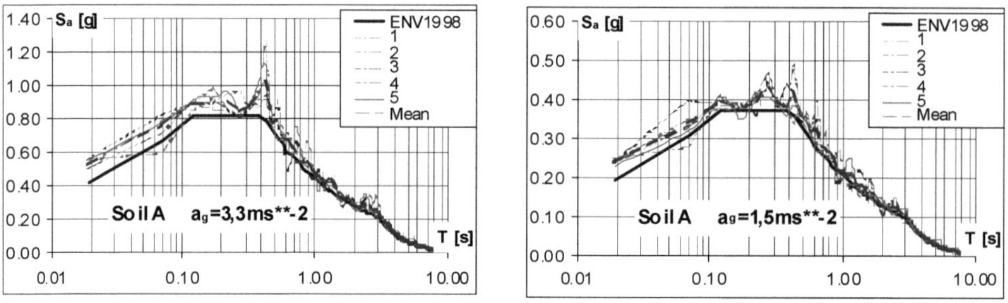

Figure 4: Response spectra (Ground type A)

5 Alternatives for strengthening

The freely suspended pipe bundles move horizontally independently and they can bounce on each other. To avoid this, the five bundles are connected in the longitudinal and transverse directions.

Five alternatives for strengthening were proposed. We will describe two of them.

5.1 Discrete dampers of the pipe bundles, casing fixed in the lower plane (Model 3)

The description of the concept of seismic isolation for earthquake protection and a review of the basic elements of a modern isolation system is given in [6]. We used discrete dampers shown on Figure 5. These are the so called high-capacity lock-up devices (Taylor) 2 x Model 600 kips, with the following parameters: damping constant min C=5.33 MN/(m/s), damping force $F_d = C\dot{v}^{0.3}$, where $\dot{v}$ is velocity, maximum velocity 0.9m/s, stroke 0.30m.

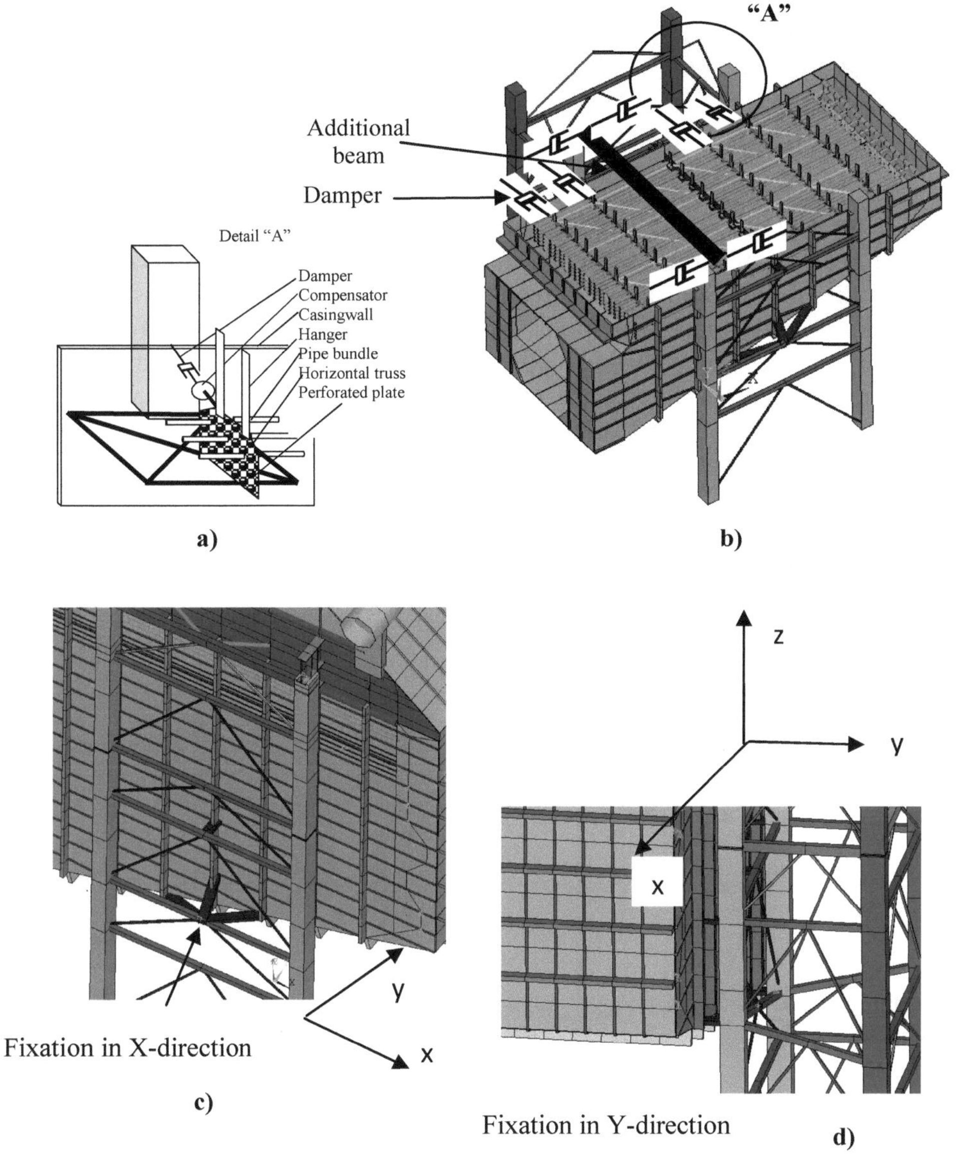

Figure 5: Discrete dampers on the pipe bundles (a) (b), casing fixed to the frame structure in the lower plane (c) (d)

These elements are connected in the longitudinal direction at the level +16.4m with the pipe bundles. Figure 5 shows the position of the dampers. Four dampers are effective in x-direction (longitudinal) and four others in transverse direction. The dampers are connected to the pipe bundle in the x-direction in the middle of the casing via an additional strong beam. This beam passes through the casing wall, where compensators should be placed. In the transversal direction the dampers are connected with the pipe bundle by means of a beam that passes through a compensator.

5.2 Truss bandage at the bumper level - casing swings freely (Model 4)

In this case a horizontal truss bandage is designed at the bumper plane level (Figure 6). The clearance between the truss and the pipe bundle is 10mm. This is needed because of large expansion due to temperature. The bandage is effective only in the Y-direction, the X-direction remains free. Nevertheless, the bandage also reduces the torsion oscillations, which can otherwise be very strong. The truss bandage is connected with the supporting frame at two places in the Y-direction where the horizontal forces are transmitted. When the bundles vibrate in the transverse direction, the perforated plates activate the bumpers and these forces are transmitted to the truss bandage.

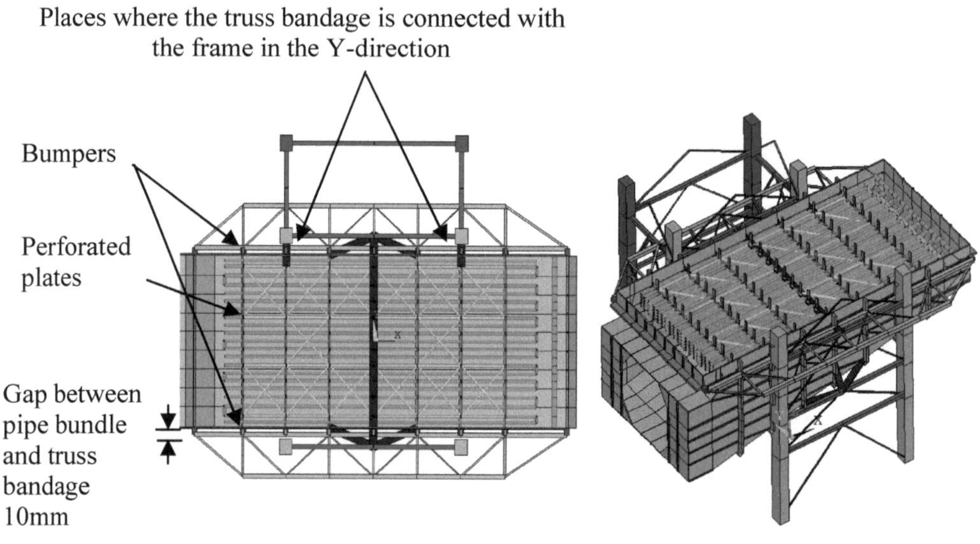

Figure 6: Truss bandage at the bumper level

6 Seismic response

To obtain a general idea about dynamic characteristics the eigenfrequencies and mode shapes are depicted on Figures 7 and 8. It is evident that the pipe bundles vibrate individually and bounce on each other (frequency No. 18, 1.46Hz, on Figure 7). The whole structure has a great mass that is located on the bearing columns in a large height.

The time-history analyses ([7], [8]) were performed for ground types A to C and with different peak accelerations a_g =1.5ms^{-2} and 3.3ms^{-2}. Six sets of results were so obtained each was calculated for five different combinations of accelerations.

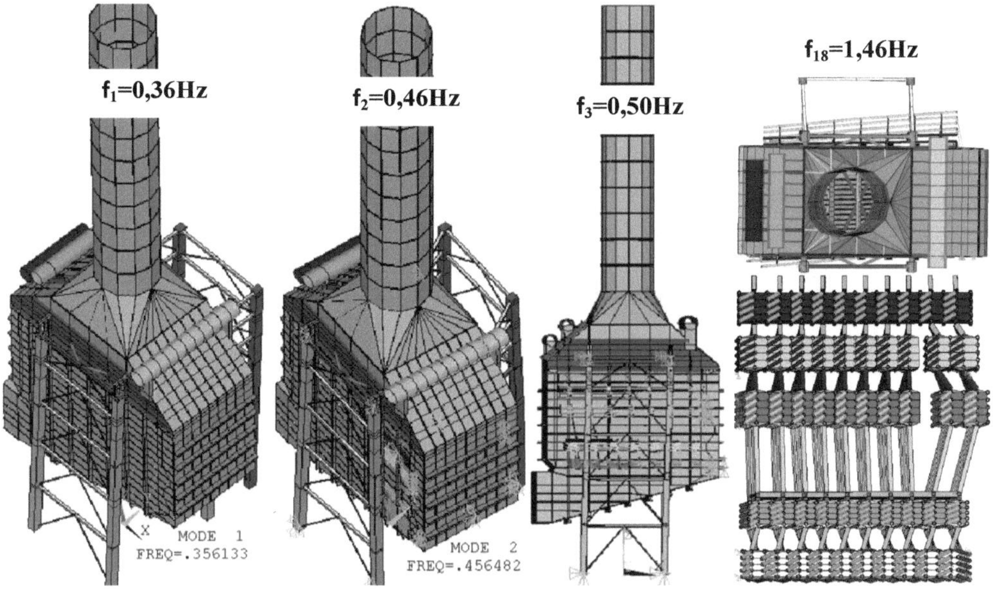

Figure 7: Eigenfrequencies and mode shapes of Model 3

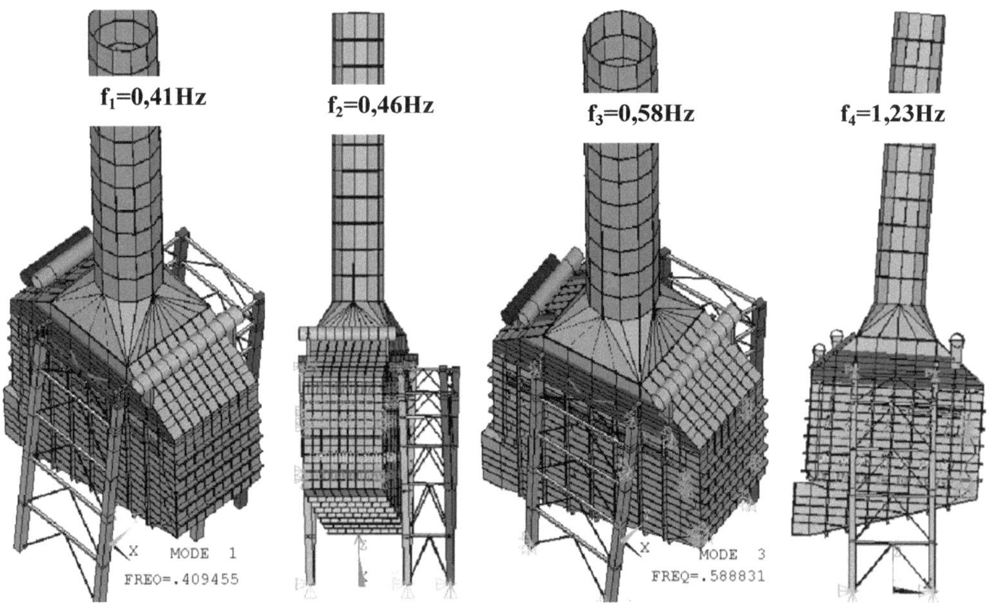

Figure 8: Eigenfrequencies and mode shapes of Model 4

We will present only sample results for ground type B, and 3.3ms^{-2} acceleration and first artificially generated acceleration time-history (B_33_01).

According to the time-history calculations the maximal X-deflection of the pipe bundle was 17cm (Figure 9a) and the maximal Y-deflection of the pipe bundle 32cm (Figure 9b).

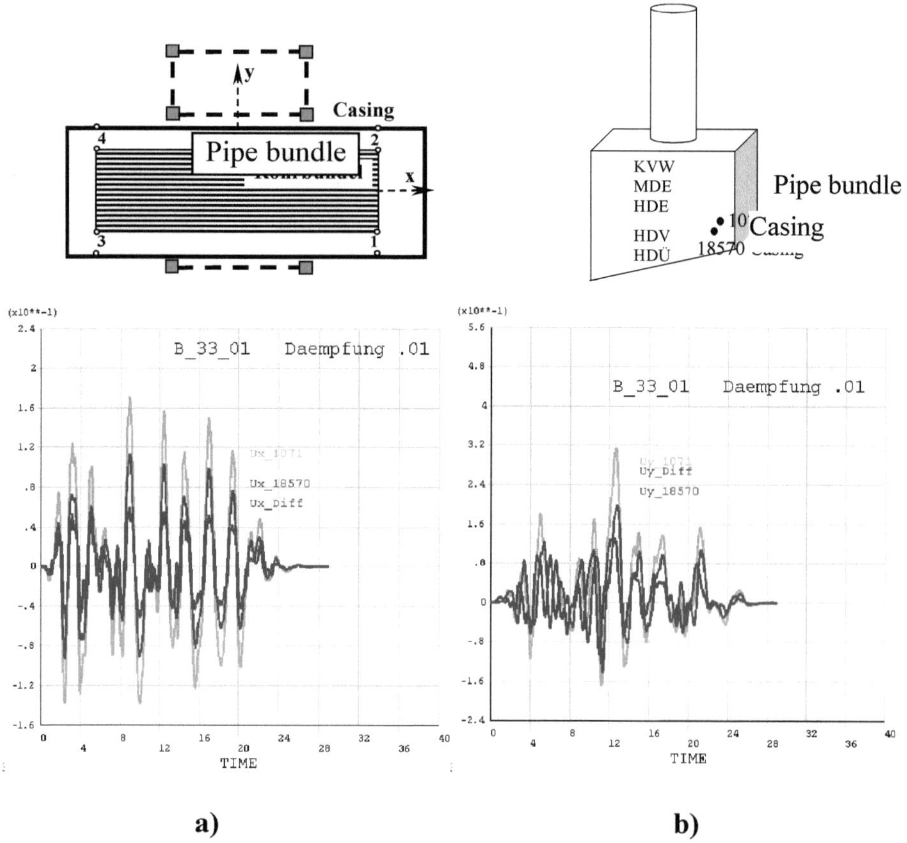

Figure 9: Seismic response of Model 3

The relative displacement (Uy_Diff) in Y-direction between the pipe bundle and casing was at the lower level 20cm (at the upper 8.5cm).

The relative displacement in Y-direction between the pipe bundle and the casing was for model 4 at the lower level 7,8cm (Figure 11) (at the upper 2,9cm) which was considerably smaller than for model 3.

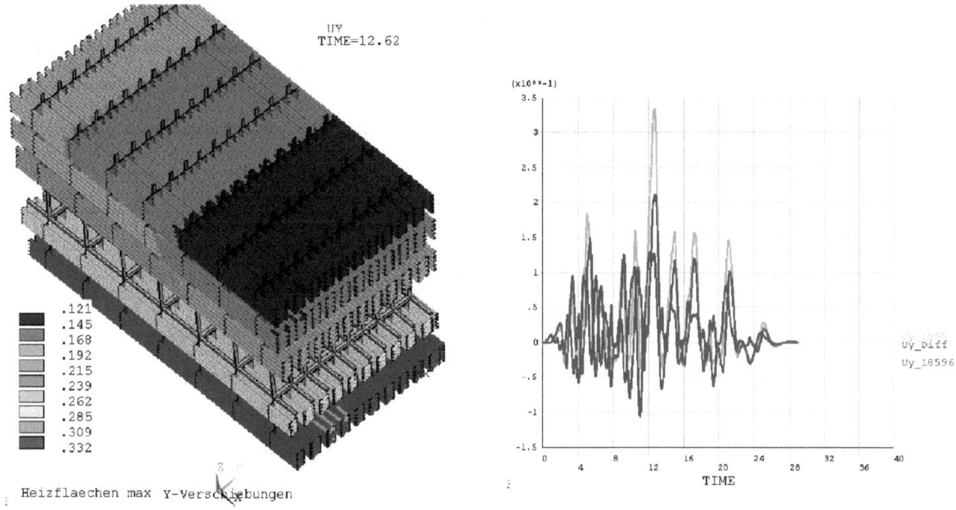

Figure 10: Pipe bundle displacements (Model 3)

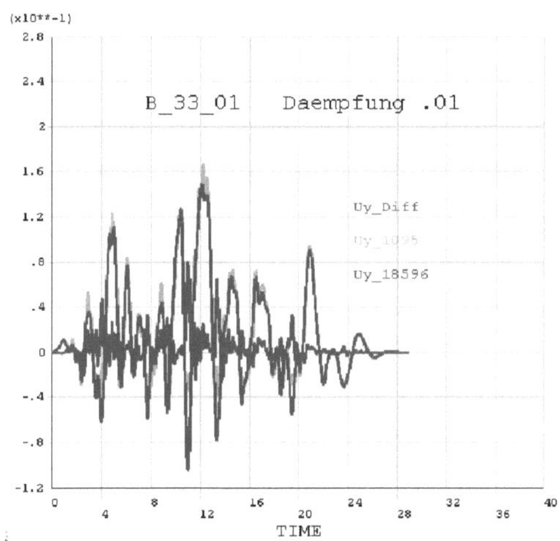

Figure 11: Pipe bundle displacements (Model 4)

7 Conclusion

Both stiffening alternatives decrease considerably the seismic effects which would otherwise reach up-to 50cm relative displacement between casing and pipe bundles. Table 1 shows the displacements of casing and pipe bundle in more detail.

The relative displacements between the pipe bundle and casing reached values up to 25cm according to the ground type and ground acceleration (Figure 12).

Table 1: Comparison of displacements between model 3 and model 4

X	Displacements					
	Pipe bundles		Casing		relative	
	lower	upper	lower	upper	lower	upper
	[m]	[m]	[m]	[m]	[m]	[m]
Model 3	0,187	0,171	0,126	0,134	0,065	0,042
Model 4	0,174	0,168	0,138	0,159	0,038	0,033

Y	Displacements					
	Pipe bundles		Casing		relative	
	lower	upper	lower	upper	lower	upper
	[m]	[m]	[m]	[m]	[m]	[m]
Model 3	0,292	0,217	0,153	0,153	0,182	0,076
Model 4	0,187	0,165	0,177	0,155	0,071	0,039

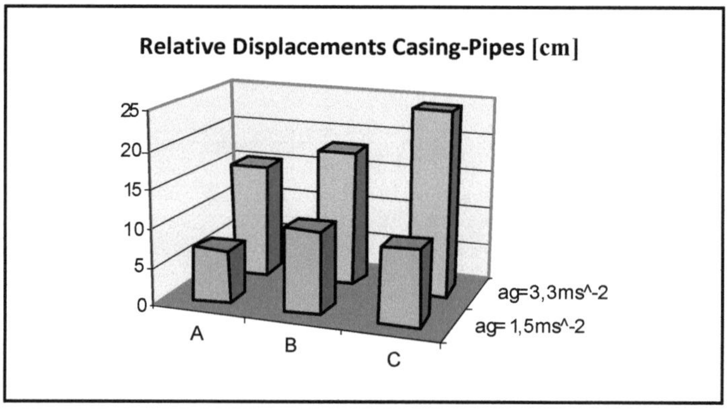

Figure 12: Relative displacement Casing-Pipes (Model 4)

Stiffening of the pipe bundles between the hangers (Model 4) decreases the relative displacements by more than 60%. The additional loading from the seismic actions in the columns is up to 80% (according to the ground type and ground acceleration) (Figure 13).

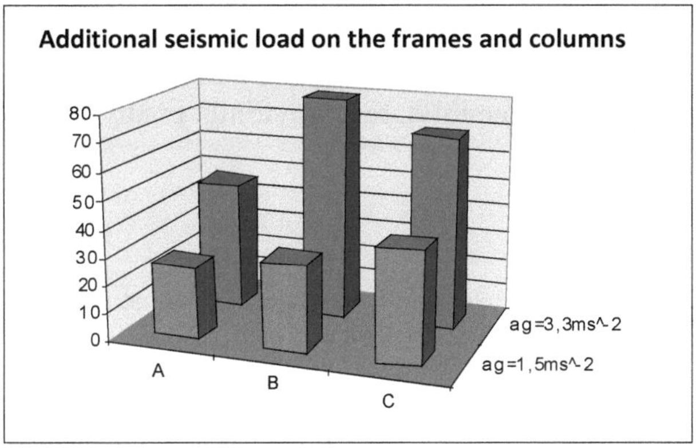

Figure 13: Additional seismic load on the frames and columns

The stresses of the horizontal grid increase by 55% (according to the ground type and ground acceleration) (Figure 14).

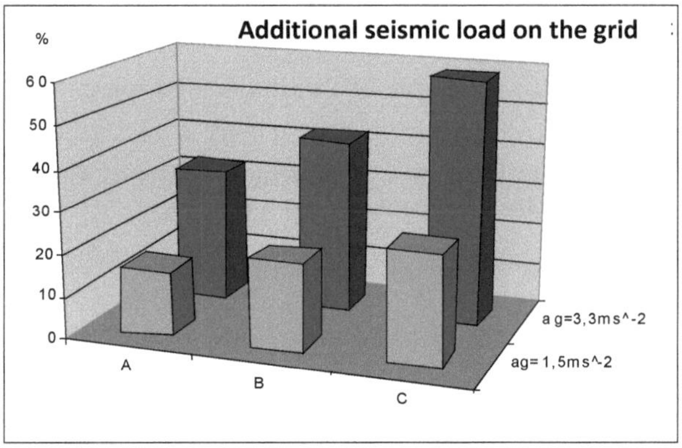

Figure 14: Additional seismic load on the grid

Also a static study was performed to model the impact of the pipe bundle on the web of the casing. As a loading of the casing a 12cm imposed deformation in the transverse direction (y direction) was considered. Especially the stiffeners were heavily loaded and the equivalent von Misses stress was exceeded almost twice.

Model 3 and Model 4 have a similar effectiveness concerning the reduction of relative displacements casing-pipe bundles. However the second one is more efficient because the cost intensive lock-up devices are not needed. The increased effect on the horizontal grid and the framing can be covered by local strengthening measures.

8 Acknowledgement

Authors thank the Grant agency of the Ministry of Education, Science, Research and Sports of the Slovak Republic for providing grant from research program VEGA Nr. 1/1119/11.

REFERENCES

[1] Sezen, H. and Whittaker, A.: Seismic Performance of Industrial Facilities Affected by the 1999 Turkey Earthquake., J. Perform. Constr. Facil., 20(1), 2006, pp. 28-36

[2] Kohnke, P.C.: Ansys, Eng. System, Theoretical Manual, Swanson Analysis System, 1989

[3] Flesch: Baudynamik, praxisgerecht, Band I, Berechnungsgrundlagen, Wiesbaden -Berlin, Bauverlag GmbH. 1993, 543 pp.

[4] EN 1998-1. Design of structures for earthquake resistance - General rules, seismic actions and rules for buildings. Brussels. 2005

[5] STN EN 1998-1/NA. Design of structures for earthquake resistance - General rules, seismic actions and rules for buildings. Slovak National Annex. SÚTN. 2009 Bratislava

[6] Buckle, I.G. and Mayes, R.L.: Seismic Isolation: History, Application, and Performance-A World View. Earthquake Spectra: May 1990, Vol. 6, No. 2, pp. 161-201

[7] Clough, R.W., Penzien, J.: Dynamics of Structures. Singapore, McGraw-Hill. 1993, 648 pp.

[8] Argyris, J., Mlejnek, H.P.: Dynamics of Structures. Texts on Computational Mechanics. Vol. 5. Amsterdam, Elsevier Science Publ. 1991, 606 pp.

Part V

Innovative Seismic Protection Systems

International Fusion Reactor Tokamak Complex Seismic Isolation

Stéphane Cazadieu[1], **Laurent Patisson**[2], **Sébastien Diaz**[3]

[1] EGIS Industries
4 rue Dolorès Ibarruri
TSA 50012
93188 Montreuil cedex - France
stephane.cazadieu@egis.fr

[2] ITER Organization
Route de Vinon sur Verdon
13115 Saint Paul Lez Durance - France
laurent.patisson@iter.org

[3] NUVIA Travaux Spéciaux
92 Cours Vitton
69006 Lyon - France
sebastien.diaz@nuvia-ts.com

ABSTRACT:

The International Fusion Reactor – ITER – is being designed and constructed with a high level of safety as an essential requirement.

In order to meet the safety and performance objectives of the French regulatory authorities and of the ITER Organization requirements, the Tokomak Complex has been isolated from the potentially highly damaging effects of the hazard seismic loading by employing seismic isolation bearings.

The Tokamak Complex seismic base isolation system and the Tokamak Complex structure have been designed by EGIS Industries as a member of the Architect-Engineer team ENGAGE.

The design, manufacturing, qualification and installation of the seismic isolation bearings have been carried out by NUVIA Travaux Spéciaux.

Keywords: Base isolation – Seismic isolation bearing – Low damping laminated elastomeric bearing

S. Klinkel et al. (Eds.), *Seismic Design of Industrial Facilities*,
DOI 10.1007/978-3-658-02810-7_13, © Springer Fachmedien Wiesbaden 2014

1 Introduction

Seismic isolation is an approach to earthquake-resistant design that is based on the concept of reducing the seismic demand rather than increasing the resistance capacity of the structure.

Application of this technology leads to the improved performance of structures, systems and equipment since they will remain essentially elastic during major seismic events.

The functional and operational requirements of the Tokamak machine and its associated systems require the Tokamak Complex to be protected from the damaging effects of seismic loading.

The high spectral accelerations of the design basis earthquake horizontal spectrum of the Cadarache site (0.739 g peak ground acceleration and 0.315 g zero period acceleration) offer a potential for seismic isolation implementation.

The Tokamak Complex measures 118.0m x 80.0m x 69.5m and is seismically isolated by 493 seismic isolation bearings as shown in Figure 1.

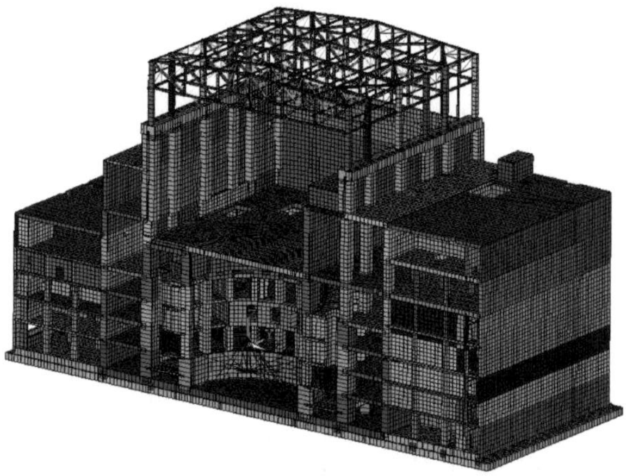

Figure 1: Tokamak Complex finite element model – Longitudinal cross-section – ANSYS model

The Tokamak Complex seismic isolation system consists of seismic isolation bearings supported on reinforced concrete plinths.

The seismic isolation system is located between the 1.50 m thick Tokamak Complex base mat (Upper base mat) and the 1.50 m thick Tokamak Complex seismic isolation structure base mat (Lower base mat) as shown in Figure 2.

The spatial arrangement of the seismic isolation bearing assemblies in the Tokamak Complex seismic isolation structure is shown in Figure 3.

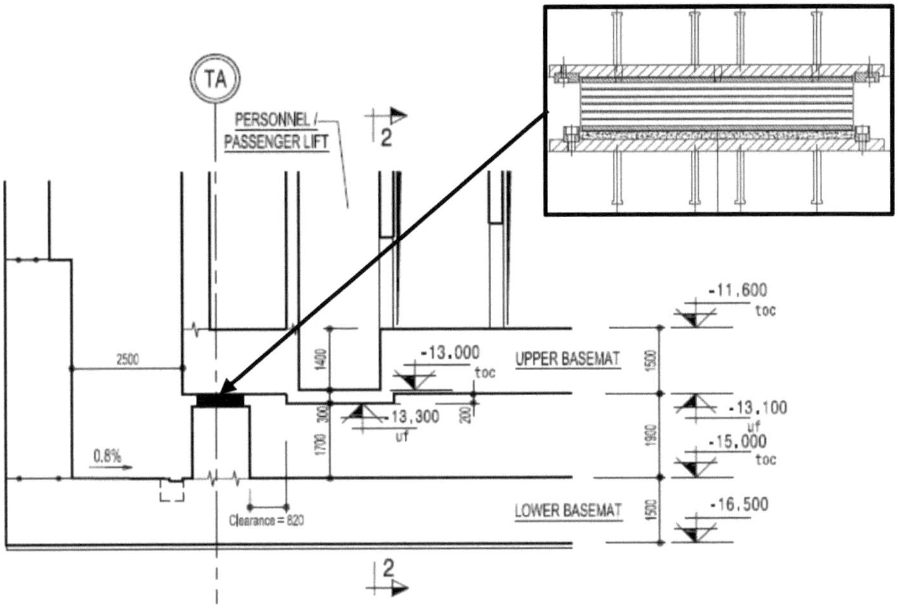

Figure 2: Seismic isolation system

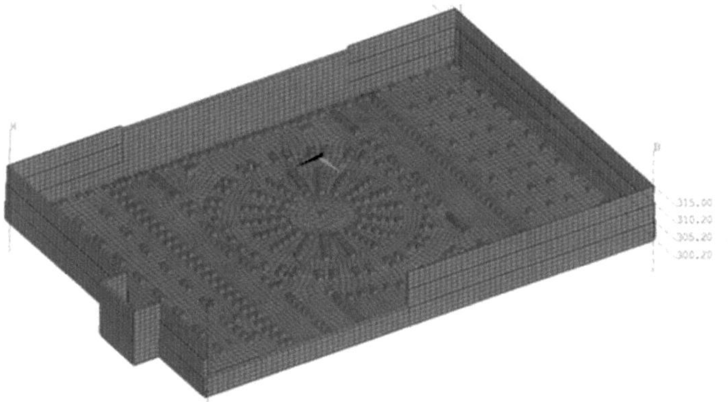

Figure 3: Seismic isolation bearing assembly arrangement inside the Tokamak Complex seismic isolation structure – ANSYS model

The total permanent gravity load supported by the seismic isolation bearings is approximately 3 180 000 kN.

The dimensions of the laminated elastomeric bearings are 900 mm x 900 mm x 181 mm thick, made of six layers of 20 mm thick chloroprene rubber, of five 5 mm thick inner reinforcing plates and of two 15 mm thick outer steel reinforcing plates.

The geometry of the seismic isolation bearing assembly is shown in Figure 4.

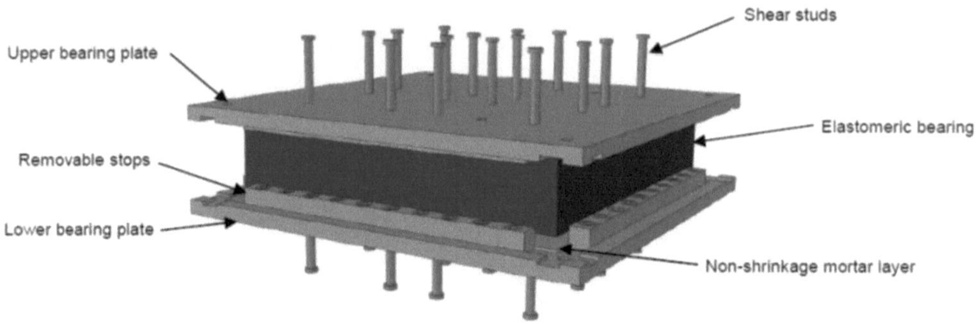

Figure 4: Seismic isolation bearing assembly

2 Design criteria

2.1 Seismic isolation

The seismic isolation system is designed in such a way as to perform its function in the expected conditions and according to the design requirements throughout the projected 70 years design life of the Tokamak Complex.

The seismic isolation system is seismically classified SC1(S) – [3].

Consequently, the seismic isolation system must remain fully operational after the design basis earthquake event (Seismic level 2 – SL-2) – [4].

The seismic isolation system must comply with the onerous requirements of ITER Structural Design Code – [4] – and the requirements of NF EN 1998-1 – [6].

The ITER Structural Design Code requirements that have to be met are – [4]:

- The inspection, maintenance and replacement of any seismic isolation bearing shall be possible at any time – (Requirement 1),
- The seismic isolation bearings shall be located immediately under or in the close vicinity of the Tokamak complex gravity-load resisting system – (Requirement 2),
- The horizontal distance between the center of gravity of the seismic isolation bearing stiffness and the center of gravity of the Tokamak complex shall be as low as practicable (Requirement 3),
- The minimum compressive stress on any seismic isolation bearing shall be 1.00 MPa under the seismic load combination at ultimate limit state (Requirement 4),
- The maximum compressive stress on any seismic isolation bearing shall not be more than 120 % of the average compressive stress under the quasi-permanent load combination at serviceability limit state (Requirement 5),

- Correspondingly, the minimum compressive stress on any seismic isolation bearing shall not be less than 80 % of the average compressive stress under the quasi-permanent load combination at serviceability limit state (Requirement 6),
- At least 90 % of the seismic isolation bearings meet the above two criteria (Requirement 7).

2.2 Seismic isolation bearings

The low-damping laminated elastomeric bearings must comply with the requirements of NF EN 1998-1, NF EN 15129 and NF EN 1337-3 – [6]-[7]-[8].

The mechanical performances required for the seismic isolation bearings are summarized in the following table:

Table 1: Required mechanical performances

Characteristics	Parameters	
Static parameters		
Static shear	Modulus	G_s = 0.97 MPa
Static compression	Stiffness	K_{vs} = 5200 MN/m
Dynamic parameters		
Dynamic shear	Modulus	At design displacement d_{bd}[1] and at a frequency of 0.55 Hz G_d = 1.10 MPa
Dynamic shear	Damping	At design displacement d_{bd}[1] and at a frequency ranging from of 0.50 Hz to 0.70 Hz $\xi_{ds} > 6$ %
Dynamic compression	Stiffness	At a frequency greater than 3.0 Hz and at a compressive force ranging from 0,5 to 1,50 of average compressive force N_{sd}[2] K_{vd} = 5200 MN/m
Dynamic compression	Damping	At a frequency greater than 3.0 Hz $\xi_{dc} > 6$ %

[1] d_{bd} = 112 mm / [2] N_{sd} = 6.32 MN

It should be noted that for the determination of the design displacement d_{bd}, the recommended value of the reliability factor γ_x of 1.20 in NF EN 1998-1 – [6] – is replaced by 1.00 according to ITER Structural Design Code – [4].

3 Design process

3.1 Seismic isolation

The arrangement, location and number of the seismic isolation bearings have been determined to satisfy the requirements by successive design iterations.

The iterative process that has been developed is:

- Verify the compliance of the selected arrangement with the requirements for inspection, maintenance and replacement of the seismic isolation bearings (Requirements 1 and 2 - Step 0),
- Develop a three-dimensional finite element model that includes the Tokamak Complex seismic isolation structure, the Tokamak Complex and the Tokamak Complex seismic isolation bearings for the selected arrangement (Step 1),
- Verify the compliance of the selected arrangement with the requirements on horizontal distance between the center of gravity of the seismic isolation bearing stiffness and the centre of gravity of the Tokamak complex (Requirement 3 – Step 2),
- Verify the compliance of the selected arrangement with the requirements on maximum and minimum compressive stress under quasi-permanent load combinations at serviceability limit state (Requirements 5, 6 and 7 – Step 3),
- Determine the seismically-induced forces and displacements on the seismic isolation bearings for the 24 combinations of the three orthogonal components of the seismic action (Step 4),
- Verify the compliance of the selected arrangement with the requirements on minimum compressive stress under seismic load combinations at ultimate limit state (Requirement 4 - Step 5),
- Verify the compliance of the seismic isolation bearings (Step 6),
- Revise arrangement, location and number of seismic isolation bearings until full compliance with the requirements is achieved.

Depending on the analysis, from six to eight iterations have been made to achieve full compliance with the requirements.

Of all the requirements, those relating to the maximum and minimum compressive stress under quasi-permanent load combinations at serviceability limit state have been, by far, the most difficult requirements to satisfy.

When, these requirements are satisfied the remaining ones are easily satisfied.

In this iterative process, the selection of the initial arrangement, location and number of the seismic isolation bearings is crucial.

In this initial selection, standardization of the reinforced concrete plinths and of the supported seismic isolation bearing assemblies has been introduced for an easy-to-build and cost-effective construction.

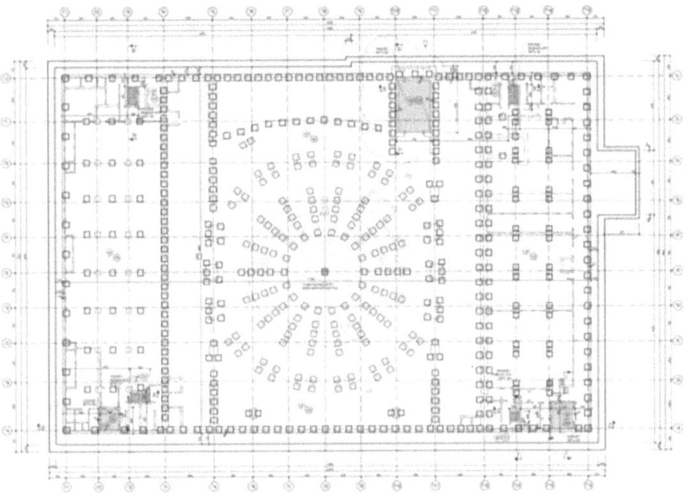

Figure 5: Tokamak Complex seismic isolation structure – Seismic isolation bearing assembly arrangement

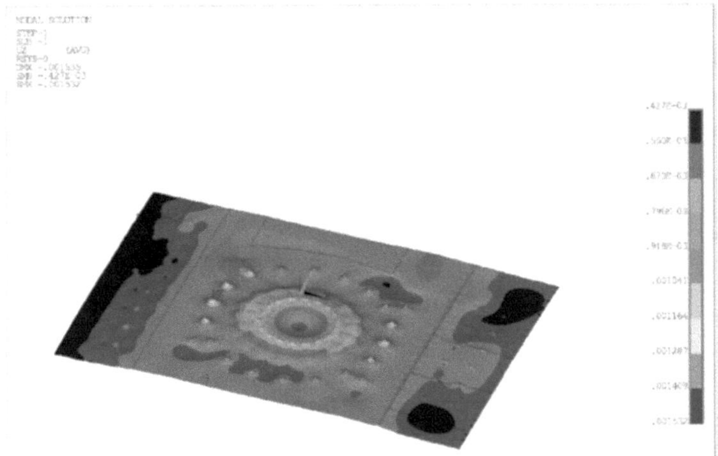

Figure 6: Tokamak Complex base mat (Upper base mat) – Vertical displacement under vertical component of the seismic action – m – ANSYS

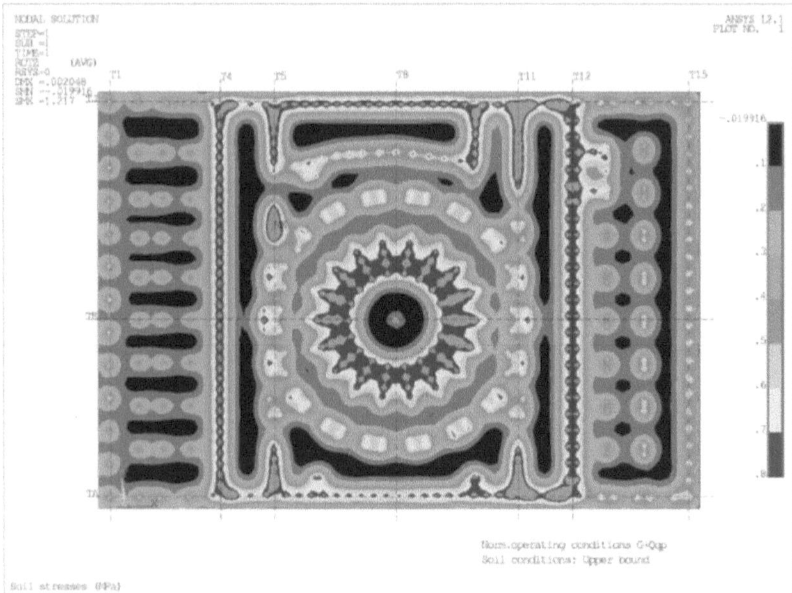

Figure 7: Tokamak Complex seismic isolation structure base mat (Lower base mat) – Ground bearing pressures – MPa – ANSYS

3.2 Anti-seismic bearings

The adequacy of the low damping laminated elastomeric bearing for their intended purpose has been demonstrated by ensuring that the requirements defined in NF EN 15129 – [6] – and NF EN 1337-3 – [7] – are complied with.

The total design strain $\varepsilon_{t,d}$ defined as the sum of the design strain due to the compressive load $\varepsilon_{c,d}$, of the design strain due to translatory movements $\varepsilon_{q,d}$ and design strain due to angular rotation $\varepsilon_{q,d}$ must not exceed the maximum permissible strain $\varepsilon_{u,k} / \gamma_m$.

$$\varepsilon_{t,d} = \varepsilon_{c,d} + \varepsilon_{q,d} + \varepsilon_{\alpha,d} \leq \frac{\varepsilon_{u,k}}{\gamma_m} \tag{1}$$

$$\varepsilon_{c,d} = \frac{N_{Ed}}{G_d \cdot \left(a' + b'\right) \cdot \left(1 + \frac{1}{2} \cdot \left(\frac{a'.b'}{\left(a' + b'\right).t_i}\right)^2\right) \cdot \left(1 - \frac{v_{d,x}}{a'} - \frac{v_{d,y}}{b'}\right) . t_i} \tag{2}$$

$$\varepsilon_{q,d} = \frac{\sqrt{v_{Ed,x}^2 + v_{Ed,y}^2}}{n.t_i + 2.t_e} \tag{3}$$

$$\varepsilon_{\alpha,d} = \frac{a'^2 . \alpha_{Ed,x} + b'^2 . \alpha_{Ed,y}}{2.n.t_i^2} \tag{4}$$

$$\varepsilon_{u,k} = 7 \qquad (5)$$

Where:

$a' =$	Width of the reinforcing plate,
$b' =$	Length of the reinforcing plate,
$n =$	Number of inner layers of chloroprene rubber,
$t_i =$	Thickness of an inner layer of chloroprene rubber,
$t_e =$	Thickness of upper and lower chloroprene rubber coating,
$N_{Ed} =$	Compressive force under seismic combination at ultimate limit state,
$v_{d,x} =$	Horizontal relative displacement in the direction of the width of the bearing under quasi-permanent load combination at serviceability limit state,
$v_{d,y} =$	Horizontal relative displacement in the direction of the length of the bearing under quasi-permanent load combination at serviceability limit state,
$v_{Ed,x} =$	Horizontal relative displacement in the direction of the width of the bearing under seismic load combination at ultimate limit state,
$v_{Ed,y} =$	Horizontal relative displacement in the direction of the length of the bearing under seismic load combination at ultimate limit state,
$\alpha_{Ed,x} =$	Relative angular rotation across the width of the bearing under seismic load combination at ultimate limit state,
$\alpha_{Ed,y} =$	Relative angular rotation across the length of the bearing under seismic load combination at ultimate limit state.

The reinforcing plate thickness t_s must be greater than the minimum reinforcing plate thickness $t_{s,min}$

$$t_s \geq t_{s,min} = Max\left(\frac{2.6.t_i.\gamma_m.N_{Ed}}{a'.b'.\left(1-\frac{v_{d,x}}{a'}-\frac{v_{d,y}}{b'}\right)}; 2\ mm\right) \qquad (6)$$

It should be noted that in Equation (1) the recommended value of the partial safety factor for the elastomer material γ_m of 1.00 given in NF EN 15129 – [7] – and NF EN 1337-3 – [8] – is taken as 1.15 in accordance with ITER Structural Design Code – [4].

It should be noted that in Equation (6) the recommended value of the partial safety factor for the elastomer material γ_m is taken as 1.00 as a design change formally managed via a Project Change Request (PCR) and instructed via a Service Order (SO).

Additional criteria regarding buckling stability, roll-over stability and minimum compressive stress for the seismic isolation bearings have also been verified – [7]-[4].

Full compliance has been demonstrated for each of the 493 seismic isolation bearings for each of the 24 combinations of the three orthogonal components of the seismic action – [9].

Figure 8: Fan-shaped arrangement of the seismic isolation system under the cryostat and Tokamak machine supporting structure

4 Conclusion

The stringent requirements of ITER Structural Design Code have made the design of the Tokamak Complex seismic isolation particularly challenging.

However, the final design solution has resulted in a cost effective structural arrangement with excellent seismic resistance capabilities.

5 Disclaimer

The views and opinions expressed herein do not necessarily reflect those of the ITER Organization.

References

[1] Nuclear safety authority safety rules – RFS 2001-01 – Seismic risk.

[2] Nuclear safety authority guides – ASN/GUIDE/2/01 – Considering seismic risk for nuclear facility civil works design – May 2006.

[3] Safety requirements for buildings – ITER Organization – PATISSON L. and STEWART P. – March 2013.

[4] ITER Structural Design Code for Buildings (I-SDCB) – Part 1: Design criteria – ITER Organization – BUET A., PATISSON L. and STEWART P. – June 2012.

[5] Load specifications for buildings with safety requirements – ITER Organization – BUET A., PATISSON L. and STEWART P. – March 2013.

[6] NF EN 1998-1 – Second issue – October 2010 – Design of structures for earthquake resistance – Part 1: General rules, seismic actions and rules for buildings.

[7] NF EN 15129 – First issue – January 2010 – Anti-seismic devices.

[8] NF EN 1337-3 – Second issue – September 2005 – Structural bearings – Part 3: Elastomeric bearings.

[9] Tokamak Complex – Construction design – Anti-seismic bearing assessment for updated finite element model configuration – EGIS Industries on behalf of ENGAGE – ENG 50 CR 11 0136 CW V2.0 – April 2013.

Strategies for the Seismic Protection of Power Plant Equipment

Peter Nawrotzki[1], **Daniel Siepe**[2]

[1] GERB Schwingungsisolierungen GmbH & Co.KG
Roedernallee 174-176, Berlin, Germany
Peter.Nawrotzki@gerb.de

[2] GERB Engineering GmbH
Ruhrallee 311, Essen, Germany
Daniel.Siepe@gerb.de

ABSTRACT:

The present paper shall give some ideas to protect power plant machinery against seismic demands. The elastic support of turbine foundations, fans, boiler feed pumps and coal mills is a well-accepted strategy for the dynamic uncoupling from their substructures and for the vibration isolation. If the corresponding bearing systems are combined with certain strategies an efficient earthquake protection for the important machinery can be achieved. Seismic control may be obtained by increasing the fundamental period or increasing the damping or changing the shape of the fundamental mode of a structure. A combination of these measures could lead to an optimum seismic protection system as described in this contribution. Here, the first step consists of the choice of the required stiffness properties of the flexible support. Helical steel springs possess the possibility of providing a three-dimensional flexibility. Thus, it is possible to obtain a vertically and horizontally acting protection system. Depending on the seismic input the spring properties could be chosen in a specific range. The system frequency can be decreased and simultaneously, the damping ratio can be increased by incorporating viscous dampers at different locations of the spring supported structure. Internal stresses of important members, acceleration amplification as well as deformations due to seismic excitation can be decreased compared to a structure without any precautions. The possible damage after a severe earthquake can be reduced significantly, and the behaviour of the structural members could remain in the elastic range. Details of executed projects and corresponding results of numerical analyses document the effectiveness of the presented seismic protection strategies. Selected pictures demonstrate the general applicability of the applied systems.

Keywords: Earthquake Protection, Passive Control, Flexible Support, Damping

DOI 10.1007/978-3-658-02810-7_14, © Springer Fachmedien Wiesbaden 2014

1 Introduction

The elastic support of machines or equipment has become "state of the art" to achieve an efficient vibration isolation. Vibration control could be defined as active, when the dissipation of vibrations from machines into the surrounding is prevented. On the other hand a passive vibration control protects machinery or equipment against vibrations from outside sources. For both approaches it is possible to use elements with helical steel springs. Each element may contain one or more springs. Concerning the type of the single spring a wide variety is available. Depending on the project requirements vertical natural frequencies of the elastic supported systems are in a range from 7 Hz down to 1 Hz.

Beneath the vertical flexibility the springs provide also a horizontal elasticity. In addition to the elastic support system often viscous dampers are installed to add damping to the system. The dampers limit the displacements of spring supported systems while they pass resonance during periodic excitation or when the systems are subjected to shock or random excitations.

The aspect of earthquake protection and the consideration of the load case "Seismic" are getting more and more important over the past years, surely influenced by the devastating seismic events during the last years. Therefore, it is a great advantage that the same elements, as described above, can be used to protect structures against earthquake by taking into account additional design criteria. Selecting the right properties of the elastic elements and of the dampers can lead to an optimum improvement of the structural behaviour due to seismic loading. Details of the required layout strategies are presented in this contribution.

After a brief outline of the fundamentals of some strategies for the seismic protection, two project examples for the earthquake protection of power plant machinery will be discussed.

2 Protection Strategies

The main objective of seismic control is the modification of the response of a structure due to seismic loading. This modification could be achieved by different methods:

- Modifying the shape of the fundamental mode,
- Increasing the fundamental period,
- Increasing the damping.

A structure like a machine together with its foundation can be dynamically uncoupled from the sub foundation or soil using an elastic support system. Usually, the machine and the foundation can be considered as one rigid mass, even if the machine itself is elastic. This assumption is valid if the supporting system is much more flexible than the machine and its foundation. This one mass system will possess six low natural frequencies and corresponding rigid-body mode shapes.

This change of the mode shapes leads to smaller internal deflections of the structure itself compared to a structure with a rigid base. The first natural frequency of a rigidly connected structure usually belongs to a mode shape including bending deformations. The change of the mode shape leads to less internal deformations and consequently smaller internal stresses.

Typically, the seismic demands for a project are defined by the description of the design response spectrum. Horizontal ground motions as well as vertical ground motions have to be considered. The vertical excitation should not be neglected as done in many current design codes. Depending on the frequency range of the highest induced accelerations (plateau area) a second protection strategy becomes possible. The elastic support could lead to a reduction of the predominant frequency (= increasing fundamental period) from the plateau values down to lower acceleration levels. As an example, it is assumed that the plateau area starts at 2.5 Hz and that the dominant frequency of the unprotected systems is nearly around this value. If the horizontal frequency of a system with a passive elastic control system is about 1 Hz the seismic demands could be reduced by about 60 %.

The third measure in using passive seismic control systems is the increase of damping. The corresponding reduction of the induced structural responses by the increase of viscous damping can be taken from different national and international earthquake standards. Eqs. (1) shows the formula of the Eurocode 8 also published by the DIN [1].

$$\eta=\sqrt{\frac{10}{5+\xi}} \geq 0.55 \qquad (1)$$

Here, the viscous damping ratio ξ of the structure should be expressed in per cent. According to Eqs. (1), an increase of structural damping from 5 % to 15 % causes a reduction of input acceleration, structural stress, strain and displacement in a range of about 30 %.

An optimum adjustment of frequencies and damping ratio by the use of a passive control system could lead to significant improvement of the seismic behaviour of the protected structure. For every project the specific requirements have to be considered during the layout of the control system. A very low frequency, for example, may lead to very low seismic accelerations, but may yield larger displacements of the supported structure. Here, it is important to find an optimum between earthquake protection and boundary conditions.

Helical steel springs and viscous dampers are one type of passive control devices that are suitable for the described mitigation measures. An example of these devices is shown in Figure 1.

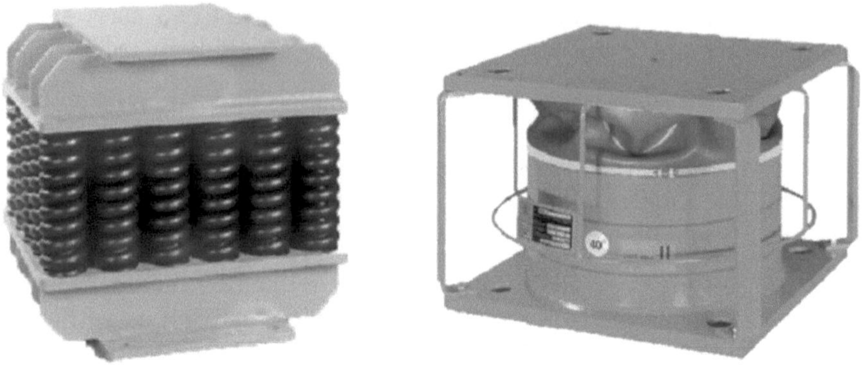

Figure 1: Spring element and viscous damper

It is well known that springs are acting in the axial direction, but they also possess a horizontal flexibility and corresponding load bearing capacity. The mechanical properties can be characterized by a linear elastic behaviour in horizontal and vertical directions. The viscous damper provides linearly velocity-dependent forces in all three spatial directions. The system behaviour can be described by the general equation of motion with constant coefficients. Due to the linearity it is easily possible to determine the system behaviour by standard procedures in regard to system frequencies, critical damping ratios and seismic effects.

The application of springs and dampers leads to a three dimensional seismic protection system. The combination of reduced frequencies and increasing structural damping yields efficient seismic protection of a structure. Accelerations and hence internal stresses are significantly reduced. Theoretical and experimental investigations with shaking-table tests, as shown by Rakicevic et al. [2], approved these positive effects. To distinguish the protection system from well-known base-isolation systems, where e.g. rubber bearings or friction pendulum systems are used, it is entitled as Base Control System (BCS). Beneath other advantages a BCS can reduce the effects of horizontal ground motions and of vertical ground motions as shown by Chouw [3].

It is possible to adjust the parameters of the BCS in regard to the requirements of the project, as the elements vary especially in the bearing capacity, in the horizontal and vertical stiffness properties, in the ratio between horizontal and vertical stiffness and in the damping. In this context, two example projects will be introduced in the following sections. One example describes the application of an elastic support system for an emergency diesel generator set in a high seismic zone. The other example presents the effects of different support systems for a turbo generator deck.

3 Project Example: Diesel Generator Set

Diesel engines are used for many purposes – in trains and ships as well as for local power generation. A typical situation for emergency diesel generator sets (EDGs) in nuclear power plants is shown in Figure 2.

Figure 2: Spring supported diesel engine

These systems are very important in regard to the safety of a nuclear power plant. In case of failure of external power they supply power for all safety related systems. Thus, the layout of an elastic support system, providing vibration isolation and seismic protection, is an ambitious task requiring special attention.

In the range of seismically significant frequencies this type of machine can be regarded as somehow “rigid”. Having a look at the entire system, subsoil conditions are often responsible for structural frequencies within the highest level of seismic amplification. Thus, the improvement of the seismic resistance can be achieved by changing the support conditions. For this project, located in a high seismic zone in Turkey, helical steel springs are used for vibration isolation purposes. A horizontally flexible layout and additional viscous dampers significantly improves the seismic performance of the supported structure.

Here, the stiffness properties of the spring devices are chosen so that structural frequencies in the horizontal directions can be found in a range between 0.8 and 1.4 Hz. A viscous damper is integrated in the used elements, leading to damping ratio for these modes in a range of 15 % and 20 %. The reduced frequencies and the increase of damping yield a significant reduction of accelerations at the machine. Concerning the low frequency of the system it is very important to have a close look at the corresponding seismic displacements. For diesel engine sets the displacements at the coupling and/or at the turbo charger connections are limited in order to avoid damage. At the same time, the vertical flexibility of the steel springs has to be chosen in order to provide sufficient vibration isolation efficiency.

4 Project Example: Turbine Generator System

The project Anpara-D is an extension of the existing thermal power station at Anpara in Uttar Pradesh, India. The two new turbine units from BHEL (Bharat Heavy Electricals Ltd.) provide a capacity of 2x500 MW. The site is located in a high seismic zone with a peak ground acceleration of about 0.22 g, thus the design and layout of the turbine deck had to consider seismic effects. The seismic behaviour of a conventional, rigidly supported turbine deck was compared with the behaviour of a spring supported deck during a case study including a seismic calculation of the whole structure. For the calculations a three-dimensional model of the system is used. This model consists of the T/G-Deck, the spring devices and the substructure. Altogether three different systems are investigated:

- System without spring devices,
- System with spring devices type 1,
- System with spring devices type 2.

The ratio of the vertical to the horizontal stiffness plays an important role in regard to the seismic behaviour of the structure. Therefore, two different type of spring devices are used for the calculations. The devices of type 2 possess a higher ratio than the devices of type 1. The sketch of the finite-element model is presented in Figure 3.

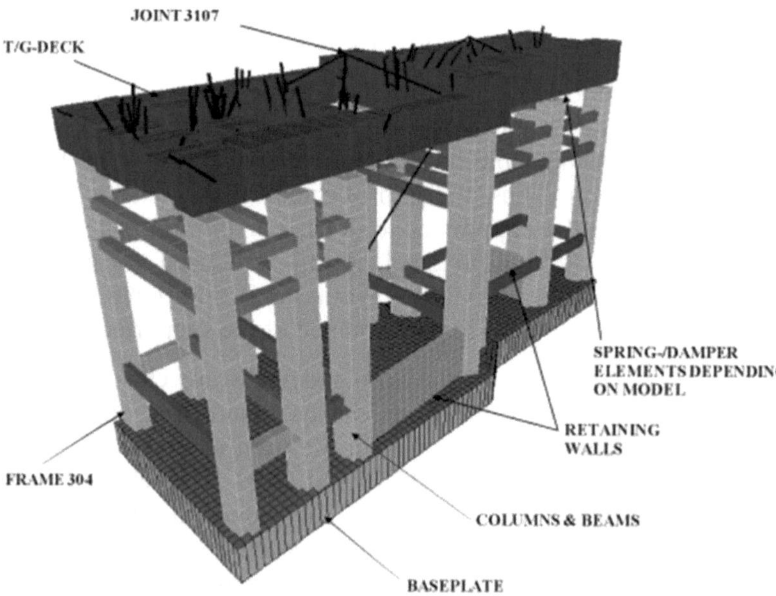

Figure 3: FE-Model of spring supported TG foundation

The introduced seismic protection strategies are considered by modifying the mode shape due to a spring support of the turbine deck. The spring devices reduce the frequencies and the used viscous dampers increase the damping. The efficiency of the elastic support systems becomes obvious, when the spectral accelerations of the three different systems are incorporated into the plot of the elastic design response spectrum for a damping of 5 % as shown in Figure 4.

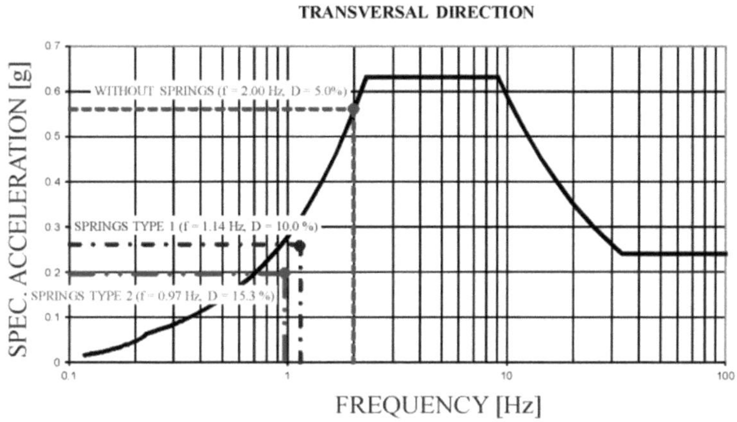

Figure 4: Effects of frequency reduction and increase of damping

The frequency range with the highest induced acceleration starts at about 2.3 Hz. As an example, the results of the transversal direction are presented. Using the first type of spring devices the frequency of the first fundamental mode could be reduced to about 1.14 Hz with a damping ratio of 10 %. Herewith, the induced demands are reduced to about 0.27 g in comparison to 0.57 g for the original system without any elastic devices.

In a second step, spring devices with a higher ratio between vertical and horizontal stiffness are implemented. In addition to the lower frequency of about 1 Hz the damping ratio is increased up to approximately 15 %. This leads to a further reduction in spectral acceleration to about 0.2 g. The same efficiency can be found for the longitudinal direction of the system.

The results of response spectrum analyses of the three systems verify the protection strategy. The values of the model without elastic devices are used as reference values (100%). The ratio between these values and the results of the model with spring devices is expressed in per cent. The output is listed in Table 1.

Accelerations as well as internal stresses are significantly reduced. Due to the positive effects finally the spring devices type 2 are applied for this project.

For the described project the direct substructure below the passive control system was not integrated into the structure of the surrounding machine house. If possible, this integration can provide several further advantages as shown by Basu et al. [4]. Space and construction time could be saved beneath the improvement of the seismic behaviour of the structures.

Table 1: Results of system without springs used as reference value (100%)

Excitation in transverse direction	Without springs	Springs type 1	Springs type 2
Abs. acceleration at shaft level (Joint 3107)	100 %	55.4 %	40.5 %
Bending moment at column (Frame 304)	100 %	69.2 %	47.4 %
Shear force at column (Frame 304)	100 %	49.8 %	33.8 %

5 Conclusion

After a brief outline of several seismic protection strategies, two practical examples for elastic supported machines are discussed. Optimizing of the parameters of the elastic devices, used already for providing vibration isolation, leads to a Base Control System, consisting of helical steel springs and viscous dampers. This system yields efficient earthquake protection of a structure by reducing accelerations and hence internal stress and strain values.

The consideration of seismic effects will play an increasingly important role for different projects, so that effective protection systems will be required. The presented strategies have already been proven in many completed projects worldwide and could be used for new projects in the future.

REFERENCES

[1] DIN Deutsches Institut für Normung e.V.: DIN EN 1998-1 Eurocode 8: Design of structures for earthquake resistance – Part 1: General rules, seismic actions and rules for buildings; German version EN 1998-1:2004 + AC:2009; Beuth Verlag GmbH, Berlin, 2010

[2] Rakicevic, Z.; Jurukovski, D.; Nawrotzki, P.: Analytical modeling of dynamic behavior of a frame structure with and without Base Control System; in: Proceedings of the 4th World Conference on Structural Control and Monitoring; 2006

[3] Chouw, N.: Reduction of the effects of strong vertical ground motions on structural responses; in: Proceedings of the 7th U.S. National Conference on Earthquake Engineering; 2002

[4] Basu, A.K.; Nawrotzki, P.; Siepe, D.: Spring supported turbo-generator with steel supporting columns integrated into the steel structure of the turbine building; in: Proceedings of POWER-GEN India & Central Asia; 2011

MARMOT – A Certified Seismic Monitoring System for Vulnerable Industrial Facilities

Andreas Stiegler[1], Hans-Jürgen Nitzpon[1], Werner Bolleter[1]

[1] SYSCOM Instruments SA
Rue de l'Industrie 21, 1450 Sainte-Croix, Switzerland
stiegler@syscom-instruments.com

ABSTRACT:

The MARMOT seismic monitoring and trip system perfectly responds to the increasing safety demand in vulnerable industries such as Nuclear Power Plants (NPP), Nuclear Storage Facilities, Liquid Natural Gas Storage (LNG), Refineries and many more. The system measures and analyses systematically tremors that occur at different locations in a facility and quickly recognizes dangerous patterns. With its distributed intelligence it guarantees dependable alarms for automatic shutdown (trip) information impacted by earthquakes on the structures. MARMOT complies with all relevant standards (e.g. IEC 61508, IEC 60780, and IEC 60880) applicable in these industries, fully tested and certified by the "TÜV-Rheinland" organization.

This paper presents requirements and the corresponding MARMOT solution regarding seismic monitoring for industrial facilities.

Keywords: Safety, Seismic monitoring, Earthquakes, Trip System

1 Introduction

When the massive March 2011 earthquake and Tsunami badly damaged the Fukushima-Daiichi Nuclear Power Plant in Japan (Figure 1), the safety issue was again the focus of industry, government and public attention.

Responsible authorities and experts defined and conducted stress tests, which analyzed the nuclear power plants worldwide. Many recommendations were made on overall plant design, operation and procedures. The mechanical design of components and the operation of seismic monitoring systems were also considered for events larger than design basis events.

Seismic instrumentation systems have been successfully utilized, primarily for structural monitoring, to determine whether the effect of an earthquake has exceeded the plant design specifications. On a smaller scale, seismic systems have also been used as "trip" systems for automatic plant shutdowns under specified conditions.

S. Klinkel et al. (Eds.), *Seismic Design of Industrial Facilities*,
DOI 10.1007/978-3-658-02810-7_15, © Springer Fachmedien Wiesbaden 2014

Figure 1: Fukushima Daiichi after March 2011

Tests and analysis carried out after the Fukushima event identified considerable optimization potential in seismic instrumentation. Three areas have been examined:

- Maintenance Parts Management: Many original system components are no longer available and many plants are still operating with systems that can no longer be supported. Repairs require similar available components. The performance quality and ongoing reliability of such obsolete systems may be degraded.
- New Technical Requirements: International regulations (IEC) have evolved and now specify solutions that reflect the current state of the art. In particular, the requirements for accuracy, recording time and electromagnetic immunity have increased. Technology developments can now provide both system reliability and performance that meets and exceeds the most rigorous standards.
- Software: As in hardware design, quality requirements for instrument firmware and system software are now specified in a detailed fashion.

The following sections describe the seismic monitoring system MARMOT as a perfect solution for industrial facilities, which fully meets all of today's requirements for both seismic monitoring and safety systems. It provides detailed information about the qualification process and related quality assurance.

2 Seismic Monitoring

A seismic instrumentation system monitors the impact of an earthquake at critical locations in an industrial facility, and at a free field location unaffected by the buildings. It records the structural vibration at each location and promptly reports whether the structural response has exceeded specified levels in both time and frequency domain. Class-A “Safety System” components can be added to the system to provide signals for the automatic shutdown of critical operations (e.g. reactors, gas turbines).

2.1 Tasks of Seismic Monitoring and Seismic Safety Systems

Earthquake monitoring and safety systems perform four major tasks:

- Recording of earthquakes at free field location: Free-field recording in three orthogonal axes is used to determine the vibration excitation (i.e. input ground acceleration unaffected by the building structures) as well as calculation of seismic intensity (e.g. Cumulative Absolute Velocity).
- Recording seismic events in the structure: Event records in three orthogonal axes at several points in the structure allow engineers to assess the impact of the earthquake and the amplification or attenuation of the vibration at critical points in the plant. They are also utilized to identify whether critical thresholds have been exceeded in the time domain.
- Providing input for automatic plant shutdown: Qualified Class-A Seismic Safety Systems can provide a signal to a reactor trip (shutdown system) when a critical acceleration threshold has been exceeded. Ideally, these systems should be integrated into the monitoring system to record the events and provide traceability. Extremely high reliability is required for trip systems.
- System management and evaluation of earthquake records: The system manager software must assure that: all events are recorded synchronously; real events (earthquakes) are recorded on all recorders (system voting logic); and, the monitoring system is working properly (state of health monitoring and reporting). The system maintains comprehensive operating logs, transfers recorded event data to a computer, which analyzes the data and provides required spectral analysis and seismic intensity reports to engineers within minutes of any event.

2.2 The MARMOT earthquake monitoring system

This system was named after the marmot, an animal known for one of nature’s most effective warning systems. The MARMOT is based on a highly reliable distributed and redundant recording system design concept coupled with solid-state sensors. The newly developed and fully-certified MARMOT system is based on SYSCOM's 25 years of experience in the development and production of strong motion instrumentation primarily for the nuclear industry. It meets highest level requirements for any seismic monitoring or safety system application.

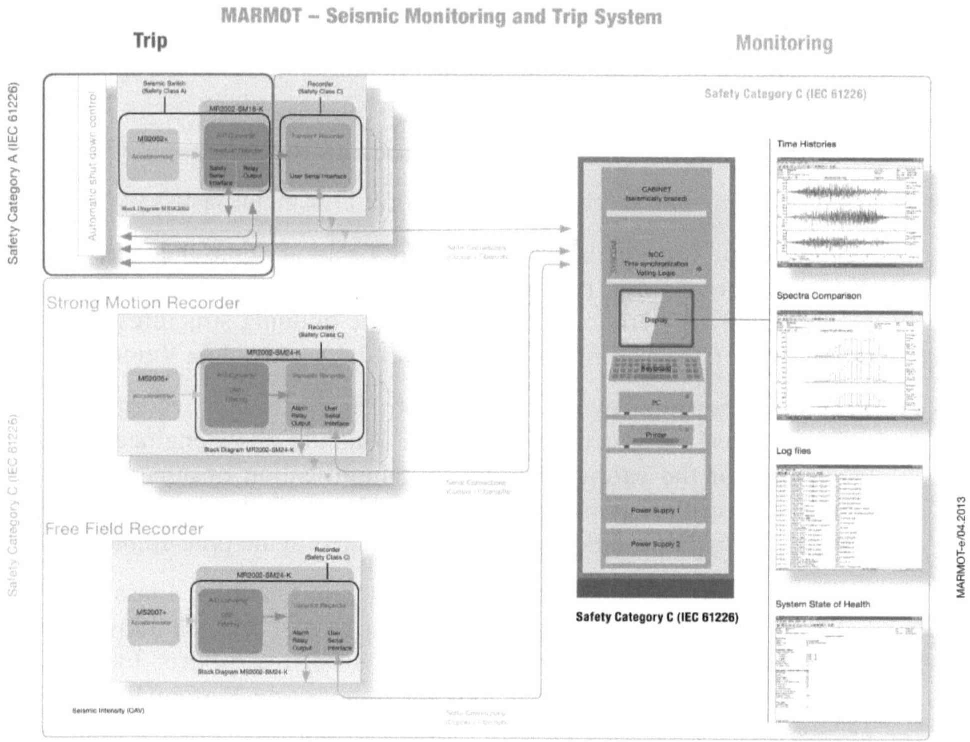

Figure 2: MARMOT system diagram

Three types of recorders are shown on the left side of Figure 2:

- Safety Class-A seismic switch / strong motion recorder (top).
- structural strong motion recorder (middle);
- free field strong motion recorder (bottom);

An industrial facility may contain multiple strong motion recorders and seismic switches in the structure in addition to the free field station. All stations record seismic events simultaneously; the Safety Class-A seismic switches additionally provides local alarm relay contacts for the automatic shutdown control system.

The Safety Class-A seismic switch / strong motion recorder, typically used in triple redundancy to reach the required safety integrity level (SIL3), serves primarily to trigger an alarm in case of critical threshold exceedance in safety critical areas, at the same time to record the event for further analysis.

The structural strong motion and free field recorders capture seismic events and can provide alarms in case of threshold exceedances for less safety critical areas.

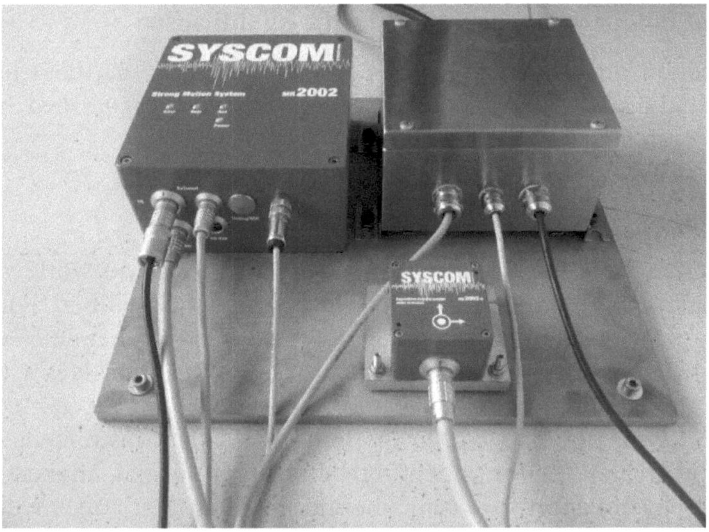

Figure 3: Strong motion recorder (MR2002), Accelerometer (MS2002) and stainless steel junction box mounted on a platform

All instruments are connected to the Network Control Center (NCC), which acts as a system manager that is built into a seismically braced cabinet. Either, or both, fiber optic or copper cable can be used to connect the instruments to the NCC. When obsolete seismic systems are replaced, the existing copper cables may be utilized. The remote stations may be powered from central station power supplies and/or local sources. An industrial PC monitors the NCC, automatically uploads data, analyses the recorded events and sends a report to the printer in a timely manner (time histories, spectral response comparisons and seismic intensity). The MARMOT system runs periodic self-tests (both continuous and programmable) and reports system state of health problems immediately.

2.3 Qualification and Quality Management

The seismic monitoring of critical facilities requires a maximum of quality and reliability of the monitoring system.

SYSCOM highly emphasized these aspects already during the design and development phase of its new MARMOT system.

Each step has been witnessed and certified against rigorous standards by independent experts and accredited laboratories.

The qualification of the newly developed system was conducted in several consecutive steps.

2.3.1 Plant independent, product specific qualification

The qualification process covered the complete product with both hardware and software including all interfaces. Test categories were established based on the definition of system safety requirements.

CLASSIFICATION

In the past there were many classification models developed worldwide, all with the same goal: to define needed test measures and confirmations. For the nuclear industry today, there is an internationally accepted standard that unifies all previous efforts. IEC 61226 describes Safety categories A, B, C and NC. Category A is required for automatic reactor protection measures, (e.g. reactor trip systems). Category C is typically used e.g. in the field of accident instrumentation and has informative character for the recording before, during and after an earthquake. Category NC dos not apply to seismic monitoring in nuclear power plants.

In the industrial field, IEC61508 has been successfully applied based on defined Safety Integrity Levels (SIL). Chemical plants, for example, have more stringent requirements regarding earthquake safety. These guidelines can also be used for nuclear applications. They offer sound approaches for the qualification of the software and the definition and testing of the reliability as well. SIL3 certification levels with instrumentation redundancy are required for safety systems (automatic shut down or trip systems). The SIL2 standard applies for seismic monitoring systems.

The MARMOT system is designed to be used in both nuclear and industrial applications. It is based on full compliance with both IEC 61226 [4] and IEC 61508 [1] standards.

SOFTWARE

For safety reasons, the newly developed software has been qualified considering requirements specified in IEC 60880 [2], IEC 61508 [1], IEC 60780 [5] and other guidelines. These requirements have been specified in a Safety Requirement Specification document, accompanied by a Safety Plan and a Verification and Validation Plan (V&V)

MARMOT distinguishes Safety Class-A (trip system) functionality and Safety Class-C (seismic monitoring) functionality. They are handled by different processors, interacting via a unidirectional communication link to exclude any backlash. The switch/recorder automatically monitors its own performance periodically and reports failures to the NCC. Redundancy was built into the system where requested (SIL3).

Extensive functional and fault insertion tests have been conducted by both the manufacturer and the institute conducting the test qualification program.

FUNCTIONAL TESTS / TYPE TESTS

After all documentation has been reviewed in accordance with the V&V plan, each element of the system was subject of a large functional type test. The developers emphasized not to do only black box tests with checking the possible combinations of input and output signals. Even more important were the tests from the view of soft- and hardware such as specified in the data sheet together with safety requirements of the product.

Therefore specific tests were performed under both normal and extreme environment conditions (e.g. IEC 61180, IEC 60439 and IEC 60068), following the procedures as described in a System Qualification Plan.

ELECTROMAGNETIC COMPATIBILITY (EMC)

The European EMC-guideline defines EMC as following: "the ability of a device, a construction or a system to work satisfactorily in an electromagnetic environment and without producing electromagnetic disturbances, which are unacceptable for the devices, constructions or systems working in the same ambience." Both EMI emission and immunity are considered. The requirements and test criteria are specified in the guidelines series IEC 61000 [6].

Emission tests assure that the electromagnetic radiation is sufficiently low. Both Conducted Radio Frequency and Radiated Radio Frequency Emissions are considered. Immunity tests assure that specified external effects have no negative impacts on the functionality of the components. Tests include: Damped Oscillatory Wave, Fast Transient Burst, Radiated Radiofrequency Electromagnetic Field, Electrostatic Discharge, Surge Immunity, Common Mode Radio Frequency, Power Frequency Magnetic Field, Pulses of Magnetic Field, Oscillatory Damped Wave, Conducted Common Mode Voltage and Damped Oscillation of Magnetic Field.

AGEING

Nuclear and chemical industry customers require a minimum system lifetime of 20 years. The question about the lifetime of a component or system has to be answered even though newly developed products have no extended operational experience. The Arrhenius equation does not provide useful results for complex electronic devices. In the nuclear industry, a procedure was established to simulate an ageing process, which includes such elements as repeated and prolonged operation, mechanical vibration and fast temperature variation in dry and damped heat. By experience these procedures can reasonably simulate and assure a lifetime of more than 30 years, however without exact mathematical evidence. After and during the ageing process, the equipment had to be checked for correct functionality. And finally, after all ageing procedures, it had to prove itself in a seismic test sequence.

Seismic Tests

Seismic tests are obviously the core of the seismic instrumentation system qualification. The Seismic tests were performed with the components that have already passed the ageing program. At the beginning, an envelope for required test response spectra has been defined starting from the floor response spectra in nuclear power plants. Then, the tests have been conducted in accordance with IEC 60980 [3] and IEEE 344 [7] on an appropriate 3-axis shaker. In the first step, resonance frequencies were identified, followed by subsequent, OBE (Operating Basis Earthquake) and SSE (Safe Shutdown Earthquake) tests. Before and after each test sequence, the correct functionality of each component has been checked.

3 Conclusion

Operators of nuclear power plants and other critical industrial facilities can be assured that the MARMOT seismic monitoring and safety system meets the latest and most rigorous international requirements regarding qualification and quality assurance. The implementation of the trip functionality with its permanent state of health monitoring is unique.

For more information and demonstration, please visit SYSCOM at the exhibition.

4 Acknowledgements

The MARMOT certification work has been carried out in cooperation with SIEMENS Erlangen, guided by an independent expert. We thank both, Mr. Guenther Soutschek and his team from SIEMENS for his valuable contributions and Mr. Kurt Zeck for his helpful expertise.

References

[1] IEC 61508 Functional Safety of Electrical/Electronic/Programmable Electronic Safety-related Systems (E/E/PE, or E/E/PES).

[2] IEC 60880 Software for computers in the safety systems of nuclear power stations

[3] IEC 60980 Nuclear power plants - Instrumentation and control systems important to safety - Software aspects for computer-based systems performing category A functions

[4] IEC 61226 Nuclear power plants - Instrumentation and control important to safety - Classification of instrumentation and control functions

[5] IEC 60780 Nuclear power plants - Electrical equipment of the safety system - Qualification

[6] IEC 61000 Electromagnetic compatibility

[7] Standard IEEE 344 Recommended Practice for Seismic Qualification of Class 1E Equipment for Nuclear Power Generating Stations

[Web-1] http://www.syscom-instruments.com

Automatic or Manual Safe Shutdown of Industrial Facilities on Earthquake Signal, Guidelines to Meet the New French Regulation: Seismological and Instrumental Aspects

Fabrice Hollender[1], Jean-Philippe Girard[2], Didier Girard[3], Sébastien Sauvignet[4] and the AFPS working group for the Guidelines "Automatic or manual safe shutdown…"

[1] Commissariat à l'Energie Atomique et aux Energies Alternatives
Centre de Cadarache, DPIE/SA2S, Bât. 352
F-13108, Saint Paul lez Durance Cedex, France
fabrice.hollender@cea.fr

[2] Association Française de Génie Parasismique
AFPS - 15, rue de la Fontaine au Roi
F-75127 Paris Cedex 11, France
jpg.aix@gmail.com

[3] Bluestar Silicone
BP 22, 1-55 avenue des Frères Perret Address
F-69191 Saint Fons Cedex, France
didier.girard@bluestarsilicones.com

[4] Adisseo France SAS
Etablissement Les Roches-Roussillon
Avenue Berthelot, Saint-Clair-du-Rhône
F-38556 Saint Maurice l'Exil, France
sebastien.sauvignet@adisseo.com

Abstract

The French regulation has been updated in 2010, and now explicitly requires that equipment of high hazard industrial facilities (outside nuclear field) do not lead to unacceptable consequences under the highest earthquake of the seismic zone where the facility is located.

As well the seismic zones have been re-evaluated, considering four levels for the French metropolitan territory.

To meet this new requirement AFPS has been asked to draft a guide that defines a strategy to stop the facility on detection of the earthquake. This specific guide is part of a set of documents that will support operators in the different design

S. Klinkel et al. (Eds.), *Seismic Design of Industrial Facilities*,
DOI 10.1007/978-3-658-02810-7_16, © Springer Fachmedien Wiesbaden 2014

requirements that could be implemented to demonstrate compliance with the regulation.

The guide explains how automatic or manual actuators could isolate the dangerous inventory inside the facility to prevent or limit the impact. As well mitigation of indirect effects is considered. Earthquake phenomenology, threshold to trigger the safe shutdown, principle to demonstrate compliance with the regulation, logic, hardware & software requirements, qualification and in service inspection are described together with real case study.

The present paper focuses on earthquake phenomenology, detection strategies and threshold to trigger the safe shutdown.

As far as the threshold level is concerned a possible -very low- default threshold that would prevent long diagnostic analysis of the weaknesses of the equipment will be discussed.

Keywords: regulation, automatic shutdown, PGA, accelerometer

1 Introduction

The French regulation has been updated in 2010, and now explicitly requires that equipment of high hazard industrial facilities (outside nuclear field) do not lead to unacceptable consequences under the highest earthquake of the seismic zone where the facility is located.

To meet this new requirement AFPS has been asked to draft a guide that defines a strategy to stop the facility on detection of the earthquake.

2 Phenomenology

When an earthquake occurs, the released energy will spread as elastic waves. It is mainly these waves will cause the surface ground motion.

There are several kinds of wave:

- the body waves that can spread throughout the earth volume,
- the surface waves, which are guided by the surface of the earth, and which is formed by conversion of energy from body waves.

In the category of body waves, we drew a distinction between the compression waves (or P-waves) generating a movement parallel to the direction of propagation, and secondly shear waves (or S-waves) which generate a movement perpendicular to the direction of propagation. There are also different surface waves, but it does not seem necessary to detail these in this guide.

Body waves are faster than surface waves. Similarly, the P-waves are faster than S-waves At a given site, P-waves are the first that will be felt (it is this property that “named” these waves, the "P" corresponding to "Primary") that arrive the S-waves (the "S" corresponding to secondary), and finally the surface waves.

Moreover, the propagation velocity of body wave decreases gradually as they approach the surface (due to the gradual decompression and weathering of geological material). This phenomenon implies refraction that involves a “verticalization” of the propagation direction. Therefore, the waves arrive at the surface with an incident angle substantially perpendicular to the surface ("vertical incidence") or at least close to the vertical.

Therefore, P-waves generate essentially vertical movements, whereas S-wave generate essentially horizontal movements (see Figure 1). Adding this feature to the fact that the P-wave amplitude is generally lower than that of S-wave amplitude, we can conclude that the P-waves are less damaging for buildings than S-waves.

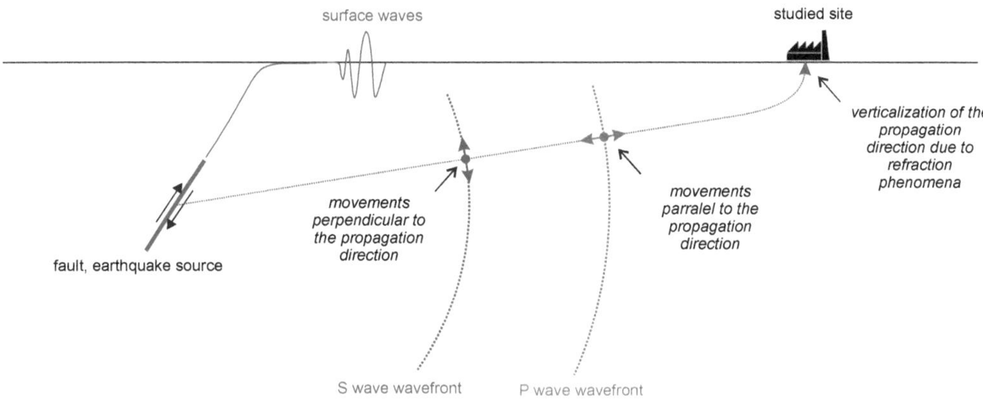

Figure 1: Wave propagation form earthquake source to the studied site. Differences between P, S and surface waves and associated polarization

The fact that the P-waves (less damaging) arrive before the S-waves (with the strongest destructive potential) may be used in some cases as part of strategies for safe shutdown procidure (see Figure 2).

In French seismic zones 1 to 4 (metropolitan area), the considered earthquakes in the framework of the seismic risk regulation have moderate to medium magnitude. Such earthquakes can have destructive effects within a few tens of kilometers up around the epicentral area.

To set orders of magnitude, the time required for the P-wave to travel 35 km will be approximately 7 seconds, the time required for the S-wave to travel the same distance will be approximately 10 seconds. The difference in arrival time between the P-wave and S-wave is around of 3 seconds in this example.

This shows that these different orders of magnitude correspond to short duration. This should be kept in mind and compared to the durations required for shutdown.

In seismic areas 5 (Guadeloupe, Martinique), we can consider thrust earthquakes with higher magnitudes that are likely to create damage at a greater distance from the epicentral area. The orders of magnitude given above are adapted to suit the distance considered (up to 80 km).

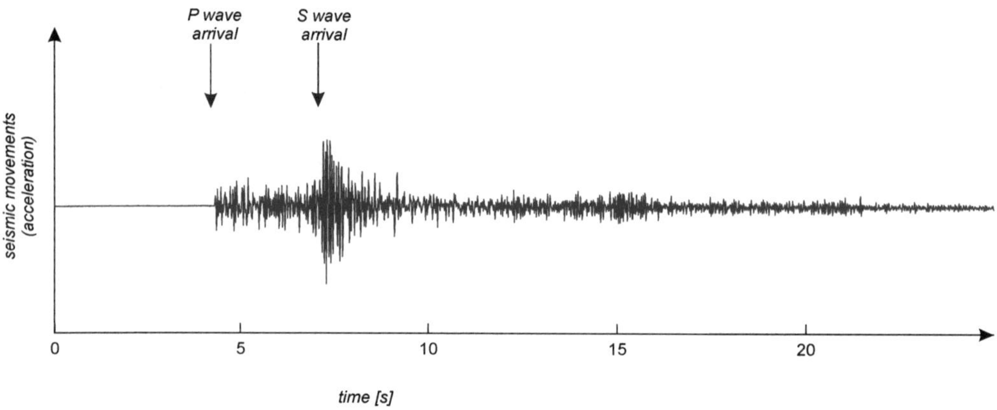

Figure 2: Ground motion (here: acceleration) recorded at a given site (distance between source and site: around 15 km)

3 Shutdown triggering strategies

Depending on the time that is available to make the installation safe shutdown, different triggering strategies are possible.

3.1 Triggering on strong movements

The first strategy consists in triggering the shutdown when the ground motion measured at seismic sensor(s) seismic(s) is already strong and reaches amplitudes that may involve acceleration near or above the safety threshold. Therefore, this approach triggers the shutdown when the most harmful waves ("S-waves") already affect the considered site.

This approach does not allow anticipation. Its choice implies that the action of shutdown can occur even when the system suffers or has suffered the most severe seismic load. It also means that any effect (e.g. release of pollutants) that may occur between the beginning of the seismic loading and the completion of the action of safe shutdown is acceptable.

However, this approach has the advantage of using relatively high trigger threshold, which limits the risk of false alarm of the security system.

3.2 Delayed triggering on strong movements

The second strategy is a variant of the previous strategy. The triggering is also made on the strongest ground motion phase, but it does not directly involve the shutdown. However, it initiates a temporization (for a period to be determined) that will involves the shutdown latter if an operator does not cancel the order during the temporization.

This is the strategy that is most appropriate to avoid the risk of false triggering. However, it is also the one that maximizes the time between the occurrence of the strongest motions and the completion of the safety procedure. It implies that any effect (e.g. release of pollutants) that may occur between the beginning of the seismic loading and the completion of the action of safety procedure is acceptable. It is the evaluation of these consequences that allows better defining the duration of the temporization.

3.3 Anticipated triggering on low threshold (called strategy "P wave")

The third strategy valorizes the arrival time delay between P-wave and S-wave. One uses here the fact that the P-wave amplitude is less than that of S-waves and S-waves have a greater destructive potential due to their orientation. The shutdown action is triggered by using an acceleration threshold relatively low, corresponding to a fraction of the acceleration threshold beyond which the shutdown is expected.

This approach has the advantage of giving a reaction time between the shutdown start and the arrival of the most damaging waves. The gain is low and therefore this only useful for very fast shutdown (<1 s) or shutdown that have to be initiated before the arrival of S-wave (even if the shutdown action is not fully completed before the S-wave arrival).

The main drawback of this strategy is a higher rate of false alarms, inherent to the choice of a low threshold. Moreover, it should be noted that although this risk is statistically low, the P-wave may already have high and damaging amplitude. It should also be noted that even if P-waves are weaker than S-waves, the facility is already subject to seismic loading between the arrival of the P-wave and S-wave.

3.4 Early triggering by remote instrumentation

A final triggering strategy is mentioned here as a reminder. It seems relatively poorly adapted to French contexts. Here, the seismic instrumentation is located close to the potential seismic sources. One tries to detect the seismic movements as earlier as possible, before the waves have reached the site to be protected.

While this strategy has the advantage to produce longer reaction delays, it has however a number of disadvantages:

- it requires placing instrumentation outside the concerned facility,
- it creates the need to maintain and demonstrate reliability and availability of systems transmitting information between remote sensors and facility,
- it implies that seismic hazard sources are well known and localized in order to identify the area to be instrumented.

Note also that, given the orders of magnitude of time provided above, this strategy seems irrelevant in France because the potential gains are very small.

4 Physical value to measure

Earthquake engineering studies may use different indicators, more or less complex. One of the most commonly used parameter is the response spectrum expressed in acceleration that produces a value of spectral acceleration at different frequencies.

In the framework automatic safety actions, however, it is difficult and unreliable to evaluate complex indicators in real-time. It seems more appropriate to analyze directly the ground acceleration, instantly felt.

We therefore propose to use the maximum instantaneous acceleration (for "free field measurement", the maximum instantaneous acceleration is the "Peak Ground Acceleration" or PGA, this notion also corresponds to the "Zero Period Acceleration" or ZPA).

The typical frequency band of seismic movements which could damage buildings and equipment is bound by 0.1 Hz and 35 Hz

The sensors to be used (accelerometers) usually allow filtering the received signal in a frequency band that rejects some unwanted signals. This filtering frequency band may be smaller than that mentioned. However, we should not choose excessive filtering.

5 Choice of sensor number and location

The choice of location for installing seismic sensors (accelerometers) and the definition of the shutdown threshold are not independent.

A given component (valve, oven ...) is generally placed in a building or a structure. However, these last may modify the seismic movement that would be recorded "in free field" (that is to say the surface, without any disturbance of buildings). On the one hand, the soil-structure interaction (effect of the building on the ground) modifies the seismic motion. On the other hand, the building or structure seismic response also changes the seismic movements (typically the upper floors of a building are subject to amplified seismic movements by comparison with the lowest floors).

Depending on the location to be chosen for placing seismic sensors, these changes in the movement should be taken into account.

We begin by presenting possible solutions for the location of seismic sensors, and then we present the proposed methodology to determining triggering threshold.

5.1 Sensor position

Different solutions are possible to place seismic sensors. Three main solutions can be developed:

- In the open field, that is to say, outside of buildings and structures (or possibly in small buildings without floors and with a small surface) and far enough form other buildings in order to avoid their effects on ground motion (typically at a distance of 2 times the height of buildings). When "free field" sites are well chosen, this solution has the advantage of minimizing the risk of false triggering due to human disturbance. A single installation (that could implement several accelerometers to allow a “2 out of 3” logical triggering strategy) may be mutualized for all of the actions that will secure the installation. However, it requires a more stringent implementation (sites outside of buildings, length of wiring ...) and involves taking into account the effects of soil structure interaction and buildings themselves (see below).
- Installation of accelerometers in the lower parts of the facility (slab, basement in direct contact with the geological formations ...). This solution is a good compromise between ease of implementation (inside the building) and reduced risk of false alarms. Such an approach also allows the instrumentation sharing for all shutdowns. However, it requires the appropriate consideration of the behavior of structures or buildings.
- Installation of accelerometers close to the component that motivates the shutdown action. This approach eliminates the need of soil / structure interaction evaluation and building behavior computation. Nevertheless, the risk of false triggering is higher (noisy area, unwanted movements amplified by structures) and this strategy does not allow the sharing of instrumentation for various components.

To increase the availability of the seismic sensor installation while limiting the risk of false alarm, it is advisable to use several seismic sensors installed in different places, associated to a triggering strategy (2 out of 2, 2 out of 3, ...). Indeed, this approach to preserve a shutdown triggering function even in case of failure of one sensor (diagnosed or not) and also to avoid false alarm if one sensor is affected by a purely local acceleration (not affecting other sensors) due to a cause other than earthquake.

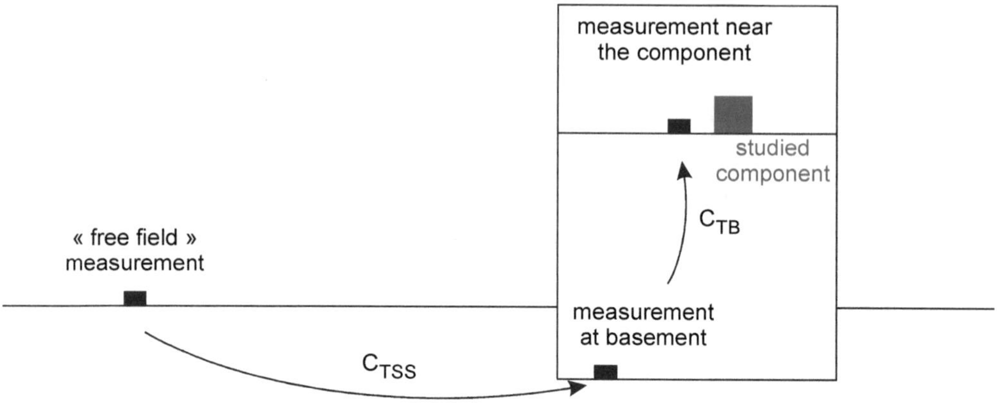

Figure 3: Different possible position of sensors

Similarly, the sensors may be either "uniaxial" (movement measurement in one space direction) or triaxial (movement measurement in three directions of space). It is advisable to use triaxial accelerometers (also called "three components"), particularly in the context of positioning sensors in open field or on basement. In the case of the use of a uni-axial accelerometer, the choice of the orientation of the sensor must be motivated. We consider that a tri-axial accelerometer exceeds a given threshold when one of the three components has exceeded this threshold. In this sense, the so-called "2 out of 3" logic is defined as "two accelerometers out of three" (with each triggered by at least one component) and not "2 out of 3 components."

5.2 Triggering threshold determination

The definition of the threshold that will trigger the shutdown procedure if the acceleration of ground motion exceeds it is a key issue of the overall shutdown procedure We propose two approaches: a simplified one and an optimized one.

The *simplify approach* implies to trigger the shutdown if the instantaneous measured acceleration exceeds 0.01 g (when accelerometers are placed in “free field” or on the basement of the facility) or 0.05 g (when accelerometers are placed near the equipment that motivated the shutdown approach). This second value takes into account the possible amplification due to the building behavior. The simplify approach cannot by applied with an anticipated P-wave strategy.

These values may appear very low, but one could be confident in the fact that if an earthquake remains below these values, no damage will occur in the facility. Moreover, if we take care in implementing an instrumental device that place sensors in free-field or basement, associated with a "2 out of 3" triggering logic, the risk of false alarm will remain very low, even at low triggering levels. Finally, in low to moderate seismicity area, like metropolitan France, the probability that and earthquake implies acceleration in a given location that exceeds 0.01 g remains acceptable. The exceedance probability of 0.01 g is usually associated to return period of several tens of years in most regions in France

Alternatively, the guide introduces the possibility of an *optimized approach* to define the shutdown threshold. This approach implies that we may be able to compute the seismic acceleration at which the different equipment (that may produce inacceptable consequence if they failed) will lose their integrity. It also implies to be able to compute the overall building behavior and also the soil-structure interaction. This approach may allow defining threshold significantly higher that the one proposed in the simplified one, but needs more studies and knowledge concerning existing buildings.

6 Conclusion

The French regulation has been updated in 2010, and now explicitly requires that equipment of high hazard industrial facilities (outside nuclear field) do not lead to unacceptable consequences under the highest earthquake of the seismic zone where the facility is located.

The new regulation, could in most cases, easily be met through robust design of equipment, but a large number of existing industrial facilities may benefit of shutdown procedure, mainly base on seismic instrumental devices, that may prevent high cost modifications or an anticipated closure which would lead to serious economic consequences.

Experimental Study on Seismic Behaviour and Vibration Control of Wind Turbine and Electrical Transmission Tower

Bin Zhao, Taixiu Cui, Zhuang Xu, Yilong Cao

State Key Laboratory of Disaster Reduction in Civil Engineering,
Tongji University,
1239 Siping Road, Shanghai 200092, China
binzh@tongji.edu.cn

ABSTRACT:

In order to research the seismic behaviour and effective vibration control strategy for the wind turbine tower and electrical transmission tower, shake table tests on reduced-scale wind turbine tower model and electrical transmission tower model were carried out at the State Key Laboratory of Disaster Reduction in Civil Engineering, Tongji University. Tuned Mass Damper (TMD) systems were applied for reducing the seismic responses of the model towers. Experimental results of both model tests are presented in this paper. The test results indicate that the TMD systems are remarkable in seismic responses reduction for the wind turbine tower and electrical transmission tower, and can be widely used for engineering application.

Keywords: seismic behaviour, vibration control, shake table test, wind turbine tower, electrical transmission tower

1 Introduction

With the rapid development of economy, large demand for electricity generation and transmission exists in the world, especially for country like China. On the other hand, the recent earthquake experiences, such as Wenchuan earthquake in 2008, show that the electrical facilities suffered serious damages from the strong ground motions [1]. In this study, shake table test technology was applied to research the seismic behaviour and effective vibration control strategy for the wind turbine tower and electrical transmission tower. Reduced-scale wind turbine tower model and electrical transmission tower model were designed and the shake table tests were carried out at the State Key Laboratory of Disaster Reduction in Civil Engineering, Tongji University. For vibration control, Tuned Mass Damper (TMD) systems were used for reducing the seismic responses of the model towers. The test

DOI 10.1007/978-3-658-02810-7_17,

results of both model tests are summarized as well as some discussions are presented in this paper.

2 Experimental design for wind turbine tower

2.1 Prototype and test model design

Shown as Figure 1, the wind turbine tower prototype was 96.52 m high [3]. The scale factor for the model tower was defined as 1/13, while the height of the model tower became 9.934m including the blades, which were simulated by 3 uniformed cantilever beams. The tower body structure was divided into 4 sections from bottom to top. The diameter and thickness of the bottom tower were 300 mm and 4 mm respectively, while the diameter and thickness of the top tower were 196 mm and 3 mm respectively. The length, width and height of the nacelle were 920 mm, 760 mm and 460 mm respectively. The mass of the nacelle was 551 Kg. The length of the model blade with a rectangular hollow section was 2400 mm, the size of the section was 80mm×40mm×3mm. The material of the tower model was Q345D.

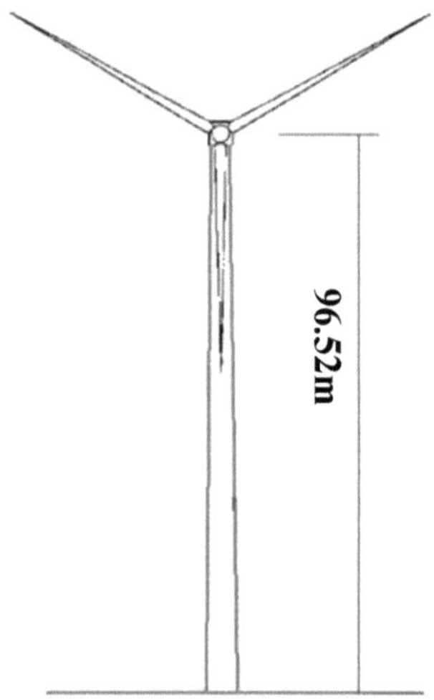

Figure 1: The facade of wind turbine tower prototype

2.2 TMD system and its parameters

A bidirectional TMD system was used to reduce the seismic responses in both horizontal directions of the model tower. The TMD mass was connected to the outer frame via a spring device, while the outer frame was rigidly connected to the top of the wind tower by bolts. The stiffness of the TMD can be modified by adding or reducing the number of springs. By changing the mass and spring of the

TMD, desired mass ratio and desired frequency ratio between TMD system and main structure can be obtained. For this test, the TMD mass was taken as 23.5 Kg, and the spring stiffness was decided by add and minus the spring number to make sure its frequencies equalled to the first and second frequencies of the model tower.

2.3 Description of the loading program

El-Centro wave, Chichi wave, Kobe wave and Wolong wave were chosen as the input motions of the shake table tests. During the test, the rotating speed of the wind blades was set to 0 rpm, 15 rpm and 30 rpm respectively for each test case. The test without TMD was conducted first, and then TMD was set up at the top of the nacelle and conducted the test with TMD. In order to obtain the mode characteristics of the model tower, the white noise sweep test was performed at the beginning of each test phase.

3 Test results of the wind turbine tower model

3.1 Measured mode parameters

Through the white noise tests, the measured first and second natural frequencies of the model tower are 1.327 and 6.657 Hz without TMD, and first frequency is reduced to 1.15 Hz when the model equipped with TMD.

3.2 Influence of blade rotation on the wind tower's response

During the shake table tests, blade speed was set to 3 levels of 0 rpm, 15 rpm and 30 rpm in order to research the effects of blade rotation on the wind tower response under different seismic wave. Under the inputs of different acceleration amplitude of El Centro wave, Chichi wave, Kobe wave and Wolong wave, the comparison results of relative displacement at the tower top are shown in Figure 2. It can be seen that the blade rotation can reduce the displacement response during the seismic events.

3.3 Vibration Control effect of TMD system

The time history comparison of the displacement at the top of the model tower between with or without TMD, while the blade speed was set to 0 rpm, is shown as Figure 3. The control efficiency of the TMD system, for top displacement and acceleration, is listed in Table 1 and 2, respectively. From these figure and tables, it can be found that the TMD system is very remarkable in reducing the displacement and acceleration responses of the model tower for different blade speeds and different seismic events. The best control efficiency even reaches 46.5% for displacement and 53.1% for acceleration, respectively.

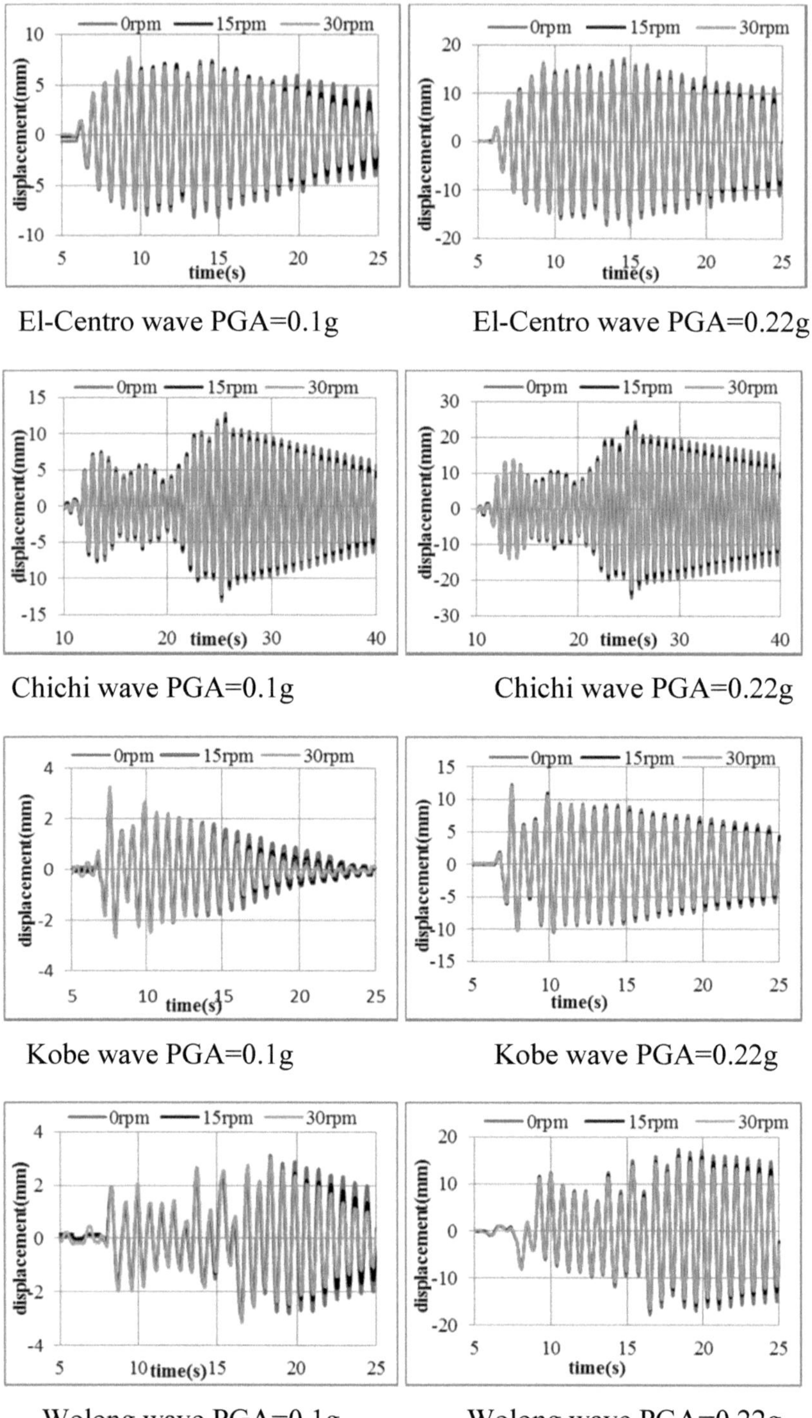

Figure 2: Time-history comparison of the top displacement of the model tower

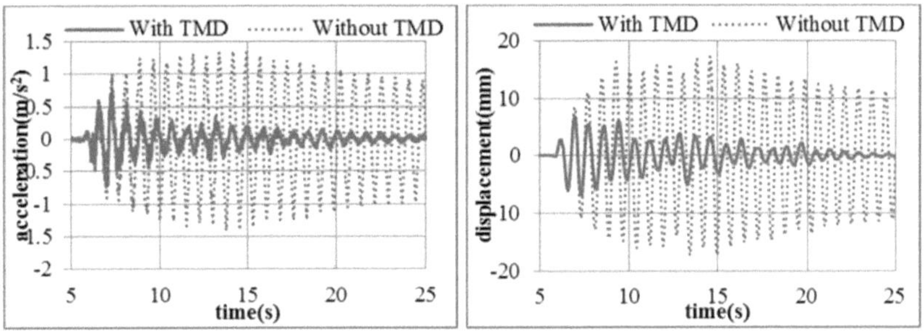

El-Centro wave

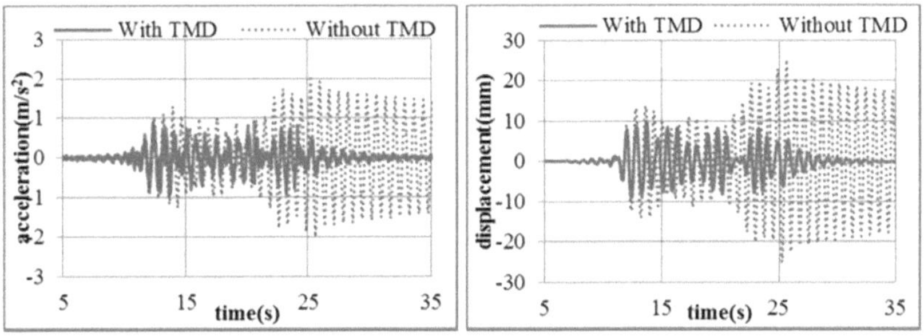

Chichi wave

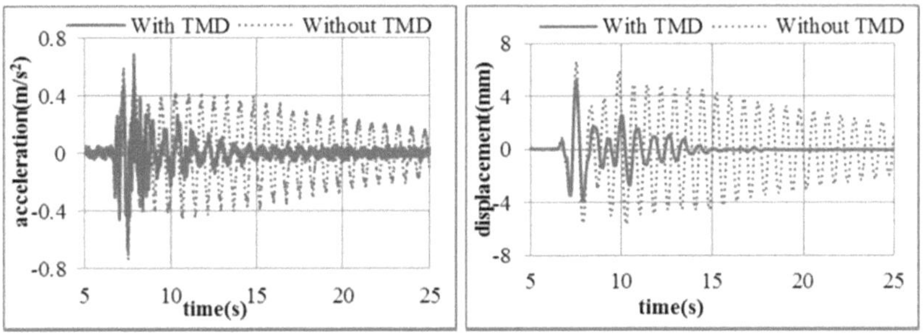

Kobe wave

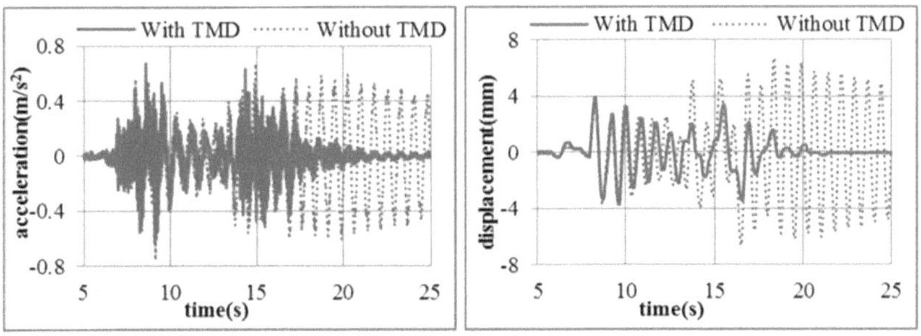

Wolong wave

Figure 3: Time-history comparison of the top acceleration and displacement of the model tower

Table 1: Control efficiency of the top displacement responses

Rotating speed (rpm)	**Seismic inputs**							
	El-Centro		Chichi		Kobe		Wolong	
	0.1g	0.22g	0.1g	0.22g	0.1g	0.22g	0.1g	0.22g
0	33.5%	42.6%	31.4%	42.6%	10.9%	20.2%	28.4%	40.9%
15	40.1%	46.5%	33.2%	46.5%	18.0%	25.4%	32.0%	41.3%
30	30.8%	45.5%	25.5%	45.5%	11.2%	13.2%	27.2%	37.6%

Table 2: Control efficiency of the top acceleration responses

Rotating speed (rpm)	**Seismic inputs**							
	El-Centro		Chichi		Kobe		Wolong	
	0.1g	0.22g	0.1g	0.22g	0.1g	0.22g	0.1g	0.22g
0	32.4%	48.8%	33.7%	32.9%	8.3%	15.3%	13.6%	10.6%
15	38.2%	53.1%	42.4%	36.5%	14.4%	22.8%	20.9%	13.2%
30	20.0%	22.0%	14.7%	18.0%	8.2%	8.3%	14.7%	7.9%

4 Experimental design for electrical transmission tower

4.1 Prototype and test model design

The prototype of transmission tower was a standard angle steel tower with height of 64.7 m, which is shown as Figure 4, and the level span of the line was 460 m. The geometric scale factor for the test model was taken as 1/8, and Q235 steel was used as the model material.

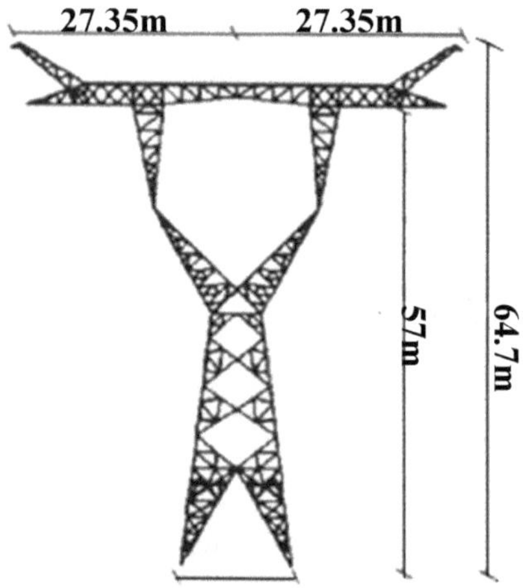

Figure 4: The facade of prototype tower

4.2 TMD system and its parameters

As the main effort of the test was focus on reducing the lateral seismic responses of the model tower, a unidirectional TMD system was applied to shake table test. The TMD mass was taken to 16.5Kg, while the spring stiffness was decided by add and minus the springs to make sure its frequency equalled to the first frequency of the model tower.

4.3 Description of the loading program

El-Centro wave, Chichi wave, Wenchuan wave, Kobe wave and SHW2 (Shanghai artificial wave which is defined by Shanghai local seismic design code) wave were used as the input motions for the shake table test. The input peak value was adjusted to 0.14g, 0.4g and 0.8g, respectively, and compare tests were conducted for two cases: model tower with or without TMD.

5 Test results of the electrical transmission tower model

5.1 Measured mode parameters

From the white noise tests, the measured natural frequencies of the model tower along the X and Y direction are 5.30 and 5.39 Hz with TMD, while the frequencies of the model tower along the X and Y direction without TMD are 5.45 Hz and 5.12 Hz.

5.2 Maximum displacement responses of the transmission tower model

The comparison of maximum displacement along the height 2m, 3.875m, 5.5m, 7.125m of the tower between with and without TMD is shown as Figure 5. For Chichi wave, the peak displacements at the top of the transmission tower are reduced by 27.45%, 26.8% and 19.38% under the input peak acceleration of 0.14g, 0.4g and 0.8g, respectively, and the damping effect was very obvious after imposing TMD system on the tower model.

5.3 Maximum acceleration responses of the transmission tower model

The comparison of maximum acceleration amplification factors along the height 2m, 3.875m, 5.5m, 7.125m of the tower between with and without TMD is shown as Figure 6. For Chichi wave, the peak accelerations at the top of the transmission tower are reduced by 40.5%, 41.3% and 37.44% under the input peak acceleration of 0.14g, 0.4g and 0.8g, respectively.

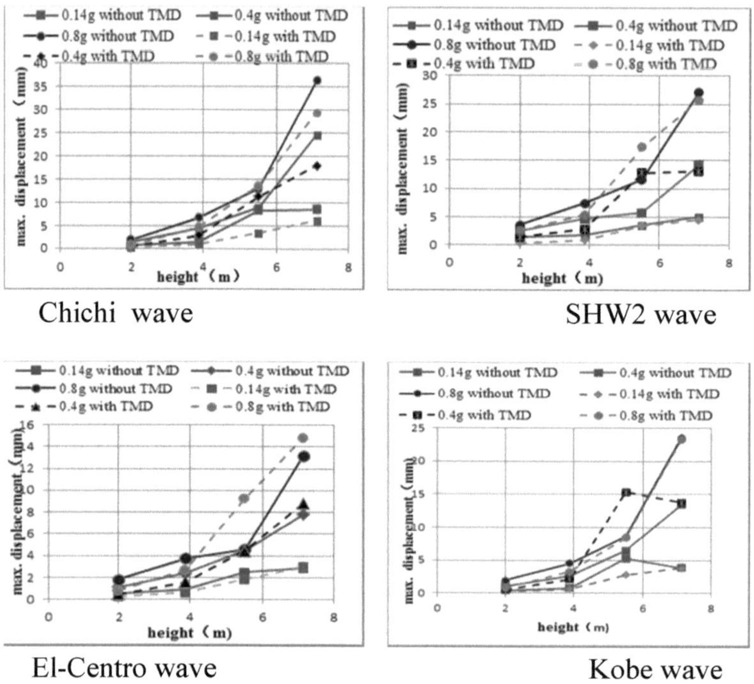

Figure 5: Comparison of max. displacement along the height of the model tower

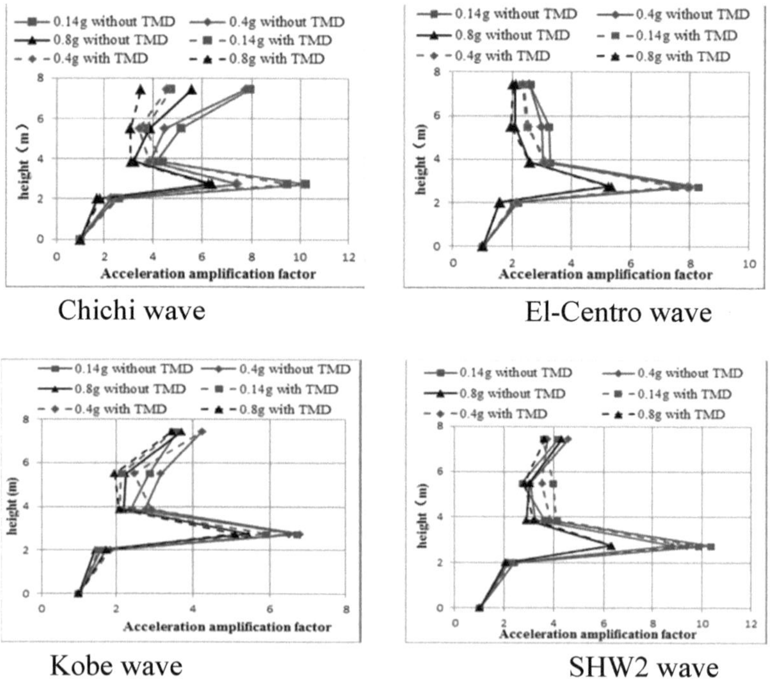

Figure 6: Maximum amplification factors of acceleration along the height of the model tower

5.4 Time history comparison

Taken the tests of Kobe wave as an example, the comparison results of absolute acceleration and relative displacement responses with TMD and without TMD under the condition of different acceleration levels are shown as Figure 7. One can find that applying TMD can significantly reduce the seismic responses of the model tower.

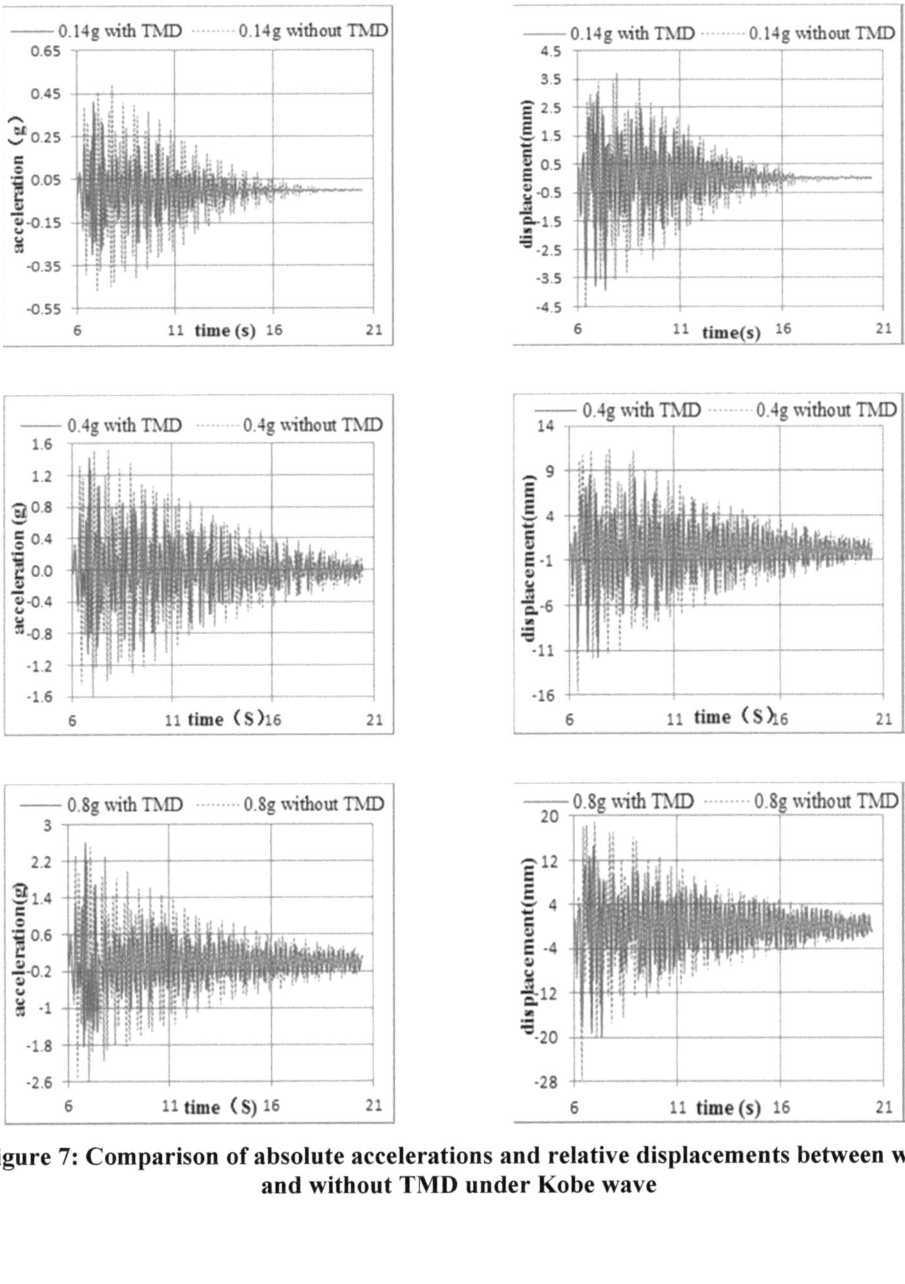

Figure 7: Comparison of absolute accelerations and relative displacements between with and without TMD under Kobe wave

6 Conclusion

Shake table results of the reduced-scale wind turbine tower model and electrical transmission tower model are presented in this paper. The comparison of the results of both models with TMD and without TMD under the condition of different earthquake and different seismic levels indicate that the control strategy of using TMD system can significantly reduce the seismic responses and suitable for widely using for engineering application.

7 Acknowledgements

The research is supported by the Ministry of Science and Technology of China with Grant No. SLDRCE 09-B-13.

References

[1] Zhao Bin; Taucer Fabio: Performance of Infrastructure during the May 12, 2008 Wenchuan Earthquake in China; Journal of Earthquake Engineering, v.14, n. 4, 2010, p. 578-600

[2] Bai Xueyuan: Stochastic Optimal Control Analysis and Experimental Studies on Offshore Wind Turbine System; A Dissertation Submitted to Tongji University in Conformity with the Requirements for the Degree of Master of Engineering. 2011

[3] Zhao Bin; Ma Fei; Chen Jianbing: Shake Table Experimental Study on Vibration Control of Wind Turbine Tower Using TMD; China Civil Engineering Journal, v 45, n. s, 2012, p.142-145+157

Part VI

Seismic Design of Secondary Structures

Systemic Seismic Vulnerability and Risk Analysis of Urban Systems, Lifelines and Infrastructures

Kyriazis Pitilakis[1], **Sotiris Argyroudis**[2]

[1] Department of Civil Engineering, Aristotle University of Thessaloniki
P.O.B. 424, 54124, Thessaloniki, Greece
kpitilak@civil.auth.gr

[2] Department of Civil Engineering, Aristotle University of Thessaloniki
P.O.B. 424, 54124, Thessaloniki, Greece
sarg@civil.auth.gr

ABSTRACT:

The basic concepts and some representative results of the work carried out within the European collaborative research project SYNER-G (http://www.syner-g.eu) are presented in this paper. The overall goal is to develop an integrated methodology for systemic seismic vulnerability and risk analysis of urban systems, transportation and utility networks and critical facilities. SYNER-G developed an innovative methodological framework for the assessment of physical as well as socio-economic seismic vulnerability and loss assessment at urban and regional level. The built environment is modeled according to a detailed taxonomy into its components and sub-systems, grouped into the following categories: buildings, transportation and utility networks, and critical facilities. Each category may have several types of components. The framework encompasses in an integrated way all aspects in the chain, from regional hazard to vulnerability assessment of components to the socioeconomic impacts of an earthquake, accounting for relevant uncertainties within an efficient quantitative simulation scheme, and modeling interactions between the multiple component systems in the taxonomy. The prototype software (OOFIMS) together with several complementary tools are implemented in the SYNER-G platform, which provides several pre and post-processing capabilities. The methodology and software tools are applied and validated in selected sites and systems in urban and regional scale. Representative results of the application in the city of Thessaloniki are presented here.

Keywords: systemic analysis, earthquakes, vulnerability, risk, socioeconomic loss, buildings, lifelines, infrastructures, interactions

1 Introduction

So far seismic vulnerability and risk assessment are performed at system level. i.e. bridge, building, water network etc. The losses are estimated at "element at risk" or at the best at system level. Then they are somehow integrated at urban or regional level to account in an elementary way the socio-economic impact. However, in reality the different systems composing an urban or industrial system are strongly interconnected to each other. For example the transportation system with the medical care system or the production and supply chain; the electrical power with almost all other systems. The real losses, physical, economic and human, are normally higher or much higher when we account the interaction among systems.

The aim of SYNER-G [1] is to tackle this issue and develop for a first time in Europe and in certain degree worldwide, a methodology to analyse systems in case of earthquakes considering inter and intra-dependencies. The goal is to establish an integrated methodology for systemic seismic vulnerability and risk analysis of buildings, different lifelines (transportation and utility networks) and critical facilities. The methodology, which is implemented in an open source software tool, integrates within the same framework the hazard, the physical vulnerability and the social consequences/impact at a system level. It is applied and validated in selected case studies at urban and regional scale: the city of Thessaloniki (Greece), the city of Vienna (Austria), the harbor of Thessaloniki, the gas system of L'Aquila (Italy), the main electric power network in Sicily, a roadway network in South Italy and a hospital facility again in Italy. In the present paper we present only some examples from the application in Thessaloniki.

Systemic studies commonly address the following two phases: a) *emergency*: short-term (a few days/weeks) at the urban/regional scale, b) *economic recovery*: medium to long-term, at the regional/national scale. SYNER-G focuses mainly on the first phase with emergency managers and insurances being the main reference stakeholders. The goal is to forecast before the strong earthquake event the expected impact for the purpose of planning and implementing risk mitigation measures. We present herein the basic concepts of the methodology and several representative results.

2 SYNER-G methodology

The goal of the SYNER-G general methodology is to assess the seismic vulnerability of an infrastructure of urban/regional scale, accounting for inter- and intra-dependencies among infrastructural components, as well as for the uncertainties characterizing the problem. The goal has been achieved setting up a model of the infrastructure and of the hazard acting upon it, and then enhancing it with the introduction of the uncertainty and of the analysis methods that can evaluate the system performance accounting for such uncertainty.

The infrastructure model actually consists of two sets of models: the first set consists of the physical models of the systems making up the infrastructure. These models take as an input the hazards and provide as an output the state of physical/functional damage of the infrastructure. The second set of models consists of the socio-economic models that take among their input the output of the physical models and provide the socio-economic consequences of the event. The SYNER-G methodology integrates these models in a unified analysis procedure. In its final form the entire procedure is based on a sequence of three models: a) seismic hazard model, b) components' physical vulnerability model, and c) system (functional and socio-economic) model.

For illustration purposes, with reference to the two socio-economic models identified and studied within SYNER-G (the SHELTER and HEALTH-CARE models), Figure 1 shows in qualitative terms the integrated procedure that leads from the evaluation of the hazard to that of the demands on the shelter and health-care system in terms of Displaced Population and Casualties, down to the assessment of social indexes like the Health Impact and the Shelter Needs. For more details the reader is referred to SYNER-G Reference Reports 1 [2] and 5 [3].

The conceptual sketch in Figure 1 can be practically implemented by developing:

- A model for the spatially distributed seismic hazard.
- A physical model of the infrastructure.
- Socio-economic models.

Development of the hazard model has the goal of providing a tool for: a) sampling events in terms of location (epicenter), magnitude and faulting type, according to the seismicity of the study region; b) predicting maps of seismic intensities at the sites of the vulnerable components in the infrastructure. These maps, conventionally conditional on M and epicenter, should correctly describe the variability and spatial correlation of intensities at different sites. This is important because systems are extended in space. Further, when more vulnerable components exist at the same location and are sensitive to different intensities (e.g. acceleration, velocity, strains and displacement), the model should predict intensities measures (IM) that are consistent at the same site.

Development of the physical model starts from the SYNER-G Taxonomy and requires: a) for each system within the Taxonomy, a description of the functioning of the system under both undisturbed and disturbed conditions (i.e. in the damaged state following an earthquake); b) a model for the physical and functional (seismic) damageability of each component within each system; c) identification of all dependencies between the systems; d) definition of adequate performance indicators (PI) for components and systems, and the infrastructure as a whole.

Development of the socio-economic model starts with an interface to outputs from the physical model in each of the four domains of SYNER-G (i.e., buildings, transportation systems, utility systems and critical facilities). Thus, four main performance indicators - Building Usability, Transportation Accessibility, Utility Functionality and Healthcare Treatment Capacity - are used to determine both direct and indirect impacts on society. A similar layout could be established at an industrial complex level. Direct social losses are computed in terms of casualties and displaced populations. Indirect social losses are considered, for the moment, in two models - Shelter Needs and Health Impact - which employ the multi-criteria decision analysis (MCDA) theory for combining performance indicators from the physical and social vulnerability models.

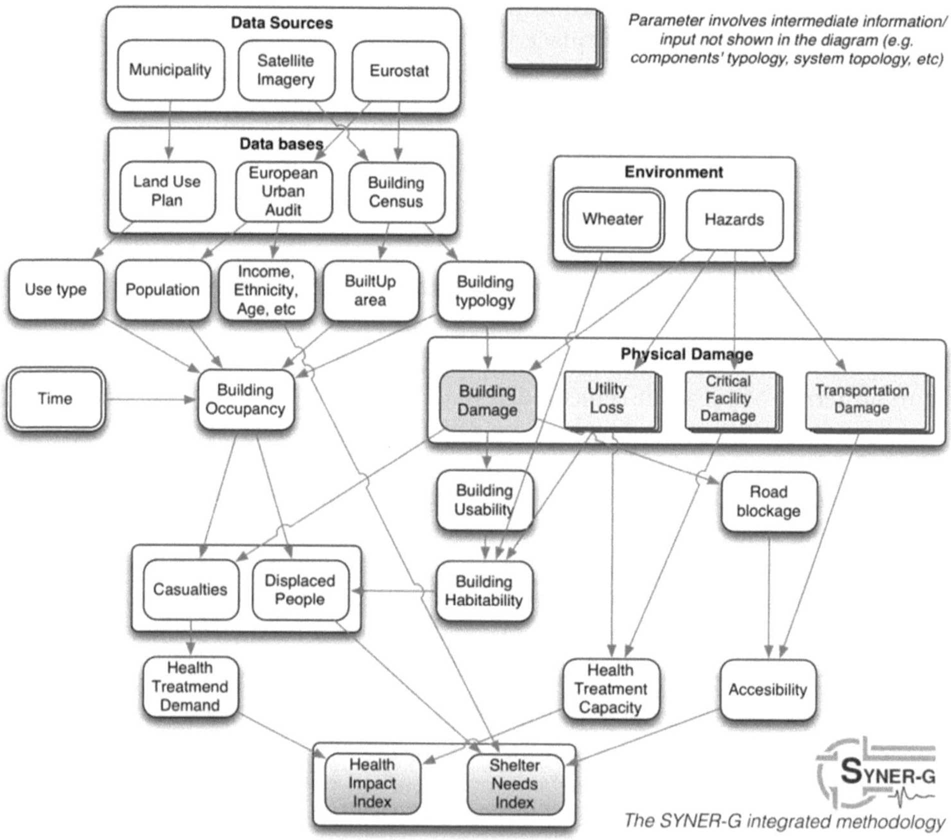

Figure 1: Integrated evaluation of physical and socio-economic performance indicators [2]

In order to tackle the complexity of the described problem the object-oriented paradigm (OOP) has been adopted. In abstract terms, within such a paradigm, the problem is described as a set of objects, characterized in terms of attributes and methods, interacting with each other [2]. Objects are instances (concrete realizations) of classes (abstract models, or templates for all objects with the same set of properties and methods). Figure 1 provides a general view of the methodological diagram.

3 SYNER-G Taxonomy

It is an essential step in urban earthquake risk assessment to compile inventory databases of elements at risk and to make a classification on the basis of pre-defined typology/taxonomy definitions. Typology definitions and the classification system should reflect the vulnerability characteristics of the systems at risk, e.g. buildings, lifeline networks, transportation infrastructures, etc., as well as of their elements at risk and sub-components in order to ensure a uniform interpretation of data and risk analyses results. Within SYNER-G a detailed taxonomy of a set of systems, sub-systems and components (elements) was identified and described, in an homogeneous way, based on all available databases and national practices in Europe and if necessary at international level. This taxonomy has been the guidance for the proposed fragility models and the modelling of systems in the next steps. The SYNER-G taxonomy is the first homogeneous ontology and taxonomy in Europe for all systems exposed at seismic risk. For more details the reader is referred to SYNER-G Reference Report 2 [4].

4 Seismic Hazard

The definition of seismic scenarios requires the development of a precise methodology for characterising the hazard input in a manner that is appropriate for application to the analysis of multiple and spatially distributed infrastructures. For novel applications such as the present one, conventional approaches for the estimation of seismic hazard are insufficient to characterise the properties of ground motion, and spatial variability, that are most relevant for each infrastructure. In accordance to the fragility models, an extensive literature review was undertaken to identify initially the best means of determining the most appropriate intensity measures (IM) for a given element, and then identifying the most efficient intensity measure for each element or collection of elements within an infrastructure [5].

For the definition of the seismic input itself, a Monte Carlo simulation methodology was developed, which has been integrated within the general methodology for systemic vulnerability analysis and the aforementioned OOFIMS prototype software. The methodology, called herein ***"Shakefield"*** approach, aims to take into account both the spatial correlation in ground motion for each intensity measure, as well as the cross-correlation and spatial cross-correlation between multiple intensity measures (Figure 2). This is a development that allows for a more direct generation of the ground motion inputs that have been identified as most efficient for each infrastructure. The spatial correlation and cross-correlation is captured via co-simulation of correlated fields of Gaussian variants, representing the residual term of the ground motion prediction equation (GMPE).

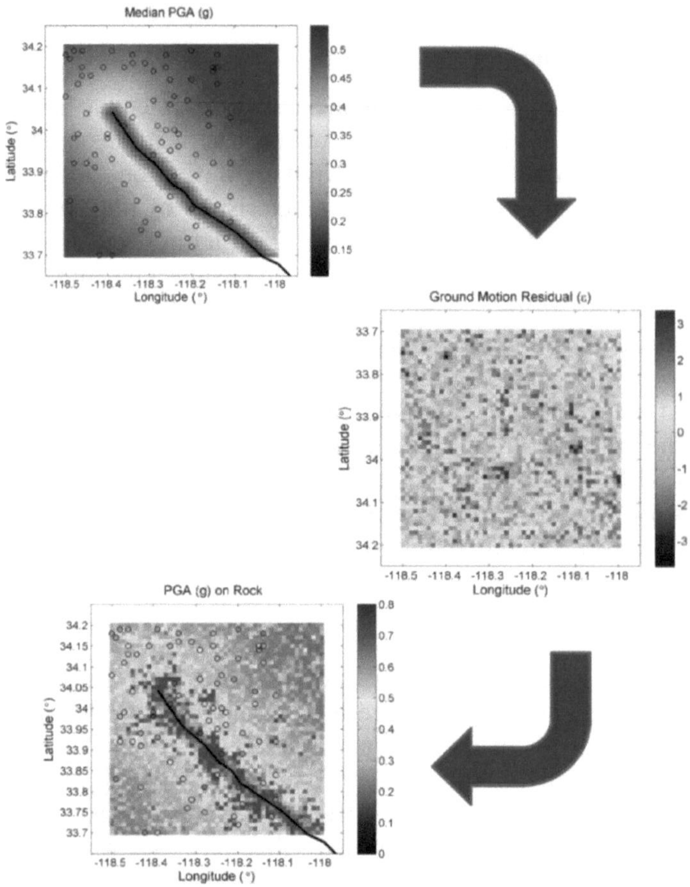

Figure 2: Overview of the "Shakefield" methodology, including the attenuation of ground motion from an event and the generation of correlated Gaussian fields as a means of simulating spatial correlation and cross-correlation in the GMPE residual term [2]

For utility systems (water and gas pipeline systems) as well as for similar systems with linear elements, fragility models are generally given in terms of permanent ground displacement (PGD), as they are most vulnerable to the permanent displacement of the ground (i.e. liquefaction or landsliding induced displacements) rather than transient shaking. To this extent "Shakefield" was further extended to incorporate geotechnical type hazards, including of course site amplification, but also liquefaction, co-seismic slope displacement and transient strain. This extension is inspired from HAZUS [6] software, with its corresponding probability definitions now interpreted in a stochastic context. However, several elements of the HAZUS model that relate the expected PGD to the strong seismic shaking, have been updated using recent empirical models that better constrain uncertainty in these terms. These new models, are also implemented in a stochastic context, while new site amplification factors will be implemented in the near future [7, 8].

5 Fragility Curves

Fragility curves constitute one of the key elements of seismic risk assessment. They relate the seismic intensity to the probability of reaching or exceeding a level of damage (e.g. minor, moderate, extensive, collapse) for each element at risk. For buildings and bridges the level of shaking can be quantified using different earthquake intensity parameters, including peak ground acceleration/velocity/ displacement, spectral acceleration, spectral velocity or spectral displacement. For other elements at risk other forms and IMs are used (i.e. repair ratio per km for pipelines correlated to PGV or PGD). They are often described by a lognormal probability distribution function, although it is noted that this distribution may not always be the best fit. Several approaches can be used to establish the fragility curves that can be grouped under empirical, judgmental, analytical and hybrid. The key assumption in the vulnerability assessment of buildings and lifeline components is that structures having similar structural characteristics, and being in similar geotechnical conditions, are expected to perform in the same way for a given seismic loading. Within this context, damage is directly related to the structural properties of the elements at risk. Typology is thus a fundamental descriptor of a system, derived from the inventory of each element.

One of the main contributions of SYNER-G is the compilation of the existing fragility curves/functions and development of new functions for all the system elements based on the proposed taxonomy. A literature review on the typology, fragility functions, damage scales, intensity measures and performance indicators has been performed for all the elements. The fragility functions are based on new analyses and collection/review of the results that are available in the literature. In some cases, the selection of the fragility functions has been based on validation studies using damage data from past and recent earthquakes mainly in Europe. Moreover, the damage and serviceability states have been defined accordingly. Appropriate adaptations and modifications have been made to the selected fragility functions in order to satisfy the distinctive features of the presented taxonomy. In other cases, new fragility functions have been developed based on numerical analyses (i.e. tunnels, road embankments/cuts, bridge abutments) or by using fault tree analysis together with the respective damage scales and serviceability rates in the framework of European typology and hazard [9].

A "***Fragility Function Manager Tool***" has been developed for buildings and bridges and is connected with the SYNER-G software platform. This tool is able to store, visualize, harmonise and compare a large number of fragility functions sets. For each fragility function set, the metadata of the functions, representative plots and the parameters of the functions can be visualized in an appropriate panel or window. Once the fragility functions are uploaded, the tool can be used to harmonise and compare the curves. The harmonisation module allows one to harmonise the curves using a target intensity measure type and a number of limit states of reference. After the harmonisation, the comparison module can be used to plot together and to compare different functions, which can then be extracted and the mean and dispersion of the parameters of the curves can be calculated. The reader may consult for more information the SYNER-G Reference Report 4 [9].

6 Socio-Economic Impact Models

The current state-of-the-art in earthquake engineering produces reasonably accurate estimates of physical damage to single elements at risk like buildings and infrastructure systems, as well as reasonable estimates of the repair and replacement costs associated with this type of damage. However, poor linkages between damage to physical systems and resultant social and economic consequences remain a significant limitation in existing loss estimation models.

A unified approach for modelling shelter needs and health impacts caused by earthquake damage, which integrates social vulnerability into the physical systems modelling approaches has been developed in SYNER-G. These two kinds of impacts have been selected as being among the most important in crisis period for the society. Figure 3 illustrates the integrated procedure that leads from the hazard to the evaluation of the demands on the shelter and health-care system, leading to the computation of two key parameters: Displaced Population (DP) and Casualties. The shelter needs and health impact models brings together the state-of-the-art social loss estimation models into a comprehensive modelling approach based on multi-criteria decision support, which provides decision makers with a dynamic platform to capture post-disaster emergency shelter demand and health impact decisions.

The focus in the ***shelter needs model*** is to obtain shelter demand as a consequence of building usability, building habitability and social vulnerability of the affected population rather than building damage alone. The shelter model simulates households' decision-making and considers physical, socio-economic, climatic, spatial and temporal factors in addition to modelled building damage states (Figure 4). The ***health impact model*** combines a new semi-empirical methodology for casualty estimation with models of health impact vulnerability, transportation accessibility and healthcare capacity to obtain a holistic assessment of health impacts in the emergency period after earthquakes. A group of socio-economic indicators were derived based on an in-depth study of disaster literature for each of the shelter, health and transport accessibility models, and harmonized based on data available for Europe from the EUROSTAT Urban Audit Database. For more details the reader may consult the SYNER-G Reference Report 5 [3].

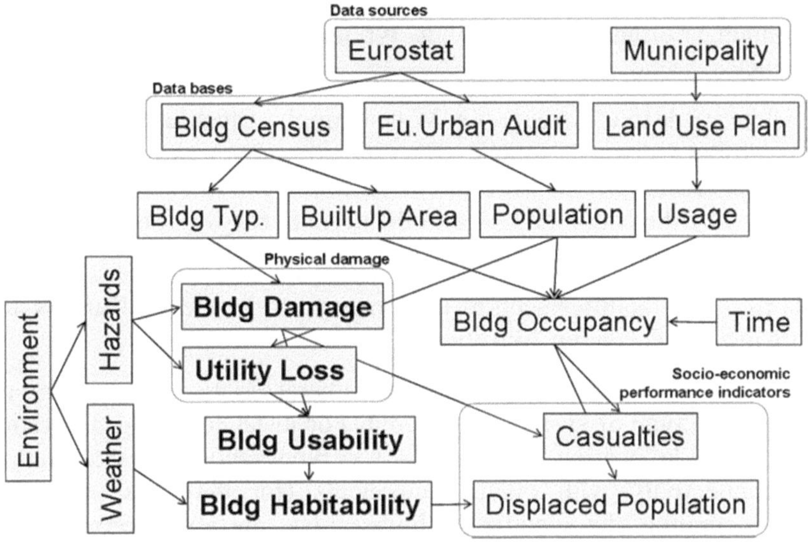

Figure 3: Integrated evaluation of physical and socio-economic performance indicators [3]

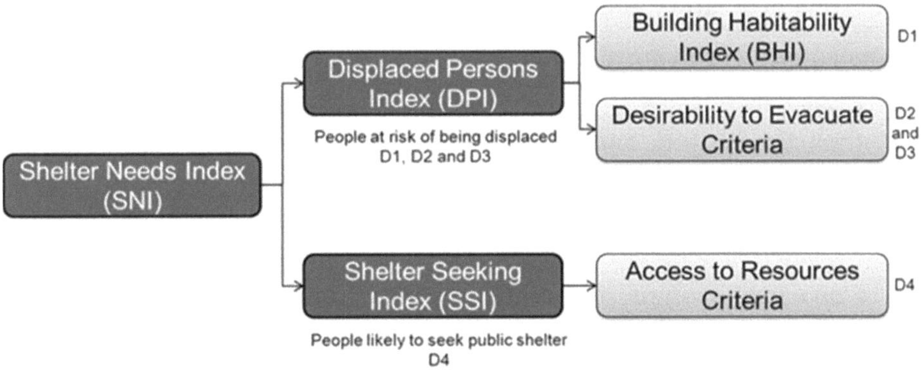

Figure 4: Multi-criteria decision model for computing Shelter Needs Index [3]

7 Systemic Analysis

Based on the SYNER-G methodology, each of the four systems considered (buildings and aggregates, utility networks, transportation networks and critical facilities) has been specified according to the following three main features [10]:

7.1 Taxonomy of components within each system

Each class of systems is composed of sub-classes that are used to describe the various types of components, based on the geographical extent and their function within the system:

- Cell classes are used to define inhabited areas (i.e. Buildings System) and contain information on buildings typologies, population or soil occupation policy.
- All network-like systems (i.e. Water Supply, Electric Power, Gas Network and Road Network) contain two types of sub-classes (Edges and Points), which are further sub-divided in specific classes, according to the role played by the component within the system: network nodes can be stations, pumps, reservoirs, sources, distribution nodes, etc.
- For critical facilities such as components of the Health-Care System, they are modelled as point-like objects.

Each of the sub-classes is specified with their characteristic attributes and methods, depending on the type of system considered. For instance, initial properties of the objects may include location, area, length, soil type, typology, associated fragility, capacity, connectivity with other components (for networks), etc. Once the simulation is running, the specific methods update the object properties, such as damage states, losses within each cell or remaining connectivity.

7.2 System evaluation and performance indicators

Three main types of solving algorithms are considered in the SYNER-G approach:

- *Connectivity analysis:* this approach removes the damaged components from the network and it updates the adjacency matrix accordingly, thus giving the nodes or areas that are disconnected from the rest of the system. This approach is used for all utility networks (water, electricity, gas) and the road transportation system.
- *Capacitive analysis:* for utility networks, graph algorithm can be used to optimize capacitive flows from sources (e.g. generators, reservoirs) to sinks (i.e. distribution nodes), based on the damages sustained by the network components (from total destruction to slight damages reducing the capacity).
- *Fault-tree analysis:* this type of approach aims to evaluate the remaining operating capacity of objects such as health-care facilities. The system is broken up into structural, non-structural or human components, each one of them being connected with logic operators.

The evaluation of ***Performance Indicators*** at the component or the system level depends on the type of analysis that is performed: connectivity analysis gives access to indices such as the connectivity loss (measure of the reduction of the number of possible paths from sources to sinks). On the other hand, capacitive modelling yields more elaborate performance indicators at the distribution nodes (e.g. head ratio for water system, voltage ratio for electric buses) or for the whole system (e.g. system serviceability index comparing the customer demand satisfaction before and after the seismic event).

7.3 Interdependencies

Three types of interactions between systems are considered within SYNER-G:

- *"Demand" interactions*: they correspond to a supply demand from a given component to another system. For instance, the presence of densely populated cells in the vicinity of a given distribution node (e.g. from a water supply or electric power system) will generate a substantial demand on the supply system. Another example could be the number of casualties that will put a strain on the treatment capacity of health-care facilities.
- *Physical interactions:* they are associated with exchanges of services or supplies between systems, like the supply of water to inhabited cells, the supply of transportation capacities by roads or the supply of power to various network facilities (e.g. water pumps) by electric generators.
- *Geographical interactions:* they are involved when two components are located in the same area and when the damage of one of them is directly influencing the physical integrity of the second one. For instance, the collapse of buildings in city centres can induce the blockage of adjacent roads due the debris accumulation.

8 SYNER-G Software Tools

A comprehensive tool box has been developed **(EQvis)** containing several pre and post-processing tools as well as other plug-ins such as the prototype software (OOFIMS), the Fragility Manager Tool, the MCDA software for modelling shelter needs and health impact (Figure 5). The product EQvis (European Earthquake Risk Assessment and Visualisation Software) is an open source product that allows owners, practicing engineers and researchers the realistic risk assessment on systemic level (Figure 6). It has been based on the similar pre and post-processing modules of MAEviz [11].

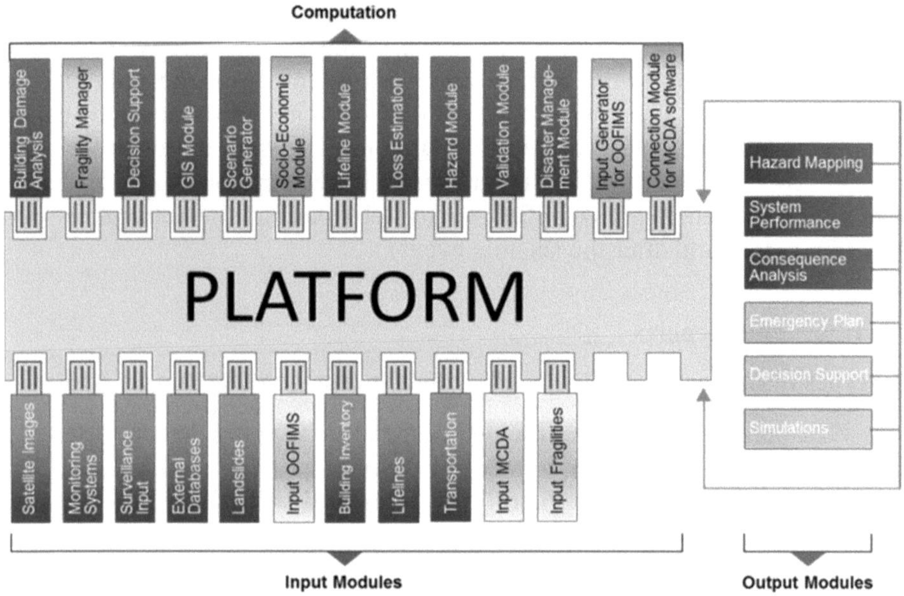

Figure 5: The plug-in based structure of the software

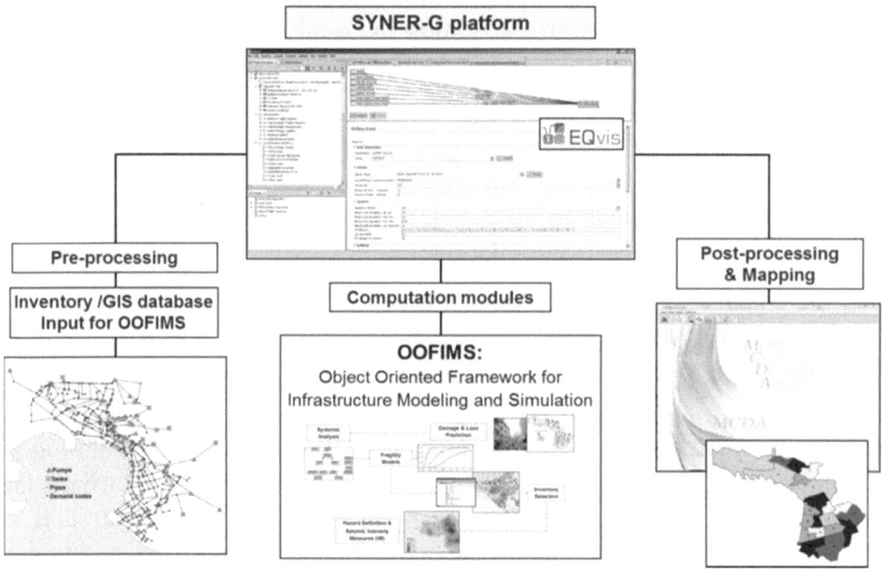

Figure 6: Layout of the SYNER-G platform

9 Application to Thessaloniki

To demonstrate the SYNER-G methodology and its tools we present in the following some representative results for the application in Thessaloniki, Greece, which is located in a high seismicity area and disposes a very good data base of all element at risk and geotechnical conditions. The study area covers the municipality of Thessaloniki, which is divided in 20 Sub City Districts as defined by Eurostat and Urban Audit approach. The case study presented herein includes the following elements: building stock (BDG), road network (RDN), water supply system (WSS) and electric power network (EPN). The networks comprising the main lines and components cover the wider Metropolitan area. The internal functioning of each network is simulated and a connectivity analysis is performed. Moreover, specific interdependencies between systems are considered: EPN with WSS (electric power supply to pumping stations), RDN with BDG (road blockage due to building collapses), BDG with EPN and WSS (displaced people due to utility loss).

A Monte Carlo simulation (MCS) has been carried out (10,000 runs) based on the methods and tools developed in SYNER-G. Each sampled event represents a single earthquake ("Shakefields" method) and all systems are analysed for each event. The results are then aggregated all over the sampled events. In this way, all the characteristics of each event (e.g., spatial correlations) are accounted for and preserved for the systemic analysis. For each system, selected Performance Indicators (PI's) are calculated based on the estimated damages and functionality losses of the different components.

The overall performance of each network is expressed through the Mean Annual Frequency (MAF) of exceedance and the moving average μ and moving standard deviation σ of the PIs. Thematic maps showing the distribution of expected damages/ losses are produced for selected events. Moreover, the significant elements for the functionality of each system are defined through correlation factors to the system PIs. An accessibility analysis to hospital facilities and shelter areas considering the damages in RDN is also performed and a shelter demand analysis based on a multi-criteria approach is applied.

9.1 Fragility curves

New fragility curves have been developed for buildings (masonry, R/C) and bridges of Thessaloniki [9, 12]. Three-dimensional finite element analysis with a nonlinear biaxial failure criterion was used to derive fragility curves for masonry buildings that consider in-plane and out-of-plane failure. Fragility curves for RC buildings that account for shear failure and consider model uncertainties and the scatter of material and geometric properties were also produced following the assessment method of EC8. Analytical fragility curves were developed for specific bridge typologies in the Thessaloniki study area, based on the available information about their geometry, materials and reinforcement. Older bridges are likely to experience damage for low to medium levels of earthquake excitation (e.g., Figure 7a). On the other hand, modern bridges are less vulnerable (e.g. Figure 7b).

For other elements (road pavements, pipelines etc.), appropriate fragility functions are developed based on the fragility models and IMs suggested in SYNER-G [9].

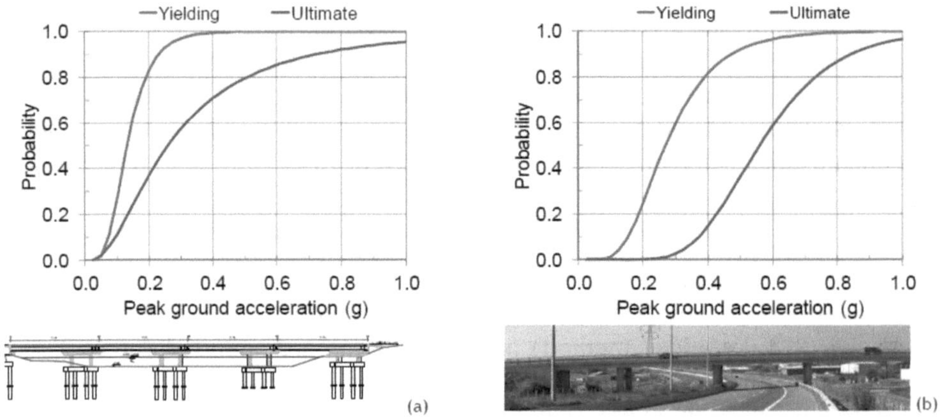

Figure 7: Example of fragility curves for Thessaloniki application (a) a bridge with the deck supported on bearings, constructed in 1985 with the old seismic code and (b) an overpass with monolithic deck-pier connection, constructed in 2003 with the new seismic code

9.2 Seismic Hazard

Five seismic zones are selected for the seismic hazard input, obtained by SHARE European research project [13]. Following the specification provided in SYNER-G the ground motion prediction equation (GMPE) introduced by Akkar and Bommer [14] is applied for the estimation of the ground motion parameters on rock basement, while the spatial variability is modelled using appropriate correlation models. For each site of the grid the averages of primary IM from the specified GMPE are calculated, and the residual is sampled from a random field of spatially correlated Gaussian variables according to the spatial correlation model. The primary IM is then retrieved at vulnerable sites by distance-based interpolation and finally the local IM is sampled conditional on primary IM.

To scale the hazard to the site condition, the current EC8 [15] amplification factors are used. For the liquefaction hazard the modelling approach proposed in HAZUS [6] is adopted for the estimation of PGD at the vulnerable sites. A detailed description of the entire hazard model adopted in the methodology can be found in Franchin et al. [16] and Weatherhill et al. [17].

9.3 Electric Power Network

Figure 8 shows the moving average (mean) curve for Electric power Connectivity Loss (ECL) as well as the mean+stdv and mean-stdv curves. The jumps present in the plots are located in correspondence of simulation runs/samples in which at least one demand node is disconnected, leading ECL to yield values greater than 0. At the end of the analysis (10,000 runs) the moving average is stabilized. The MAF of exceedance for ECL is also shown in Figure 8. The ECL with mean return period Tm=500 years (λ=0.002) is 24%. Functional and non-functional components (transmission substations and demand nodes-WSS pumping stations) for a seismic event (#6415) corresponding to the specific return period of ECL are shown in Figure 9.

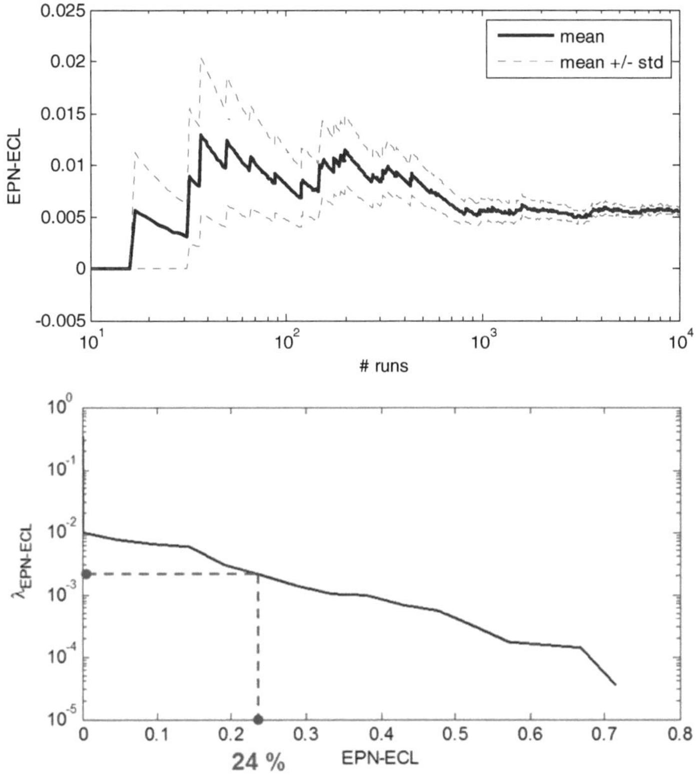

Figure 8: Moving average μ, μ+σ, μ-σ (up) and MAF (down) curves for ECL

Figure 10 shows the level of correlation between the ECL and non-functional transmission substations. In this way the most critical components of the network can be identified in relation with their contribution to the connectivity loss of the network. The majority of substations present high levels of correlation near or over 35%. This can be mostly attributed to the low level of redundancy of the network in combination to the substations vulnerability and distribution of PGA in average over all runs of the simulation.

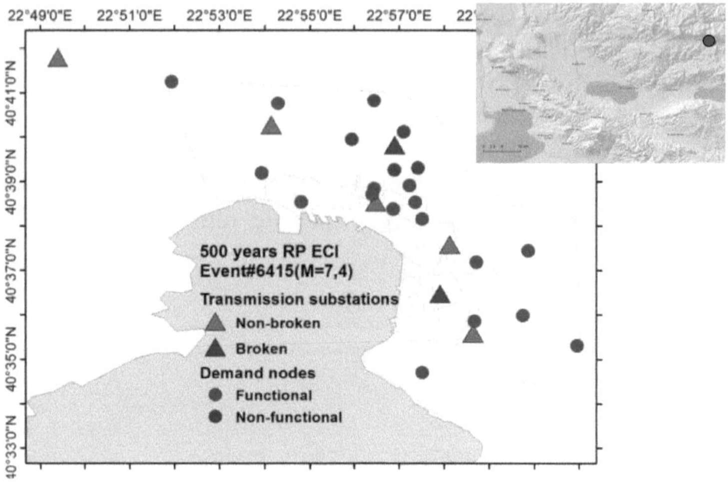

Figure 9: Electric power network damages for an event (#6415 M=7.4, R=40km) that corresponds to ECL with Tm=500 years

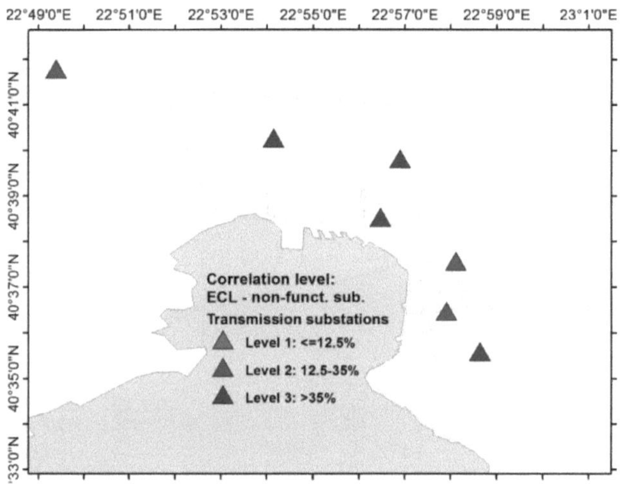

Figure 10: Correlation of non-functional transmission substations to electric power network connectivity loss

9.4 Water Supply System

Figure 11 shows the moving average (mean) curve for Water Connectivity Loss (WCL) as well as the mean+stdv and mean-stdv curves. The jumps present in the plots are located in correspondence of simulation runs/samples in which at least one node is disconnected, leading WCL to yield values greater than 0. At the end of the analysis (10,000 runs) the moving average is stabilized. Figure 11 shows the

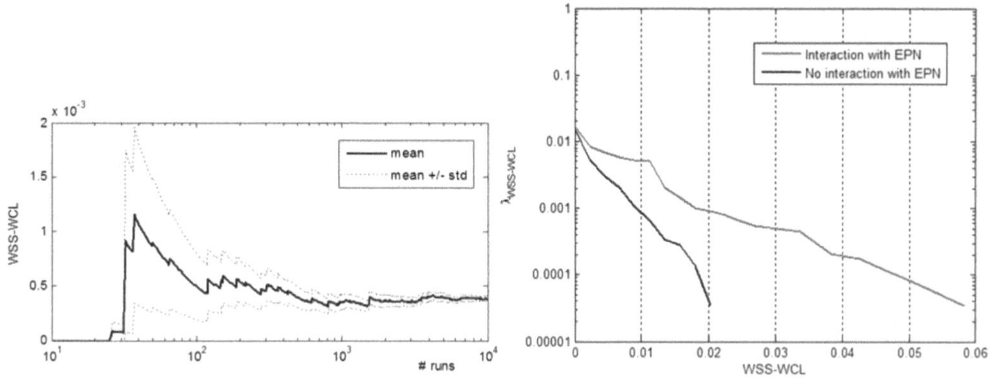

Figure 11: Moving average μ, μ+σ, μ-σ curves for WCL (left) and MAF curves with and without interaction with electric power netwrok (EPN) (right)

MAF of exceedance for WCL. In the same figure, the estimated MAF of exceedance curve for WCL when the interaction with electric power network is not considered in the analysis is compared. The interaction can be important; as an example the connectivity loss is increased from 1% to 1.8% for λ=0.001 (Tm= 1000 years) when the connections of water pumping stations to EPN are included in the analysis.

Figure 12 shows the level of correlation between the WCL and damages in pipelines as well as the non-functional EPN substations supplying the water

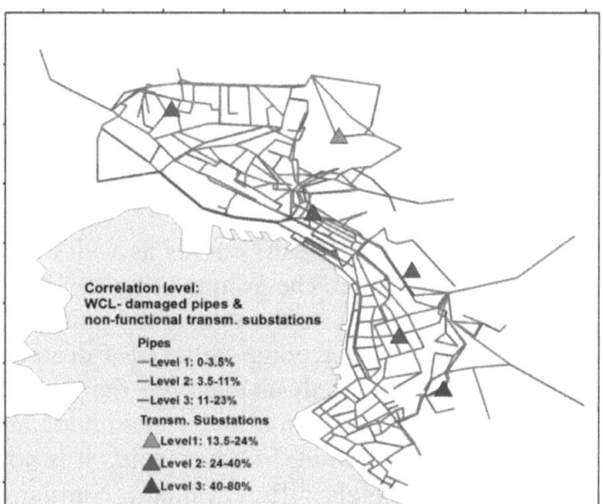

Figure 12: Correlation of damaged pipes and non-functional EPN transmission stations to water network connectivity

pumping stations. The most correlated pipelines are concentrated along the coast where the liquefaction susceptibility is high and therefore damages due to permanent ground displacement are expected. Interestingly, a higher level of correlation is estimated for the EPN transmission substations. The highest value of 80 % is attributed to component in the S-E part of the city, where several pumping stations (connected to EPN) are located. Figure 13 shows an example of the expected distribution of damages for an event that corresponds to connectivity loss (WCL=1.4%) with mean return period Tm=500 years. Only few broken pipes are observed, while the majority of non-functional pumping stations and not-connected demand nodes are accumulated at the S-SE part of the city.

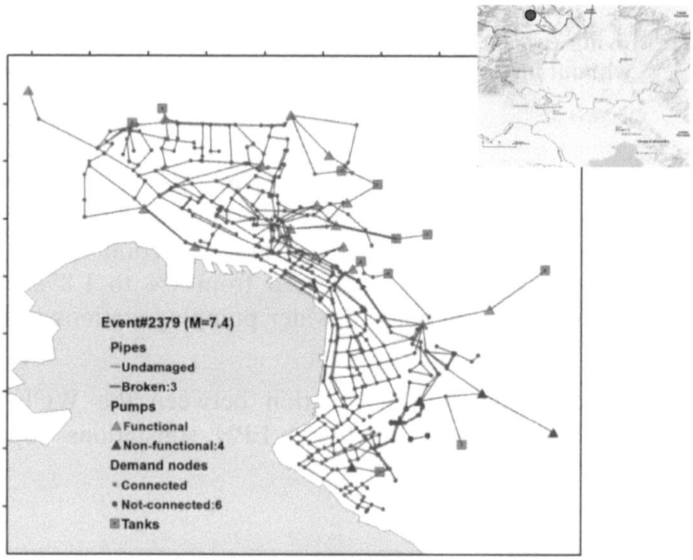

Figure 13: Water supply system damages for an event (#2379, M=7.4, R=72km) that corresponds to WCL with Tm=500 years

9.5 Buildings

Figure 14 shows the moving average (mean) curves as well as the mean+stdv and mean-stdv curves for expected deaths. The values are given as percentages of the total population (790,824 inhabitants). At the end of the analysis (10,000 runs) the moving average is stabilized with an average value of 4 deaths. This low fatality rate is reasonable in this case as the analysis averages the results over all possible magnitudes and epicentral distances, and the lower magnitude and longer distance events are certainly controlling the output. In other words it is not a scenario-based event, which will produce a completely different image. Similar curves and results are derived for injuries and displaced people (in bad and good weather conditions).

Figure 14 also shows the MAF of exceedance curves for deaths (as percentages of the total population). The expected deaths for λ=0.002 (return period Tm=500 years) are 201. The distribution of building damages for an event that corresponds to this return period of deaths is shown in Figure 16. Similar maps can be obtained for casualties and displaced people. For this event, the estimated losses are: 2,248 collapsed and 16,634 yielding buildings, 201 deaths, 492 injuries, 180,000 (in good weather) and 288,000 (in bad weather) displaced people. Figure 15 shows the level of correlation between the damaged WSS and EPN components and the displaced people. It is observed that the correlation is higher with the EPN substations, which highlights the importance of the interaction between EPN loss and habitability.

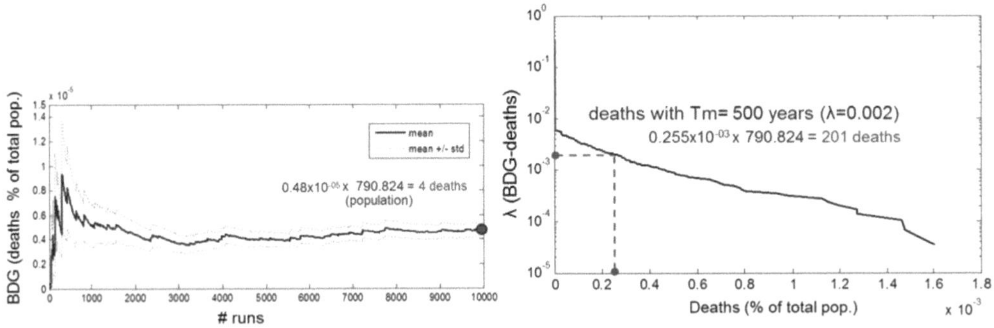

Figure 14: Moving average μ, μ+σ, μ-σ (left) and MAF curve for deaths (right)

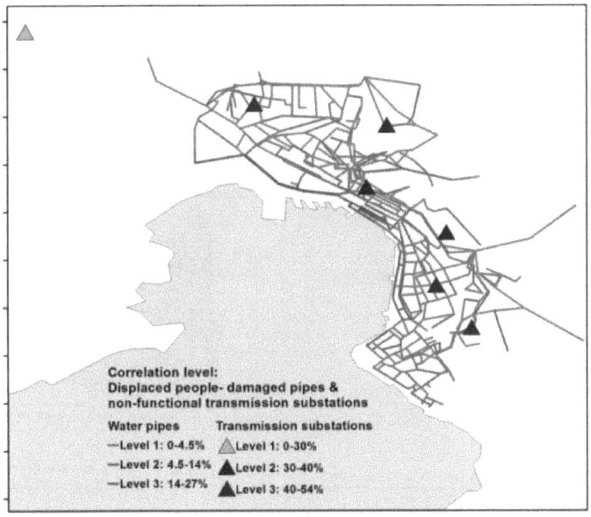

Figure 15: Correlation of damaged EPN and WSS to displaced people

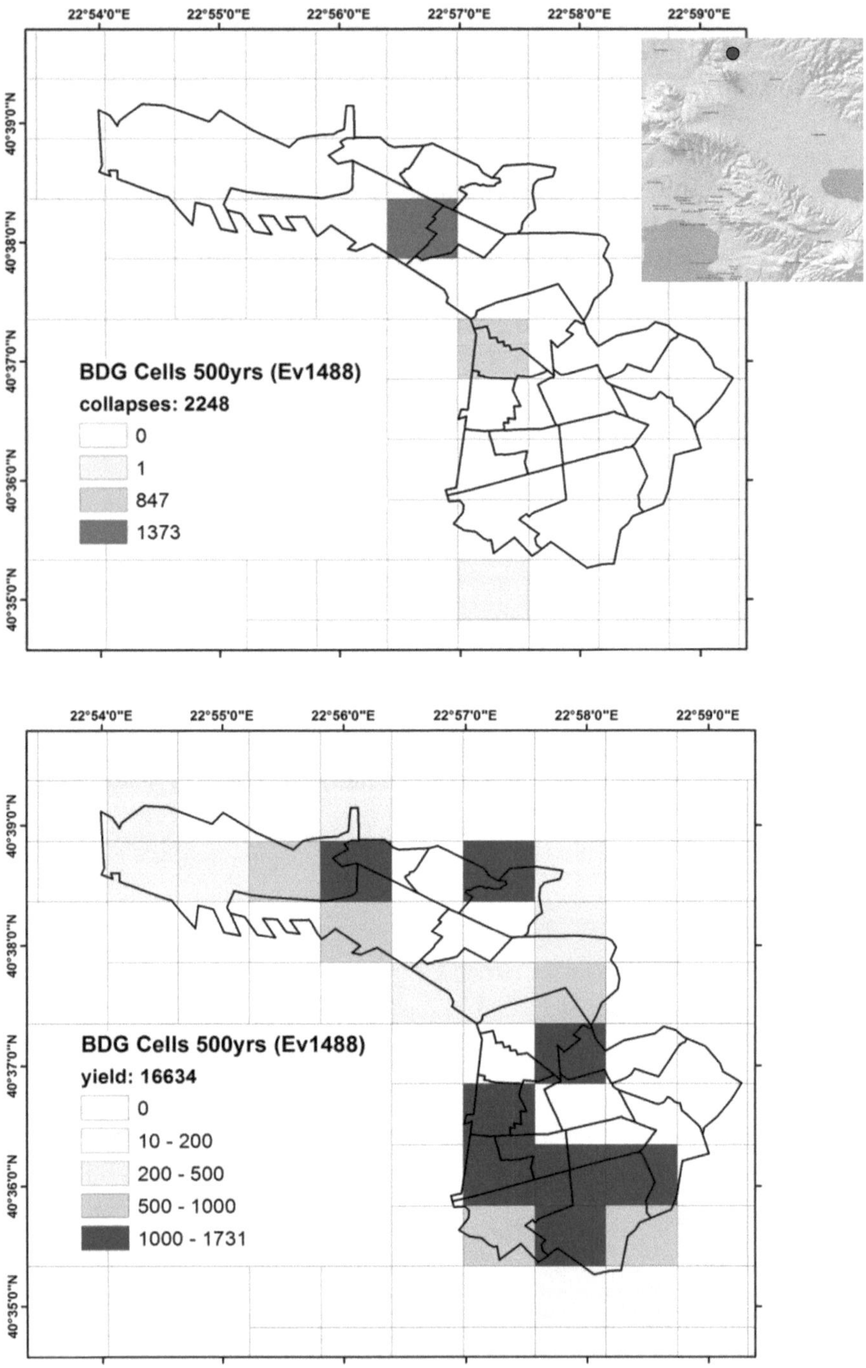

Figure 16: Distribution of estimated damages (collapsed and yielding buildings) into cells of the study area for an event (#1488, M=5.5, R=24 km) that corresponds to death rate with Tm=500 years

9.6 Road Network

Figure 17 shows the moving average (mean) curves for Simple Connectivity Loss (SCL) and Weighted Connectivity Loss (WCL), as well as the mean+stdv and mean-stdv curves for the two PIs. The figures indicate that the expected value of connectivity loss given the occurrence of an earthquake is higher for WCL than for SCL, as expected. This is because WCL takes into account not only the existence of a path between two Traffic Analysis Zones (TAZs), but also the increase in travel time due to the seismically induced damage suffered by the RDN. The jumps present in the plots are located in correspondence of simulation runs/samples in which at least one TAZ node is disconnected, leading SCL and WCL to yield values greater than 0. At the end of the analysis the moving average is stabilized.

Figure 18 shows the MAF of exceedance curves for SCL and WCL. As expected, weighting the computation of connectivity loss with the path travel times yields higher values of exceedance frequency. The same figure compares the estimated MAF of exceedance curve for SCL and WCL when the road blockage due to collapsed building is not considered in the analysis. The interaction with building collapses can be important especially for mean return periods of WCL higher than 500 years (λ=0.002). As an example the WCL is increased from 20% to 33% for λ=0.001 (Tm= 1000years) when the building collapses are included in the analysis.

Figure 19 and Figure 20 show the level of correlation between the WCL and the distribution of damages in bridges and road blockages respectively. In this way the most critical segments can be identified in relation with their contribution to the connectivity loss of the network. These bridges present a high risk of failure due to their vulnerability (old, simple span bridges) and the high values of PGA. The most correlated blocked roads are mainly in the historical centre of the city, where the vulnerability of buildings (mostly build with the oldest seismic code of 1959) is

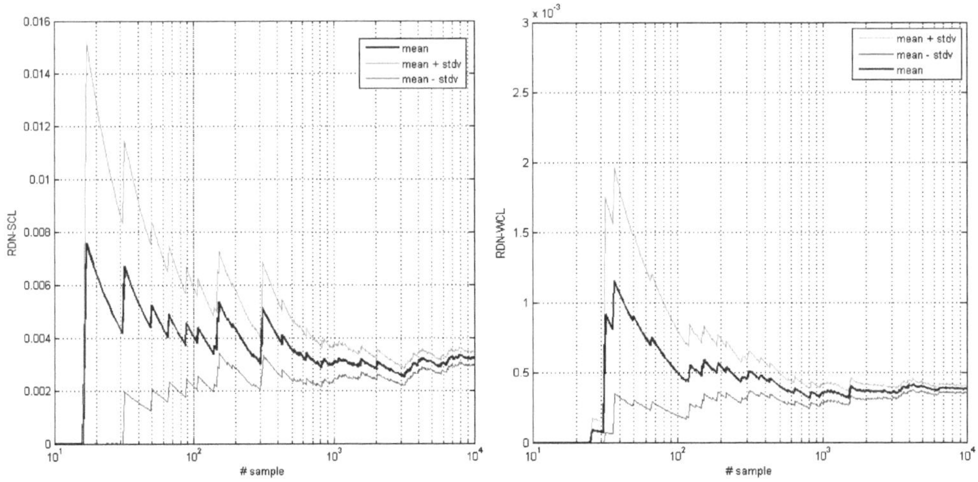

Figure 17: Moving average μ, $\mu+\sigma$, $\mu-\sigma$ curves for SCL (left) and WCL (right)

higher and the road to building distance is shorter. Several road segments in the city centre and the SE part of the study area present a medium correlation due to building collapses. Few roads near the coast which are subjected to ground failure to liquefaction are also highly correlated to the network connectivity.

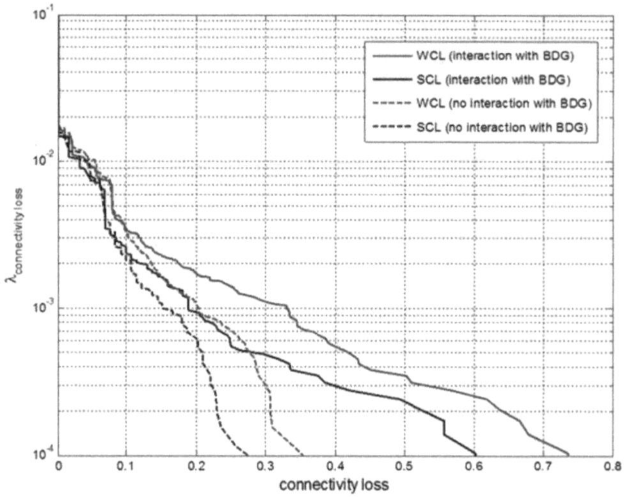

Figure 18: MAF curves for simple (SCL) and weighted (WCL) connectivity loss with and without interaction with building collapses

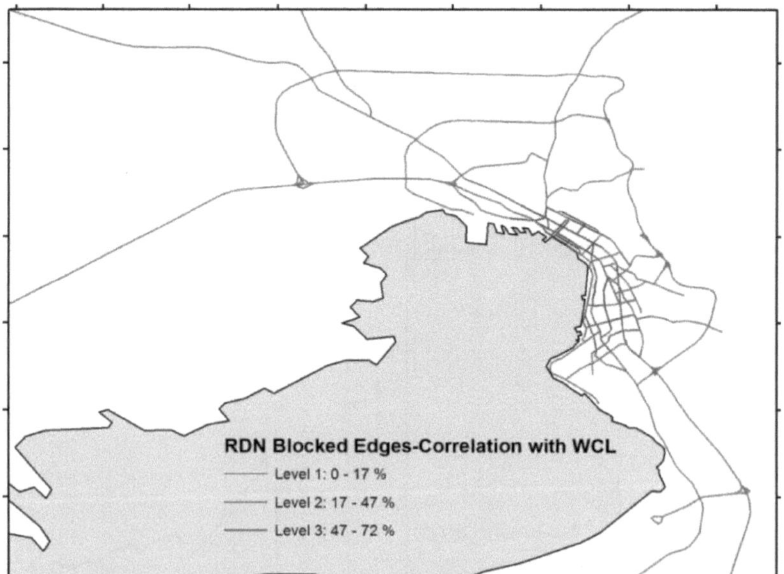

Figure 19: Correlation of blocked by buildings edges to road network connectivity (PI=WCL)

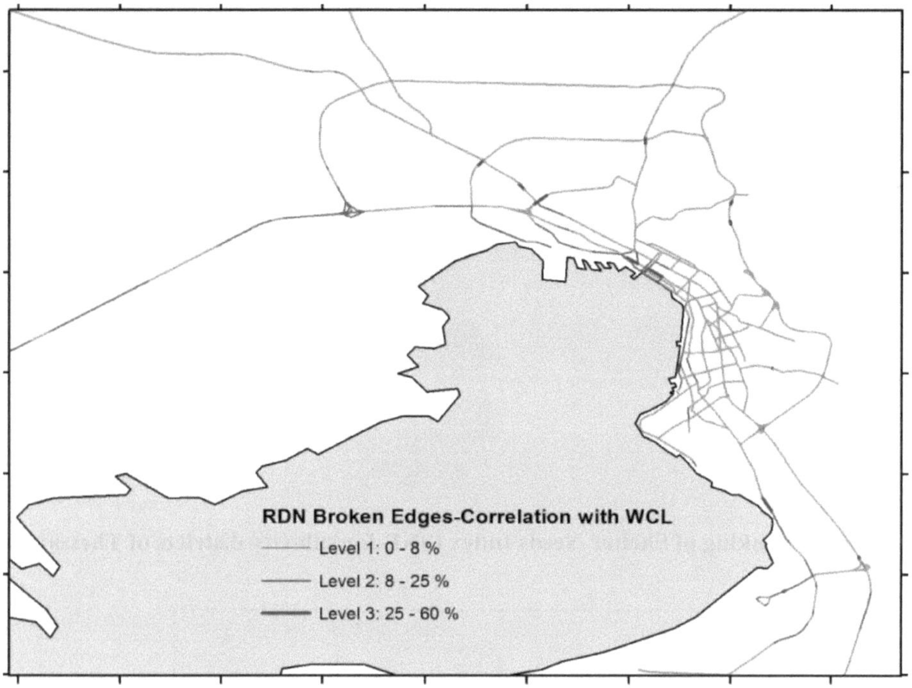

Figure 20: Correlation of broken edges (bridges) to road network connectivity (PI=WCL)

9.7 Shelter Needs and Accessibility Analysis

The estimated damages and losses for buildings, utility and road networks are used as input to the integrated shelter need model developed in SYNER-G (section 7). In particular, a Shelter Needs Index (SNI) is estimated for each one of the 20 Sub City Districts (Figure 21) based on: a) the displaced people estimates for bad and good weather conditions, which are a function of the building damages (BDG) and the utility losses (WSS and EPN), b) the desirability of people to evacuate and c) their access to resources. Criteria b) and c) are evaluated based on indicators from the Urban Audit survey (e.g. age, family status, unemployment rate, education level etc). In this way the Hot Spots" for shelter needs are identified using an interactive decision-support tool.

The estimated damages and losses of the road network provided input for the accessibility modelling to shelters and hospital facilities using isochrone-based and zone-based techniques. An example is given in Figure 22, where the accessibility to health facilities is estimated using the results of RDN over all runs.

Figure 21: Ranking of Shelter Needs Index (SNI) for sub-city districts of Thessaloniki

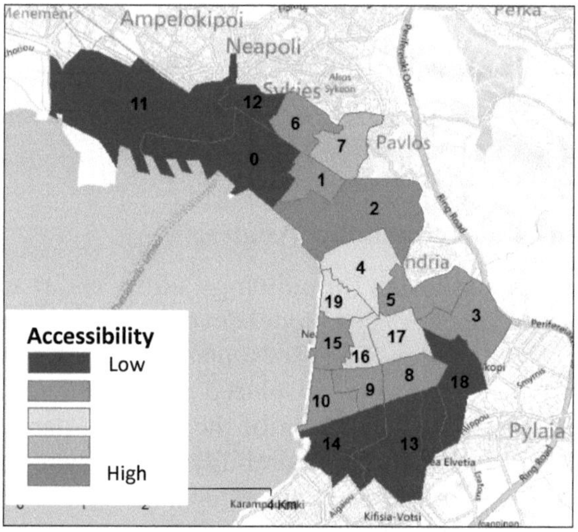

Figure 22: Accessibility to hospitals for Thessaloniki SCDs (zone based technique)

10 Conclusions

SYNER-G has developed a highly innovative and powerful methodology and tool for modern and efficient seismic risk assessment and management of complex urban or regional systems, lifelines and infrastructures. The basic idea is to account in the vulnerability and risk assessment the interdependencies and intra-dependencies (synergies) among various systems and networks, which is finally producing higher damages and losses. It is probably the first time that so many important components of this complex problem have been put together in a comprehensive and scientifically sound way. The whole methodology and tools have been applied and validated in different case studies of variable typology and complexity.

Several sources of epistemic and aleatory uncertainties are inherent in the analysis, which are related among others to the seismic hazard and spatial correlation models, the fragility assessment or the functionality thresholds of each component. The next step of the SYNER-G development is to tackle this issue and to make the whole software package more friendly and easily usable by end users.

11 Acknowledgements

This work has been developed in the framework of the research project SYNER-G funded from the European Community's 7th Framework Programme (FP7/2007-2013) under grant agreement no 244061.

REFERENCES

[1] SYNER-G: Systemic Seismic Vulnerability and Risk Analysis for Buildings, Lifeline Networks and Infrastructures Safety Gain. European Collaborative research Project (FP7-ENV-2009- 1-244061) http://www.syner-g.eu.

[2] Franchin P. (ed.): Methodology for systemic seismic vulnerability assessment of buildings, infrastructures, networks and socio-economic impacts. SYNER-G Reference Report 1, Publications Office of the European Union, (2013), ISBN 978-92-79-28975-0.

[3] Khazai B. (ed.): Guidelines for the consideration of socio-economic impacts in seismic risk analysis. SYNER-G Reference Report 5, Publications Office of the European Union, (2013), ISBN 978-92-79-28968-2.

[4] Hancilar U., Taucer F. (eds): Guidelines for typology definition of European physical assets for earthquake risk assessment. SYNER-G Reference Report 2, Publications Office of the European Union, (2013), ISBN 978-92-79-28973-6.

[5] Weatherhill G., Crowley H., Pinho R.: D2.12- Efficient intensity measures for components within a number of infrastructures. Deliverable of SYNER-G EC project, (2011), available in www.syner-g.eu.

[6] National Institute of Building Sciences (NIBS): HAZUS-MH: Users' Manual and Technical Manuals. Report prepared for the Federal Emergency Management Agency, Washington, D.C, (2004).

[7] Pitilakis K., Riga E., Anastasiadis A.: Design spectra and amplification factors for Eurocode 8. Bulletin of Earthquake Engineering, 10 (2012), 1377-1400.

[8] Pitilakis K., Riga E., Anastasiadis A.: New code site classification, amplification factors and normalized response spectra based on a worldwide ground-motion database. Bulletin of Earthquake Engineering, (2013), doi: 10.1007/s10518-013-9429-4.

[9] Kaynia A.M. (ed.): Guidelines for deriving seismic fragility functions of elements at risk: Buildings, lifelines, transportation networks and critical facilities. SYNER-G Reference Report 4, Publications Office of the European Union, (2013), ISBN 978-92-79-28966-8.

[10] Gehl P., Desramaut N., Monfort-Climent D., Argyroudis S.: D5.1-Systemic vulnerability and loss for building aggregates in urban scale, Deliverable of SYNER-G EC project, (2011), in www.syner-g.eu.

[11] Schäfer D., Bosi A. (2013): Systemic seismic vulnerability assessment: Software users manual. SYNER-G Reference Report 7, Publications Office of the European Union, (2013), in press.

[12] Fardis M.N., Papailia A., Tsionis G.: Seismic fragility of RC framed and wall-frame buildings designed to the EN-Eurocodes, Bulletin of Earthquake Engineering, 10 (2012), 1767-1793.

[13] Arvidsson R., Grünthal G. and the SHARE Working Group on the Seismic Source Zone Model: Compilation of existing regional and national seismic source zones, Deliverable of SHARE EC project (2010).

[14] Akkar S., Bommer J.J.: Empirical equations for the prediction of PGA, PGV and spectral accelerations in Europe, the Mediterranean Region and the Middle East. Seismological Research Letters, 81 (2010), 195-206.

[15] EC8 Eurocode 8: Design of Structures for Earthquake Resistance, Brussels, Belgium, European Committee for Standardisation, The European Standard EN 1998–1, (2004).

[16] Franchin et al.: D2.1- General methodology for systemic vulnerability assessment, Deliverable of SYNER-G EC project, (2011), available in www.syner-g.eu.

[17] Weatherhill G., Crowley H., Pinho R., Franchin P., Cavalieri F. Iervolino I., Esposito S.: D2.13-A review and preliminary application of methodologies for the generation of earthquake scenarios for spatially distributed systems, Deliverable of SYNER-G EC project, (2011), available in www.syner-g.eu.

SeDIF Conference

International Conference on
Seismic Design of Industrial Facilities
2013, RWTH Aachen University

Floor Response Spectra Considering Influence of Higher Modes and Dissipative Behaviour

Marius Pinkawa[1], **Benno Hoffmeister**[1], **Markus Feldmann**[1]

[1] Institute of Steel Construction, RWTH Aachen University
Mies-van-der-Rohe-Str. 1, 52074 Aachen, Germany
m.pinkawa@stb.rwth-aachen.de

ABSTRACT:

Seismic design forces of nonstructural components are commonly obtained by application of floor response spectra. This method is usually applied using estimated modal shapes and periods of the main structure; it allows for a separated design of components and their anchorages by the producers of equipment. Simplified formulas for determination of floor response spectra are provided by current codes such as Eurocode 8. All of them follow the assumption of the first fundamental elastic mode governing the acceleration values at the floors. These approaches do not take into account effects of higher modes, topology, ground response spectrum and plastification of supporting structures.

Floor response spectra of four different building frames, one typical for an industrial 5-storey steel supporting structure and other three representing 5-, 10- and 15-storey regular steel buildings, were investigated using nonlinear incremental dynamic analyses. The results were compared to current code provisions revealing large discrepancies which have impact on safety as well as on economy of the design.

Three aspects were identified and qualified:

- Application of ground response spectrum values instead of peak ground acceleration as basic input variable
- Importance of higher modes
- Impact of plastification of the main structure and the components

It could be shown that all three parameters have a significant influence on the acceleration values, on the dimensioning of the anchorages and on the ductility demand for components designed to dissipate energy.

Keywords: Floor Response Spectra, Nonstructural Components, Secondary Structures, Incremental Dynamic Analysis, Seismic Design

S. Klinkel et al. (Eds.), *Seismic Design of Industrial Facilities*,
DOI 10.1007/978-3-658-02810-7_19, © Springer Fachmedien Wiesbaden 2014

1 Introduction

Secondary structures are mechanical, electrical or architectural components usually attached to primary supporting structures. Such nonstructural components are frequently found in industrial facilities and are primarily designed for functionality rather than for seismic resistance. Nevertheless, due to dynamic response of the supporting structure during an earthquake they can be subjected to high accelerations at their attachment points. Therefore and also owing to their usually high investment costs and/or risk potential special attention should be paid to their seismic design. However, nonstructural components often suffered severe damage in recent earthquakes, resulting in threat for lives and in high economic losses.

To determine seismic force demands on secondary structures different types of analyses can be applied. Time history analyses using a combined model of supporting structure and attached secondary structure provide the most accurate results. However, it is the most complex type of analysis and therefore more practical methods are often used in current practice like the floor response spectrum method. This cascaded approach has advantages over the use of a combined model, since the analysis of primary and secondary structures are separated. This is particularly preferable due to the fact that design processes of components and supporting structures are usually not only partitioned between different design teams but also often take place at different time stages. A drawback of this two-step procedure is that neglecting dynamic interaction effects can lead to unreal high accelerations of the component if its mass is not negligible in relation to its supporting structure's mass.

Simplified formulas to determine design forces for nonstructural components separated from the supporting structure's design are contained in current code provisions like Eurocode 8 [1] and ASCE 7 [2]. They are very similar in their approach: the peak ground acceleration serves as basic input and is amplified on the one hand through the supporting structure's vibration, i.e. amplification from ground to attachment point, and on the other hand through the vibration of the component itself, i.e. amplification from attachment point to centre of mass of the secondary structure. The first amplification effect is reflected by a linear increase of accelerations up to the top of the supporting structure, approximating the fundamental mode. The second effect is accounted for by a constant factor of 2.5 for flexible components (ASCE 7) or a given resonance function depending on the ratio of component period T_a to fundamental period of the supporting structure T_1 (Eurocode 8). Energy dissipation by the component's inelastic behaviour is taken into account by response modification factor R_p (ASCE 7) or behaviour factor q_a (Eurocode 8) respectively. In contrast energy dissipation by the supporting structure's inelastic behaviour is fully neglected. Also ground response spectra available in current codes are not taken into account as basis for simplified formulas. The simplified approach of Eurocode 8 is described in more detail below.

2 Simplified Eurocode 8 approach

For important or hazardous components the force demands have to be determined by a realistic model of the relevant structures and shall be based on appropriate response spectra derived from the response of the supporting structural elements of the main seismic resisting system; i.e. the generation of floor response spectra is prescribed. In other cases properly justified simplifications are allowed. Such a simplification is given in Eurocode 8 with a formula for the determination of seismic force demand F_a on nonstructural components shown in Eq. (1).

$$F_a=\frac{S_a \cdot W_a \cdot \gamma_a}{q_a} \quad \text{with} \quad S_a=\frac{a_g}{g}\cdot S\cdot\left[\frac{3\cdot\left(1+\frac{z}{H}\right)}{1+\left(1-\frac{T_a}{T_1}\right)^2}-0,5\right]\geq\frac{a_g}{g}\cdot S \qquad (1)$$

Here W_a is the weight, γ_a the importance factor and q_a is the behaviour factor of the component, which takes into account its energy dissipation capacities. The basis of the formula is the seismic coefficient S_a, which assumes a linear increase of floor accelerations along the building's height and resonance in the case when the period of the component T_a approaches the fundamental period of the supporting structure T_1. The value z corresponds to the height of the component's attachment point above ground whereas H is the total height of the supporting structure. The seismic coefficient has to be at least the peak ground acceleration, which is normalized to the gravitation constant. This equals the product of design ground acceleration a_g normalized to the gravitation constant g and the soil factor S.

The seismic coefficient comprises the two different amplification effects which were mentioned above: (I) the amplification of acceleration from ground to floor which is caused by the response of the primary structure; (II) the amplification of acceleration from floor to the component's centre of mass which is caused by the response of the component itself. The approximations of the first and the second amplification effects are shown in Figure 1 (a) and (b) respectively; the combined amplification factor, which equals the term in parentheses in Eq. (1), is shown in Figure 1 (c).

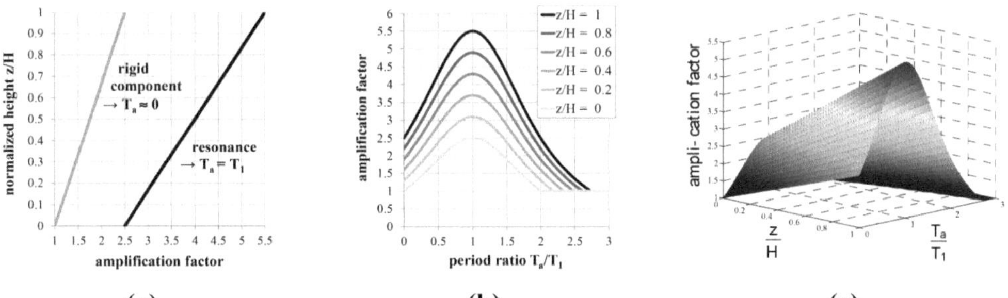

Figure 1: Eurocode 8 formula (a) Linear approximation of floor accelerations along supporting structure height; (b) approximation of resonance phenomenon when component period T_a approaches fundamental period of supporting structure T_1; (c) resulting amplification factor as a function of attachment height and fundamental period ratio

3 Numerical investigations

Nonlinear time history analyses were carried out on four moment resisting plane steel frames: a 5-storey slightly irregular frame with 1 bay, a 5-, 10- and 15-storey regular frame with each 3 bays, covering a meaningful range of supporting structures with increasing relevance of higher modes. These investigations were a significant extension of studies performed before using single-degree-of-freedom (SDOF) models only [3]. The geometries of the investigated frames are shown in Table 1 along with some important properties.

Table 1: Properties of investigated moment resisting steel frames

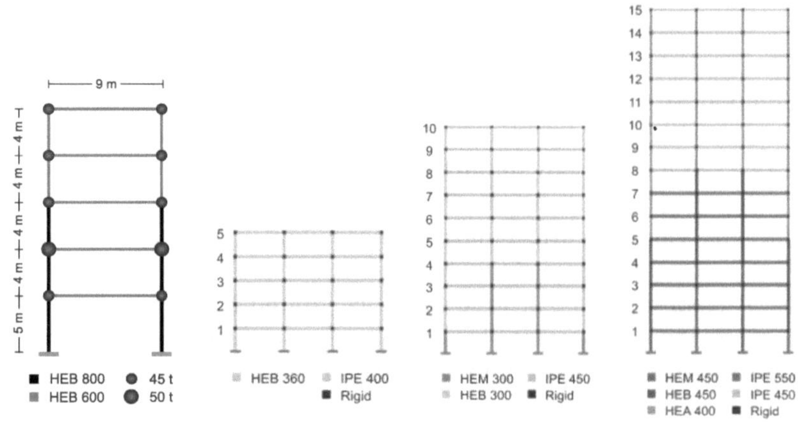

	5-storey 1-bay	5-storey 3-bay	10-storey 3-bay	15-storey 3-bay
Bay width	9 m	7 m each	7 m each	7 m each
Storey height	5 / 4 / 4 / 4 m	3.5 m each	3.5 m each	3.5 m each
Steel grade	S235	S355	S355	S355
Lumped masses at connections	As shown in sketch	786.5 kg	786.5 kg	786.5 kg
Distributed masses (regular / roof)	0 / 0	3065 kg/m / 494 kg/m	3065 kg/m / 494 kg/m	3065 kg/m / 494 kg/m

The transient dynamic analyses were conducted by well-proven in-house software DYNACS [6] taking into account geometric and material non-linearity. The columns and beams were modelled by fibre elements, thus allowing for distributed plasticity and M-N interaction directly. Panel zones were modelled as rigid. Important dynamic characteristics for the first three modes of the investigated structures are contained in Table 2, which are the period T, the fraction of effective modal mass m_{eff} of total mass m_{tot} and the damping values ξ. The latter results from an assumed stiffness and mass proportional Rayleigh approach with a damping of 5% in the first and second mode. The shapes of the first three natural modes are very similar among all buildings, as can be seen in Figure 2. As input a set of 7 artificially generated accelerograms was used, which matched a specific

Eurocode 8 elastic response spectrum (type 1, soil type B, a_g=0.25g, importance factor γ_I=1, 5% damping ratio). Taking into account the soil factor S=1.2 this resulted in a peak ground acceleration PGA of 0.3g. The 3-bay structures were designed to this specific spectrum according to Eurocode 8 in [4], whereas the 1-bay structure is a modified example found in literature [5]. The spectral acceleration value S_{ag} obtained from the ground response spectrum of each earthquake at the corresponding period is also given in Table 2 as mean value of all 7 accelerograms.

Table 2: Dynamic characteristics

Steel frame	Mode	T [s]	m_{eff}/m_{tot} [-]	ξ [%]	S_{ag} [m/s²]
5-storey 1-bay	1	1.05	0.83	5.0	3.5
	2	0.32	0.12	5.0	7.2
	3	0.16	0.04	8.4	7.3
5-storey 3-bay	1	1.12	0.81	5.0	3.3
	2	0.34	0.11	5.0	7.3
	3	0.18	0.04	7.9	7.4
10-storey 3-bay	1	2.03	0.78	5.0	1.8
	2	0.68	0.11	5.0	5.6
	3	0.39	0.04	7.2	7.6
15-storey 3-bay	1	2.36	0.72	5.0	1.3
	2	0.85	0.15	5.0	4.5
	3	0.49	0.04	7.2	7.3

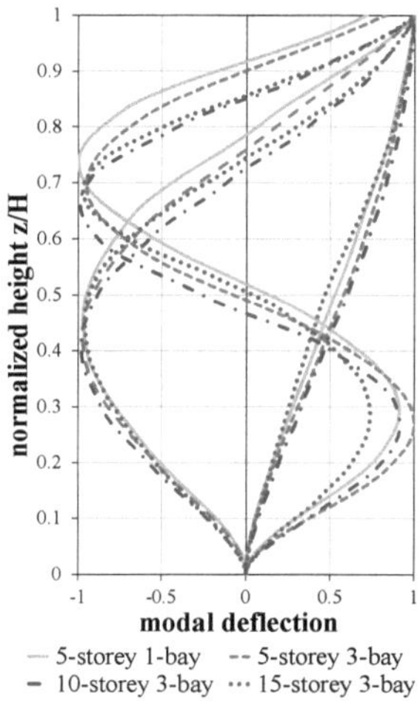

Figure 2: First 3 natural mode shapes

In order to investigate the impact of inelastic behaviour of the supporting structure as well as of the component incremental dynamic analyses were conducted. At first for each earthquake the seismic level was determined, at which the building's global behaviour was still elastic. This characteristic state was defined when a cross section's moment reached the plastic section modulus taking into account M-N interaction according to Eurocode 3 [7]. The magnitude of shear forces showed to be negligible for moment bearing capacity. The scaled accelerogram which satisfied this condition first was assigned the earthquake intensity "1" label. Higher seismic levels which yielded an inelastic behaviour of the supporting structure were obtained by scaling the intensity 1 accelerogram. Therefore the seismic level entitled as intensity measure IM k is the k-scaled accelerogram which first reached the plastic section modulus of an arbitrary cross section.

3.1 Peak floor accelerations

First of all the peak accelerations of rigid components were investigated, which equal the peak accelerations of the floors where they are attached. According to [2] rigidity is assumed if the fundamental period of a structure is less than 0.06s. The

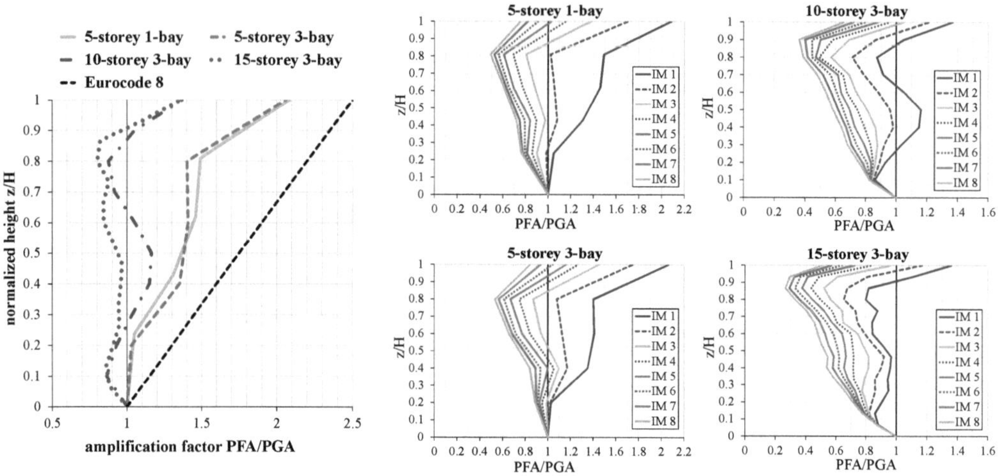

(a) (b)

Figure 3: Amplification factor PFA/PGA for frames behaving elastically in comparison to Eurocode 8 approach (a) and for eight earthquake intensity measures IM for each frame (b)

peak floor accelerations of all four supporting structures are shown in Figure 3 (a) for earthquake intensity 1, thus elastic global behaviour. For reasons of comparability the peak floor accelerations PFA are normalized to the corresponding earthquake's peak ground acceleration PGA. As in all following diagrams the mean values of all 7 accelerograms are shown. As can be seen, the linear approach of Eurocode 8 fits reasonably well for the 5-storey buildings, whereas the PFA in higher buildings are significantly overestimated. The impact of primary structure's dissipating behaviour at higher earthquake intensity measures IM is shown in Figure 3 (b). In all cases the acceleration amplification values PFA/PGA are significantly reduced. This beneficial effect is the strongest at the first few intensity increments, whereas further increase in earthquake intensity at already high intensities results in less reduction.

3.2 Elastic floor response spectra

If the component is not rigid but flexible, the amplification of the attachment point's acceleration considering the component's response has to be taken into account. Therefore the secondary structure was idealized by a SDOF system and its demands were determined by a decoupled approach, i.e. floor response spectra were computed from the obtained floor acceleration time histories. A component damping ratio of 3% was applied for all presented spectra.

Exemplary mean floor response spectra are shown in comparison to the Eurocode 8 approach in Figure 4. The peak component acceleration PCA is normalized to PGA and the period axis is normalized to the supporting structure's fundamental period

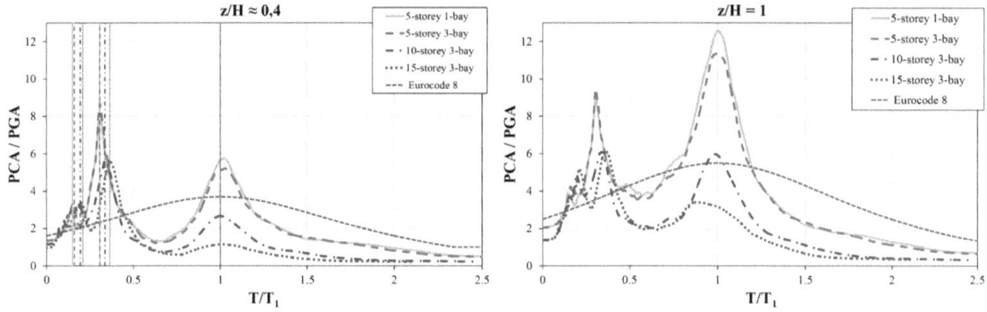

Figure 4: Floor response spectra normalized to peak ground acceleration PGA for all frames and two different relative heights: 40% (left) and roof level (right)

T_1 for each building. The positions of higher modes are marked by vertical lines in the left diagram. The resonance effects with higher modes are clearly identifiable and in some cases they exceed the values obtained for the fundamental mode. Therefore neglect of higher modes as done in Eurocode 8 approach can lead to unsafe results.

Concerning the prediction of component acceleration values by the Eurocode 8 formula the relevance of assumed component damping in computation of floor response spectra is highlighted in Figure 5. Unlikely high damping values are needed to comply with the Eurocode 8 formula in this example. Thus the amplification effects in case of resonance are underestimated with the Eurocode 8 approach. Keeping in mind the overestimation of peak floor accelerations, this underestimation is even more relevant. Counteraction of both aspects – overestimation of the first amplification effect and underestimation of the second one – yields reasonable results for the 10-storey frame. Nevertheless both effects represented in the Eurocode 8 formula are not well balanced.

The scatter in Figure 4 between different buildings of varying height is very large. The uniform Eurocode 8 approach neglecting building properties beside the fundamental period T_1 is not able to reflect these differences among building topologies. To eliminate the variation of different peak floor accelerations a further amplification factor peak component acceleration PCA normalized to peak floor acceleration PFA was considered. Although diminished the scatter was still high. The most suitable amplification factor was shown to be the peak component acceleration PCA normalized to spectral acceleration S_{ag} obtained from the 3%-damped ground response spectrum at the component's

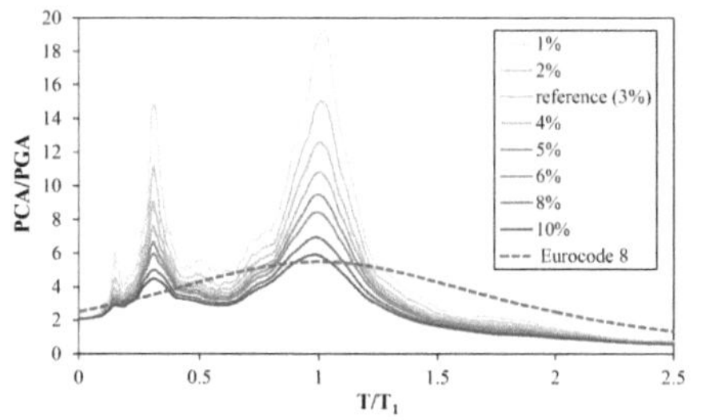

Figure 5: Influence of component damping at roof for 5-storey 1-bay frame

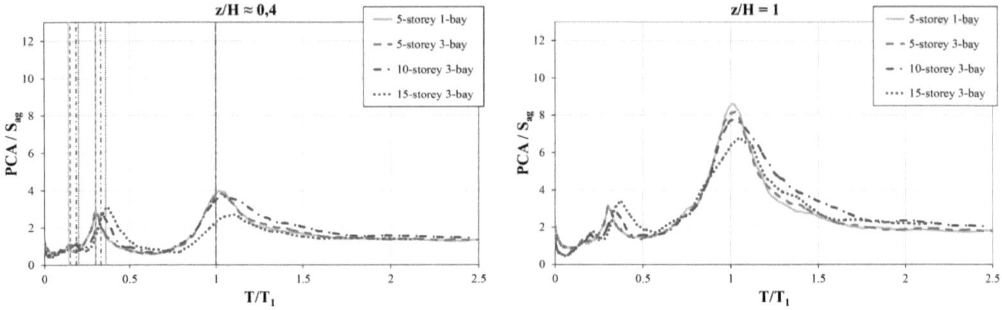

Figure 6: Floor response spectra normalized to ground response spectrum values S_{ag} at component period at two different relative heights: 40% (left) and roof level (right)

period. This equals the normalization of the floor response spectrum to the ground response spectrum. This amplification factor is shown in Figure 6 in the same fashion as in Figure 4. The scatter among various frames is noticeably reduced, because such an approach takes into account the amount of energy which the earthquake contains at modes of the supporting structure and thus indirectly includes the supporting structure's properties.

The magnitude of amplification effects compared between various floors showed a big influence of height of attachment point and a direct proportionality with the modal shapes. This means if for example the modal shape's deflection of the second mode was zero at a specific floor no amplification effects appeared at this floor with this mode. Thus the consideration of amplification due to resonance with specific modes should take into account the shape of considered natural mode.

The influence of energy dissipation through the supporting structure on amplification factor PCA/S_{ag} is shown in Figure 7. Especially the severe amplification at the fundamental mode is significantly reduced by the inelastic behaviour. The biggest decrease takes place at the first few intensity increments, when entering into the nonlinear range, whereas at already high intensities the

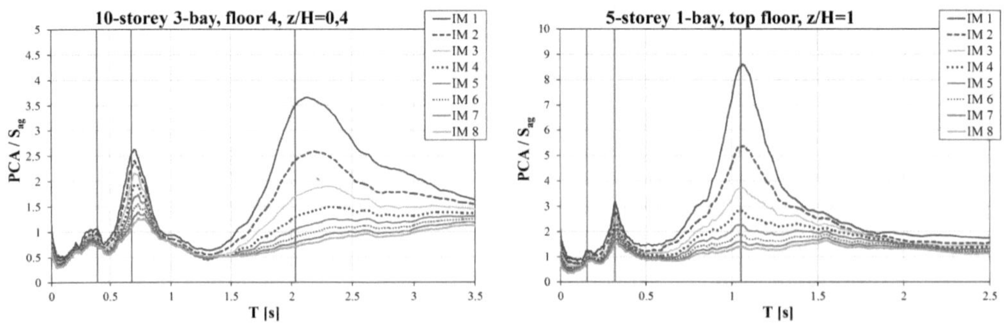

Figure 7: Impact of inelastic frame behaviour on amplification factors PCA/S_{ag} for two of the frames at different relative heights

additional benefit is smaller. On the contrary very flexible and stiff components as well as components with a period between relevant modes of supporting structure are slightly affected by plastification.

The in most cases beneficial effect of supporting structure plastification is not accounted for in the above mentioned current design provisions ASEC 7 and Eurocode 8. The most critical aspect in including this aspect is the overstrength of the main system, which consideration would be crucial. Otherwise a force decrease in the component would be anticipated whereas the force in fact would increase, which could lead to severe differences in capacity and demand of the component. However, the incorporation of primary system plastification seems difficult when no detailed information on the supporting structure's overstrength is given.

3.3 Inelastic floor response spectra

Inelastic floor response spectra where calculated assuming an ideal elastic-plastic force-displacement relationship of the SDOF system. The force at yield was set to the maximum force which the elastic SDOF system had experienced at earthquake intensity 1, thus when the supporting structure behaved globally still elastic. These maximum forces were obtained from the floor response spectra at intensity 1 as a function of considered earthquake, floor and component period. Therefore the components – as well as the frames – plastified at higher intensities than 1. With the maximum force acting in the component set to an upper limit, demands on the deformations were investigated. As suitable parameter the ductility demand μ was determined, which is defined as ratio of maximum absolute displacement during the time history to the displacement at onset of yielding. Ductility demand has to be lower than ductility capacity, which is an inherent property of the specific component and its anchorages. If ductility demands are too high, a limitation to these forces would not be justified in the components design.

Ductility demands for components mounted on structures behaving elastically were investigated first. Consequently the net effect of plastification of just the component could be studied. For this purpose the floor acceleration time histories at higher earthquake intensities were extrapolated from intensity 1. Some exemplary ductility demand floor response spectra for ideal elastic behaving supporting structures are shown in Figure 8. It should be noted that at earthquake intensity k the forces which would act on an ideal elastic SDOF were k times higher than the forces actual acting in the inelastic SDOF system. Thus at high intensities the forces are strongly reduced.

Ductility demands for stiff components are extremely high. In general hysteretic damping as well as viscous one is not suitable to reduce demands in very stiff systems. For flexible components, having a period larger than the fundamental one of the supporting structure, the ductility demands were approximately proportional or lower to the earthquake intensity increase. Ductility demands are noticeably the lowest when the component is in tune with the fundamental period of supporting

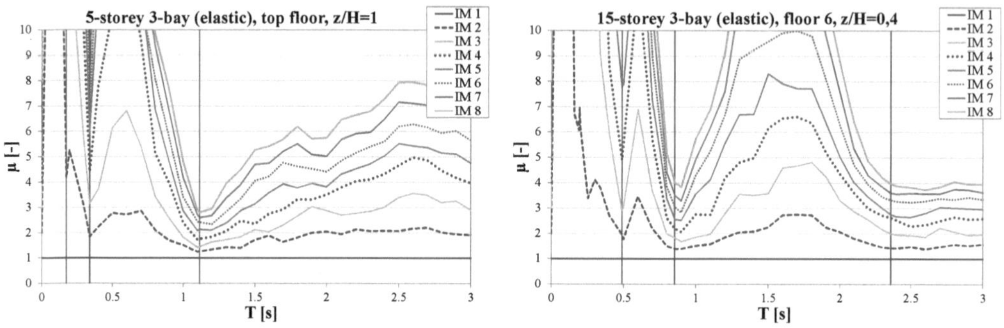

Figure 8: Ductility demands assuming elastic supporting structure behaviour

structure. They are also lower when in tune with the second mode. In contrast when component period is between two relevant frame mode periods demands are very high. For components in tune the low ductility demands can be explained by the design of the component for high loads corresponding to the peaks of the floor response spectra. Due to inelastic behaviour the component detracts from classical resonance and thus high demands. In other words a strong system experiences a relative decrease in demands, compared to the elastic system. The opposite is observed for components which lacked resonance with the supporting structure's modes and were designed for low elastic forces. Due to plastification these components can shift to resonant regions of floor response spectra and thus weak systems exhibit a relative increase in demands. In general it can be stated that ductility demands for components are extremely sensitive to the initial location in the floor response spectrum and thus to period estimation. Therefore if floor response spectrum approach is used the reduction of design forces by behaviour factors is risky if the component does not match the fundamental period of the building. Too low design forces predicted by simplified formulas in current code provisions through application of high behaviour factors could lead to very large and unfeasible ductility demands. Of course peak broadening and enveloping techniques as done in practice for nuclear power plants, as well as use of an averaged response spectrum as done for ordinary structures, would at least reduce such unfavourable effects.

Moreover the combined effect of primary as well as of secondary structure plastification was investigated. Figure 9 shows ductility demands corresponding to Figure 8, but this time taking into account the supporting structure's plastification. Thus the actual computed floor acceleration time history records where used rather than extrapolated ones. As can be seen, beside at small periods, ductility demands are further reduced. Especially at the fundamental but also at the second mode periods they can be extremely low, justifying the high reduction of force demands. Also the valleys are broadened as compared to elastic behaving supporting structures, thus affecting a wider range of components near in tune with ones of the relevant supporting structure's modes.

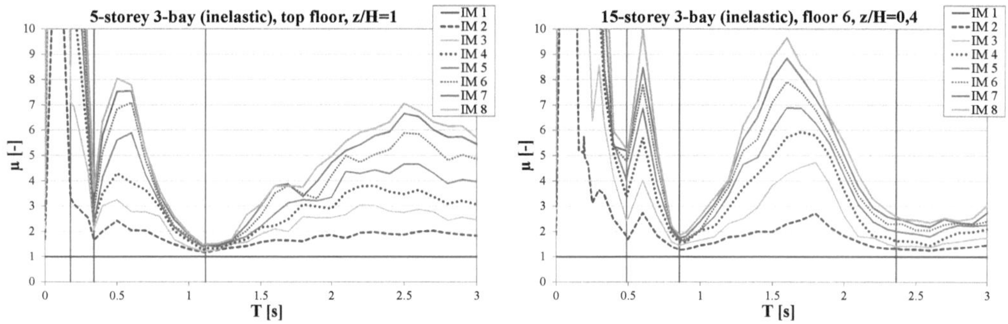

Figure 9: Ductility demands for inelastic supporting structure behaviour

4 Conclusion

Floor response spectra obtained by time history analyses on four moment resisting steel frames showed a strong influence of three investigated main aspects: (a) use of ground response spectrum values as suitable normalization basis for floor response spectra; (b) resonance with higher modes; (c) energy dissipation through inelastic behaviour of the primary, the secondary as well as both structures combined.

Ground response spectrum values corresponding to the component period showed to be the most suitable input variable for implementation in a simplified approach for determination of component accelerations. Compared to peak ground acceleration used as reference in current codes, component accelerations normalized to ground response spectrum values yielded the lowest scatter between all investigated building frames. Especially if supporting structure's topology is not taken into account, the consideration of ground response spectrum in simplified formulas revealed to be favourable.

Resonance effects with higher building modes were very important. On lower floors resonance with higher modes could lead to more severe accelerations than resonance with the fundamental mode, especially for higher buildings. Thus neglecting higher modes, as done in the simplified Eurocode 8 formula, can lead to unsafe design of components.

The inelastic behaviour of the supporting structure lowered in general the component force demands. This beneficial reduction was especially high when the component's period matched one of the building's relevant periods, because classical resonance effect was diminished by plastification. However, to take into account the beneficial effect of energy dissipation of the primary structure for design purposes, the consideration of its overstrength would be crucial.

Ductility demands for ideal elastic-plastic components assuming elastic behaviour of the supporting structure were low to acceptable for components in tune with the first and second building mode. Caution should be paid to components with a fundamental period between relevant periods of the supporting structure, if they are designed to forces obtained from the raw floor response spectrum and which are diminished by a response modification factor. The inelastic behaviour can shift components to a resonant region of the spectrum, and thus components initially designed for low forces would exhibit very large ductility demands. Peak broadening techniques and spectra enveloping or averaging as done in practice would minimize such unfavourable effects.

Finally, the combined effect of supporting structure and component plastification yielded considerable lowered force demands accompanied by very small ductility demands for tuned components, which would be highly loaded in case of elastic behaviour of building and component. For flexible components not in tune with building modes moderate ductility demands were determined. Thus reasonable consideration of energy dissipation of building and component could lead to more economic design of nonstructural components.

In conclusion further effort is needed to enhance current code provisions and to deduce more reliable but simple formulas for the determination of seismic force demands on nonstructural components. Ideally they should take into account: (a) the supporting structure's modal properties of relevant natural modes; (b) the energy input for the corresponding mode; here the ground response spectrum should be used as input parameter instead of the peak ground acceleration; (c) beneficial dynamic interaction effects as function of mass and period ratio, in order to not obtain over-conservative results and (d) possible demand reductions due to energy dissipating behaviour by plastification of primary as well as of secondary structure.

References

[1] EN 1998-1:2010-12, Eurocode 8: Design of Structures for Earthquake Resistance – Part 1: General Rules, Seismic Actions and Rules for Buildings, European Committee for Standardization, Brussels, 2010.

[2] ASCE/SEI 7-10, Minimum Design Loads for Buildings and Other Structures, American Society of Civil Engineers (ASCE), Reston, 2010.

[3] Hoffmeister, B., Gündel, M., Feldmann, M., Floor Response Spectra for Dissipative Steel Supports of Industrial Equipment, COMPDYN 2011, Corfu, Greece, 26-28 May 2011.

[4] Gündel, M., Zuverlässigkeitsanalysen zur Kapazitätsbemessung von Stahl-rahmen, Institute of Steel Construction, RWTH Aachen, 2013 (in progress).

[5] Meskouris, K., Hinzen, K-G., Butenweg, C., Mistler M., Bauwerke und Erdbeben. Grundlagen – Anwendung – Beispiele, Kapitel 7: Bauwerke und Komponenten im Anlagenbau, Vieweg+Teubner Verlag, 2011.

[6] Kuck, J., Hoffmeister, B.: User Manual for DYNACS – A Program for Dynamic Nonlinear Analysis of Composite and Steel Structures, Institute of Steel Construction, RWTH Aachen, 1999.

[7] EN 1993-1-1:2010-12, Eurocode 3: Design of Steel Structures – Part 1-1: General Rules and Rules for Buildings, European Committee for Standardization, Brussels, 2010.

Application and Distinction of Current Approaches for the Evaluation of Earthquake-Response of Secondary Systems

Wehr, Franziska[1, 2], Bach, Andreas[1], Zahlten, Wolfhard[2]

[1] Schüßler-Plan Ingenieurgesellschaft mbH
St.-Franziskus-Str. 148, Düsseldorf, Germany
abach@schuessler-plan.de

[2] Institute for Structural Mechanics and Numerical Methods
Bergische Universität Wuppertal
Pauluskirchstr. 7, Wuppertal, Germany
fwehr@uni-wuppertal.de
zahlten@uni-wuppertal.de

ABSTRACT:

The generation of synthetic earthquakes is an important point in earthquake design in order to have representative earthquake time histories for a given response spectrum. Different possibilities like amplitude modification and wavelet modification exist to match the synthetic earthquake with the target response spectrum as closely as possible and hence to allow for a design of secondary systems using direct time integration methods. However, for the design of secondary systems within a building against earthquake excitation, other methods are also applicable. Besides a calculation using direct time integration with a complete FE-model of the building including the secondary system, the floor response spectrum method or a simplified method, as given e.g. within the EC 8, may also be used. The applicability of these approaches, however, depends on their validity compared to the direct time integration method. This paper compares and discusses the known methods for the generation of earthquake time histories and checks the design methods for secondary systems within a large reinforced concrete structure to enable a reliable design of secondary systems against earthquake excitation.

Keywords: Synthetic Earthquakes, Dynamic Analysis, Engineering Methods, Secondary Systems

1 Introduction

Earthquakes may not only cause damage to the buildings themselves, but also to parts of the buildings infrastructure, such as pipes, machinery and electronic equipment. Especially for buildings with a high degree of technical infrastructure, the loss of these secondary systems can lead to a breakdown of the building performance. This may cause severe societal and financial losses. Therefore, appropriate approaches for the evaluation of the earthquake response of secondary systems are necessary. Examples are power plants for which the value of the building's infrastructure (machinery, turbines, pipes etc.) is much higher than the value of the structural components. Therefore in addition to structural integrity, serviceability under earthquake excitation becomes an issue, as the loss of parts of the building infrastructure may cause the loss of its functionality and therefore cause serious financial, as well as societal losses. Therefore it is necessary to describe the influence of earthquake excitations on the secondary systems for large structures adequately. This leads to the establishment of design procedures for their secondary systems which guarantee serviceability of the building under earthquake excitation.

For the design of the structural components three approaches may be chosen. On the one hand there are advanced methods using direct time integration whereby the structure is excited with a ground motion, on the other hand are simplified methods as presented in EC 8.

The numerical effort of these methods varies considerably. Especially an advanced design with direct modelling of the secondary systems is a challenge, as the full dynamic data of the secondary systems (eigenfrequencies, stiffness etc.) is often unknown.

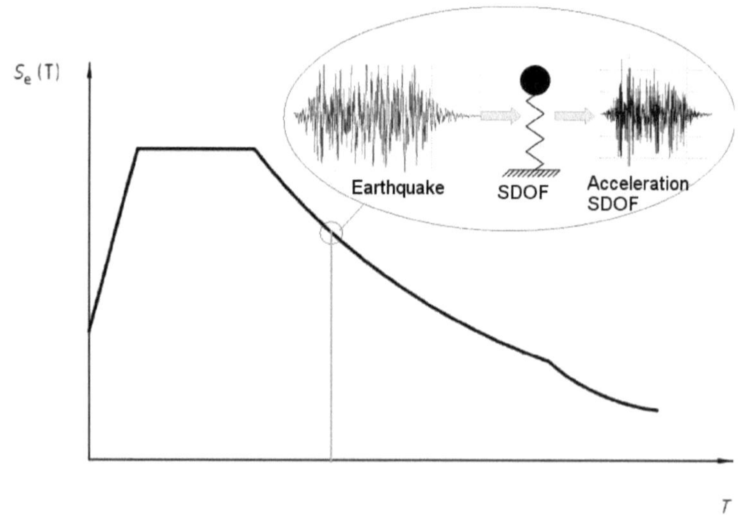

Figure 1: Generation of a response spectrum from an earthquake time record

However, all methods are based on a chosen target response spectrum for the design which defines the earthquake excitation, e.g. in Eurocode 8 [1].

Fig. 1 shows how one point of a response spectrum is gained from an earthquake time record. It can be seen that one value of the response spectrum only represents the maximum response of a one single degree of freedom system (short: SDOF system) to an earthquake. For multiple earthquakes each earthquake excites certain frequencies stronger than others. The response spectrum may therefore be described as an envelope of possible earthquake excitations for a given probability of occurrence. That means that the earthquake design of buildings with the response spectrum method is on the safe side, which is also true for a time domain simulation if the synthetic earthquake covers the entire response spectrum.

2 Generation of synthetic earthquakes

2.1 General remarks

Earthquake time histories are generated on the basis of a given target response spectrum. The target spectra can be found in literature, e.g. Eurocode 8 [1]. This is the official framework for earthquake design in Europe and it also defines certain rules that need to be followed when generating synthetic earthquakes. The synthetic earthquake needs to be generated such that the difference of the maximum acceleration of a SDOF system is not larger by ten percent than the corresponding spectral ordinate if the natural period of the SDOF oscillator lies between 0.2 T_1 and 2 T_1. T_1 denotes the first natural period of the building to be designed.

2.2 Generation of time histories

A common tool for the generation of earthquakes is the program *SIMQKE* which was developed by Gasparini and Vanmarcke. The program is based on the theory of random oscillations. Lestuzzi [2] describes the operating mode of this program.

The earthquake is hereby described by Eq. (1).

$$x(t) = \sum_{i=1}^{N_f} A_i\,(\omega_i)\sin(\omega_i t + \Phi_i) \qquad (1)$$

The amplitudes A_i can be evaluated from the spectral density function and the frequencies ω_i, but the phase angle Φ_i is chosen randomly for each frequency i. The random phase angles cause all generated earthquakes to be different from each other, even though their amplitudes are the same.

The quality of the generated time history can be checked by computing its corresponding response spectrum and comparing it to the target spectrum. An example is shown in Fig. 2. It can be easily seen that the spectrum of the synthetic earthquake diverges considerably from the target spectrum.

In order to reduce the divergence from the target spectrum the earthquake needs to be modified. One method is to modify the amplitudes of the harmonic terms in Eq. (1).

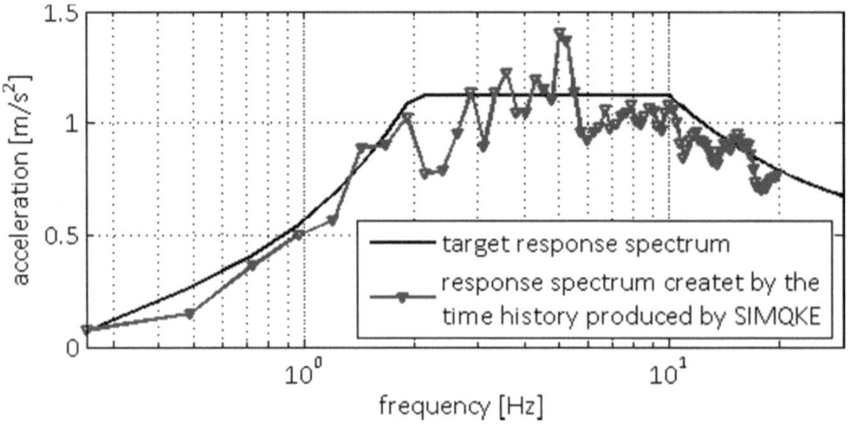

Figure 2: Acceleration response spectrum evaluated by SIMQKE and the target response spectrum

2.3 Adjustment of earthquake time histories by modifying the amplitudes

Lestuzzi developed in 2002 the computer program *SimSeisme* at the École Polytechnique Fédérale de Lausanne. It can be used to modify the amplitudes of an earthquake produced by *SIMQKE*.

In this program the difference between the value of the response spectrum of the synthetic earthquake $S_{a,Target}$ and the value of the target response spectrum S_a is determined at a certain number of frequencies ω_i. In order to minimize the difference between the two response spectra, the amplitude A_i for each considered frequency is modified by Eq. (2).

$$A(\omega)_{i,new} = A(\omega)_i \frac{\frac{S_{a,Target}(\omega)_i}{S_a(\omega)_i} + \left(\frac{S_{a,Target}(\omega)_i}{S_a(\omega)_i}\right)^2}{2} \tag{2}$$

It is not possible to match a given response spectrum exactly by this modification method. The problem is illustrated in Fig. 3. It shows the response of an SDOF oscillator with an eigenfrequency of 10 Hz to an earthquake time history in the frequency domain.

It is obvious that the SDOF system not only reacts with its natural frequency, but also for excitation frequencies nearby, in particular for lower frequencies. For this reason a change in the amplitude of a nearby frequency has also effects on the response of the oscillator, as shown in Fig. 3.

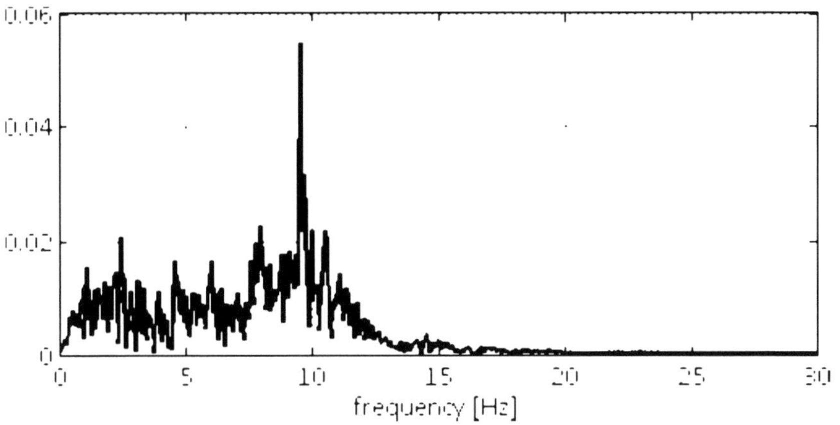

Figure 3: Response of an SDOF with 10 Hz to an earthquake time history

The program *SimSeisme* stops approximating the target spectrum when the differences at the considered frequencies are smaller than a defined value which holds for all points.

2.4 Modification of earthquake time histories in time domain

A more precise possibility to adjust a synthetic earthquake to a given target response spectrum is the modification in time domain. Similar to the amplitude modification the adjustment is made at chosen frequencies. The modification is achieved by adding functions in the form of corrected tapered cosine wavelets to the earthquake time history. The frequency of the wavelets is equal to the chosen modification frequencies. The modification frequencies of the wavelets do hereby not need to coincide with the frequencies of the Fourier series of the earthquake. However, for simplicity these may usually be chosen. With this modification Eq. (1) is extended with an additional wavelet term a_j which is shown in Eq. (3).

$$a_j(\mathrm{t}) = \cos[\omega_j'(t - t_j')]\,\mathrm{e}^{-|t-t_j'|\Psi_j} + [c_1(t - t_j') + c_2]\mathrm{e}^{-|t-t_j'|5\Psi_j} \qquad (3)$$

With:

- ω_j': damped eigenfrequency
- t_j': time when the largest acceleration of the corresponding SDOF oscillator occurs.
- ψ_j: adapts the length of the wavelet.
- c_1 and c_2: adapt the wavelet to achieve zero displacement, velocity and acceleration at the beginning and end of the wavelet.

The amplitudes of the wavelets are adapted such that the difference between the target response spectrum and the response spectrum gained from the synthetic earthquake equals zero. This modification is described in detail by Hancock [3] and Wehr [4].

2.5 Comparison of the presented modifications

A modification by wavelets has less influence on the neighbouring frequencies than the amplitude modification. This behaviour is illustrated in Fig. 4. The original earthquake time history has a greater similarity to the modified time history when using wavelets. This method achieves better results for the approximation than the amplitude modification for the chosen frequencies. However, the time needed for calculation until convergence is usually greater for the wavelet method.

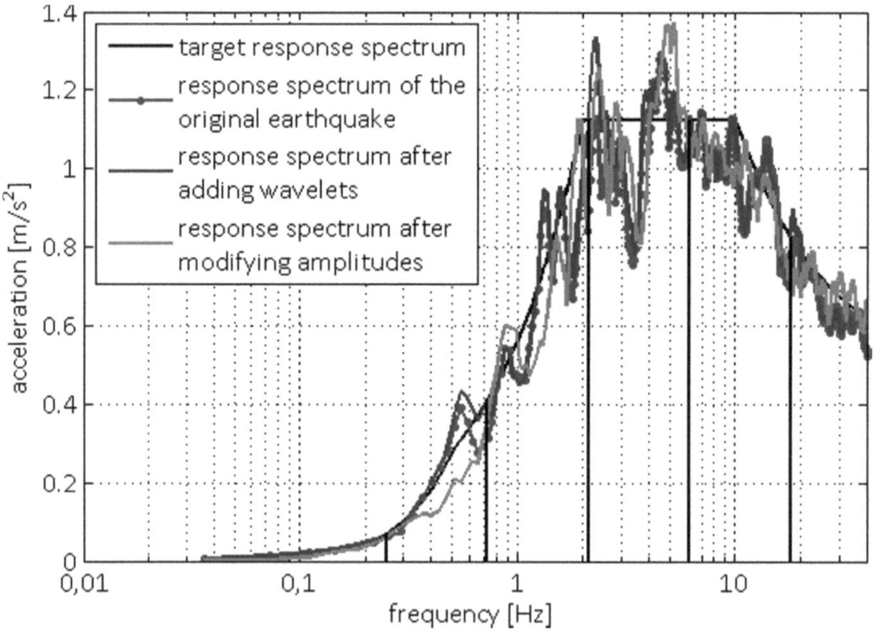

Figure 4: Comparison of the modifications for approximating 5 frequencies

3 Dynamic analysis

Apart from the evaluation of the earthquake time histories it is important to choose a suitable design method for the description of the dynamic responses of secondary systems. In this chapter three different methods for determining the acceleration of secondary systems due to earthquakes are compared. The synthetic earthquakes are created as described in chapter 2. The limits for a divergence from the target spectrum are hereby in accordance with EC 8 [1].

The analysis is performed for a turbine building of a power plant. The reinforced concrete structure is 49.5 m high and has a ground plan area of 48.0 m x 90.5 m (Fig. 5). Two positions for non-bearing elements have been analysed. One point is located at the top floor of the building and another at mid-height. At these points, marked in Fig. 5, SDOF oscillators with different eigenfrequencies are placed. The chosen frequencies are the first two horizontal eigenfrequencies of the building in each direction (x-direction: 1.94 Hz and 4.60 Hz, y-direction: 1.15 Hz and 2.71 Hz) and 0.5 Hz, 1.0 Hz, 1.5 Hz, 2.0 Hz, 3.0 Hz, and 4.0 Hz. In addition their masses are also varied between 0.001 t, 1.0 t, 100 t, 1000 t, and 5000 t.

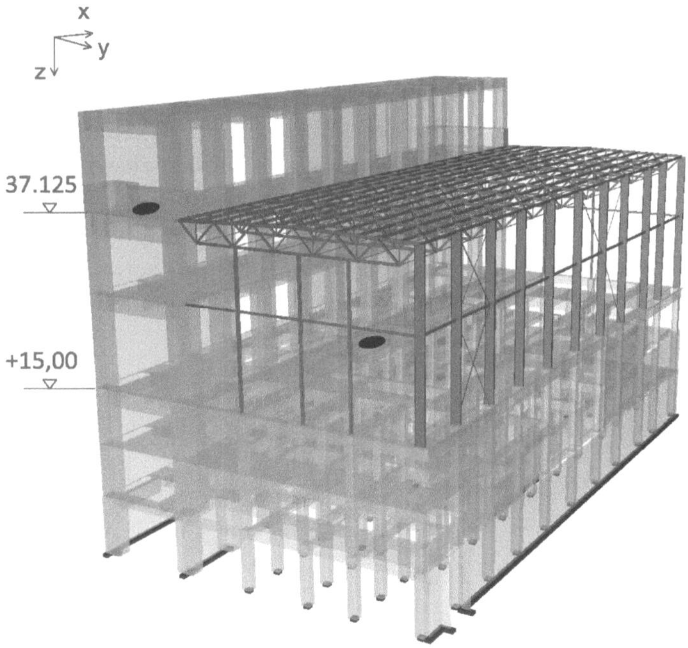

Figure 5: Position of the fictive non-bearing elements in a machine building

The displacements and accelerations for the SDOF systems are computed by the following methods:

- simplified method from EC 8 [1],
- floor response spectrum as shown by Holtschoppen [5],
- direct time integration with a complete FE-Model.

The results for each method, i.e. a comparison of the computational methods, are shown in the following chapters.

3.1 Simplified method from EC 8 [1]

This method computes a horizontal static equivalent load that acts in the centre of mass of the secondary system. The size of the load depends on the vertical position z/H of the non-bearing system in the building and the ratio between its natural period T_a and the natural frequency of the building T_1.

For a better comparison with the other methods an equivalent acceleration is computed in Eq. (4).

$$a_a = a_g \times S \times \left[\frac{3 \times \left(1 + \frac{z}{H}\right)}{1 + \left(1 - \frac{T_a}{T_1}\right)^2} - 0.5 \right] \times \frac{\gamma_a}{q_a} \geq \frac{a_g \times S \times \gamma_a}{q_a} \quad (4)$$

3.2 Floor response spectrum

Fig. 6 explains the generation of floor response spectra. The building is excited with a synthetic earthquake time history. As a result the acceleration at a given point within the building is obtained. This acceleration is further used to excite various SDOF systems with different eigenfrequencies to obtain a response spectrum. This response spectrum is called the floor response spectrum since it is valid not for the entire structure but only for the point, i.e. the floor, for which it has been computed.

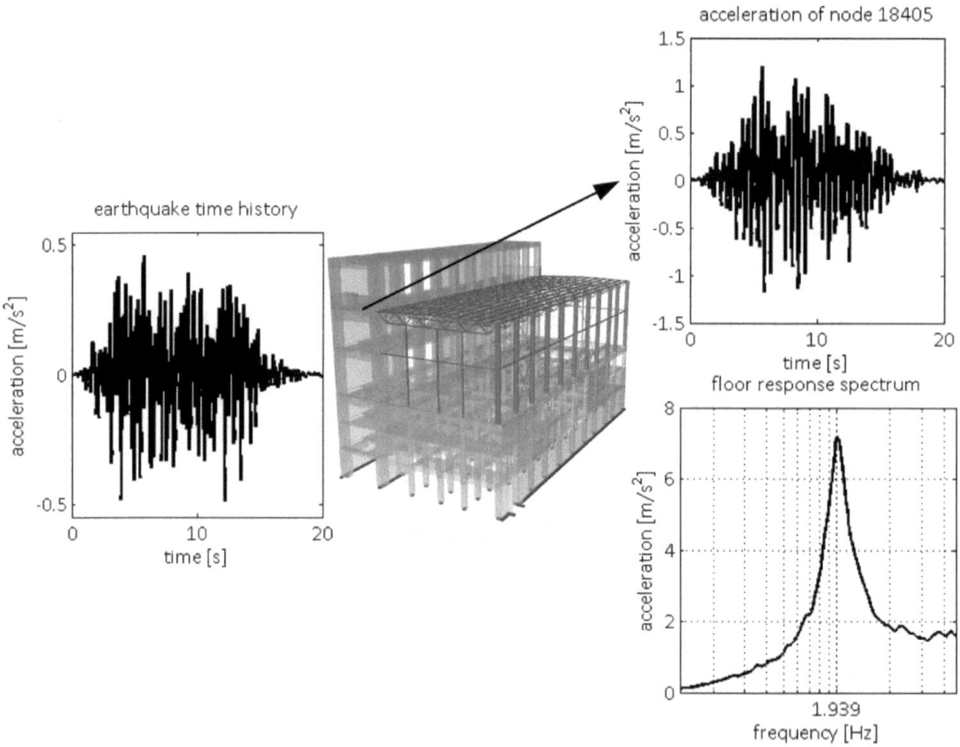

Figure 6: Generation of a floor response spectrum

As the building acts as a complex filter for the earthquake excitation which transmits the vibration to the secondary system, a floor response spectrum has peaks at the dominating resonance frequencies of the primary structure, depending on its location within the building. If multiple resonance frequencies dominate, the spectrum may have multiple peaks as shown in Fig. 7 (black line).

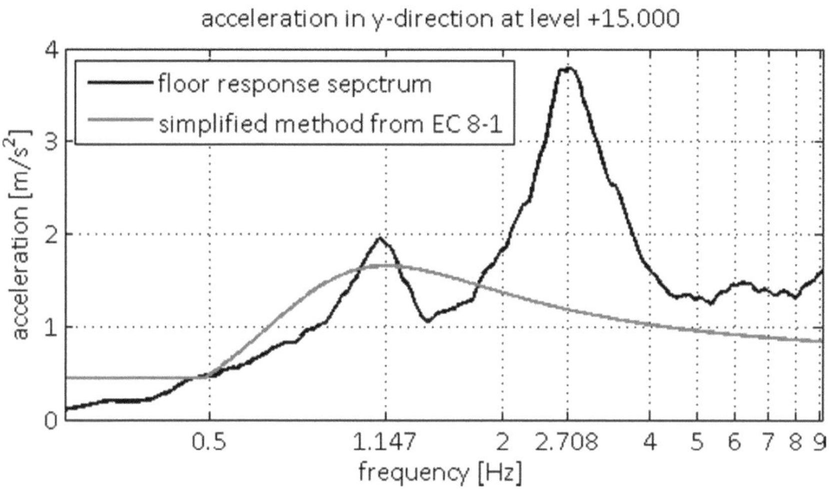

Figure 7: Comparison of the simplified method from EC 8-1 and the floor response spectrum

For small frequencies the value of the floor response spectrum converges to zero, for large frequencies the acceleration corresponds to the maximum acceleration of the floor.

3.3 Direct time integration

For the computation of the acceleration of the secondary systems that are situated in the building, it is necessary to incorporate these in the FE-model. So a possible interaction between the oscillation of the secondary system and the building during the earthquake excitation is taken into account. The computation has been performed using the Newmark method with constant average accelerations.

3.4 Comparison of direct time integration and floor response spectrum

For small masses of the secondary system the natural frequencies of the building are not influenced, therefore the results for the acceleration of the SDOF oscillators using the floor response spectrum method and the direct time integration method are almost identical. The comparison is inter alia pictured in Fig. 8. For higher masses of the secondary systems the interaction causes a small decline of the acceleration of the secondary system. Therefore the results from the direct time integration provide lower values than the floor response spectrum. This behaviour can be observed in the given example for masses from 1000 t and more. For secondary systems with such high masses, such as e.g. a turbine, the response spectrum may overestimate the acceleration up to 40 percent in the horizontal direction (100 t – 1 Hz). By using floor response spectra, however, the resulting static equivalent load leads to a design on the save side.

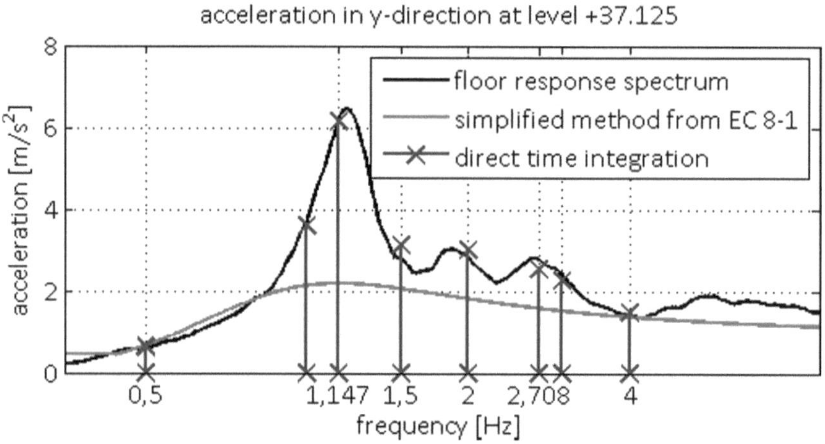

Figure 8: Comparison of the acceleration of the secondary systems gained from the different methods

3.5 Comparison of the simplified method from EC 8 and the floor response spectrum

Fig. 8 also shows the results from the simplified method from EC 8-1. It is apparent that the accelerations, especially in the range of the first resonance frequency, differ considerably compared to the other methods. Similar results are observed for the other directions and also for the point on level +15.00 m, as can be seen in Fig. 7.

The large difference between the simplified and the advanced methods can be explained by analysing the theoretic foundation of the equation for the simplified method (Eq. (4)). This formula has been developed for framed structures where it is sufficient to include only the first resonance period of the building for the considered direction. For a simple, regular, symmetric building the equation gives reasonable results, but for a building with irregularities regarding stiffness and/or mass it leads to unsafe results as higher frequencies as well as torsional effects may dominate the response. This issue is presented within Fig 7. If a secondary system of the same eigenfrequency as the second eigenfrequency of the building is excited, the simplified method is not capable to describe this influence.

Holtschoppen represented in [5] an additional design formula for the consideration of the 2nd eigenfrequency, whereby a value of 1.6 S (≈ 1.8 m/s²) may be evaluated as a conservative value for the given building. This upper limit is however not conservative for the given structural system, see Fig. 7. This can be reasoned by the irregularity of the mass and stiffness distribution.

4 Conclusions

For an evaluation of the responses of secondary systems within larger structures, as can be found e.g. in power plants, synthetic earthquakes have been generated based on commonly used methods. Starting from a first approximation by harmonic series, matching procedures in the frequency domain (amplitude modification) and the time domain (wavelet superposition) have been implemented. Numerical tests showed a good convergence to the target response spectrum. The time domain matching procedure proved to be superior to the frequency domain approach. Therefore this method is recommended for the generation of synthetic earthquakes.

For the computation of the influence of earthquakes on secondary systems the floor response spectrum is a good choice. Only for heavy non-bearing systems the accelerations are too far on the safe side (for the given structure more than 1000 t). For those massive secondary components a complete modelling with an FE-program of the entire structure, including the secondary systems, is recommended. Especially when it is not obvious whether the secondary systems influence the behaviour of the building the direct time integration is the better choice.

Additionally, it could also be shown that the simplified method from EC 8-1 should not be used for buildings with irregularities in the distribution of stiffness and mass. In these cases the simplified method tends to underestimate the accelerations, since more than one eigenfrequency for each direction and torsional effects influence the behaviour of the building. Therefore it should be carefully investigated beforehand if the simplified method is applicable for the given case. A remark pertaining to the limitations of the simplified method should be included in future versions of EC 8.

REFERENCES

[1] Normenausschuss Bauwesen (NABau) im DIN, Eurocode 8: Auslegung von Bauwerken gegen Erdbeben – Teil 1: Grundlagen, Erdbebeneinwirkungen und Regeln für Hochbauten; Deutsche Fassung EN 1998-1: 2004 + AC: 2009; 2010

[2] Lestuzzi, P.: Dynamisch plastisches Verhalten von Stahlbetontragwänden unter Erdbebeneinwirkung, Dissertation, Institut für Baustatik und Konstruktion, 2000

[3] Hancock, J.; Watson-Lamprey, J.; Abrahamson, N.: An Improved Method of Matching Response Spectra of Recorded Earthquake Ground Motion Using Wavelets; Journal of Earthquake Engineering; Vol. 10 (2006), Pages 67-89

[4] Wehr, F.: Auslegung nichttragender Strukturen gegenüber Erdbeben, Masterthesis, Bergische Universität Wuppertal, 2012, unpublished

[5] Holtschoppen, B.: Beitrag zur Auslegung von Industrieanlagen auf seismische Belastungen, Dissertation, Rheinisch-Westfälische Technische Hochschule Aachen, 2009

Seismic Design of Mechanical and Electrical Components According to Safety Standard KTA 2201 of the German Nuclear Safety Standards Commission

Matthias Wacker[1]

[1] WK Consult Hamburg Ingenieure für Bauwesen VBI (inspection engineers) and WKC Hamburg GmbH Planungen im Bauwesen (design office)
Tempowerkring 1B, 21079 Hamburg, Germany
wacker@wk-consult.com; www.wk-consult.com

ABSTRACT:

The KTA safety standards not only apply to nuclear power plants but also to other nuclear facilities. The experience gained from retrofitting of structural and non-structural components in nuclear power plants can be applied to other areas where KTA standards are required. When designing and executing according to these standards best practices taken from conventional design often cannot be used. In many cases engineers and contractors are not aware of the additional expenditures involved. Differences between conventional design and design according to KTA standards are shown in the following areas: planning objectives, sources of information, changes to basis of design, probability levels, and, as an example, anchoring of a cable tray support to concrete.

Keywords: Nuclear power plant, Seismic design of components, Concrete anchor, Cost estimation

1 Introduction

Since most areas of Germany can be considered of low seismicity, many engineers and contractors are not aware of the particularities of seismic design and the additional expenditures involved. In many Universities, seismic design is not part of the curriculum. In-service training programs are few.

The German Nuclear Safety Standards Commission (KTA) has issued 93 safety standards. Another 13 are in preparation [Web-1]. They are in effect not only at nuclear power plant sites but also at interim storage facilities and permanent disposal sites such as the Konrad mine in Germany, a repository for low-level and intermediate-level radioactive waste. When designing and executing according to KTA standards best practices taken from conventional design often cannot be used. Expenditures increase further and estimating engineering or construction cost and

DOI 10.1007/978-3-658-02810-7_21, © Springer Fachmedien Wiesbaden 2014

the preparation of adequate quotes become a challenge. This article intends to highlight differences between conventional design and design according to KTA standards, and their effects on engineering and execution.

2 Experience gained from retrofitting of nuclear power plants

The WK-Consult engineering office, established in 1942, is involved in planning of nuclear power plants since the 1970's. In the last 10 years, more than 60 individual projects have been completed, several with a multi-annual scope.

A four-year project included the assessment, recalculation, and retrofit of working platforms. Some of these working platforms could be retrofitted; others had to be replaced by new structures. For instance one working platform required the installation of 202 heavy-duty concrete anchors.

In a further two-year project the exchange of anchors was accompanied and supervised. In 200 anchor plates of component supporting structures 325 anchors were removed and 734 anchors were installed.

Figure 1: Because of obstructed access, pipes had to be shut off and removed

The main challenges to be dealt with were:

1. Anchor plates where anchors were exchanged required a change request to the building authorities. This was particularly challenging if it occurred during execution e.g. when reinforcement was hit. During the time from preparing the change request until receiving permission the work had to be suspended and on occasion time slots for the work passed (see point 3).

2. In some places, the high reinforcement ratio of the substructure made it difficult to find a suitable anchor location. Although areas of dense reinforcement such as columns and lower surfaces of beams were avoided, the remaining areas often had multiple-layer reinforcement, too.
3. Access to some supporting structures was greatly obstructed. Pipes had to be shut off and removed, as shown in Figure 1. Before that could be done, applications had to be submitted and permissions had to be obtained. They include design of temporary support structures and verification of the required degree of redundancy of security relevant systems. The redundancy constraints determined where, when, and for how long work could be done. Where shut-off was not possible, alternative component supports were installed.
4. Admittance to some areas was not possible because of remaining radiation.

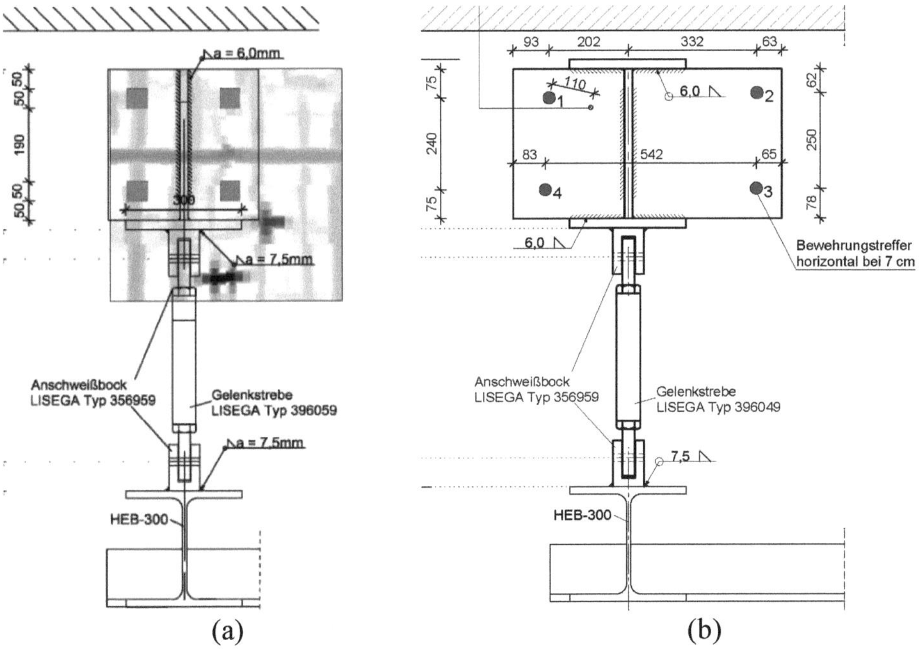

Figure 2: (a) Anchor plate as originally designed and the subsequent reinforcement scan (b) As-built drawing of redesigned anchor plate

Not all anchors could be installed according to the drawings. These cases may be categorized as follows:

1. Even with careful planning and scanning anchor locations before installation, collisions with reinforcement bars were encountered when drilling the holes. Besides training and choice of equipment, reasons are the limits of

the reinforcement detection methods (Taffe et al. [1]). While detection of reinforcement close to the surface is reliable, it becomes increasingly difficult with multiple-layer reinforcement and where the concrete cover is greater than the bar spacing. The incomplete boreholes were documented and assessed by the inspection engineer. If only the surface of the reinforcement bar was touched, the aborted borehole was filled with high strength mortar and a new hole was drilled, observing the required minimum distance. The affected anchor plate was then redesigned with a new anchor pattern or with a new anchor plate shape. An example can be seen in Figure 2. In this example, another bar was hit during the installation of the redesigned anchor plate.

2. In some instances, because of obstructions, there was no alternative than drilling through reinforcement bars or the damage to the bar was significant. Structural verification had to be provided showing that the reinforcement requirements of the substructure were still fulfilled without the drilled-through bar. An example, where changing the shape of the anchor plate was not possible, is shown in Figure 3.

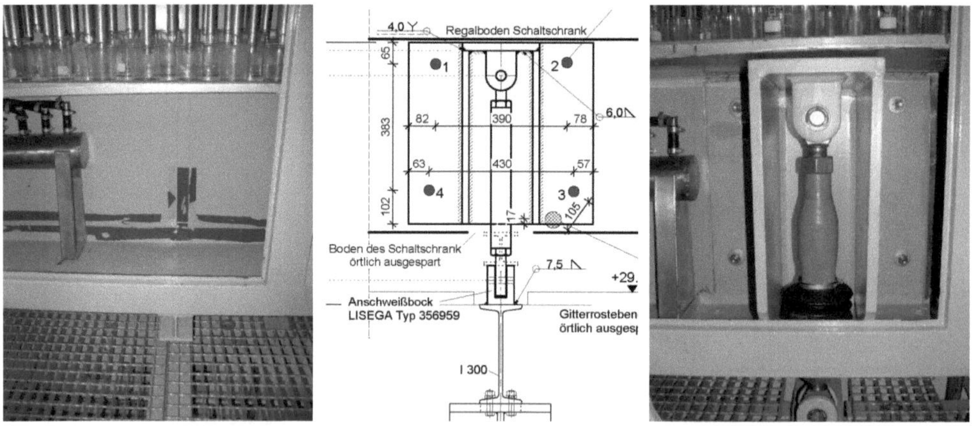

(a) (b) (c)

Figure 3: (a) proposed location of new anchor plate within a control cabinet (b) anchor plate design (c) completed anchor plate at the proposed location

In Table 1 two KTA-projects are analyzed. Column 2 shows the amount of anchors that could be installed according to the original design without changes or further verifications. Redesign of anchors and their anchor plates was necessary because reinforcement was detected prior to installation (column 3, see also Figure 2) or because the installation as built deviated from the design more than the allowance made in the design (column 4). Some reinforcement collisions were found despite reinforcement detection and redesign (column 5, see also Figure 3). In most of these cases the bars had to be drilled through and further verifications of the structure were necessary. The figures show the high degree of redesign required although the original design was carefully performed.

Table 1: Analysis of two KTA-projects

	1	2	3	4	5
	Total of installed anchors	**Installed according to original design**	**Redesign required because of rebar detection before installation**	**Redesign required because of installation deviations**	**Reinforcement collisions after redesign**
Project 1: new anchors	202	106	34	48	14
		52%	17%	24%	7%
Project 2a: anchors exchanged in place	70	53	9	1	7
		76%	13%	1%	10%
Project 2b: new anchors	664	593	26	18	27
		89%	4%	3%	4%

3 Differences between conventional design and design according to KTA standards

The following section describes selected differences between conventional design and design according to KTA standards. The mentioned areas are planning objectives, sources of information for design, changes to basis of design, probability levels, and, as an example, anchoring of a cable tray support to concrete. It is acknowledged that industrial facilities may vary in the expected level of design and detailing depending on the infrastructural importance or economic and environmental consequences of an accident. Therefore, in some cases strategies and requirements similar to those found in KTA standards and related documents have been introduced.

3.1 Planning objectives

Conventional Design: Structural design as a consequence of an investment decision is influenced, among others, by the following factors:

1. The market determines what products can be sold. Developments of energy and raw material prices open up new markets or render existing markets unprofitable.

2. Plant engineering determines what is needed to serve the market or demand. It may mean adjustments to electrical and mechanical systems, to software and logistics. This may or may not mean changes to existing structures or building new ones.
3. The decision to build new structures or to remodel existing ones is made.
4. Additional measures depending on the decision where to build are necessary. These include seismic design, provisions for high wind or high snow events.

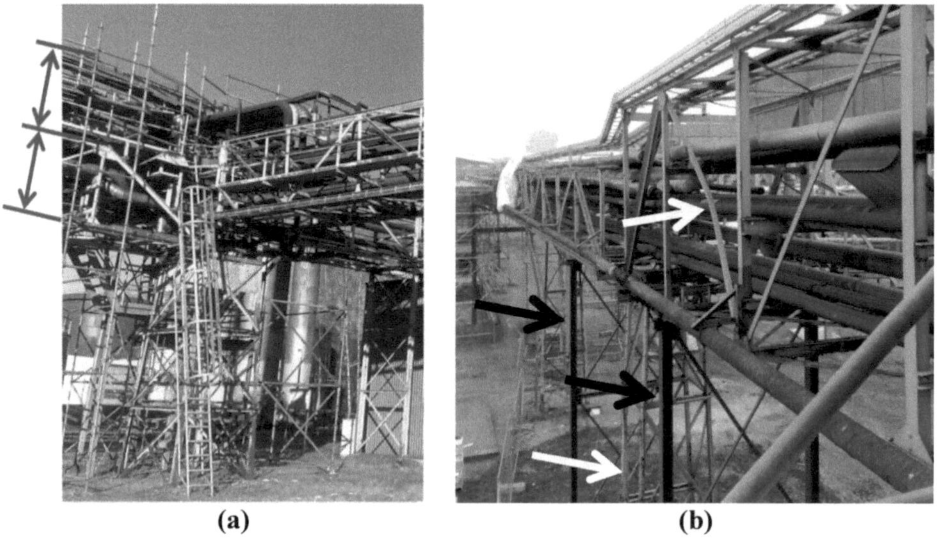

Figure 4: (a) Accumulation of pipes and cables on top of an existing utilities bridge. (b) After members have failed (white arrows), temporary supports are installed (black arrows)

When decisions are made based on this sequence the following effects have been observed:

1. Before the decision to build is made, all other options are exploited. Figure 4 shows a common situation with utility bridges. Initially pipes and cables are added with every plant modification. After that, if members have failed due to corrosion or overload, temporary supports are installed.
2. The clients desire to minimize the time from the decision to build until realization. Otherwise market opportunities may have passed. However, during the first two steps mentioned above there is no involvement of structural or civil engineers which limits the available time for geological site survey, search for unexploded ordnance, the design, and to prepare

applications for building permission or permission according to emission control acts.

3. There is only little tolerance to project costs increase. Such increase may result from difficulties encountered during construction, such as disposal of unexploded Second World War ordnance, but also from foregone profits due to production interruptions. Sometimes the latter means replacing structures in full operation of the plant, as can be seen in Figure 5. Construction cost may easily double under such circumstances. Since profit margins are small that in turn could mean that the whole marketing project and thus the construction project is endangered or will be transferred to another plant or continent.

Design According to KTA standards: One determining factor for projects where KTA standards apply is the long time span between idea and realization. For the Konrad mine repository, located about 100 km SE of Hannover, Germany, it took more than 30 years from the first examinations to reach building permission and to settle objections [Web-2]. Since then, six years have passed and another six years of construction can be expected until completion. During such a long time span, initial support of the public may change into opposition. As a result, high emphasis is put on formal correctness. Additionally, such time spans pose a challenge to office organization and personnel development in engineering offices. It may mean reactivating retired staff and finding ways to hand over design results through changes in personnel as well as in computer hard- and software.

Figure 5: Replacement of a utilities bridge in full operation

3.2 Sources of information

Conventional Design: The actions on structures such as dead, live, snow, and ice loads, as well as wind, temperature, accidental, and seismic actions are usually determined by the structural engineer when establishing the basis of design. They are taken from building standards, the geological site survey, and from manufacturers' specifications of components. For projects abroad, establishing the basis of design may additionally mean for the engineer to interpret local data and to design according to a double set of technical standards, on the one hand the minimum client requirements, such as the Eurocodes [2] or the IBC [3], and on the other hand the local building regulations.

Design According to KTA standards: Additional actions have to be considered resulting from external events (EVA) such as high water, air craft crash, and pressure wave from explosions as well as internal events (EVI) such as differential pressures, jet impingement forces, plant internal flooding, and load crash. It may also include "hardening" measures to discourage or fend off intrusions and sabotage attempts (razor wire barriers, exterior wall height and thickness, design of wall openings). Such actions are normally not determined by the structural engineers but are subject of technical reports and expert opinions. Whenever such reports and opinions are updated during the design phase, it usually means redesigning at least part of the structure. The number of experts and engineering offices involved increases drastically compared to conventional design of industrial facilities. The interaction between those requires extra time and coordination effort.

3.3 Changes to basis of design

Conventional design: Basis of design is the generally accepted codes of practice, which principally means following the technical standards in effect at the time of construction. After that, apart from few exceptions, the works continue to be accepted though building standards may change.

Design According to KTA standards: Basis of design is the state of science and technology. Whenever this state is updated e.g. through new findings or through failure modes observed in other nuclear facilities, the facility under consideration has to be examined and updated to the current state. Figure 6 shows the replacement of anchors in a nuclear power plant. Though the removed anchors had a technical approval at the time of their installation, they were replaced to accommodate the present state of science and technology.

3.4 Probability levels

Conventional design: For ordinary buildings, the generally accepted level of risk/protection of the seismic action is a probability of exceedance of 10% over the assumed building life time of 50 years. This results in a return period of 475 years.

Depending on the hazard potential or the post-earthquake-importance of an industrial site, higher return periods may be stipulated by the operational license. Return periods of up to 2475 years (2% probability of exceedance in 50 years) are used, which lead to an increase of the design seismic action of about 50%.

Figure 6: Replacement of Liebig Anchors, technical approval of 1975 (black arrow), by Hilti HDA Anchors, technical approval of 2008 (white arrow)

Design according to KTA standards: According to KTA 2201.1, the design seismic action is determined for a return period of 100000 years [4]. All known seismic events have to be considered, including historic events. An example is the catalog of earthquakes in Germany and adjacent areas between 800 AD and 2008 by Leydecker [5].

In Table 2 a comparison is shown between the two design standards for the Konrad mine, located about 100 km SE of Hannover. Eurocode 8 does not require design for seismic action in this area [6]. Therefore, the horizontal acceleration was taken from Grünthal et al. [7]. The value was picked based on the used color scheme. It can be seen that design according to KTA standards leads to seismic actions that are multiples higher than required by conventional design. Even areas where seismic activity is not known by the public may have considerable design accelerations for nuclear facilities.

3.5 Anchoring of a cable tray support to concrete

Conventional design: All varieties of anchors are used to connect components to concrete structures. Cable tray supports may serve as an example of items to be anchored. The black arrow in Figure 7 points to a support system commonly used in conventional design. It consists of prefabricated cold formed steel profiles to be installed using one or two anchors. Most of these anchors and support systems have not been tested for their behavior in seismic action and may perform poorly in the event of an earthquake with the associated adverse implications.

Table 2: Comparison of two design standards for the Konrad mine

Source	Return Period	Horizontal acceleration
Peak ground acceleration according to GSHAP Region 3 Map [7].	475 years	~0,25 m/s²
Design earthquake according to Leydecker and Kopera [8].	100000 years	1,12 m/s²

Design according to KTA standards: Only anchors with a technical approval for the use in nuclear facilities may be used, like the Hilti HDA [9] and the Fischer Zykon FZA [10] undercut anchors. The white arrow in Figure 7 points to a support designed to resist horizontal and vertical accelerations, and to allow for a certain degree of positional deviation of the anchors. Instead of picking a prefabricated system from a catalog, this support was verified by a structural calculation and reviewed by inspection engineers.

The main difference between conventional design and design according to KTA standards in this area is not of technical nature. Rather, in so many cases of conventional design the need to provide suitable seismic design for components and their supports is not respected. Economical losses in an earthquake often do not result from damaged primary structural members but from failed non-structural elements and components. The complexity and the restrictions involved in proper seismic design of concrete anchor connections must not be underestimated even in conventional design.

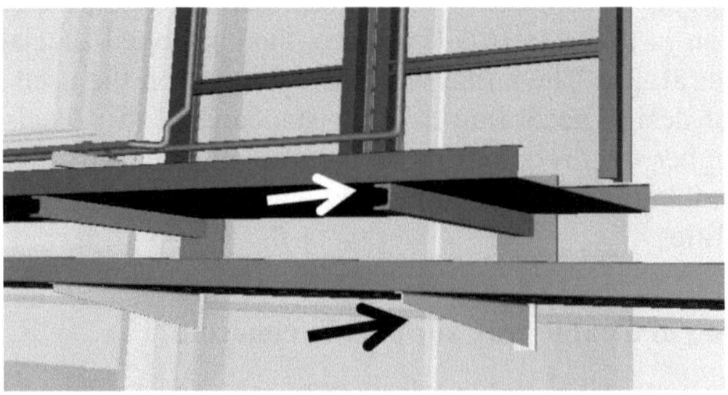

Figure 7: Cable tray supports for conventional design (black arrow) and for design according to KTA standards (white arrow)

4 Trends

Although conventional design and design according to KTA standards differ in many areas as highlighted in section 3 of this article, some basic needs of both worlds lead to a similar approach: suitable interfaces between the engineering disciplines and between engineering and construction have to be defined.

As an example, for some nuclear facilities, catalogs of different cast-in anchor plates with headed studs have been issued, creating an interface. Components can be installed observing the permissible loads of these anchor plates. The degree of interaction and redesign, e.g. because of reinforcement collisions, is reduced. Besides anchor plates, cast-in channels as shown in Figure 8 have been used based on an approval for the individual case. A trend from post-installed anchors toward cast-in elements has been observed. It is hoped that more structural elements of such kind with a general technical approval will be available in time.

When designing conventional industrial facilities, interfaces allow the structural design to move forward even though plant engineering is still in progress. Identifying possible interfaces in early stages of design is vital for the success of a project and constitutes an advantage on the market.

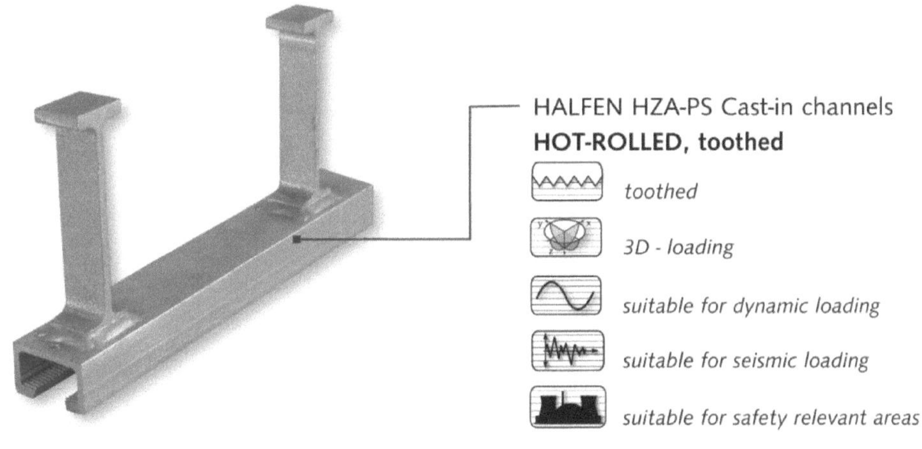

Figure 8: Cast-in channel. © 2013 Halfen GmbH, Germany

5 Conclusion

When preparing quotas for the seismic design of mechanical and electrical components or the execution according to KTA standards, many engineers and contractors are not aware of the increased complexity compared to conventional design of industrial facilities. Instead of selecting component support systems available on the market, they have to be verified by a structural engineer. The increased number of involved engineering offices and experts require extra time and coordination.

Seismic actions are multiples higher. A high degree of redesign can be expected. Finally, long time spans between idea and realization pose challenges to office organization and personnel development.

REFERENCES

[1] Taffe, A. et al.: Schulung zur zuverlässigen Ortung von Bewehrung in Stahlbetonbauteilen von Kraftwerken; in: Tagungsband zur DGZfP Jahrestagung 2011; 2011

[2] Structural Eurocodes Program, European Standards EN 1990 to EN 1999, European Committee for Standardization, Brussels, 2002-2007

[3] International Building Code 2012, 2 Volumes, International Code Council, 2011, ISBN 978-1-60983-040-3

[4] KTA 2201.1 Design of Nuclear Power Plants against Seismic Events; Part 1: Principles, Version 2011-11, Kerntechnischer Ausschuss (KTA), Köln, 2011

[5] Leydecker, G.: Erdbebenkatalog für Deutschland mit Randgebieten für die Jahre 800 bis 2008, Stuttgart, 2011, ISBN 978-3-510-95989-1

[6] DIN EN 1998-1/NA National Annex - Nationally determined parameters - Eurocode 8: Design of structures for earthquake resistance - Part 1: General rules, Seismic actions and rules for buildings, Berlin, 2011

[7] Grünthal, G. et al.: Seismic Hazard Assessment for Central, North and Northwest Europe: GSHAP Region 3, Annali de Geofisica, Vol. 42, No. 6 (1999), Pages 999-1011, ISSN 2037-416X

[8] Leydecker, G.; Kopera, J.: Seismological hazard assessment for a site in Northern Germany, an area of low seismicity, Engineering Geology, Vol. 52 (1999), Pages 293-304, ISSN 0013-7952

[9] National Technical Approval Z-21.1-1696 for Hilti undercut anchor HDA for accidental actions, Deutsches Institut für Bautechnik (DIBt), Berlin, 2011

[10] National Technical Approval Z-21.1-1646 for fischer-Zykon-Anker FZA-K, FZA, FZA-I for accidental actions, Deutsches Institut für Bautechnik (DIBt), Berlin, 2009

[Web-1] http://www.kta-gs.de/, publisher: Kerntechnischer Ausschuss (KTA), data gathered on 25. Jan. 2013

[Web-2] http://www.endlager-konrad.de/cln_321/nn_1073268/DE/Themen/Historie/__node.html?__nnn=true, publisher: Bundesamt für Strahlenschutz, data gathered on 25. Jan. 2013

Seismic Qualification of Equipment in Industrial Facilities

Carsten Block[1], Thomas Bauer[1], Fritz-Otto Henkel[1]

[1] WÖLFEL Beratende Ingenieure GmbH + Co. KG
Max-Planck-Str. 15, 97204 Höchberg, Germany
block@woelfel.de

ABSTRACT:

This paper summarizes general aspects for the seismic qualification of equipment in industrial facilities. In the first part of the paper a guideline for the seismic qualification of equipment is described. The purpose of the guideline is to assist engineers in addressing the seismic design requirements for the equipment. Important points that need to be addressed are design philosophy, seismic performance goals, scope of application, methods of qualification, applicable codes and the description of seismic input. In the second part of the paper the seismic qualification of equipment for a facility located in China is described. The development of the seismic loads according to the applicable Chinese code is shown and for selected components the seismic qualification using numerical analysis and qualification by analogy is demonstrated.

Keywords: seismic qualification, design guideline, equipment

1 Introduction

Industrial facilities located in seismically active areas have to meet the national building codes including the seismic requirements. Besides the building itself, equipment inside the facilities may need to be seismically qualified as well. The seismic requirements may range from safety of people and functional capability after the earthquake for safety relevant equipment to investment protection.

For a seismic assessment of equipment, a systematic approach is necessary, where in an ideal case all required information is available in the beginning of the project. Especially if several different components have to be qualified within a project a seismic design guideline is beneficial. The purpose of such a guideline is to put all information required for the seismic qualification of the equipment in one document and to assist engineers in addressing the seismic design requirements.

S. Klinkel et al. (Eds.), *Seismic Design of Industrial Facilities*,
DOI 10.1007/978-3-658-02810-7_22,

2 Seismic design guideline

2.1 General

The more information, requirements and specifications are included and described in the guideline the more straightforward and cost effective the actual seismic qualification can be performed. Of course, it might be sufficient in some cases to just specify the applicable codes. However, a summary of the most important aspects in one document avoids misinterpretations, simplifies the coordination between different involved parties and helps in discussions with the customer, design authorities, etc.

In the following, the main points that need to be addressed in such a guideline are described in more detail.

2.2 Design philosophy and seismic performance goals

The guideline has to define the main objectives of the seismic design requirements for the industrial facility. For example safe evacuation of the building, avoid release of hazardous material, maintain electrical supply for critical components during and/or after the earthquake, investment protection.

Often the equipment in the facility is classified in different seismic categories for which seismic performance goals are defined. The three typical performance goals for components with increasing requirements are:

- Structural Stability
  The component shall be designed to maintain its structural respectively position stability. It has to meet certain stress limits and be protected against falling over / down or impermissible slipping.
- Leak Tightness
  The component shall be designed to maintain its integrity and not to leak out e.g. any fluid.
- Functional capability
  The component shall be designed to remain operable during and after a specified earthquake.

In addition, different design earthquake levels may be specified for the seismic categories.

2.3 Scope of application

It has to be clearly identified for which equipment a seismic assessment has to be performed and the guideline has to be applied to.

The guideline needs not to be limited to one project / site / location but may specify the minimum requirements which shall be met e.g. at all potential sites worldwide.

An equipment list should be prepared and included specifying each single component to be qualified with an assignment of the applicable performance goal, earthquake level, etc.

2.4 Method of qualification

The following qualification methods may be applied either individually or in combination with each other:

- Qualification by analysis

  A seismic analysis is commonly used when the component is required to maintain its structural stability and / or leak tightness. The analysis may be performed using equivalent static methods or dynamic response spectrum or time history methods. For many applications the equivalent static method is sufficient applying a constant acceleration at the centre of gravity of the component.

- Qualification by testing
  Seismic qualification by testing is the best or even only option when functional capability of the component is required during or after the earthquake. In most cases dynamic shake table tests are performed.

- Qualification by analogy (similarity)
  The component may be qualified by referencing to results from an analytical or experimental qualification of a similar or type-identical component.

The seismic guideline should specify any preferred or exclude methods not permitted.

2.5 Applicable codes

The seismic guideline shall identify the applicable codes and specifications including the edition (year).

To simplify the work for the engineer a summary of the applicable design criteria from the codes to achieve the desired performance goal is beneficial. The decisive safety factors and parameters should be specified. If dynamic analyses shall be performed the specification of the damping factor for the components is of special importance.

2.6 Seismic input

The seismic input for the qualification of equipment may be static load coefficients (e.g. force or acceleration), response spectra or time histories.

For components often the equivalent static method is used where the static loads is applied at the centre of gravity of the component. The determination of a static load coefficient is exemplarily shown for the Chinese Code for Seismic Design of Buildings [1] in section 3.2.

Beside the seismic input the guideline should also specify the approach for the combination of excitation directions and the combination with other load cases. Even if this information is included in the specified codes the implementation in the guideline avoids any misinterpretation.

In case other loads have to be investigated the corresponding data has to be specified as well. For operational loads this could be for example pressure and temperature data.

3 Seismic qualification of equipment for an industrial facility

3.1 General

For an industrial facility located in China in total more than 40 components had to be seismically qualified. Before the actual qualification a seismic design guideline was developed covering the topics described above. This guideline helped to qualify the components in a highly efficient way.

In the following the main aspect are summarized. The development of the applicable seismic loads according to the Chinese Code for Seismic Design of Buildings [1] is described and for selected components the seismic qualification using numerical analysis and qualification by analogy is demonstrated.

The main objective of the seismic design requirements is to safeguard against major failure of the equipment and loss of life, not to limit damage or maintain function. The performance goal for the equipment is to maintain structural stability.

3.2 Calculation of earthquake loads

All equipment in the facility with a net weight of > 180 kg must withstand an earthquake of intensity level 7 with a basic ground acceleration of 0.10 g, according to the Chinese Code for Seismic Design of Buildings [1].

In order to supply the manufacturer of the equipment and the engineer in charge with the earthquake load required for a safe design of the support structures of the equipment and its anchorage in the building, the following data were specified in the guideline:

- Earthquake loads in form of equivalent forces, which act in the centre of gravity of the equipment.
- Support forces at the anchorage of the equipment to the civil structure.

The earthquake loads are derived for the equipment as non-structural components by use of the equivalent force method according to section 13.2.3 of [1]. The value of the horizontal seismic action F_{Eh} at the gravity centre of the component is calculated by following equation:

$$F_{Eh} = \gamma \cdot \eta \cdot \zeta_1 \cdot \zeta_2 \cdot \alpha_{max} \cdot G \qquad (1)$$

G is the weight of the equipment and γ the importance factor, which is 1.0 for the regarded components according to [1].

The response modification factor η accounts for the global ductility capacity of the lateral force resisting system, e.g. support structure. According to Table 13.2.3-2 in the appendix to [1], the range of the factor is between 0.6 (e.g. for cabinet supports) and 1.2 (e.g. for water tank and cooling tower supports).

The amplification factor ζ_1 accounts for resonance amplification of flexible structures. According to [1] it should be 2.0 for cantilever components, any equipment whose bearing point is below its center point and flexible systems, while it may be 1.0 for other cases (e.g. quasi rigid systems). The factor $\zeta_1 = 2.0$ is specified for all investigated components.

The position factor ζ_2 accounts for the attachment elevation of the equipment within the civil structure. It is 1.0 at the bottom of the structure and 2.0 at the top of the structure. The factor for elevations between is linearly distributed. It is calculated as follows:

$$\zeta_2 = 1 + (H_i - H_{bot})/(H_{top} - H_{bot}) \qquad (2)$$

H_i is the elevation of attachment of the component within the civil structure, H_{bot} the elevation of the base of the civil structure and H_{top} the elevation of the top of the civil structure.

The maximum earthquake impact coefficient for horizontal earthquake α_{max} is defined in Table 5.1.4-1 of [1]. To accomplish a save and conservative design of the equipment a coefficient of $\alpha_{max} = 0.30$ is specified in the guideline. This is about the mean value of the coefficients for high-chance and low-chance earthquake.

Vertical seismic action need not be regarded for intensity levels less than 8. Nevertheless it is recommended to consider vertical seismic actions for cantilevering or suspended equipment. According to section 5.3.3 of [1] 10 % of the gravity load should be applied as vertical earthquake load at long cantilever and big span structures for intensity level 8.

$$\text{vertical seismic action} \quad F_{Ev} = 0.1\ G \qquad (3)$$

For the equipment which is supported on the floor by vertically acting structures no vertical seismic loads need be regarded.

The specification [1] gives no information about the superposition of the earthquake excitations in the two horizontal directions. The guideline proposes to superpose the different earthquake directions as square root of the sum of squares (SRSS rule).

3.3 Load superposition

The basic combination of the earthquake loads S_E with the dead load S_G is as follows:

$$S = \gamma_G S_G + \gamma_E S_E \quad (4)$$

The factor γ_G shall be taken as 1.2 if dead load increases the actions by earthquake load and as 1.0 if dead load reduces the actions by earthquake load. The earthquake load S_E shall be regarded with changing sign and with a factor $\gamma_E = 1.3$.

To prevent the anchoring from failure before the structure's ductility can develop the seismic design force according to equation (1) and (3) is specified to be increased by an over-strength factor $\Phi = 1.3$ for steel failure of bolts or welding seams and $\Phi = 2.0$ for concrete failure of anchors in concrete.

3.4 Qualification by analysis

For the qualification by analysis generally 3D-finite-element-models of the frame structures are developed. The stresses in the frames and the anchorage forces are calculated for dead load and earthquake loads. The allowable stress for the load combination including earthquake loads is the yield stress. The analysis of welds and bolts is performed via hand-calculations.

As the flexibility of the regarded components themselves has no impact on the structural stability only the decisive supporting structures are modelled in detail. The components are included as point-masses.

In Figures 1 and 2 two typical examples are illustrated. Figure 1 shows a component with dimensions 2.76 m x 1.35 m x 2.55 m and a total mass of 10400 kg. The steel frame is made of S355 steel material and bolted to the floor. The horizontal earthquake load F_{Eh} is 56 kN, which corresponds to an acceleration of 0.54 g.

The maximum von Mises stress in the beam elements is 213 MPa, which is well below the yield strength of 355 MPa. Hand-calculations show that the weld and bolt stresses are within the allowable limits.

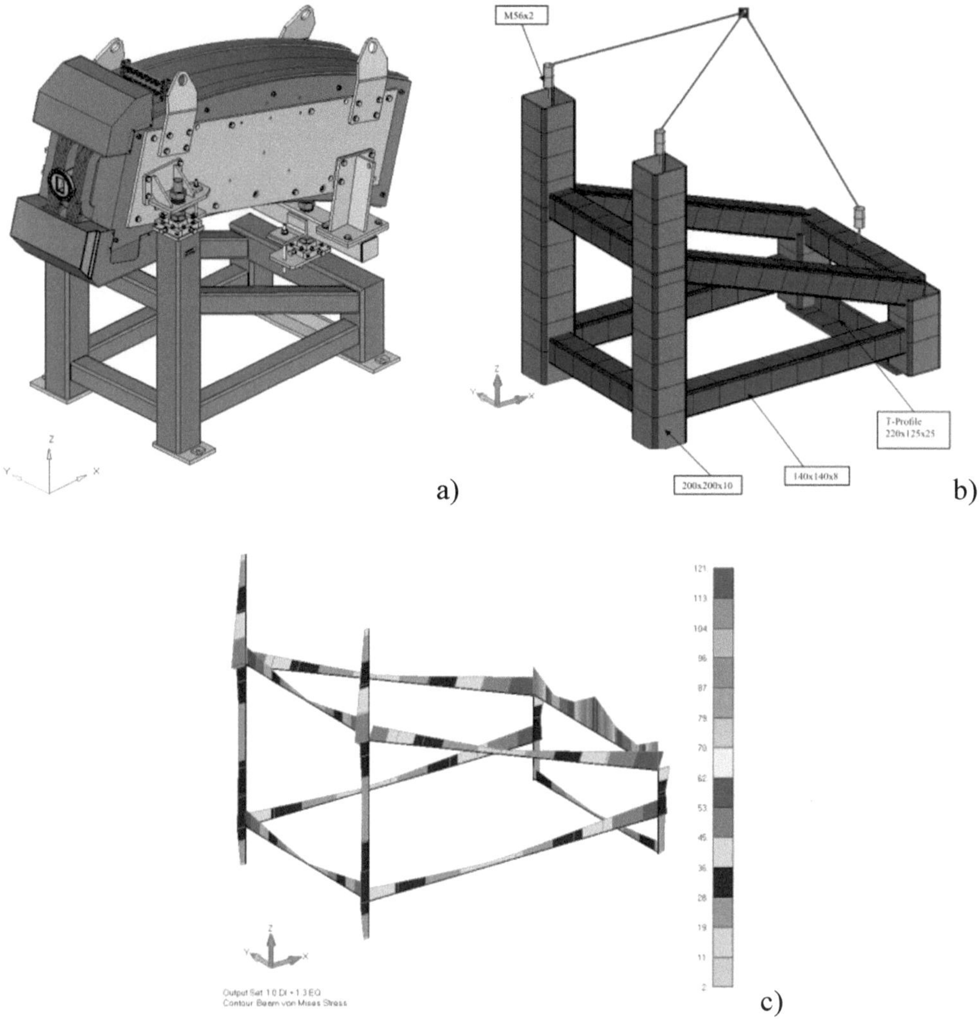

Figure 1: a) CAD model, b) FE-model, c) von Mises stresses [MPa]

Figure 2 shows a component with dimensions 8.87 m x 1.14 m x 7.47 m and a total mass of 16670 kg. The supporting frame is made of S355 steel material and bolted to the floor at four locations. As for the component before the corresponding seismic acceleration is 0.54 g.

To include the effect of the local load distribution in the area of the cantilever, beam elements are combined with shell elements to model the connection plates in more detail. The maximum von Mises stress in the beam elements is 121 MPa which is well below the yield strength of 355 MPa. Local analyses show that the welded and bolted connections are acceptable.

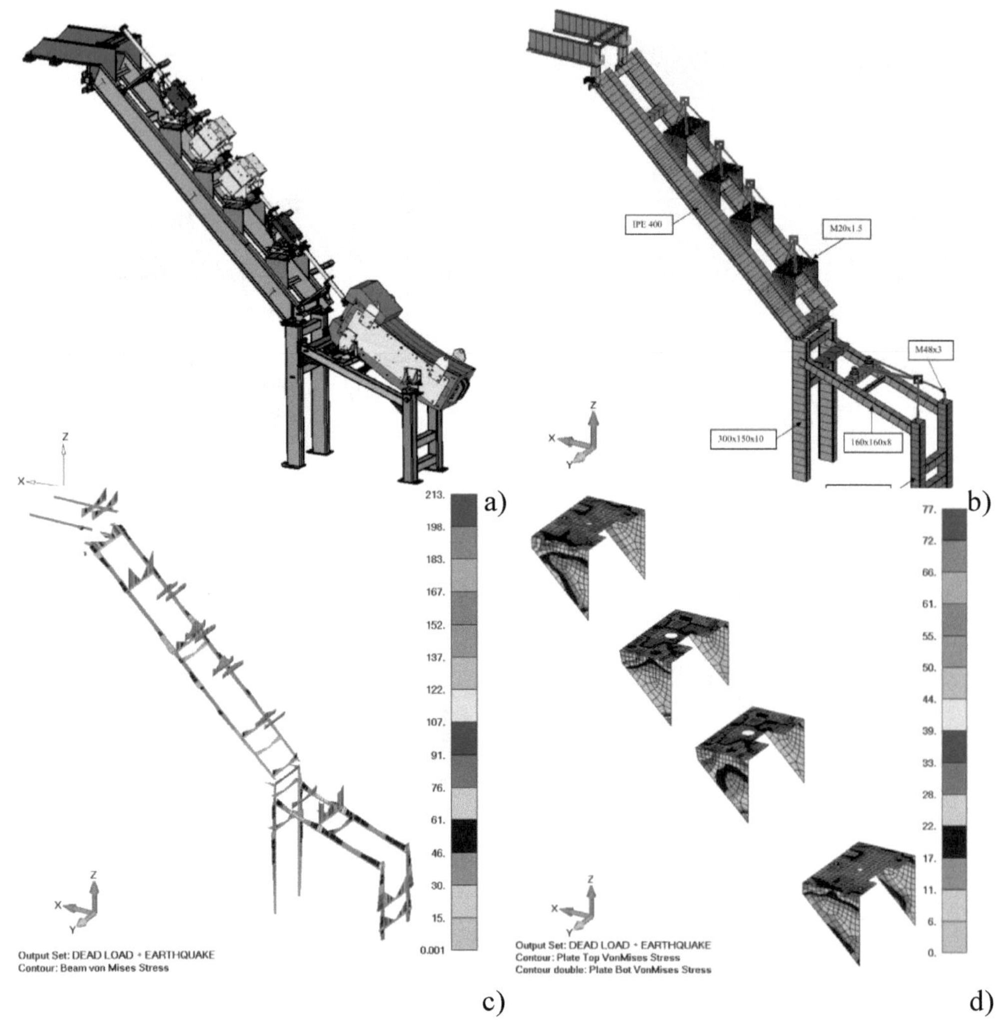

Figure 2: a) CAD model, b) FE-model, c) von Mises stresses [MPa] in beam elements, d) von Mises stresses in connection plates

3.5 Qualification by analogy

Some of the regarded components and their supporting structures are very similar to components located at a different site. As these components have already been qualified by analysis a qualification by analogy is performed by comparing the constructions, dimensions, anchor bolts, materials, masses and forces from dead load and earthquake. For modified details of the supporting structures additional stress analyses are performed.

Figure 3 shows a typical example. Table 1 compares the key attributes of the two components to show the similarity. With similar construction and mass, but lower earthquake loads no further investigations are necessary for the new component.

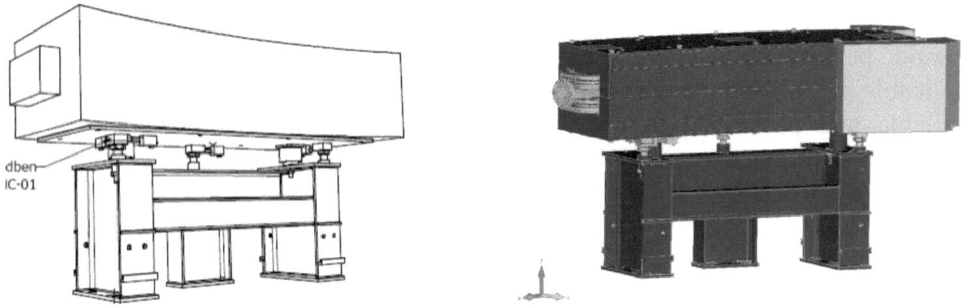

Figure 3: Previously assessed component (left) and new component (right)

Table 1: Comparison of similar components at two facilities

Item		OLD	NEW
Cx	mm	1650	1654
Cy	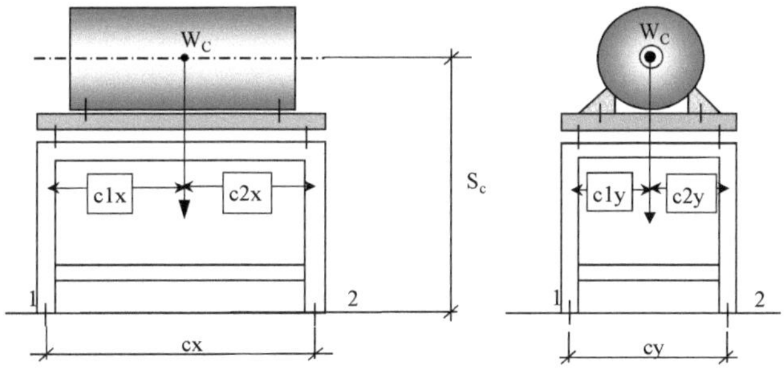	320	320
c1x		825	841
c2x		825	813
c1y		160	151
c2y		160	169
s_c		1246	1253
Mass	kg	9355	9600
dead load	kN	93.6	96.0
Earthquake (hor. res.)	kN	93.6	68.5

4 Conclusion

For a seismic qualification of equipment it is beneficial to have all required information available in the beginning of the project in one comprehensive document. This paper summarized the most important points that should be included in such a seismic design guideline. Namely these are design philosophy, seismic performance goals, scope of application, methods of qualification, applicable codes and the description of seismic input. If all this information is available and processed in a structured and comprehensive way even a large number of components can be seismically qualified in a straight forward and time and cost effective way.

References

[1] National Standard of the People's Republic of China GB 50011 – Code for Seismic Design of Buildings, Version 2008, including Appendix

Shake Table Test on the 1:30 Model Structure of A Large Cooling Tower for Fire Power Plant

J. W. Dai[1], X. R. Weng[1] and Y. Hu[1]

[1] Key Laboratory of Earthquake Engineering and Engineering Vibration, China Earthquake Administration
29 Xuefu Road, Harbin, 150080 China
jwdai@iem.cn

ABSTRACT:

For understanding the seismic behaviour of extra-large scale cooling tower with dimension of 220 meters high and 188 meters in diameter, the shake table tests for its' 1:30 (length ratio) tower model were carried out to simulate the structural response to potential earthquake impacts. The model structure was excited by three dimensional white noise and different intensity of earthquake motions from PGA=0.04g to PGA=0.40g in considering of four different site conditions from soft soil to hard rock (I~IV). Through the tests, the dynamic responses and damage patterns of the cooling tower under different three-dimensional seismic excitations were studied.

Keywords: cooling tower, damage pattern, shake table test, earthquake

1 Introduction

With the rapid demand of the fire power plant, extra-large indirect-air-cooling tower (1000MW) will be constructed in the high seismic risk areas with PGA 0.2g and higher in China such as mid-north and west-north regions. The dimension of the studied huge tower structure reaches up to 220 meters high and 195 meters in diameter. It's constructed with X type R/C column supported hyperboloid shell and the X column's length-width ratio can reach up to 1:40. It's really a challenge but very necessary and urgent to know the seismic behavior and design weak points of the huge tower under strong earthquake attacks.

In 2005, S. Sabouri-Ghomi and M.H.K. Kharrazi took a study on the reinforced concrete column supported hyperboloid cooling tower stability assessment for seismic loads. In their study, finite element analyses have been performed to obtain the stress concentration, nonlinear behavior, stability or safety factor of the R_C_ tower due to earthquake loads. Outcomes of their study show that considerable plastic hinges were created in the X shape long columns of the R/C hyperboloid

S. Klinkel et al. (Eds.), *Seismic Design of Industrial Facilities*,
DOI 10.1007/978-3-658-02810-7_23, © Springer Fachmedien Wiesbaden 2014

cooling tower due to seismic loads, which resulted in a significant decrease in the stability safety factor. According to W.S. Guo's introduction, R.Harte and U.Montag performed a study on computer simulations and crock-damage evaluation for the durability design of the world-largest cooling tower shell (200m high and 152m span) at Niederaussem power station (1000MW grade). But as we all know, Germany is not located in the seismic region and the Niederaussem power station is not exposed to severe earthquake risk. The study on the 200m high and 152m span cooling tower can't provide useful reference to the seismic design for the world largest 220m high and 188m span cooling tower in China.

This paper provides test results about the seismic behaviour of the world-largest R/C hyperboloid cooling towers with very long X shape supporting columns. The shake table tests for its' 1:30 (length ratio) model were carried out to simulate the earthquake impacts. A new model material simulation method is developed to fulfil the goal of shaking table test. Specially treated lead sand is used as one of the main aggregates of the model construction micro–aggregate concrete. The earthquake resistant capacity of the tower as well as its' critical element, the support X-type columns were inspected and studied carefully.

2 Length Ratio 1:30 Model Similitude Design

2.1 Describe of the Prototype Cooling Tower

The huge prototype R/C hyperboloid cooling tower has a total height of 220 m, a span of 188 m in diameter on the foundation, a span of 169 m in diameter at the transition of columns to shell, a span of 107 m at the throat section and a span of 110 m in diameter at the top. The total elevation from the grade for the X shaped column is 28.7 m. The columns have a dimension of 1.6 m by 0.9 m and the thickness does not vary throughout the height. They were built on the concrete supporting piers with the dimension of 4.0m High by 4.0m wide by 3.5m thick. The thickness of the shell varies from 1.7 m close to the columns top end to 0.45 m at an elevation of 38.6 m. From there, it decreases to 0.4 m at the elevation of 165.3 m, and the keep 0.4 m to the elevation of 214.6 m, then it increase to 0.65 m at the top. The cooling tower is built on a ring strip foundation, which is 4.0 m below grade and with a width of 14.0 m and an average height of 2.0 m. A concrete stiffening ring (or transient ring) with a thickness and width of 0.40 m and 1.7 m is built together with the upper tower shell at the top of the X shaped columns. As well, a concrete stiffening ring (or top ring) with a thickness and width of 0.45 m and 1.8 m, respectively, has been built at the top of the cooling tower. Fig. 1 shows the elevation plan of the R/C cooling tower.

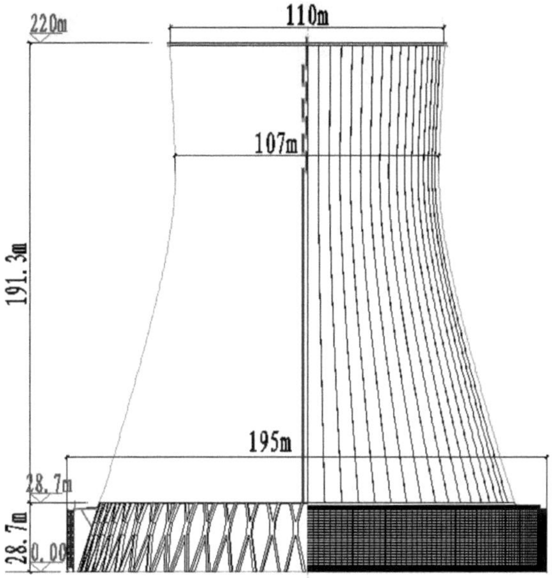

Fig. 1: Elevation of the prototype cooling tower

2.2 Design and Construction of the Model Structure with Length Ratio 1:30

All structural elements are scale down to 1:30 of the prototype tower in geometric dimension. Due to the special shape and structure of the hyperboloid shell tower, it's difficult to add the artificial mass on the model's shell during the dynamic earthquake simulation test. For solving this problem, in design and construction of the tower model, a kind of specially treated lead sand was used as one of the main aggregates of the model micro-concrete. Correspondingly, in the dimensional analysis of the similitude law for dynamic test, the equivalent density ratio, length ratio as well as the efficient elastic modulus ratio can be set as the basic variables, and other variables such as acceleration, frequency and time etc. could be derived from the dynamics formulation easily, shown in Table 1.

Table 1: Similitude relationship used in dynamic test and finite element analysis for model structure

Physical parameters	Similitude ratio		
	Lower excitation	Medium excitation	Large excitation
Length	l_r	l_r	l_r
Equivalent modulus	E_{r0}	$E_{r1} = \frac{E_{r0} f_1^{\,2}}{f_0^{\,2}}$	$E_{ri} = \frac{E_{r0} f_i^{\,2}}{f_0^{\,2}}$
Density	ρ_r	ρ_r	ρ_r

Stress	$\sigma_r = E_{ro}$	$\sigma_r = E_{ro}$	$\sigma_r = E_{ro}$
Time	$t_r = l_r \left(\frac{E_{r0}}{\rho_r}\right)^{-0.5}$	$t_r = l_r \left(\frac{E_{r1}}{\rho_r}\right)^{-0.5}$	$t_r = l_r \left(\frac{E_{ri}}{\rho_r}\right)^{-0.5}$
Deformation	$r_r = l_r$	$r_r = l_r$	$r_r = l_r$
Velocity	$v_r = \left(\frac{E_{r0}}{\rho_r}\right)^{0.5}$	$v_r = \left(\frac{E_{r1}}{\rho_r}\right)^{0.5}$	$v_r = \left(\frac{E_{ri}}{\rho_r}\right)^{0.5}$
Acceleration	$a_r = \frac{E_{ro}}{l_r \rho_r}$	$a_r = \frac{E_{r1}}{l_r \rho_r}$	$a_r = \frac{E_{ri}}{l_r \rho_r}$
Frequency	$v_r = l_r^{-1} \left(\frac{E_{r0}}{\rho_r}\right)^{0.5}$	$v_r = l_r^{-1} \left(\frac{E_{r1}}{\rho_r}\right)^{0.5}$	$v_r = l_r^{-1} \left(\frac{E_{ri}}{\rho_r}\right)^{0.5}$

3 Shake Table Test and Loading Sequence

3.1 Locations of Sensors

Considering the axial symmetry of the cooling tower, along x and y direction of shake table, one each measure point was selected in the transient ring of the hyperboloid shell (the meridional and hoop direction of the top of X shaped columns), the throat (the meridional and hoop direction), and the top ring (hoop direction of the inside and outside); at corresponding positions, one each X shaped column was selected to be fixed with measure points, as Fig. 2 shows. 66 strain gages were used in all.

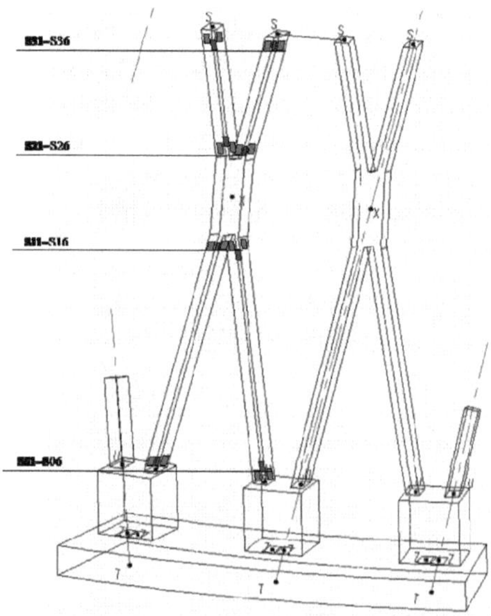

Fig. 2: Strain gage's location of the X shaped columns

As shown in Fig. 3, along x and y direction of shake table, measure points were fixed on shake table and the outer surface of the hyperboloid shell. There were ten measure points altogether. At each measure points x, y and z direction acceleration transducers were located, consequently the total number of acceleration transducers is $10 \times 3 = 30$.

As shown in Fig. 4, in order to get the absolute displacement of the cooling tower, ten measure points were fixed on the same place as the ones for acceleration measurement, and at each measure points x and y direction absolute displacement sensors were located; to obtain the relative deformation (x and y direction) of the tower, four relative displacement sensors were located as shown in Fig. 4.

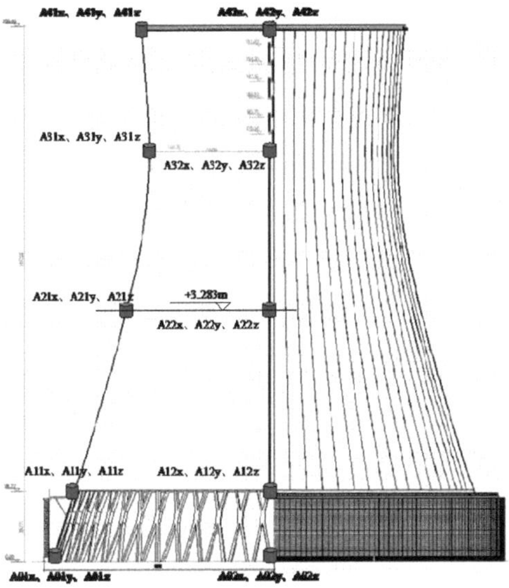

Fig. 3: location of the acceleration transducers

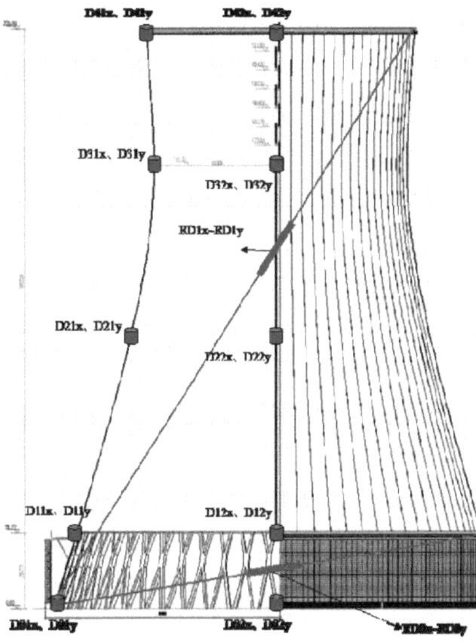

Fig. 4: location of the displacement sensors

3.2 Input Seismic Excitation

To study the seismic performance of the cooling tower in four different site conditions (I~IV), 12 seismic acceleration records, 3 (2 actual seismic records and 1 artificial simulated acceleration record) for each site condition, were selected as input seismic excitation. The actual seismic records were selected from the "16 most unfavourable seismic records", which could be used in time history analysis, listed in General Rule for Seismic Design of Buildings by Lili Xie.

The test contained 72 cases that the model structure was orderly excited by minor earthquake under seismic fortify intensities of VII, minor earthquake under seismic fortify intensities of VIII, moderate earthquake under seismic fortify intensities of VII, major earthquake under seismic fortify intensities of VII (minor earthquake under seismic fortify intensities of VIII), major earthquake under seismic fortify intensities of VIII in four different site conditions (I~IV). Meanwhile, white noise excitation tests were inserted before and after each condition to real-time monitor changes of model structure's natural vibration characteristics. Duration of the earthquake waves was compressed according to similitude relationship, PGA was determined by Code for Seismic Design of Buildings and similitude relationship; to the codes, the ratio of three-dimensional acceleration peak should be adjusted to 1 (principal horizontal direction x):0.85 (second horizontal direction y):0.65(vertical direction z).

Table 2 Input Seismic Excitation

Site condition	Seismic wave number	Name of seismic wave
I	Wave01	1992, LANDERS-JUNE 28, AMBOY
	Wave02	1999, Chi-Chi earthquake, TCU046
	Wave03	Artificial ground motion, Class1
II	Wave04	1979,Imperial Valley CA, El Centro ,Array #10
	Wave05	1999, Chi-Chi earthquake, TCU070
	Wave06	Artificial ground motion, Class2
III	Wave07	1979, Imperial Valley, CA, Meloland Overpass FF
	Wave08	1999, Chi-Chi earthquake, TCU052
	Wave09	Artificial ground motion, Class3
IV	Wave10	1995, Kobe, Osaka
	Wave11	1976,Tianjing Hospital, Tangshan Aftershock
	Wave12	Artificial ground motion, Class1

4 Test Phenomenon

In minor earthquake under seismic fortify intensities of VII and VIII, the cooling tower basically remained intact, without cracks obvious; in moderate earthquake under seismic fortify intensities of VII, at the top of X-shaped columns appeared a few transverse cracks (Fig. 5-a), at the top and bottom of the columns cracks gradually increased and expanded with the increase of acceleration peak, finally leading to crush of concrete at the top of some X-shaped columns (Fig. 5-b) and transverse cracks appearing at the middle of some X-shaped columns' limbs (Fig. 5-c);the sway of upper shell got larger with the increase of acceleration peak, in major earthquake under seismic fortify intensities of VII (minor earthquake under seismic fortify intensities of VIII) in the minus x direction at the height of 4.5m outside the hyperboloid shell the concrete crushed and fell off, appearing a 60cm long transverse crack (Fig. 5-d), in major earthquake under seismic fortify intensities of VIII, the crack expanded and eventually became a crack throughout the second quadrant (Fig. 5-e), at the same time, on the other side of the hyperboloid shell the upper concrete fell off, appearing two transverse crack because of tension (one 4.5m long, the other 6m) (Fig. 5-f); by the end of the test, the cooling tower had not collapsed (Fig. 5-g).

(a)

(b)

(c)

(d)

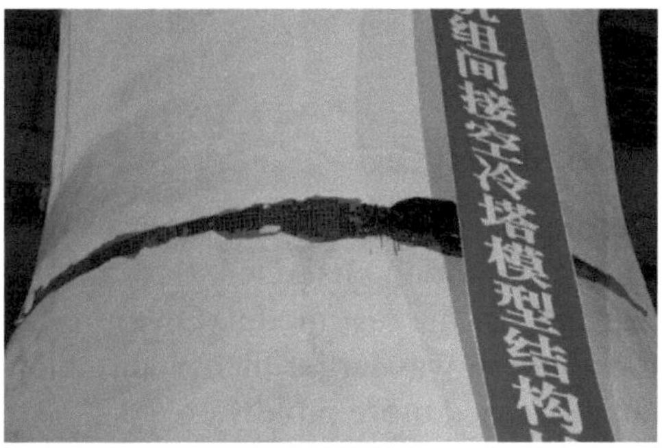

(e)

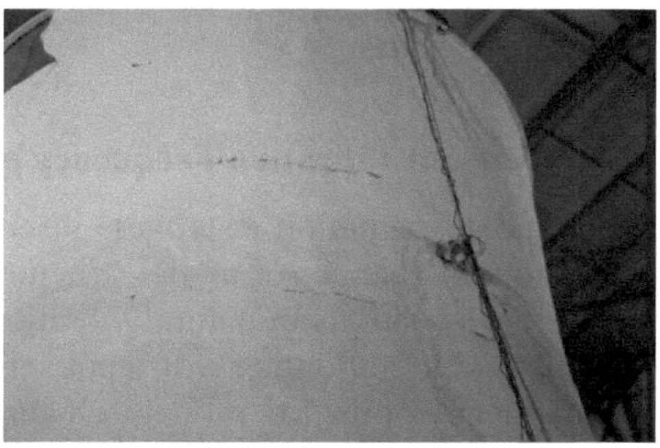

(f)

Fig. 5: Damage situation of support X shaped columns and hyperboloid shell

(g)

Fig. 5: *(continued)*

5 Test Results

5.1 Natural Vibration Frequency of Model Structure

White noise excitation tests were inserted before and after each condition to real-time monitor changes of model structure's natural vibration characteristics. Table 3 shows the variations in natural vibration frequency of the test model. Owing to X-direction acceleration peak of input seismic excitation was larger than Y-direction, the frequency of the test model in X-direction decreased more rapidly.

Table 3 Model Structure's Natural Vibration Frequency

Earthquake level	Initial state	After Minor of VII (0.04g)	After Minor of VIII (0.07g)	After Moderate of VII (0.10g)	After Major of VII (0.20g)	After Major of VIII (0.40g)
X-direction	9.8HZ	8.3HZ	7.7HZ	7HZ	5HZ	5HZ
Y-direction	9.2HZ	8.7HZ	7.5HZ	7.5HZ	7.5HZ	7.5HZ

5.2 Dynamic Response of Model Structure

Due to limited space, the test results of 4 seismic waves selected from 12 waves are listed below, and the 4 seismic waves are wave01, wave05, wave08 and wave11, which separately belong to Site I~IV.

5.2.1 Acceleration Response

The acceleration response under seismic excitation can be measured by acceleration transducers, which are fixed at ten measure points from top of the tower to the shake table along x and y direction. Fig. 6 is the diagram of acceleration amplification coefficient in different intensity earthquakes. As shown in Fig. 6, under small earthquake the top of the tower and the top of the X shaped columns are the positions where acceleration is relatively larger. With the acceleration peak of input earthquake waves increasing, the acceleration of the X shaped columns' top changes not much, the acceleration of the tower's top also changes little or even decreases, while the acceleration of the throat increases obviously, leading to the throat becoming the position where acceleration is relatively larger.

As the test results show, with the increasing of the acceleration peak value of input earthquake motions, stiffness of the structure degrades. After some failures happen, the acceleration amplification coefficient decreases.

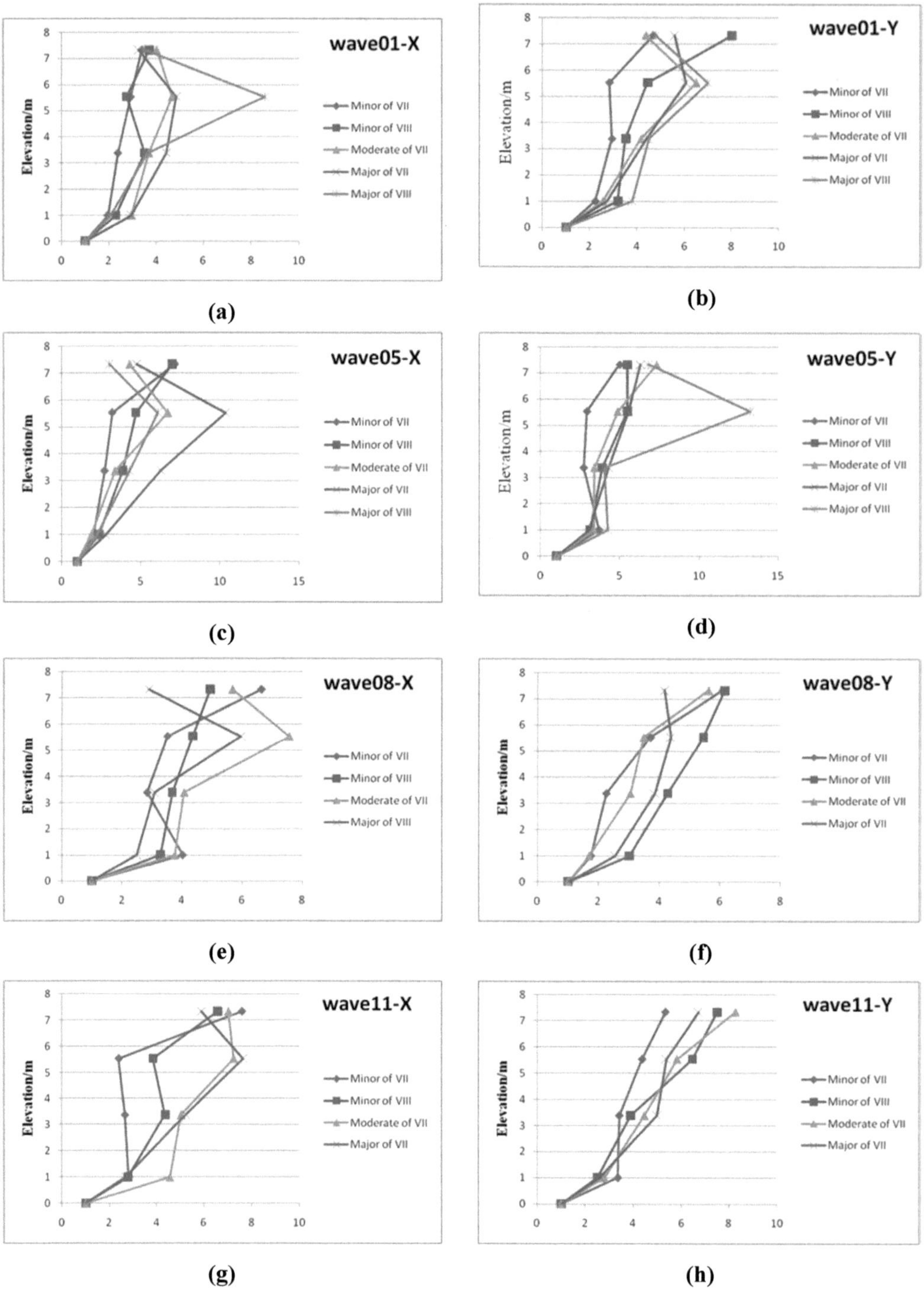

Fig. 6: Acceleration amplification coefficient

5.2.2 Displacement Response

As shown in Fig. 7, the displacement responses under different seismic excitations are not all the same. In X, Y direction under wave01 and in Y direction under wave11, the displacement of the middle of the shell is larger and the displacement of each measure point increases with the increase of acceleration peak of input earthquake waves; in X direction under wave05 and wave11, the displacement of the middle of the shell is larger under small earthquake, while the displacement of the top of the shell is larger under big earthquake; In X, Y direction under wave08 and in Y direction under wave05, the displacement of the middle of the shell is larger under small earthquake, then with the increase of acceleration peak of input earthquake waves, the displacement of the top of the shell increases and the top becomes the position where acceleration is relatively larger, at last when acceleration peak of input earthquake waves is big, the displacement of the top decreases and the displacement of the middle increases, the middle becomes the position where acceleration is relatively larger again.

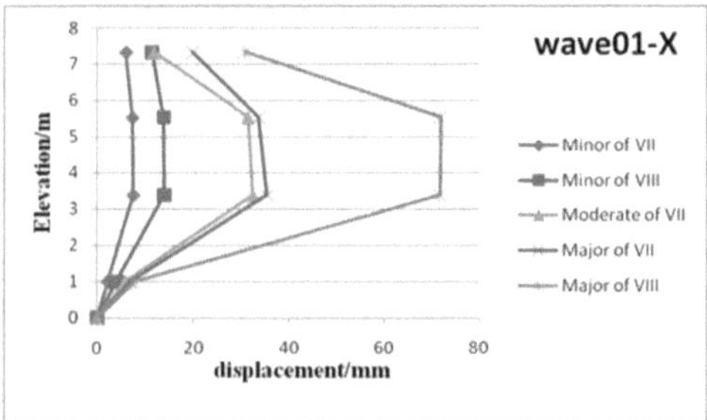

(a)

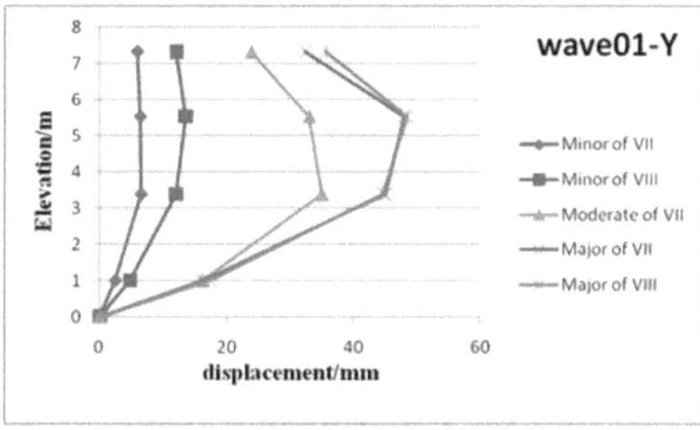

(b)

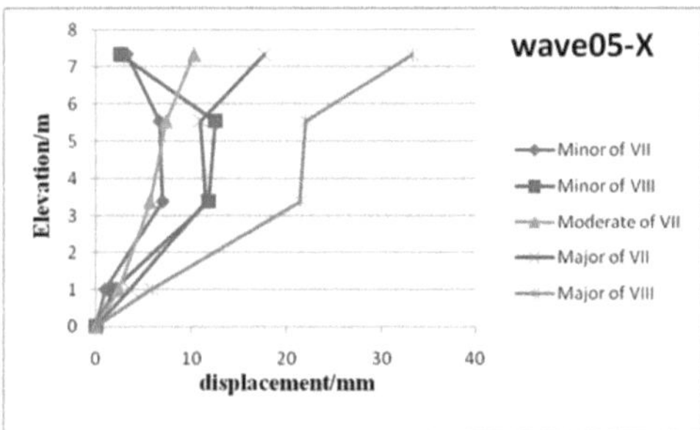

(c)

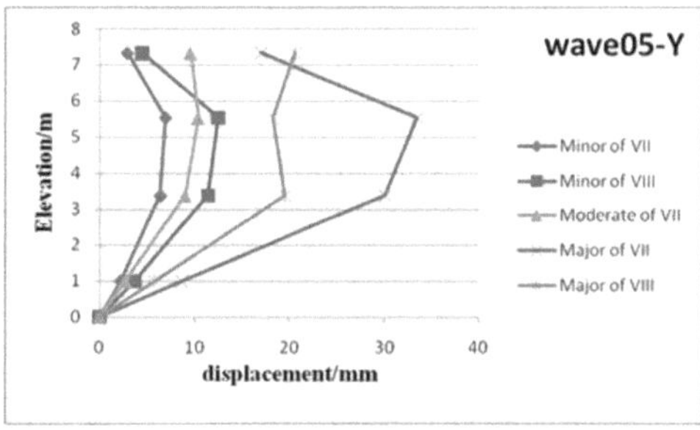

(d)

Fig. 7: Displacement

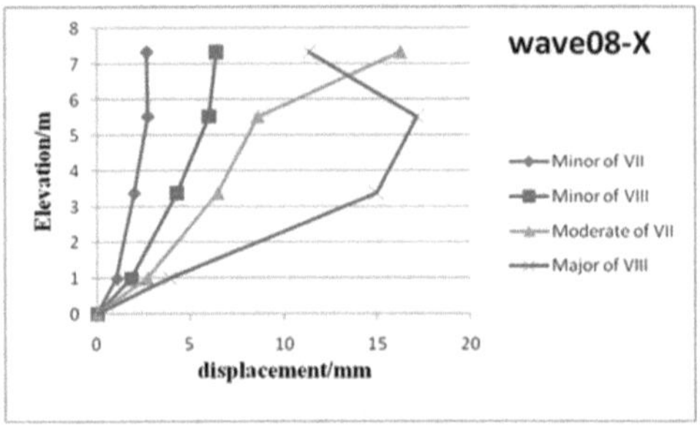

(e)

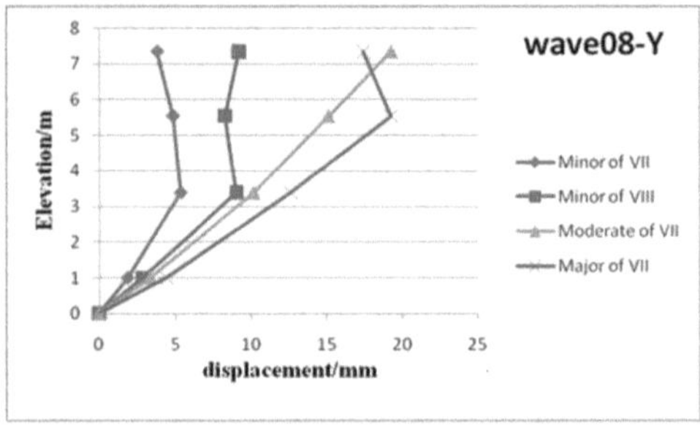

(f)

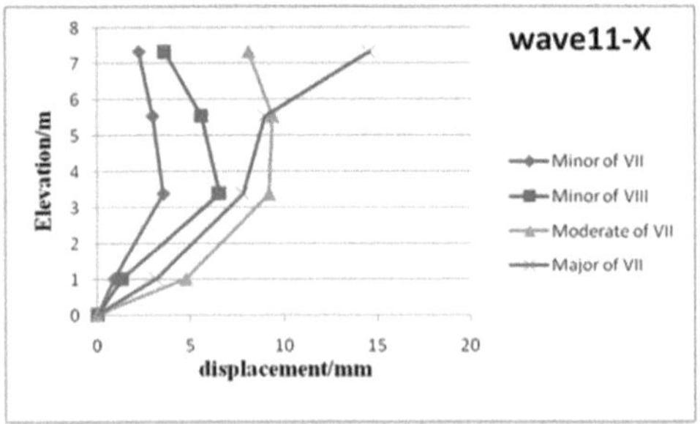

(g)

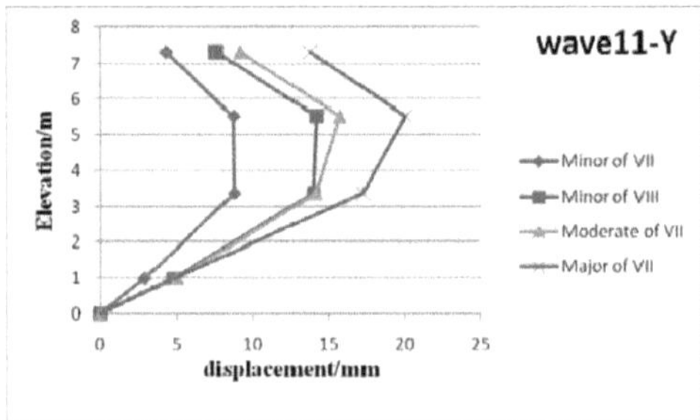

(h)

Fig. 7: *(continued)*

6 Conclusion

(1) From the experimental results, it can be seen that the aseismic capacity of the cooling tower can satisfy the demands of the local seismic design intensity VIII (PGA=0.4g) in different soil sites.

(2) Both the bottom and top end of the X shaped column are the weak points to earthquake impacts, where the failure appears earlier than all other damages. Because of whiplash effects and local vibration superposition, the upper part above the throat of the cooling tower is also vulnerable to shaking impacts, where concrete may flake away under major earthquakes.

(3) Due to the influence of multiple adjacent high modes coupled vibration, as well as the global response to the vertical components of the ground motion, severe hoop damage may appear at the thinnest wall where close to the throat part of the tower shell.

ACKNOWLEDGEMENT

This paper is jointly sponsored by the National Science and Technology Supporting Program (2012BAK15B02) and the National Natural Science Foundation program (50938006). The investigation cannot be conducted without these financial supports.

REFERENCES

[1] J. Y. LI; C. L. REN; Z. L. HUANG: Test and FE Analyses for Natural Ventilation Cooling Tower; Mechanics Quarterly; 28(3) (2007), 443-447.

[2] M. WANG; Z. L. Huang; L.Q. LI: FE Analysis for Natural Ventilation Smoke-Cooling Tower; Mechanics and Practice; 28(4) (2006), 64-67.

[3] Aksu T.: A Finite Element Formulation for Column Supported Hyperboloid Cooling Towers; Computers & Structures; 59(5) (1996), 965-974.

[4] Castiau T.; Gaurios R. : The Design of Cooling Towers in Extremely Severe Earthquake Conditions; Engineering Structure; 13(1991).

[5] Chen H.F.; Salip A.F. : Constitutive Equations for Materials of Concrete and Soil (1st ed.); China Architecture & Building Press.

[6] Chen H.F.; and Salip A.F.: Elasticity and Plasticity (1st ed.). Beijing; China Architecture & Building Press.

[7] Chen, Y.J.: The Seismic Responses and Experiment Research on the Models of Cooling Tower; Zhejiang University.

[8] GB50011-2010, Code to Seismic Design of Building; China Architecture & Building Press.

[9] Institute of Engineering Mechanic; China Earthquake Administration; Institute of Earthquake Engineering; CABR.; Harbin Institute of Technology: General Rule for Performance-based Seismic Design of Buildings. Beijing; China Planning Press.

Seismic Qualification of Electrical Cabinets

Marcus Ries[1], Helmut Kennerknecht[1], Philipp Moor[2], Fritz-Otto Henkel[1]

[1] Wölfel Beratende Ingenieure GmbH + Co KG
Max-Planck-Str. 15, 97204 Höchberg, Germany
ries@woelfel.de

[2] GUTOR by Schneider Electric
Hardstrasse 72 – 74, 5430 Wettingen, Switzerland
Philipp.Moor@schneider-electric.com

ABSTRACT :

The seismic qualification of electrical cabinets can be established by different methods like analysis, test and proof by analogy. This contribution gives two examples of seismic qualification; the first example shows the qualification of a cabinet with respect to stability and functionality. Stability is proved by analysis using a finite element model of the cabinet and performing an RSMA-calculation. By comparing the von-Mises comparative stress against permissible values, the stability of the cabinet is assessed. Functionality is proved by separate component tests. To define the test loading for the uniaxial component tests, the calculated maximum accelerations are used. The second example shows how to make use of a successful seismic qualification of a reference cabinet to qualify a similar cabinet with respect to stability. For this purpose the method of 'proof by analogy' (similarity) is used.

Keywords: cabinet, seismic qualification, analysis, test, analogy

1 Introduction

When installing (safety related) electrical cabinets in a plant, their seismic qualification often is requested by different regulations (e.g. IEEE 344 [2], KTA 2201.4 [4], IEEE 693 [3]) or dedicated specifications. Generally for electrical cabinets two objectives of qualification are found: stability and functionality during and after the seismic event. To achieve these qualification objectives typically three methods – or combination of these – are applicable: test, analysis and proof by analogy.

Seismic qualification by test is the only option for a functional qualification of components built into the cabinet. Such a qualification can be done by testing the built-in components individually or by testing the whole cabinet. Testing whole cabinets with a mass of several tons is a common but resource consuming method

DOI 10.1007/978-3-658-02810-7_24, © Springer Fachmedien Wiesbaden 2014

to prove stability and functionality simultaneously. The option to prove the functionality of components built into a cabinet by testing the components individually e.g. by uniaxial sine-sweep tests, is much less expensive. For such tests the in-cabinet mounting conditions as well as the in-cabinet earthquake loading has to be regarded. Once successfully tested, the components can mostly be used in similar cabinets for earthquakes of similar or lower levels. By such component tests the functionality and stability of the individual built-in component is considered, the stability of the whole cabinet has to be proved separately e.g. by analysis or proof by analogy. In [1] seismic qualification procedures for electrical cabinets or built-in modules can be found.

This paper will give examples to explain and clarify the above addressed methods of seismic qualification of electrical cabinets by analysis (stability) in combination with component tests (functionality) as well as proof by analogy.

2 Qualification of a Cabinet by Analysis and Component Test

The cabinet that has to be seismically qualified with respect to stability and functionality is shown in Figure 1. The cabinet has overall dimensions of (WxDxH) 2150x800x2200 mm. The total mass is 2.4 tons. The cabinet consists of three sections which are connected by screws.

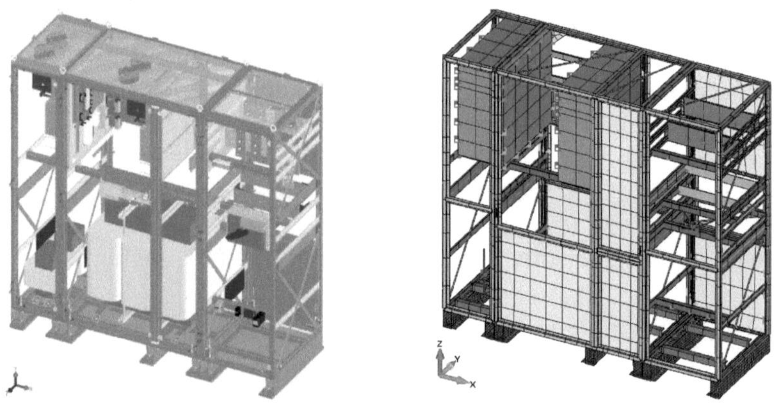

Figure 1: CAD- (left) and FE-Model of cabinet

2.1 Loading

The seismic loading is given by response spectra depicted in Figure 2. Two spectra are shown representing the seismic load case; the horizontal response spectrum has to be applied simultaneously in each horizontal direction together with the vertical spectrum. If a cabinet is supposed to be erected at different installation locations an envelope of all horizontal spectra as well as of all vertical spectra is usually used. The results of the seismic loading in each direction are typically combined by the square root of sum of squares (SRSS) rule. This is a typical seismic loading and results combination situation found in the nuclear field ([2], [4]). In conventional plant engineering other seismic loading conditions have to be applied, e.g. in Eurocode 8 [5] no vertical loading is requested for cabinets (non-structural elements), the horizontal seismic loading is given by a resulting acceleration. Besides seismic loading also gravitational loading has to be regarded. The results of both loadings have to be combined.

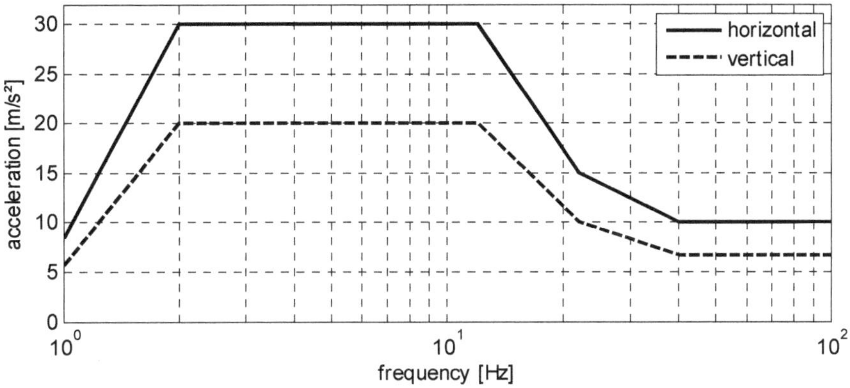

Figure 2: seismic loading

2.2 Finite Element Model

Based on the available CAD-data a Finite Element Model (FE-model) of the cabinet is created – see Figure 1.

As the global frame structure is responsible for the stability of the cabinet and the highest stressing of this structure is typically caused by the excitation of the first global eigenmodes with high modal masses, the FE-model must best estimate the behaviour of these first global modes. Thus the focus during the modelling process is to get those modes in the frequency range up to the cut-off frequency right. For this purpose use of beam, plate or structural elements is sufficient to get an appropriate model representing the relevant dynamical behaviour of the real cabinet.

The least effort is needed by creating a linear elastic model. This can be used for further static (gravitational load, quasi-static method) and (mode based) dynamic analyses like eigenfrequency analysis, response spectrum modal analysis (RSMA), time history modal analysis (THMA).

2.3 Analysis of Cabinet

As a static analysis with equivalent seismic loads will not give in-cabinet accelerations and the loading / stressing of dynamic subsystems could be judged significantly wrong, the RSMA-method is the best choice for the seismic analysis of the cabinet with respect to effort and quality of results.

After the calculation of the gravitational load case (dead load), i.e. -1 g in vertical direction, the modal extraction step is performed followed by the RSMA analysis. The first eigenfrequency of the considered cabinet is determined at 15 Hz. The corresponding mode is related to a global vibration of the cabinet in horizontal y-direction (front-back). The second mode at 16.4 Hz shows a global motion in x-direction (side-side). No significant vertical motion exists in the amplification range of the spectrum. Figure 3 shows the shape of the first eigenmode and modal parameters of the first four modes.

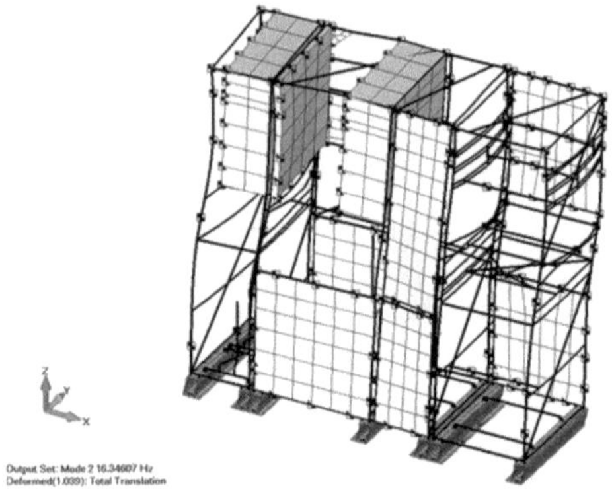

mode	eigen-frequency	modal mass		
		x	y	z
	[Hz]	[kg]	[kg]	[kg]
1	14.98	142.21	554.8	0.08
2	16.35	1198.9	100.8	0.41
3	20.54	198.34	10.3	0.14
4	24.19	13.26	0.96	0.03
…	…	…	…	…

Figure 3: modal results: 2nd eigenform (left) and modal parameters (right)

In the RSMA analysis the contribution of the eigenmodes up to 50 Hz are combined by the Complete-Quadratic-Combination (CQC) rule for each excitation direction. The contribution of the higher modes is regarded by considering the residual modes (missing mass). These resultant response values of each of the three spectra are combined by the SRSS rule.

The result values of dead load and seismic load are combined by the formula 'dead load ± seismic load'. To judge the internal stressing of the cabinet, the comparison stress, e.g. Von-Mises stress is compared against its permissible value. This value depends on the cabinet's material and the regulation used for the proof. For the considered cabinet a stress of 250 N/mm² is permissible for the cabinet's frame elements.

Results of the seismic analysis are depicted in Figure 4. The Von-Mises stress is shown. A maximum stress of 26 N/mm² is achieved for load case 'deadload' in the supporting frames of the transformer, earthquake loading gives a maximum value of 229 N/mm². This value can be found in the vertical struts (frame). The combination of both load cases yields a local maximum value of 231 N/mm² in the frame. This stress is below the permissible value of 240 N/mm².

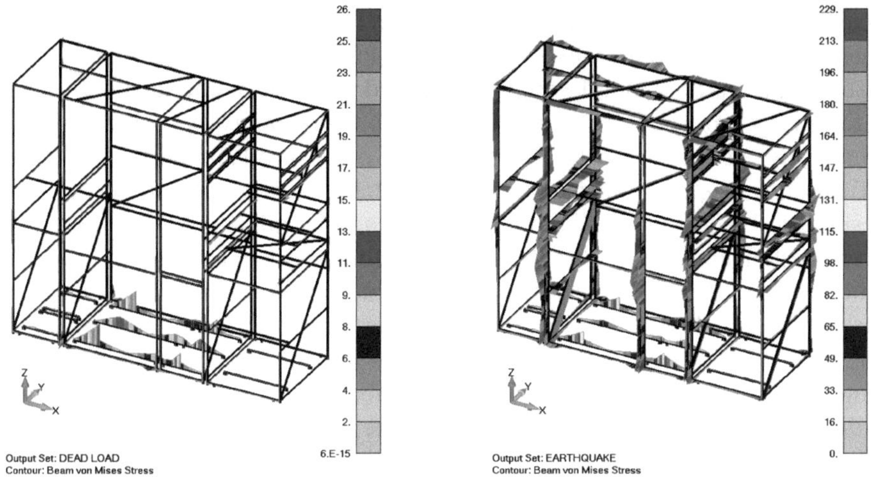

Figure 4: Von-Mises comparison stress in the frame elements of the cabinet: deadload (left), earthquake loading (right)

By proving that the stressing values of the elements of the FE-model as well of the connecting elements (e.g. anchors) are below the permissible value the seismic qualification of the cabinet is achieved with respect to stability.

Besides the stressing of the cabinet, also the resulting accelerations due to earthquake loading can be determined by such an analysis. Figure 5 shows the calculated maximum absolute accelerations exemplarily at different vertical frame levels.

	accelerations [m/s²]			
	hor. x	hor. y	vert. z	res.
Frames, level 1/2	47.6	30.9	7.1	57.2
Frames, level 2/3	44.7	34.3	7.3	56.8
Frames, level 5/6	46.2	36.1	7.6	59.1
Frames, top	43.5	38.0	7.7	58.3

Figure 5: Maximum acceleration at different frame levels

These accelerations can be used to determine loading parameters of functionality tests for the built-in components.

2.4 Functionality-Test of Components

After having proved the stability of the cabinet structure against earthquake loading, the stability and functionality of the built-in electrical components can be qualified by separate tests. Therefore the dynamic loading at the in-cabinet mounting points of the built-in components has to be determined. By defining the mounting points at the interconnection of the component (including its supporting structure) with the vertical frames, the determined accelerations in Figure 5 can be used to setup the test parameters for functionality tests. As the components and their supporting structure are usually of small size and weight compared with the whole cabinet, there is no need for a large multiaxial shaking table which is not easily available having a long planning time and high costs. Thus the use of a simple uniaxial shaker for the qualification test is more attractive with respect to availability and cost. Therefore the test has to be designed to be performed on a uniaxial shaker. The three principal axes are tested consecutively. Per axis a resonance search test followed by the seismic test is performed. For both resonance search and seismic test a sine-sweep loading is chosen. The sweep rates are set to be 1 octave per minute. The amplitude for the resonance search is set to be 0.2 g, the excitation amplitude of the seismic loading of the individual axes is determined for single frequency excitation according to [4]. Firstly the spectra at the mounting points of the built-in components are determined by the 'substitution method' (see [4]) based on the location with the resulting maximum acceleration (59.1 m/s², see Figure 5) at 'Frames, level 5/6'. The maximum accelerations in each direction for this frame level are equal to the Zero-Period-Acceleration of the mounting points (tertiary responses) for the corresponding direction. By directly applying a spectra amplification factor of $V = 8.2$ for $D_1 = 7\,\%$ (damping ratio of the bolted cabinet) and $D_2 = 3\,\%$ (damping ratio of the built-in components) the spectral maxima of the tertiary responses can be conservatively estimated by the 'substitution method'. As the test will be conducted uniaxial, the resulting spectrum is determined by taking the x-, y and z- axis into account. By assuming, that the eigenfrequencies of the components are typically in the ZPA-range of the cabinet spectra (see Figure 2), the factor for measuring the relative shares of several natural vibrations k_i is set to be 1. The excitation specific amplification factor is determined to be $\ddot{U} = 1/2/D_2$. By applying the above mentioned values, a maximum excitation amplitude of 29 m/s² per axis is determined. An additional test of the rigid body acceleration is performed per axis with the resulting ZPA-value of 59.1 m/s² (see Figure 5).

According to the above values and the capabilities of the test facility, the test loading described in Table 1 were conducted with each electrical component of the cabinet. Also seismic tests with lower loading were planned to be performed.

Table 1: loading per axis for component tests

description	waveform	excitation level		
		0 – 8 Hz	**8 – 35 Hz**	**35 – 50 Hz**
resonance search	sine sweep	0.8 mm	0.2 g	0.2 g – 0.0 g
seismic loading, 1.5g	sine sweep	5.8 mm	1.5 g	1.5 g – 0.0 g
seismic loading, 2 g	sine sweep	7.8 mm	2 g	2 g – 0.0 g
seismic loading 3 g	sine sweep	11.6 mm	3 g	3 g – 0.0 g
seismic loading, ZPA	shock	6 g		

During the tests, the input motion as well as structural responses of the components were acquired. The proper functioning of the component during and after the seismic loading was recorded and checked, too.

The stated uniaxial sine-sweep tests are assessed to be quite demanding in comparison to corresponding simultaneous multiaxial time-history tests.

As a vertical hydraulic cylinder was used for excitation, the specimen's orientation had to be adjusted for excitation of its three principal axes. Figure 6 depicts the mounting of the specimen.

Figure 6: Orientation of specimen during tests for excitation of: vertical (left), horizontal-x (middle) and horizontal-y (right) direction

Combining the two methods analysis of the cabinet's stability (see previous chapters 2.2 and 2.3) and testing the functionality of the cabinet's components is an attractive procedure to seismically qualify whole cabinets.

3 Proof by Analogy

Seismic qualification by means of proof by analogy is only possible if a test or an analysis of a similar cabinet (reference cabinet) is available. It has to be shown, that the cabinet to be qualified can resist its expected earthquake loading and fulfil its safety related function based on the available reference results. The concept of proof by analogy can be found e.g. in KTA 2201.4 [4], in IEEE 344 [2] named 'extrapolation for similar equipment' or in conventional regulations like IEEE 693 [3] named 'qualifying equipment by group' or in many dedicated technical specifications where the term 'similarity' is used. For the purpose of proof by analogy the cabinet to be qualified and the reference cabinet have to be compared with respect to: e.g. (mechanical) design, total mass, mass distribution, stiffness of the supporting structure, materials and earthquake loading. Based on this comparison it has to be shown quantitatively, that the cabinet to be qualified can fulfil its safety-related function during/after earthquake loading.

Based on the results in chapter 2.3 a cabinet is exemplarily proved by analogy. Figure 7 shows layout drawings of the reference cabinet, the cabinet to be qualified (cabinet B) and a comparison of the floor response spectra of both cabinets.

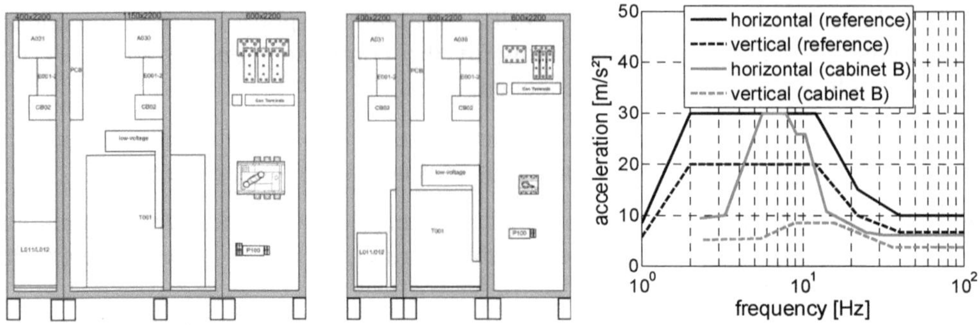

Figure 7: Layout drawings of reference cabinet (left) and cabinet B (cabinet to be qualified, middle) and comparison of floor response spectra (right)

Both cabinets are manufactured with same framework profiles (identical cross-sections, interconnections, material, etc.) by using the same seismic resistant design, i.e. bracings and other seismic reinforcements. Both cabinets consist of three single cubical sections. Due to a big transformer, the middle section of the reference cabinet is wider than the middle section of cabinet B and it has an additional vertical strut and an additional base profile. With respect to stiffness cabinet B is assessed to offer nearly the same global stiffness k_i in each horizontal direction $i=\{x, y\}$ as the reference cabinet. By comparing the masses – see Table 2 – it can be seen, that the total mass m_{total} of cabinet B is 47% of the reference cabinet's total mass. As the global horizontal eigenfrequencies of the two cabinets are significantly influenced by the mass in the upper area of the cabinet, the comparison of these masses (see Table 2) shows, that cabinet B's top level mass m_{top} is about 28% lower than the top level mass of the reference cabinet.

Table 2: Comparison of cabinet masses

parameter	reference cabinet	cabinet B
total mass m_{total} [kg]	2372	1125
top level mass m_{top} [kg] (≥50% of height, supported by vertical struts)	587	422

As the global stiffness k_i is judged nearly equal the stiffness of the reference cabinet and in view of a 28% lower top level mass, the (horizontal) eigenfrequencies f_i are expected to be 18% higher than the eigenfrequencies of the reference cabinet $f_{i,ref}$:

$$f_i=\sqrt{k_i/m_{top}} = \sqrt{1/0.72} \cdot f_{i,ref} = 1.18 \cdot f_{i,ref} \qquad (1)$$

Conservatively it is assumed, that cabinet B has the same eigenfrequencies as the reference cabinet. With the lowest mode at 15 Hz (see chapter 2.3) it can be seen from the floor response spectra in Figure 7 that the seismic loading for cabinet B in the range 15 Hz and above is considerably lower than the loading of the reference cabinet. The lower seismic loading in combination with lower cabinet masses will give lower stresses in the frame structure than in the reference cabinet (see Figure 4). For this reasons cabinet B is seismically qualified by analogy.

4 Conclusion

To seismically qualify electrical cabinets different approaches are viable. In this paper the different qualifications are presented by example: analysis, test (of components) and proof by analogy.

If only the stability of a cabinet has to be seismically qualified, the most effective way of qualification is proof by analogy provided an analysed or tested similar cabinet with appropriate seismic loading is available. If proof by analogy is not viable, proof by analysis is the next best choice. Analysis provides an additional insight in the global and local stress distribution of the cabinet's structure, offering the possibility to improve the cabinet's strength in critical areas. Even if a test is demanded for seismic qualification, a preceding analysis is helpful to find the cabinet's weak spots, strengthen them and thus to prepare a successful test.

If stability and functionality of a cabinet is requested, the afore made statements regarding stability still hold. The additional functional demands can be met by successful separate functional tests of the cabinet's electrical components. If no previous sufficient component test results are available, new component tests have to be conducted or the whole cabinet including its built-in components has to be tested, which is the most direct way of proving stability and functionality.

No matter which method is chosen as qualification procedure for a cabinet, the prescribed regulations and specifications have to be met. This may limit the number of the qualification possibilities.

REFERENCES

[1] Henkel, F.-O. et.al: Methods of Qualifying electrical cabinets for the load case Earthquake; in Proceedings of 18th International Conference on Structural Mechanics in Reactor Technology (SmiRT 18); 2005.

[2] IEEE Standard 344-2004: Recommended Practice for Seismic Qualification of Class 1E Equipment for Nuclear Power Generating Stations.

[3] IEEE Standard 693-2005: Recommended Practice for Seismic Design of Substations.

[4] KTA 2201.4, Issue 2012-11: Design of Nuclear Power Plants against Seismic Events, Part 4: Components.

[5] DIN EN 1998-1: Eurocode 8: Design of structures for earthquake resistance, Part 1: General rules, seismic actions and rules for buildings.

Seismic Design of Fastenings with Anchors in Nuclear Power Plants

Rüdiger Meiswinkel[1], Franz-Hermann Schlüter[2]

[1] TU Kaiserslautern, Structural Mechanics
Paul-Ehrlich Straße 14, 67663 Kaiserslautern, Germany
ruediger.meiswinkel@bauing.uni-kl.de

[2] SMP, Consulting Engineers
Stephanienstr. 102, 76133 Karlsruhe, Germany
fh.schlueter@smp-ing.de

ABSTRACT:

For the connection of steel structures and mechanical components like steel platforms, piping systems or vent pipes to concrete structures fastenings with metal anchors will be used. In nuclear power plants safety related fastenings require an adequate seismic design which is based on nuclear specific standards like the German safety standard series KTA 2201 "Design of Nuclear Power Plants against Seismic Events". This safety standard series defines demands on determining the design basis earthquake as well as the design requirements of components and building structures including dynamic analysis procedures.

The different demands on safety-related fastenings with anchors have been established in the German DIBt-guideline with the specifications for the technical approval of metal anchors and for the design of anchor connections. This guideline considers extraordinary action effects like earthquake actions. For example the load bearing of anchors has to be guaranteed in cracked concrete structures with large crack openings considering cyclic loading typically for seismic events.

In addition to the DIBt-guideline the status report KTA-GS-80 presents a review about safety related fastenings in nuclear power plants. This report comprehends the essential information about the design and safety concept of those anchor fastenings regarding the interface between mechanical and constructional engineering. In this context the high demands on the limitation of deformations represents an important design criterion for the components with the assumption of a rigid connection to concrete structures.

Keywords: anchor, fastening, anchors, seismic design, nuclear power plants

1 Introduction

For nuclear power plants seismic events belong to that group of design basis accidents that requires preventive plant engineering measures against damage. The basic requirements of these preventive measures are dealt in the German safety standard series KTA 2201 compromised of six parts [1 – 6]. The first four parts KTA 2201.1 to KTA 2201.4 [1 - 4] represent the design basis for safety-related components and building structures including fasteners.

For the fastening of safety-related mechanical components to concrete structures adequate fasteners as so called post-installed anchors have to be applied. Safety related fastenings also are needed for the anchoring of structural members or components which could detrimentally affect the functions of safety related components or building structures. With regard to the safety related fastenings in nuclear power plants specifications for the approval of metal anchors and for the design of anchor connections are given in the guideline of the German institute for constructional engineering called DIBt-guideline [7].

A complete review of the application of safety-related fastenings with anchors will be presented in the status report KTA-GS-80 [8]. This report clarifies the safety aspects and the interface between mechanical and constructional engineering as well as the design and safety concept of those fastenings.

2 Fastenings with post-installed metal anchors

2.1 Anchor systems

In nuclear power plants different steel structures and mechanical components like steel platforms, piping systems or vent pipes must be connected to concrete structures using steel anchor plates with welds for the connection between the plates and the components. Especially for modification and retrofitting measures so-called post-installed metal anchors can be used for such fastening (see Fig. 1). They have to guarantee the transfer of axial and shear forces (N, V) and the limitation of the corresponding deformations (δ_N and δ_V).

Three types of such post-installed metal anchors seem to be suitable for the application in nuclear power plants (see Fig. 2):

- expansion anchor: anchor with friction connection for the anchoring of axial tensile forces,
- undercut anchor: anchor which develops its tensile resistance from the mechanical interlock provided by undercutting of the concrete at the embedded end of the fastener,
- bonded anchor as bond expansion anchor (chemical anchors): threaded bar embedded in the bore holes by an adhesive mortar.

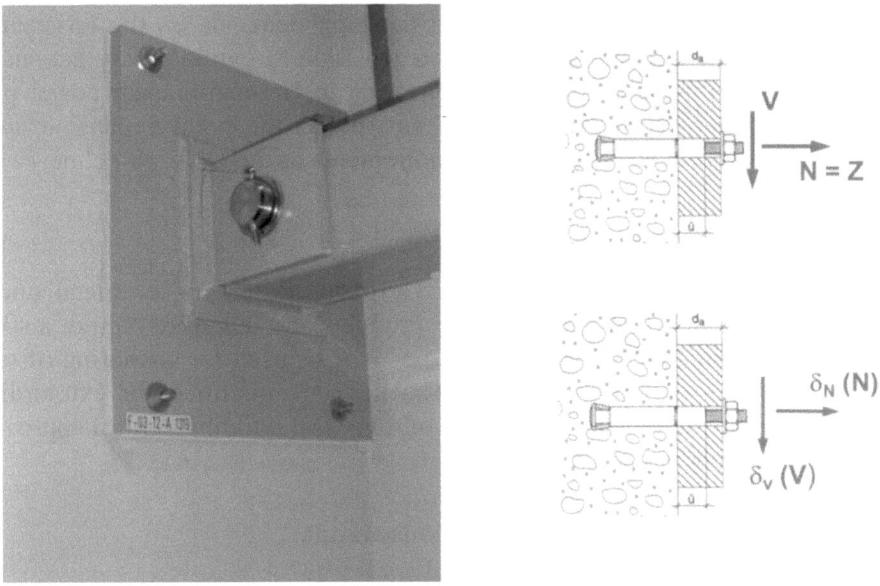

Figure 1: Anchoring with metal anchors (acting forces N and V with corresponding displacements δ_N and δ_V)

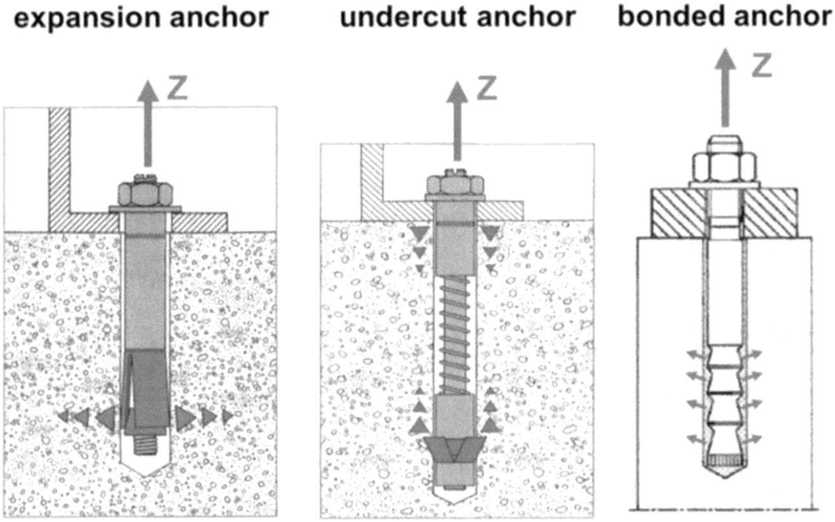

Figure 2: Different types of metal anchors: expansion anchor, undercut anchor, bond expansion anchor

First of all expansion anchors like "Liebig-safety anchor" have been used in German nuclear plants. Later on undercut anchors have been preferred due to the possible load transfer in cracked concrete. In addition to undercut anchors also bond expansion anchors are able to fulfil the high demands on the resistance in concrete with cyclic opening and closing cracks due to extraordinary actions like seismic actions. So nowadays for the application in German nuclear power plants the undercut anchor "Hilti HDA KKW" as well as the bond expansion anchor "MKT VMZ" is approved based on the requirements of the DIBt-guideline.

2.2 Qualification of anchor types

With regard to European countries like Germany fastenings or metal anchors represent so called non-regulated building products and therefore require a special applicability confirmation. Only anchor types will be used for anchoring of safety related components or building structures which are qualified for extraordinary actions like seismic actions in the context of a general building control approval or an approval in individual case (see Fig. 3).

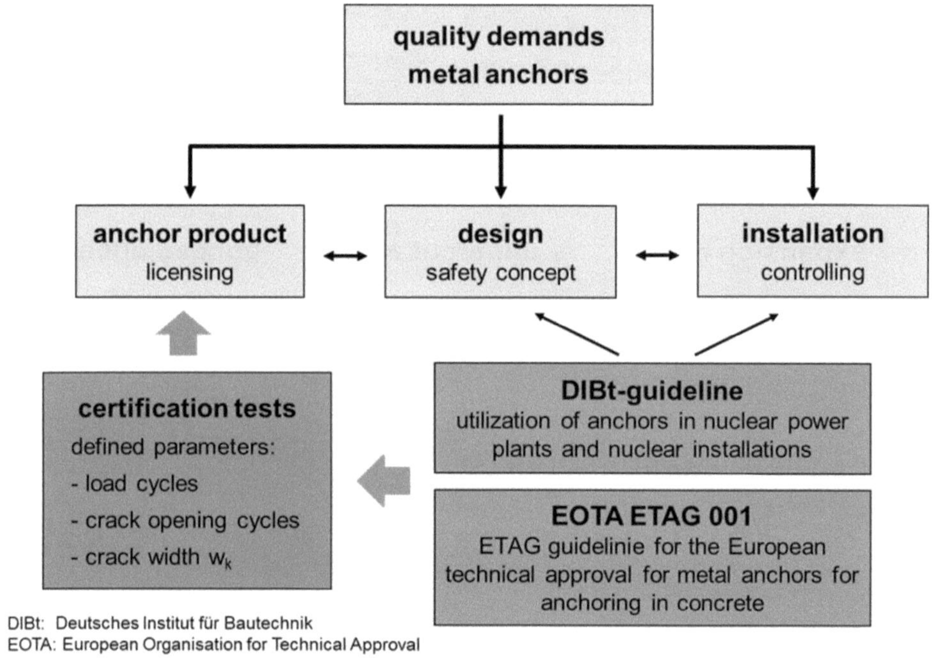

Figure 3: German licensing process for metal anchors

The technical approvals contain information with regard to the manufacture and installation of anchors as well as characteristic values for the design and product specific design procedures. They are currently based on the criteria referred to in the ETAG 001 [9] and which cover the requirements of the general building construction. However the demands on the safety related fastenings in nuclear power plants exceed those criteria given in the guideline of ETAG 001. These beyond demands are summarized in the DIBt-guideline.

Regarding extraordinary actions like seismic actions or actions due to an airplane crash specified verification tests are needed in addition to adequate design concepts. Especially in view of seismic actions the DIBt-guideline defines the anchor tests including the number of load cycles and crack opening cycles which have to be considered by an assumed maximum crack width.

3 Verification procedure for seismic events

3.1 Seismic actions

For the verification of safety-related fastenings extraordinary design situations and in particular a design basis earthquake according to a nuclear safety standard like the German safety standard KTA 2201.1 have to be applied. This first part of the safety standard series KTA 2201 specifies demands for determining the design basis earthquake and provides fundamental requirements for the following five parts.

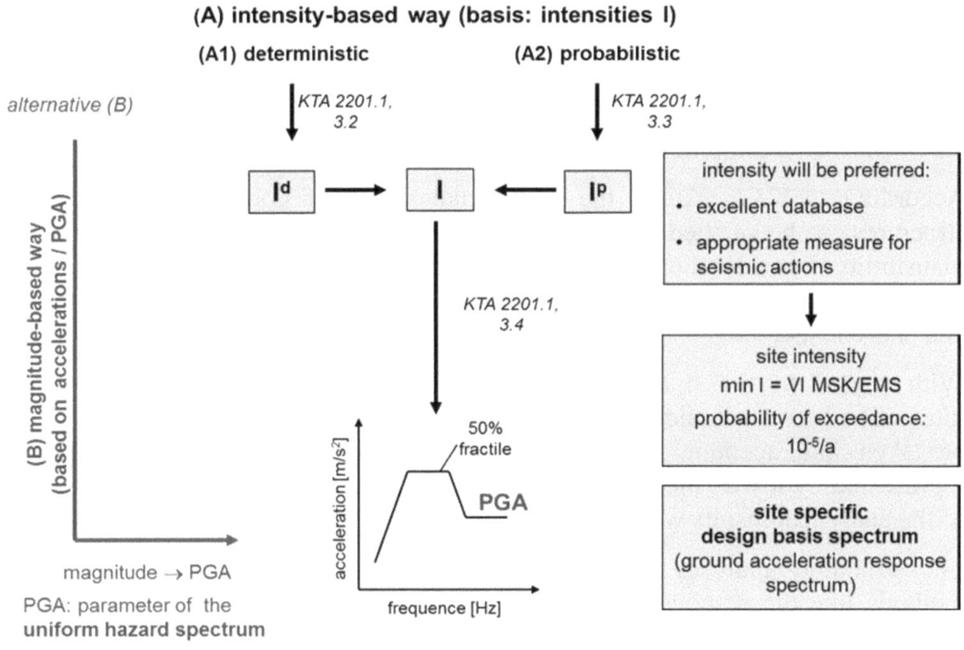

Figure 4: Determining the design basis earthquake (KTA 2201.1)

In KTA 2201.1 the intensity will be used as a characteristic parameter for the design basis earthquake because in opposite to the magnitude the intensity represents a robust measure for the expected seismic actions. Moreover the alternative of using intensities is justified by the excellent database for European countries, especially for Germany. The design basis earthquake will be specified by evaluating deterministic seismic hazard assessment (DSHA) as well as probabilistic seismic hazard assessment (PSHA). DSHA and PSHA result in the site specific intensity with the corresponding ground acceleration response spectrum (see Fig. 4).

In the context of the deterministic determination the design basis earthquake is the seismic event with the maximum intensity at a specific site which, according to scientific knowledge, may occur at the site or within a larger radius of the site (up to approx. 200 km from the site). The probabilistic approach to specifying the design basis earthquake is based on a probability of exceedance of 10^{-5} per annum (10^{-5}/a) regarding seismic actions which may be specified for the 50 %-fractile value (median values).

As a result of the deterministic and the probabilistic determinations the actions of the earthquake can be described by seismo-engineering parameters, in particular, by the ground response spectra with the corresponding rigid-body accelerations (PGA: peak ground acceleration) and the strong-motion duration. However KTA 2201.1 demands a minimum intensity of VI with respect to the target of a basic protection against seismic events.

3.2 Seismic design of building structures and components

According to KTA 2201.1 the earthquake safety of components and building structures can be verified analytically, experimentally, or by analogy (similarity) or plausibility considerations (experience based). However generally analytical verification procedures will be applied which require adequate structural models and analytic methods.

With regard to the dynamic behavior of the structure the influence of the interaction between building structure and subsoil (soil–structure interaction) must be taken into account, varying the soil characteristics in a reasonable range represented by lower, medium and upper soil stiffness (see also [2]). The envelope of the analytical results with different soil stiffness's must then be determined.

The structural analyses can be carried out using the usual dynamic analytic methods like the response spectrum methods, linear and non-linear time history methods or frequency response methods. Also the quasi-static method as a simplified method can be applied in special cases. The resulting method to be used in performing structural analyses and verifications of building structures or components with respect to the different parts of KTA 2201 is shown in Figure 5.

Generally for the anchoring of components the component structures will be analysed by the response spectrum method using tertiary responses or tertiary spectra. The variation of the soil parameters and the evaluation of the spectra indicate the conservative determination of the structural response (see also [8]). For the supports of the components additional analyses will be carried out which consider rigid building connections fundamentally. The resulting forces of the building connections represent the forces of the anchor plates considering a rigid connection of the fixing points.

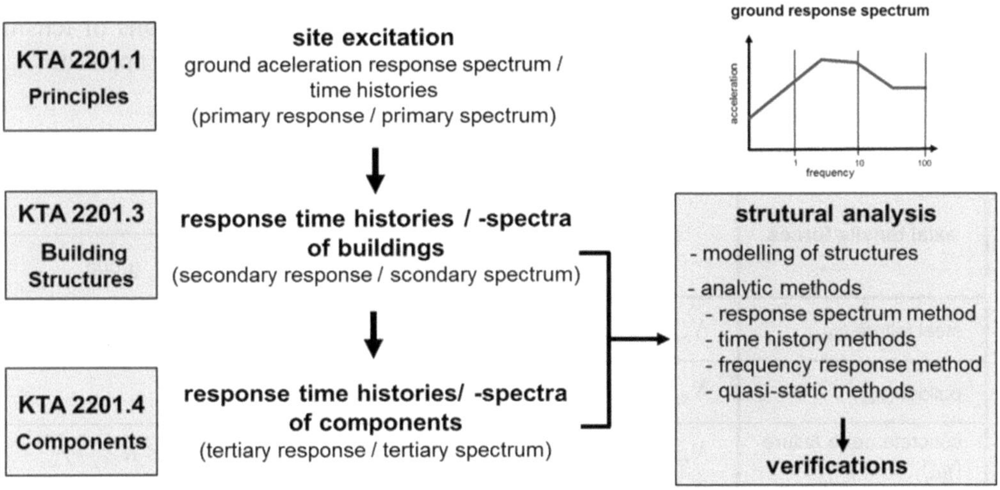

Figure 5: Earthquake analysis and verifications (KTA 2201 series)

4 Design and safety concept of anchors

4.1 Partial safety concept

The design of anchor fastenings in nuclear power plants is based on a partial safety concept according to the European standards. This safety concept requires the non-exceeding of the design value of the resistance R_d for the design value of the actions S_d for ultimate limit states (ULS) and the limit states of serviceability (SLS):

$$S_d = \gamma_E \cdot S_k \leq R_d = R_k / \gamma_M \qquad (1)$$

Values for the partial safety factors γ_E together with the necessary load combination factors ψ are given in DIN 25449 [10]. The safety factors γ_M of the structural resistance depend on the different design limit states. For SLS the safety coefficients of the structural resistance may be assumed to 1.0. For ULS the consideration of internal and external incidents requires three different design requirements given by the categories A1 (combinations of actions of permanent and temporary design situations), A2 (combinations of accidental actions) and A3 (combinations of accidental actions with a minor probability of occurrence,

occurrence rate $\leq 10^{-4}$ /year), defined in DIN 25449. For seismic actions according to KTA 2201.1 category A3 has to be considered.

With respect to the requirement categories the safety factors γ_M for the verifications of metal anchors will be subdivided in partial safety factor for concrete failure (γ_{Mc}), failure due to splitting (γ_{Msp}), failure due to pullout (γ_{Mp}) and steel failure (γ_{Ms}). These partial safety factors are quantified in the DIBt-guideline and have to be considered for the verifications of the tensile axial forces (Table 1) as well as for the verifications of the shear forces (Table 2). For simultaneous actions of tensile forces and shear forces interactions diagrams of ETAG 001 have to be applied.

Table 1: Verification for tensile forces N

axial tensile forces	single anchor	anchor group	
		anchor with peak action effect	group
steel failure (γ_{Ms})	$N_{Sd} \leq N_{Rk,s} / \gamma_{Ms}$	$N^h_{Sd} \leq N_{Rk,s} / \gamma_{Ms}$	
pullout (γ_{Mp})	$N_{Sd} \leq N_{Rk,p} / \gamma_{Mp}$	$N^h_{Sd} \leq N_{Rk,p} / \gamma_{Mp}$	
concrete cone failure (γ_{Mc})	$N_{Sd} \leq N_{Rk,c} / \gamma_{Mc}$		$N^g_{Sd} \leq N_{Rk,c} / \gamma_{Mc}$
splitting (γ_{Msp})	$N_{Sd} \leq N_{Rk,sp} / \gamma_{Msp}$		$N^g_{Sd} \leq N_{Rk,sp} / \gamma_{Msp}$

Table 2: Verification for shear forces V

shear forces	single anchor	anchor group	
		anchor with peak action effect	group
steel failure, shear force without level arm (γ_{Ms})	$V_{Sd} \leq V_{Rk,s} / \gamma_{Ms}$	$V^h_{Sd} \leq V_{Rk,s} / \gamma_{Ms}$	
steel failure, shear force with level arm (γ_{Ms})	$V_{Sd} \leq V_{Rk,s} / \gamma_{Ms}$	$V^h_{Sd} \leq V_{Rk,s} / \gamma_{Ms}$	
concrete pryout failure (γ_{Mc})	$V_{Sd} \leq V_{Rk,p} / \gamma_{Mc}$		$V^g_{Sd} \leq V_{Rk,p} / \gamma_{Mc}$
concrete edge failure (γ_{Mc})	$V_{Sd} \leq V_{Rk,c} / \gamma_{Mc}$		$V^g_{Sd} \leq V_{Rk,c} / \gamma_{Mc}$

4.2 Displacements

Generally for the supports of components like piping systems the anchor forces result from an analysis with the assumption of rigid building connections. This assumption can be justified for small displacements up to approximately 3 mm [7].

So technical approvals of anchors according DIBt-guideline consider the displacements depending on the permitted axial tensile forces or shear forces.

For determining the anchor displacements for extraordinary actions like seismic actions where larger cracks in the concrete cannot be excluded the DIBt-guideline defines the test conditions. Particularly due to seismic actions anchor displacements will be influenced by concrete crack width and crack opening cycles. So the tests are carried out in special crack test specimens with controllable single-axis parallel crack assuming a crack width of 1.0 mm (if no location-specific lower crack widths are given).

Fig. 6 shows the typical behavior of undercut anchors in the tests according the DIBt-guideline, which have to be carried out to determine the displacement in the cyclical opening and closing crack. The axial deformation is defined as the deformation value after 5 crack opening cycles. The displacements corresponding to shear force have to be determined by tests with 10-times alternating shear forces.

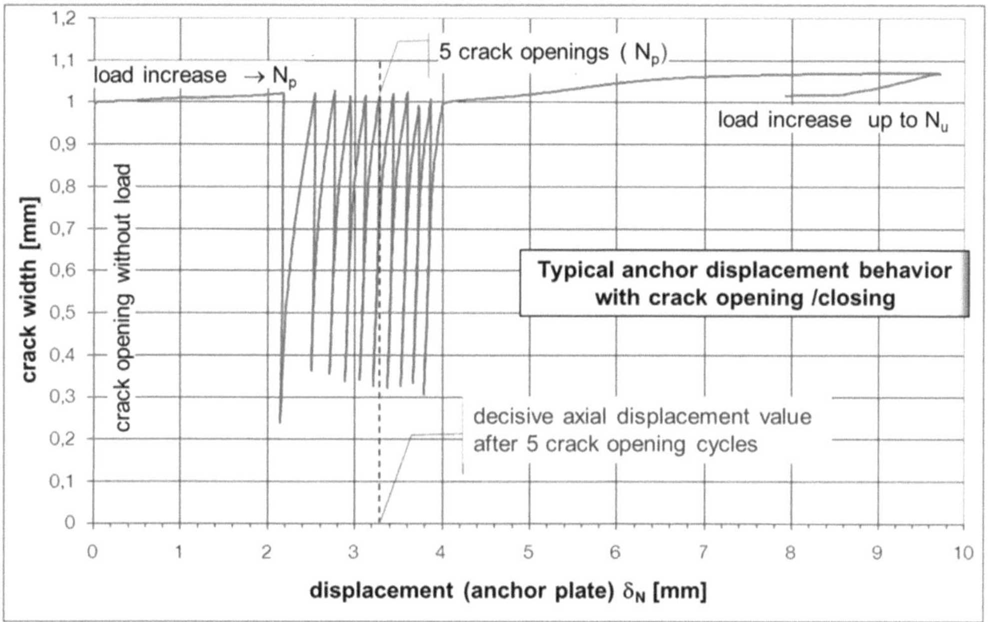

Figure 6: Typical displacement behaviour – tensile forces

5 Conclusion

Safety related anchor connections in nuclear power plants have to guarantee the load transfer of components or structural elements in the load bearing structure regarding the actions of normal operation as well as extraordinary actions like those due to seismic events. Demands for determining the design basis earthquake and the design requirements for components and building structures are provided in the German safety standard series KTA 2201.

With regard to seismic induced crack widths, crack opening and load cycles the DIBT-guideline defines high demands on the anchor technical approval, design and installations. As a result anchoring of components and structural elements fulfil all requirements for seismic deign in nuclear power plants in order to meet the protective goals of controlling reactivity, cooling fuels assemblies, confining radioactive substances and limiting radiation exposure (see [8]).

References

[1] KTA 2201.1 (2011): “Design of nuclear Power Plants against Seismic Events, Part 1: Principles”, Safety Standard of the Nuclear Safety Standard Commission (KTA), Salzgitter

[2] KTA 2201.2 (2012-11). “Design of Nuclear Power Plants against Seismic Events; Part 2: Subsoil”, Safety Standard of the Nuclear Safety Standard Commission (KTA).

[3] KTA 2201.3 (2013-11, prepared). “Design of Nuclear Power Plants against Seismic Events; Part 3: Building Structures”, Safety Standard of the Nuclear Safety Standard Commission (KTA).

[4] KTA 2201.4 (2012-11). “Design of Nuclear Power Plants against Seismic Events; Part 4: Components”, Safety Standard of the Nuclear Safety Standard Commission (KTA).

[5] KTA 2201.5 (2012, in revision). “Design of Nuclear Power Plants against Seismic Events; Part 5: Seismic instrumentation”, Safety Standard of the Nuclear Safety Standard Commission (KTA).

[6] KTA 2201.6 (2012, in revision). “Design of Nuclear Power Plants against Seismic Events; Part 6: Post-seismic measures”, Safety Standard of the Nuclear Safety Standard Commission (KTA).

[7] DIBt-Guideline (2010): “Guideline for Fastenings with Anchors in Nuclear Power Plants and Nuclear Facilities”, Deutsches Institut für Bautechnik, Berlin

[8] KTA-GS-80 (2013): “Application of Fastenings with Anchors in Nuclear Power Plants”, Status Report of the Nuclear Safety Standard Commission (KTA), Salzgitter,

[9] ETAG 001 (2006): “ETAG 001 guideline for the European technical approval for metal anchors for anchoring in concrete , Brussels

[10] DIN 25449 (2008): Reinforced and prestressed concrete structures in nuclear facilities – Safety concept, actions, design and construction, Beuth Verlag, Berlin

New European Seismic Regulations Provide Guidance for the Qualification and Design of Post-installed Anchoring

J. Gramaxo[1], M. Di-Sario[2]

[1] BU Anchors, Hilti Corporation, Liechtenstein,

[2] Engineering Manager, Hilti Italy

ABSTRACT:

Under seismic loading, the performance of a connection in a structure is crucial either to its stability or in order to avoid casualties and major economic impacts, due to the collapse of non-structural elements. In the United States the anchor seismic resistance shall be evaluated in accordance with ACI 318 Appendix D. Created in accordance with the ACI 355.2 regulated testing procedures and acceptance criteria ICC-ES AC193 and AC308, pre-qualification reports provide sound data in a proper design format. With the release of the ETAG 001 Annex E in 2013, the seismic pre-qualification of anchors became regulated in Europe. Anchors submitted to these new test procedures will now also incorporate in the ETA (European Technical Approval) all the required technical data for seismic design. Until the release of the EN 1992-4, planned for 2015, EOTA TR045 (Technical Report) will set the standard for the seismic design of steel to concrete connections. Therefore, the design framework for the seismic design of anchors is already available through both the U.S. and European regulations.

Keywords: Anchors, Seismic design, Seismic qualification

1 Background and Recommendations

In all parts of the world, seismic design methodologies not only for primary structures, but also including equipment, installation and other non-structural element supports have significantly gained in importance over the past years. This does not apply solely to "classical" earthquake regions, but also to Central Europe where, for example, the threat from earthquakes has been underestimated in the past. As the 1756 Düren earthquake and the seismicity distribution shown in Fig. 1, large earthquakes in Europe are not just historical references.

DOI 10.1007/978-3-658-02810-7_26, © Springer Fachmedien Wiesbaden 2014

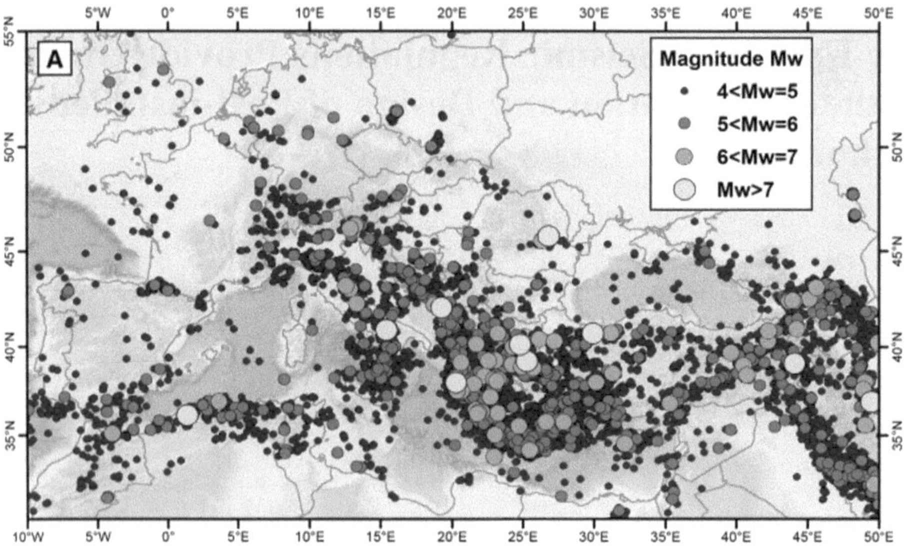

Figure 1: European seismicity distribution for the 1976–2009 period (Source: NEIC catalog)

In fact the economic and social costs associated with the failure or interruption of certain services and equipments such as water, energy or telecommunication supply systems and traffic lines are of comparable magnitude to the costs associated with structural failures, if not greater.

As post-installed anchors are often used to fix structural members and non-structural components, their adequate design and selection is of crucial importance to guarantee safety and minimize costs associated with seismic events. The connections should then be clearly detailed during design phase in order to allow a common understanding of the project specifications by contractors and building inspectors. Ultimately, this practice avoids the high risk of leaving the responsibility to subcontractors.

1.1 Influence of earthquake resulting cracks in concrete base material

As a structure responds to earthquake ground motion it experiences displacement and consequently deformation of its individual members. This deformation leads to the formation and opening of cracks in the concrete members. Consequently all anchorages intended to transfer earthquake loads should be suitable for use in cracked concrete and their design should be predicted on the assumption that cracks in the concrete will cycle open and closed for the duration of the ground motion.

Parts of the structures may however be subjected to extreme inelastic deformation as exposed in Fig. 2. In the reinforced areas yielding of the reinforcement and cycling of cracks may result in cracks width of several millimetres, particularly in regions of

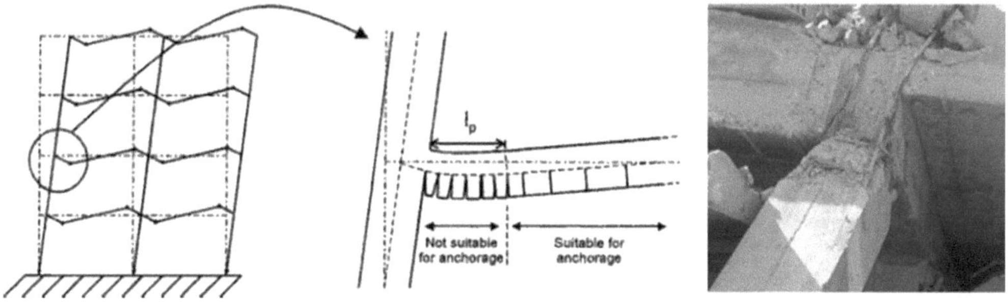

Figure 2: Member cracking assuming a strong-column, weak girder design (lp = plastic hinge length)

plastic hinges. Qualification procedures for anchors do not currently anticipate such large crack widths. For this reason, anchorages in these regions where plastic hinging is expected to occur should be avoided unless apposite design measures are taken.

1.2 Suitability of anchors under seismic loading

An anchor suitable (approved) to perform in a commonly defined cracked concrete, about 0.3 mm, is not consequently suitable to resist seismic actions, it's just a starting point.

During an earthquake cyclic loading of the structure and fastenings is induced simultaneously. Due to this the width of the cracks will vary between a minimum and a maximum value and the fastenings will be loaded cyclically. Specific testing programs and evaluation requirements are then necessary in order to evaluate the performance of an anchor subjected to seismic actions. Only the anchors approved after the mentioned procedure shall be specified for any safety relevant connection.

Anchors generally suitable for taking up seismic actions are those which can be given a controlled and sustained pre-tensioning force and are capable of re-expanding when cracking occurs. Also favourable are anchors which have an anchoring mechanism based on a keying (mechanical interlock) as it is the case for undercut anchors. Furthermore, some specific chemical anchors have also been recognized good performance to resist seismic actions. Displacement controlled expansion anchors should be avoided considering that their performance under seismic is proven unsuitable.

The following Table 1 provides a rough overview of the suitability of various types of anchors to resist seismic actions. This suitability depends to a great extent on how badly the concrete has cracked and how large the cracks are in the event of an earthquake. The classifications presented are based on a generic assessment of the anchor types not reflecting a particular evaluation of any product or anchor manufacture.

Table 1: Suitability of anchors under seismic loading (- unsuitable, + limited suitability, ++ very suitable)

Type of anchor		Displacement controlled expansion anchors	Adhesives anchors	Concrete screws	Torque-controlled expansion anchors	Adhesive-expansion anchors	Sleeved torque-controlled expansion anchors	Undercut anchors
Cracked concrete with crack width, w	small (w < 0.5mm)	-	++	++	++	++	++	++
	medium (0.5 ≤ w ≤ 1.0mm)	-	+	+	+	+	++	++
	large (w > 1.0mm)	-	-	-	-	+	+	++

Note that the precise understanding of an anchor ability to tackle seismic loading should always be checked by consulting the anchor approvals being Table 1 a guidance for a general understanding of the different anchor type capacities and limitations.

1.3 Influence of annular gaps in the anchorage resistance under shear loading

Under shear loading, if the force exceeds the friction between the concrete and the anchoring plate, the consequence will be slip of the fixture by an amount equal to the annular gap. The forces on the anchors are amplified due to a hammer effect on the anchor resulting from the sudden stop against the side of the hole (Fig. 3a). This justifies the new European seismic design guideline recommendation for annular gaps between the anchors and the fixture to be avoided in seismic design situations.

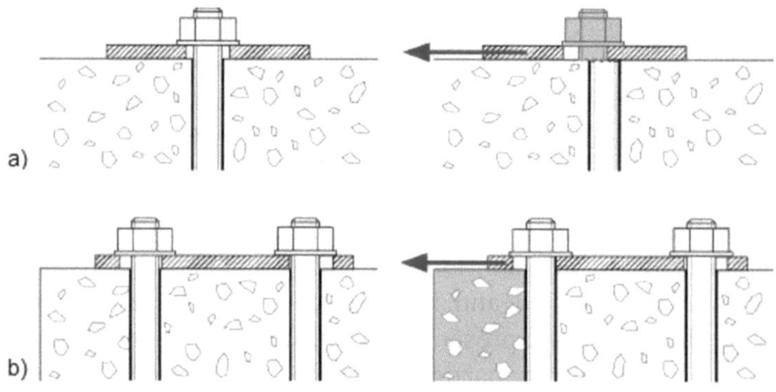

Figure 3: Mains consequences possibility resulting from annular gaps

Moreover, where multiple-anchor fastenings are concerned, it must be assumed that due to play of the hole on the steel plate a shear load may not be distributed equally among all anchors. In an unfavourable situation, when anchor fastenings are positioned near to the edge of a building member, only the anchors closest to the edge should be assumed loaded and this could result in failure of the concrete edge before the anchors furthest from the edge can also participate in the load transfer (Fig. 3b).

By eliminating the hole play, filling the clearance hole with an adhesive mortar e.g., the effects mentioned above are controlled with great benefit to the anchorage performance. The use of Hilti Dynamic Set (Fig. 4) will ensure a professional approach for a controlled filling of the annular gaps as well as it will prevent the loosening of the nut since it also comprehends a lock nut, effect that also complies with a European seismic design guideline clear recommendation. Also according to the same guideline, in case it can be ensured that there is no hole clearance between the anchor and the fixture, the anchor seismic resistance for shear loading is doubled compared to connections with hole clearances.

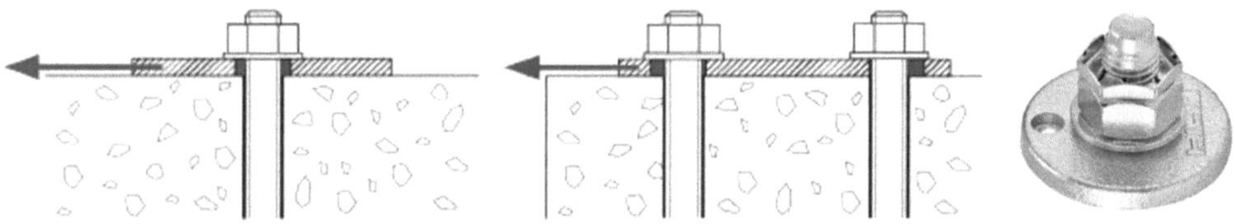

Figure 4: Benefits of filled annular gaps and Hilti Dynamic Set: Filling washer, conical washer, nut and lock-nut

2 United States and European Seismic Regulations

For a sound seismic design of a post-installed anchorage the first step begins with the correct definition of the acting loads. In the United States ASCE 7 establishes the provisions for the definition of the seismic action and the anchor performance shall be evaluated in accordance with ACI 318 Appendix D and AC308 in case of chemical anchors. Pre-qualification reports, created in accordance with published testing procedures and acceptance criteria, (ACI 355.2 with ICC-ES AC193 and AC308) provide sound data in a proper format for design.

Following the same design flow, in Europe the action definition is available through the EN 1998:2004 (Eurocode 8). Until the release of the EN 1992-4, planned for 2015, an EOTA TR (Technical Report) will set the standard for the seismic design of steel to concrete connections. This regulation is in full alignment with ETAG 001 Annex E, the new European guideline for the anchor's seismic pre-qualification testing. As such, the European framework is also already harmonized in order to allow the design of a post-installed anchorage under seismic conditions.

Table 2: Seismic design framework for fastenings in concrete

	United States	Europe
Load definition	ASCE 7	EN 1998-1:2004
Design resistance	ACI 318 Appendix D AC308	EOTA TR
Technical data	ICC-ESR	ETA
Pre-qualification criteria	ACI 355.2 with ICC-ES AC193/AC308	ETAG 001, Annex E

As an overview, Table 2. displays the application ranges of the different guidelines or codes mentioned above. The presented design codes represent the state of the art for the testing of fasteners and the design of fastenings in concrete worldwide. Note that even if not all, most of the countries in the world refer to one of these frameworks for the design of anchors.

2.1 Seismic load definition

The starting point for the definition of the seismic actions is the seismic design spectrum. In the case of the US a seismic design category (SDC) is endorsed and the seismic design spectrum is obtained by the mapped maximum (short period, 0.2s) and 1.0s period acceleration whereas in Europe the seismic hazard is defined by the peak ground acceleration (PGA) and no SDC is established. There is however a clear definition for low and very low seismicity, based on the design ground acceleration, and in case of very low seismicity no specific seismic provisions need to be observed.

The influence of the soil type is considered in both codes by a site coefficient which is based on matching ground classifications, considering the shear wave velocity limits and soil descriptions. Based on the risk of an eventual improper seismic performance, the categorization of buildings is placed in the same way by both codes and the correspondent importance factor is assigned with similar values (even if at different phase in the design flow).

Considering the above mentioned, the equations to derive the seismic design spectrum are expected to be different between the codes but, considering equivalent importance class and ground type, the resulting shape and spectral acceleration are very much similar. In simple terms, it can be said that mathematically the two codes are just pointing different coordinates of the design spectrum (Fig. 5). Note that the design response spectrum according to ASCE7 does not contemplate the influence of the building importance (being considered later in the design) and as such the comparison was made considering the resulting spectrum accordantly scaled by this factor.

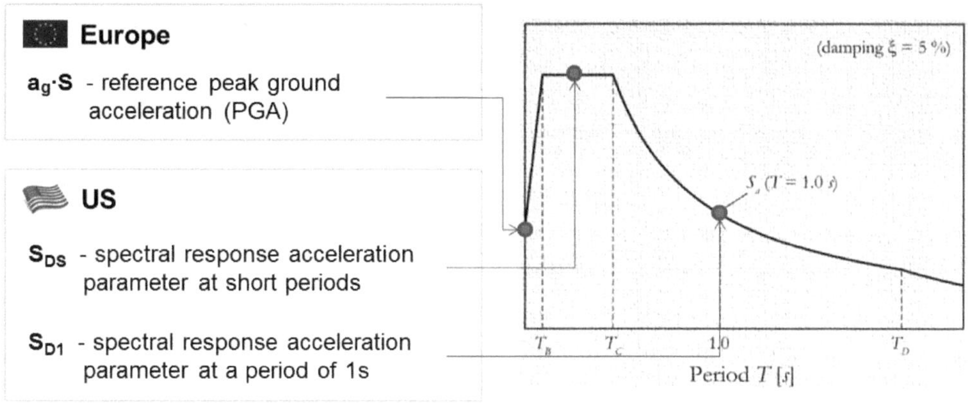

Figure 5: Design response spectrum according to Eurocode 8 and ASCE 7

A comparison was also established between the seismic base shear force using the EN1998-1:2004 and the ASCE7. Evaluating the different expressions as well as some practical applications of the codes we can conclude that the values are decidedly coincident. From the seismic base shear force different well-known methods can be used to determine the load acting at each level of the structure.

As such, comparing the resulting seismic design spectrums with equivalent importance classes and ground types (S being the soil factor), it's possible to correlate the European seismicity rating with the United States seismic design category, as expressed in Table 3.

As the only yet important exception to the Table 3, in case of a building with an importance class IV and a seismicity rating of low or above the corresponding seismic design category is C or above. This means that in the case of buildings that in the event of a failure could pose a substantial hazard to the environment or community (e.g. hospitals, fire stations, power plants) the design should consider all the seismic specific provisions.

Table 3: European seismicity rating relation to seismic design category (SDC) for importance class I, II, III

EN 1998-1:2004 (Eurocode 8)		ASCE7	
Seismicity rating	Design repercussion	SDC	Design repercussion
Very low $ag \cdot S \leq 0.05 \cdot g$	No seismic specific provisions need to be observed	A	No seismic specific provisions need to be observed
Low $ag \cdot S \leq 0.1 \cdot g$	Reduced or simplified design procedures may be used	B	
$ag \cdot S > 0.1 \cdot g$	Seismic design must be attend to all the elements	C to F	Seismic design must be attend to all the elements

2.2 Anchors seismic design resistance

Design provisions for the anchor seismic design are provided by the ACI 318 Appendix D or the recent EOTA TR045. Both design regulations work with the CC-method (concrete capacity method) to calculate the characteristic resistances of fastenings. Differences between the codes occur in the basic assumptions for the design equations which partially result in different factors. According to the CC-method the design resistances are calculated for tension loading and shear loading considering all possible failure modes.

All discussed safety concepts calculate resistance and actions based on partial safety factors. The main requirement for design of the discussed codes is that the factored action E shall be smaller or equal to the factored resistance R (Eqn. 2.1.). All codes factor the characteristic action E_k with partial safety factors γ (Eqn. 2.2.).

$$E_d \leq R_d \tag{2.1}$$

$$E_d = E_k \cdot \gamma \tag{2.2}$$

For the characteristic resistance there is a conceptual difference since the European codes divide the characteristic resistance R_k by a partial safety factor γ (Eqn. 2.3.) whereas the United States codes factor the characteristic resistance R_k with a strength reduction factor ϕ (Eqn. 2.4.). The effect of these factors is however the same reducing the characteristic value to design level. The design resistance R_d is generally very similar for all the evaluated failure modes independently on the adopted code.

$$R_d = R_k / \gamma \tag{2.3}$$

$$R_d = \phi \cdot R_k \tag{2.4}$$

As per the new European design guideline, EOTA TR045, the design incorporates three design approaches which are described below. Note that all three of these approaches are acceptable within their application conditions. Table 4 provides an overview of these different design options.

Note that the ACI 318 also considers thee design approaches that are conceptually the same as the ones presented by the EOTA TR045. The main difference, that nevertheless has the same background intention, comes from the fact that the "Elastic design" defined as per European guideline has a different approach in the U.S. regulations. In the ACI 318 this design option consider the loads resulting from a regular seismic design (not elastic) and introduces a reduction factor (recommended as 0.4) directly applied on all concrete failure modes. It is the authors' opinion that the new European regulations have made the different design approaches more clear compared to the ACI 318 interpretation.

Table 4: Seismic design options per European seismic guideline

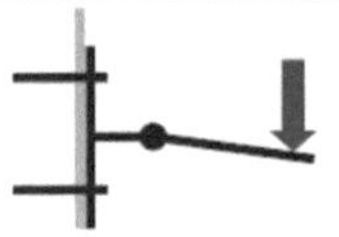	**a1) Capacity design** The anchorage is designed for the force corresponding to the yield of a ductile component or, if lower, the maximum force that can be transferred by the fixture or the attached element.
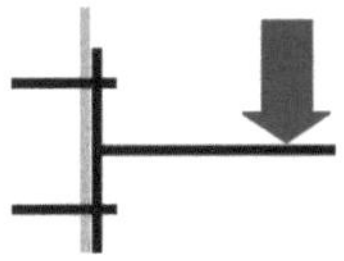	**a2) Elastic design** The fastening is designed for the maximum load assuming an elastic behaviour of the fastening and of the structure.
	**b) Design with requirements on the ductility of the anchors** This design for ductile steel failure requires an anchor classified as ductile. Additionally, this approach is applicable only for the tension component and some provisions require to be observed in order to ensure that the cause of failure is steel failure.

2.3 Evaluation of the anchor seismic performance

For testing of fastenings in concrete three different basic guidelines must be considered. In the United States ACI 355.2 covers testing of post-installed mechanical anchors under static and seismic loading and prescribes testing programs and evaluation requirements for post-installed mechanical anchors intended for use in concrete under the design provisions of ACI 318. This guideline is the basis for the acceptance criteria AC193 and AC308 by the International Code Council (ICC). While AC193 covers testing of mechanical anchors, AC308 covers testing and design of adhesive anchors.

Referring the main testing procedures, the anchors are installed in a closed crack that then is open to 0.5mm. The anchors under testing are afterwards subject to the sinusoid varying loads specified, using a loading frequency between 0.1 and 2Hz as exposed in Fig. 6. The maximum seismic tension and shear test load is equal to 50% of the mean capacity in cracked concrete from reference tests.

After the simulated seismic-tension and seismic-shear cycles have been run, the anchors are tested to failure in static-tension and static-shear. The mean residual tension and shear capacities shall be assessed according the guideline defined limits.

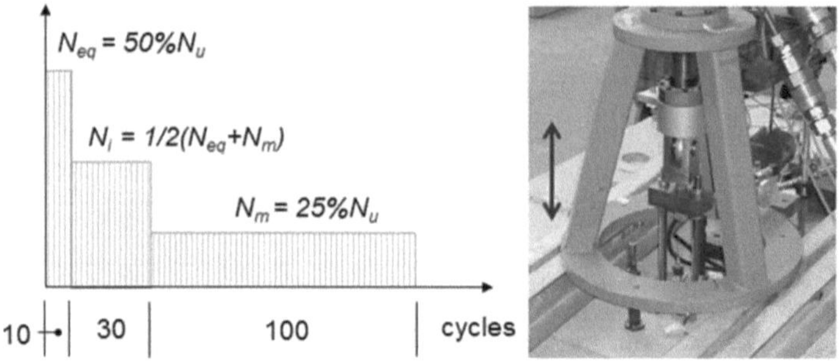

Figure 6: Loading pattern for simulated seismic tension tests according to ACI355.2

In Europe the ETAG 001 is valid for testing of post- installed mechanical (Part 1 to Part 4) and bonded anchors (Part 5). With the release of the ETAG 001 Annex E in 2013, the seismic pre-qualification of anchors became regulated in Europe. Two different testing programs are presented to assess the anchor's suitability to seismic loading resulting in two seismic performance categories classified as follows:

- Seismic category C1 - similar to the US seismic pre-qualification procedure and only suitable for non-structural applications.
- Seismic category C2 - very demanding seismic crack movement tests classify an anchor as suitable for structural and non-structural applications.

While seismic category C1 is identical to the U.S. seismic pre-qualification procedure, seismic category C2 involves a set of quite more demanding seismic load and/or crack cycle tests especially considering that for assessing the tension seismic performance one of the tests involves the cycling of the cracks until a width of 0.8mm.

In practical terms, according to the EOTA TR045 and for $a_g{\cdot}S$ above 0.05g, anchors intended for connections between structural elements of primary or secondary seismic members should always have a seismic category C2. For anchors used in the attachment of non-structural elements, if the acceleration $a_g{\cdot}S$ is between 0.05g and 0.10g then a seismic category C1 can be used. Please note that these are the generic recommendations that member states can locally adjust.

3 Conclusion

Considering all the exposed above, the design framework for the seismic design of anchors is already available through both the U.S. and European regulations. This means that there it is no longer a need for an engineering judgement on the use of U.S. anchor performance provisions along with the European seismic action definition, solution suggested by one of the authors during the last years in the absence of European seismic regulations to assess and design anchors.

It's now the responsibility of the anchoring manufactures to provide designers with seismic design data according to the new European testing procedures. Hilti has already approved for seismic category C2HY200+HIT Z and HST for the ETA C1&C2 seismic approval.

REFERENCES

[1] European Organization for Technical Approvals (2013). EOTA TR045, Design of Metal Anchors Under Seismic Actions. Brussels. Belgium.

[2] European Organization for Technical Approvals (2013). ETAG 001 Annex E, Assessment of metal anchors under seismic actions. Brussels. Belgium.

[3] Hilti Corporation (2013). Build a future safer from earthquakes: New EU guidelines. Schaan. Liechtenstein.

[4] Gramaxo, J. (2012). Engineering judgement on the use of American's anchor performance provisions along with the European seismic action definition. 15th World Conference Earthquake Engineering. Portugal.

[5] ICC Evaluation Service, Inc. (2011). AC 193, Acceptance Criteria for Mechanical Anchors in Concrete Elements. United States of America.

[6] ICC Evaluation Service, Inc. (2011). AC 308, Acceptance Criteria for Post-Installed Adhesive Anchors in Concrete Elements. United States of America.

[7] Hilti Corporation (2011). Build a future safer from earthquakes. Schaan. Liechtenstein.

[8] Tonning, K. (2009). Design of High-rise Onshore Steel and Reinforced Concrete Structures for Earthquake Resistance. University of Stavanger. Norway.

[9] Applied Technology Council (2008). ATC-69, Reducing the risks of nonstructural earthquake damage. California. United States of America.

[10] American Concrete Institute (2008). ACI 318-08 Appendix D, Building Code Requirements for Structural Concrete - Anchoring to Concrete. United States of America.

[11] American Concrete Institute (2007). ACI 355.2, Qualification of Post-Installed Mechanical Anchors in Concrete. United States of America.

[12] Eligehausen R., Mallee, R. and Silva, J.F. (2006). Anchorage in Concrete construction. Ernst & Sohn. Berlin.

[13] Höhler, M, S. (2006). Behavior and testing of fastenings to concrete for use in seismic applications. Doctor Thesis. University of Stuttgart.

[14] American Society of Civil Engineers (2005). ASCE 7-05, Minimum Design Loads for Buildings and Other Structures. United States of America.

[15] European Committee for Standardisation (2004). EN 1998:2004, Eurocode 8: Design of structures for earthquake resistance. Brussels. Belgium.

[16] Hilti Corporation (2004). Guideline for earthquake resistant design of installations and non-structural elements. Schaan. Liechtenstein.

S e D I F Conference

International Conference on
Seismic Design of Industrial Facilities
2013, RWTH Aachen University

Fastenings for Use in Concrete – Seismic Actions

Wolfgang Roeser[1]

[1] H+P Ingenieure GmbH & Co KG (Hegger and Partner)
Kackertstr. 10
52072 Aachen / Germany
wroeser@huping.de
www.huping.de

Abstract

Fastenings like headed studs and post-installed mechanical or chemical anchors for use in concrete are often used in Industrial Facilities. This paper deals with fastenings that are used to transmit seismic actions by means of tension, shear, or a combination of tension and shear,

- between connected structural elements
- between non-structural attachments and structural elements

Although the majority of fastenings up to now are designed and tested for use in non-seismic environments, they are commonly used for applications in structures in earthquake regions. Fasteners will be subjected to both crack cycling and load cycling at dynamic rates during an earthquake (Fig. 1 [1, 2]). Therefore in future special requirements for the use of fasteners in seismic regions are demanded.

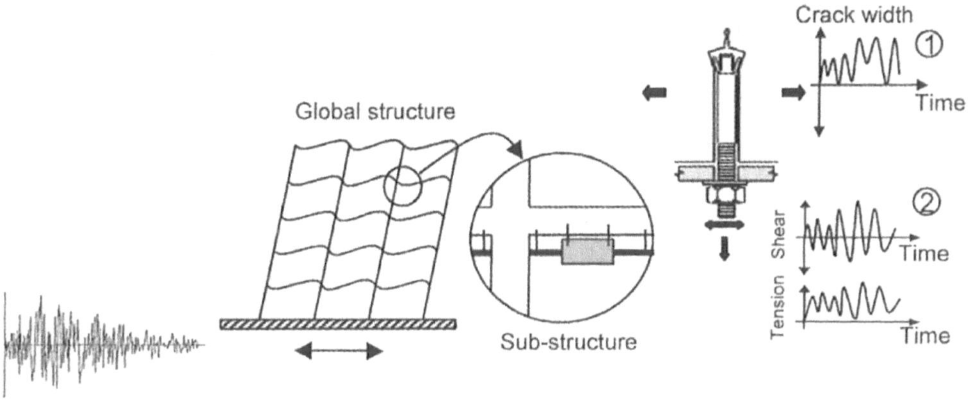

Fig. 1: Actions acting on a non-structural anchorage under earthquake loading acc. to Eligehausen et al. [1] / Hoehler [2]

DOI 10.1007/978-3-658-02810-7_27, © Springer Fachmedien Wiesbaden 2014

Metal anchors used to resist seismic actions shall meet the requirements of ETAG 001 Annex E: “Assesment of Metal Anchors under seismic action” [3]. The document deals with the preconditions, assumptions, required tests and assessment for metal anchors under seismic action. The design value of the effect of seismic actions acting on the fixture shall be determined according to Eurocode 8 [4]. Furthermore the EOTA “Technical report TR 045 – Design of Metal Anchors under Seismic Actions” [5] and the draft of Eurocode 2 part 4 “Design of Fasteners for Use in Concrete” [6] gives further requirements regarding the design of fasteners under seismic actions in addition to Eurocode 8.

Keywords: fastening, fasteners, metal anchors, chemical anchors, headed studs, earthquake

1 Introduction

In General the following types of connections are distinguished:

Type A: Connection between structural elements of primary and/or secondary seismic members.

Type B: Attachment of non-structural elements. According to Eurocode 8, 4.3.5.1 non-strutural elements may be parapets, cables, antennae, mechanical appendages and equipement, curtain walls, partitions, railings and so on.

Fasteners used to transmit seismic actions have to be tested according ETAG 001, Annex E in order to achieve a European Technical Approval (ETA) that defines the resistance in the case of seismic action. In dependence of the testing method it is divided between the **performance categories C1** and **C2**.

The design of the fasteners shall be in accordance with the rules given in the Technical Report TR 045, which are nearly similar to Eurocode 2, part 4, 7th draft, chapter 9 and annex C. In the design of fastenings one of the following options a1), a2) or b) should be satisfied.

Option a1) **Capacity design**: The anchors are designed for the maximum load that develops a ductile yield mechanism in the fixture or the attached element.

Option a2) **Elastic design** – the fastenings is designed for the maximum load assuming an elastic behaviour of the fastening and of the structure.

Option b) Design with requirements on the **ductility of the fasteners.** The tension steel capacity of the fastener shall be smaller than the capacity of the attached elements. Therefore sufficient elongation capacity of the anchors is required.

The general matrix in Table 1 shows that Option a1) and a2) may be used for structural elements (Type A) and non-structural elements (Type B) as well, and that the performance category of the anchor may be C1 or C2 in dependence of seismicity level. On the other hand Option b) should be used only for non-structural elements and the performance category C2 is required.

Table 1: General Matrix for the performance category

	Type A: Structural Elements	**Type B: Non-Structural Elements**
Option a1)	C1 or C2	C1 or C2
Option a2)	C1 or C2	C1 or C2
Option b)	Not recommended	C2

2 Performance Categories C1 and C2 according to Annex E of ETAG 001

For the evaluation of the performance of anchors subjected to seismic loading two seismic performance categories C1 and C2 are distinguished, with C2 being more stringent than C1. Annex E of ETAG 001 deals with the required tests in order to achieve an ETA that defines the characteristic resistance in the performance category C1 and C2. The performance category C1 is in agreement with the requirements according to ACI 318 Annex D [7].

Performance category C1 provides anchor capacities in terms of strength (forces), while performance category C2 provides anchor capacities in terms of both strength (forces) and deformations. In both cases the effect of concrete cracking is taken into account. The maximum crack width considered in C1 is $\Delta w = 0{,}5$ mm and in C2 it is $\Delta w = 0{,}8$ mm.

Qualification of anchors for category C1 comprises tests under pulsating tension load and tests under alternating shear load. Qualification of anchors for category C2 includes reference tests up to failure, tests under pulsating tension load, tests under alternating shear load as well as under crack cycling. Based on the respective load histories and crack widths the design information's for C1 contains values of tension and shear resistance of the anchor, while for C2 it contains values of tension and shear resistance as well as anchor displacement.

Table 2 relates the anchor seismic performance categories C1 and C2 to the seismicity and the building importance class. The designer shall use Table 2 unless a different national requirement is recommended.

Table 2: Performance Categories according to ETAG 001, Annex E

Seismicity level[a]		Importance Class acc. to EN 1998-1:2004, 4.2.5			
Class	$a_g \cdot S$ [c]	I	II	III	IV
Very low[b]	$a_g \cdot S \leq 0{,}05\ g$	No additional requirement			
Low[b]	$0{,}05\ g < a_g \cdot S \leq 0{,}1\ g$	C1	C1 [d] or C2 [e]		C2
> low	$a_g \cdot S > 0{,}1\ g$	C1	C2		

[a] The values defining the seismicity levels are may be found in the National Annex of EN 1988-1.
[b] Definition according to EN 1998-1:2004, 3.2.1.
[c] a_g = design ground acceleration on Type A ground (EN 1998-1:2004, 3.2.1),
S = soil factor (see e.g. EN 1998-1:2004, 3.2.2).
[d] C1 for Type 'B' connections (see 5.1)
[e] C2 for Type 'A' connections (see 5.1)

3 Design according to Option a)

In General fastenings used to resist seismic actions shall meet all applicable requirements for non-seismic applications. Only fasteners qualified for cracked concrete and seismic applications shall be used. The design method does not apply to the design of fastenings in plastic hinge zones (critical sections) of the concrete structures, which is defined in Eurocode 8.

In **option a1) capacity design** the fastening is designed for the maximum load that can be transmitted to the fastening based either on the development of a ductile yield mechanism in the attached steel component or in the steel base plate. Strain hardening or material over-strength of the attached element has to be taking into account.

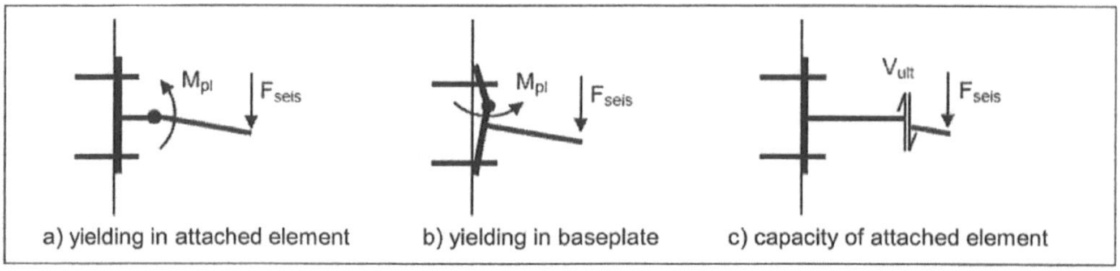

Fig. 2: Seismic design by option a1) capacity design

In **option a2) elastic design** the fastening is designed for the maximum load obtained from the design load combinations according to Eurocode 8 assuming an elastic behaviour of the fastening and the structure. Therefore the behaviour factor is $q = 1{,}0$ for structural elements (Type A) and $q_a = 1{,}0$ for non-structural elements (Type B). If action effects are derived in accordance with a simplified approach they shall be multiplied by an amplification factor.

For structural connections (Type A) the vertical component of the seismic action shall be taken into account if the vertical design ground acceleration is greater than 2,5 m/sec^2.

For non-structural elements (Type B) subjected to seismic actions, any beneficial effects of friction due to gravity should be ignored. In Addition to Eurocode 8 some further equations are given in TR045, in order to determine the horizontal and vertical effects on non-structural elements.

For the seismic design situation the verifications $F_{Sd,seis} \leq R_{d,seis}$ shall be performed for all loading directions (tension, shear, combined) as well as failure modes (steel, pull-out, cone, splitting, pry-out, edge failure) (Table 3).

Table 3: Required verifications

	failure mode	single anchor	anchor group: most loaded anchor	anchor group: anchor group
tension	steel failure	$N_{Sd,seis} \leq N_{Rd,s,seis}$	$N^h_{Sd,seis} \leq N^h_{Rd,s,seis}$	
tension	pull-out failure	$N_{Sd,seis} \leq N_{Rd,p,seis}$	$N^h_{Sd,seis} \leq N^h_{Rd,p,seis}$	
tension	combined pull-out and concrete cone failure [1]	$N_{Sd,seis} \leq N_{Rd,p,seis}$		$N^g_{Sd,seis} \leq N^g_{Rd,p,seis}$
tension	concrete cone failure	$N_{Sd,seis} \leq N_{Rd,c,seis}$		$N^g_{Sd,seis} \leq N^g_{Rd,c,seis}$
tension	splitting [3]	$N_{Sd,seis} \leq N_{Rd,sp,seis}$		$N^g_{Sd,seis} \leq N^g_{Rd,sp,seis}$
shear	steel failure, shear load without lever arm [2]	$V_{Sd,seis} \leq V_{Rd,s,seis}$	$V^h_{Sd,seis} \leq V^h_{Rd,s,seis}$	
shear	concrete pry-out failure	$V_{Sd,seis} \leq V_{Rd,cp,seis}$		$V^g_{Sd,seis} \leq V^g_{Rd,cp,seis}$
shear	concrete edge failure	$V_{Sd,seis} \leq V_{Rd,c,seis}$		$V^g_{Sd,seis} \leq V^g_{Rd,c,seis}$

[1] Verification for bonded anchors only.
[2] Steel failure for shear loads with lever arm is not covered in this Technical Report (see Section 5.1).
[3] Verification is not required if cracked concrete is assumed and reinforcement resists the splitting forces.

The seismic design resistance of a fastener is given by equation (1).

$$R_{d,seis} = R_{k,seis} / \gamma_{m,seis} \tag{1}$$

With: $R_{k,seis} = \alpha_{gap} \cdot \alpha_{seis} \cdot R^0_{k,seis}$

α_{gap} = Reduction factor to take into account inertia effects due to an annular gap between anchor and fixture in case of shear loading
= 1,0 in case of no hole clearance between anchor and fixture
= 0,5 in case of connections with hole clearance according to table 4

α_{seis} = Reduction factor to take into account the influence of large cracks and scatter or f load displacement curves, according to Table 5

$R^0_{k,seis}$ = basic charateristic seismic resistance according to ETA

$\gamma_{m,seis}$ = The partial safety factor should be identical to the corresponding values for static loading according to ETAG 001, Annex C or EOTA TR 029

Table 4: Diameter of clearance hole in the fixture

external diameter d or d_{nom} [1]	[mm]	6	8	10	12	14	16	18	20	22	24	27	30
diameter d_f of clearance hole in the fixture	[mm]	7	9	12	14	16	18	20	22	24	26	30	33

1) diameter d if bolt bears against the fixture; diameter d_{nom} if sleeve bears against the fixture

Table 5: Reduction factor α_{seis}

Loading	Failure mode	Single anchor [1]	Anchor group
tension	Steel failure	1,0	1,0
	Pull-out failure	1,0	0,85
	Combined pull-out and concrete failure	1,0	0,85
	Concrete cone failure • undercut anchors with the same behaviour as cast-in headed fasteners [2] • all other anchors	 1,00 0,85	 0,85 0,75
	Splitting failure	1,0	0,85
shear	Steel failure	1,0	0,85
	Concrete edge failure	1,0	0,85
	Concrete pry-out failure • undercut anchors with the same behaviour as cast-in headed fasteners [2] • all other anchors	 1,0 0,85	 0,85 0,75

1) In case of tension loading single anchor also addresses situations where only 1 anchor in a group of anchors is subjected to tension.
2) Undercut anchors with the same concrete cone capacity in cracked concrete as cast-in headed fasteners, i.e. at least $N^0_{Rk,c} = 8.0 \cdot (f_{ck,cube})^{0,5} \cdot (h_{ef})^{1,5}$; given in the relevant ETA.

The interaction between tension and shear forces shall be determined according to equation (2).

$$(N_{sd}/N_{Rd,seis}) + (V_{Sd}/V_{Rd,seis}) \leq 1 \quad (2)$$

With: $N_{sd}/N_{Rd,seis} \leq 1$

$V_{Sd}/V_{Rd,seis} \leq 1$

4 Design according to option b)

According to TR 045 the use of **option b) ductility of the fastener** is limited as follows:

- The approach is applicable only for the tension component of the load acting on the anchor. Shear loads should be resisted by additional elements.
- The fastening shall not be used for energy dissipation unless proper justification is provided by a non-linear time-history dynamic analysis and the hysteretic behaviour of the anchor is provided by an ETA.
- The fastening may not be suitable for primary seismic members. Therefore it is recommended to use option b) only for the use of secondary seismic members.
- The anchor is qualified for seismic performance category C2
- There are special requirements regarding the stretch length, the ultimate strength, the ratio between yield strength to ultimate strength, the rupture elongation and using a reduced section of the fastening.

The definition in option b) is in discrepancy to Eurocode 8. In option b) the fastening should be ductile on the one hand but may not be hysteretic on the other hand. (If otherwise the benefit of energy dissipation and the hystertic behaviour is taken into account a non-linear time-history dynamic analysis should be performed and the hysteretic behaviour has to be provided by an ETA). According to Eurocode 8, 2.2.2 (2) the ductility and the energy dissipation (hysteretic behaviour) are deeply connected by the behaviour factor [8] and can't be separated as optional proposed in option b). Because of the given limitations, option b) is not further described in this paper.

5 The use of fastenings in German earthquake regions

In Germany the seismicity level is low or very low. Therefore some simplified design rules have been proposed in agreement with the german authorities (Ministerium Baden Württemberg, Deutsches Institut für Bautechnik), the University of Stuttgart (IWB, Prof. Hofmann) and the consulting engineers (Dr. Schlüter, SMP, and the author):

- If the behaviour factor is $q \leq 1{,}5$ there are no special requirements on the fastener. The fastenings shall be designed for static or quasi-static loading regarding changing load-directions due to earthquakes.
- If the behaviour factor of the global structure is $q > 1{,}5$ the fastening should fulfil performance category C1. The use of performance category C2 is not required.
- The fastenings should not be placed in plastic hinge zones (critical sections) of the concrete structures.

References

[1] Eligehausen, R.; Mallee, R.; Silva, J.F.: Anchorage in Concrete Construction; Ernst & Sohn, Berlin, 2006

[2] Hoehler, M.S.: Behavior and Testing of Fastenings to Concrete for use in Seismic Applications; Dissertation, IWB, University of Stuttgart, 2006

[3] EOTA: ETAG 001- Guideline for European Technical Approval of Metal Anchors for use in Concrete – Annex E: Assesment of Metal Anchors under seismic actions; European Organisation for Technical Approvals, Edition October 2012, Brussels

[4] Eurocode 8: Design of structures for earthquake resistance – part 1: General rules, seismic actions and rules for buildings; European Standard, EN 1998-1, December 2004, Brussels

[5] EOTA: Technical Report TR 045 for ETAG 001 - Guideline for European Technical Approval of Metal Anchors for use in Concrete Design of Metal Anchors under Seismic Actions; European Organisation for Technical Approvals, Draft February 2013, Brussels

[6] CEN TC 250: Design of Fastenings for Use in Concrete; EN 1992-4:2012/prA7:2013, 7th draft, Brussels

[7] ACI 318-Building Code Requirement for Structural Concrete (ACI 318-11) and Comentary; ACI Standard, reported by ACI Comitee 318, American Concrete Institute, Michigan, USA, 2011

[8] Butenweg, C.; Roeser, W.: Erdbebenbemessung von Stahlbetonbauwerken nach DIN EN 1998-1; in: Goris, A.; Hegger, J.: Stahlbetonbau aktuell 2013 – Praxishandbuch, Beuth Verlag, Berlin, 2013

Part VII

Seismic Design of Primary Structures

Reliability Analysis on Capacity Design Rules for Steel Frames

Max Gündel[1], Benno Hoffmeister[1], Markus Feldmann[1]

[1] Institute of Steel Construction, RWTH Aachen University
Mies-van-der-Rohe-Str. 1, 52074 Aachen, Germany
guendel@stb.rwth-aachen.de

ABSTRACT:

Capacity design rules are applied to ensure the intended plastic behaviour of structures subjected to earthquakes. They need to cover scattering due to the seismic action and material strength. Recent evaluations of data of structural steel from European producers show for low steel grade an overstrength, which is higher than covered by the recommended value in DIN EN 1998-1. In this paper results of reliability analysis performed on this topic are presented. For this purpose a stochastic model for the seismic action was derived, which enables to perform the investigations via push-over analysis instead of time-consuming non-linear time history analysis. The main findings are discussed in the view of adjustments of design rules in DIN EN 1998-1 and European product standards, respectively.

Keywords: steel frame, capacity design, reliability analysis, probabilistic seismic load model

1 Introduction

Steel frames resist earthquake very efficiently by means of energy dissipation due to plastic deformation. However, the design according to current seismic standards is usually carried out by elastic analysis with equivalent static loads based on the elastic response spectra and the effective mass of the building. Here, energy dissipation due to plastic deformation is considered by force reduction factors (named q in DIN EN 1998-1), where the elastic response spectra are reduced in relation to the plastic deformation behaviour and capacity of the structure. To guarantee the expected and required plastic behaviour capacity design rules are applied in the design procedure. These contain for steel frames following verifications (Figure 1):

(i) Weak storey failure has to be prevented and the desired global plastic mechanism should form. To guarantee this, the moment capacity of columns needs

S. Klinkel et al. (Eds.), *Seismic Design of Industrial Facilities*,
DOI 10.1007/978-3-658-02810-7_28, © Springer Fachmedien Wiesbaden 2014

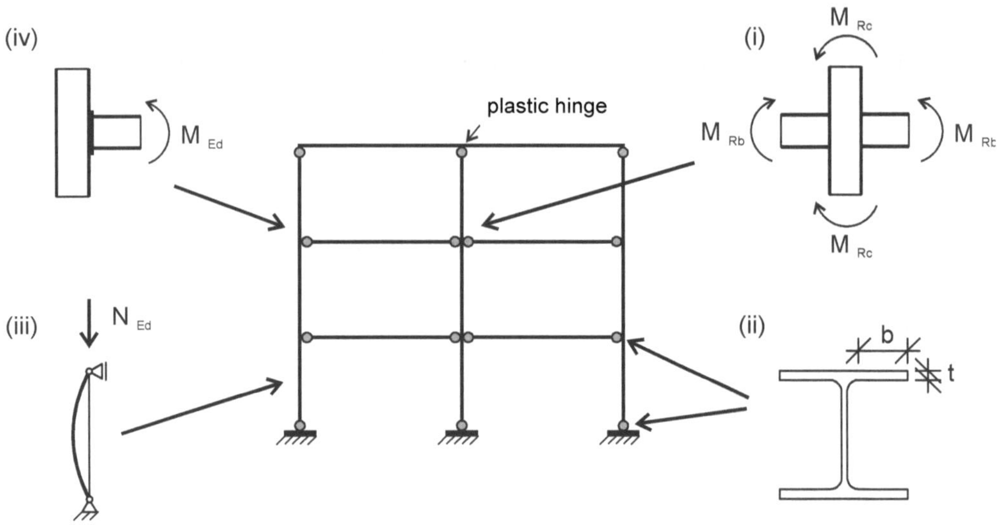

Figure 1: Capacity design of steel frames acc. to DIN EN 1998-1

a sufficient overstrength compared to the moment capacity of the adjacent beams at each node (DIN EN 1998-1):

$$\sum M_{Rc} / \sum M_{Rb} \geq COF \qquad (1)$$

(ii) Dissipative elements (structural members where plastic hinges are intended) need a sufficient rotation capacity to ensure the considered plastic moment capacity even under strong plastic deformations and repeated loading. For this purpose sections need to be semi-compact for medium dissipative design ($1.5 < q \leq 4$) and compact for high dissipative design ($4 < q \leq 6.5 \cdot \alpha$, where α considers the global structural overstrenght).

(iii) Buckling of columns (non-dissipative elements) has to be prevented, as this failure mode provides only little deformation capacity and the failure consequences are considerable. Hence, design forces resulting from seismic actions are increased by an amplification factor considering structural and material overstrength:

$$N_{Ed} = N_{Ed,G} + 1{,}1 \cdot \gamma_{ov} \cdot \Omega \cdot N_{Ed,E} \qquad (2)$$

(iv) Finally, failure of non-dissipative connection has to be prevented to achieve the desired global plastic deformation capacity. Therefore, joints are designed with a sufficient overstrength considering the plastic section capacity of the attached structural member and expected material overstrength:

$$R_{Ed} \geq 1{,}1 \cdot \gamma_{ov} \cdot R_{fy} \qquad (3)$$

Material scattering and overstrength is considered in the capacity design rules for steel structures in DIN EN 1998-1 by the material overstrength factor $1.1 \cdot \gamma_{ov}$,

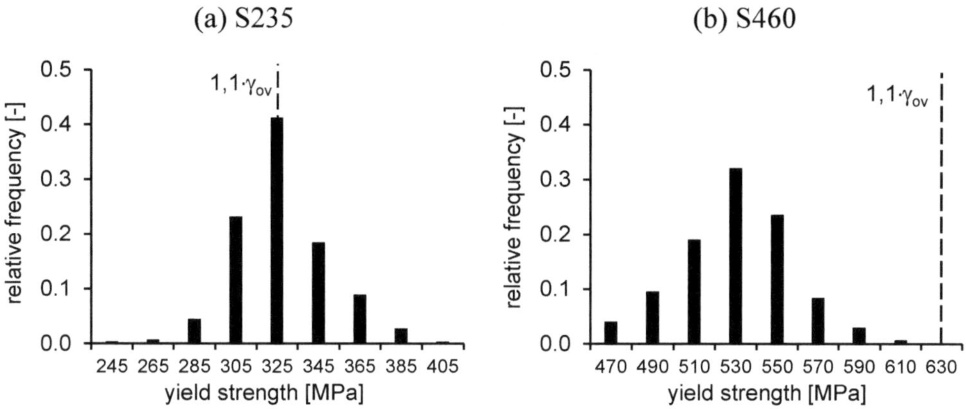

Figure 2: Distribution of yield strength of structural steel [1]

where the recommend value of γ_{ov} is 1.25. However, recent statistic evaluations of material data from European steel producers show that for steel grades with low nominal strengths even the mean value of yield strength is higher than the recommended material overstrenght factor (Figure 2 left). On the other hand, the overstrength of high strength steel grades is very small (Figure 2 right).

The influence of material strength distribution representative for current structural steel production in European on the capacity design rules for steel frames according to DIN EN 1998-1 has been investigated by seismic reliability analysis. Methods and results of this study are presented in this paper and are discussed in view of possibly necessary adjustments of production and/or seismic design standards.

2 Probabilistic seismic load model

2.1 Basics

The structural behaviour of steel frames subjected to earthquakes can be simulated by non-linear time history analysis with very high accuracy. The scattering of seismic actions is considered by repeating the calculations with a number of (scaled) recorded or artificial accelerograms which fulfil the design response spectra representative for the location of the building. However, in reliability analysis - even with advanced methods - hundreds of calculations are required to determine the failure probability. In this context non-linear dynamic calculations are disadvantageous, as they are rather time consuming. Furthermore, there are difficulties to derive direct relationships between seismic action and structural performance, as the scattering of the action is considered inherently by a number of accelerograms. Simplified and more efficient reliability methods specifically developed for seismic design (e. g. the SAC2000/FEMA procedure [2]) are inadequate for this study, as they do not consider the influence of material scattering on the structural performance (demand) as well as system failure (system of limit state functions).

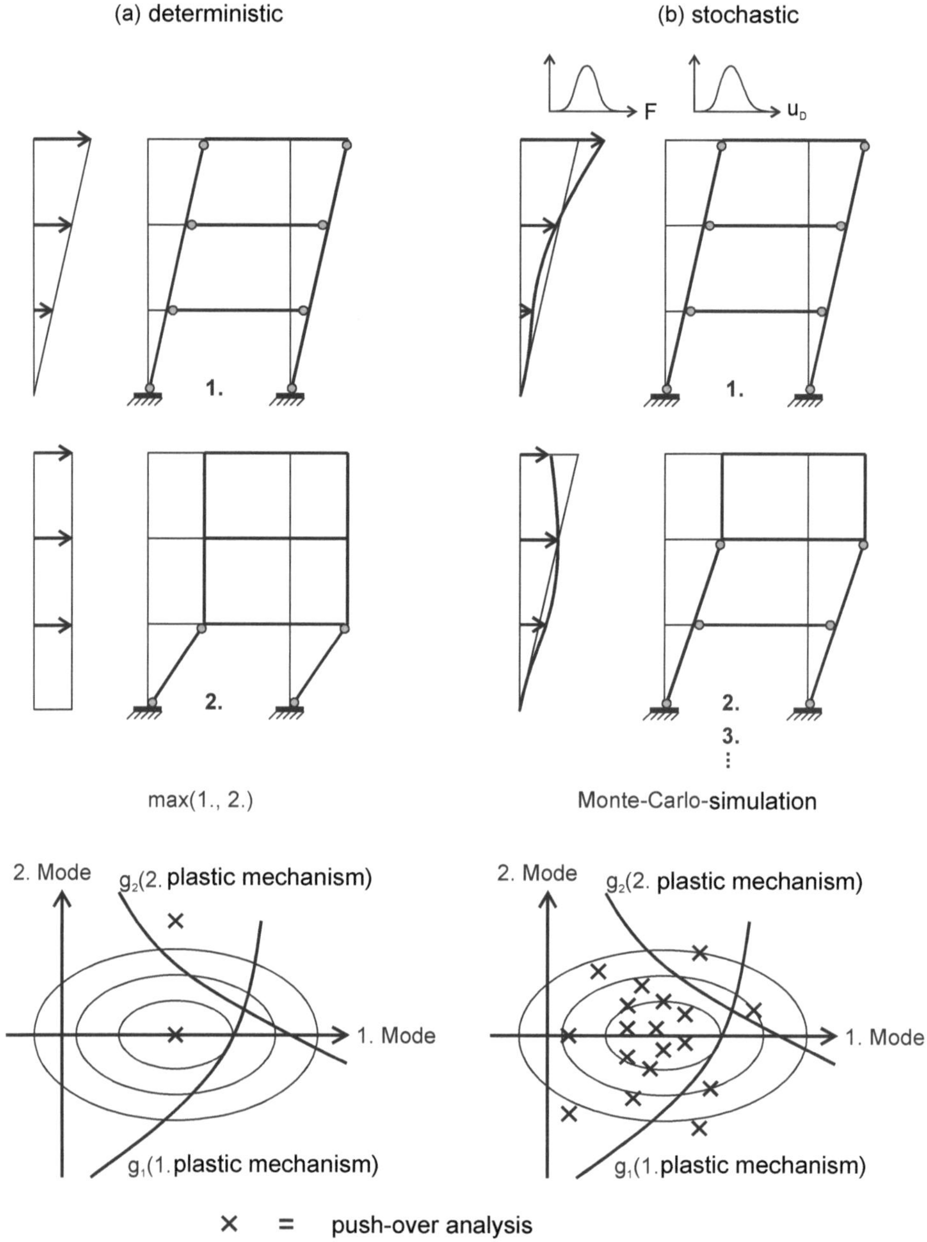

Figure 3: Schematic description of deterministic and probabilistic push-over analyses procedure

Therefore, non-linear static analyses (push-over analysis) were used in the reliability analysis instead of non-linear dynamic analyses. Push-over analyses enable to evaluate the plastic behaviour of structures with a sufficient accuracy and are much less time consuming than time history analysis. In this method horizontal forces at each storey are increased until the target displacement of the reference point (usually the top of the building) is reached, while gravity loads are kept constant. The target displacement is determined on the basis of an equivalent single degree of freedom system (SDOF) and the displacement response spectra. Higher mode effects are considered in DIN EN 1998-1 by repeating the calculation with two different distributions of the horizontal forces over the height of the building: one distribution is based on the fundamental mode shape and one is based on a constant displacement over the height of the building. Obviously, this procedure does not consider the scattering of the seismic actions is in a probabilistic way; a probabilistic model for the horizontal seismic forces is not available so far.

2.2 Stochastic model for distribution of horizontal seismic forces

The description of seismic actions consists of the probability of occurrence of earthquakes with a specific intensity as well as the variability of the acceleration time history itself. The first part is described by a hazard functions, which connects the mean value of the target displacement with the probability of occurrence of a seismic action with a specific intensity. The corresponding distribution of the horizontal forces over the height of the building is related to the fundamental modal shape and period. The variation of acceleration time history yields to scattering of the target displacement and variations of the horizontal forces due to higher modes.

The probabilistic model for horizontal seismic forces is derived based on non-linear and linear time history analyses for a number of case studies and 20 artificial accelerograms each. For this purpose, analyses with elastic material behaviour are beneficial, as there is a direct relationship between deformation and horizontal forces as well as models can be derived by the linear random processes theory. The evaluation of non-linear (with nominal material properties) and linear time history analysis in the critical time step shows that the distribution of plastic hinges is very similar (Figure 4). Non-linear material behaviour does not lead to a considerable redistribution of internal forces and different plastic mechanism. Therefore, it can be assumed that the distribution of the horizontal forces in structures with elastic and plastic material behaviour is very similar and the stochastic model can be derived based on linear time history analyses.

The evaluation of horizontal deformations of the steel frame in the critical time step for different accelerograms shows that the scattering is very small and close to the fundamental eigenmode (Figure 5 left). By subtracting the deformation of the fundamental eigenmode from the total deformation, the influence of higher modes becomes visible (Figure 5 right).

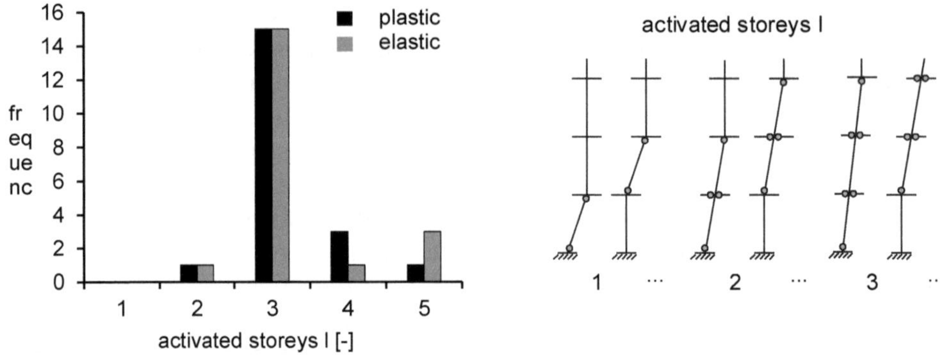

Figure 4: Distribution of plastic hinges in non-linear and linear time step analyses

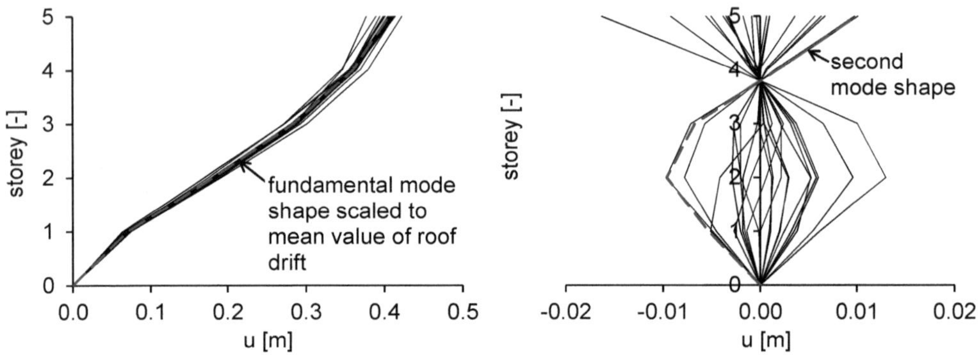

Figure 5: Horizontal deformation in the critical time step (left) deformation of higher modes (right)

To separate the deformations coming from the fundamental and from higher modes, linear time-step analyses with equivalent SDOF for each relevant eigenmode are carried out. The time step of the maximum displacement of the frame is consistent with that of the SDOF for the first eigenmode (Figure 6 left, displacement of SDOFs without participation factors). The first eigenmode represents the mean value and the corresponding target displacement is the ordinate of the displacement spectra (Equ. 4). The standard deviation of the first eigenmode between different accelerograms is zero at the time step of the maximum displacement.

The amplitude of higher eigenmodes at the time step of the maximum displacement is random, if the eigenmodes can be considered as stochastically independent $T_i/T_j \leq 0.9$). The standard deviation of the amplitude of higher modes can be determined by linear random process theory, as the displacement history of SDOFs can be described by a stationary Gauss process. The mean value of higher modes is

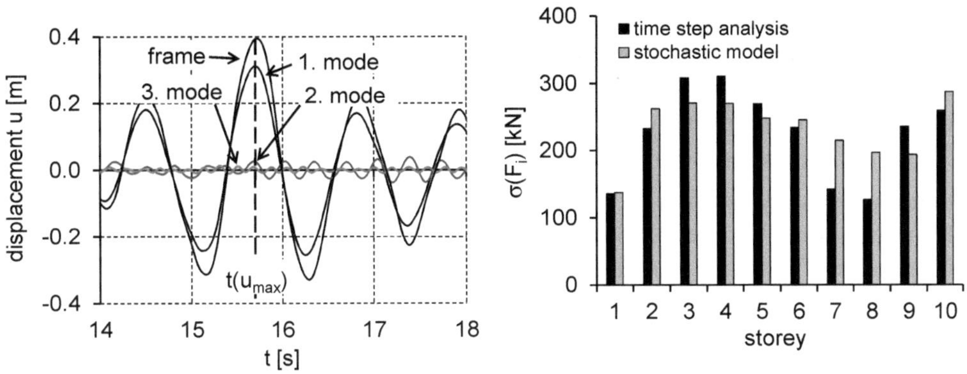

Figure 6: Displacement histories of equivalent SDOFs (left); resulting standard deviation over the height of the building (right)

zero and the standard deviation results from the ordinates of displacement spectrum divided by the peak factor (Equ. 5). The peak factor represents the ratio between maximum value and standard deviation of a random process and is about 3 for earthquakes [3].

$$\mu(\mathbf{u}) = |\Gamma_1| \cdot S_d(T_1) \cdot \boldsymbol{\phi}_1 \tag{4}$$

$$\sigma(\mathbf{u_n}) = |\Gamma_n| \cdot \frac{S_d(T_n)}{r_n} \cdot \boldsymbol{\phi}_\mathbf{n} \approx |\Gamma_n| \cdot \frac{S_d(T_n)}{3} \cdot \boldsymbol{\phi}_\mathbf{n} \tag{5}$$

With $\boldsymbol{\phi}_\mathbf{n}$ the eigenvector, Γ_n participation factor, $S_d(T_n)$ ordinate of the displacement response spectra at eigenmode n. The horizontal forces can be determined by the condensed stiffness matrix or alternatively directly by the acceleration response spectra:

$$\mu(\mathbf{F}) = |\Gamma_1| \cdot \mathbf{M} \cdot \boldsymbol{\phi}_1 \cdot S_a(T_1) \tag{6}$$

$$\sigma(\mathbf{F_n}) = |\Gamma_n| \cdot \mathbf{M} \cdot \boldsymbol{\phi}_\mathbf{n} \cdot \frac{S_a(T_n)}{r_n} \approx |\Gamma_n| \cdot \mathbf{M} \cdot \boldsymbol{\phi}_\mathbf{n} \cdot \frac{S_a(T_n)}{3} \tag{7}$$

With $S_a(T_n)$ ordinate of the acceleration response spectra at eigenmode n, $\mathbf{M}$ mass matrix. The total standard deviation resulting from all higher modes is determined with the SRSS-rule. A very good correlation of the standard deviation of the horizontal forces based on time-step analysis and the probabilistic model (Equ. (6) to (7)) can be observed (Figure 6 right).

2.3 Stochastic model for target displacement

Besides the scattering of distribution of horizontal forces also the target displacement itself is subjected to stochastic variations. The deviations result from higher modes (elastic part), which can be determined by Equ. 8, and from non-linear material behaviour (plastic part). The latter is investigated by non-linear time history analysis with different accelerograms. In Figure 7 left the mean value of maximum roof displacement over intensity level (scaled to the ductility μ respectively deformation ductility μ_d) is shown. The behaviour of different structures is very similar. Plastic material behaviour reduces the maximum deformation due to energy dissipation. At the same time the variation of the maximum roof displacement increases (Figure 7 right). This can be described by following empiric linear equations:

$$u_{D,\max} = S_d(T_1) \cdot \Gamma_1 \qquad \text{for } \mu \le 1{,}0 \cdot \Omega_{ov} \tag{8}$$

$$u_{D,\max} = S_d(T_1) \cdot \Gamma_1 \cdot \frac{0{,}7 \cdot \mu + 0{,}3 \cdot \Omega_{ov}}{\mu} \qquad \text{for } 1{,}0 \cdot \Omega_{ov} < \mu \le 8{,}0 \cdot \Omega_{ov} \tag{9}$$

$$\sigma(u_{D,\max}) = 0 \qquad \text{for } \mu \le 1{,}0 \cdot \Omega_{ov} \tag{10}$$

$$\sigma(u_{D,\max}) = u_{D,\max} \cdot \left(0{,}06 \cdot \frac{\mu}{\Omega_{ov}} - 0{,}06\right) \qquad \text{for } 1{,}0 \cdot \Omega_{ov} < \mu \le 6{,}0 \cdot \Omega_{ov} \tag{11}$$

$$\sigma(u_{D,\max}) = u_{D,\max} \cdot 0{,}30 \qquad \text{for } 6{,}0 \cdot \Omega_{ov} > \mu \tag{12}$$

Global structural overstrength Ω_{ov} delays yielding and leads to higher mean target displacement and less scattering.

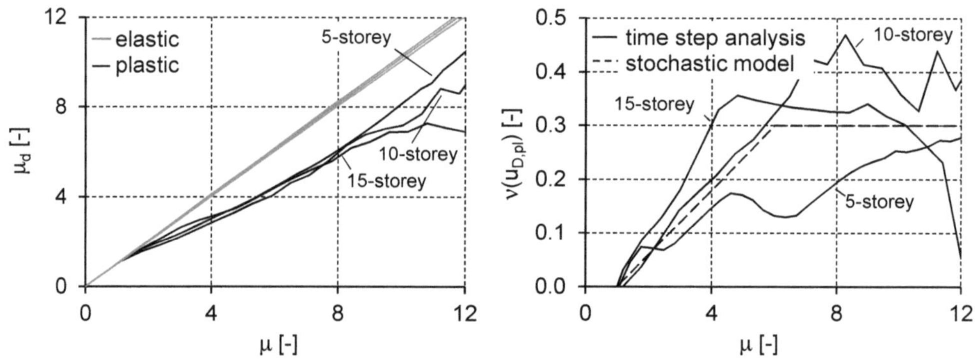

Figure 7: Mean value (left) and coefficient of variation (right) for target displacement over ductility

3 Reliability analysis

Based on the probabilistic model for seismic actions presented in section 2 reliability analyses on the capacity design rules for steel frames are carried via push-over analysis. For the yield strength following probabilistic model is used [1]:

$$\mu(f_y) = 0.83 \cdot f_{y,nom} + 125N/mm^2 \qquad \sigma(f_y) = 25N/mm^2 \qquad (13)$$

Reliability analyses on the buckling of columns are not performed, as non-linear time history analyses have shown only negligible influence of variations caused by the seismic action as well as the material strength scattering. The reference case is a 5-storey-3-bay steel frame in steel grade S355 designed for a PGA of 0.25 g.

3.1 Weak storey failure

The plastic performance of steel frames is investigated by evaluating the safety index of each possible plastic mechanism by first order reliability methods (FORM) analyses. Obviously, the overdesign of columns (COF) shifts the probability of occurrence (equal to the inverse of reliability index) from storey mechanism to global mechanism (Figure 10). Therefore, the required COF is defined in such a way that the probability of storey failure has to be smaller than the probability of global mechanism. The results for different steel grades and standard deviation of seismic action (mainly influenced by the first eigenperiod) on the COF_{req} is shown in Figure 9. The influence of the steel grade is small, while the influence of the scattering of seismic action is dominant. The determined required COF with values up to 4 would govern the design and would lead to uneconomic structures. Hence, in the following investigations COF is kept to 1.3 (recommended value acc. to DIN EN 1998-1), even if the probability for storey mechanism increases the rotation demand in plastic hinges of beams and columns.

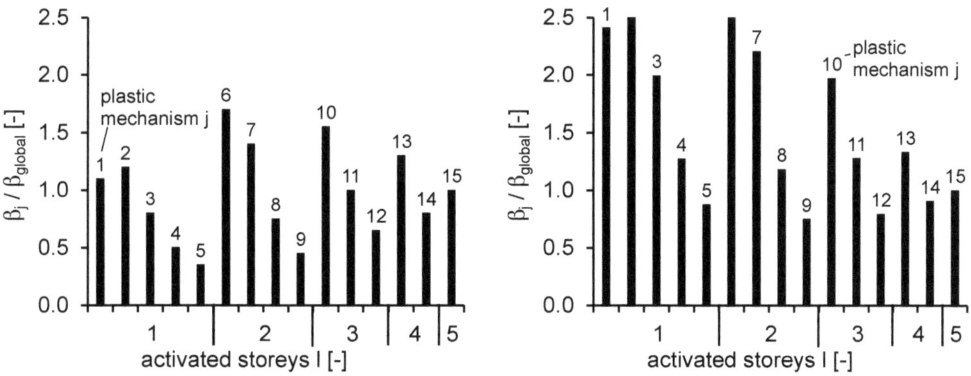

Figure 8: Reliability index β_j of plastic mechanisms of a steel frame: COF = 1.0 (left) and COF = 1.3 (right)

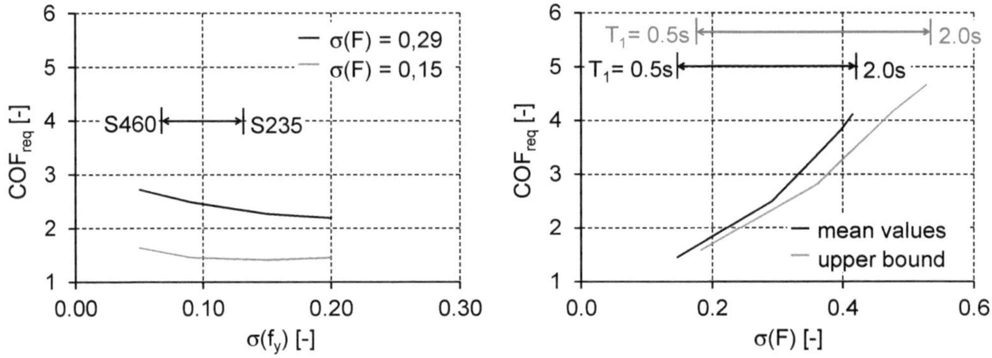

Figure 9: Required overdesign factor COF for steel frames under seismic action: variation of yield strength (left) and of action (right)

3.2 Rotation capacity

The available rotation capacity is evaluated by mechanical models presented by Feldmann (moment-rotation curve) and Ibarra-Medina-Krawinkler (cyclic degradation). The coefficient of variation is rather constant for different sections and steel grades (COV = 0.3). Failure is defined, if the rotation capacity is smaller than the maximum rotation in on of the outer columns or both internal columns or of all beams in one of the storeys. The rotation capacity check is analysed by fragility curves. Repeating fragility analysis for different rotation capacities yields the required rotation capacity required to reach the target reliability index of 2.0 (following [2]).

Figure 11 shows that a number of design parameters related to the seismic action and the type of the structure has a strong influence on the required rotation capacity. In contrast, the influence of the steel grade is rather small. It has to be

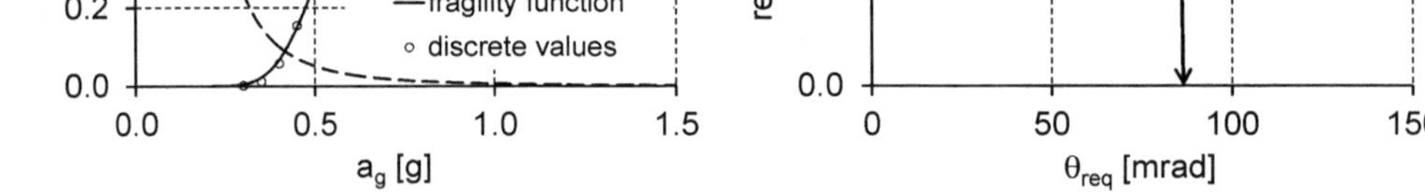

Figure 10: Fragility function (left), det. of required rotation capacity (right)

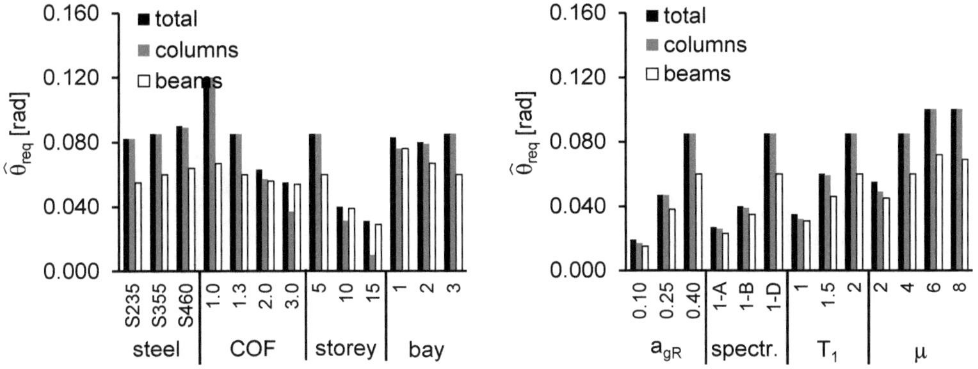

Figure 11: Required rotation capacity for various design parameter

mentioned that in the case studies the storey drift limitation according to DIN EN 1998-1 are not meet. The correlation between this design requirement and rotation demand is poor, which is an indication for the necessity to improve design rules for rotation capacity checks.

3.3 Non-dissipative connections

Statistical evaluation of tests on various components of connections leads to the conclusion that joints with failure of bolts in tension are the most crucial, brittle connections. Hence, each connection is described by a random variable representative for this failure mode ($\mu = 1.38$, $\sigma = 0.14$). Failure is defined, if the strength of the connection of any beam, of one of the outer column base or both internal column bases is smaller than its connection force. The fragility curve of non-dissipative joints is considerably different than fragility curves of the rotation capacity check (Figure 13). While for rotation capacity the failure probability increases continuously with the intensity level, for non-dissipative connections a

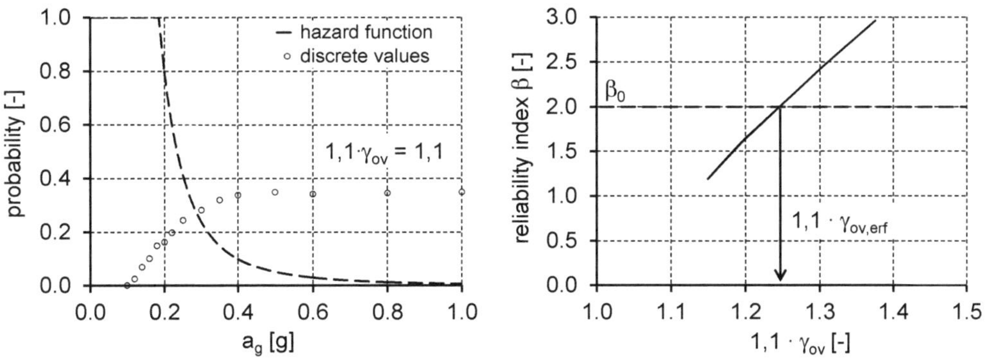

Figure 12: Fragility function (left) and required joint overstrength (right)

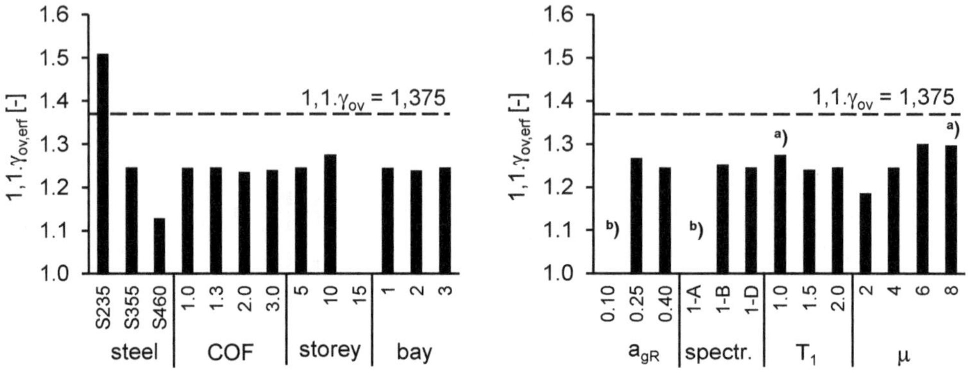

Figure 13: Required joint overstrength for various design parameter

plateau of failure probability is reached, which corresponds to the plastification of the adjacent member. The consequence is that all design parameters related to the seismic action has only a small influence on the reliability level. Hence, for non-dissipative connections the steel grade is the governing parameter.

4 Conclusions

In this paper results of reliability analysis on the capacity design rules in DIN EN 1998-1 for steel frames are presented. For this purpose a stochastic model for the seismic action in push-over analysis was derived. The probabilistic model for yield strength is based on current data from European steel. The main findings are:

- The type of plastic mechanism as well as the rotation demand is mainly influenced by the variation of the seismic action and not by the material strength scattering.
- The design of non-dissipative joints is mainly governed by the material strength distribution; the influence of the seismic action is negligible.

The recommend values to consider material overstrength in the design of non-dissipative connections are 1.05 for S460, 1.20 for S355 and 1.40 for S235.

REFERENCES

[1] Gündel, M.: Zuverlässigkeitsanalysen zur Kapazitätsbemessung von Stahlrahmen, PhD-thesis, Institute for Steel structures, RWTH Aachen, 2013

[2] FEMA 350: Recommended Seismic Design Criteria for New Steel Moment-Frame Buildings. Federal Emergency Management Agency, 2000

[3] Pinto, P. et al: Seismic reliability analysis of structures. Pavia: IUSS press, 2004

Dissipative Devices for Vulnerability Reduction of Precast Buildings

Marco Mezzi[1], Fabrizio Comodini[2], Leonardo Rossi[3]

[1] Dept. of Civil and Environmental Engineering, University of Perugia
via Duranti 93, 06125 Perugia, Italy
marco.mezzi@unipg.it

[2] University eCampus
via Isimbardi 10, 22060 Novedrate (CO), Italy
fabrizio.comodini@uniecampus.it

[3] PhD Student, University of Perugia
via Duranti 93, 06125 Perugia, Italy
leonardo.rossi1@studenti.unipg.it

ABSTRACT

The paper presents the development of a study on low cost seismic protection devices to put in place at the joints of prefabricated structural systems with the aim of improving their seismic response. In particular, this phase of the research focuses on the optimisation of protection devices used on two-dimensional mono- and multi-storey frames. A comparative analysis of the seismic response of the systems varying the mechanical characteristics of the devices was developed. The friction-type protective devices adopted were installed at the beam-column and column-foundation interfaces. The performed analyses show a significant improvement in seismic response, in terms of both reduction of stresses and increase of dissipative capacity.

Keywords: precast building, beam–column joints, seismic capacity

1 Introduction

It is now established that, in regard to seismic actions, the poor performance of prefabricated reinforced concrete buildings is mainly due to the inadequate capacity of the joints between the structural elements to dissipate energy and to ensure appropriate connection. This deficiency in some cases has contributed to the spread of mixed systems resulting from the coupling of prefabricated elements with cast-in-place structures. To the latter is usually assigned the task of supporting the majority of the seismic demand. An improvement of the seismic capacity of these prefabricated systems can be achieved by inserting suitable dissipation devices at

DOI 10.1007/978-3-658-02810-7_29, © Springer Fachmedien Wiesbaden 2014

the beam-column and column-foundation joints. The aspects to be analysed are of two types. The first concerns the optimisation of the calibration of the devices, to maximise the global dissipation capacity of the system; the second aspect relates to the alteration of the degree of constraint that the devices generate in the joints of the structural elements and the resulting redistribution of the stresses on them. Moreover, seismic protection devices can be also installed on existing structures allowing an improvement of seismic performance. This paper reports the main results of the research devoted to the study of the behaviour of rotational friction dissipators and their effectiveness on prefabricated systems in areas of high seismicity.

2 Friction based connections

The seismic response of a prefabricated structure is closely related to the type of beam-column and column-foundation joints. With the new Italian design codes for prefabricated structures, also known as NTC2008 [1] and Circolare n°617 [2], the need to understand the mechanical characteristics of the joints has increased; in fact, they play a key role in ensuring the development of ductile mechanisms. There are many types of connection devices currently proposed in literature and available on the market, as shown by Comodini & Mezzi [3]. The present study hypothesizes the use of a rotational friction device installed at the beam-column and at the column-foundation joints. The device adopted in the performed studies is a rotational friction dissipator capable of providing a semi-fixed joint constraint and energy dissipation (Morgen & Kurama [4]). The devices can be installed locally on the extrados of the beams at the beam-column joints and laterally to the columns at the column-foundation joints. The devices are activated through the relative rotations that are generated between the interconnected structural elements as a result of the lateral deformation of the system. The magnitude of the energy dissipated is thus closely related to the floor drift. The device showed in Figure 1 consists of five components in cast steel, with four friction interfaces, obtained with discs of lead-bronze inserted in the middle. The friction interfaces are pre-stressed by a spring clamp.

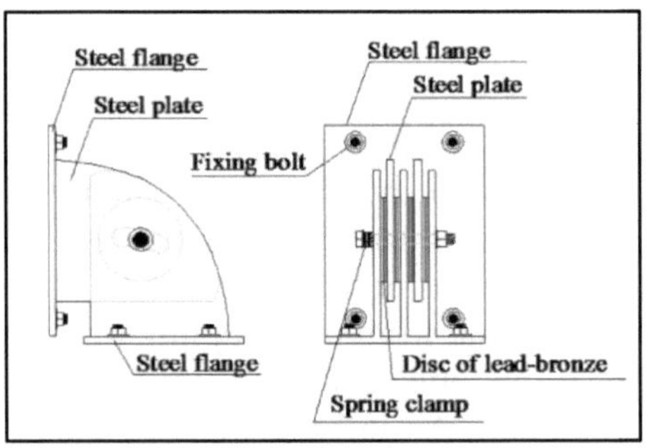

Figure 1: Rotational friction devices

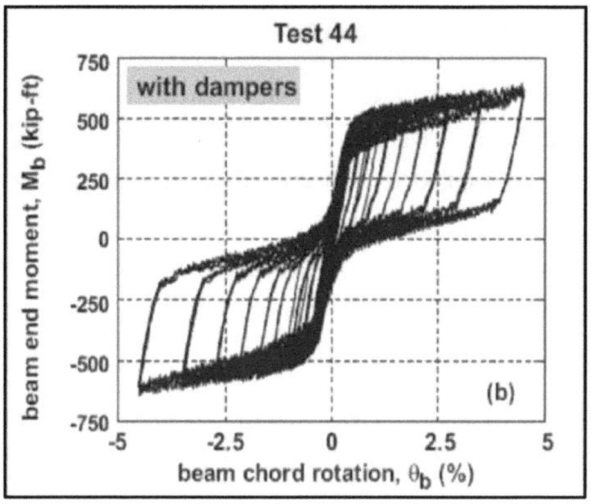

Figure 2: Experimental hysteretic loops of the friction devices (Morgen et al. 2004)

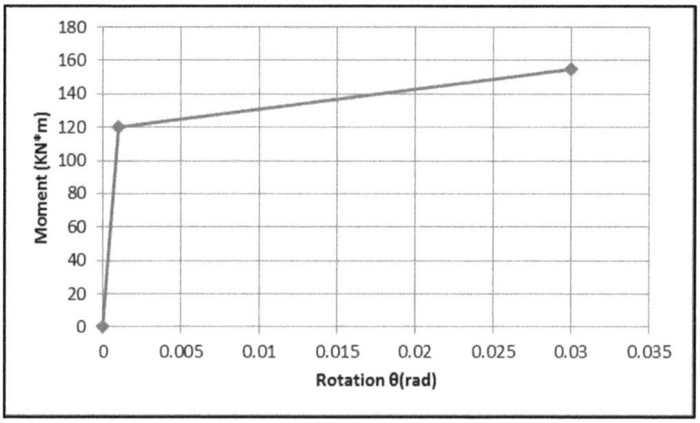

Figure 3: Moment-rotation relationship for friction devices

The seismic improvement of the prefabricated structure is pursued using passive supplemental dissipation of energy. Experimental tests (Morgen & Kurama [4]) carried out on a single device determined the hysteretic loop in terms of the moment-rotation relationship, as shown in Figure 2. The hysteretic loop of the device taken as the baseline for the first parametric analysis was traced back to an equivalent elastic-plastic model with hardening (Figure 3), characterised by the following parameters: M_y = 120 kN·m (yield moment); M_u = 155 kN·m (ultimate state moment); Θ_y = 0.1% (yield rotation); Θ_u = 3.0% (ultimate state rotation); K_{el} = 120000 kNm/RAD (elastic stiffness); K_p /K_{el} = 0.01 (post-elastic stiffness ratio).

The choice of the activation threshold and of the elastic stiffness of the device influences the transfer of bending stresses to the columns and the equivalent viscous damping associated with the loops of hysteresis described by the rotational-frictional behaviour of the devices.

3 Analysis of a bi-dimensional frame

The frame studied (shown in Figure 4) is made from monolithic 12.86 m high columns with variable section from 1300·1300 mm to 1300·1200 mm, pre-stressed hollow-core slabs with a constant 340 mm high section supplemented by a cooperative structural slab in reinforced concrete and prefabricated T and L-shaped beams in reinforced concrete. The optimisation procedure of the devices involves estimation of the rotations activation limit and the determination of the optimal values of elastic stiffness and plastic threshold of the same.

A basic condition for a correct operation of the devices is that the rotation activation limit must be higher than the nodal rotations induced by the vertical and horizontal loads of service. The variation of the constitutive model of the devices produces stress distributions that are different from those associated with a scheme of isostatic columns and beams simply recumbent. It is also clear that the calibration of the devices is closely related to the dynamic characteristics of the structure. Improper calibration of the devices can compromise the benefits of the installation of the same.

A first phase of numerical processing consisting of non-linear static analysis for vertical and horizontal loads was performed, considering three different configurations of multi-storey frame, two of which with dissipative devices. The goal of this first step was to estimate floor drift and the nodal rotations required to meet the seismic demand respectively for the damage limit state and the safeguarding of life with different structural configurations. Looking at Figure 5 and Figure 6, an important difference can immediately be seen. In the model without devices (Figure 5), the curve has a perfectly elastic-plastic trend due to the formation of plastic hinges at the base of the columns, which produces a fragile floor mechanism. This problem is completely avoided in the model with the devices (Figure 6).

The technology with which the devices are made makes it possible to modify the elastic stiffness and the plastic threshold increasing the pre-stressing force of the horizontal retaining screw. For devices to be inserted in the beam-column joints of the three-level frame, 21 constitutive models were generated; starting from the reference values, the stiffness and the plastic threshold were respectively: +50%, +100%, +150%, +200% and ±10%, ±20% and 35% (Figure 7 and Figure 8).

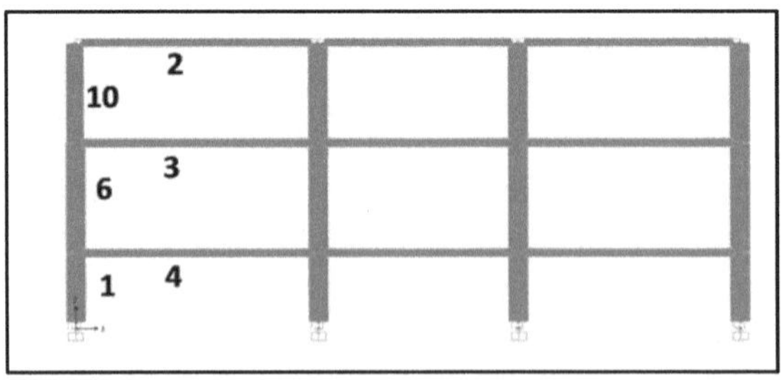

Figure 4: Sample model analysed

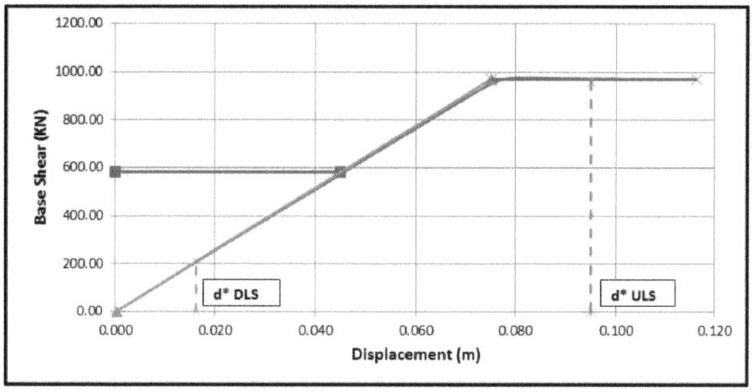

Figure 5: Capacity curve of the model without devices

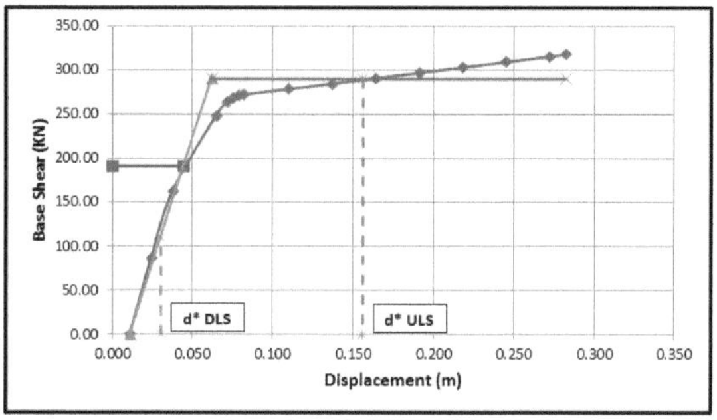

Figure 6: Capacity curve of the model with devices in all joints

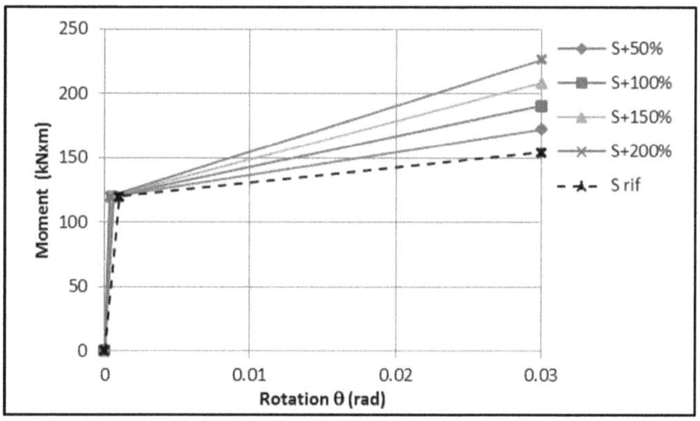

Figure 7: Moment-rotation relationship of the friction device for different stiffness

As for the devices inserted at the column-foundation joints, the following mechanical parameters were assumed: M_y = 360 kN·m (yield moment); M_u = 392 kN·m (ultimate state moment); Θ_y = 0.3% (yield rotation); Θ_u = 3.0% (ultimate state rotation); K_{el} = 120000 kNm/RAD (elastic stiffness) and K_p/K_{el} = 0.01 (post-elastic stiffness ratio). The latter were defined in order to postpone the activation of

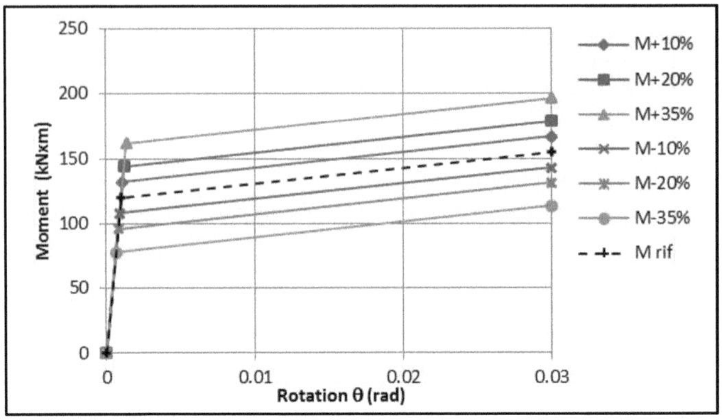

Figure 8: Moment-rotation relationship of the friction device for different elastic limits

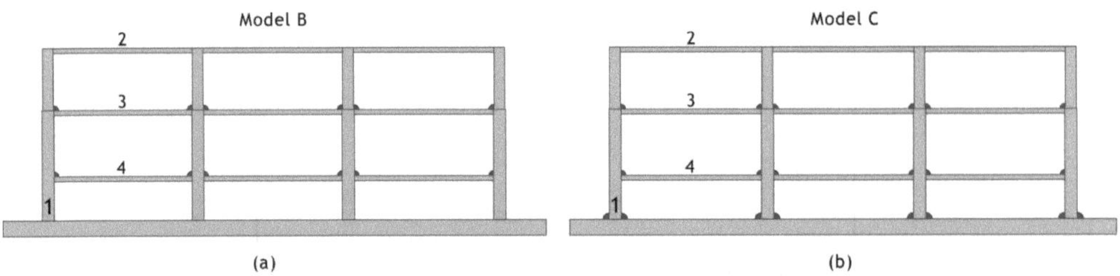

Figure 9: Schemes of models Type B and Type C

the base devices with respect to those of the beams and to ensure an adequate level of horizontal elastic stiffness. One of the objectives of the optimization procedure consists in the study of the distribution of stresses between the columns and beams for different values of the mechanical characteristics of the devices and also the identification of the optimum cycle of operation of the devices, in order to ensure greater energy dissipation. Two configurations were analysed (Figure 9): the first involved the insertion of the devices only at the beam-column joints (Model B) and the second also at the column-foundation joints (Model C).

The results obtained from models B and C were compared with those obtained for a reference model without devices (called Model A). From the results of numerical processing, carried out by varying the mechanical parameters of the devices, it was possible to compare the evolution of the bending stress in the columns and beams and the rotations required of the devices according to floor drifts. In particular, Figure 10 shows the trend of the maximum bending moment of column 1 for the three different configurations of the frame with variations in parameter K_{el}. The Figure 11 shows a detail which concerned only the results for column 1 of Model C to better understand the differences in the results.

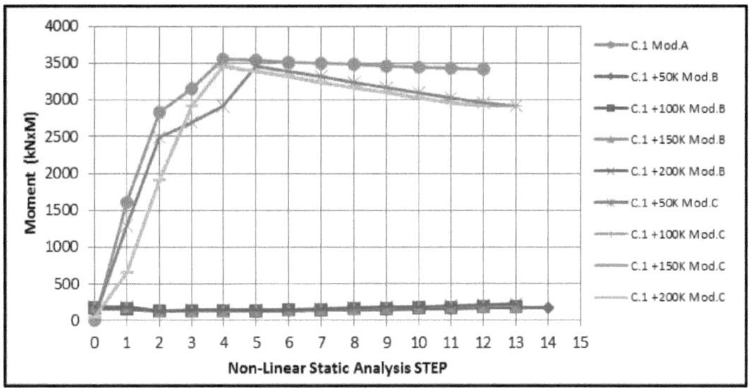

Figure 10: Moment-step diagram of the column n.1 for model A, B and C

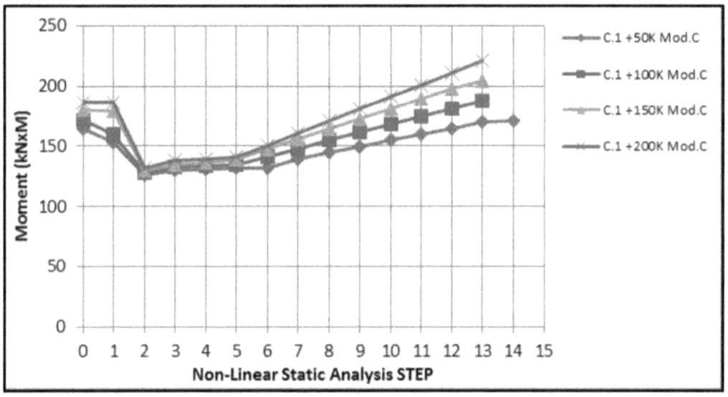

Figure 11: Moment-step diagram of the column n.1 for model C

Figure 12 and Figure 13 show the maximum values of the bending moment of beam 2 respectively for the left end and the right end. Even in this case, the results are parametrised according to elastic and post-elastic stiffness and refer only to models B and C, because the beams are hinged in model A.

In absolute values, there is a clear variation in the maximum bending stress. This is due to two main factors: the first is related to the presence of the devices at the beam-column joints that produce a semi-fixed joint constraint at the end of the beams, reducing the bending action on the columns. The second factor is associated with the insertion of the devices at the foot of the columns such as to reduce the degree of rotational constraint. Therefore, the model with the base device has a greater lateral deformability and its modes of vibration are characterised by greater periods. Model C will therefore be subjected to a base seismic shear that is smaller than in models A and B. The devices with elastic stiffness increased by 200% are able to override the negative rotations produced by static loads of service and to fully develop the assigned hysteresis loop. The definition of the plastic threshold appears to be a secondary concern because the increases and decreases of it do not produce significant changes in the bending moment. On the basis of the obtained

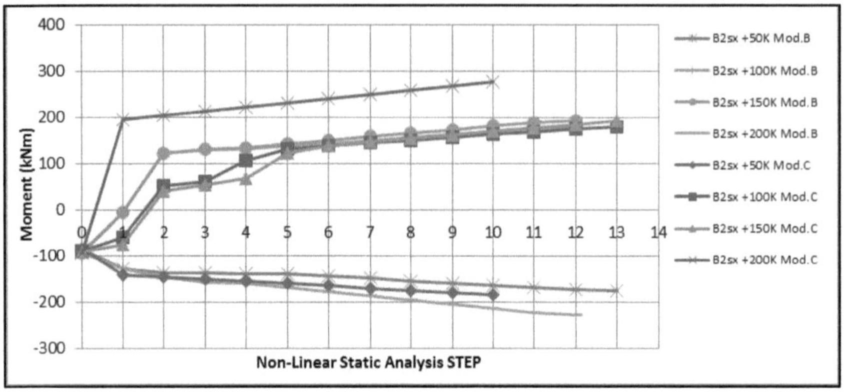

Figure 12: Moment-step diagram of the left end of beam n.2 for model B and C

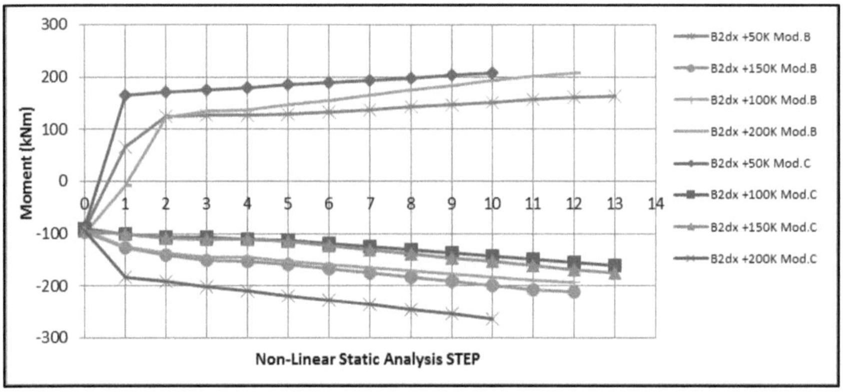

Figure 13: Moment-step diagram of the right end of beam n.2 for model B and C

results, it can be concluded that the optimum cycle must be characterised by a high elastic stiffness and a plastic threshold defined according to the strength characteristics of the beam and the column. The optimal configuration is that in which the devices are put in place also at the base of the structure, because this produces a greater absolute reduction of bending actions in the columns.

4 Non-Linear dynamic analysis

The non-linear dynamic analysis was conducted for models A and C, with the aim of verifying the effective cyclic operation of the devices and the benefit provided by them in terms of energy dissipation and stress reduction at the base of the columns. The non-linear dynamic analyses were performed using the software SAP2000NL [5]. The hysteretic loops of the devices adopted were modelled using a "Plastic Wen" type NLLink element. The mechanical parameters that characterize the non-linear element for the beams are the following: M_y = 162 kN·m (yield moment); M_u = 268 kN·m (ultimate state moment); Θ_y = 0.045% (yield rotation); Θ_u = 3.0% (ultimate state rotation); K_{el} = 360000 kNm/RAD (elastic stiffness) e K_p/K_{el} = 0.01 (post-elastic stiffness ratio). The constitutive model adopted earlier for parametric analyses remains unchanged for the columns. The choice to assign a plastic threshold lower than that of the columns to the devices of the beams is motivated by the opportunity to ensure their early activation, preventing the transfer of high bending stresses to the columns. The increase of the equivalent viscous damping associated with the behaviour defined for the dissipative element is equal to 35.0% of the critical value.

5 Results for the multi-storey frames

The non-linear dynamic analysis highlighted the cyclic operation of the rotational friction devices. The passive energy dissipation introduced by the devices produced a significant reduction of the bending stresses in the columns. The analyses show a different functioning between the devices inserted at the beam-column joints and those inserted at the base of the columns. In fact, the devices at the base, with higher plastic threshold and lower elastic stiffness are activated by greater rotations and produce hysteresis loops of lower amplitude than the devices placed on the beams. This result is in line with the original project. With the maximum stresses obtained from the non-linear dynamic analysis, sizing of sections and reinforcement respectively for the model without devices and for the model with type C devices setup was conducted. A considerable difference in terms of geometrical dimensions and reinforcement is evident. This means that it is possible to reduce the stresses in the columns without need to use mixed structural systems.

6 Single storey frames case study

It is interesting to evaluate the use of friction devices in typical structural schemes for industrial buildings. For this reason, a preliminary study of application to a single-storey frame with precast elements in reinforced concrete is conducted. The structural scheme of the frames comprises two isostatic columns fixed to the ground and a beam placed between them. Two calculation models, one equipped with a friction device, were prepared using SAP2000NL software [5]. The friction devices schematised in frame (b) of the Figure 14 are modeled through a "Plastic Wen" type "NLLink" element. For this, an elastic-plastic hardening type diagram is assumed, with hardening ratio K_p/K_{el} equal to 0.01, yield rotation of 0.045% and ultimate rotation Θ_u of 3.0% RAD. The stiffness K_{el} is equal to 360000 kNm/RAD and the yield moment M_y is 162 kN·m, and the ultimate state moment Mu is 268 kNm.

Figure 15 shows how the system works. For moderate values of seismic demand, the stresses on the columns are lower compared with the cantilever diagram, as a result of the semi-rigid behaviour of the beam-column joint. At high values of seismic demand, the elastic deformation limit is exceeded in the devices and dissipative cycles are completed.

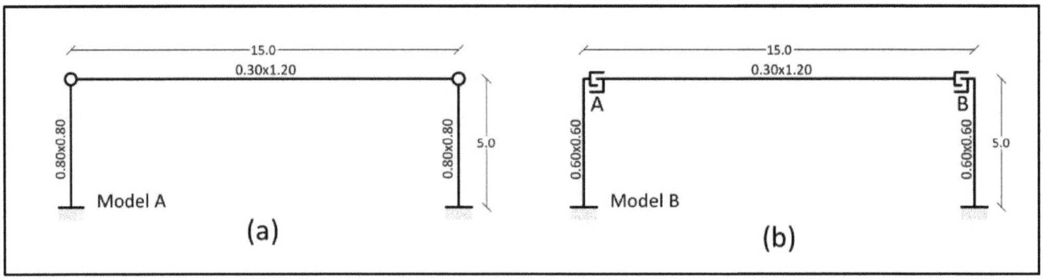

Figure 14: Geometric schemes; (a) beam with hinges; (b) system with devices

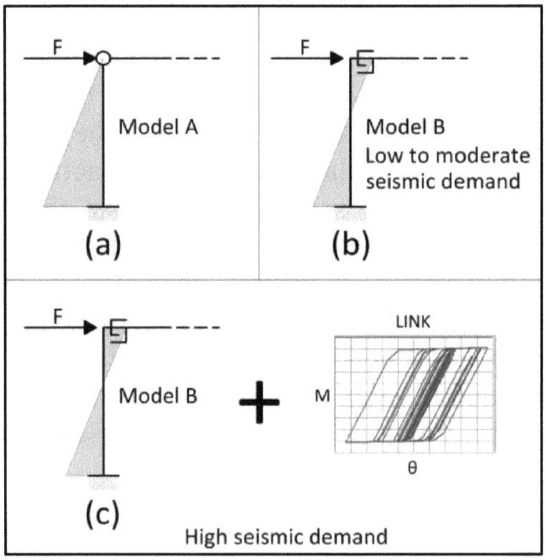

Figure 15: How the system works

For the initial sizing of the structural elements, dynamic analysis with response spectrum was used, according to the state of the art design principles (see Parducci [6]) and Italian design guidelines (NTC2008 [1] and Circolare n°617 [2]). On models case study was performed nonlinear dynamic analysis with direct integration, using three spectrum compatible accelerograms (Iervolino, Maddaloni and Cosenza [7]). The reference site for the design is L'Aquila (Italy). The design value of the bedrock acceleration is equal to 0.261 g. The amplification factor S for the site is equal to 1.33. Rotation-moment cycles on friction devices (see the Figure 16) were obtained from non-linear dynamic analysis. The graph shows that the devices trigger the plastic phase, forming dissipative cycles.

The bending stresses on the columns are reduced by about 50% where devices are installed. When hinge constraints are present on the beams, the stresses are such as

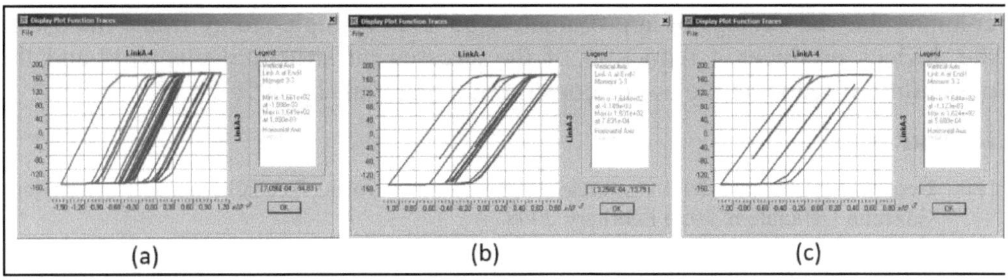

Figure 16: Rotation-bending moment diagrams for the link A; (a) dynamic analysis 1; (b) dynamic analysis 2; (c) dynamic analysis 3

to require a side section of 800 mm with reinforcement ratio $\rho = 1.18\%$ for the columns. In the model with friction devices, it is deemed sufficient to adopt a 600 mm side section with reinforcement ratio $\rho = 1.17\%$. In this case, the use of friction devices makes it possible to reduce the volume of concrete required for the columns by about 43% and the quantity of steel by about 44%.

7 Conclusions

The objectives of this work are optimisation of the mechanical parameters of rotational friction devices and evaluation of the effectiveness of the inclusion of them also at column-foundation joints for the seismic protection of precast buildings. With reference to a bi-dimensional frame consisting of only prefabricated elements, a first phase was to study the behaviour of individual devices, parametrising elastic stiffnesses and plastic thresholds. From the comparison of the results it was possible to identify the parameters that define the optimal cycle of operation of the devices. In the second phase, a study of the global response of the bi-dimensional frame was conducted using non-linear dynamic analysis, comparing the results of the frame without devices and the frame with type C devices setup. The non-linear dynamic analyses performed on the structure showed that it is possible to reduce the stresses in the columns through the insertion of devices at the base of them, with an appropriate definition of hysteresis loops differentiated from those of the beams. The structure with devices at the base is more deformable for seismic actions, and although the natural consequence is the increase of energy dissipated by the devices, to keep the floor drift below the thresholds allowed by the regulations it is necessary to calibrate the cycles of the devices also in relation to the latter parameter. The insertion of these rotational dissipators produces a benefit in terms of reduction of repair costs and non-use costs following a seismic event. As regards the construction of a new building, the devices assumed are low-cost, their operation is of mechanical type and their installation and maintenance do not require special precautions. Expeditious calculations of comparison for assessing the reduction in the cost of construction due to the installation of the devices were carried out: the comparison related to the columns showed savings of approximately 35%. In a single-storey frame the use of the devices produces a reduction of the stresses on structural elements and the possibility of reducing the use of construction materials by 40% circa.

References

[1] Decreto Ministeriale del 14 gennaio 2008 - “Nuove Norme Tecniche per le Costruzioni”, Gazzetta Ufficiale n. 29, 4 February 2008

[2] Circolare 2 febbraio 2009, n. 617 C. S. LL. PP. -”Istruzioni per l'applicazione delle «Nuove norme tecniche per le costruzioni» di cui al decreto ministeriale 14 gennaio 2008”. Gazzetta Ufficiale n. 47, 26 February 2009

[3] Comodini F., Mezzi M., Ricerca di sistemi di protezione appropriati per schemi prefabbricati ricorrenti, Eda-esempi di architettura on-line, 2012, ISSN 2035-7982

[4] Morgen B.G. and Kurama Y., A friction damper for post-tensioned precast concrete beam-to-column joints, 23th World Conference on Earthquake Engineering Vancouver, B.C., Canada, August 1-6, 2004

[5] CSI, Computers and Structures Inc., Analysis Reference Manual Berkeley, California, USA, 2005

[6] Parducci A., Progetto delle costruzioni in zona sismica, Liguori Editore, 2005

[7] Iervolino I., Maddaloni G., Cosenza E., Accelerogrammi Naturali Compatibili con le Specifiche dell’OPCM 3274 per l’Analisi Non Lineare delle Strutture. RELUIS Rete dei Laboratori Universitari di Ingegneria Sismica, 2006

Seismic Performance of Concrete-Filled Steel Tubular (CFST) Structures

Lin-Hai Han*, **Wei Li**

Department of Civil Engineering, Tsinghua University
Beijing, 100084, China
* lhhan@tsinghua.edu.cn

ABSTRACT:

Concrete-filled steel tube (CFST) consists of outer steel tube and concrete in-filled, which combines the merits of steel and concrete. This kind of composite member has various advantages, i.e., high strength and high ductility, favorable cyclic behaviour, high fire resistance and excellent constructability, have been recognized all over the world. Nowadays CFST has been widely used in construction, including many industrial facilities. This paper gives a brief review on the investigations of seismic behaviour of CFST members, joints, planar frames, hybrid walls and high-rise buildings, especially in China. The development of concrete-filled steel tubular members' family is introduced. Some industrial projects utilizing CFST members are also presented.

Keywords: Concrete-filled steel tube, Members, Joints, Structural systems, Seismic behaviour

1 Introduction

In concrete-filled steel tubular (CFST) members, steel and concrete are used such that their natural and most prominent characteristics are taken advantage of. The behaviour of the composite member will be better than the simple combination of two materials. In addition, there is no need for the use of shuttering during concrete construction, and the construction cost and time are therefore reduced. These advantages have been widely recognized and have led to the extensive use of CFST structures [1].

Fig. 1(a) shows a typical CFST cross-section, where the concrete is filled in a circular hollow section (CHS). The square hollow section (SHS) and the rectangular hollow section (RHS) are also widely used in construction. Other cross-sectional shapes have also been used for esthetical purposes, such as polygon, round-ended rectangular and elliptical shapes. Besides the common concrete-filled steel tubes, there are other types of "general" member designation in the CFST family. Some of them are shown in Fig. 1 as follows: concrete-filled double skin

S. Klinkel et al. (Eds.), *Seismic Design of Industrial Facilities*,
DOI 10.1007/978-3-658-02810-7_30, © Springer Fachmedien Wiesbaden 2014

steel tube (CFDST) (Fig. 1 (b)) [2], concrete-encased concrete-filled steel tube (Fig. 1 (c)) [3], reinforced concrete-filled steel tube (Fig. 1 (d)). Besides being used as single elements in construction, various combinations of concrete-filled steel tubular members are also used. For instance, the hollow steel tubes can be used to form a latticed member, as shown in Fig. 1 (e) [4]. Moreover, due to architectural or structural requirements, inclined, tapered or non-prismatic members have been used [5]. Research results for these columns have shown that the steel tube and the concrete can work together well, despite the inclined angle, the tapered angle or the curvature of the member.

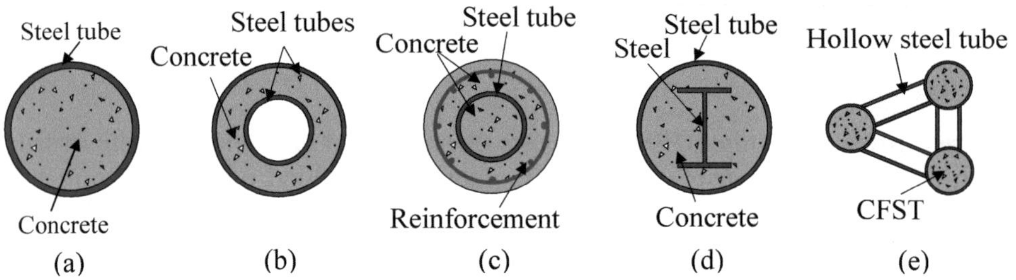

Figure 1: General CFST cross sections

Some recent research work on the seismic performance of CFST structures in China is summarized in this paper. The investigations on members, joints, planar frames, hybrid walls and high-rise buildings are reviewed. Some examples of industrial facilities using CFST structures are also presented.

2 Seismic performance of CFST structures

2.1 Members

Numerous investigations have been conducted for concrete-filled steel tubes under cyclic loading, and several state of the art reports or papers were also published on CFST structures [1]. It has been demonstrated that this kind of composite member has excellent ductility and energy dissipating capacity.

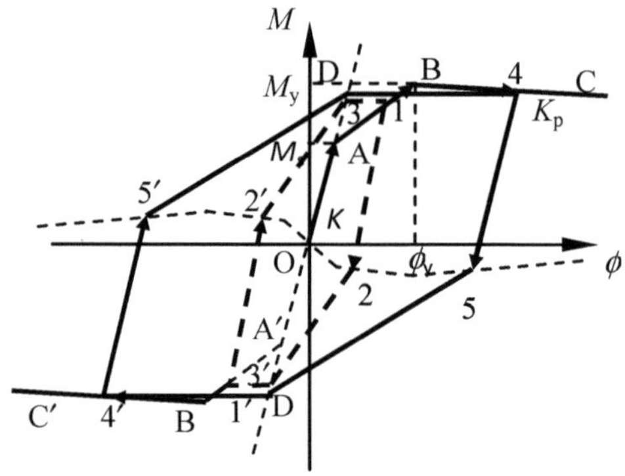

Figure 2: Moment versus curvature relationship for circular CFST [7]

For the moment versus curvature response and the lateral load versus lateral displacement relationship, hysteretic models were proposed for the cyclic response based on parametric studies, as shown in Fig. 2. Key parameters such as axial load level, steel ratio, slenderness ratio and material strength were studied. The results from theoretical models showed a good agreement with the test ones (with a difference less than 12%) [6][7].

As the general CFST members are used in structures in seismic regions, studies on the cyclic behaviour of concrete-encased CFST members, FRP-concrete-steel members, and CFDST members were also conducted [3][8][9]. In general these columns exhibited a good ductility and favorable energy dissipation capacity.

2.2 Joints

A proper connection details plays an important role in the structural system. The "weak beam-strong column" concept is adopted in various seismic design codes in different countries. In the past, some research has been conducted on steel beam to CFST column joints, which involved experimental studies to assess the elasto-plastic behaviour of the composite joints, and were reviewed by Han and Li [10].

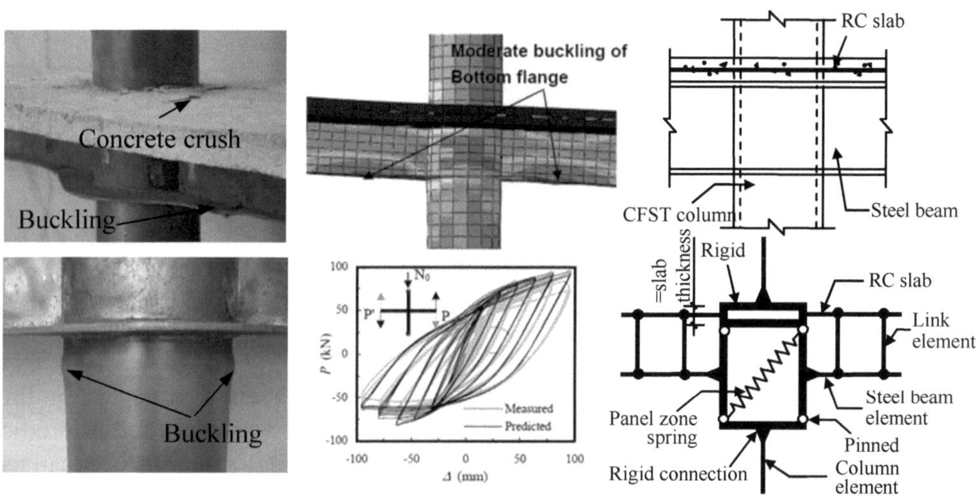

Figure 3: Research on CFST joint [10][11][12]

For composite joints consisted of circular CFST columns and steel beams, Han and Li [10] conducted experimental investigation on the joint seismic behaviour where the reinforced concrete slab was attached, as shown in Fig. 3. Experiments were carried out on the composite joints with constant axial load on the top of column and reverse cyclic loading at the ends of beams. The results showed that the load versus deflection curves were plump, and stable strength and stiffness degradations were observed under cyclic loading. Nonlinear finite element analysis (FEA) was also conducted [11]. The accuracy of the FEA model was verified by extensive experimental results. The failure modes, force transfer mechanism, force versus deformation relations of the composite joints were analyzed by the FEA model.

For the macro joint model used in the structural system analysis, Li and Han [12] proposed a joint macro model for the CFST column to beam joint with RC slab, as shown in Fig. 3. A shear versus shear deformation hysteretic relation for the panel zone was established based on the parametric analysis, and then it was implanted in this macro model. It is concluded that the proposed hysteretic relation and the joint macro element had a favorable accuracy when compared with the FEA and experimental results.

2.3 Planar Frame

Composite frames using CFST columns are being used more and more popularly in building structures, which is owed to the excellent earthquake-resistant and fire-resistant properties of the column. The CFST column is usually connected to the steel beam in the structural system (named CFST frame in this paper).

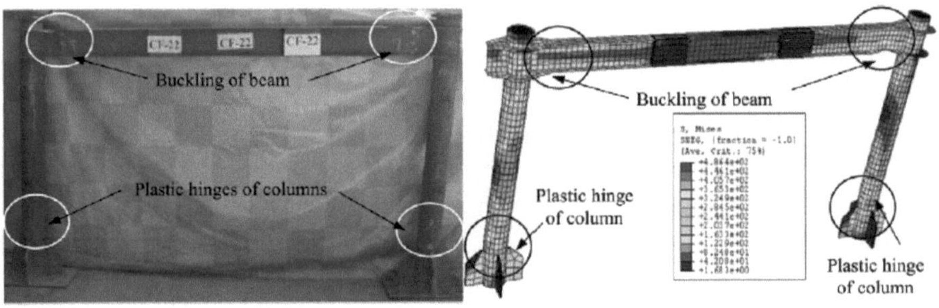

Figure 4: Failure modes of CFST frame (Adopted from Han et al. 2011[13])

Experimental as well as numerical investigations have been conducted for this particular kind of composite frame. An accurate FE model was proposed to predict the frame behaviour under lateral load, and the experimental research was carried out to study the frame behaviour under the cyclic loading, as shown in Fig. 4 [13]. The results showed that this kind of composite frame had an excellent seismic resistance and the beam failure mode was expected when the weak-beam-strong-column design criteria was used. The lateral load-carrying capacity, ductility coefficient and the energy dissipation capacity decreased when the column axial load level increased from 0.07~0.6. Simplified hysteretic models for lateral load versus lateral displacement relationship were also proposed for composite frames, which will be useful in the dynamic analysis of CFST structures [14][15].

2.4 Hybrid shear wall

In the CFST hybrid structural systems, the shear walls can also be built after the composite frame was established. The CFST frame served as the outer boundary of the shear wall panels. The overturning moment can be resisted by the frame and the cracking development of RC wall can be restrained. Besides, the CFST columns can still resist part of the lateral load and considerable axial load after the RC wall deteriorated.

Tests results showed that the hybrid shear wall exhibited a shear-dominant failure mode, as shown in Fig. 5 [16]. The measured load versus deformation hysteretic curves showed an obvious pinch effect due to the deterioration of the RC shear wall. The deformation capacity of this hybrid structures could meet the Chinese code's requirements for seismic design, and the CFST columns and RC shear wall can work together well by using U-shaped connectors.

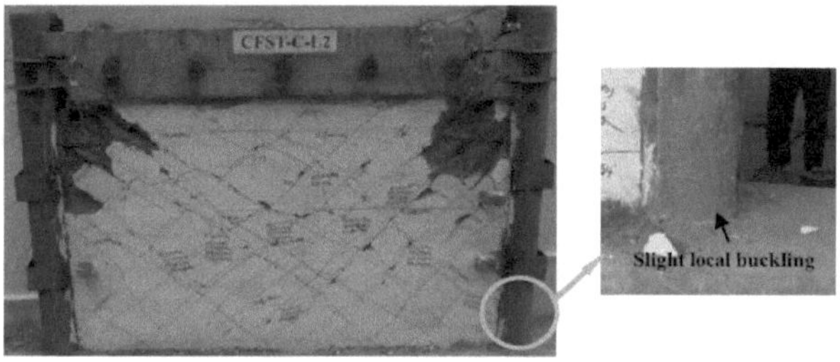

Figure 5: Failure mode of Hybrid wall. (Adopted from Liao et al. 2009 [16])

2.5 Hybrid structural system

In high-rise buildings or super high-rise buildings, the CFST composite frame structures are often combined with other lateral load resisting systems such as reinforced concrete or steel shear walls or core tubes. The frame using concrete-filled steel tubular columns integrates high stiffness and high ductility, and works well with the shear walls or core tubes in hybrid structural systems. The RC core walls can be built several storeys before the frame installation to accelerate the construction speed. Shaking table tests have been performed for the CFST frame and RC core tube hybrid system, as shown in Fig. 6. Two building models with 30 storeys were tested under various earthquake excitations [17]. Each building model had 20 CFST columns, and the difference between two models was the cross-sectional type of CFST columns, i.e. circular and square respectively. The results showed that the first order damping ratios of the building models range from 3.0% to 3.5% before the earthquake excitations. The second order damping ratio is about 2.5%~3%. The first order damping ratios range from 3.5% to 4% after 0.6g earthquake excitations. The

Figure 6: Shaking table tests of CFST hybrid systems [17]

frames using circular and square CFST columns both exhibited the excellence of high stiffness and outstanding ductility, and cooperated well with the core wall in the high-rise hybrid structural system.

3 CFST used in industrial facilities

Concrete-filled steel tubular columns have been used in China for almost 50 years. They have been used in numerous buildings, bridges and other structures, including many industrial facilities. The high resistance, high stiffness and favorable dynamic behaviour of CFST members met the requirements of heavily loaded industrial facilities. When compared to steel structures, less steel can be used for CFST structures, and the fire resistance will be better. When compared to reinforced concrete structures, the fast-built construct ability of CFST structures can save the time as well as the cost.

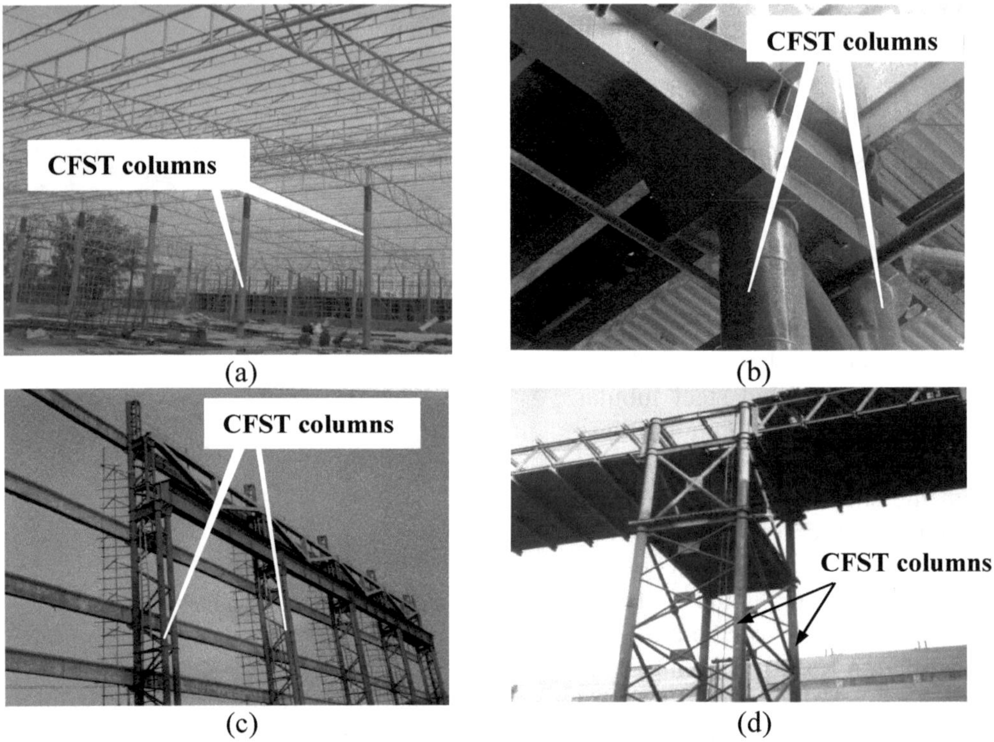

(a) (b) (c) (d)

Figure 7: CFST members used in industrial facilities [1]

The concrete-filled steel tube has been used in industrial facilities in the north of China since 1970s. The column usually resists axial load and bending in workshop or industrial buildings. If the single column is applied, the column is an eccentrically-loaded one. Therefore built-up CFST members are popular in workshop buildings. Each column in the built-up member is close to an axially-loaded member. The number of the longitudinal column elements depends on the load resistance requirement.

Fig. 7 (a) shows a workshop using the single CFST column as the main support. Fig. 7 (b) shows the latticed CFST columns used in a power plant workshop. The hollow steel tubes were used as the lacing strut. The steel used in CFST column was only 55% of the pure steel column in similar workshops. Fig. 7 (c) shows a shipyard under construction, where triangular latticed CFST columns were used. Fig. 7 (d) shows a photo of coal trestle using CFST members, where latticed members with four longitudinal column elements and lacing bars were used.

On the other hand, it is well known that industrial facilities may be subjected to other aggressive environmental conditions such as the corrosion. Therefore the structural life-cycle performance should also be taken consideration of during the design. Some primary research has been conducted, and it was important to include all the loading and environmental conditions in the analysis [18].

4 Concluding remark

The scope of "concrete-filled steel tube" has been extended greatly by researchers and engineers. In general, the concrete-filled steel tubular structures have favorable ductility and energy dissipation capacities, and are suitable for the structures in seismic regions. Simplified hysteretic models for load-deformation relationships were developed for CFST members, joints and frames. When compared to reinforced concrete and steel structures, the CFST structures have their own advantages, and could be used in industrial facilities in earthquake-prone areas.

5 Acknowledgements

The research reported in the paper is part of Projects 51178245 and 51208281 supported by National Natural Science Foundation of China (NSFC), as well as the Tsinghua Initiative Scientific Research Program (No. 2010THZ02-1).

References

[1] Han, L.H.: Concrete-filled steel tubular structures-theory and practice (Second edition); China Science Press; 2007 (in Chinese).

[2] Zhao, X.L.; Han, L.H.: Double skin composite construction; Progress in Structural Engineering and Materials; Vol. 8, Issue 3 (2006), Page: 93-102.

[3] Han, L.H.; Liao, F.Y.; Tao, Z.; Hong, Z.: Performance of concrete filled steel tube reinforced concrete columns subjected to cyclic bending; Journal of Constructional Steel Research; Vol. 65, Issue 8-9 (2009), Page: 1607-1616.

[4] Han, L.H.; He, S.H.; Zheng, L.Q.; Tao, Z.: Curved concrete filled steel tubular (CCFST) built-up members under axial compression: Experiments; Journal of Constructional Steel Research; Vol.74, Issue 7(2012), Page: 63-75.

[5] Han, L.H.; Ren, Q.X; Li, W.: Tests on inclined, tapered and STS concrete-filled steel tubular (CFST) stub columns; Journal of Constructional Steel Research; Vol. 66, Issue 10 (2010), Page: 1186-1195.

[6] Han, L.H.; Yang, Y.F.; Tao, Z.: Concrete-filled thin walled steel RHS beam-columns subjected to cyclic loading; Thin-Walled Structures; Vol. 41, Issue 9 (2003), Page: 801-833.

[7] Han, L.H.; Yang, Y.F.: Cyclic performance of concrete-filled steel CHS columns under flexural loading; Journal of Constructional Steel Research; Vol. 61, Issue 4 (2005), Page: 423-452.

[8] Han, L.H.; Tao, Z.; Liao, F.Y.; Xu, Y.: Tests on cyclic performance of FRP–concrete–steel double-skin tubular columns; Thin Walled Structures; Vol. 48, Issue 6 (2010), Page: 430-439.

[9] Han, L.H.; Huang, H.; Zhao, X.L.: Analytical behaviour of concrete-filled double skin steel tubular (CFDST) beam-columns under cyclic loading; Thin-Walled Structures; Vol. 47, Issue 6-7 (2009), Page: 668-680.

[10] Han, L. H.; Li, W.: Seismic performance of CFST column-to-steel beam joint with RC slab: experiments; Journal of Constructional Steel Research; Vol. 66, Issue 11 (2010), Page: 1374-1386.

[11] Li, W.; Han, L. H.: Seismic performance of CFST column-to-steel beam joints with RC slab: analysis; Journal of Constructional Steel Research; Vol. 67, Issue 1 (2011), Page: 127-139.

[12] Li, W.; Han, L. H.: Seismic performance of CFST column-to-steel beam joint with RC slab: joint model; Journal of Constructional Steel Research; Vol. 73, Issue 6 (2012), Page: 66-79.

[13] Han, L.H.; Wang, W.D.; Zhao, X.L.: Behaviour of steel beam to concrete-filled SHS column frames: finite element model and verifications; Engineering Structures; Vol. 30, Issue 6 (2008), Page: 1647-1658.

[14] Wang, W.D.; Han, L.H.; Zhao, X.L.: Analytical behavior of frames with steel beams to concrete-filled steel tubular column. Journal of Constructional Steel Research; Vol. 65, Issue 3 (2009), Page: 497-508.

[15] Han, L.H.; Wang, W.D.; Tao, Z.: Performance of circular CFST column to steel beam frames under lateral cyclic loading; Journal of Constructional Steel Research; Vol. 67, Issue 5 (2011), Page: 876-890.

[16] Liao, F.Y.; Han, L.H.; Tao, Z.: Seismic behaviour of circular CFST columns and RC shear wall mixed structures: Experiments; Journal of Constructional Steel Research; Vol. 65, Issue 8-9 (2009), Page: 1582-1596.

[17] Han, L.H.; Li, W.; Yang, Y.F.: Seismic behaviour of concrete-filled steel tubular frame to RC shear wall high-rise mixed structures; Journal of Constructional Steel Research; Vol. 65, Issue 5 (2009), Page: 1249-1260.

[18] Li, W.; Han, L.H.; Zhao, X.L.: Numerical investigation towards life-cycle performance of concrete filled double skin tubular (CFDST) columns; in: First Conference on Performance-based and Life-cycle Structural Engineering; 2012, Page 17 (Full paper in flash disk).

System Identification of Industrial Steel Building Based on Ambient Vibration Measurements and Short Time Monitoring

Sergey Churilov[1], **Simona Markovska**[1], **Elena Dumova-Jovanoska**[1], **Goran Markovski**[1]

[1] Ss. Cyril and Methodius University, Faculty of Civil Engineering-Skopje
blvd. Partizanski odredi 24, P.O. Box 560, Skopje, Macedonia
curilov@gf.ukim.edu.mk

ABSTRACT

This paper presents a case study for identification of the dynamic characteristics of an industrial building with flexible steel moment resisting frame system based on ambient vibration measurements and short time monitoring. The field tests were conducted after detection of damage on non-structural separation masonry wall in the building with the intention to identify the properties of the building and detect possible sources of extreme operational conditions that lead to appearance of cracks in the walls. The accelerations, displacements and variations of temperature inside the building were monitored for 5 days via real time online monitoring system. The results revealed presence of continuous, but low level accelerations with different intensities throughout the building, significant variations of the relative displacements of the cracked wall in relation to the floor system and negligible variation of temperature.

Keywords: system identification, ambient vibration, monitoring, in-situ test, cracks

1 Introduction

The authorities of a industrial facility that operates in the fields of emission control technologies for the automobile industry detected certain damage in the non-structural elements in one of their production halls. The preliminary on-site visual inspection revealed cracks in the infill walls of the structural system without any indicative crack orientation. It was found out that the cracks appeared in the relatively large aerated concrete infill walls, with significant existence in the central part of the walls, as well as in the contact regions of the walls and the structural system. The main structural system of the industrial hall is flexible steel moment resisting frame. Visible cracks were also detected in the floor structure on the ground floor near the supporting columns in the west part of the facility.

S. Klinkel et al. (Eds.), *Seismic Design of Industrial Facilities*,
DOI 10.1007/978-3-658-02810-7_31,

With the intention to identify the sources that lead to occurrence of cracks and to establish a level of structural safety and stability, an experimental programme was launched. Considering the operational conditions in the facility and the limitations imposed by the management staff not to disturb the weekly established production cycle, it was decided that it was not possible to perform destructive testing. A solution to diagnose the problem and to understand the actual structural behaviour was to apply dynamic structural health monitoring of the building.

2 Industrial facility building

The existing structure of the facility is located in Skopje industrial area. The whole industrial complex was build in 2009 and consists of several production halls and supporting buildings and covers production area of about 18.000 m^2. The main structural system is steel moment resisting frames with lightweight concrete panels, steel and aluminium sandwich panels for glazing and façade walls and steel sandwich panels on the roof.

All buildings are structurally and functionally independent and have different geometry parameters in plan and height. Due to their plan dimensions some of the buildings are additionally divided in a number of structural parts. A subject of the performed investigations was one structural part of a three part production hall, see Figure 1.

The structural system is spatial steel frame structure, composed from steel columns, beams and plane braces. The columns have complex shape, created by welding two wide flange cross-sections 2xHEA600A at an angle of 90^0.

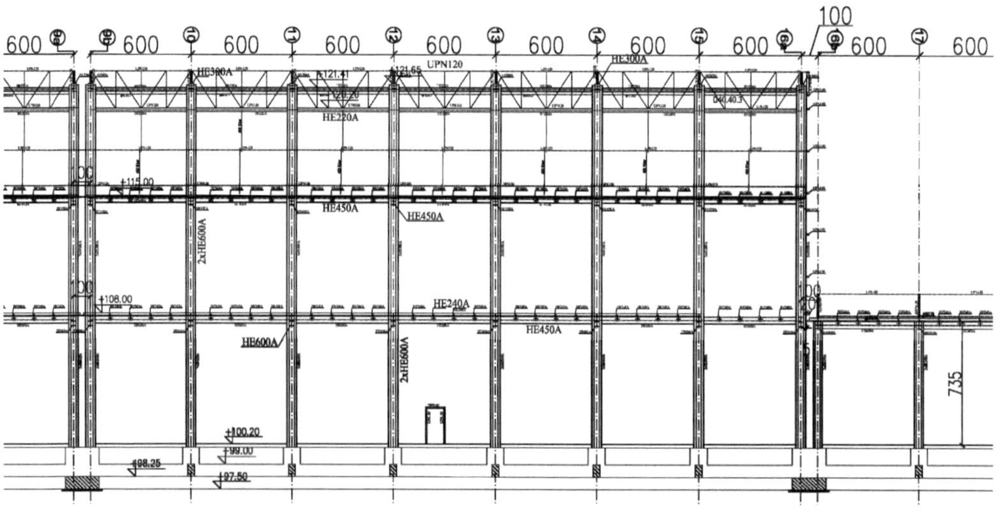

Figure 1: Lateral section of the production hall

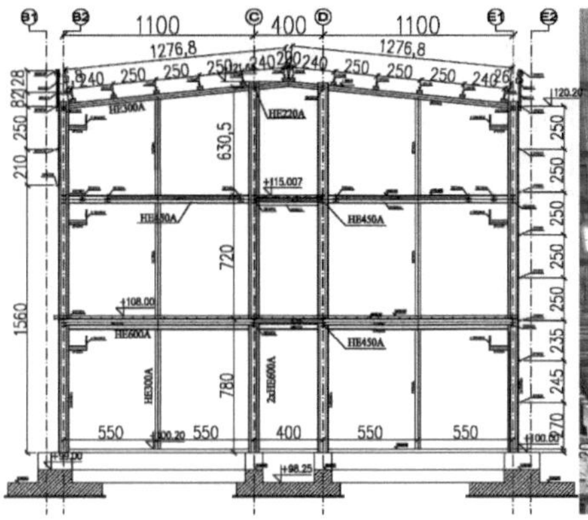

Figure 2: Transversal section of the production hall

Figure 3: Top floor view of the inspected building

The primary beams in the transversal direction and longitudinal section are HE600A and HE450A. The greatest part of the secondary beams are constructed with cross-section HE360A, see Figure 2. The floor system on the first floor is created by overlying reinforced concrete slab with total thickness of 200 mm over corrugated trapezoidal sheets. The second floor system is constructed differently by laying plain steel plates with thickness of 12 mm directly on the steel beams. The roof structure is constructed from roofing sheets, see Figure 2.

Few structural spans are enclosed by infill walls to create separate rooms. The infill walls were constructed by aerated concrete blocks and cement-lime mortar.

The building in the longitudinal direction is composed of 7 modules with spans of 6000 mm, while in transversal direction the frame has variable spans of 11000/4000/11000 mm.

The foundation system is a combination of strip foundations along the longitudinal direction and tie beams in the transversal direction which directly support the ground floor system.

3 Preliminary inspection and damage description

The initial inspection in the building was performed by visual identification of the detected damage. Neither photographic documentation nor mapping of the cracks was performed due to the strict policy rules of the facility.

The visual inspection revealed damage in the non-structural walls between the structural axes 14 and 15, see Figure 4. The damage was identified as low with presence of thin cracks on the walls that during the construction were plastered with cement-lime mortar as architectural finish. The cracks were located in the masonry walls and near the edges of the masonry walls and structural steel elements. According to the information given by the facility authorities and in line

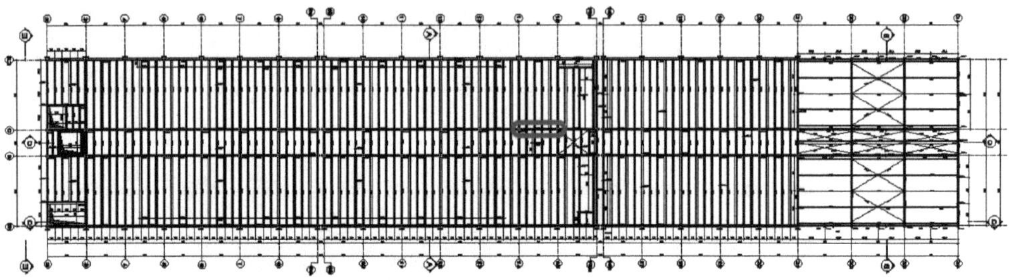

Figure 4: Location of the damaged infill wall

with some indications perceived during the visual inspection of the building, it was assessed that during the construction works no additional measures were taken to ensure better quality of the connection between the masonry and steel elements.

During the inspection, certain levels of vibration were easily felt by the inspection team that originated from the production process technology. It resulted mainly from random starting of the mixers located in affected part of the building.

4 Experimental tests

With the intention to find out the reasons for the occurred damage, as well as to monitor the development of the cracks in time, several short-time and continuous experimental tests were performed [1]. They consist in: monitoring the development and the size of the existing cracks, monitoring the vertical deformations in the damaged wall, ambient vibration testing, and measurement of accelerations, relative displacements and ambient temperature near the damaged wall.

4.1 Monitoring the cracks and the vertical deformations

With the aim of determining the development of the cracks in time, in the period of 4 months, at seven previously decided locations, seven non-destructive control measurements were performed. The measurements were done with a deformation measuring instrument of type Hugenberger that operates with a precision of 1/1000 mm. During the same period, at another location, a deflection meter with a precision of 1/100 mm was positioned, but due to the production process, available testing space and constant vibrations of the structural system, it was not possible to perform this measurement with acceptable quality.

Table 1: Measured strains at several crack locations

Pos.	D01	D21	D22	D23	D41	D51	D52
Strain (‰)	0.164	0.320	0.472	-0.328	0.092	0.240	0.120

According to the obtained results from the control measurements of the cracks, strains that developed in the inspected time period were calculated and are given in Table 1. The obtained strains show certain crack growth in six test locations, except in one location were decrease of the crack size was detected.

4.2 Ambient vibration testing

From the available possibilities of the in-situ tests, one of the most useful procedures is experimental modal identification of the structural system by ambient vibration testing. This procedure assesses the global properties of the structure and allows identification of the dynamic properties of the buildings, their natural frequencies, mode shapes and damping ratios. With the intention to relate this parameters with parameters calculated in the design project documentation, ambient vibration test were conducted at the beginning of October 2012. These tests were executed to measure the dynamic response in 12 different points, with the excitation being associated to environmental loads and to the production process. It should be noted that the production process was not interrupted during the tests, so the vibration level associated with the production process was captured as well.

Figure 5 shows a schematic representation of the sensor layout. Since a maximum of 16 channels were available for testing and three channels were held stationary for reference measurements, a series of three set-ups was used to cover the 12 measurement points of Figure 5.

The tests were conducted using 16 channels, 24 bits resolution Digitexx PDAQ Premium portable system with 5 tri-axial Digitexx MEMS accelerometer sensors. The sensors were connected with a high quality conductor cables to the data acquisition system. For each channel, the ambient time histories, in terms of accelerations, were recorded for 184 s at intervals of 0.005 s, which resulted in a total of 36,800 data points per channel.

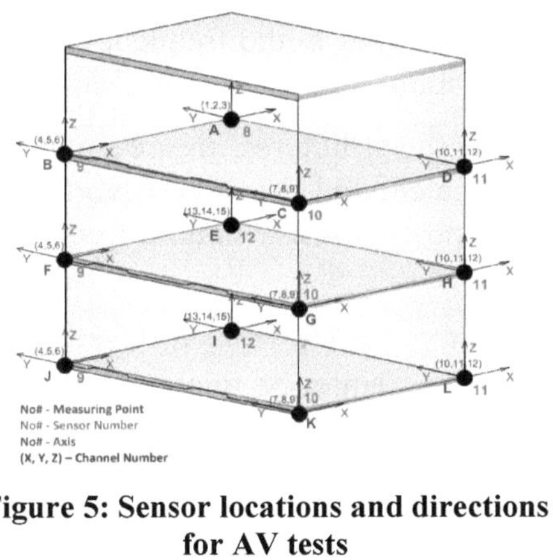

Figure 5: Sensor locations and directions for AV tests

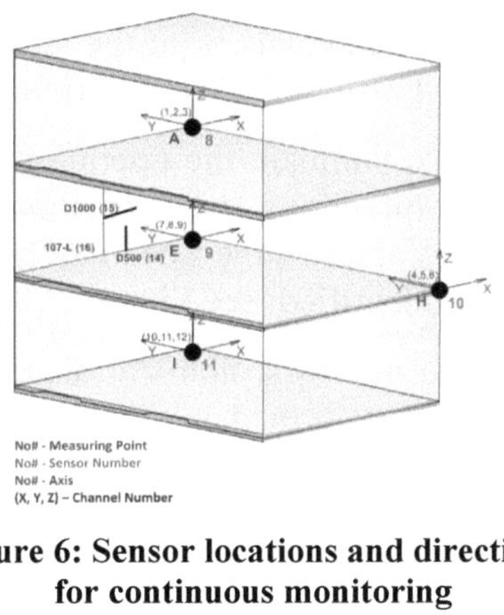

Figure 6: Sensor locations and directions for continuous monitoring

4.2.1 Qualification and analysis of experimental data

Prior to further data analysis, qualification of the recorded acceleration time-histories was completed. In order to define the appropriate analysis procedure the experimental data was classified, validated and edited.

- Data Classification

The correct analysis, as well as interpretation of random data is strongly influenced by the basic stationarity characteristics of the data. Therefore the performed data classification covered the following three important properties: (1) stationarity of data, (2) presence of periodic components [2].

The stationarity test of the measured response data was accomplished by the nonparametric approach: reversed arrangements test. Each time record was divided into 20 equal time intervals and for each interval the mean square value was computed. The reversed arrangement test of the sequence of mean squared values showed that the recorded accelerations are nonstationary. The hypothesis of stationarity was rejected at 5% level of significance.

In order to reveal any periodic components the autospectral densities of the raw measured data were visually inspected. No significant peak was detected.

- Data validation and editing

The steps of data validation were performed by careful visual inspection of the raw measured time-histories. Some of the potential anomalies that could be eliminated in this manner are: excessive instrumentation noise, signal clipping, noise spikes, spurious trends and signal dropouts. For the process of raw data editing the statistical software Minitab 16 [3] was used. The digital data samples were transformed to a new set of values that have zero mean value. Furthermore, any spurious trends were removed by fitting a low-order polynomial to the digital data samples.

- Modal analysis

The recorded signals from the structure were transformed in the frequency domain using the Fast Fourier Transformation (FFT) algorithm.

In order to simulate the operational behaviour of the structures, the responses for different directions were combined. From the calculated Averaged Normalized Singular Values of the Spectral Density Matrices no conclusive natural frequencies could be identified, see Figure 7.

Having in mind the non-stationarity of the data it is obvious that further analysis should be performed taking into account the time-frequency relationship of the recorded acceleration signals.

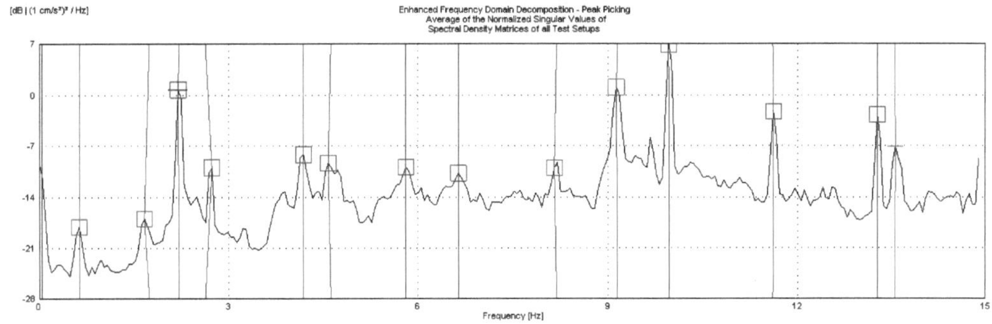

Figure 7: Spectral Density Plot for all Test Setups

The frequency domain analysis could not provide sufficient conclusions for the dynamic parameters of the structure.

Furthermore, the absence of knowledge of the operational frequencies and working pattern of the active mixers during the on-site experiment limited the identification of the present harmonics in the recorded accelerations.

4.3 Measurement of accelerations, relative displacements and ambient temperature

Since the appeared cracks in the infill wall could not be directly related to any particular reason it was decided to perform 5 days continuous monitoring of the affected structural part of the building. The primary aim was to identify the accelerations, the relative displacements and the ambient temperature that are applied to the structural system from the production process. Since the production process involves rotation of several masses in the mixers with different frequencies, but also certain heat from the moulding process it was suspected that these action might be the reason for damage in the wall.

The accelerations were monitored in 4 measurement points (A, E, H, I), while the relative displacements were recorded in two orthogonal directions between the structural frame system surrounding the wall and the damaged wall itself. In the same time, temperature changes were monitored and recorded at a measurement point in the vicinity of the damaged wall, see Figure 6. The location of the measurement points was determined according to the disposition of the structural system and the position of the non-structural elements were damage was detected. The relative displacements were monitored with LVDT sensors with maximal capacity of 25 mm. The vertical LVDT was fixed with the floor structure on the first floor and a point on the wall. The horizontal LVDT was fixed between a steel column and the measuring point. The variations of the ambient temperature were recorded with a temperature sensor.

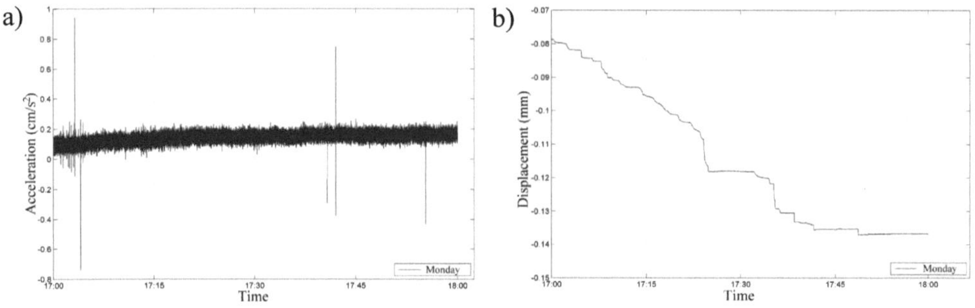

Figure 8: a) Acceleration time history of the point A in vertical direction; b) Vertical displacement time history of the measured point

The selection of the measurement points and the duration of the recording allowed overview of the structural behaviour during one work week of the facility. The monitoring produced large set of data. For each channel, the acceleration, displacement and temperature time histories were continuously recorded at frequency of 200 Hz, which resulted in a total of 84,240,000 data points per channel. The recorded data was stored for each hour in separate files for easier manipulation. During the recording there were no interruptions of the production process which was maintained with the usual routine. This set-up let real insight view of the structural behaviour in operational conditions during one work week.

4.3.1 Data processing

The recorded accelerations, displacements and variations of the ambient temperature were read from the individual files and processed with a custom Matlab [4] code. First, the data was filtered in real time domain using a Lowpass Butterworth filter of order 8 and cut-off frequency of 40 Hz.

All recorded signals were visually inspected in order to check their quality and possible spurious trends and peaks. Due to the large data stored in the files, additional processing by decimation with factor 10 was performed. All linear trends and offsets of the recorded time histories were previously removed and eliminated. Typical illustration of the acceleration time history at one measurement point and relative displacement time history at the measuring point are presented in Figure 8.

4.3.2 Results from the continuous 5 days monitoring

Accelerations

The accelerations recorded in the measurement points show certain variations of the amplitudes. The most distinguished level of accelerations was identified in

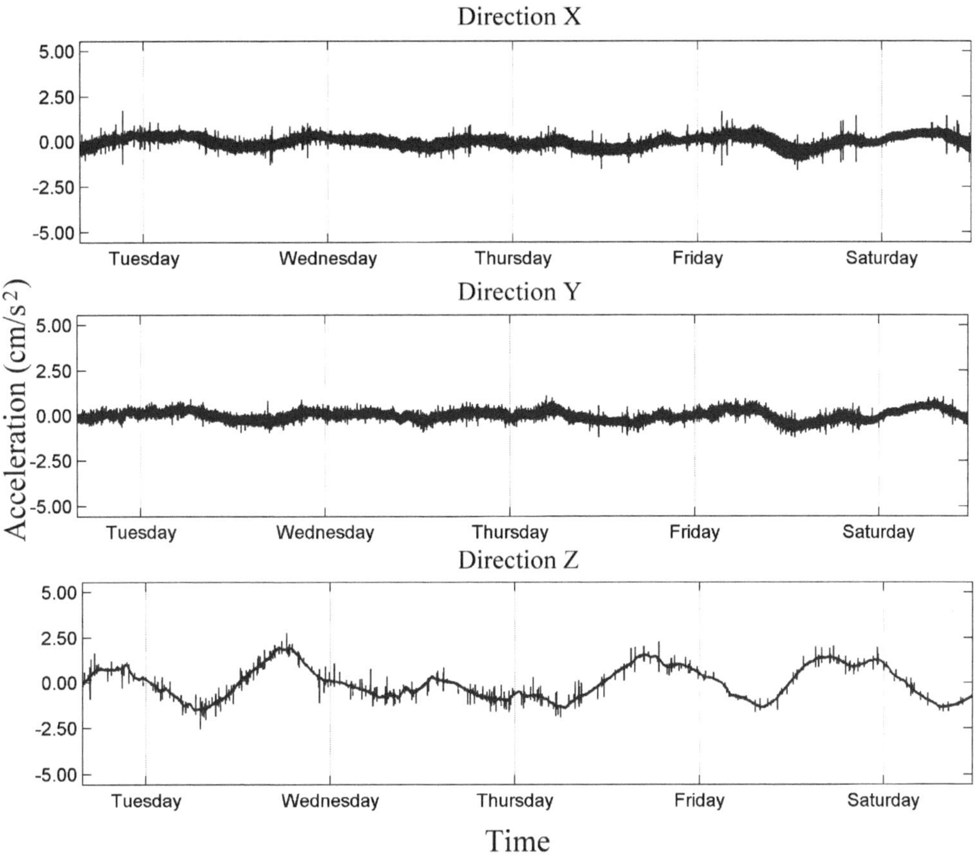

Figure 9: Component acceleration time histories in point E

point E. The accelerations in the vertical direction (Z) change with certain frequency with approximate periodicity of 24 hours, see Figure 9.

Also, the acceleration variations in the Z direction are twice greater than the variations recorded in the horizontal directions X and Y. The variations of the accelerations detected in Z directions range from -2 to 2 cm/s^2, while in X and Y directions their range is from -1 to 1 cm/s^2. The point E was located on the lower edge of the cracked wall. The levels of acceleration recorded in the ground floor, point I, showed significantly lower amplitudes than on other levels, as expected.

Relative vertical displacements

Figure 10 shows the relative vertical displacements recorded between the measurement point on the cracked wall and the floor structure. The time history plot shows increasing relative displacements overt the time, with significant jumps

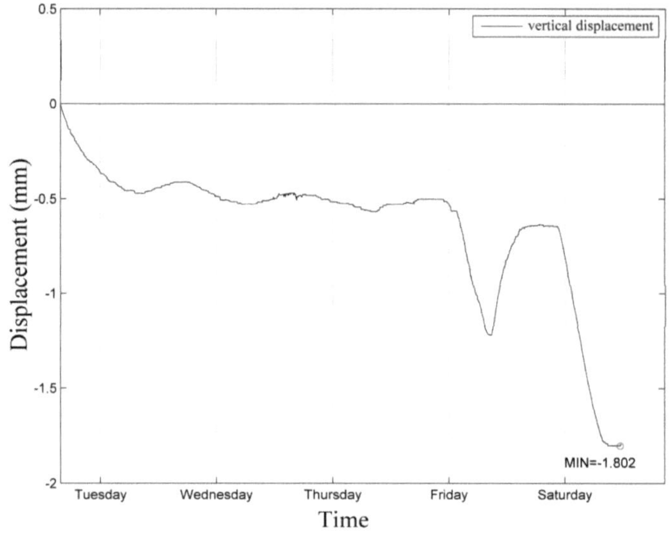

Figure 10: Vertical displacement time history

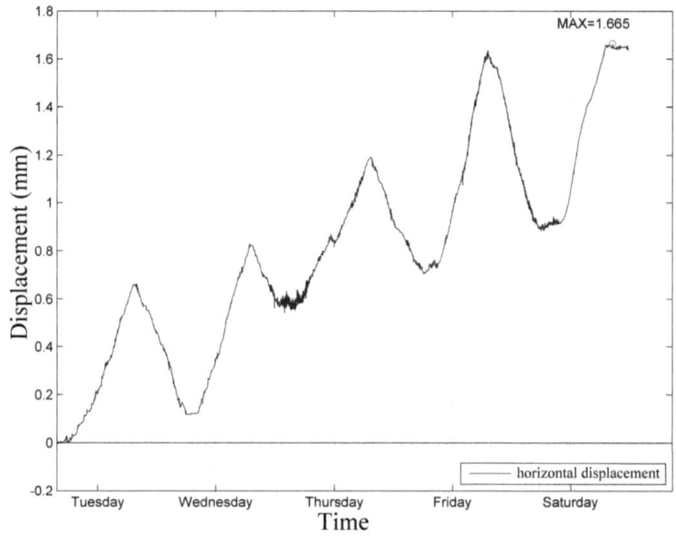

Figure 11: Horizontal displacement time history

in the last two days of the monitoring interval. Two extreme values have been found in those two days, with a maximal displacement of 1.802 mm in the last day.

Relative horizontal displacements

Figure 11 presents the obtained results from the continuous monitoring of the relative horizontal displacements of the cracked wall with respect to the steel

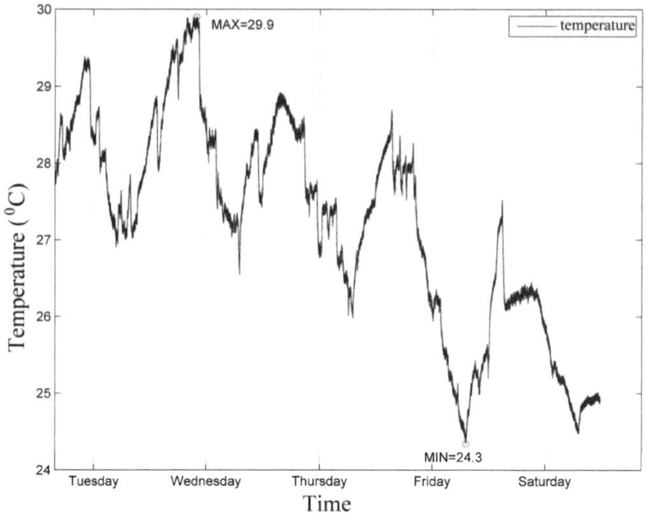

Figure 12: Temperature changes during the monitoring period

column that frames the infill wall. Theses displacements show clear tendency for increasing the horizontal displacements with considerable displacement jumps that correlate to the 24 hour production sequence in the facility. The peaks in the displacements have been found at 8:00 in the morning when the operating machines are put in operation at full capacity. The maximal increase of the relative horizontal displacements measured in the 5 days monitoring period was 1.665 mm.

Temperature variations

Figure 12 shows the temperature variations that occur in the surrounding area of the cracked wall. As can be seen, the changes in temperature show tendency for repeating in a 24 hour cycles. It was found out that the maximum recorded temperature is 29.9^{0}C and the minimum temperature is 24.3^{0}C. The temperature difference of 5.6^{0}C could not be the reason for the appeared cracks in the wall.

5 Conclusions

From the obtained results and performed analysis of the affected part of the structural system of the industrial hall it can be concluded that due to the non-stationarity of the recorded accelerations, no reliable data about the natural frequencies could be obtained. Therefore, no identification of the structural system could be performed in the given operational conditions of the facility.

The technological production process generates vibrations in all orthogonal directions, with the most noticeable vibrations being in vertical direction. The vertical vibration component is more distinct in the north-west part of the building which probably results from non-uniform mass distributions of the equipment and/or the nature of the production process. Moreover, the vertical vibration component has highest amplitudes in the first floor where the cracked wall is located. All recorded vertical accelerations show certain periodicity with an order of 24 hours.

The vibrations that originate from the production process and the equipment excite the non-structural infill walls as well. With respect of the stiffness differences in the built materials, a certain level of relative displacements occurs between the supporting structural elements and the masonry walls which cause development of cracks in the brittle materials.

The experimental investigations identified the reasons for cracking in the walls. The cracks appeared due to usage of inappropriate material for the infill walls in environment with constant vibrations conditions, but also due to the inadequate construction technique. It is recommended that the walls made from heavy masonry blocks should be replaced with partition walls from lighter material. In order to control and monitor the structural behaviour and to prevent possible damage of the structural elements caused by fatigue or relaxation of the assembled joints it is recommended to perform continuous or periodical long term monitoring.

REFERENCES

[1] Markovski G. et al: Analysis for determination of the reasons for cracking in certain parts of the industrial facility, Faculty of Civil Engineering-Skopje, 2013 (in Macedonian).

[2] Bendat, J. Allian,P.: Random Data: Analysis and measurements procedures, Wiley Series in Probability and Statistics; 2010, SBN: 978-0-470-24877-5.

[3] Meet Minitab 16. Users Manual, Minitab Inc; 2010.

[4] MATLAB. 2012b. Users Manual, The Mathworks, Inc.

Collapse Simulation of Building Structures Induced by Extreme Earthquakes

Xinzheng Lu[1], Xiao Lu[1] and Linlin Xie[1]

[1] Department of Civil Engineering / Tsinghua University
Beijing, China
luxz@tsinghua.edu.cn

ABSTRACT:

Research development has demonstrated that numerical simulation is becoming one of the most powerful tools for collapse analysis of building structures in addition to the conventional laboratory model tests and post-earthquake investigations. In this paper, a finite element (FE) method based numerical model encompassing fiber-beam element model, multi-layer shell model and elemental deactivation technique is proposed to predict the collapse process of buildings subjected to extreme earthquake. The potential collapse processes are simulated for several different types of buildings. The analysis results indicate that the proposed numerical model is capable of simulating collapse process of buildings by identifying potentially weak components of the structure that may induce collapse. The study outcome will be beneficial to aid further development of optimal design philosophy.

Keywords: fiber-beam element model, multi-layer shell model, elemental deactivation technique, collapse simulation, super-tall building

1 Introduction

Collapse is a critical ultimate state for structures during extreme earthquakes. Only the collapse process is understood clearly, the structural collapse can be effectively prevented. Research development has demonstrated that numerical simulation is becoming one of the most powerful tools to study the collapse process and mechanism. Despite the development of many numerical models, such as the Discrete Element Method [1] and Applied Element Method [2] in simulating structural collapse, and some important progresses have been made, these methods still have a long way before they can be used to simulate complicated real super-tall buildings.

In view of this, the present study aims to develop a simulation model that is based on a well-developed finite element (FE) framework to provide a feasible collapse simulation methodology for practical application. In the FE models, fiber-beam

S. Klinkel et al. (Eds.), *Seismic Design of Industrial Facilities*,
DOI 10.1007/978-3-658-02810-7_32,

element and multi-layer shell element are adopted to simulate the frame beams/columns and the shear walls respectively. And the efficiency and accuracy is verified by many literatures [3-6]. Time-history analyses are carried out to simulate the entire collapse process. Three numerical examples including two actual high-rise RC frame-core tube buildings and an actual super-tall building are analysed to demonstrate the applicability and efficiency of the proposed collapse simulation method.

2 Elemental deactivation

During the collapse simulation, the key issue is how to simulate the phenomenon that the whole structure changes from a continuum system into discrete parts through structural fracturing and element crushing. In this paper, elemental deactivation technique is adopted to simulate this process, where the failed elements are deactivated when a specified elemental-failure criterion is reached. Since both elemental models (fiber-beam element model and multi-layer shell model) are based on material stress-strain relations, corresponding material-related failure criterion must be adopted to monitor the failure of structural elements. For the fiber-beam element model, each element has at least 36 concrete fibers and 4 steel rebar fibers and each fiber has 3 Gauss integration points. Similarly, for the multi-layer shell model, each element has at least 10 layers (the number of layers depends on the specific situation of the actual reinforcement) and each layer has 4 Gaussian integration points. If the strain at any integration point in a fiber or layer (either concrete or steel) exceeds the material failure criterion, the stress and the stiffness of this fiber/layer are deactivated, which means that the fiber/layer no longer contributes to the stiffness computation of the whole structure. If all fibers/layers of an element are deactivated, the element is considered fully deactivated from the model.

3 Collapse simulation of 18-story building

Shown in Figures 1a is the FE model of an existing high-rise building which has 18 stories above the ground and a 4-story basement with a total height of 74.8 m. The core-tube is made up of four sub-tubes connected by coupling beams. The thickness of the shear wall changes from 500mm (at the bottom story) to 350mm (at the top story). The columns and beams are simulated by the fiber-beam element model, and the RC shear wall and coupling beams are simulated using the multi-layer shell model. More details of this structure are described in Lu et al. [6,7].

The fundamental period of this structure T_1=1.55 s. El-Centro EW Ground Motion that is scaled to PGA=1500 gal is used as an earthquake input to the structure along the X-axis. Figure 1 clearly displays the potential collapse process of this high-rise building under El-Centro ground motion. The ground story is identified to be the weakest part of the building due to its much larger height than the other stories. As can be seen from Figure 1b, the failure of the shear wall starts from the outer flange of the core-tube in the ground floor, which is caused by the gravity load of the building and the over-turning effect of the seismic load. Note that in the outer flange of the core-tube, the compressive load is much larger than the shear force. Therefore, the failure of the shear wall is dominated by concrete crushing induced by the axial load and bending moment. Subsequently, significant force redistribution occurs in the ground floor. This in turn results in a steady increase in the vertical and horizontal forces in the columns thereby leading to buckling of the columns (as shown in Figure 1c). With an increase in time, collision occurs between the basement and the upper stories (Figure 1d) which in turn results in a total collapse of the ground floor and subsequently the whole building.

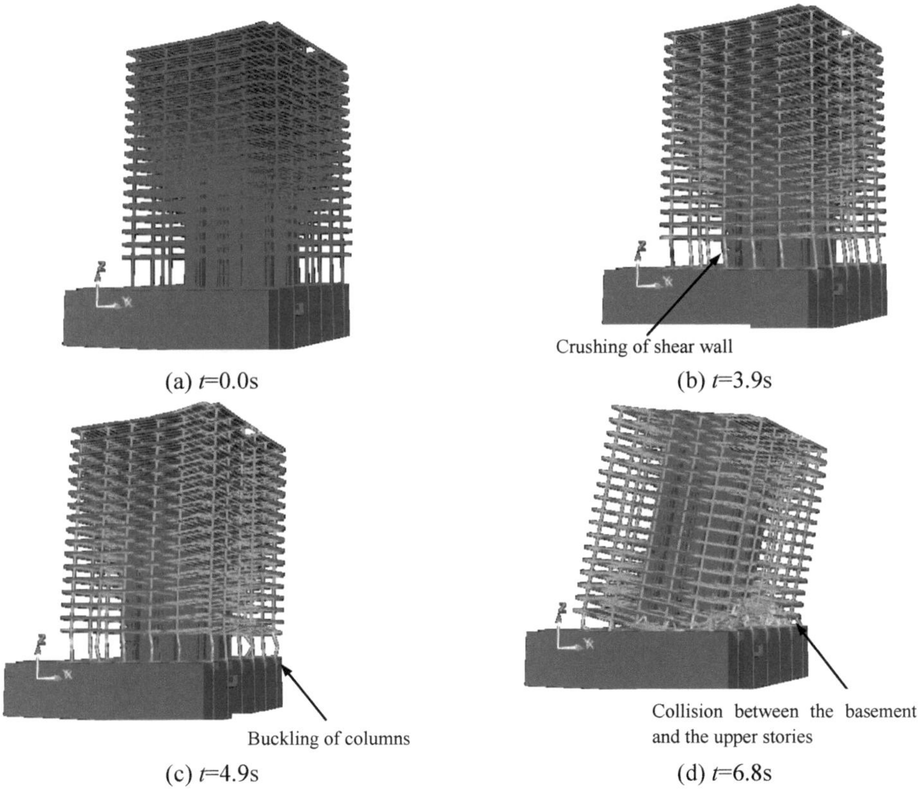

(a) t=0.0s (b) t=3.9s (c) t=4.9s (d) t=6.8s

Figure 1: Collapse process of the 18-story frame-core-tube building

4 Collapse simulation of 20-story frame-core tube building

This structure is a 79.47 m tall, 20-story office with a 4-story skirt building. The finite element model is shown in Figure 2a. The lateral-force-resisting system of the building consists of reinforced concrete external frame and core-tube. The cross-sectional dimensions of the columns from bottom to top of the building are

800 mm×800 mm, 700 mm×700 mm, 600mm×600mm. The beam sections are 350mm×650mm in the X-direction and 350mm×600mm in the Y-direction. The thickness of the core-tube is 350mm. And more details of the structural geometries are described in Lu et al. [6, 7].

Illustrated in Figure 2 is the collapse process of this building subjected to El-Centro EW Ground Motion which is scaled to PGA=4000 gal. The shear wall at the 10th story has its concrete strength changed from C40 to C30 and the column section changes from 700mm×700mm to 600mm×600mm. This results in a sudden change in stiffness which in turn yields stress concentration. In consequence, at *t*=4.5 s, the shear wall at this story is crushed as demonstrated in Figure 2b. With propagation of the failed structural elements including buckled columns (Figure 2c), the stories above the 10th story comes down and impacts on the lower stories (Figure 2d), thereby leading to a progressive collapse of the whole building. The failure mechanism is similar to the eighteen-story frame-core tube building (Section 3) in which collapse is initiated by concrete crushing in the outer flange of the core-tube in the weak story.

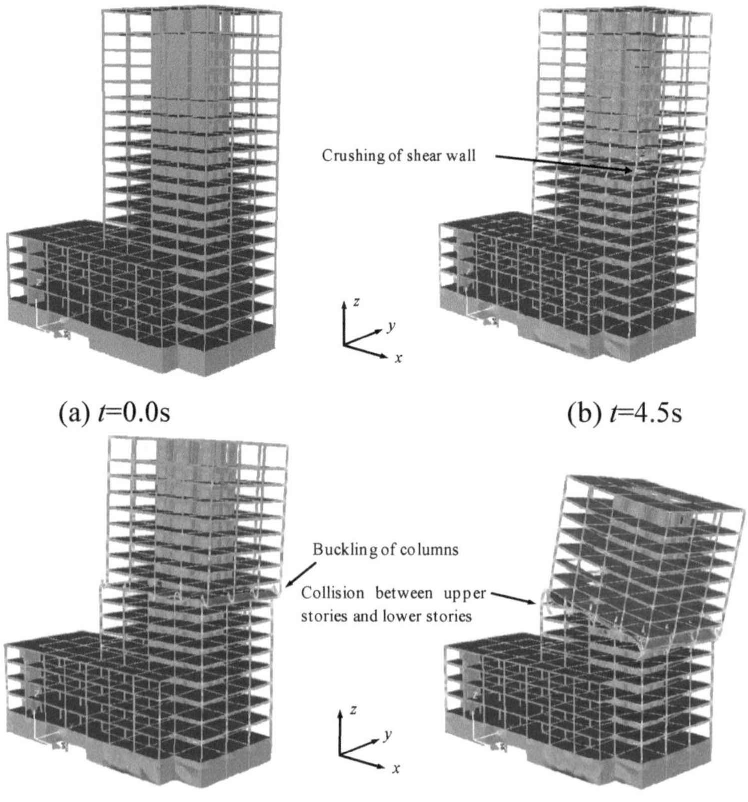

(a) *t*=0.0s (b) *t*=4.5s

(c) *t*=5.1s (d) *t*=7.5s

Figure 2: Collapse process of the 20-story frame-core tube building

5 Collapse simulation of the Shanghai Tower

The Shanghai Tower, located in Lujiazui, Shanghai, is a multi-functional office building (as shown in Figure 3). The total height of the main tower is 632 m with 124 stories. A hybrid lateral-force-resisting system referred to as "mega-column/core-tube/outrigger" was adopted for the main tower.

The main part of the core-tube is a 30 m by 30 m square RC tube. The mega-column system consists of 12 shaped-steel reinforced concrete columns with a maximum cross-sectional dimension of 5,300 mm×3,700 mm. 8 mega-columns extend from the bottom to the top of the building. The remaining 4 columns are located at each corner and only extend from the ground floor to Zone 5. The outrigger system, located at the mechanical stories, consists of circle trusses and outriggers with a total height of 9.9 m. All of the components of the outriggers are composed of H-shaped steel beams. The more details of structural properties are available in Lu et al. [8].

Figure 3: The location of Shanghai Tower

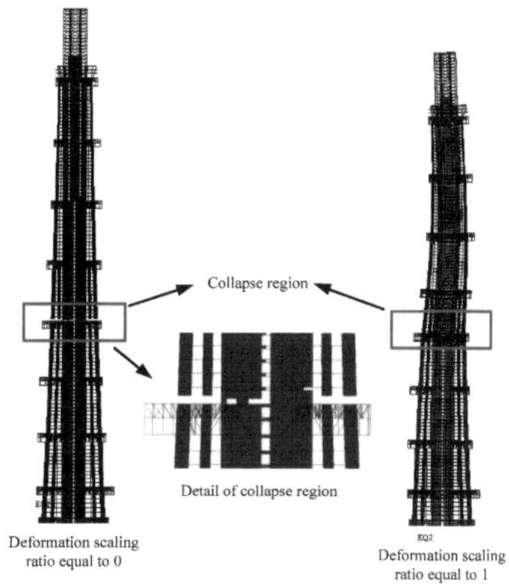

Figure 4: Collapse mode subjected to El-Centro EW ground motion

The external frames and outriggers are modelled with traditional fiber beam element and the shear walls of core-tube are simulated by multi-layer shell elements. Meanwhile, few experimental data regarding the mega-columns can be found in the literature, so a multi-layer shell element-based simplified model was proposed for the mega-columns and the parameters of the simplified model were determined based on the detailed FE model of mega-columns with solid elements. More details of these numerical models and failure criteria can be seen in Lu et al. [8].

The fundamental period of the Shanghai Tower in x direction is 9.83 s, which is far beyond the range of 6 s specified in the design response spectrum in the Chinese Code for the Seismic Design of Buildings [9]. Similar to the analysis above, the El-

Centro EW ground motion was chosen as a typical example of ground motion input. The peak ground acceleration (PGA) is scaled to 1960 gal, and then used as input for the FE model in the x direction. The final collapse mode is shown in Figure 4.

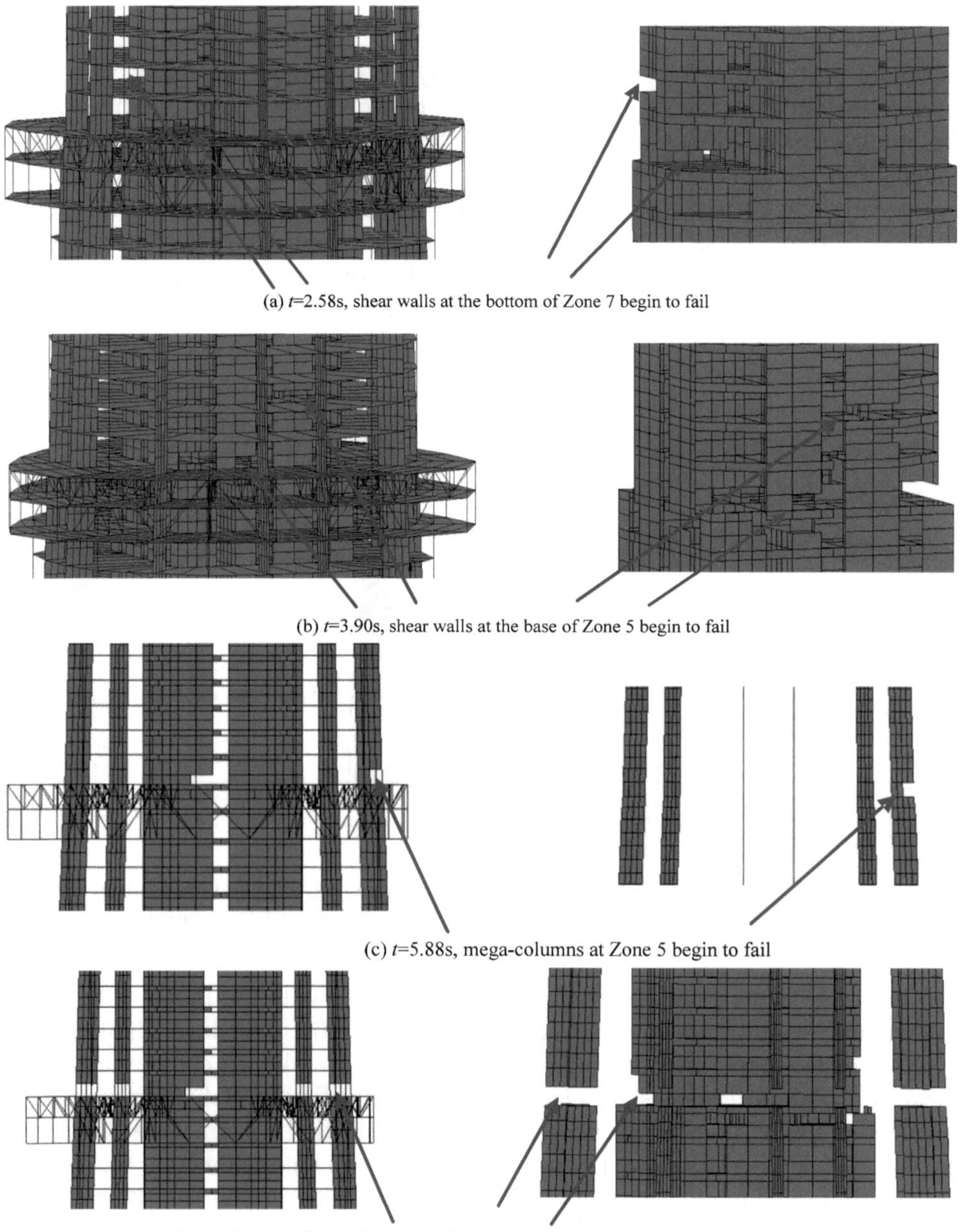

(a) t=2.58s, shear walls at the bottom of Zone 7 begin to fail

(b) t=3.90s, shear walls at the base of Zone 5 begin to fail

(c) t=5.88s, mega-columns at Zone 5 begin to fail

(d) t=6.18s, more than 50% shear walls and all mega- columns destroyed at Zone 5 the whole structure begins to collapse

Figure 5: Collapse process of the Shanghai Tower

The details of the collapse process are clearly shown in Figure 5. First, when t=2.58 s, some coupling beams in the core-tube begin to fail, and the flange wall of the core-tube at the bottom of Zone 7 is crushed. The reason for this crushing is that the layout of the openings in the core-tube changes between Zones 6 and 7, resulting in a sudden change of stiffness and stress concentration. After that, when t=3.90 s, the shear wall at the bottom of Zone 5 begins to fail because the cross section of the core-tube changes from Zone 4 to Zone 5 as shown in Figure 5b. When t=5.88 s, more than 50% of the shear walls at the bottom of Zone 5 fail, and the internal forces are redistributed to other components. The vertical and horizontal loads in the mega-columns increase gradually and reach their load capacities. The mega-columns then begin to fail. Finally, when t=6.18 s, the core-tube and mega-columns in Zone 5 are completely destroyed, and the collapse begins to propagate to the entire structure.

Obviously, when subjected to El-Centro ground motion in the x direction, the Shanghai Tower is mainly damaged in Zones 5, 6 and 7. Finally, collapse occurs in Zone 5, and the entire structure breaks into two parts. It can be clearly seen that Zone 5 is a potentially weak part, where structural collapse can be initiated.

6 Conclusions

A finite element framework to simulate the structural collapse subjected to extreme earthquakes is proposed and its application is illustrated using three actual high-rise buildings. For a given strong ground motion, the potential collapse modes and the corresponding weak parts can be predicted, which gives a better understanding of the collapse mechanism of building structures. The outcome of this study can also be used as references in engineering practice for collapse resistance design of similar building structures.

7 Acknowledgements

The authors are grateful for the financial support received from the National Nature Science Foundation of China (No. 51222804, 51178249, 51261120377) and the Fok Ying Dong Education Foundation (No. 131071).

References

[1] Lemos, JV. Discrete element modeling of masonry structures. International Journal of Architectural Heritage; 1(2007),190-213.

[2] Asprone, D., Nanni, A., Salem, H., et al. Applied element method analysis of porous GFRP barrier subjected to blast. Advances in Structural Engineering; 13(2010), 152-170.

[3] Li, Y., Lu, XZ., Guan, H., et al. An improved tie force method for progressive collapse resistance design of reinforced concrete frame structures, Engineering Structures; 10(2011), 2931-2942.

[4] Lin, XC., Lu, XZ., Miao, ZW., et al. Finite element analysis and engineering application of RC core-tube structures based on the multi-layer shell elements, (in Chinese). China Civil Engineering Journal; 3(2009), 51-56.

[5] Miao, ZW., Wu, YH., Ma, QL., et al. Seismic performance evaluation using nonlinear time history analysis with three-dimensional structural model for a hybrid frame-core tube structure (in Chinese). Journal of Building Structures; 4(2009), 119-129.

[6] Lu, X., Lu, X., Guan, H., et al. Collapse simulation of reinforced concrete high - rise building induced by extreme earthquakes. Earthquake Engineering & Structural Dynamics; 42(2013), 705-723.

[7] Lu, XZ., Ye, LP., Miao, ZW., Elasto-plastic analysis of building against earthquake, (2009), Beijing, China, China Architecture and Building Press.

[8] Lu, X., Lu, XZ., Zhang, WK., et al. Collapse simulation of a super high-rise building subjected to extremely strong earthquakes. Science China Technological Sciences; 10(2011), 2549-2560.

[9] Ministry of Construction of the People's Republic of China. Code for seismic design of buildings, Code No. GB-50011, (2010), Beijing: China Architecture and Buildings Press.

Part VIII

Seismic Design of Silos, Tanks and Vessels

The Eurocode Approach to Seismic Design of Liquid-Filled Steel Storage Tanks

Margarita Chasapi[1]

[1] SDA-engineering GmbH
Kaiserstr. 100, TPH III-B, 52134 Herzogenrath, Germany
chasapi@sda-engineering.de

ABSTRACT:

The seismic analysis and design of liquid-filled storage tanks is an engineering problem connected with a significant degree of complexity, due to the liquid-structure-soil interaction that defines the dynamic response and determines the design of the tank. The implementation of adequate design rules is essential for ensuring the continuous operation of tanks after strong earthquakes and avoiding significant property loss and environmental damage. The European Standard Norms provide guidelines for steel shells, which can be implemented to investigate seismically excited liquid-filled tanks against buckling. This paper addresses the thematic area of shell buckling for the condition of seismic loading relevant for liquid-filled tanks, as defined in Eurocode 8 Part 4. The stress design approach described in Eurocode 3 Part 1-6 is examined, discussing the estimation and influence of the buckling relevant boundary conditions, geometrical tolerances and resistances. The concept is thereafter implemented by means of a parameterized design tool applicable for cylindrical anchored tanks based on rigid foundations. The investigation of a typical storage tank, the characteristic damage forms and their influence on the design are finally presented and evaluated. As the stress design procedure proposed in Eurocode 3 Part 1-6 remains the main option for the engineering practice, its implementation and results are examined in regard to their contribution to a cost-effective and earthquake resistant design.

Keywords: storage tank, earthquake, shell buckling, design, Eurocode

1 Introduction

The consequences of damages in storage tanks after strong earthquakes can be disastrous for the population and the environment. Deficiencies in the structural design can lead to significant property loss, lifeline damages, exposure to toxic substances and eventually contaminations, fires or explosions. Sufficient analysis and design procedures are therefore essential to ensure the safe and continuous operation of tanks.

DOI 10.1007/978-3-658-02810-7_33, © Springer Fachmedien Wiesbaden 2014

Engineering practitioners face though certain challenges regarding the seismic evaluation and design of tanks. The accurate estimation of the seismic response of the tank under consideration of the liquid-shell-soil interaction is a complex issue that has been intensely investigated during the last decades (Rammerstorfer et al. [1]). As the main damage forms observed in storage tanks after earthquakes are attributed to buckling, a further understanding of the structural behaviour of shells in regard to stability under seismic loads is necessary for an adequate design. The application and evaluation of existing associated standards can contribute significantly to a better understanding of the problem.

Thereafter, an insight into the thematic area of seismically induced buckling of liquid storage tanks will be provided. The suggestions of Eurocode 8 Part 4 [2] for the estimation of the seismic response of tanks will therefore be considered, in regard to their contribution to the structural behaviour of the shell. Furthermore, the stress based design of Eurocode 3 Part 1-6 [3] will be examined discussing the estimation and influence of buckling relevant parameters. Finally, the concept will be applied on a typical water tank located in Germany. The seismic evaluation will be based on the finite element method in combination with a parameterised tool, which provides a reliable estimation of the seismic response and elaborate design results. The implementation of the procedure and its results will be examined and discussed in terms of their contribution to an effective seismic design.

2 Seismically induced buckling of tanks

Liquid-filled storage tanks in normal operating conditions are subjected to permanent loads such as self-weight and hydrostatic pressure of the liquid in storage. The behaviour of the shell is in this case quite simple, characterised by circumferential tension due to internal pressure. A risk of buckling can occur with the appearance of additional wind loads, resulting in an interaction of axial and circumferential compressive stresses critical for the upper thinner areas of the tank. However, shear stresses are usually of minor importance for the conventional tank design.

In the event of seismic actions, liquid filled tanks can develop a more complex behaviour, characterised by a three-axial stress state implying buckling risk along the shell height. Moreover, as a practical difficulty is connected with the estimation of the seismic load for the shell-liquid system, insufficient tank design and subsequent buckling-induced damage can be attributed to underestimation of the seismic forces. EC 8 Part 4 [2] provides informative guidelines for the definition of seismic actions as equivalent static loads on liquid filled tanks, by means of three independent hydrodynamic pressure components:

- Convective (sloshing) component describing the vibration of the liquid due to horizontal earthquake excitation

- Impulsive rigid component describing the vibration of the rigid shell due to horizontal and vertical earthquake excitation
- Impulsive flexible component describing the interaction vibration between flexible steel shell and liquid due to horizontal and vertical earthquake excitation

The combined vibration of shell and liquid under earthquake excitation is defined by an iterative procedure, where the liquid is applied as added-mass on the 'dry' shell. After the application of the pressure distributions as static loads, the seismic analysis is based on the elastic response spectrum under consideration of different damping characteristics for the convective and impulsive flexible component (0.5% and 5% suggested in EC 8 Part 4 [2]). The application of an appropriate behavioural factor for the impulsive flexible component is also possible, which leads to a significant reduction of the overall seismic action and influences accordingly the design. Under consideration of the axial symmetry of the tank, only one horizontal and the vertical component have to be considered as coexistent (EC 8 Part 4 [2]).

Different combination rules can be applied for the superposition of the pressure components in each direction (e.g. SRSS) as well as for the combination of the seismic resultants with other permanent or variable actions when available (e.g. linear superposition) as described by Meskouris et al. in [4]. The resulting design stress condition is characterised by high axial compressive and shear stresses concentrated at the bottom of the tank, due to the overturning moment and base shear force of the horizontal seismic component, in combination with circumferential tensile stresses due to the combined action of internal pressure and the vertical seismic component. This state can lead to plastic axial buckling characterised as 'elephant-foot buckling'. Elastic axial buckling characterised as 'diamond-shaped buckling' can occur when the stabilising hydrostatic pressure is reduced by the action of the vertical seismic component. Finally, areas of negative pressure towards the upper part of the shell can lead to circumferential buckling in cases of inadequate wall thickness.

3 Stress based design according to Eurocode 3 Part 1-6

After the successful estimation of the seismic forces on liquid storage tanks, the need for an effective yet practical and directly applicable design procedure leads to the traditional stress based design of EC 3 Part 1-6 [3].The verification concept is based on a linear elastic (LA) analysis, assuming a linear shell-bending theory, perfect shell geometry and linear material properties. A direct advantage of the method is therefore that it can be easily linked to a commonly practised finite element analysis. An algebraic estimation of the ideal elastic buckling stresses and their reduction to the buckling resistances under consideration of the buckling relevant parameters can thereafter be achieved for typical geometries and loading conditions.

3.1 Boundary conditions

The buckling relevant boundary conditions are defined under consideration of the translational and rotational degrees of freedom at each end of the investigated cylinder. Three basic boundary condition types can therefore be defined, corresponding to a fixed, jointed and free support. Applicable combinations of the bottom and top condition for typical storage tanks are summarised in Table 1 .

Table 1: Boundary conditions for Tanks, EC 3 Part 1-6 [3]

Unanchored tank with roof	BC2-BC2
Anchored tank with roof	BC1-BC2
Anchored tank with open top	BC1-BC3
Cylinder segment between stiffeners	BC2-BC2

The influence of the boundary conditions is included in the determination of the ideal elastic buckling stresses by means of an amplification factor C. Typical geometries of storage tanks correspond to middle-length cylinders, which results in an amplification factor C = 1 for the axial and shear component regardless of the boundary condition type. In the case of circumferential buckling, the amplification factor depends on the chosen boundary condition. For a typical anchored tank with roof a 25% amplification of the critical circumferential buckling stress is defined, in contrast to a 40% decrease in the case of an anchored tank with open top. For cylinders with variable wall thickness an overall factor equal to C = 1 has to be applied regardless of the chosen boundary conditions.

3.2 Geometrical tolerances

Imperfection sensitivity is taken under consideration for the estimation of the buckling strength with the application of different fabrications qualities coupled with associated tolerance levels. Three different quality classes are therefore defined (Table 2) representing the following buckling relevant geometrical tolerances:

- Out-of-roundness tolerance of the shell
- Accidental eccentricity tolerance at joints in the shell wall
- Pre-buckling (dimple) tolerance of the shell wall

An elastic imperfection reduction factor α is defined in accordance with the quality class for each component (axial, circumferential, shear). For axially compressed cylinders with internal pressure, the factor is further modified to account for the stabilising role of internal pressure. For the upper areas of the shell, the coexistence of internal pressure is though doubtful when combined with the hydrodynamic pressure components. To avoid the risk of overestimating the stabilising role, elaborate information regarding the pressure distribution over the shell length should be available.

Table 2: Definition of quality classes, EC 3 Part 1-6 [3]

Fabrication tolerance quality class	Description	Out of roundness $U_{r,max}$ for d≥1.25m	Accidental eccentricity $U_{e,max}$	Pre-buckling $U_{o,max}$
A	Excellent	0.007	0.14	0.006
B	High	0.010	0.20	0.010
C	Normal	0.015	0.30	0.016

3.3 Buckling resistance

The first step towards the estimation of the buckling strength is the calculation of the elastic critical buckling stresses ($\sigma_{x,Rcr}$, $\sigma_{\theta,Rcr}$, $\tau_{x\theta,Rcr}$) for the perfect shell. Three basic cases of cylinder buckling under uniform stress (axial compression, external pressure and uniform torsion) are therefore provided. The available expressions in Annex D of EC 3 Part 1-6 [3] are applicable for unstiffened cylinders with constant and stepwise variable wall thickness.

$$\lambda_x = \sqrt{f_{yk}/\sigma_{x,Rcr}} \qquad \lambda_\vartheta = \sqrt{f_{yk}/\sigma_{\vartheta,Rcr}} \qquad \lambda_\tau = \sqrt{\frac{f_{yk}/\sqrt{3}}{\tau_{x\vartheta,Rcr}}} \tag{1}$$

The next steps are the definition of the relative slenderness of the shell relating the elastic critical buckling stress and the yield strength (Eq. 1), and the determination of the buckling reduction factor χ_x, χ_θ, χ_τ for each component as a function of the relative slenderness (Eq. 2 - 4), representing the capacity curve of the shell and accounting for imperfection sensitivities and plasticity effects.

$$\chi = 1 \qquad when \quad \lambda \le \lambda_o \tag{2}$$

$$\chi = 1 - \beta\left(\frac{\lambda-\lambda_o}{\lambda_p-\lambda_o}\right)^n \qquad when \quad \lambda_o \le \lambda \le \lambda_p \tag{3}$$

$$\chi = \frac{\alpha}{\lambda^2} \qquad when \quad \lambda_p \le \lambda \tag{4}$$

The buckling reduction factors are applied for the deduction of the characteristic buckling stresses (Eq. 5).

$$\sigma_{x,Rk} = \chi_x \cdot f_{yk} \qquad \sigma_{\vartheta,Rk} = \chi_\vartheta \cdot f_{yk} \qquad \tau_{x\theta,Rk} = \chi_\tau \cdot f_{yk}/\sqrt{3} \tag{5}$$

The allowable design buckling stresses (Eq. 6) are expressed with the application of the partial safety factor $\gamma_{M1} = 1.1$ as defined for tanks in the associated standard EC 3 Part 4-2 [5].

$$\sigma_{x,Rd} = \sigma_{x,Rk}/\gamma_{M1} \qquad \sigma_{\theta,Rd} = \sigma_{\theta,Rk}/\gamma_{M1} \qquad \tau_{x\theta,Rd} = \tau_{x\theta,Rk}/\gamma_{M1} \tag{6}$$

Finally the buckling strength verification is conducted for the three buckling-relevant stress components as well as for the combined stress state (Eq. 7 - 8).

$$\sigma_{x,Ed} \leq \sigma_{x,Rd} \qquad \sigma_{\theta,Ed} \leq \sigma_{\theta,Rd} \qquad \tau_{x\theta,Ed} \leq \tau_{x\theta,Rd} \tag{7}$$

$$\left(\frac{\sigma_{x,Ed}}{\sigma_{x,Rd}}\right)^{k_x} - k_i \left(\frac{\sigma_{x,Ed}}{\sigma_{x,Rd}}\right) \cdot \left(\frac{\sigma_{\theta,Ed}}{\sigma_{\theta,Rd}}\right) + \left(\frac{\sigma_{\theta,Ed}}{\sigma_{\theta,Rd}}\right)^{k_\theta} + \left(\frac{\tau_{x\theta,Ed}}{\tau_{x\theta,Rd}}\right)^{k_\tau} \leq 1 \tag{8}$$

The application of the interaction buckling verification is required for a set of stress components that are present at each point of the structure excluding areas adjacent to the boundaries.

4 Example

The evaluation of the seismic design will be based on the simulation and analysis of a typical cylindrical storage tank by means of a parameterised routine by Cornelissen [6], which estimates the interaction between the liquid and flexible shell on the basis of the 'added-mass' concept of EC 8 Part 4 [2]. The buckling verifications according to EC 3 Part 1-6 [3] will be carried out with the design tool SHEND (Chasapi [7]), based on the finite element model and stress condition of the analysis. This approach requires coupling between mathematical software (Maple [8]) and finite element program (ANSYS [9]) for the analysis, as well as elaborate input for the design (accurate pressure and stress resultants for each element of the FE model) and provides thus adequate accuracy.

The investigated tank has a total cylindrical height of L = 12.0 m and a filling height of H = 11.4 m, which corresponds to a filling grade of 95%. The radius of the tank is R = 10 m, resulting in a tank slenderness of γ = H / R = 1.14 m. The storage content is water with a density of ρ = 1.0 t/m^3. The tank is anchored on a concrete foundation and has a base plate with a thickness of t = 8 mm. The stiffening at the top of the cylindrical shell is achieved by a circumferential girder with a section L 90x9. The tank has a floating roof, and will therefore be considered as open top for the design. The material is S235 steel with yield stress $f_{yk} = 2.35 \cdot 10^5$ kN/m^2 and module of elasticity $E = 2.1 \cdot 10^8$ kN/m^2. The steel shell is divided in 5 courses with a height of h = 2.4 m each and a stepwise variable wall thickness of t = 6 – 10 mm decreasing from bottom to top. An additional investigation of the cylindrical shell under consideration of intermediate stiffening girders with a section L 45x5 will also be examined and evaluated in terms of its influence on the buckling design. The cylinder is simulated with shell elements and is fixed at its base in three spatial directions. The stiffeners are simulated with beam elements.

The seismic action is defined by the elastic response spectrum for the location Friedrichshafen, Germany according to EC 8 Part 1/NA [10] (Table 3).

Table 3: Response spectrum input values, EC 8 Part 1/NA [10]

Seismic zone 2	$a_{gh} = 0.6\ m/s^2$ | $a_{gv} = 0.3\ m/s^2$
Ground class C-S	S = 0.75
Control periods horizontal (s)	$T_A = 0$ | $T_B = 0.1$ | $T_C = 0.5$ | $T_D = 2.0$
Control periods vertical (s)	$T_A = 0$ | $T_B = 0.05$ | $T_C = 0.2$ | $T_D = 2.0$
Importance value	1.0
Damping convective pressure	0.5%
Damping imp. flexible pressure	5%

4.1 Analysis

An iterative procedure for the estimation of the interaction vibration between the flexible shell and liquid results in a fundamental period for the impulsive flexible component equal to $T_{if,h} = 0.116$ s in horizontal direction and $T_{if,v} = 0.165$ s in vertical direction. Both periods correspond to the plateau area of the response spectrum, indicating a maximum spectral acceleration of $a_{if,h} = 1.125\ m/s^2$ and $a_{if,v} = 0.9\ m/s^2$ accordingly.

The resulting seismically induced pressure distributions in horizontal and vertical direction (Figure 1) indicate that the impulsive flexible components in horizontal and vertical direction contribute significantly to the overall seismic response, whereas the contribution of the convective component is relatively minor.

The hydrodynamic pressure components are applied on the FE model as static loads in each direction and combined through SRSS superposition. The combination of the resultant seismic action in horizontal and vertical direction with the permanent actions is thereafter achieved through linear superposition. As a result of the horizontal hydrodynamic action on the tank wall, the overturning moment M = 12320.6 kNm and base shear force F = 2325.9 kN induce maximum axial compressive and shear stresses at the tank bottom (Figure 2-Figure 3). The combined action of hydrostatic and hydrodynamic pressure components leads to circumferential tension over the shell height.

4.2 Design

A boundary condition type BC1-BC2 will be chosen for the unstiffened cylinder, which corresponds to a radially and rotation restrained support provided by the anchorage and a radially restrained and rotation free support provided by the top stiffening girder accordingly. The amplification factor will be set to C = 1 as

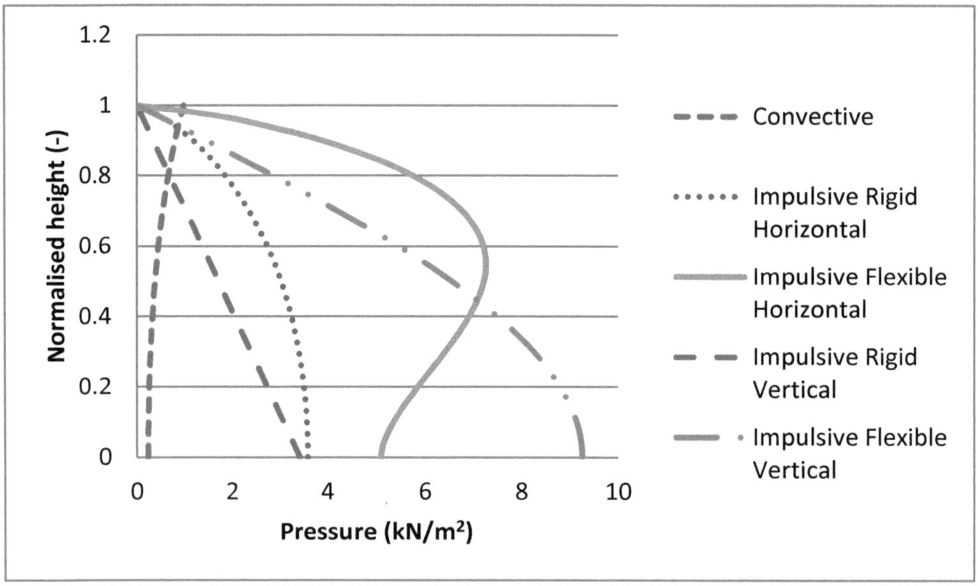

Figure 1: Hydrodynamic pressure components

defined for cylinders with variable wall thickness. High fabrication tolerance quality class "B" will be applied, with an imperfection factor a = 0.65 for circumferential and shear buckling and a fabrication quality parameter Q = 25 for axial buckling. The partial factor for resistance of the shell wall to stability will be set to $\gamma_{M1} = 1.1$ as indicated in EC 3 Part 4-2 [5].

Due to the stepwise variable wall thickness of the tank, a transformation of the stepped cylinder into an equivalent three-course cylinder is required for the estimation of the ideal critical buckling stress in circumferential and shear direction according to sections D.2.3 and D.2.4 of EC 3 Part 1-6 [3]. The resulting equivalent parameters amount to course length l_a = 6 m and thickness t_a= 6.2 mm for the upper course, and $l_b = l_c$ = 3 m with a thickness t_b = 7.6 mm and t_c = 9.6 mm for the intermediate and bottom course accordingly. The effective length of the equivalent cylinder results in l_{eff} = 8.57 m with a thickness t_a = 6.2 mm.

The decisive combination of maximum horizontal and maximum vertical pressure coordinates responsible for plastic buckling ('elephant-foot buckling') leads to maximum axial stresses with a utilisation of 22-23% for the lower two courses. Under consideration of the stabilising action of the internal pressure, the utilisation is less than 10% at the tank bottom. The influence of the stabilising internal pressure is significant especially for the lower courses (Figure 2).

A decisive factor for the design of the tank is the shear component. The allowable shear buckling stresses of the equivalent cylinder are exceeded by 17% at the lowest course (Figure 3). It has to be pointed out, that the equivalent cylinder procedure is defined as applicable for the shear component in EC 3 Part 1-6 [3] in

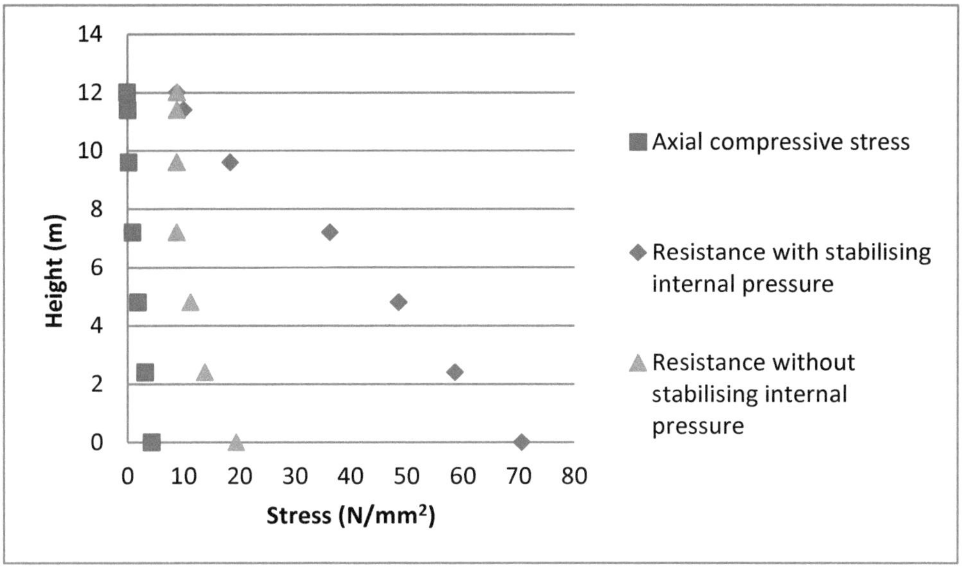

Figure 2: Axial buckling of unstiffened cylinder

analogy with the circumferential component, in contrast to previous regulations such as DIN 18800-4 [11], where the topic of shear buckling was not covered for cylinders with variable wall thickness. The procedure results in buckling resistances which increase from bottom to top corresponding to a buckling stress state induced by constant external pressure. For variable forces over the shell length, a modification of the membrane stresses under consideration of the maximum membrane force and the corresponding wall thicknesses is suggested in D.2.3.2 of EC 3 Part 1-6 [3], which results in a constant utilisation ratio over the shell height (Figure 3). The application of an alternative approach according to EC 3 Part 4-1 [12] results in an 'average' shear resistance over the shell length with a utilisation of 91% for the bottom course. The equivalent cylinder procedure underestimates the allowable stresses at the bottom of the tank compared to the results of the alternative approach (Figure 3). The opposite can be observed for the upper courses, where the alternative approach results in more conservative values.

The interaction of axial and shear component under consideration of the coexistent stresses at each point of the shell excluding the parts adjacent to the boundaries leads to a utilisation of 85% for the lowest course according to the alternative approach, in comparison to the equivalent cylinder procedure that leads to a 10% exceedance of the allowable value.

The cylinder is additionally investigated for the given seismic action under consideration of intermediate stiffeners. The wall thickness are in this case reduced to t = 5.25 - 3mm achieving thus an optimal utilisation. The seismic action remains though unchanged, as the fundamental period of the impulsive flexible component depending on the wall thickness remains on the plateau area of the spectrum. The

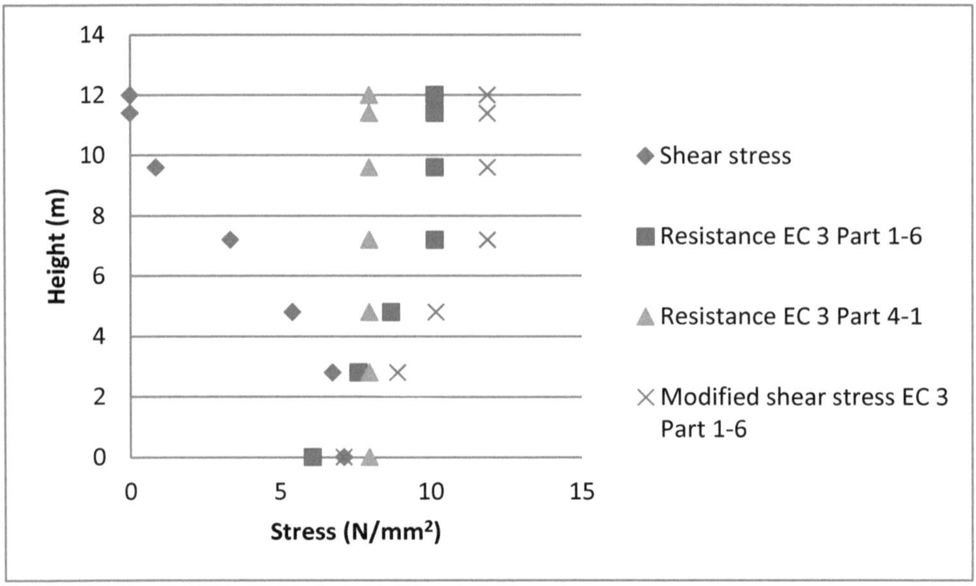

Figure 3: Shear buckling of unstiffened cylinder

main difference for the design is the boundary condition and the estimation of the shear buckling resistance. A boundary condition type BC2-BC2 will be chosen for the cylinder segments, which corresponds to a radially restrained and rotation free support provided by the girder at both ends of the cylinder segments. Each cylinder section is considered as an equivalent cylinder with constant wall thickness supported at both ends by stiffeners. The utilisation for axial buckling under consideration of the stabilizing internal pressure is equal to 70% at the lowest course. The shear component leads to 91% utilisation at the tank bottom. The interaction of axial and shear stresses leads to a utilisation of 93%.

5 Conclusion

The stress based design of EC 3 Part 1-6 can be applied to investigate the seismic behaviour of tanks against buckling. An accurate estimation of the stress and pressure resultants over the shell length is necessary for an adequate design, and can be achieved with the application of the provisions of EC 8 Part 1 and appropriate computational tools. Especially the contribution of the interaction vibration between shell and liquid as well as the stabilising internal pressure is significant for the overall design. The design procedure for unstiffened cylindrical shells with variable wall thickness can lead to conservative values regarding the shear component of the seismic action, whereas the consideration of stiffeners influences significantly the design. Further studies with nonlinear computer assessments can contribute to the validation of the behaviour under seismic loads and the development of improved practical solutions in the future.

REFERENCES

[1] Rammerstorfer, F.G. et al.: Storage Tanks under earthquake loadings, Applied Mechanics Review, Vol. 43, No 11, Pages 261-279, Nov. 1990.

[2] Eurocode 8: Design of structures – Part 4: Silos, Tanks and Pipelines, German Version EN 1998-4:2006.

[3] Eurocode 3: Design of steel structures - Part 1-6: Strength and stability of shell structures, German Version EN 1993-1-6:2007+AC: 2009.

[4] Meskouris et al.: Bauwerke und Erdbeben: Grundlagen – Anwendung – Beispiele, 3. Auflage, Vieweg & Teubner Verlag, 2011, ISBN 978-3-8348-0779-3.

[5] Eurocode 3: Design of steel structures - Part 4-2: Tanks, German Version EN 1993-4-2:2007 + AC:2009.

[6] Cornelissen, P.: Erarbeitung eines vereinfachten impulsiv-flexiblen Lastansatzes für die Berechnung von Tankbauwerken unter Erdbebenlast, Diplomarbeit, Lehrstuhl für Baustatik und Baudynamik, RWTH Aachen, 2010.

[7] Chasapi, M.: Seismic analysis and design of cylindrical liquid storage tanks according to DIN EN 1998-4, Master Thesis, Chair of Structural Statics and Dynamics, RWTH Aachen, 2011.

[8] Maple v. 13, Maplesoft, 2009.

[9] ANSYS v.14.0, ANSYS Inc. and ANSYS Europe Ltd, 2012.

[10] National Annex – Nationally determined parameters – Eurocode 8: Design of structures for earthquake resistance – Part 1: General rules, Seismic actions and rules for buildings, January 2010.

[11] DIN 18800: Stahlbauten – Teil 4: Stabilitätsfälle, Beuth Verlag, Berlin, 2008.

[12] Eurocode 3: Design of steel structures - Part 4-1: Silos; German Version EN 1993-4-1:2007 + AC:2009.

Lateral Free Vibration of Liquid-Storage Tanks

Konstantinos Mykoniou, Britta Holtschoppen

Chair of Structural Statics and Dynamics, RWTH Aachen
Mies-van-der-Rohe-Str.1, 52074, Aachen, Germany
konstantinos.mykoniou@lbb.rwth-aachen.de
holtschoppen@lbb.rwth-aachen.de

Abstract

Liquid filled tanks play an important role in the infrastructure of many industrial facilities assuring the supply with raw material needed for the production process or serving as storage for intermediate products. Due to their oftentimes large dimensions in diameter and height the stored fluid develops high seismic loads to the tank shell induced by the vibration of the liquid and the interaction of shell and liquid. In the design of tank shells the determination of the seismically induced pressure to the tank shell and the resulting overturning moments pose some challenges in engineering practice, especially with respect to the impulsive load component (interaction of shell and liquid). The following paper presents two different methods to calculate the eigenperiod, the eigenmode and the associated hydrodynamic pressure distribution for thin cylindrical liquid storage tanks for the circumferential wave number m=1 (lateral ground excitation). The first method includes an improved variation of the added-mass-iteration scheme: It employs a Rayleigh quotient's formulation of the liquid-shell free vibration and repetitively manipulates the distribution of the kinetic energy of the fluid until convergence occurs. The second method involves the calculation of the added mass matrix directly from the appropriate expression for the work done by the liquid-shell interface forces on the basis of the radial displacement shape functions. The closed form solution of the governing matrix equation of motion of the shell enables the computation of higher eigenmodes and no iterative procedure is required.

Keywords: Liquid-shell interaction, flexible tank, added mass matrix, Rayleigh quotient

1 Introduction

It is well known that for design purposes the hydrodynamic effects of a tank can be evaluated as the sum of two parts: A convective part, which represents the action of the portion of the liquid that experiences sloshing motion and an impulsive part, which represents the action of the portion of the liquid that moves in unison with

DOI 10.1007/978-3-658-02810-7_34, © Springer Fachmedien Wiesbaden 2014

the tank. Analytic studies and post-earthquake observations have manifested that the hydrodynamic impulsive forces induced by seismic ground motion in flexible tanks may be appreciably higher than those in rigid tanks of the same dimensions. Therefore, the interaction between the liquid and the elastic container should be taken into account in the seismic design of flexible tanks. This fact necessitates the solution of the eigenvalue problem and the computation of the associated hydrodynamic impulsive pressures. The integration of the latter in conjunction with the seismic motion results to the design impulsive tank forces and moments applied to the tank and the foundation.

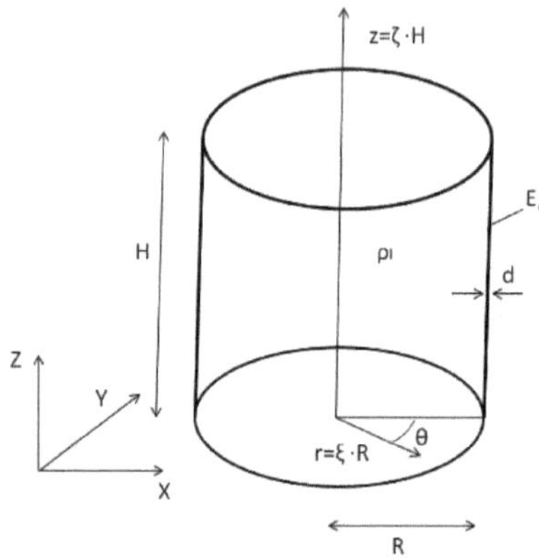

Figure 1: Parameters and geometry of the liquid-storage tank

The objective of this study is threefold: 1. to provide an insight into the current standard proposed method [1] for the determination of the impulsive fundamental period for a beam-like mode and settle the limits of its applicability; 2. to propose an improved variation of the latter method; and 3. to present results from a closed-form solution for the virtual mass of the fluid, which are useful for the design of flexible broad tanks.

The concept of an added hydrodynamic mass to represent the inertial influence of the liquid interacting with the structure is utilized throughout this paper, which circumvents efficiently the necessity of cumbersome fully coupled fluid-structure models. The added mass matrix was incorporated on the nodes of the wet-surface of an empty shell, which has the stiffness provided by the conventional finite element modeling. In order to emphasize the simplicity and competence of the proposed procedures, 4-nodes flat elements were used for the shell model.

2 System and assumptions

The system investigated is shown in figure 1. The tank wall is considered to be of uniform thickness, completely filled with liquid and clamped to the base. The shell of radius R and height H is regarded as a perfect cylinder, thus only the anti-symmetrical mode (m=1) may be excited by a lateral excitation. Modes of m>1 correspond only to imperfect shells [2]. The liquid is presumed to be homogeneous, incompressible, inviscid, free at its upper surface, the flow field is irrotational and only small amplitude, undamped oscillations are investigated. The impulsive and convective components of the response are considered uncoupled due to the fact that the significant liquid sloshing modes and the combined liquid-elastic tank vibrational modes have well-separated frequency ranges [3].

3 Fundamental equations of the oscillating liquid

For an irrotational flow the potential function Φ of an incompressible, non viscous fluid satisfies the Laplace equation:

$$\frac{\partial^2 \Phi}{\partial \xi^2} + \frac{1}{\xi} \cdot \frac{\partial \Phi}{\partial \xi} + \frac{1}{\xi^2} \cdot \frac{\partial^2 \Phi}{\partial \theta^2} + \frac{1}{\gamma^2} \frac{\partial^2 \Phi}{\partial \zeta^2} = 0 \tag{1}$$

where $\Phi = (\zeta, \xi, \theta, t)$ is the velocity potential, which must satisfy the proper kinematic boundary conditions and $\gamma = H/R$. The velocity vector of the liquid is the gradient of the velocity potential and consequently the boundary conditions of the flexible tank impulsive vibration are: At the tank wall the radial velocity component of the fluid must be equal to the corresponding component of the ground motion; therefore:

$$\frac{1}{R} \cdot \frac{\partial \Phi}{\partial \xi} = -\frac{\partial w}{\partial t} \quad \text{at } \xi = 1 \tag{2}$$

At the tank bottom the vertical velocity of the fluid must be zero; therefore:

$$\frac{1}{H} \cdot \frac{\partial \Phi}{\partial \zeta} = 0 \quad \text{at } \zeta = 0 \tag{3}$$

On the free surface of the liquid the pressure is assumed zero; therefore:

$$\frac{\partial \Phi}{\partial t} = 0 \quad \text{at } \zeta = 1 \tag{4}$$

The solution for this case is given by [4]:

$$\Phi_m(\xi, \zeta, \theta, t) = \sum_{n=0}^{\infty} 2 \cdot R \cdot \frac{I_m\left(\frac{\nu_n}{\gamma} \cdot \xi\right)}{\frac{\nu_n}{\gamma} \cdot I_m'\left(\frac{\nu_n}{\gamma}\right)} \cdot \cos(\nu_n \cdot \zeta) \cdot \cos(m\theta) \int_0^1 \dot{w}_{m,n}(\zeta, t) \cdot \cos(\nu_n \cdot \zeta)\, d\zeta \tag{5}$$

where I_m is the modified Bessel function of the first kind of order m, I_m' its derivative, $\gamma = \frac{H}{R}$, $\dot{w}_{m,n}$ is the radial velocity of the shell for the mth circumferential wavenumber and n^{th} eigenmode and $\nu_n = 0.5 \cdot \pi \cdot (2 \cdot n + 1)$. This solution will be subsequently utilized for the constitution of the added mass matrix. Primary the expressions for the potential and kinetic energy of the empty shell will be formulated, which are essential for the solution of the governing matrix equation of the free vibration of the liquid-filled shell. This is the purpose of the following section.

4 Equations governing the shell motion

In terms of calculus of variations, Hamilton's principle is defined as:

$$\delta \int_{t_1}^{t_2} (T - U + W)dt = 0 \quad (6)$$

where T is the kinetic energy, U the potential energy, W is the work done by external loads and δ a variational operator taken during the specified time interval.

Kinetic Energy of the Shell

For the finite element model with the shape functions provided by the model, the translational kinetic energy is written as:

$$T(t) = \frac{1}{2}\sum_{e=1}^{N_{elem.}} \rho_e \int_{t_1}^{t_2}\int_{s_1}^{s_2}\int_{r_1}^{r_2} \left([N]\dot{\mathbf{U}}_{\mathbf{e}}\right)^T \cdot \left([N]\dot{\mathbf{U}}_{\mathbf{e}}\right)\det J \cdot dr \cdot ds \cdot dt = \frac{1}{2}\dot{\mathbf{q}}^T[M_s]\dot{\mathbf{q}} \quad (7)$$

where ρ_e is the density of element e and $[M_s]$ is the mass matrix of the assembled system:

$$[M_s] = \sum_{e=1}^{N_{elem.}} \rho_e \int_{t_1}^{t_2}\int_{s_1}^{s_2}\int_{r_1}^{r_2} [N]^T \cdot [N]\det J \cdot dr \cdot ds \cdot dt \quad (8)$$

$[N]$ is a 3x12 matrix, which contains linear shape functions, r,s,t are the natural coordinates, J is the Jacobian operator relating the natural coordinate derivatives to the local coordinate derivatives and $\dot{\mathbf{q}} = \sum_{\mathbf{e=1}}^{\mathbf{N_{elem.}}} \dot{\mathbf{U}}_{\mathbf{e}}$ is the assemblage velocity nodal vector.

Potential Energy of the Shell

The strain energy of the model is:

$$U(t) = \frac{1}{2}\sum_{e=1}^{N_{elem.}} \int_{t_1}^{t_2}\int_{s_1}^{s_2}\int_{r_1}^{r_2} ([B]\mathbf{U}_{\mathbf{e}})^T \cdot [D]_e \cdot ([B]\mathbf{U}_{\mathbf{e}})\det J \cdot dr \cdot ds \cdot dt = \frac{1}{2}\mathbf{q}^T[K]\boldsymbol{q} \quad (9)$$

where $[K]$ is the stiffness matrix of the assembled system:

$$[K] = \sum_{e=1}^{N_{elem.}} \int_{t_1}^{t_2}\int_{s_1}^{s_2}\int_{r_1}^{r_2} [B]^T \cdot [D]_e \cdot [B]\det J \cdot dr \cdot ds \cdot dt \quad (10)$$

$[B]$ is the strain-diplacement matrix, obtained by appropriately differentiating and combining rows of the matrix $[N]$. $[D]_e$ is the stress-strain matrix of each element and $\mathbf{q} = \sum_{\mathbf{e=1}}^{\mathbf{N_{elem.}}} \mathbf{U}_{\mathbf{e}}$ is the assemblage displacement nodal vector. The element kinememática allow for both finite bending and membrane strains. Having established the above matrices, the solution steps of the tank's free vibration problem on the basis of an iteration scheme will be presented in the following section.

5 Iteration scheme for the antisymmetric vibration

5.1 General concept-implementation

The iteration scheme for the determination of the fundamental frequency of cylindrical above-ground liquid storage tanks for an antisymmetric vibration was first proposed by Fischer and Rammerstorfer [5] and was included in the current standard provisions [1]. The method consists of an initial assumption of the first radial eigenmode, which is used to estimate the normalized dynamic impulsive pressure, the latter being subsequently transformed to an equivalent density distributed over the height of the shell. This density, together with the material density of the shell, is ascribed to the „dry" shell and the eigenvalue problem is solved resulting to a new eigenmode, which is used for an improved assumption of the deformation figure. The iteration process is continued until two successive eigenmodes are practically the same.

The following equations for the assessment of the virtual fluid density should be regarded as part of an arbitrary iteration step i. In the subsequent formulations, quantities are associated with two subindices: The first index refers to the circumferential wavenumber and the second to the axial wavenumber. The resultant of the pressure distribution per unit length in the axial direction is:

$$R_{1,1}(\zeta, t) = \int_0^{2\pi} p_{1,1}(\xi = 1, \zeta, \theta, t) \cdot \cos\theta \cdot R d\theta \tag{11}$$

where the pressure $p_{1,1}$ can be obtained from equation (5) as follows:

$$\begin{aligned} p_{1,1} &= \rho_l \frac{\partial \Phi_1}{\partial t} \\ &= \sum_{n=0}^{\infty} 2 \cdot R \cdot \rho_l \cdot \frac{I_1\left(\frac{\nu_n}{\gamma} \cdot \xi\right)}{\frac{\nu_n}{\gamma} \cdot I_1'\left(\frac{\nu_n}{\gamma}\right)} \cdot \cos(\nu_n \cdot \zeta) \cdot \cos\theta \int_0^1 \ddot{w}_{1,1}(\zeta, t) \cdot \cos(\nu_n \cdot \zeta)\, d\zeta \end{aligned} \tag{12}$$

The first modal radial acceleration maximum value is:

$$\ddot{w}_{1,1}(\zeta, t) = \psi_{1,1}(\zeta) \cdot \Gamma_{1,1} \cdot S_{a,x,1}^{rel.}(t) \tag{13}$$

where $\Gamma_{1,1}$ is the modal participation factor, $\psi_{1,1}$ is the radial eigenmode and $S_{a,x,1}^{rel.}(t)$ is the horizontal relative spectral acceleration. The mass that is attached to the shell at height ζ experiences the same acceleration as the shell itself is determined by:

$$m_{1,1}(\zeta) = \frac{R_{1,1}(\zeta,t)}{\ddot{w}_{1,1}(\zeta,t)} = \frac{\pi \cdot R^2 \cdot \rho_l \cdot \bar{p}_{1,1}(\xi=1,\zeta)}{\psi_{1,1}(\zeta)} \tag{14}$$

where $\bar{p}_{1,1}$ is the normalized pressure defined as [6]:

$$\bar{p}_{1,1} = \frac{p_{1,1}}{R \cdot \rho_l \cdot \cos\theta \cdot \Gamma_{1,1} \cdot S_{a,x,1}^{rel.}} \tag{15}$$

Accordingly, the shell is attributed with a virtual additional mass density $\rho(\zeta)$, which due to the discretization of the model in the axial direction in $k = 1..N_{elem.}^{Z}$ elements can be expressed as:

$$\rho_{1,1}(\zeta) = \frac{R \cdot \rho_l \cdot [\bar{p}_{1,1}(\xi=1,\zeta_{k+1}) + \bar{p}_{1,1}(\xi=1,\zeta_k)]}{2 \cdot d(\zeta) \cdot [\psi_{1,1}(\zeta_{k+1}) + \psi_{1,1}(\zeta_k)]} \tag{16}$$

The work done by the distributed inertia force $F_w = \rho_{1,1} \cdot \ddot{W} = \rho_{1,1} \cdot \ddot{w} \cdot \cos\theta$ is:

$$\delta W(t) = -\sum_{e=1}^{N_{elem.}} \rho_{1,1} \int_{t_1}^{t_2} \int_{s_1}^{s_2} \int_{r_1}^{r_2} ([N]W_e)^T \cdot ([N]\ddot{W}_e) \det J \cdot dr \cdot ds \cdot dt = \delta \mathbf{q}^T [M_A] \ddot{\mathbf{q}} \tag{17}$$

where $[M_A]$ is the added mass matrix and $\rho_{1,1}$ the density obtained by equation (16). Inserting equations (7), (9) and (17) into (6), Hamilton's principle is rewritten as:

$$\int_{t_1}^{t_2} (\delta \dot{q}^T [M_s] \dot{q} - \delta q^T [K] q - \delta q^T [M_A] \ddot{q}) dt = 0 \tag{18}$$

The integration by parts in time of the first term of equation (18) leads to the governing matrix equation of the free lateral, undamped vibration of the liquid-storage tank since the time-boundary term vanishes according to the application conditions of Hamilton's principle. This equation is given by:

$$([M_s] + [M_A])\ddot{q} + [K]q = 0 \tag{19}$$

It is evident from equation (14) that large values of the virtual density appear near the tank base due to the fixed-base conditions (figure 2a). However, the solution of equation (19) shows that the singularity has a minor influence on the fundamental frequency and the first eigenmode of slender tanks with $H/R \geq 2$, independent on the ratio R/d, and the iteration procedure can be therefore followed readily leading to precise results, as can be shown in table 1 given in Appendix, where the dimensionless coefficient $c_{1,1} = H \cdot \omega_{1,1} \cdot \rho/E$ of the angular frequency $\omega_{1,1}$ is listed for different proportions of steel tanks. Nevertheless, within the framework of the iteration routine, the free vibration of broad liquid-filled tanks (H/R<2) becomes a formidable problem, since in that case the centroid of the total mass moves towards the base and the singularity becomes dominant leading to unstable iterative steps and finally loss of convergence.

5.2 Rayleigh's quotient of the Euler-Bernoulli beam

In order to compensate with this singularity, which has no physical meaning, the Rayleigh's quotient is employed for a system with distributed mass. The concept is based on keeping constant the virtual mass below a specific heigt ζ_c (figure 2c) during the iteration and distributing the cutted off portion of the mass linearly along the height of the shell (figure 2b). In order to eliminate the arbirtrariness of this manipulation the Rayleigh's quotient between the initial and the modified dynamic system is enforced to be equal and therefore the linear distributed added mass can be determined. The development of this method, assuming that the shell is behaving as a slender beam, was proposed by Rammerstorfer et al. [7]. In this paper, a similar scheme has been initially carried out, but some novel observations are made regarding the applicability of the method for broad tanks.

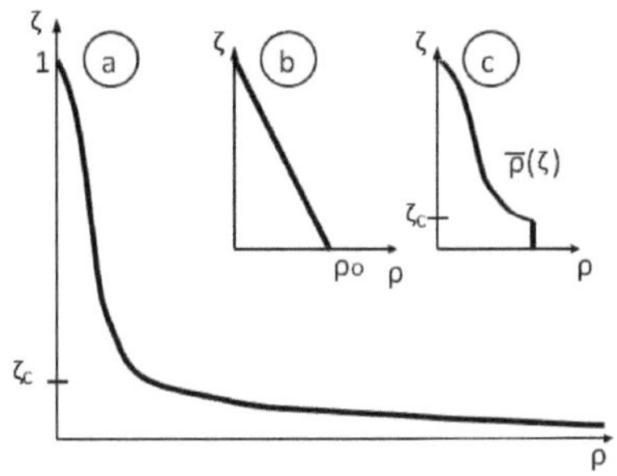

Figure 2: Distribution of the virtual fluid density without (a) and with (b),(c) use of the Rayleigh's quotient

For the uniform Euler-Bernoulli beam of length L and constant cross-section the Rayleigh's quotient can be promptly obtained by:

$$S_b = \omega^2 = \frac{V_{max}}{T_{max}} = \frac{\int_0^1 EI[\psi''(\zeta)]^2 d\zeta}{\int_0^1 \rho(\zeta) A[\psi(\zeta)]^2 d\zeta} \tag{20}$$

where ζ is a dimensionless length coefficient, $\psi(\zeta)$ is an assumed shape function and EI, A, $\rho(\zeta)$ the stiffness, surface area and density respectively. $\psi(\zeta)$ and $\rho(\zeta)$ correspond to $\psi_{1,1}(\zeta)$ and $\rho_{1,1}(\zeta)$ respectively.

After truncating the density up to a dimesionless height ζ_c and distributing the still unknown density ρ_o axialsymmetrically and linearly along the height, the Rayleigh's quotient is formulated as follows:

$$S_b = \frac{EI \int_0^1 [\psi''(x)]^2 dx}{A \int_0^1 [\psi(x)]^2 (1-x)\rho_o + [\psi(x)]^2 \bar{\rho}(x) dx} \tag{21}$$

By equating equations (20) and (21) and splitting the integrals at height ζ_c, the following equation for the density ρ_o is obtained :

$$\rho_o = \frac{\int_0^{\zeta_c} \Delta\rho(\zeta)[\psi(\zeta)]^2 d\zeta}{\int_0^1 [\psi(\zeta)]^2 (1-\zeta) d\zeta} \tag{22}$$

where $\Delta\rho$ stands for the truncated portion of the density. In this study, a minimum height ζ_c was estimated exclusively on the basis of the criterium of convergence for the iterative procedure. This is believed to be justified as it furnishes greater values of ζ_c as the slenderness of the tanks decreases: In this case the singularity spreads out over a greater relative height of the shell due to the gradual shift to the bottom of the maximum value of deflection, especially for higher values of the ratio R/d.

Even though the application of the Rayleigh's quotient for a tank, assuming that it behaves as a slender beam, generally results in convergence of the fundamental eigenmode, it is obvious from table 2 that by increasing the broadness of the tank, notable discrepancies from the eigenvalues reported in the literature appear. It has to be mentioned, that for broad tanks and lower values of the ratio R/d convergence can be only ensured by sufficiently large values of ζ_c, which leads to spurious eigenmodes, since the distribution of the virtual mass tends to be necessarily linear along the height, which violates the validity of equation (16). This approach reaches its limits for the cases bounded by the ratios $\gamma \leq 0.8$ and $R/d \geq 1000$. Within them convergence cannot be reached independent of the choice of ζ_c.

It can be clearly stated that this method underestimates the density ρ_o, resulting almost monotonically in larger eigenvalues when the slenderness of the tank decreases. The latter conclusion may be confirmed by the closeness of the values obtained by the present study (table 2) and the corresponding values resulting from the equation proposed in [1], which emanates from a similar study [7].

5.3 Proposal of Rayleigh's quotient for the coupled shell-liquid vibration

In this study, focusing on disposing of the aforementioned singularity, a more precise approach is proposed, which takes into account the coupled free vibration of the fluid-structure. By using the orthogonality relations of wet modes obtained by Zhu [8] and neglecting free surface waves, the Rayleigh's quotient for the coupled free vibration of a inviscid, irrotational, incompressible liquid and the cylindrical shell is formulated as follows [9]:

$$S_T = \omega^2 = \frac{\iint_\Omega \mathbf{U} \cdot \mathrm{M}(\mathbf{U}) dS}{\rho d \iint_\Omega \mathbf{U} \cdot \mathbf{U}\, dS + \rho_l \cdot \iiint_V \nabla\Phi \cdot \nabla\Phi dV} \tag{23}$$

where M denotes a partial differential operator, **U** the displacement vector of the middle surface, Ω the mean surface of the structure which coincides with the wetted area S, and Φ the fluid deformation potential. The latter is given by:

$$\Phi_m(\zeta, \theta) = 2 \cdot R \cdot \sum_{n=0}^{\infty} A_m \frac{I_m\left(\frac{\nu_n}{\gamma}\right)}{\frac{\nu_n}{\gamma} \cdot I_m'\left(\frac{\nu_n}{\gamma}\right)} \cdot \cos(\nu_n \cdot \zeta) \cdot \cos(m\theta) \tag{24}$$

where $A_m = \frac{1}{\pi} \int_0^{2\pi} \int_0^1 w_m(\zeta, \theta) \cdot \cos(\nu_n \cdot \zeta) \cos(m\theta) d\zeta d\theta$

The radial displacements W can be expressed as:

$$W_m(\zeta, \theta) = \cos(m\theta) \cdot \psi(\zeta) \tag{25}$$

If the Green's thorem for the kinetic energy of the fluid is applied, the Rayleigh's quotient can be written as:

$$S_T = \omega^2 = \frac{\iint_S \mathbf{U} \cdot \mathrm{M}(\mathbf{U}) dS}{\rho d \iint_S \mathbf{U} \cdot \mathbf{U}\, dS + d \iint_S \bar{\rho} \cdot W^2\, dS + \iint_{S_c} \Phi \cdot \frac{\partial\Phi}{\partial n} dS} \tag{26}$$

where $\bar{\rho} = \bar{\rho}(\zeta)$ is the virtual density and S_c is the wetted surface bounded from the tank bottom and the height ζ_c. The above equation presumes that the kinetic energy associated with the curtailed virtual mass constitutes the total kinetic energy of the fluid contained up to the dimesionless height ζ_c. This is justified since this portion of the mass is substantially larger than the constant complementary part. The

Rayleigh's quotient, after cutting off the density at a height ζ_c and distributing axialsymmetrically and linearly along the height the still unknown density ρ_o, is formulated as follows:

$$S_T = \omega^2 = \frac{\iint_\Omega \mathbf{U}\cdot\mathbf{M}(\mathbf{u})dS}{\rho d \iint_\Omega \mathbf{U}\cdot\mathbf{U}\, dS + d\iint_\Omega (\bar{\rho}_o + \bar{\rho})\cdot W^2\, dS} \tag{27}$$

where $\bar{\rho}_o = (1-\zeta)\cdot\rho_o$. Taking into account the boundary condition on the tank wall and equating (26) and (27) one obtains for m=1:

$$\rho_o = \frac{T_l}{T_l^o} \tag{28}$$

where :

$$T_l = 2\cdot R^2\cdot\rho_l\cdot H\cdot\pi\cdot\sum_{n=0}^{\infty}\frac{I_1\left(\frac{\nu_n}{\gamma}\right)}{\frac{\nu_n}{\gamma}\cdot I_1{}'\left(\frac{\nu_n}{\gamma}\right)}\cdot F_{1,1} \tag{29}$$

$$F_{1,1} = \int_0^{\zeta_c}\psi_{1,1}(\zeta)\cdot\cos(\nu_n\cdot\zeta)\,d\zeta\int_0^1\psi_{1,1}(\zeta)\cdot\cos(\nu_n\cdot\zeta)\,d\zeta$$

$$T_l^o = d\cdot H\cdot R\cdot\pi\int_0^{\zeta_c}(1-\zeta)\cdot\left[\psi_{1,1}(\zeta)\right]^2 d\zeta \tag{30}$$

This approach assured the convergence of the iteration procedure for all tanks examined in few loop steps and furnishes results of high accuracy as can be seen from table 3.

6 Computation of a closed-form added mass matrix

Another way of treating the free vibration problem is to derive the closed-form added mass matrix directly from the appropriate expression for the work done by the liquid-shell interface forces. In this method, the recurrent evaluation of a large number of Bessel functions terms in the equation of pressure (eq. 12) is avoided.

The work done by the liquid pressure through an arbitrary virtual displacement $\delta w_m\cdot\cos(m\theta)$ can be written as:

$$\delta W = \int_0^H\int_0^{2\pi} p_m(R,z,\theta,t)\cdot\delta w_m(z,\theta,t)\cdot R\cdot d\,\theta dz \tag{31}$$

The proper substitution of (12) into (31) for m=1 results to:

$$\delta W = -\sum_{n=0}^{\infty}\frac{2\cdot\pi\cdot R\cdot\rho_l\cdot I_1(\nu_n\cdot R)}{H\cdot\nu_n\cdot I_1{}'(\nu_n\cdot R)}\left(\int_0^H\delta w\cdot\cos(\nu_n\cdot z)dz\right)\left(\int_0^H\ddot{w}\cdot\cos(\nu_n\cdot z)dz\right) \tag{32}$$

In order to compute the added mass matrix, the integrals of (32) are expressed in terms of the nodal displacement vector, which coincides with the corresponding vector of an axisymmetric shell, modelled with cylindrical elements, since the integration in (31) is analytically performed in the circumferential direction. However, due to the symmetry of the structure, the implemented shell model consisting of flat elements can still be used if the added mass matrix is incorporated to shell nodes proportionally to the size of each element in the circumference.

The second integral of equation (32) can be representatively discretized as:

$$\int_0^H \ddot{w}(z,t)\cdot\cos(\nu_n\cdot z)dz = \sum_{k=1}^{N^Z_{elem.}} \mathbf{d}_\mathbf{k}^\mathbf{T}\cdot\ddot{\mathbf{w}}(\mathbf{t}) = \mathbf{D}^\mathbf{T}\ddot{\mathbf{q}}(\mathbf{t}) \tag{33}$$

where:

$$\mathbf{d}_\mathbf{k}^\mathbf{T} = \int_0^{L_e} \mathbf{N}(\bar{\mathbf{z}})^\mathbf{T}\cdot\cos(\nu_n\cdot(\bar{z}+(e-1)\cdot L_e)d\bar{z} \tag{34}$$

$\mathbf{N}(\bar{\mathbf{z}})$ is the vector of linear shape functions of the elements in the axial direction, L_e the length of each element and $\bar{z}$ the local element coordinate.

Equation (31) can be subsequently formulated as:

$$\delta W = -\delta\mathbf{q}^T\left(\sum_{n=0}^{\infty}\frac{2\cdot\pi\cdot R\cdot\rho_l\cdot I_1(\nu_n\cdot R)}{N^Z_{elem.}\cdot H\cdot\nu_n\cdot I_1{}'(\nu_n\cdot R)}\mathbf{D}\cdot\mathbf{D}^\mathbf{T}\right)\cdot\delta\ddot{\mathbf{q}} = -\delta\mathbf{q}^T[M_A]\ddot{\mathbf{q}} \tag{35}$$

where $[M_A]$ is the added mass matrix of the elements with same circumferential coordinates. Inserting equations (7), (9) and (35) into (6), the equation of the free lateral vibration of the tank is deduced similarly to the paragraph 5.1.

Having solved the eigenvalue problem, the tank forces may be evaluated by integration of the impulsive pressures. For design purposes the base shear force and the overturning moments immediately above and below the tank base can be evaluated by multiplication of the quantities $m_{1,n}$, $m_{1,n}\cdot h_{1,n}$ and $m_{1,n}\cdot\Delta h_{1,n}$ respectively with spectral accelerations provided in standard codes [4]. The quantity $m_{1,n}$, represents the nth modal mass of the shell-liquid system, and $h_{1,n}$, and $\Delta h_{1,n}$ heights at which this mass must be concentrated to furnish the correct modal components of base moments. In table 4 the obtained results are presented in normalized form for a range of typical broad tanks. Satisfactory agreement was realized by comparing these values with the results published in [10].

7 Conclusions

In this study the drawbacks of the iteration scheme proposed by the current standard provisions for the determination of the fundamental period of the liquid-shell system were highlighted and an improved approach was proposed. An alternative, efficient method for the solution of the eigenvalue problem was developed, which enables the assessment of higher eigenfrequencies.

8 Acknowledgement

The study was carried out under a research grant of the German Research Council (DFG). The financial support of the DFG is very much appreciated. Furthermore we thank Dr. Ing. Christoph Butenweg at the Chair of Structural Statics and Dynamics for his encouraging and helpful discussions.

REFERENCES

[1] Eurocode 8: Design of structures for earthquake resistance-Part 4: Silos, tanks and pipelines.

[2] A. Kalnins; D. A. Godfrey: Seismic analysis of thin shell structures; Nuclear Engineering and Design; 1974, 68-76, 27.

[3] M. A. Haroun: Dynamic analyses of liquid storage tanks, Reserarch Report, Earthquake Engineering Research Laboratory, California Institute of Technology, 1980.

[4] J. Habenberger: Beitrag zur Berechnung von nachgiebig gelagerten Behältertragwerken unter seismischen Einwirkungen, Dissertation, Bauingenieurwesen der Bauhaus-Universität Weimar, 2001.

[5] D. F. Fisher; F. G. Rammerstorfer: The stability of liquid-filled cylindrical shells under dynamic loading; Proceedings of a State-of-the-Art Colloquium; 1982, 569-597.

[6] K. Meskouris; K.G. Hinzen; C. Butenweg; M. Mistler: Bauwerke und Erdbeben, Viewer & Teubner, 2011.

[7] F. G. Rammerstorfer; K. Scharf; F. D. Fischer; R. Seeber: Collapse of earthquake ecxited tanks; Res Mechanica, 1988, 129-143, 25.

[8] F. Zhu: Orthogonality of wet modes in coupled vibration; Journal of Sound and Vibration; 1991, 439-338, 146.

[9] F. Zhu: Rayleigh quotients for coupled free vibrations; Journal of Sound and Vibration; 1994, 641-649, 171.

[10] Y. Tang: Studies on the dynamic response of liquid-storage tanks, Ph.D. Thesis, Rice University, Houston, Texas, 1986.

Table 1: Dimensionless coefficient $C_{1,1} = H \cdot \omega_{1,1} \cdot \rho/E$ for a fully filled shell with liquid (ν=0.3, ρ_l/ρ=0.127, m=1,n=1). Use of standard added-mass concept. In parantheses the relative % difference from the references

	R/d = 1000				R/d = 500			
H/R	Present Analysis	Ref. 10 (Tang)	Ref. 4 (Habenberger)	Ref. 1 (EC 8 part 4)	Present Analysis	Ref. 10 (Tang)	Ref. 4 (Habenberger)	Ref. 1 (EC 8 part 4)
2.0	0.095	0.0896 (6%)	0.0895 (6%)	0.095 (0%)	0.133	-	0.125 (6%)	0.135 (1%)
2.5	0.087	0.0848 (3%)	0.0838 (4%)	0.089 (2%)	0.121	-	0.118 (3%)	0.125 (3%)
3.0	0.080	0.0792 (1%)	0.0780 (3%)	0.082 (2%)	0.111	-	0.110 (0%)	0.116 (4%)
3.5	0.072	-	0.0735 (2%)	0.076 (5%)	0.101	-	0.104 (3%)	0.107 (6%)
4.0	0.067	-	0.0690 (3%)	0.070 (4%)	0.093	-	0.097 (6%)	0.098 (5%)

Table 2: Dimensionless coefficient $C_{1,1} = H \cdot \omega_{1,1} \cdot \rho/E$ for a fully filled shell with liquid (ν=0.3, $\rho l/\rho$=0.127, m=1,n=1). Use of Rayleigh's quotient of a slender beam vibration. In parantheses the relative % difference from the references

	R/d = 1000				R/d = 500			
H/R	Present Analysis	Ref. 10 (Tang)	Ref. 4 (Habenberger)	Ref. 1 (EC 8 part 4)	Present Analysis	Ref. 10 (Tang)	Ref. 4 (Habenberger)	Ref. 1 (EC 8 part 4)
0.6	no conv.	0.0763	0.0763	0.1	0.148	-	0.112 (32%)	0.142 (4%)
0.8	0.111	0.0829 (34%)	0.0830 (34%)	0.104 (7%)	0.154	-	0.118 (31%)	0.147 (5%)
1.0	0.111	0.0875 (27%)	0.0870 (28%)	0.105 (6%)	0.155	-	0.122 (27%)	0.149 (4%)
1.2	0.109	0.0903 (20%)	0.0900 (21%)	0.105 (4%)	0.152	-	0.125 (22%)	0.148 (3%)
1.4	0.106	0.0915 (16%)	0.0915 (16%)	0.103 (3%)	0.147	-	0.128 (15%)	0.146 (1%)
1.6	0.102	0.0917 (11%)	0.0917 (11%)	0.101 (1%)	0.143	-	0.128 (12%)	0.143 (0%)
1.8	0.098	0.09090 (8%)	0.0909 (8%)	0.098 (0%)	0.138	-	0.127 (9%)	0.139 (1%)

Table 3: Dimensionless coefficient $C_{1,1} = H \cdot \omega_{1,1} \cdot \rho/E$ for a fully filled shell with liquid (ν=0.3, ρ_l/ρ=0.127, m=1,n=1). Use of Rayleigh's quotient of the shell-liquid coupled vibration. In parantheses the rel. % diff. from the references

	R/d = 1000				R/d = 500			
H/R	Present Analysis	Ref. 10 (Tang)	Ref. 4 (Habenberger)	Ref. 1 (EC 8 part 4)	Present Analysis	Ref. 10 (Tang)	Ref. 4 (Habenberger)	Ref. 1 (EC 8 part 4)
0.6	0.070	0.0763 (8%)	0.0763 (8%)	0.100 (30%)	0.102	-	0.112 (9%)	0.142 (28%)
0.8	0.087	0.0829 (5%)	0.0830 (5%)	0.104 (16%)	0.126	-	0.118 (7%)	0.147 (14%)
1.0	0.092	0.0875 (5%)	0.0870 (6%)	0.105 (12%)	0.133	-	0.122 (9%)	0.149 (11%)
1.2	0.093	0.0903 (3%)	0.0900 (3%)	0.105 (11%)	0.133	-	0.125 (6%)	0.148 (10%)
1.4	0.093	0.0915 (2%)	0.0915 (2%)	0.103 (10%)	0.132	-	0.128 (3%)	0.146 (10%)
1.6	0.096	0.0917 (4%)	0.0917 (4%)	0.101 (5%)	0.129	-	0.128 (1%)	0.143 (10%)
1.8	0.094	0.0909 (3%)	0.0909 (3%)	0.0980 (4%)	0.126	-	0.127 (1%)	0.139 (9%)

Table 4: Dimensionless coefficients in the expressions of impulsive tank forces and moments (ν=0.3, ρ_l/ρ=0.127, R/d=1000). Use of the closed form added mass matrix

H/R	$\frac{m_{1,1}}{m_l}$	$\frac{m_{1,2}}{m_l}$	$\frac{m_{1,1} \cdot h_{1,1}}{m_l \cdot H}$	$\frac{m_{1,2} \cdot h_{1,2}}{m_l \cdot H}$	$\frac{m_{1,2} \cdot \Delta h_{1,1}}{m_l \cdot H}$	$\frac{m_{1,2} \cdot \Delta h_{1,2}}{m_l \cdot H}$
0.6	0.36	0	0.14	0	0.29	0
0.8	0.46	0	0.19	0	0.22	0
1.0	0.54	0	0.22	0	0.16	0
1.2	0.60	0.01	0.26	-0.010	0.12	0
1.4	0.65	0.02	0.29	-0.010	0.08	0.01
1.6	0.66	0.04	0.30	-0.013	0.06	0.02
1.8	0.68	0.05	0.33	-0.020	0.05	0.02

Seismic Design of Spherical Pressure Vessels

Matthias Wieschollek[1], **Marius Pinkawa**[1], **Benno Hoffmeister**[1] **and Markus Feldmann**[1]

[1] Institute of Steel Construction, RWTH Aachen University
Mies-van-der-Rohe-Str.1, 52074 Aachen, Germany
wieschollek@stb.rwth-aachen.de

ABSTRACT

Spherical pressure vessels are globally used for storage of pressurized liquids or gases of different hazard classes. An adequate seismic design of these structures must consider their particular structural behaviour and consequences of possible damage or failure. A study of the current standard situation for seismic design of pressure vessels revealed significant gaps and missing design rules, in particular for spherical pressure vessels. Within the European Research Project INDUSE the seismic performance and applicability of existing European and American codes to pressure vessels with cylindrical and spherical shape were investigated. This paper describes the results of a study on different examples of spherical pressure vessels which were selected to be representative for the current practice. The study comprised numerical investigations as well as simplified models for the estimation of the dynamic properties of the vessel structures. It is shown, which failure modes and stress concentrations areas are crucial in the event of an earthquake. In addition engineering calculation methods to determine fundamental periods and internal forces for braced and non-braced spherical pressure vessels were developed and compared to results of numerical simulations. The applicability of behaviour factors is discussed based on proposals made by European and American codes in comparison to own results. Recommendations for the behaviour factor of spherical pressure vessels with different dimensions were developed based on push over analyses and non-linear incremental dynamic analyses. Furthermore the influence of sloshing effects in spherical vessels, for which no specific rules are given in the codes, was investigated according to the current state of the art.

Keywords: Spherical pressure vessel, behaviour factor, fundamental period, non-linear static pushover analysis, failure mode

S. Klinkel et al. (Eds.), *Seismic Design of Industrial Facilities*,
DOI 10.1007/978-3-658-02810-7_35,

1 Introduction

Within the European Research Project INDUSE "Structural safety of industrial steel tanks, pressure vessels and piping systems under strong seismic loading" guidelines for seismic design and analyses of industrial pressure vessels were developed [1]. The considered types of pressure vessels were vertical pressure vessels on skirt supports, horizontal pressure vessels on saddle supports and spherical pressure vessels supported by an even number of braced or non-braced columns with circular hollow sections. In particular the determination of fundamental periods, the application of adequate behaviour factors and the determination of seismic forces as well as the definition of limit states (failure modes) and dimension limits for the pressure vessels and the supporting structures are in the focus of these guidelines. The recommendations and seismic design rules are also illustrated by means of design examples of pressure vessels with various geometries as well as parametric studies given within the background document of these guidelines [2]. The main results of the design examples are already discussed in [3]. This paper focuses on the guidelines and recommendations, which comprise common rules valid for any type of pressure vessel as well as rules depending on the type of the pressure vessel. They are meant to be considered in addition to existing seismic design rules for steel structures which - to large extend - remain valid for pressure vessels and in particular for their supporting structures. The application of these rules is given in [2] and [3] based on two design examples as well as by means of a parametric study with 78 different geometries.

2 General

The following recommendations and design rules refer to spherical pressure vessels with two different supporting systems as shown in Figure 1 (non-braced and braced columns) and even number of columns. The columns are made of circular hollow sections with hinged connections to the foundation, whereas for the bracings only the cross sectional areas are considered. The recommendations are based on investigations of spherical pressure vessels with a mean diameter of the sphere of $15\ m \leq d_s \leq 25\ m$ and a number of columns $n_c = 4, 8, 12$; they are mainly related to the supporting structure and to the connection of the columns to the spherical shell of the vessel (shell-column connection). They aim at achieving a response of the spherical vessel governed by the supports, minimising the influence of the deformations of the shell and of the local behaviour of the connections [1].

3 Limit values for dimensions

In order to obtain stable results for seismic analyses and a pressure vessel system providing a pronounced ductile behaviour as well as to obtain realistic results by using the simplified beam model given below the following conditions for the dimensions of the whole system should be observed [1]:

1. For columns at least cross-sectional class 3 is required and at least cross-sectional class 2 is recommended

2. The relative slenderness of the columns λ_c shall not exceed the following limit:

$$\lambda_c = \frac{l_{c,ef}}{i_c} \leq 65 \tag{1}$$

with: $l_{c,ef} = \beta_c \cdot l_c \qquad (\beta_c = 2.0)$

$$i_c = \frac{1}{2}\sqrt{r_{c,e}^2 + r_{c,i}^2}$$

Note: The buckling length factor for braced columns depends on the position of the column and the resulting elongation stiffness c_b as well as the position of the attached bracings $l_{c,1}$; however the factor is in the range of $2/3 \leq \beta_c \leq 2.0$, whereby $\beta_c = 2.0$ corresponds to $c_b = 0$ and is always on the safe side. The limit value of $\lambda_c \leq 65$ is only valid for $\beta_c = 2.0$.

3. The ratio of the cross-sectional areas of columns and bracings $\psi_{c,b}$ fulfils the following condition:

$$\psi_{c,b} = \frac{A_c}{A_b} \geq 8 \tag{2}$$

In addition in order to exclude as far as possible the influence of the shell structure on the elastic behaviour as well as to avoid plastification of the shell structure under high seismic loads the following conditions are recommended [1]:

4. For braced or non-braced 4-columns systems the dimensions of the sphere are within the following limit:

$$d_s / t_s \leq 265 \tag{3}$$

The mean diameter of the sphere should not exceed $d_s \leq 20m$.

5. For braced or non-braced 8-columns systems the dimensions of the sphere are within the following limit:

$$d_s / t_s \leq 265 \tag{4}$$

The mean diameter of the sphere should not exceed $d_s \leq 25m$.

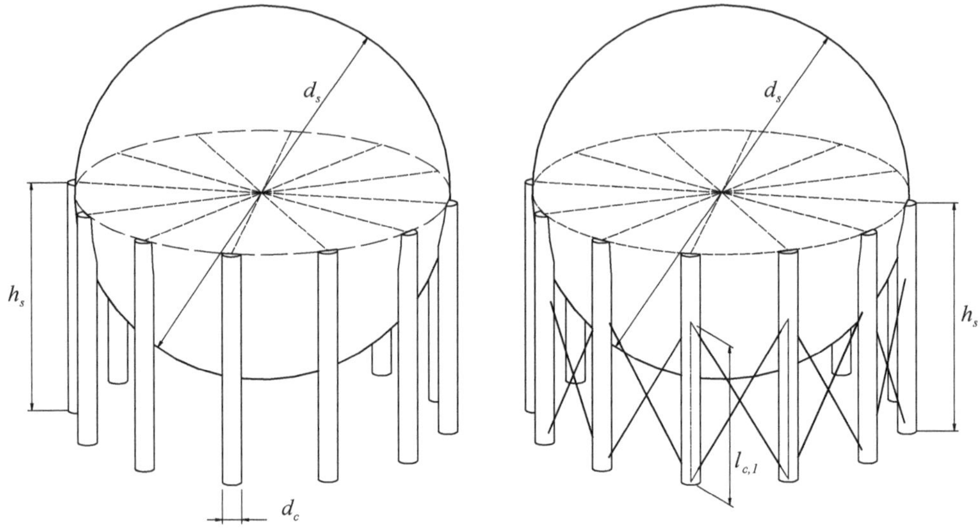

Figure 1: Geometry of spherical pressure vessels supported by non-braced columns (left) and braced columns (right) [1]

4 Simplified beam model

4.1 Shell-column connection

In accordance with the above mentioned conditions spherical pressure vessels supported by columns may be represented by simplified beam models. Figure 1 show the two different supporting systems, which are covered by these guidelines. The effective length of the columns l_c to be used in such model shall be determined considering the shell-column connection in Figure 2; it is given by equation (5). Using this effective length the elastic stiffness of the system and the formation of plastic zones can be reproduced with sufficient accuracy [1].

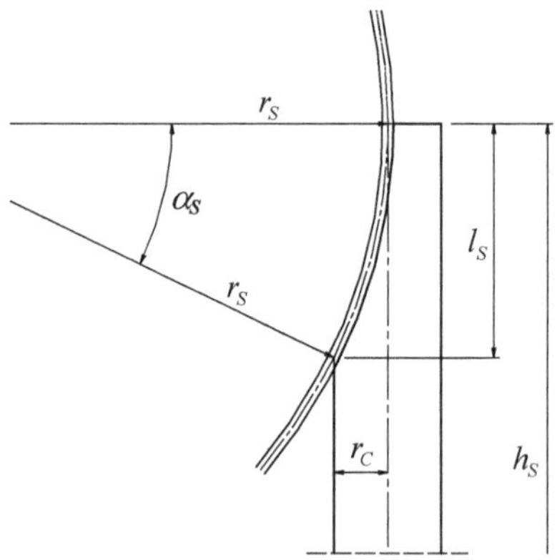

Figure 2: Geometry of shell-column connection [1]

$$l_c = h_s - 0.45 \cdot l_s \qquad (5)$$

with: $l_s = r_s \cdot sin(\alpha_s)$ $\alpha_s = arccos\left(1 - \frac{r_c}{r_s}\right)$

4.2 Global elastic stiffness of non-braced systems

The global elastic stiffness of spherical pressure vessels supported by non-braced columns k_{pv} is independent of the load direction and can be calculated using equation (6). Figure 3 (right) shows schematically the simplified beam model of the whole braced or non-braced system [1].

$$k_{pv} = n_c \cdot \frac{3EI_c}{l_c^3} \tag{6}$$

with: $I_c = \frac{\pi}{4} \cdot \left(r_{c,e}^4 - r_{c,i}^4 \right)$

4.3 Global elastic stiffness of braced systems

The global elastic stiffness of spherical pressure vessels supported by braced columns k_{pv} is also independent of the load direction and can be calculated using equation (7) by neglecting bracings in compression and equation (8) by consideration of bracings in compression and tension. It should be noticed, that this expression does not consider imperfections (pre-curvatures) of bracings, so that the calculation with consideration of bracings in compression and tension leads to global elastic stiffness which is generally too high and furthermore in consequence to higher seismic actions. Thus it is recommended to neglect bracings in compression [1].

$$k_{pv} = \frac{n_c}{2} \left[k_1 + \frac{k_1 \cdot k_2}{b\left(a^2 - 3\right) + k_2} \right] \tag{7}$$

$$k_{pv} = n_c \left[\frac{k_1 \cdot k_2}{b\left(a^2 - 3\right) + k_2} \right] \tag{8}$$

with: $k_1 = \frac{3EI_c}{l_c^3}$ $\qquad k_2 = \frac{4 \cdot k_1}{c_b \cdot a^2} + 12 - 8a$

$a = l_{c,1} / l_c$ $\qquad b = 3 - a^2$

The above mentioned expressions refer to a simplified single column-bracing system as shown in Figure 3 (left), which is directed parallel to the load direction. By means of this system also the elongation stiffness of a single bracing c_b can be calculated as follows [1]:

$$c_b = \frac{EA_b}{l_b} \cdot cos^2 \alpha_b \quad \text{or} \quad c_b = \frac{EA_b}{l_b'} \cdot cos^3 \alpha_b \tag{9}$$

with: $\alpha_b = arctan\left(\frac{l_{c,1}}{l_b'} \right)$ $\qquad l_b' = d_s \cdot sin\left(\frac{\beta}{2} \right)$ $\qquad \beta = \frac{360°}{n_c}$

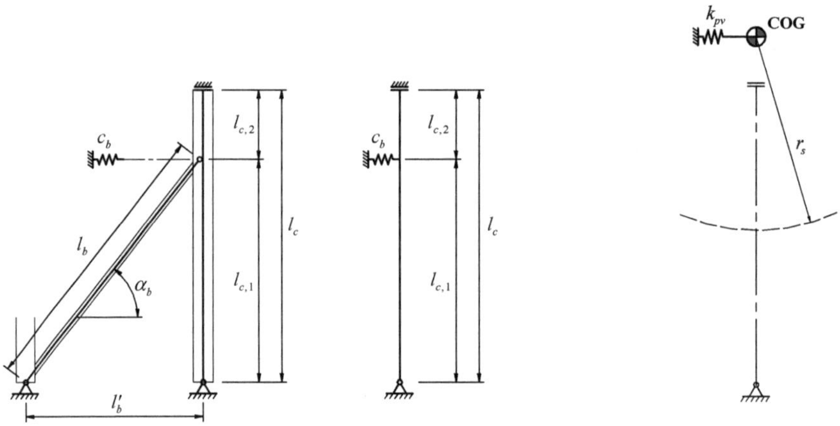

Figure 3: Simplified single column-bracing system (left) and simplified beam model of the whole braced or non-braced system (right) [1]

5 Calculation of fundamental period

Alternatively to a modal analysis the calculation of fundamental period may be determined using the above described simplified beam model, which represent an inverted pendulum and behave like a single degree of freedom system (SDOF). Thus the fundamental period T_1 results to [1]:

$$T_1 = \frac{1}{f_1} \qquad (10)$$

with: $$f_1 = \frac{\omega_1}{2\pi} \qquad \omega_1 = \sqrt{\frac{k_{pv}}{m_t}}$$

The total operating mass m_t consists of the mass of the empty pressure vessel m_{pv} and the mass of the filling m_f and can be determined in simplified terms by equation (11), whereby m_f depends on the filling level k_f or height h_f respectively.

$$m_t = m_{pv} + m_f \qquad (11)$$

with: $$m_{pv} = \rho_{st} \cdot \pi \cdot \left[\frac{1}{6}\left(d_{s,e}^3 - d_{s,i}^3\right) + n_c \cdot h_s \cdot \left(r_{c,e}^2 - r_{c,i}^2\right) + n_b \cdot l_b \cdot \left(r_{b,e}^2 - r_{b,i}^2\right)\right]$$

$$m_f = \rho_f \cdot \frac{\pi \cdot h_f^2}{3} \cdot \left(3 \cdot r_{s,i} - h_f\right) \qquad h_f = k_f \cdot d_{s,i}$$

The conducted parametric numerical study of braced and non-braced pressure vessels with various geometries described in [2] showed an excellent agreement between the proposed simplified calculation of fundamental periods and the significantly more complex modal analysis. The deviations were below 2% in average and maximum 8.0% for braced systems (neglecting bracings in compression) and also below 2% in average and maximum 6.2% for non-braced systems, whereby the maximum deviations referred to systems where the recommended limit values for the dimensions (see chapter 3) were not complied with. By using a modal analysis for the calculation of fundamental periods of braced systems it should be mentioned, that even though consideration of pre-curved bracings the structure behaves too stiff, which leads unnecessarily to higher seismic actions (see [2] and [3]). With regard to the performed pushover analyses in [2] it was observed that all bracings in compression buckled very early in the elastic range and failed immediately by reaching the yield load in terms of plastic hinges. Therefore it is recommended to neglect bracings in compression. For calculating the horizontal seismic base shear force including effects of liquid sloshing a simple and efficient method is given in [4], which is compatible with the corresponding methods in existing specifications for liquid storage tanks, and can be used for the seismic design of spherical pressure vessels [1].

6 Behaviour factor

The behaviour factor q used for steel structures represents their capacity to dissipate energy by means of plastifications caused by an earthquake. It is used for determination of design response spectra taking into account and allowing for non-linear behaviour of the structure. The parts of Eurocode 8 (EN 1998-1 [5] and EN 1998-4 [6]) provide no clear regulations for the behaviour factor of spherical pressure vessels. However the conducted parametric numerical study [2] showed that pressure vessels supported by braced or non-braced columns behave like an inverted pendulum, so that the behaviour factor can be assumed to $q = 2.0$ according to EN 1998-1, Table 6.2 [5]. On the other hand the American standard ASCE/SEI 7-05 [7] distinguishes between building and non-building structures and Table 15.4-2 includes precise declaration for non-building elevated pressure vessels, so that the behaviour factor can be assumed with $q = 2.0$ for non-braced systems and $q = 3.0$ for braced systems [1].

In order to determine exemplary the behaviour factor, for the design examples given in [2] and [3] incremental non-linear dynamic analyses were performed, resulting in behaviour factors above $q = 2.0$. A behaviour factor of $q = 2.7$ was found for the non-braced system, which was limited by dynamic instability phenomena. For the braced system a behaviour factor of $q = 2.3$ was identified under the condition that the plastification shall be limited to the bracings only. Otherwise higher values than $q = 3.0$ were determined, if additional plastifications of the columns were accepted. Moreover the parametric numerical study (pushover analyses) [2] showed that by compliance with the limit values for the dimensions of the vessels (see chapter 3) the displacement ductility $\mu = e_{ult}/e_y$ for different braced and non-braced systems with 4, 8 or 12 columns remains relatively constant. For the non-braced systems slightly higher values than $\mu = 2.0$ were found, whereas for the braced systems the displacement ductility showed to be higher than $\mu = 4.0$. This led to the following recommendations [1]:

- $q = 2.0$ for spherical pressure vessels supported by non-braced columns
- $q = 3.0$ for spherical pressure vessels supported by braced columns

However these recommendations apply only for spherical pressure supported by an even number of braced or non-braced columns with circular hollow sections as well as only under the condition that the limit values for the dimensions of the vessels given in chapter 3 are complied with.

7 Seismic design

The determination of the seismic loads shall be done using seismic actions provided by the relevant codes (e.g. response spectra). Generally a simple model representing the spherical vessel as a SDOF system with its fundamental period is sufficient for the determination of the seismic base shear force. The design values may be obtained by considering the behaviour factors as mentioned before. The verification of the supporting columns, anchorages, foundations and bracings are to be done according to the rules provided by the codes for steel structures. The crucial point remains the verification of the shell-column connection which can be done either by simplified approaches or by FE-analysis, which may however be limited to the detail only [1].

Within the parametric study in [2] it was observed that the fundamental periods are in the range of $0.4\ s \leq T_1 \leq 1.5\ s$ for braced systems and $0.8\ s \leq T_1 \leq 4.4\ s$ for non-braced systems; consequently the fundamental periods of braced and non-braced pressure vessels are in general in the decreasing range or in the plateau of the design response spectrum depending on the ground type according to EN 1998-1

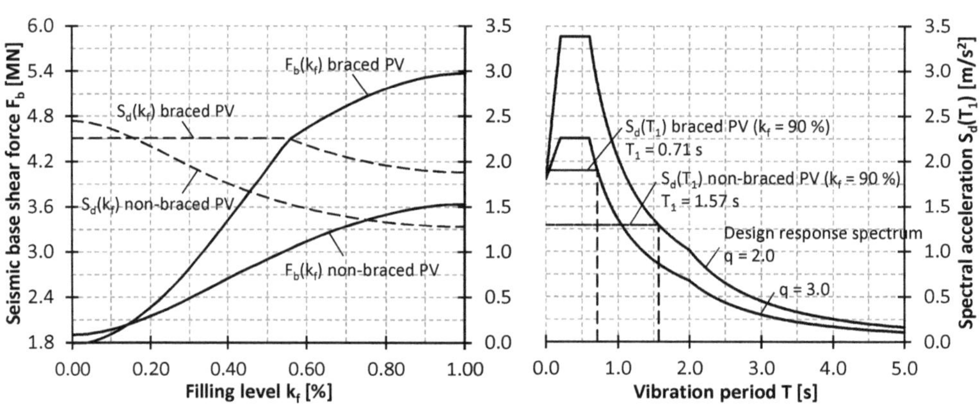

Figure 4: Seismic base shear force F_b and spectral acceleration S_d depending on the filling level k_f (left) and design response spectrum ($a_g = 0.24\ g$, type 1, ground type C) showing the resulting spectral acceleration $S_d(T_1)$ (right) both using the design examples [1]

[5]. With regard to the seismic design and the seismic base shear force respectively the maximum filling level is decisive. Using the design examples given in [2] and [3], Figure 4 shows that even though the fundamental periods for different filling levels are in the decreasing range of the design response spectrum, the maximum filling level is governing (for both design examples the maximum filling level is $k_f = 90\ \%$) [1].

8 Limit states (failure modes)

8.1 Non-braced spherical pressure vessels

The non-linear behaviour of non-braced pressure vessels as well as the corresponding failure modes (limit states) may be described generally as follows, provided that the geometry of the vessel is in compliance with the limits given in chapter 3. The first plastifications occur directly below the shell-column connections of the columns which are loaded under the greatest compressive stress due to the overturning moment. Thereafter up to a displacement equal to the ultimate state all columns successively plastified also directly below the shell-column connections. By reaching the ultimate state first local buckling was observed in the same sequence and locations as the plastifications. Finally in the case of further displacements all other columns buckle locally, whereby for the columns loaded in tension due the overturning moment the location of local buckling shifts in the range of compressive strains due to bending in the columns. Furthermore it should be mentioned that generally no plastifications occur in the sphere up to the collapse of the whole system [1].

8.2 Braced spherical pressure vessels

Analogous to non-braced systems the non-linear behaviour of braced pressure vessels may be described generally as follows. At the time of reaching the yield point the first plastifications occur in the most loaded bracings, which are directed parallel to the load direction and therefore loaded under the highest tension stress. Directly after reaching the yield point all bracings in compression fail suddenly by formation of plastic hinges. By further displacements up to the ultimate state the highest loaded bracings in tension fail due to reaching of the ultimate elongation, which leads in consequence to the first plastifications in the columns loaded in compression due the overturning moment. In the case of further displacements after reaching the ultimate state all bracings in tension failed as well as all columns successively plastified directly below the shell-column connections. Since all bracings in tension and compression are failed as well as all columns are plastified locally, the post critical behaviour is equal to non-braced systems. First local buckling occurs in the columns which are loaded most in compression due to the overturning moment and finally all other columns buckle locally. Furthermore it should be mentioned that generally no plastifications occur in the sphere up to the collapse of the whole system [1].

9 Conclusion

Within the European Research Project INDUSE the seismic performance and applicability of existing European and American codes to spherical pressure vessels were investigated. The seismic behaviour of braced and non-braced systems was analysed using pushover and incremental dynamic analysis techniques. Based on two design examples as well as a parametric study with 78 different geometries described in [2], design rules and recommendations were developed. In particular a simplified beam model for the calculation of global elastic stiffnesses and fundamental periods is proposed. By compliance of the recommended limit values for the dimensions of the vessels an excellent agreement between the simplified method and the conducted modal analyses was observed, the deviations were below 2% in average. For the determination of seismic base shear forces including effects of liquid sloshing it is referred to a simple and efficient method described in [4] and it was shown that in general the maximum filling level is decisive, even if the fundamental period is in the decreasing range of the design response spectrum.

For the two design examples given in [2] and [3] the behaviour factors were determined and showed to be higher than $q = 2.0$ suggested by EN 1998-1 [5]. A behaviour factor of $q = 2.7$ was found for the non-braced system and for the braced system higher values than $q = 3.0$ were determined, if additional plastifications of the columns were accepted. Considering the results of the performed pushover analyses within the parametric study, the different braced and non-braced vessels with 4, 8 or 12 columns showed an appropriate ductile behaviour, so that the displacement ductility $\mu = e_{ult}/e_y$ for the different systems remains relatively constant. For the non-braced systems slightly higher values than $\mu = 2.0$ were found, whereas for the braced systems the ductility showed to be higher than $\mu = 4.0$. This led to the judgement, that for non-braced systems a behaviour factor of $q = 2.0$ and for braced systems a behaviour factor of $q = 3.0$ according to the American standard ASCE/SEI 7-05 [7] is applicable.

10 Nomenclature

n_c = number of columns
l_c = effective length of columns
$l_{c,1}$ = partial length of columns (bracing to column connection)
$l_{c,ef}$ = buckling length of column
β_c = buckling length factor of columns
λ_c = relative slenderness of columns
i_c = inertia radius for circular hollow section
r_c = mean radius of column cross-section
$r_{c,e}$ = external radius of column cross-section
$r_{c,i}$ = internal radius of column cross-section
A_c = cross-sectional area of columns

I_c	=	moment of inertia for circular hollow sections
$\psi_{c,b}$		ratio of cross-sectional areas of columns and bracings
h_s	=	equator height of sphere (total length of columns)
d_s	=	mean diameter of sphere
$d_{s,e}$	=	external diameter of sphere
$d_{s,i}$	=	internal diameter of sphere
r_s	=	mean radius of sphere
t_s	=	shell thickness of sphere
l_s	=	overlapping length of shell-column connection
α_s	=	overlapping angle of shell-column connection
n_b	=	number of bracings
c_b	=	elongation stiffness of single bracing
α_b	=	inclination angle of bracings
l_b	=	length of bracings
l'_b	=	projected length of bracings
$r_{c,e}$	=	external radius of bracings cross-section
$r_{c,i}$	=	internal radius of bracings cross-section
A_b	=	cross-sectional area of bracings
β	=	angle of bracings related to the base area of the pressure vessel
k_1	=	elastic stiffness of single column
k_2	=	relative stiffness of simplified single column-bracing system
a	=	length ratio of columns
b	=	non-dimensional intermediate factor
k_{pv}	=	global elastic stiffness of spherical pressure vessel
T_1	=	fundamental period of spherical pressure vessel in [s]
f_1	=	natural frequency in [Hz]
ω_1	=	circular natural frequency in [rad/s]
m_t	=	total operating mass in [t]
m_{pv}	=	mass of empty spherical pressure vessel in [t]
m_f	=	mass of filling in [t]
ρ_{st}	=	steel density
ρ_f	=	filling density
h_f	=	filling height
k_f	=	filling level
q	=	behaviour factor
μ	=	displacement ductility
e_y	=	displacement by reaching the yield load
e_{ult}	=	displacement by reaching the ultimate load

11 Acknowledgments

This work was carried out with a financial grant from the Research Fund for Coal and Steel of the European Commission, within INDUSE project: “Structural safety of industrial steel tanks, pressure vessels and piping systems under strong seismic loading”, Grant No. RFSR-CT-2009-00022.

REFERENCES

[1] Deliverable D6.2: “Design guidelines for seismic analysis and design of industrial pressure vessels”, RFCS project INDUSE, Grant No.RFSR-CT-2009-00022

[2] Deliverable D6.1: “Background document of design guidelines for seismic analysis and design of industrial pressure vessels”, RFCS project INDUSE, Grant No.RFSR-CT-2009-00022

[3] Wieschollek, M., Kalliopi, D., Pinkawa, M., Hoffmeister, B., Feldmann, M.: “Guidelines for seismic design and analysis of pressure vessels”, in proceedings of ASME 2013 Pressure Vessels & Piping Division Conference, Paris, France, 14–18 July 2013, in press

[4] Karamanos, S.A., Patkas, L.A., Paltyrachos, M.A., “Sloshing Effects on the Seismic Design of Horizontal-Cylindrical and Spherical Industrial Vessels“, Journal of Pressure Vessel Technology, ASME, Volume 128, Issue 3, page 328-340, 2006

[5] EN 1998-1:2004: “Design of structures for earthquake resistance – Part 1: General rules, seismic actions and rules for buildings”, European Committee for Standardization, Brussels, 2004

[6] EN 1998-4:2006: “Design of structures for earthquake resistance – Part 4: Silos, tanks and pipelines”, European Committee for Standardization, Brussels, 2006

[7] ASCE/SEI 7-05: “Minimum Design Loads for Buildings and other Structures”, American Society of Civil Engineers, 2006

Seismic Isolation of Cylindrical Liquid Storage Tanks

Julia Rosin[1], Thomas Kubalski[1], Christoph Butenweg[1]

[1] Chair of Structural Statics and Dynamics, RWTH Aachen University
Mies-van-der-Rohe-Str. 1, 52074 Aachen, Germany
Julia.Rosin@lbb.rwth-aachen.de

ABSTRACT:

Seismic excited liquid filled tanks are subjected to extreme loading due to hydrodynamic pressures, which can lead to nonlinear stability failure of the thin-walled cylindrical tanks, as it is known from past earthquakes. A significant reduction of the seismically induced loads can be obtained by the application of base isolation systems, which have to be designed carefully with respect to the modified hydrodynamic behaviour of the tank in interaction with the liquid. For this reason a highly sophisticated fluid-structure interaction model has to be applied for a realistic simulation of the overall dynamic system. In the following, such a model is presented and compared with the results of simplified mathematical models for rigidly supported tanks. Finally, it is examined to what extent a simple mechanical model can represent the behaviour of a base isolated tank in case of seismic excitation.

Keywords: Liquid-filled tank, base-isolation, fluid-structure-interaction

1 Introduction

Tanks are preferably designed as cylindrical shells, because the geometry is able to carry the hydrostatic pressure from the liquid filling by activating membrane stresses with a minimum of material. In combination with the high strength of steel this leads to thin-walled constructions, which are highly vulnerable to stability failures caused by additional axial and shear forces in case of seismic excitation. However, an earthquake-resistant design of rigid supported tanks for high seismic loading requires unrealistic und uneconomic wall thicknesses. Compared to increasing the wall thickness an earthquake protection system can be a much more cost-effective alternative. Especially a base isolation with elastomeric bearings offers advantages in terms of an earthquake-friendly tank design. But the calculation capabilities of base-isolated, liquid-filled tanks are quite limited because of the complex interaction of the seismic isolation behaviour and the combined modes of vibrations of tank and fluid. Generally accepted calculation approaches are only available for rigid supported tanks (Meskouris et al. [1]). To

S. Klinkel et al. (Eds.), *Seismic Design of Industrial Facilities*,
DOI 10.1007/978-3-658-02810-7_36, © Springer Fachmedien Wiesbaden 2014

capture the hydrodynamic loading of isolated tanks, a complete modelling of the fluid-structure interaction including the behaviour of the seismic isolation is presented in the following.

2 Calculation of anchored liquid storage tanks

2.1 Seismically induced load components of liquid filled tanks

As a result of seismic excitation hydrodynamic pressure components, produced by the movement of the fluid, appear and have to be superimposed with the hydrostatic pressure. Since the oscillation periods of the individual seismically induced pressure components are far apart, each mode of oscillation with its associated pressure distribution can be determined individually. In case of a horizontal seismic excitation the convective part of sloshing vibrations, the impulsive rigid pressure component of the rigid-body motion as well as the impulsive flexible pressure component caused by the combined interaction vibration mode of tank and liquid must be considered (Fig. 1).

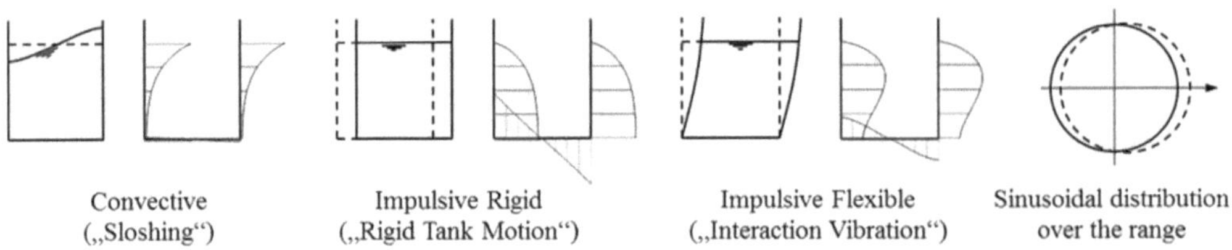

Figure 1: Horizontal seismic action: Modes of vibrations and pressure distributions (Meskouris et al. [1])

Furthermore the vertical seismic excitation must be taken into account, which leads to two additional modes of vibrations and corresponding pressure distributions. The impulsive rigid pressure is activated by the rigid-body motion of the tank and the flexible pressure component is caused by the flexibility of the tank shell (Fig. 2).

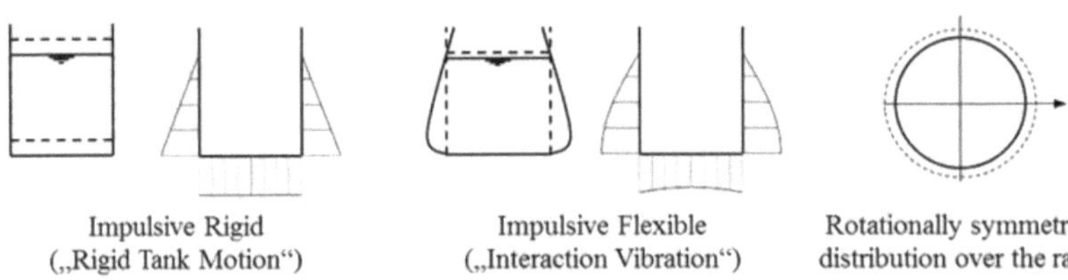

Figure 2: Vertical seismic action: Modes of vibrations and pressure distributions (Meskouris et al. [1])

2.2 Computational models for seismic excited tank structures

In the literature different approaches for modeling and calculation of seismically excited tank structures can be found. On the one hand engineering-based analytical calculation approaches are used which are usually represented by simple mass oscillators. Most of these approaches are based on the findings of Housner [2], who developed formulas for rigidly supported tanks with non-deformable walls to calculate the modes of vibrations and the corresponding dynamic pressure components. Based on these findings several approaches have been developed. A widely applied simplified method was developed by Veletsos [3] for rigid supported tanks with flexible walls. This method is fast and easy to apply, but it delivers only the seismically induced shear force and the overturning moment at the bottom of the tank. An accurate calculation of the stress distribution is not possible using such simplified approaches. DIN EN 1998-4 [4] proposes a more precise calculation that allows a three dimensional finite element analysis of the tank by applying the seismically induced pressure components as equivalent static loads on the dry shell. However, the approaches for calculating the individual seismically induced pressure components are based on the assumption of a rigid support at the tank bottom and they are not applicable to base isolated tanks. To gather the hydrodynamic loading of isolated tanks, a simulation model taking the fluid-structure interaction and the seismic isolation effects into account, is required. In the following the software LS-DYNA [5] is used for the necessary fluid dynamics calculations.

2.3 Base isolation

A base isolation is aimed at a decoupling of the building and the ground motion. Elastomeric bearings are a widely used base isolation and can optionally be installed with or without reinforcement, often in combination with a lead core (Petersen et al. [6]). However an unreinforced execution is unusual nowadays, since elastomers are subjected to high deformations up to 25% under vertical loads. These deformations cause lateral strains, by which unwanted rocking motion in case of seismic excitation can occur. Reinforced bearings can be considered as quasi-rigid in the vertical direction, so they are suitable to transfer vertical loads. Under cyclic loading, elastomers behave almost like springs. They have – depending on the material properties – a certain stiffness which causes a reset of the bearing and thus of the entire system after the release. Through the use of high-damping elastomers (addition of oils, resins, extra fine carbon black and other fillers), or a lead core, the damping capacity of the bearings can be increased significantly. In case of a distortion of 100%, high damping elastomers have damping rates from 0.1 to 0.2 during normal elastomers from 0.04 to 0.06 (Petersen et al, [6]). The use of elastomeric bearings as earthquake protection systems for tanks has already been realized by Bachmann and Wenk [7].

2.4 Example of Calculation

The following calculations are carried out for a steel tank with constant wall thickness, firmly anchored to a reinforced concrete base plate. The geometry of the tank is illustrated in Figure 3. The base plate is supported on elastomeric bearings, which properties are taken from Baumann and Boehler [8]. The calculation model considers the horizontal stiffness of the elastomeric bearings, whereas a rigid behaviour is assumed in vertical direction. The fluid is idealized as incompressible and friction-free. The material parameters of the tank and the isolation and the seismic hazard input parameter according to DIN EN 1998-1/NA [9] are given in Table 1.

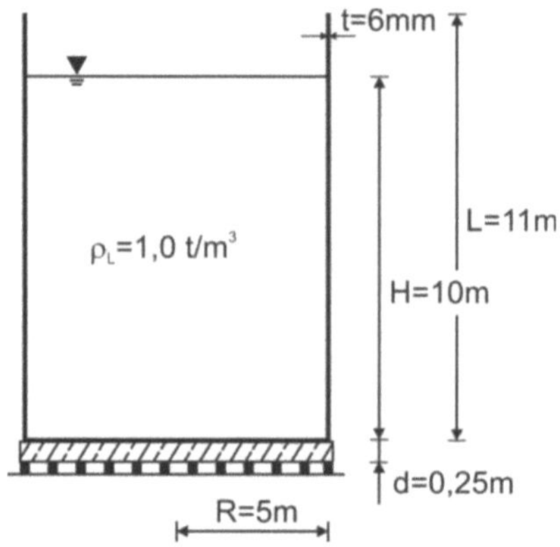

Figure 3: System of the isolated tank with elastomeric bearings

Table 1: Input parameter for the calculation

Location		Material		Base isolation	
PGA:	0,6 m/s^2	Shell:	S 235	Number:	20
Subsoil class:	CS	Foundation:	C 50/60	Stiffness:	700 kN/m
Importance Factor:	1,2			Damping:	15 %

3 Fluid-structure-interaction model

The software LS-DYNA [5] is used for the simulation of the fluid-structure-interaction of the liquid-filled tank. The software provides an explicit solver, which offers advantages especially for the solution of short-term dynamic problems. In addition, LS-DYNA provides formulations for the modeling of fluids and standardized contact formulations, which are able to represent the interaction of the tank shell and the fluid during a seismic excitation. Details of the following material, element and contact formulations can be found at the LS-DYNA manuals [10].

3.1 Material formulation

Basically both, elastic and plastic approaches are applicable for the tank shell. Since the focus is set on the reduction of the seismic loading by applying a base isolation, an elastic behaviour of the tank shell is assumed (MAT_ELASTIC). For the base plate the concrete material model MAT_CSCM is used, which is applied with the default settings. Two material formulations are investigated for the fluid: a linear (MAT_ELASTIC_FLUID) and a non-linear (MAT_NULL).

3.2 Element formulation

The tank wall and the tank bottom are idealized by Belytschko-Lin-Tsay shell elements with reduced integration, which are characterized by a high efficiency in terms of computing power required for explicit analysis. The foundation plate is idealized by 8-node solid elements and the fluid is represented by an Arbitrary-Lagrangian-Eulerian finite element formulation (ALE). This element formulation can be combined with both material models, but it cannot be used for an implicit calculation (modal analysis). When using the ALE formulation extra volume elements are generated within the scope of the freeboard up to the top edge of the tank wall, so the fluid surface can move freely (sloshing). Also, the ALE mesh must enclose the Lagrange mesh. For this reason a series of elements at the top and bottom of the tank, below the base plate and outside of the tank wall are generated. The elements are assigned to the vacuum material MAT_VACUUM, which has no physical meaning, but merely represents a region within the ALE mesh in which the fluid can move.

3.3 Contact formulation

The interaction between the tank shell and the fluid represents an important aspect of modelling. If a contact of the two parts appears, compression stresses are transferred while the transfer of tensile and shear stresses is disregarded. LS-DYNA provides essentially two different contact formulations for coupling the tank shell (Lagrange) and the fluid (ALE) with each other: ALE_FSI_PROTECTION (AFP) and CONSTRAINED_LAGRANGE_IN_SOLID (CLIS). Both formulations are well suited for fluid-structure interaction, but the latter formulation offers more configuration options, for example a separate specification of damping. Furthermore the formulation allows the evaluation of the contact forces as a result of the hydrostatic and hydrodynamic pressures on the tank shell.

3.4 Base isolation

The base isolation is simply idealized by linear spring and damper elements, which are integrated in the model between the supporting nodes and the nodes of the base plate. The elements exhibit a corresponding equivalent stiffness and damping representing the behavior of the base isolation. The stiffness and damping values are given in Table 1. The base isolation is acting in the direction of the seismic excitation, whereas in vertical and horizontally perpendicular direction fixed supports are applied.

3.5 Sequence of loading

In a first step the system is loaded with the acceleration of gravity. The load is linearly applied within a period of one second to avoid an excessive oscillation of the system. For the next half second the system is unloaded, so that the resulting oscillations subside. Then the seismic excitation is applied to the supporting nodes as a displacement-time history, artificially generated from the code spectrum according to DIN EN 1998-1/NA [9].

3.6 Results

To validate the fluid-structure interaction model the stress results of a rigidly supported tank are compared to those from an equivalent force analysis according to Eurocode 8, Part 4 [4]. In case of the equivalent force analysis the seismic induced pressure components (Fig. 1) are calculated separately and then they are superimposed to the resulting hydrodynamic pressure using the SRSS-rule. Finally the resulting hydrodynamic pressure is applied as an equivalent static load to the dry tank wall. Afterwards the hydrodynamic pressure is combined with dead load and hydrostatic pressure. The calculations are carried out for the subsoil class CS (soft soil) according to DIN EN 1998-1/NA [9]. Figure 4 and 5 show the circumferential, axial and shear stress distributions over the tank height. It has to be pointed out, that the decisive stresses for the design appear at different circumferential angles θ: circumferential stresses ($\theta = 180°$), axial stresses ($\theta = 0°$) and shear stresses ($\theta = 90°$). According to DIN EN 1998-4 [4] damping values of 2% for the tank shell and 0.5% for the fluid are applied for the rigidly-supported tank. By using these damping values the numerical simulation results of the rigidly-supported tank considering fluid-structure interaction effects show a good agreement with results according to DIN EN 1998-4 [4] for both material formulations of the fluid which is shown in figure 4 for the non-linear fluid formulation. Generally the hoop stresses of the tank wall are dominated by tension stresses due to the hydrostatic pressure of the liquid (Fig. 5). Except the upper edge of the tank wall with low hydrostatic pressures shows local compression stresses which can lead to stability problems of the thin steel sheet in the upper tank section. The axial and shear stresses comply qualitatively for both fluid material formulations with the results according to DIN EN 1998-4 [4]. The results of the nonlinear fluid formulation are consistent with the results according to DIN EN 1998-4 [4], while the results of the linear fluid formulation are somewhat less than the results of the DIN EN 1998-4 [4] calculation.

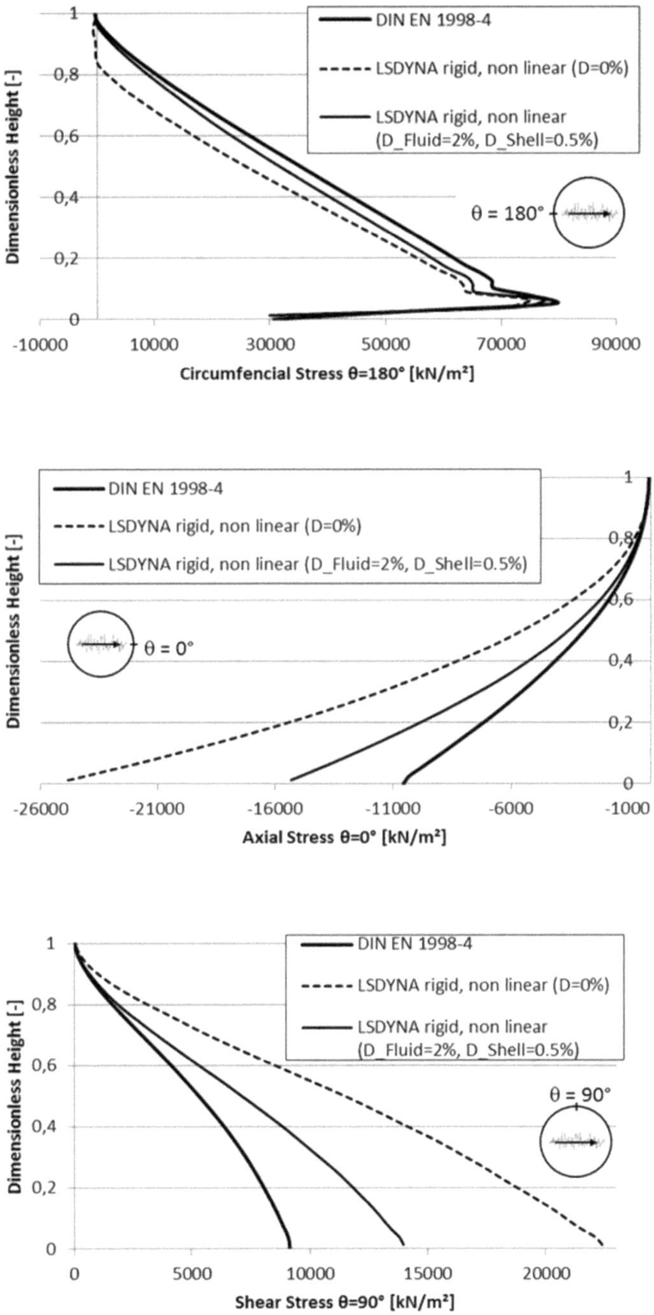

Figure 4: Stress distribution over the tank height for the non-linear fluid formulation and different damping values (rigidly supported tank)

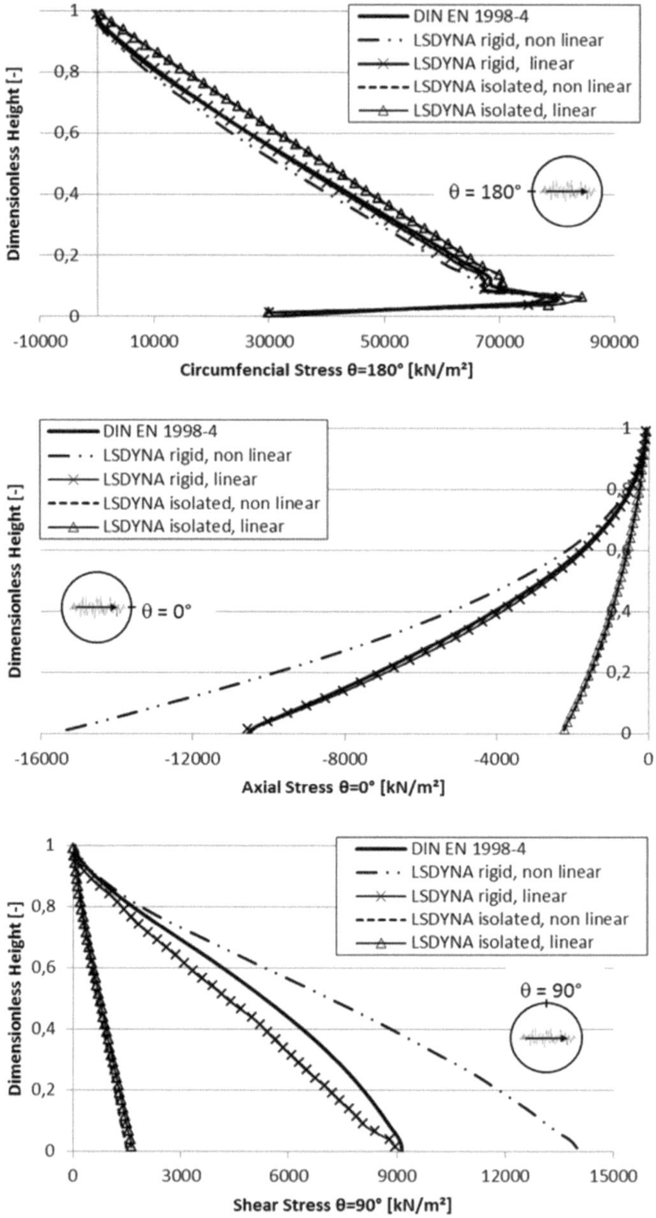

Figure 5: Stress distribution over the tank height for the rigidly supported and isolated tank

The calculations of the base isolated tank are carried out for an elastic behavior of the tank itself and a damping of the fluid of 0.5%. The results show a significant decrease of the axial and shear stresses, while the differences of the circumferential stresses are negligible because they are dominated by the tension stresses due to the hydrostatic pressure (Fig. 5).

4 Simplified mechanical model

The development of simplified mechanical models for isolated liquid-filled tanks has been studied in recent years and is particularly interesting in terms of practicality. For example, in Malhotra [11], Christovasilis and Whittaker [12] equivalent dynamic systems in the form of single-mass oscillators are derived. The base isolation is simply taken into account by introducing a horizontal degree of freedom at the base of the single-mass oscillator. In addition, the degree of freedom is combined with spring and damper elements, which reflect the characteristics of the corresponding protection system. The seismically induced pressure components are considered by single masses with certain lever arms. The masses and lengths of the lever arms correspond to those of rigidly-supported tanks, which is not really correct for the calculation of base-isolated tanks. The equivalent dynamic systems can be used for the calculation of the total shear force and the overturning moment, but they cannot be applied for the determination of the pressure distribution over the tank wall. Schäpertöns [13] investigated the influence of soil-structure interaction effects for seismically excited tanks and noticed that the normalized impulsive pressure distribution over the tank height is not significantly affected by soil-structure interaction effects. Veletsos and Tang [14] showed in their studies, that it is sufficient to consider the impulsive pressure components for the soil-structure interaction, since the changes of the convective pressure component is negligible small. These findings can be used to derive a simplified mechanical model for isolated liquid storage tanks. The starting point is the calculation of the pressure distributions for a rigidly supported tank according to DIN EN 1998-4 [4]. The integration of the impulsive flexible pressure $p_{if,h,rigid}$ normalized to the spectral acceleration of the first eigenperiod of the flexible vibration mode delivers the seismic mass in node 1 for an equivalent two-mass oscillator (Fig. 6):

$$m_{if,h} = \int_0^H \int_0^{2\pi} \left[p_{if,h,rigid}(\xi = 1, \zeta, \theta) \cdot \cos(\theta)\right] \cdot R \, d\theta \, dz \qquad (1)$$

Herein, ξ and ζ are the dimensionless coordinates for radius ($\xi = r/R$) and wetted height of the tank wall ($\zeta = z/H$) and θ is the peripheral angle. The integration assumes a sinusoidal pressure distribution in circumferential direction which is projected to the direction of excitation. With the mass $m_{if,h}$ and the natural period of the impulsive flexible vibration mode of the rigidly supported system, the stiffness k_{ifh} of the two-mass oscillator is calculated. The damping factor $c_{if,h}$ of the impulsive flexible vibration mode is set to 0.5%. The mass of the foundation plate and the base isolation (m_{Bi}) is applied in node 2. The idealization of the base isolation is performed by a combined spring-damper element, which represents the damping and stiffness properties of the base isolation. The load a(t) is applied as a synthetically generated acceleration time history in node 3 of the system.

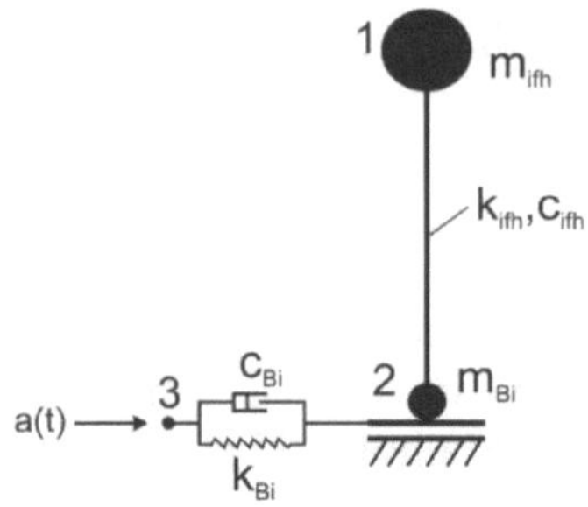

Figure 6: Simplified model of the isolated liquid storage tank

The needed result for the further calculation steps is the maximum response acceleration max $a_{if,h,iso}$ of the mass $m_{if,h}$ in node 1. This acceleration is used for scaling the normalized impulsive flexible pressure $p_{if,h,rigid}$ in Eq. 1, which leads to the impulsive flexible pressure distribution corresponding to the isolated vibration mode:

$$p_{if,h,iso}(\xi=1, \zeta, \theta) = \max a_{if,h,iso} \cdot p_{if,h,rigid}(\xi=1, \zeta, \theta) \qquad (2)$$

For the calculation of the impulsive flexible pressure component in Eq. 2, the horizontal response acceleration of the mass of $m_{if,h}$ relative to the ground has to be applied, because the impulsive rigid pressure component already includes the ground acceleration. The impulsive pressure component can be disregarded (Veletsos and Tang [14]), if the absolute acceleration is applied in Eq. 2. In this case the resulting seismically induced pressure of the isolated tank can be calculated by superposition of the impulsive flexible pressure component of the isolated tank and the convective pressure component of the rigidly supported tank. The stress calculation is carried out according to DIN EN 1998-4 [4].

4.1 Results

Figure 7 compares the axial and shear stress distributions over the tank height calculated with the simplified mechanical model and the fluid-structure interaction model with a linear and non-linear material approach for the fluid. The results show a good agreement, but it should be noted, that the results of the simplified model are more conservative.

5 Conclusion

The seismic excitation of rigidly supported liquid storage tanks activates hydrodynamic pressure components which lead to uneconomic wall thicknesses. A significant reduction of the seismic induced stresses can be obtained by the application of base isolations with elastomeric bearings. The paper introduced two calculation models for base-isolated tanks with different levels of accuracy. The simplified mechanical model is an equivalent two-mass oscillator, which is used for the calculation of modified impulsive pressure components for the base isolated tank. The more sophisticated simulation model is realized with LS-DYNA [5] and takes the fluid-structure interaction into account. The results of both models show a satisfactory agreement with the analysis results according to DIN EN 1998-4 [4]. Although the obtained results with the developed models are very promising further investigations are needed for different slenderness ratios, isolation systems and subsoil classes.

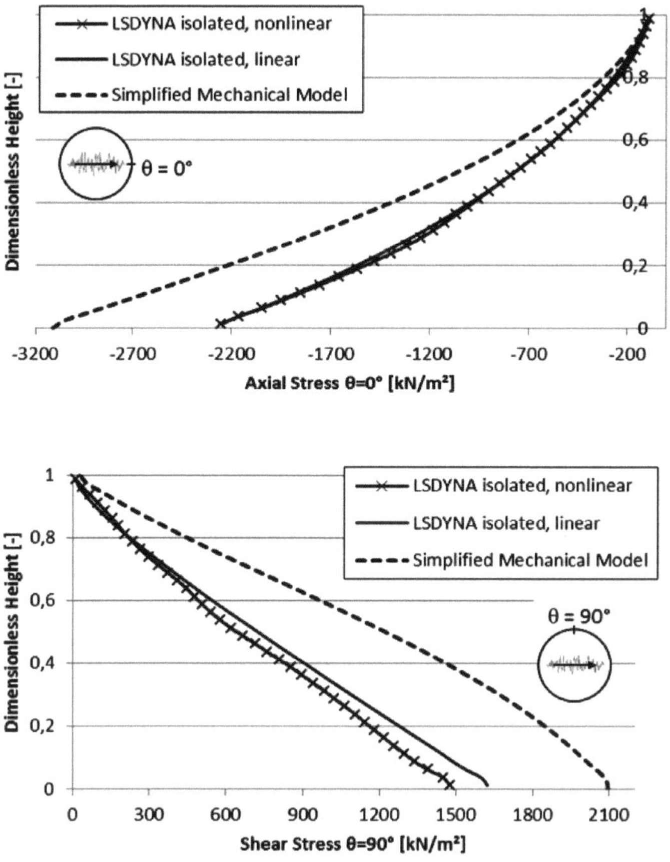

Figure 7: Axial and shear stress curve of the isolated tank

REFERENCES

[1] Meskouris, K., Hinzen, K.-G., Butenweg, C., Mistler, M.: Bauwerke und Erdbeben – Grundlagen, Anwendungen, Beispiele, 3. Auflage, Vieweg und Teubner Verlag, Wiesbaden, 2011, ISBN 978-3-8348-0779-3

[2] Housner, G. W.: The dynamic behavior of water tanks; Bulletin of the Seismological Society of America; Vol. 53, No. 2 (1963), Pages 381-387

[3] Veletsos, A. S.: Seismic effects in flexible liquid storage tanks, in: Proceedings of the 5th World Conference on Earthquake Engineering; Vol. 1, 1974, Pages 630-639

[4] DIN EN 1998-4 – Eurocode 8: Auslegung von Bauwerken gegen Erdbeben – Teil 4: Silos, Tankbauwerke und Rohrleitungen. Deutsches Institut für Normung, 2006

[5] LS-DYNA, Livermore Software Technology Corporation (LSTC), 2013

[6] Petersen, C., Beutler, H. Braun, C., Mangering, I.: Erdbebenschutzsysteme für den Hoch- und Brückenbau, In: Kuhlmann, U. (Hrsg.), Stahlbaukalender 2005; Verlag Ernst und Sohn, Berlin, 2005

[7] Bachmann, H., Wenk, T.: Schwächen statt Verstärken bei der Erdbebensanierung; Schweizer Ingenieur und Architekt; Band 118, Heft 3 (2000), Pages 18-23

[8] Baumann, T., Böhler, J.: Seismic Design of Base-Isolated LNG-Storage-Tanks; Structural Engineering International; Vol. 11, No. 2 (2001), Pages 139-144

[9] DIN EN 1998-1/NA – Eurocode 8: Auslegung von Bauwerken gegen Erdbeben – Teil 1: Grundlagen, Erdbebeneinwirkungen und Regeln für Hochbauten; Nationaler Anhang, Deutsches Institut für Normung, 2010

[10] LS-DYNA: Keyword User's Manual Volume I and II. Version 971 R6.1.0 (2012), Livermore Software Technology Corporation

[11] Malhotra, P. K.: New Method for Seismic Isolation of Liquid-Storage Tanks; Earthquake Engineering and Structural Dynamics; Vol. 26 (1997), Pages 839-847

[12] Christovasilis, I. P., Whittaker, A. S.: Seismic Analysis of Conventional and Isolated LNG Tanks Using Mechanical Analogs; Earthquake Spectra; Vol. 24, No. 3 (2008), Pages 599-616

[13] Schäpertöns, B.: Über die Wellenausbreitung im Baugrund und deren Einfluss auf das Tragverhalten von Flüssigkeitsgefüllten Behältern Dissertation, Technische Universität München, 1996

[14] Veletsos, A. S., Tang, Y.: Soil structure interaction effects for laterally excited liquid storage tanks; Earthquake Engineering and Structural Dynamics; Vol. 19 (1990), Pages 473-496

A Comparison of Piping Stress Calculation Methods Applied to Process Piping System for Seismic Design

Cheng Weimin[1], Jopp Heiko[1], Jan Pekař[2]

[1] TGE Gas Engineering GmbH
Mildred-Scheel-Straße 1, 53175 Bonn, Germany
Weimin.Cheng@tge-gas.com
Heiko.Jopp@tge-gas.com

[2] CAD-PRO, s.r.o.
4D Centrum, Kodaňská 1441/46, 100 10 Praha 10, Czech Republic
Jan.Pekar@tge-gas.com

Abstract

For design of industrial plants like LNG (liquefied natural gas) terminal the earthquake engineering for piping design is one of the most important design criteria [1]. The required calculation approaches in analyzing reactions of piping systems due to seismic events are specified in a variety of international and European codes and standards (e.g. in [2], [3] and [4]). Within these methods the simplified static equivalent method and the modal response spectra analysis are the most used in practice. From the engineering's point of view the simplified static analysis has obviously its advantages. This is why it is often used to perform some preliminary or final stress calculations. But in practice it also can be seen that this approach is even extended to the piping connected to the storage tank, where the modal response spectra analysis shall be applied according to the codes [3] and [4]. Furthermore there is no precise prediction about the results of the simplified static method in the area of piping design, neither in aspect of reliability nor in aspect of economy.

This article, based on a calculation of a typical unloading line for a new LNG storage tank – carried out by means of the CAESAR II program [7], compares the simplified static equivalent method and the modal response spectra analysis. The aim of this article is trying to set a general evaluation criterion and to give an answer to questions, under which conditions the simplified calculation method can be used. How big are the differences of the results between the two approaches?

Keywords: LNG terminal, unloading line, piping stress calculation, modal response spectra analysis, simplified static equivalent method

DOI 10.1007/978-3-658-02810-7_37,

1 Introduction

In practice of the industrial plant design the simplified static analysis is often used instead of performing a dynamic calculation. One situation is for example that the LNG terminal is not located in a very active earthquake zone. Therefore for piping design of the whole terminal the simplified static analysis is foreseen in the project specification or in the stress calculation procedure. The second situation is that at the beginning of a project execution it is usually that stress calculation engineers do not have enough input data to carry out a complete stress analysis for a given piping system, because the tank design is also at the beginning phase or it is ongoing. All information they know about seismic events is, where the terminal is located and how big the corresponding PGA (peak ground acceleration) is. But they still need to do some preliminary piping stress analysis and provide estimated support loads to the civil department so that the civil engineers are able to start their steel structure design. It is even not rare to see there are companies who only use the simplified static analysis for their seismic design.

For all these situations there is no precise prediction about the results of the simplified static equivalent analysis in the area of piping design, neither in aspect of reliability nor in aspect of economy. This article describes and compares results of the performed piping stress analysis by both methods - the simplified static equivalent method and the modal response spectra analysis for the typical unloading line of a new LNG storage tank. Some recommendations are given.

2 Design Data of the unloading line

This unloading line is designed for a 150.000 m^3 storage tank. The ship unloading rate is 12000 m^3 /h. The design data of the unloading line are listed as follows:

Line outside diameter:	813 mm
Wall thickness:	9.53 mm,
Material:	A358 TP304
Fluid density:	470 kg/m^3
Insulation thickness:	170 mm
Insulation density:	90 kg/m^3
Design temperature:	-165°C (min) and +50°C (max)
Design pressure	18.9 barg,
Installation temperature:	+21°C

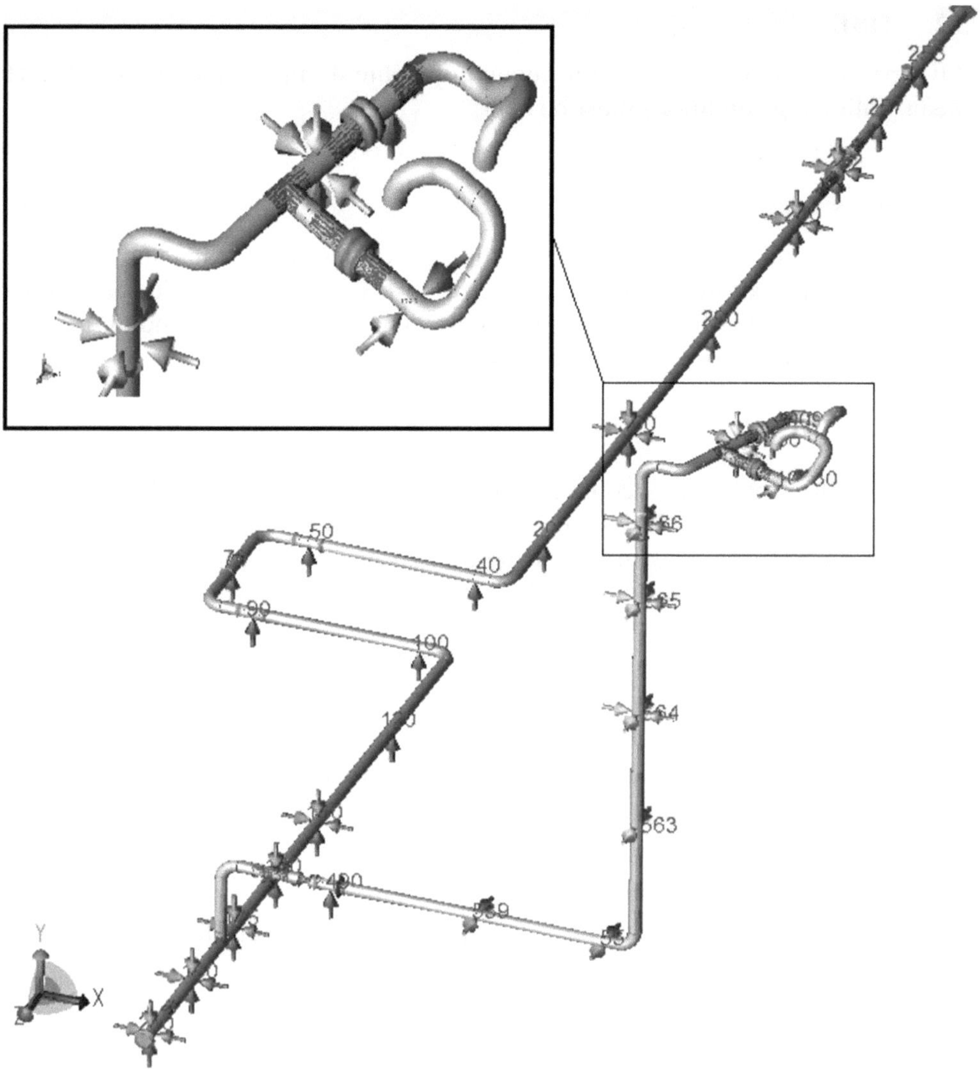

Figure 1: 3D model of the unloading line

3 Seismic Design Criteria

3.1 Overview

According to [1] the piping system should be designed to resist sustained loads, thermal loads, piping movement, snow and wind loading, and particularly some dynamic effects like earthquake and surge. For seismic design of piping system two different earthquake levels shall be considered: OBE (Operating Basis Earthquake) and SSE (Safe Shutdown Earthquake) [5] and [6].

3.2 OBE

All pipe lines shall be designed to remain operable during and after an OBE. That means following conditions must be kept:

$$S_{LO} < 1.33\ S_h \quad (1)$$

$$F_{exit} < F_{allow} \quad (2)$$

Herein S_{LO} is the existing stress from the sum of sustained loads and the seismic loads. S_h is the allowable stress according to [1]. F_{exit} is the calculated support and/or nozzle load under seismic conditions. F_{allow} is the corresponding allowable load provided by support vendor and tank designer.

3.3 SSE

After SSE the storage tank/container shall be in a safe condition. That means the design shall be such that during and after SSE there is no loss of container capacity and after an SSE the container shall be able to be emptied and inspected. That's why only tank connected piping shall be checked for this case. Similar to OBE the requirements can be formulated as follows:

$$S_{LO} < 1.2\ S_y \quad (3)$$

$$F_{exit} < F_{allow} \quad (4)$$

The condition (3) is adapted from [2]. Herein S_y is the yield strength at the metal temperature of the operating condition being considered

4 Definition of seismic loading

For LNG terminal located in south-west Europe the PGAs are given by

$$PGA_{SSE} = 0.5g \quad (5)$$

$$PGA_{OBE} = 0.15g \quad (6)$$

4.1 Seismic loads for static calculation

For a simplified static calculation the seismic loads are the horizontal acceleration a_h and the vertical acceleration a_v. From different considerations different accelerations as static loads can be defined.

1) Using PGAs directly

SSE

$$a_h = 0.5g, \quad a_v = 2/3*0.5g=0.33g \quad (7)$$

OBE

$a_h = 0.15g, \quad a_v = 2/3*0.15g=0.1g$ (8)

2) Considering coefficient 1.5 for PGAs according to [2] in case of lack of building information

SSE

$a_h = 0.75g, \quad a_v = 2/3*0.75g=0.5g$ (9)

OBE

$a_h = 0.225g, \quad a_v = 2/3*0.225g=0.15g$ (10)

3) Using plateau acceleration of the elastic response spectrum according to [3] in case of not having any soil information or assuming ground type A.

SSE

$a_h = 2.5*0.5g=1.25g, \quad a_v = 2/3*1.25g=0.833g$ (11)

OBE

$a_h = 2.5*0.15g=0.375g, \quad a_v = 2/3*0.375g=0.25g$ (12)

4) Considering coefficient 1.5 for plateau acceleration from 3)

SSE

$a_h = 1.5*1.25g=1.875g, \quad a_v = 2/3*1.875g=1.25g$ (13)

OBE

$a_h = 1.5*0.375g=0.563g, \quad a_v = 2/3*0.563g=0.375g$ (14)

5) Peak accelerations from the corresponding response spectra (see Fig. 2-5 in 4.2)

SSE

$a_h = 2.16g, \quad a_v = 2.34g$ (15)

OBE

$a_h = 0.58g, \quad a_v = 0.6g$ (16)

4.2 Seismic loads for dynamic calculation

Based on the PGAs specified in (5) and (6) and the design principles according to [3] the response spectra regarding tank roof were calculated by tank designer [8] by means of a FEM analysis in time domain. The results are shown in Figure 2 to Figure 5.

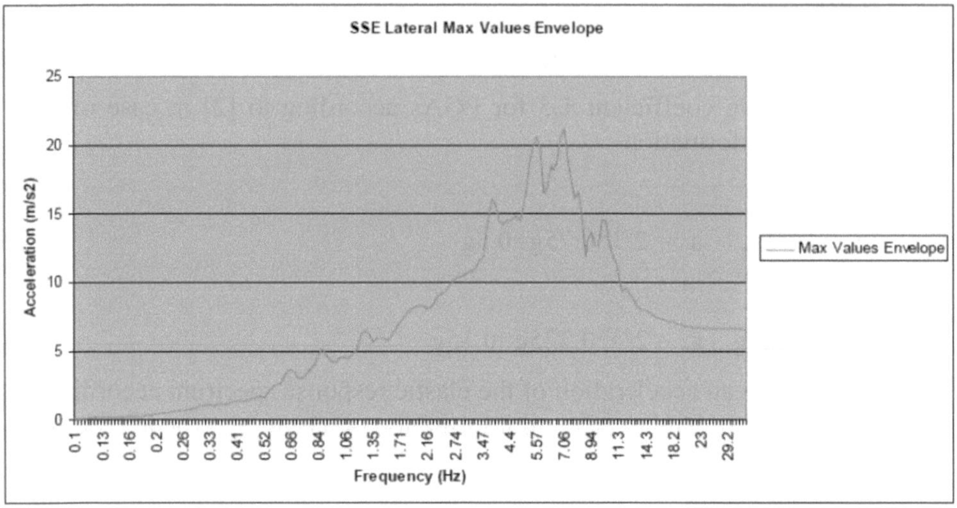

Figure 2: Horizontal response spectrum for SSE

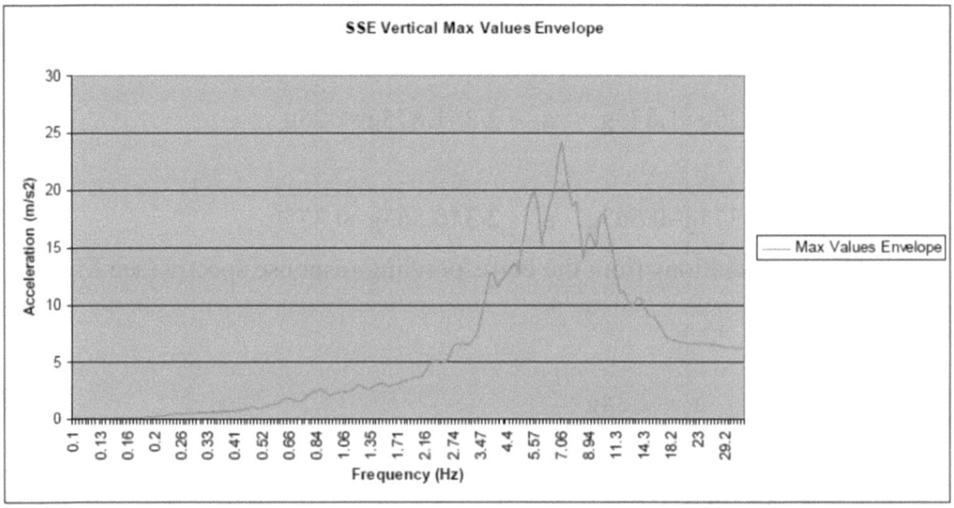

Figure 3: Vertical response spectrum for SSE

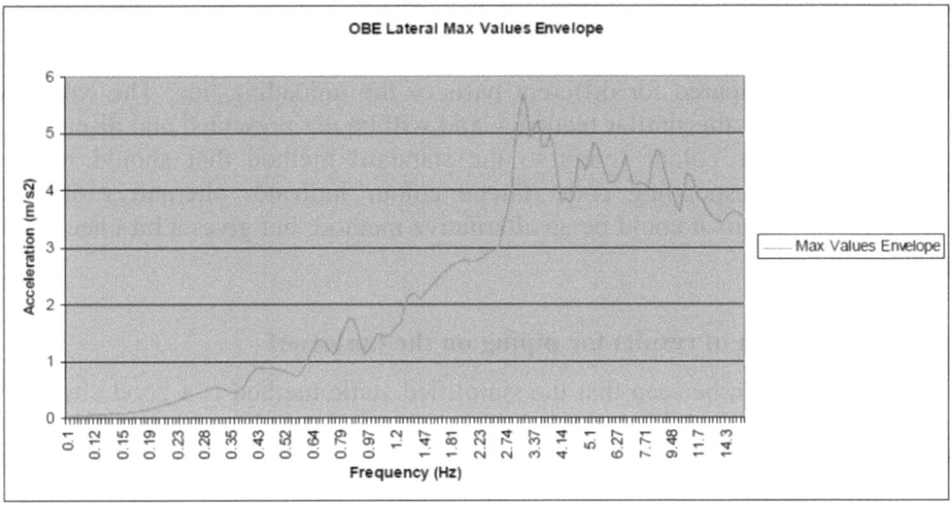

Figure 4: Horizontal response spectrum for OBE

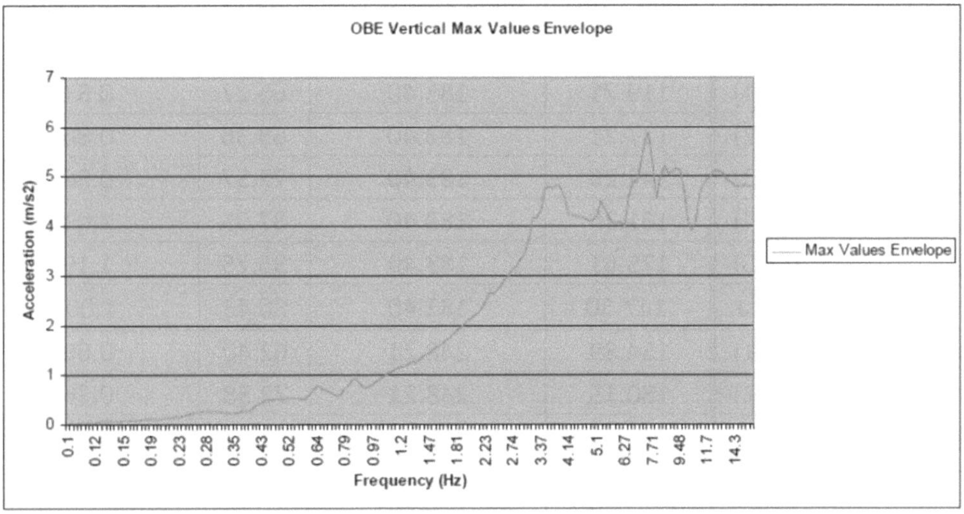

Figure 5: Vertical response spectrum for OBE

5 Results

In the following tables the calculated results with respect to the code stress are presented and compared for different parts of the unloading line. The calculated support loads show the similar tendency and will be not presented and discussed in this paper. Yellow colour refers to the standard method that should be used according to corresponding code. Green colour indicates alternative methods. Orange colour means it could be an alternative method, but gives a bit uneconomic results.

5.1 Comparison of results for piping on the tank roof

From Table 1 it can be seen that the simplified static method is a good alternative for the dynamic analysis with static load level 3) and 4). Using peak values of the response spectra for a static calculation (static 5) is not necessary and leads to uneconomic results.

Table 1: Comparison of results for piping on the tank roof

Part on the tank roof (top part)					
Earthq. Event	Seismic load	Code stress N/mm^2	Allowable stress N/mm^2	Ratio %	Comparison Static / Spectra
OBE	Static 1)	119.71	183.40	65.27	0.81
	Static 2)	127.21	183.40	69.36	0.86
	Static 3)	142.26	183.40	77.57	0.96
	Static 4)	161.30	183.40	87.95	1.09
	Static 5)	175.61	183.40	95.75	1.19
	Spectra	147.50	183.40	80.43	1.00
SSE	Static 1)	154.89	248.21	62.40	0.66
	Static 2)	180.15	248.21	72.58	0.76
	Static 3)	230.85	248.21	93.01	0.98
	Static 4)	293.95	248.21	118.43	1.25
	Static 5)	378.53	248.21	152.50	1.60
	Spectra	236.00	248.21	95.08	1.00

5.2 Comparison of results for tank riser (vertical piping)

The values from Table 2 show that for the tank riser piping we can get the same conclusion as for the tank roof piping.

Table 2: Comparison of results for the tank riser piping

Part of the tank riser					
Earthq. event	Seismic load	Code stress N/mm²	Allowable stress N/mm²	Ratio %	Comparison Static / Spectra
OBE	Static 1)	52.19	183.40	28.46	0.92
	Static 2)	53.62	183.40	29.24	0.94
	Static 3)	56.43	183.40	30.77	0.99
	Static 4)	60.76	183.40	33.13	1.07
	Static 5)	64.67	183.40	35.26	1.13
	Spectra	57.00	183.40	31.08	1.00
SSE	Static 1)	58.88	248.21	23.72	0.68
	Static 2)	66.23	248.21	26.68	0.77
	Static 3)	80.96	248.21	32.62	0.94
	Static 4)	99.36	248.21	40.03	1.16
	Static 5)	119.65	248.21	48.21	1.39
	Spectra	86.00	248.21	34.65	1.00

Table 3: Comparison of results for the pipe rack piping

Part of the pipe rack					
Earthq. event	Seismic load	Code stress N/mm²	Allowable stress N/mm²	Ratio %	Comparison others/Static 2)
OBE	Static 1)	89.75	183.40	48.94	0.85
	Static 2)	105.88	183.40	57.73	1.00
	Static 3)	141.91	183.40	77.38	1.34
	Static 4)	189.17	183.40	103.15	1.79
	Static 5)	209.10	183.40	114.01	1.98
	Spectra	117.00	183.40	63.79	1.11
SSE	Static 1)	175.26	248.21	70.61	0.83
	Static 2)	212.10	248.21	85.45	1.00
	Static 3)	313.09	248.21	126.14	1.48
	Static 4)	437.96	248.21	176.45	2.06
	Static 5)	570.82	248.21	229.97	2.69
	Spectra	333.60	248.21	134.40	1.57

5.3 Comparison of results for piping on pipe rack or sleeper

For piping on the pipe rack or sleeper it is not necessary to consider using plateau acceleration of the elastic response spectrum (static 3) or seismic load level more than that. A dynamic calculation based on the response spectra analysis confirms the rule according to [3]. That means it is sufficient enough if we consider the coefficient 1.5 for the peak ground acceleration for case of without having any building information or no dynamic analysis for the pipe rack steel structure.

6 Conclusion

The simplified static equivalent method is a useful and reliable method for seismic design of piping connected to storage tank, only if an adequate seismic load level is applied. The response spectrum analysis for pipe rack piping, based on the response spectra of tank roof could even lead to acceptable results. They are not uneconomical than the results from other static calculations.

This conclusion is based on calculation of one unloading line and should be confirmed by additional calculations.

References

[1] Process Piping
ASME B31.3 2012 edition

[2] Metallic Industrial Piping - Part 3: Design and calculation
EN 13480-3:2012

[3] Eurocode 8: Design of structures of earthquake resistance -
Part 4: Silos, tanks and pipelines

[4] Minimum Design Loads for Buildings and other Structures
ASCE 7: 2005 edition

[5] Standard for the Production, Storage, and Handling of Liquefied Natural Gas
NFPA 59A: 2013 edition

[6] Installation and equipment for liquefied natural gas - Design of onshore installations
EN 1473: 2007 edition

[7] Intergraph: CAESAR II Version 6.10

[8] TGE Gas Engineering: Tank Seismic Spectra Data for Tank T-231,
05063/TD37/CAL/1900/0523

Seismic Analysis of Pressure Vessels in Correspondence to the VCI-Guideline

Jörg Habenberger[1], **Sebastian Villiger**[1]

[1] Basler & Hofmann AG
Forchstrasse 395, CH-8032 Zürich, Switzerland
joerg.habenberger@baslerhofmann.ch, sebastian.villiger@baslerhofmann.ch

ABSTRACT:

The DIN EN 1998 standard and the VCI-Guideline provide comprehensive information for the design and verification of industrial facilities concerning earthquakes. For the fabrication and distribution of pressure equipment in the European Union the Pressure Equipment Directive (directive 97/23/EG) defines a basic framework. In Germany it was enacted by the Pressure Equipment Regulations. The requirements of the Pressure Equipment Regulations are included in harmonised standards, such as EN 13445. The application of other technical regulation is also possible in order to fulfil the Pressure Equipment Regulations. In Germany especially the AD-regulations are applied for the design of pressure vessels. They were developed by the chemical industry over many decades for a reliable and economical operation of the pressure equipment. The accordance of the AD-regulations with the Pressure Equipment Regulations has to be approved by a *Certification Body* when used in design practice. In the present paper the application of the AD-regulations in accordance with the specifications of the VCI-Guideline is explained by means of a current project for the *Evonik Industries GmbH Rheinfelden* in the earthquake evaluation of pressure vessels. The consequences for the structural models and the design procedures will be shown. The influences on safety factors and mechanical properties are discussed in detail. With the suggested procedure it is possible to achieve a consistent earthquake design for pressure equipment in accordance with the current standards and guidelines.

Keywords: Seismic Design, Pressure Vessels, Standards

1 Introduction

For the fabrication and distribution of non-portable pressure vessels within the European Union the Pressure Equipment Directive (directive 97/23/EG, [Web-1]) has to be taken in account. The demands of the Pressure Equipment Directive can be met by the application of the standard EN 13445 e. g. (unfired pressure vessels).

It is also possible to use other specifications like the AD 2000-guideline [Web-2] and ASME-Codes respectively. The accordance of these specifications with the Pressure Equipment Regulations has to be approved by a *Certification Body*. In Germany the so called *Technical Inspection Association (TÜV)* can perform such an assessment of the manufacturer of these pressure vessels.

In Germany pressure vessels have been designed and fabricated very often using the AD 2000-regulation. The AD 2000-regulation is a collection of bulletins issued by the *Working Committee Pressure Vessels* (Arbeitsgemeinschaft Druckbehälter, "AD"), which consists of seven members (AD-associates), i.e. manufacturers, operators, technical inspection and industrial injury coorporations. The German Chemical Industry Association (VCI) is also a member of the Working Committee Pressure Vessels.

The AD-Bulletins contain a large amount of practical experiences in the fabrication of pressure vessels and their use offers a very economic design. The AD-associations constitute working groups. These working groups revise existing AD-Bulletins and establish new ones (see also AD 2000-Bulletin G1).

The AD-Bulletins are based on the DIN-standards (see Bulletins G1/G2) and in general they also cover seismic actions (AD 2000-Bulletin S3/0). As far as seismic loads are concerned, the current issue of AD-Bulletin S3/0 (February 2013) refers to DIN EN 1998 [3] and the VCI-Guideline (issue 2009).

For the seismic evaluation of existing process facilities it is therefore reasonable to apply the AD 2000-regulation. Especially if they were originally designed and fabricated according to this guideline.

2 Current Standards and Safety Concepts

2.1 Eurocodes

In Germany, the series of standards DIN EN 1990-1999 (Eurocodes) were established in 2012. National related parameters and supplementary provisions are included in the respective national addenda. In principle, these standards should be considered when verifying the earthquake safety of chemical facilities.

Particularly the following parts of the Eurocode are relevant to the earthquake safety of plant constructions:

- DIN EN 1990, Basis of structural design
- DIN EN 1991, Actions on structures
- DIN EN 1993, Design of steel structures
- DIN EN 1998, Design of structures for earthquake resistance

The Eurocode design concept is based on the verification of both: ultimate and serviceability limit states. Depending on the limit state, appropriate models, calculation methods and actions shall be used.

The verification of limit states relies predominantly on the concept of partial safety factors. In order to conform to limit state demand the design reactions must not exceed the resistances on design level. Partial safety factors as well as combination coefficients are applied to the characteristic values for the calculation of design values (see DIN EN 1990).

For exceptional actions like earthquakes all partial safety factors γ are equal to 1. All variable actions are to be considered as accompanying actions. Quasi-permanent values ψ_2 shall be used for the combination of the different variable actions.

Permanent and variable actions are regulated by the code DIN EN 1991. Earthquake action is regulated by the code DIN EN 1998.

The partial safety factors for construction materials, included in DIN EN 1991-1999, have to be considered for earthquake verifications.

2.2 VCI-Guideline

The VCI-Guideline [2, 3] was first published in 2009. An updated version of the guideline, adjusted to the Eurocodes, was published in 2012.

The guideline contains information for chemical plant constructions regarding earthquake safety, as these constructions differ from common engineering structures. Especially the following topics are covered: earthquake action (importance factor γ_I, response factor q), modelling of process facilities, calculation methods and evaluation of existing plants.

For chemical plant constructions the VCI-Guideline is a commentary to DIN EN 1998.

2.3 AD 2000-Regulation

For the design of pressure vessels the bulletins of series B (calculation), series S (exceptions) and series W (construction materials) are of great significance.

Basic calculation principles and safety factors are given in bulletin B0, whereas the other bulletins of series B deal with the different construction parts of pressure vessels (bottom plate, shell, etc.).

The characteristic material parameters *K* are given in the bulletins of series W, depending on the operating temperature and material thickness.

The design of bearing and supporting constructions is regulated in bulletin series S. Similar to the former steel construction code DIN 4114, different load cases are defined in bulletin series page S3/0:

- Operation load case (BF)
- Testing load case (PF)
- Installation load case (MF)
- Exceptional load case (SF)

Earthquake is considered as an exceptional load case "SF". For exceptional load cases, stresses σ in the construction have to be calculated and compared to admissible calculation stresses f. The permissible design stresses are determined from characteristic material parameters K and a global safety factor S. The global safety factor depends on the load case, the material and the present load case. This concept is similar to that of the former steel construction code DIN 4114.

2.4 Earthquake design acc. to VCI-Guideline, Eurocode and AD 2000-Regulation

Following procedure is suggested for earthquake safety verification:

- Determination of the design actions according to EC 0, EC 1 and EC 8 with partial safety factor $\gamma=1$
- Combination of the actions according to load case "SF" given in AD 2000 (i.e. bulletin series sheet S3/0, Tafel 1)
- Calculation of stresses in the structure following AD 2000 bulletin series B and S
- Determination of material parameters according to AD 2000 bulletin series W
- Verification of stresses following AD 2000 bulletin series B and S

3 Example

The previously described design procedure will be explained by the verification of an existing pressure vessel. Only the dished tank end at the connection with the supporting columns will be examined (section 5.1 AD 2000-Bulletin S3/3).

During a preliminary seismic assessment the tank was classified to priority group III acc. to VCI-guideline [2].

3.1 Pressure vessel and model

The pressure vessel was built in 1996. It is filled with isobutene and was designed and constructed for an operating pressure of 7.7 bar according to the AD-regulations. The cylindrical tank with dished endings has dimensions of 7.2 m x 3.2 m (H x D) and a volume of approx. 50 m^3 (Figure 1).

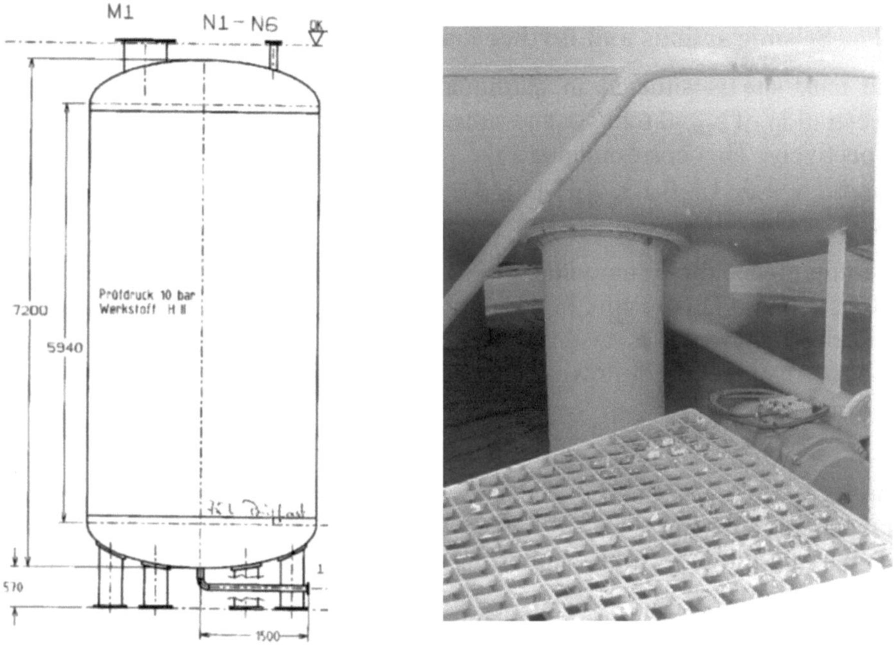

Figure 1: Drawing of the analysed Isobutene tank and connection between vessel bottom and steel support feet

The tank rests on six steel columns (ROR 323.9x7.1). The thickness of tank bottom amounts to 20.9 mm. The tank bottom is reinforced by steel plates located on top of the columns (Figure 1).

The material of the supporting steel feet is steel S235 and the material of the tank wall is steel P265GH. The vessel operates under normal temperature (T<100°C). The overall mass amounts to 42 tons.

A finite element model of the tank was established in order to calculate the natural frequencies and the stresses in the columns. The overall fluid was added as additional mass to the tank wall (instead of a combination factor ψ_2=0.6 acc. to VCI-guideline, Tab. 5.4). Hydrodynamic effects (sloshing of the liquid) were not considered.

Plane triangular shell elements were used to model the tank wall and beam elements for the columns. The columns are pin-jointed with the tank base and clamped with the base plate (see AD 2000-Bulletin S3/3).

The dominant natural frequencies of 7.5 and 34 Hz were determined by a vibration analysis. The participation factors are 0.78 and 0.21 respectively. In the first natural frequency the tank behaves like a rigid mass on a flexible supporting structure.

3.2 Seismic actions and further loads

The tank site is situated in earthquake zone 2 with an effective peak ground acceleration of a_{gR}=0.6 m/s^2. The subsoil class and geological bedrock are C and R respectively. This corresponds to the following control parameters of the response spectra: S=1.5, T_B=0.1 s, T_C=0.3 s, T_D=2.0 s (acc. to Table NA.4 DIN EN 1998, T_B acc. VCI-Guideline).

The importance factor amounts to γ_I=1.4. The structural damping is assumed to be ξ=0.02 (steel). This results in a damping correction factor of η=1.2. The value of the behaviour factor is taken equal to q=1.5 (lowest value for steel structures). Altogether we get the following horizontal spectral acceleration:

$$S_e = a_g\,\gamma_I\,S\,\eta\,\frac{2.5}{q} = 2.52\,\frac{m}{s^2} \tag{1}$$

The second relevant natural frequency is outside the constant spectral acceleration branch acc. DIN EN 1998-1 (T=0.03 s < T_B=0.1 s). A conservative estimation of the maximum forces in the supporting structure can be received, if the first natural frequency and the overall fluid mass are considered (ψ_2=1.0).

With a total mass of 42‘000 kg a resultant horizontal earthquake force of F_b=105 kN results. Vertical accelerations can be neglected acc. to the VCI-guideline. Because of the alignment of the columns in the layout, the perpendicular horizontal earthquake component does not increase the structural stresses.

Besides the dead load of the structure no other actions are considered. Snow loads are not relevant; the tank is encased.

3.3 Verification of the dished vessel end

The AD 2000-Bulletin S3/3 is used to verify the connection between the tank bottom and the supporting structure. The parameter K of the tank bottom material P265GH with t=20.9 mm at T<100°C is 255 N/mm^2 (yield strength, acc. AD 2000-Bulletin W1).

The global safety coefficient for special load cases is S_S=1.0 acc. to AD 2000-Bulletin B0, table 2 and S3/3, section 4.3.4.1 and the allowable design stress for the vessel is f_S=K/S_S=255 N/mm^2.

The safety coefficient for the supporting structure is determined within AD 2000-Bulletin B0: S=1.5. The related design stress results of AD-Bulletin S3/3 to 1.5xf = 1.5x255 N/mm^2.

The pressure vessel is simply supported by the columns. The columns are clamped into the base plate (Figure 2).

Section 5.1 of AD 2000-Bulletin S3/3 specifies the following stress proofs:

Table 1: Stress analysis acc. to AD 2000, S3/3

Nr.	Stresses (N/mm^2)	inside	outside	design σ
1	$\sigma_{mp} = R_m p/(20\ e)$	+7.7		
2	$\bar{\sigma}_{mx} = \left(\frac{N_x e}{F}\right) F/e^2$	-40.0		
3	$\bar{\sigma}_{my} = \left(\frac{N_y e}{F}\right) F/e^2$	-14.2		
4	$\sigma_{mx} = \sigma_{mp} + \bar{\sigma}_{mx}$	-31.9		
5	$\sigma_{my} = \sigma_{mp} + \bar{\sigma}_{my}$	-24.2		
6	$\sigma_{mV} = \sqrt{\sigma_{mx}^2 + \sigma_{my}^2 + \sigma_{mx}\sigma_{my}} \leq 1.5f$	28.8		≤ 255
7	$\sigma_{bx} = \left(\frac{M_x}{F}\right) 6F/e^2$	+135.1	-135.1	
8	$\sigma_{by} = \left(\frac{M_y}{F}\right) 6F/e^2$	+42.9	-42.9	
9	$\sigma_x = \sigma_{mx} + \sigma_{bx}$	+103.3	-167.0	
10	$\sigma_y = \sigma_{my} + \sigma_{by}$	+18.7	-67.1	
11	$(\sigma_m + \sigma_b)_V = \sqrt{\sigma_x^2 + \sigma_y^2 + \sigma_x\sigma_y} \leq (\sigma_m + \sigma_b)_{Vzul.}$	95.3	145.6	≤380
12	$q = \sigma_{mV}/K$			0.11
13	$z = 1.5 - 0.5\ q^2$			1.49
14	$(\sigma_m + \sigma_b)_{Vzul.} = 1.5\ z\ f$			380

Thereby the verification of the tank bottom at the connection to the columns concerning seismic loads is concluded.

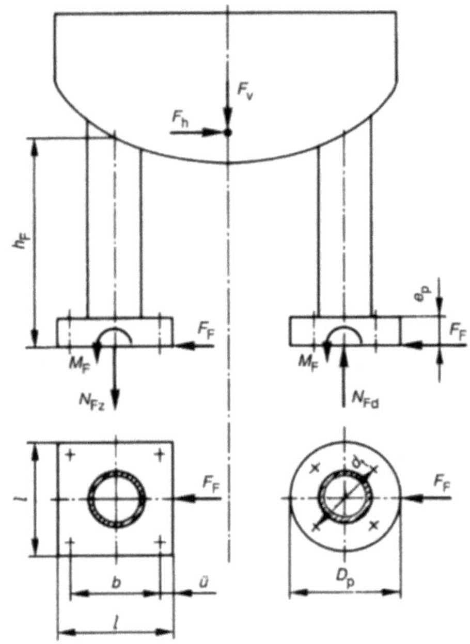

Figure 2: Geometrie and designation of the pressure vessel supporting feet (from AD 2000 –Bulletin S3/3)

4 Conclusion

Earthquake verifications of existing pressure vessels can be carried out using the AD-regulations especially if they were originally designed according to these specifications. Although AD-regulations have a different design concept than DIN EN 1998, it is possible to use them for the verification of pressure vessels due to the fact that earthquakes are exceptional actions. Furthermore, the AD-regulations contain important information about the modelling of the existing constructions and the material parameters.

5 Acknowledgements

The authors wish to thank *Evonik Industries GmbH Rheinfelden (D)* for their support to the writing of this paper.

References

[1] VCI-Guideline, "Guideline, Seismic design in process industry", German Chemical Industry Association, October 2013

[2] VCI-Guideline, "Commentary to the guideline, Seismic design process industry", German Chemical Industry Association, October 2013

[3] DIN EN 1998-1:2010-12, Eurocode 8: Earthquakes, Part 1: General rules, Seismic actions and general requirements for structures, German version

[Web-1] http://ec.europa.eu/enterprise/sectors/pressure-and-gas/documents/ped/guidelines/index_de.htm, publisher: European Commission, data gathered on 5. Feb. 2013

[Web-2] http://www.ad-2000-online.de/, publisher: Beuth Verlag GmbH, data gathered on 10. May. 2013

Seismic Assessment of Horizontal Cylindrical Reservoirs

Christos Baltas[1], Pierino Lestuzzi[1, 2], Martin G. Koller[1]

[1] Résonance Ingénieurs-Conseils SA
21 rue Jacques Grosselin, 1227 Carouge, Switzerland
christos.baltas@resonance.ch, martin.koller@resonance.ch

[2] Ecole Polytechnique Fédérale de Lausanne
EPFL Station 18, 1015 Lausanne, Switzerland
pierino.lestuzzi@epfl.ch

ABSTRACT:

Industrial facilities contain a large number of constructions and structural components. Both building and non-building structures typically can be found in an industrial/chemical plant. Above ground pressurised tanks are typical examples of non-building structures of such sites. These equipments are typically used for the storage of gas and liquid materials, e.g. chlorium, ammonia etc. The overall design of such structures, especially in low to moderate seismicity areas, has neglected any seismic loading in the past, basically due to the absence of relevant seismic requirements in previous codes. The seismic security of above ground pressurised tanks is of great importance, since failure of these structures can lead to negative impact for the environment and to economic losses. Recent codes for seismic design and construction of horizontal cylindrical reservoirs provide tools which can serve also to assess existing tanks. From experience, the seismic deficiencies of reservoirs of this type are in general concentrated in some strategic points. This paper describes the main deficiencies of such structures and the simplified methodology used for their assessment based on the guidelines presented in Eurocode 8. In addition, typical cost effective solutions for the retrofit of the tanks with these shortcomings are presented and critically discussed. The above assessment and retrofit methodology is illustrated for some examples of typical equipment.

Keywords: horizontal cylindrical reservoirs, seismic assessment, structural seismic deficiencies, retrofit

1 Introduction

Above ground pressurised tanks are typical examples of non-building structures of industrial facilities sites [1]. These equipments are typically used for the storage of

S. Klinkel et al. (Eds.), *Seismic Design of Industrial Facilities*,
DOI 10.1007/978-3-658-02810-7_39,

gas and liquid materials, e.g. chlorium, ammonia etc. The overall design of such structures, especially in low to moderate seismicity areas, has neglected any seismic loading in the past, basically due to the absence of relevant seismic requirements in previous codes. The seismic security of above ground pressurised tanks is of great importance, since failure of these structures can lead to a great negative impact for the environment and to big economic losses. Recent codes for seismic design and construction of horizontal cylindrical reservoirs provide tools which can serve also to assess existing tanks. From experience, the seismic deficiencies of reservoirs of this type are in general concentrated in some strategic points. This paper describes the main deficiencies of such structures and the simplified methodology used for their assessment based on the guidelines presented in Eurocode 8. In addition, typical cost effective solutions for the retrofit of the tanks with these shortcomings are presented and critically discussed. The above assessment and retrofit methodology is illustrated for some examples of typical equipment.

2 Type of reservoir

2.1 Dimensions, Materials

Typically, the horizontal reservoirs have a cylindrical shape with flat or spherical ends. Their volume varies from small 1 t to larger 500-800 t [Web-1]. Their main body-structure is manufactured from structural steel. Due to internally applied pressure and the static system itself, the steel thickness of these reservoirs is usually quite important. Indicatively, this can exceed a thickness of 2 cm.

Tanks of this type, in general, are supported, at the level of the ground, on reinforced concrete foundation systems.

The configuration of their supporting system, meaning the system which transfers the forces from the structure of the tank to the foundation system, depends on their weight, dimensions, the elevation height and the seismicity. For tanks constructed few meters above the ground the supporting system may include only simple elements. Indicatively, these include steel plates, bearings and bolts.

2.2 Bearings, fixed and sliding

In general, the reservoirs of this type are supported on two axes either on four legs system support, two per axis, see Figure 1, or on two saddles, one per axis, see Figure 2. At the location of these axes, the reservoir is usually strengthened with external or internal circular rings. The aforementioned bearing elements are placed on a foundation system whose details will be described later. The connection between these elements is either fixed, or partially fixed, or often in the longitudinal direction of the tank permits sliding between them, in order to prevent additional loading of the foundation system, but also of the entire structure, from temperature changes, differential settlements and other loads with similar effects. The longitudinal direction is along the elongated dimension of the reservoir. Hence, it is not rare that the bearing elements of one of the two axes are fixed on the foundation and the opposite elements are free to slide, a configuration unfavourable for a good seismic behaviour of the structure, since the seismic loading is concentrated on few elements. The sliding connection is usually achieved with guided in one direction sliding bearings. Sometimes, this special detail is omitted. Instead, the bearing element rests on the foundation system without providing any special detailing. Hence, the only mean of transferring a horizontal force is via friction which depends mainly on the roughness of the interface and the axial load which sometimes can be reduced to zero. This kind of supports does not provide a satisfactory seismic behaviour. In fact, large torsion effects can be developed for such kind of supports. Concerning fixed connections, these are constructed by fixing the bearing element on the foundation with bolts. These are able to transfer some horizontal loads mainly due to wind, but usually unable to transfer the total seismic forces.

Figure 1: Horizontal cylindrical reservoirs resting on four legs system support

Figure 2: Horizontal cylindrical reservoirs resting on saddles of support

Figure 3: Horizontal cylindrical reservoirs resting on saddles

Experience has shown that the number of the bolts as per the construction drawings is in reality smaller, as shown in Figure 3. For that reason, a very good check of the actual condition of the structural elements of the tanks is essential before any assessment procedure.

2.3 Foundations, isolated and lab on grade

The foundation system usually comprises some of the following elements: independent footings, connecting beams, slab on grade and piles. The foundation system, designed for transmitting merely static, vertical loads to the ground has relatively small dimensions in terms of area. The dimensions of the foundation are also usually limited by the dimensions of the reservoir in plan view. One reason for this configuration is that often more than one reservoir is constructed in a row, each close to each other due to space limitations. The space, below or close to the tank, required for the attachment of pipes on it is an additional reason of the limited dimensions and the configuration of the foundation elements.

On the one hand, when four legs system support is provided for the tank, then the foundation system usually comprises four column-shape elements founded on four independent footings. On the other hand, when two saddles of support are used, then the foundation system comprises two wall shape elements founded on a longer footing, perpendicular to the longitudinal axis of the tank. Sometimes the latter system is used also for the four legs system support case. Now the above elements are connected via a slab on grade or they are completely independent.

Being small in plan view, the foundations are susceptible to overturning when subjected to earthquake loading, and the column-shape and more rarely the wall-element shape structural elements described above are susceptible to flexure and shear failure.

3 Seismic loading

3.1 Seismic action

The horizontal seismic action to be used for the design of tanks should be that defined in EN 1998-1 [2]. For the case of the assessment of existing tanks, the same seismic actions may be used. EN 1998-1 provides information concerning the vertical component of the seismic action that could be used.

The importance factor γ_I taken into consideration depends on the importance class of each structure. This class depends on the potential loss of life due to the failure of the particular structure and on the economic and social consequences of failure. For example, a Class IV, as per EC8, refers to situations with high risk to life and considerable economic and social consequences of failure. Further description of each class can be found on EN 1998 and EN 1990 [3]. In case of exceptional risk to life and extreme economic and social consequences of failure, higher importance factors might be necessary. It is important that the importance class of each structure is well defined in order to set the requirements for the assessment procedure as well as for eventual retrofitting of a tank. In general, the importance factor is imposed for each country from National authorities.

The contribution of each component will be derived for the value of q and of the damping ratio considered appropriate for the corresponding component [4].

3.2 Structural response

For steel tanks, the inertia forces on the shell due to its own mass are small compared with the hydrodynamic forces, as described in the following paragraph.

3.3 Content response

The dynamic behaviour of the fluid contents of tanks can be approximated by a mechanical analogue of springs and masses, as shown in Figure 4. m_I represents that portion of the total fluid mass m_f which acts as though rigidly attached to the tank walls, and is thus subject to the same accelerations as the tank walls. This is generally referred to as the impulsive mass of the fluid. The sloshing or convective response is represented by a number of masses and springs simulating the different antisymmetrical slosh-modes of the fluid. In practice regarding one slosh-mode is normally sufficient to represent the convective forces on the walls. For the cases of rigid tanks, the total base shear can be calculated by adding the absolute maxima of the force of each component in assuming that the impulsive mass to respond with the peak ground acceleration whereas the convective mass to respond with the acceleration which corresponds to the first convective period [4].

It is normally unconservative to consider the tank as rigid (especially for steel tanks, and moreover those with L/R ratios smaller than 10 having two supports at their ends [5]). In flexible tanks, the fluid pressure is usually expressed as the sum of three contributions, referred to as: 'rigid impulsive', 'sloshing' and 'flexible' [4]. Nevertheless, simplified ways for defining an acceptable value of the total base shear have been proposed by several researchers [4]. One of these concludes that this can be done by adding the seismic force of the impulsive and sloshing component in assuming the entire impulsive mass to respond with the amplified absolute response of the flexible tank system.

It should be noted that even in the case of rigid tanks, the impulsive mass is not always responding with the peak ground acceleration, dotted line in figure 5, but with an amplified one as well, dashed line or solid thin line in Figure 5. This is due to the flexibility of soil and the support system and soil-structure interaction effects. In general, in order to calculate the response of the impulsive mass, it is advisable to use the acceleration value of the plateau of the spectrum, solid thin line in Figure 5. It could be decided then that a reduced value could be used if this can be further justified and provided that a check with the value of the plateau leads to small exceedance of the resistance of the structure. Finally, the peak ground acceleration could be used in case the reservoir is very stiff and founded on rock.

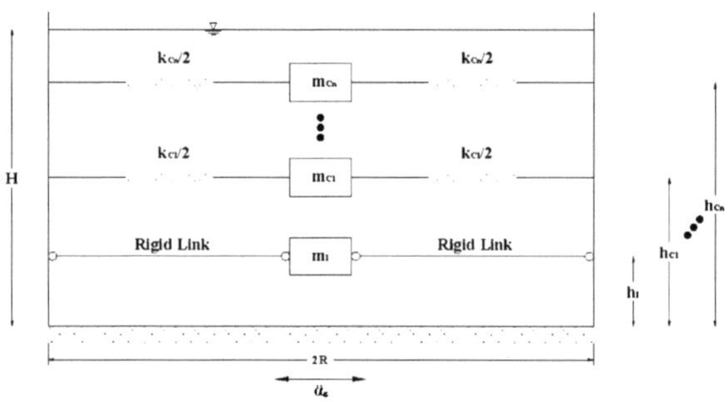

Figure 4: Mechanical analogue of response of fluid contents of tank

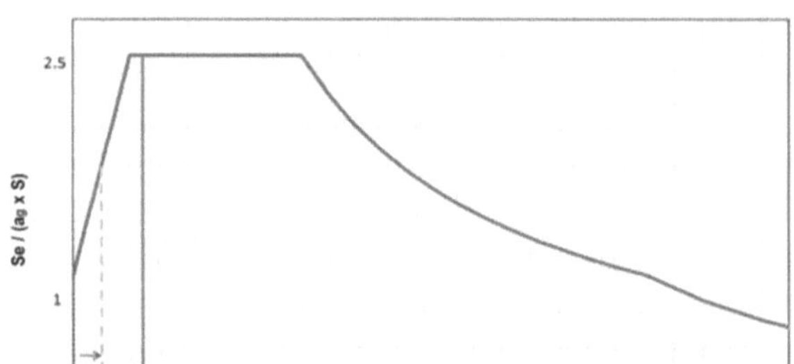

Figure 5: Typical acceleration response spectrum shape

3.4 Forces to be considered

Horizontal cylindrical tanks should be analyzed for seismic action along the longitudinal and along the transverse axis (see Figure 6 for notations).

Approximate values for hydrodynamic pressures induced by seismic action in either the longitudinal or transverse direction may be obtained by considering a rectangular tank with the same depth at the liquid level, the same dimension as the actual one and in the direction of the seismic action and third dimension (width) such that the liquid volume is maintained. The maxima result values of a sophisticated numerical model analysis by Carluccio et al. [1] of a horizontal cylindrical reservoir agreed well with the base shear computed using the combination rule described already for the impulsive and convective components. In general, given that the vertical acceleration, as per EC8 requirements, is smaller than the horizontal one and that the vertical component is reduced to 30% when this is combined with the horizontal component of the seismic excitation, the effects of the vertical component of the earthquake are negligible. Nevertheless, in the case of assessment of the foundation system of a reservoir and especially of a four legs support system on which the axial load is increasing/decreasing substantially, its contribution may be critical and should be taken into account in the transversal direction of the reservoir.

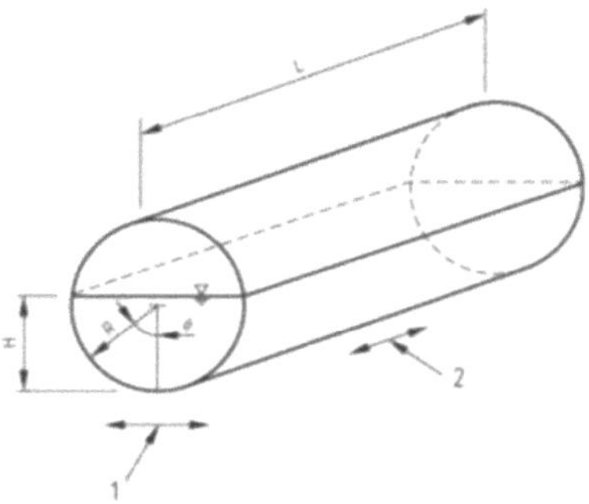

Key: 1: seismic action in transverse direction; 2: seismic action in longitudinal direction.

Figure 6: Notations for horizontal axis cylindrical tank [4]

Being the critical elements against failure due to seismic loading, the bearings of the reservoir and the foundation system usually have to be checked. Hence, the knowledge of the total base shear and the height where each component is applied is of importance.

The determination of the fraction of each of the above modes to be used is defined in EN 1998-4 in tables [4], providing hence a useful and easy way for the assessment of the tanks.

4 Simplified analysis

4.1 Verification of bearings

In order to determine the seismic demand of the bearings, simple models can be used. Being quite stiff and strong, the reservoir from experience can transmit the horizontal seismic force to its bearings. The latter, designed merely to withstand loads from the weight of the reservoir and its content as well as small horizontal loads imposed from wind loading of the reservoir, have usually insufficient capacity to withstand seismic loads, even when situated in small to moderate seismicity regions. Careful check should be performed for all its constituent components: steel plates, bolts, welds and other bearing equipment.

4.2 Verification of foundations

An additional critical control for the assessment integrity of the reservoirs is the one of the foundations. In order to assess if the foundation can transmit the seismic loads at the ground their structural capacity should be first checked. Secondly but still important, a control of overturning stability and sliding of the foundation system should be effectuated. For instance, for the case of a rectangular footing, the eccentricity of loading should be checked not to exceed the 1/3 of its width as per EC7 [6]. The soil stresses below the foundation should be checked as well. These, in general, should not exceed the bearing resistance of the foundation soil. Nevertheless, in some cases, the exceedance of the aforementioned value of the 1/3 as well as the bearing capacity of the foundation soil could be tolerated [7].

5 Typical example

5.1 Numerical example

A horizontal cylindrical steel reservoir, situated in Switzerland, in an area of relatively high seismicity, is assessed hereafter with the aid of the aforementioned simplified method. A general view of the reservoir with some general dimensions is shown in Figure 7.

A closer view of the foundation system with its dimensions is shown in Figure 8. In the same figure, the support conditions of the reservoir on the two legs of support are shown, one of each support axis. The absence of anchoring of the bearings on the second axis support yields to a sliding support of the reservoir in both horizontal directions.

The diameter of the reservoir is 1,3 m. Its length is 9,0 m approximately. The distance between the supports in the longitudinal and the transversal direction is 5,0 m and 1,9 m respectively. A four legs system support has been adopted for this reservoir. A slab on grade of approximately 25 cm thickness connects them.

Not being strong and stiff, the slab on grade cannot withstand big forces and is not able to distribute the axial load of the "columns" at all its surface. Hence, its contribution to the resistance and the stiffness of the system has been, in general, neglected in the calculations. Nevertheless, a small contribution of the foundation was taken into account by assuming a 45° degree dispersal of the axial load of the four legs inside the slab on grade.

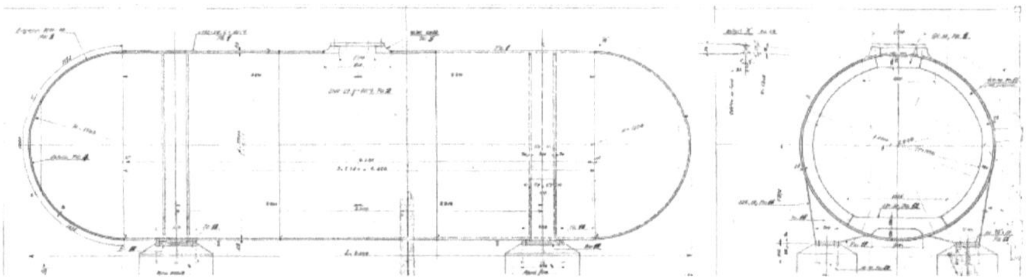

Figure 7: Construction drawing of a horizontal cylindrical reservoir including some general dimensions

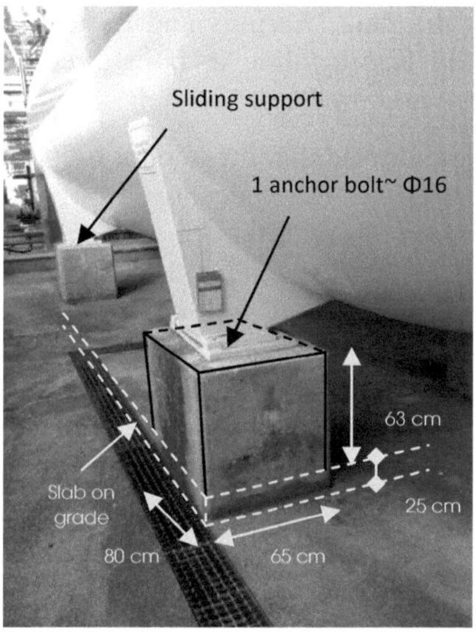

Figure 8: View and dimensions of column-shape leg-support of the reservoir

The seismic demand for the assessment of the reservoir was obtained from the spectrum as imposed in [Web-2]. This corresponds to a D soil class as defined in SIA 261 [8] and it is identical to the equivalent spectrum EC8 [2] for the same soil class. The seismic zone of the site of the reservoir is Z3a, the second highest in Switzerland, which corresponds to a ground acceleration value of a_{gR} equal to 1,3 m/s^2.

Given the relatively limited risk to the environment and people in case of failure of such a tank and given the fact that the tank is full only for a short time during the year, an importance factor of $\gamma_I = 1.4$ was considered as adequate for its control.

The mass of the liquid is 50 t, corresponding to a filling ratio of 83%, and the mass of the tank is 26 t. This filling ratio corresponds to a fluid height of around H/R = 1,52. For the excitation in the transversal direction of the reservoir, the impulsive mass corresponds to 61 t, tank mass and 70% of liquid mass, and the convective mode to 15 t, 30% of liquid mass. For the excitation in the longitudinal direction of the reservoir, the impulsive mass corresponds to 41 t, tank mass and 30% of the liquid mass, and the convective mode to 35 t, the 70% of the liquid mass. From graphs in EC8 [4], it can be easily found that the period of the convective mode is 1,58 s and 3,38 s, respectively, for the two directions. The period for the impulsive mass is assumed, conservatively, to correspond to the plateau of the spectrum.

The total seismic force in the transversal and the longitudinal direction is 320 kN and 214 kN, respectively. Indicatively, the seismic forces of each component as well as the total corresponding reactions in the columns are shown in Figure 9 for the case of the excitation in the transversal direction.

The connection between the tank bearings and the foundation has been assumed articulated at one axis and free to slide in both directions at the second axis of support. The supports were assumed fixed at their base.

The control of the bearings showed that the anchor bolts are failing to withstand the seismic forces in both directions. The eccentricity of the load, assuming that the bolts can transmit the seismic forces to the foundation, exceeds the 1/3 of the width of the footing. The same result was obtained in the case the seismic load in the transversal direction is distributed to all four columns. The resistance in sliding has been found adequate due to the connection of the four columns with the slab on grade. The soil stresses due to the large values of eccentricities proved to exceed the soil bearing resistance.

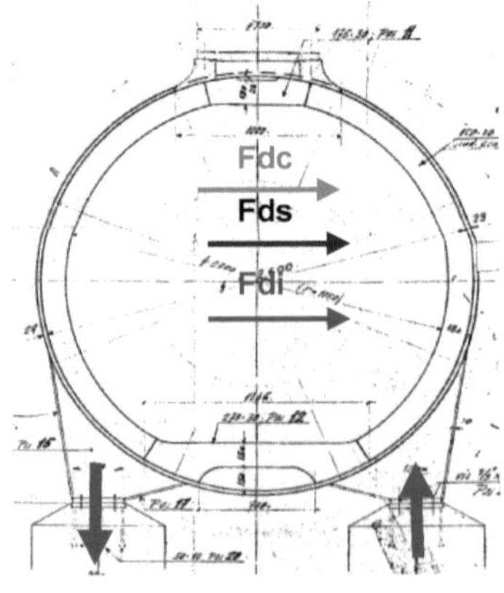

Figure 9: Seismic forces in the transversal direction

6 Retrofitting measures

6.1 Bearings, Anchorages

On the one hand, in the longitudinal direction, the seismic forces have to be transmitted only from one support axis columns, in order to leave the reservoir able to expand due to temperature changes. On the other hand, in the transversal direction, the seismic forces can be distributed to all four bearings, if in the second support axis, a bearing system which can leave the reservoir to slide in the longitudinal direction and block it in the other direction is provided. This can be achieved by fixing the bearings with bolts on the foundation providing to the connection steel plates oval holes with their larger dimension being along the longitudinal direction. The shear resistance of the bolts to be provided on the one support axis should be more than 214 kN and the shear resistance of the bolts on the one way sliding support axis more than 320 / 2 = 160 kN.

6.2 Foundations

In order to reduce the eccentricity of the load, the dimensions of each support axis foundation elements were increased. These were increased in both directions in order to avoid any permanent eccentricity from the weight of the reservoir.

6.3 Other typical retrofitting measure

Other typical example of retrofitting measure for the case of saddle support is the rigidification of the foundation system by adding a connecting beam or wall between the two independent elements, see Figure 10. This measure results in distributing the seismic forces for an excitation in the longitudinal direction of the reservoir to both support axes, especially when a sliding bearing is provided to the one of the two saddles. Consequently, shear and moment forces are reduced at each

Figure 10: Other typical retrofitting measure

support and the compliance factors from the control of the foundation system are substantially increased.

7 Conclusion

A simplified code based procedure for the seismic assessment of horizontal cylindrical reservoirs has been presented in the present study. Simple hand calculations and engineering judgement can lead to the easy detection of the most important seismic deficiencies of these tanks. Not being designed for seismic loads, the supporting system and the foundation are susceptible to shear failure and overturning, respectively. Cost effective interventions can upgrade and improve considerably the seismic behaviour of this kind of structures.

References

[1] Carluccio, A.D. et al.: Analysis of pressurized horizontal vessels under seismic excitation; in China: Proceedings of 14th Conference on Earthquake Engineering; 2008.

[2] European committee for Standardization: Part 1-General rules-seismic actions and rules for buildings; Eurocode 8, Design provisions for earthquake resistance of structures; 2004.

[3] European committee for Standardization: Basis of structural design; Eurocode 0; 2002.

[4] European committee for Standardization: Part 4- Silos, tanks and pipelines; Eurocode 8, Design provisions for earthquake resistance of structures; 2006.

[5] Karamanos, S.A. et al.: Sloshing Effects on the Seismic Design of Horizontal-Cylindrical and Spherical Industrial Vessels; Journal of Pressure Vessel Technology; Vol.128 No.3 (2006), Pages 328-340.

[6] European committee for Standardization: Part 1- General rules; Eurocode 7, Geotechnical design; 2004.

[7] Gazetas, G.: Seismic Design of foundations and soil-structure interaction; in Geneva, Switzerland: Proceedings of 1st European Conference on Earthquake Engineering and Seismology; 2006.

[8] Swiss society of engineers and architects: Actions on structures; SIA 261; Zurich, 2003.

[Web-1] http://www.michigan.gov/documents/Vol2-5UIP11Tanks_121080_7.pdf, publisher: Michigan state, data gathered on 28. Apr. 2013.

[Web-2] http://www.crealp.ch/phocadownload/Carte_MSS-Chablais_8nov11.pdf, publisher: CREALP, data gathered on 28. Apr. 2013.

Part IX

Soil-Structure Interaction: Applications

The Significance of Site Effect Studies for Seismic Design and Assessment of Industrial Facilities

Corinne Lacave[1], Martin G. Koller[1] Pierino Lestuzzi[1, 2] and Christelle Salameh[3]

[1] Résonance Ingénieurs-Conseils SA
21 rue Jacques Grosselin, 1227 Carouge, Switzerland
corinne.lacave@resonance.ch, martin.koller@resonance.ch

[2] Ecole Polytechnique Fédérale de Lausanne
EPFL Station 18, 1015 Lausanne, Switzerland
pierino.lestuzzi@epfl.ch

[3] Institut des Sciences de la Terre ISTerre
BP 53, 38041 Grenoble Cedex 9, France
salamehc@ujf-grenoble.fr

ABSTRACT:

Site effects can significantly modify the seismic motion in certain frequency domains, due to the resonance of soft deposits and subsequent amplification of the motion and / or due to the shape of the bedrock surface under soft deposits. Consequently, the shape of an appropriate elastic response spectrum might significantly differ from those proposed in building codes like EC8 based on a few soil classes. A site specific elastic response spectrum can either be lower or higher than the corresponding code spectrum or even both together, depending on the considered frequency band. Especially in the framework of assessment and reinforcement of existing industrial facilities, it might be of great importance to determine a site specific spectrum, much more adapted to account for local site effects. In some cases, such a specific spectrum makes it possible to save millions of unnecessary reinforcements. Some brief methodological aspects are presented, followed by real case examples, showing the importance of specific site effect studies and their consequences in terms of elastic response spectra for a more appropriate assessment or design of industrial facilities. In particular, the soil classification in EC8 is essentially based on Vs30 whereas site specific studies also account for the velocity contrast between the bottom of loose soil deposits and the bedrock, a parameter that can have a great influence on the amplitude of the resulting response spectrum.

Keywords: site effects, site specific spectra, seismic action, seismic design, seismic assessment of existing structures

DOI 10.1007/978-3-658-02810-7_40, © Springer Fachmedien Wiesbaden 2014

1 Introduction

It is now well known, and widely accepted amongst the earthquake engineering community, that the effects of surface geology on seismic motion exist and can be large. It has been recognised for a very long time that earthquake damage is generally larger over soft sediments than on firm rock outcrops. This is particularly important because most of modern urban and industrial settlements have occurred along river valleys over such young, soft surface deposits. Older settlements where indeed often built on firm soils on the side of the valleys because of frequent inundations; before the rivers were corrected, people rarely settled within the valleys.

The fundamental phenomenon responsible for the amplification of motion over soft sediments is the trapping of seismic waves due to the impedance contrast between sediments and the underlying bedrock. When the structure is horizontally layered (which will be referred to in the following as 1-D structures), this trapping affects only body waves travelling up and down in the surface layers. When the surface sediments form a 2-D or 3-D structure, i.e., when lateral heterogeneities such as thickness variations are present, this trapping also affects the surface waves which develop on these heterogeneities and reverberate back and forth.

The interference between these trapped waves leads to resonance patterns, the shape and the frequency of which are related to the geometrical and mechanical characteristics of the structure. While these resonance patterns are very simple in the case of 1-D media (vertical resonance of body waves), they become more complex in the case of 2-D and a fortiori 3-D structures. An illustration of 2-D effects is given in Figure 1.

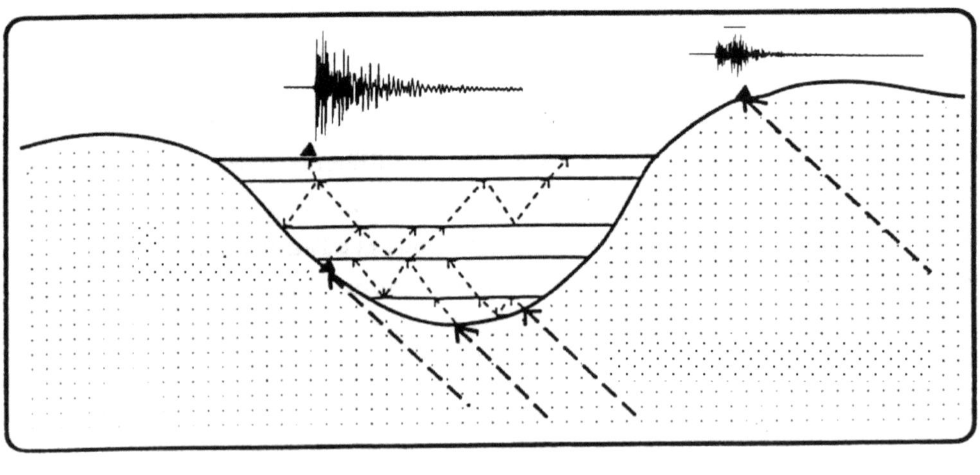

Figure 1: Illustration of site effects due to trapped waves in a sedimentary basin; ground motion within the basin is significantly stronger than on an adjacent rock site

More physical insight, together with a general overview on key factors controlling seismic hazard, is given by Bard [1], and a more detailed presentation on 2D site effects can be found in Makra et al. [2].

2 Site effect study

In order to take site effects into account, the shape of the response spectrum used to design or evaluate structures can be adapted to reflect the characteristics of the amplification of the ground motion. The first way to do this was to introduce different soil classes into the building codes. Most of the building codes include five or six different soil classes with an associated spectral shape.

Of course, these shapes are supposed to be compatible with a kind of mean behaviour of all soils belonging to one class. Looking to the shapes of real recordings, it appears that there is a great variability between sites belonging to the same class. Most of the soil classes' definitions (Eurocode 8, ASCE 7-10, etc.) do account for the mean Vs over the first 30 m of deposits; no account is taken for the total thickness of the deposits, nor for the velocity contrast at top bedrock, etc.

In order to take into account the influence of the ground structure on the resulting seismic behaviour of a site in more details, it is possible to conduct a site effect study. The goal of such a study is to define a site specific spectrum, to be used instead of the mean code spectrum.

First, an S-wave velocity profile has to be defined, based on all available geological and geotechnical information, as well as some geophysical measurements such as MASW or H/V measurements, for example. If necessary, the shape of the bedrock has to be defined also, for a 2D account of site effects. Then, some computations (1D or 2D) are conducted in order to get the seismic motion at the surface, given a regional input rock motion at the bottom of the profile (as shown on Figure 2). The account for the non linear behaviour of the soil can also be included, as well as for the uncertainties in the Vs determination, in the incidence angle in case of 2D computations, and so on.

Finally, using all resulting spectra at a given site, a site specific spectrum is defined, characterizing the soil response at this site much better than the mean spectrum of the corresponding code soil class. For example, it can be seen on Figure 2 that the shape of the resulting spectrum for the lateral shallow part of the valley is very different from the one for the deep central part.

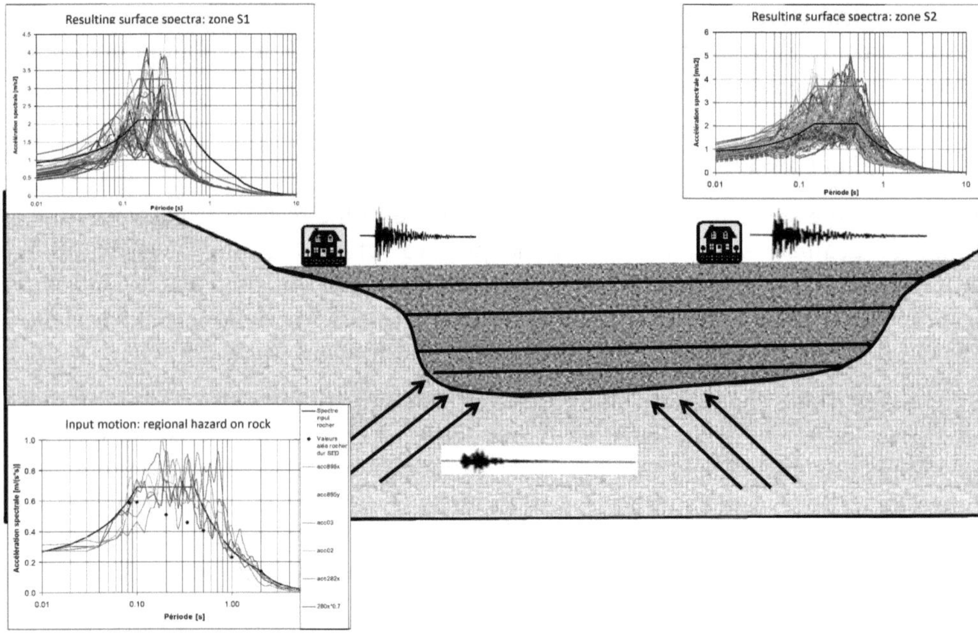

Figure 2: 2D computation of site effects with different input motions and the account for different incidence angles and uncertainties on the Vs profile

3 Influence of the velocity contrast at top bedrock

The influence of the velocity contrast at top bedrock on the response spectrum is illustrated for a simple, but realistic example. Three velocity profiles are considered; all correspond to the soil class E according to Eurocode 8 (Figure 3). All three profiles have a strictly identical gravelly sand layer of 18 m thickness, with a shear wave velocity Vs growing from 200 m/s at the surface to 400 m/s at 18 m depth. The only difference is in the underlying rock profile. In profile 1, Vs jumps to 800 m/s at 20 m depth (top bedrock), then grows smoothly, attaining 2400 m/s in about 350 m depth and remaining constant below. In profile 2, Vs jumps to 1400 m/s at 20 m depth (top bedrock), then grows smoothly, attaining 2400 m/s in 250 m depth and remaining constant below. Finally, in profile 3, Vs immediately jumps to 2400 m/s at 20 m depth and remains constant below. From 18 m to 20 m depth, in all profiles, there is a layer of weathered rock with a value for Vs that is in-between those at 18 m and 20 m depth.

Figure 4 compares the resulting response spectra, assuming that the site is located somewhere near Berne (Switzerland). All response spectra show a pronounced peak around a period of 0.2 s, corresponding to the natural period of the sandy deposit. However, the height of this peak depends on the velocity contrast below the sandy deposit: the stronger the velocity contrast, the higher the resulting amplification of the spectral acceleration around the natural period. In the opposite,

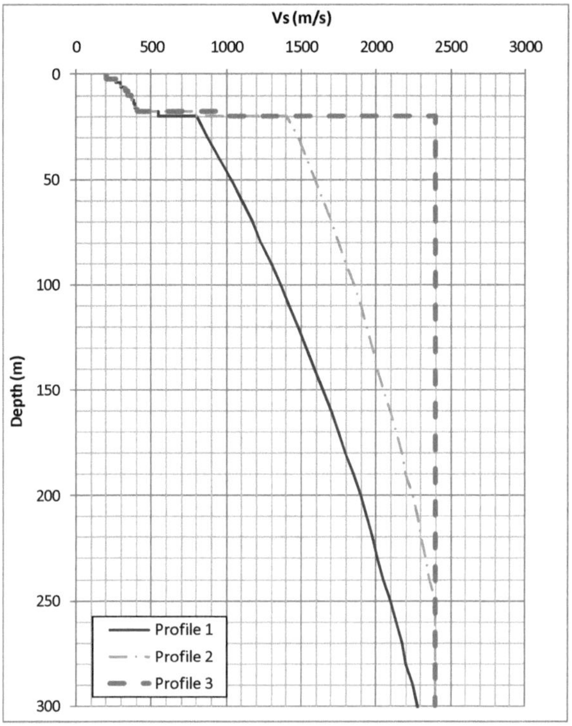

Figure 3: Three different velocity profiles, all belonging to the Eurocode 8 soil class E; the Vs profile for the first 18 m (gravelly sand) is always identical

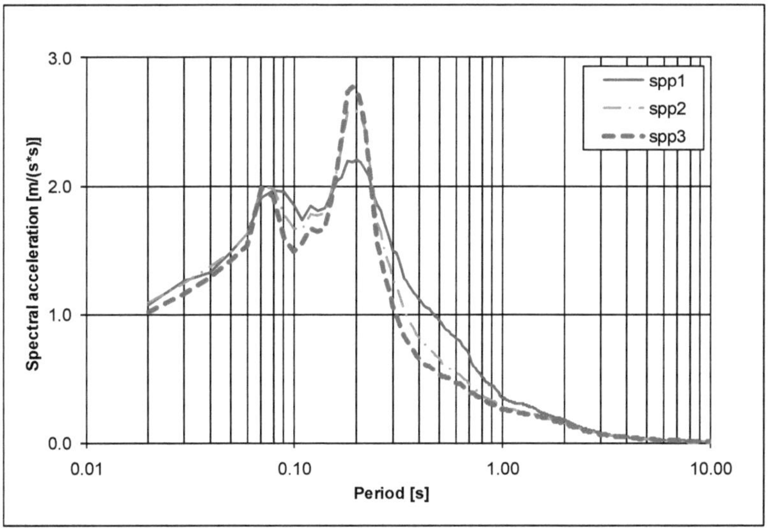

Figure 4: Resulting site specific response spectra for the soil profiles shown in Figure 3 for a regional rock hazard corresponding to Berne (Switzerland)

for a period around 0.5 s, profile 3 leads to a spectral acceleration which is only about half of the one resulting from profile 1, whereas according to Eurocode 8, the same response spectrum for soil class E would have to be used for both profiles. However, in reality, profile 3 is more unfavourable or, in the contrary, much more favourable than profile 1, depending on the (fundamental) natural period of the structure to be designed. Profile 2 is somewhere in-between the others.

4 Examples of site specific spectra

This section presents four examples of site specific spectra obtained in different areas in Switzerland, in order to illustrate the variation between the code spectra and the specific spectra, depending on the type of local geological conditions.

The first site is characterized by a 3 to 10 meter thick alluvial layer over the bedrock. The site is classified in A class where the deposits are less than 5 m thick, and in E class elsewhere. Figure 5 shows the resulting site specific spectrum, compared to the code spectra for A and E classes. It appears, in this case, that the code E spectrum can significantly be reduced, for periods longer than 0.2 s, due to the fact that the alluvial layer in not very thick (maximum 10 m).

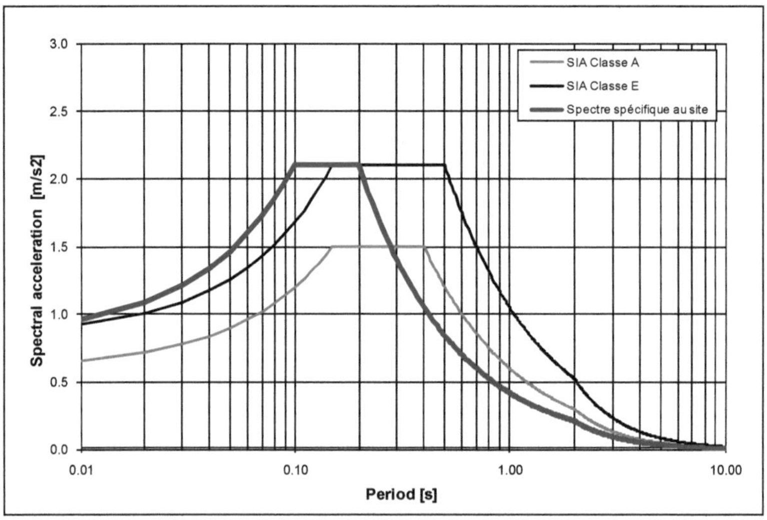

Figure 5: Site specific spectrum (in thick red), compared to A and E class spectra of the code (grey and black lines respectively)

The second example comes from a site characterized by a layer, about 40 to 50 m thick, of silty clay. Figure 6 shows the resulting proposed site specific spectrum, compared with the class C code spectrum that would have to be used without a site effect study. The plateau of the site specific spectrum is much higher than the one of the code spectrum, due to the very high velocity contrast between the clay layer and the bedrock. However, the proposed spectrum could significantly be reduced for periods longer than 0.8 s, which can be of great interest for existing flexible buildings or structures.

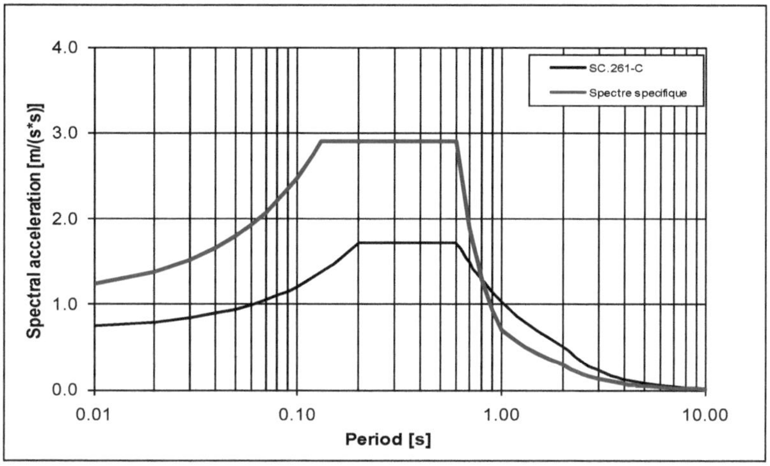

Figure 6: Site specific spectrum (in thick red), compared to C class spectrum of the code (black line)

A third example shows the case of a deep 2D sedimentary valley filled with 300 to 400 m of sedimentary deposits; the site would be class D. As shown on Figure 7, in this case, the resulting site specific spectrum is larger than the code class D spectrum over the whole frequency range.

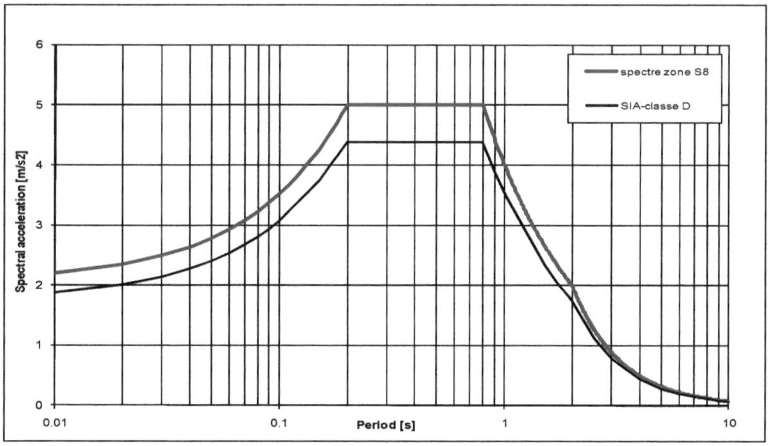

Figure 7: Site specific spectrum (in thick red), compared to D class spectrum of the code (black line)

Finally, a last example shows the case of a deep 2D sedimentary valley filled with 800 m of sedimentary deposits. As shown on Figure 8, in this case, the resulting site specific spectrum is lower or equal to the code class C spectrum over the whole frequency range, and of course then much lower than the class D spectrum. The site specific spectrum stems from a spectral microzonation study and is valid for an area that corresponds partly to soil class C, and partly to soil class D. In this case, the damping effect in the very thick sedimentary deposits is stronger than the amplification effect due to the trapping of waves in the 2D valley.

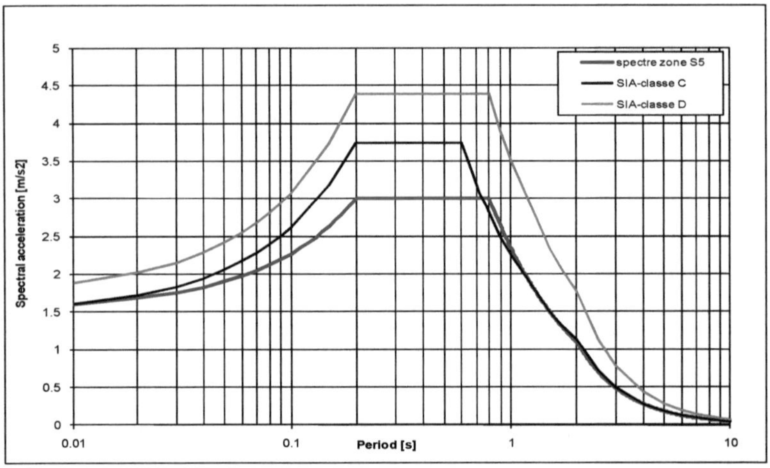

Figure 8: Site specific spectrum (in thick red), compared to C and D class spectra of the code (grey and black lines, respectively)

5 Consequences for structures

The impact of site effect studies on the seismic design or assessment depends on the dynamic characteristics of the considered structures. Generally speaking, the plateau of the site specific spectrum is significant for stiff structures such as low-rise buildings or typical industrial equipment, whereas the long period range is significant for more flexible structures such as high-rise buildings.

The consequence of the site specific spectrum of Figure 5 was that several relatively flexible buildings of a CIF did finally not need any seismic upgrade; a large amount of money could be economised.

A similar outcome was the consequence of the site specific spectrum of Figure 6. In fact, the site study had been carried out for a high-rise building with a natural period of about 2 s. For this period, the site specific spectrum turned out to be significantly lower than the corresponding code spectrum – although the site showed a particularly strong site amplification due to a strong velocity contrast at top bedrock. However, this amplification was limited to periods shorter than 0.8 s. Needless to say that for most types of industrial equipment, with natural periods shorter than 0.8 s, this spectrum would have been particularly unfavourable. Or in other words: the simple use of the code spectrum would significantly underestimate the seismic risk associated with most types of industrial equipment and 'non-high-rise' buildings at that site.

The site specific spectrum of Figure 7 leads to the same consequences for all structures. Compared to soil class D, the seismic action is slightly increased by a factor of about 1.2.

Finally, the site specific spectrum of Figure 8 is favourable for stiff structures such as low-rise buildings or most types of industrial equipment, since the plateau is clearly lower than the plateaus of the spectra for soil classes C and D. The seismic action is reduced by a factor of about 0.8 in the short and medium period range. For flexible structures, such as high-rise buildings, however, this site specific spectrum is neutral with respect to the spectrum for soil class C.

6 Conclusions

Site effects can significantly modify seismic ground motion owing to the local resonance of soil deposits and subsequent amplification of ground motion and/or owing to the shape of the bedrock topography below the soil deposits. A site specific response spectrum can either be lower or higher than the corresponding code spectrum. Often, the modification is different and in an opposite way for the plateau and for the long period range.

Compared to code spectra, seismic action can sometimes be reduced by a factor of up to 2 or, in rare cases, even more, within a limited period range. Consequently, particularly in the framework of assessment of existing industrial facilities, it might be of great interest to determine a site specific spectrum. Sometimes, a seismic reinforcement that would be necessary according to the code spectrum can be shown to be superfluous thanks to a site specific study.

REFERENCES

[1] Bard, P.-Y.: Seismic loading. Engineering characterisation of seismic ground motion and key controlling factors; European Journal of Environmental and Civil Engineering, Vol. 15, Special Issue: Hazard prevention and protecttion , some issues on seismic and gravitational risks, 2011, Pages 141-184, Hermès Editions.

[2] Makra, K.; Chavez-Garcia, F.J.; Raptakis, D.; Pitilakis, K.: Parametric analysis of the seismic response of a 2D sedimentary valley: implications for code implementations of complex site effects; Soil Dynamics and Earthquake Engineering; No.25 (2005), Pages 303-315.

Soil Liquefaction: Mechanism and Assessment of Liquefaction Susceptibility

Roberto Cudmani

Smoltczyk and Partner GmbH
Untere Waldplätze 14, Stuttgart, Germany
cudmani@smoltczykpartner.de

ABSTRACT

The basic mechanisms of earthquake-induced soil liquefaction are introduced by considering the shaking of a block on a thin granular layer, which mechanical behaviour is modelled with a hypoplastic constitutive model. If the block is founded on a dry cohesionless soil or drainage of the granular layer is fully allowed, the soil densifies and the block settles step-wise. On the other hand, if drainage is impeded pore pressure develops and effective pressure decays with increasing number of shaking cycles, until, depending on the initial density, either a quasi-stationary cyclic state is reached or the effective pressure vanishes (liquefaction). The coupled nature of dynamic problems involving soils is also shown by the results of the analyses, i.e. the motion of the block causes changes of the soil state which in turn affect the block motion. Investigations of soil liquefaction under dynamic earthquake-like excitation with a 1-g laminar box confirm the predicted behaviour. The same constitutive equation is applied to the numerical simulation of the propagation of plane waves in homogeneous and layered level soil deposits induced by a wave coming from below. Well-documented sites during strong earthquakes are used to verify the adequacy of the hypoplasticity-based numerical model for the prediction of soil response during strong earthquakes. It is concluded that liquefaction susceptibility during strong earthquakes can be reliably assessed with the proposed method. The influence of local site conditions, seismic excitation and nonlinearity of the soil behaviour on the ground response can be realistically taken into account by the model.

Keywords: liquefaction, earthquake, dynamic

1 Introduction

Reports of recent strong earthquakes present clear evidence of the devastating effects of the so-called '"soil liquefaction"' on human lives, infrastructure and buildings. Historically, the term '"liquefaction"' has been used in a broad sense to indicate a variety of phenomena involving the decay of shear resistance and

DOI 10.1007/978-3-658-02810-7_41, © Springer Fachmedien Wiesbaden 2014

excessive deformation caused by monotonic or repeated loading of saturated soils. Seed and Lee [1] defined *initial liquefaction* in a cyclic triaxial test as the state at which the pore water pressure becomes equal to the total pressure. In this sense, the word liquefaction is used here to indicate the loss of effective stresses during alternating shearing.

The mechanism leading to soil liquefaction during pure (stress-/strain-controlled) cyclic shearing is rather well-known. A dry cohesionless soil compacts under cyclic shearing. Exactly the same compaction occurs in a saturated soil if fully drainage is guaranteed. The looser is the soil initially, the faster and stronger the compaction. On the other hand, if drainage is impeded pore pressure develops, intergranular forces decay and with them also the effective pressure and the shear stiffness of the granular skeleton. According to the drained behaviour, the looser is the soil, the larger the shear strain amplitude and the higher the rate of reduction of the effective pressure.

The basic mechanism behind the earthquake-induced soil liquefaction is the same as that described above. However, it must be realized that soil shearing during an earthquake is not cyclic: The amplitude of shear stress and strain are functions of both time and space. The ground response and the degree of reduction of effective stress at a certain depth is controlled by both local site conditions (geology, material properties, fine content and state of the soil) and the characteristics of the bedrock motion (amplitudes, frequency content and duration).

Let us consider the first passage of an earthquake-induced shear wave coming from the bedrock through a homogeneous cohesionless soil stratum. The induced shearing of the soil is not homogeneous over the depth of the stratum. Thus, the reduction rate of the effective pressure, and, consequently, the reduction rate of the shear stiffness are different at different depths. The second waves lead to a further increase of the shear strain amplitude and the pore water pressure, especially at those depths, where the shear strain induced by the first waves was larger and the reduction of pore pressure stronger, which means a positive feedback. This mechanism of effective pressure reduction can lead to a localization of liquefaction in narrow zones as observed in dynamic analysis Osinov [2]. For layered soils, the dynamic ground response is much more complex, but the described feedback effect can be still observed Cudmani et al. [3].

Existing methods for evaluation of the liquefaction potential of soils can be divided in empirical and mathematical. To the first group belongs the *cyclic stress approach* proposed by Seed and Idriss [4] following the disastrous earthquakes in Alaska and Niigata in 1964. The method is based on the comparison of a *cyclic stress resistance* τ_{fd} determined from stress-controlled undrained simple shear or triaxial tests in the laboratory or estimated from in-situ tests with an *equivalent cyclic shear stress* τ_d expected to occur during an earthquake and can be estimated empirically or using elastic response analysis. For a given depth, liquefaction is said to occur if the equivalent shear stress exceeds the cyclic shear resistance.

Such approach does not properly take into account the dynamic nature of earthquake induced liquefaction: Owing to positive feedback, the shear resistance can vanish at a depth at which, according to the outlined safety criteria, liquefaction is not expected. On the other hand, an *unsafe* soil layer may be actually not endangered if liquefaction (vanishing shear stiffness) of a deeper layer impedes further shear waves to propagate upward. Despite of its limitations and mainly due to its apparent simplicity, the method has become a standard in North America and much of the world earthquake engineering practice.

In the present contribution, the mechanism leading to soil liquefaction is introduced by analysing the dynamic response of a block upon a thin horizontal soil layer under horizontal base shaking (sec. 1). The constitutive equations used to model soil behaviour in the numerical simulations are briefly described in sec. 2. In sec. 3, the results of experimental investigations of earthquake-induced soil liquefaction carried out with a 1-g laminar shake box are presented. Finally, an effective-stress ground response analyses for layered soils also based on hypoplasticity is applied to the evaluation of liquefiable soil response during past strong earthquakes (Sec. 4.).

2 Constitutive model

Two hypoplastic constitutive relations are employed in the present study. One describes the rate independent behaviour of granular soils (e.g. sand), and the other one takes into account viscous effects and is used for the modeling of clayey soils. Both relations describe plastic deformations of a solid skeleton under monotonic as well as cyclic loading for drained and undrained conditions. They incorporate the critical state concept of soil mechanics and the dependence of the stiffness on the current stress, the density and the history of deformation. For rate independent materials, the rate of the effective stress $\dot{\boldsymbol{\sigma}}'$ is determined by the rate of strain $\dot{\boldsymbol{\varepsilon}}$, the current effective stress $\boldsymbol{\sigma}'$, the void ratio e and the so-called intergranular strain tensor $\boldsymbol{\delta}$ which takes into account the influence of the recent deformation history. The constitutive equation is written as a tensor-valued function

$$\dot{\boldsymbol{\sigma}}' = \mathbf{H}(\boldsymbol{\sigma}', e, \boldsymbol{\delta}, \dot{\boldsymbol{\varepsilon}}) \qquad (1)$$

For a rate dependent material, the function involves a viscous strain rate tensor $\dot{\boldsymbol{\varepsilon}}_v$ as an additional variable:

$$\dot{\boldsymbol{\sigma}}' = \mathbf{H}(\boldsymbol{\sigma}', e, \boldsymbol{\delta}, \dot{\boldsymbol{\varepsilon}}, \dot{\boldsymbol{\varepsilon}}_v) \qquad (2)$$

As distinct from elastoplasticity theories, the description of the plastic deformation through equations (1), (2) does not require the introduction of a yield surface and a flow rule, and the decomposition of the deformation into elastic and plastic parts. A detailed description of the hypoplastic relations (1), (2) can be found in [5],[6]. The solution of a boundary value problem requires both material parameters and initial values of the state variables. The constitutive equation (1) or (2) contains 13 model parameters. They are independent of the state variables, that is, the behaviour of a given material can be modelled in a wide range of stresses and densities with the same set of parameters.

3 Shaking of a block on a thin dry and saturated soil layer

Consider the shaking of a block upon a horizontal shaking base with a thin dry granular layer in between (Fig. 1).

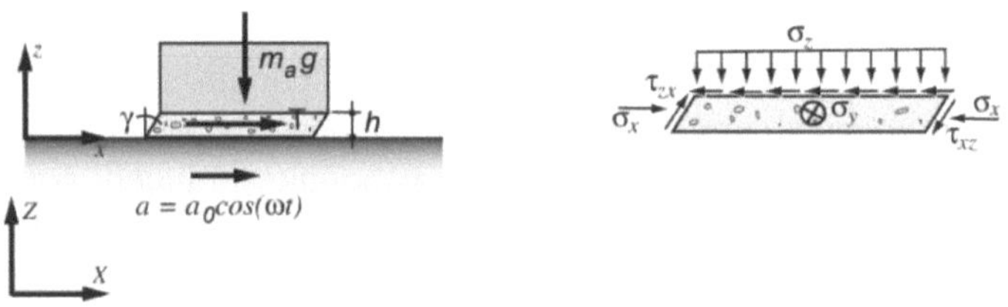

Figure 1: Block upon a horizontal shaking base with a thin granular layer

Simple shearing is assumed to occur, i.e. lateral squeezing out is prevented. In the normal direction, the initial pressure σ_z is given by the weight $m_a g$ of the block. The horizontal pressures are $\sigma_y = \sigma_x = K\ \sigma_z$ with an earth pressure coefficient K. The height h of the layer changes together with the void ratio e, its initial value e_0 may correspond to a rather loose packing. h decreases from h_0 due to shaking, $h_0/(1+e_0) = h/(1+e)$ expresses conservation of solid mass with constant grain volume. The equations of motion and its numerical solution are described in detail by Gudehus et al. [7]. Calculated displacements for harmonic base shaking are shown in 2a and b.

The horizontal block motion u_x is retarded and not harmonic, $|u_x| > |u_{base}|$ indicates amplification. Due to densification of the sand layer the block settles step-wise. Both densification and rate of increase of permanent displacements u_z decrease with further cycles. The calculated evolution of shear stress τ and void ratio e are plotted in Fig. 2c and d as a function of the shear strain γ. For the 30th cycle the layer is stiffer and the hysteresis smaller and denser than for the 2nd one. With two reversals of γ, the density goes through four reversals. In other words, one shear cycle induces nearly *two dilatancy cycles* (without change of pressure).

We consider now the case of the thin granular layer with water saturation and without drainage. The layer height h is constant, and also the total vertical pressure σ. Effective (or skeleton) pressure σ' und pore water pressure p_w are variable with $\sigma = \sigma' + p_w$. As can be seen in Fig. 3b and c, shear stiffness, shear amplitude and mean effective pressure p' decrease with the number of cycles. p' oscillates with twice the frequency of the base. After a certain number of cycles that depends on

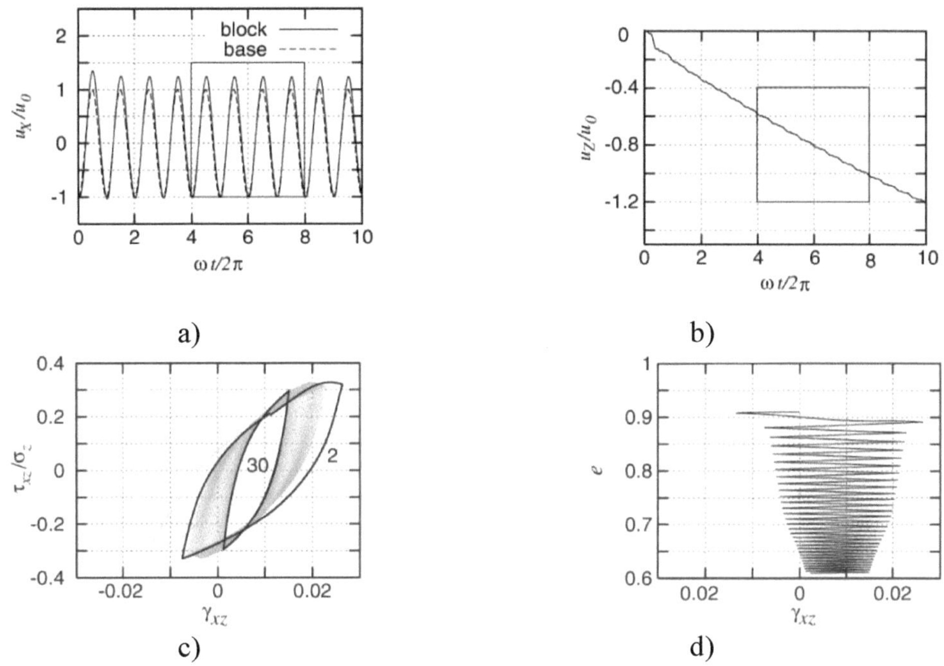

Figure 2: Results for a dry granular layer: a) horizontal and b) vertical displacements, c) shear stress vs. shear strain, d) void ratio vs. shear strain

the initial shear amplitude, the normal stress and the void ratio, the block stands almost still on a nearly liquefied layer (a). This sort of liquefaction-induced seismic isolation is indirectly substantiated by the analysis of damage on structures founded on liquefied soils.

For instance, Yoshida et al. [8] observed after the Adapazari 1999 Kocaeli earthquake in Turkey that buildings founded on liquefiable soils settled or tilted, but did not suffer severe inertia-induced damage, whereas numerous buildings in the non liquefied areas were considerable damaged or collapsed during the earthquake.

For a block on a thin saturated layer of soil with soft particles base shaking leads to a quantitatively different response which may be briefly indicated without further drawings. The shear amplitudes under a block like in Fig. 1 are bigger due to lower stiffness and as the effective pressure p' is only moderately reduced. As OCR increases with the decrease of p' the response after one or two strong cycles becomes nearly hypoelastic. Close to a certain frequency the oscillation amplitude is markedly amplified.

The effects outlined in this section may only qualitatively be transferred to field situations. Drainage is negligible during a group of strong shocks if the

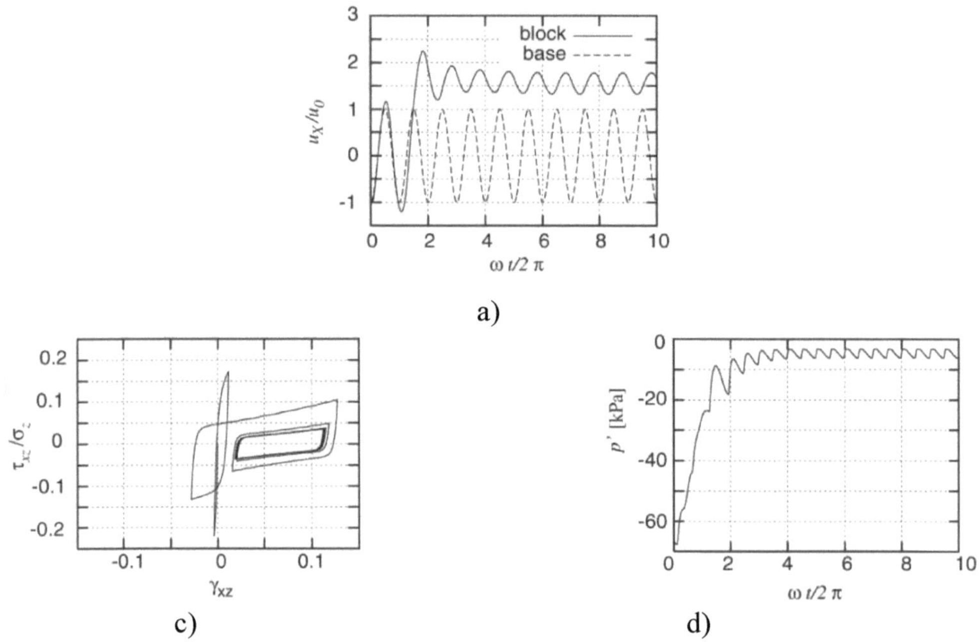

Figure 3: Results for a saturated granular layer: a) horizontal c) shear stress vs. shear strain, d) effective mean pressure vs. shear strain

permeability is below ca 10^{-4} m/s, but plays a stabilising role afterwards. Effective stress redistribution (which was excluded in the cases of Fig. 3) can lead to stabilization, in case of soft particles also a long time after the strong shocks. Any quantitative assessment of all the named effects requires the numerical solution of more sophisticated boundary-value problems via finite difference or finite element method.

4 Experimental investigation with a laminar box

In order to investigate the dynamic response of soil under dynamic earthquake-like excitation, model tests have been carried out with a 1-g laminar box shown in Fig. 4 (length 400 mm, width 300 mm, height 500 mm) at the University of Karlsruhe. The lamellas of the laminar box are allowed to translate and rotate. Opposite lamellas are constrained to undergo exactly the same motion by hinged bar connectors. Parallel to the shaking direction, the box consists of smooth steel walls, which are rigidly connected to the base. The dynamic base excitation is generated by means of springs attached to the base of the box. The spring forces are activated by enforcing an initial displacement to the base and fixing it in the new position. Shaking is initiated by a manual release mechanism. Displacements of the lamellas, settlements of the surface, and pore pressures at the bottom can be recorded, processed and consequently analysed.

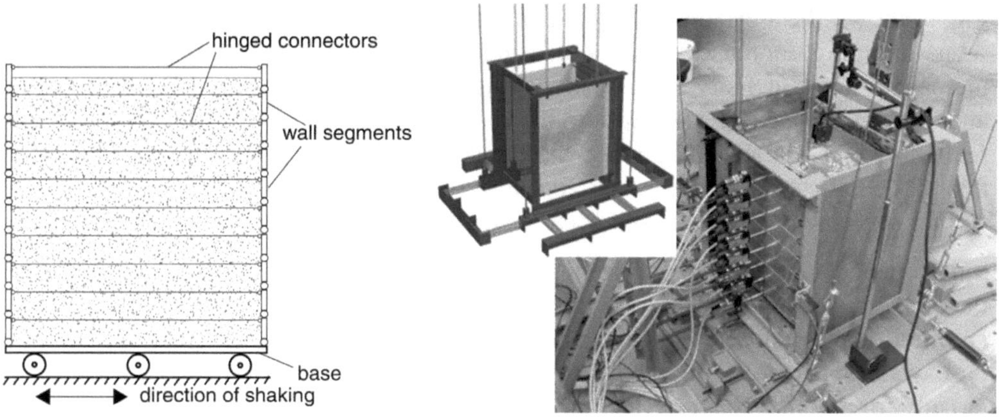

Figure 4: Laminar box at the University of Karlsruhe

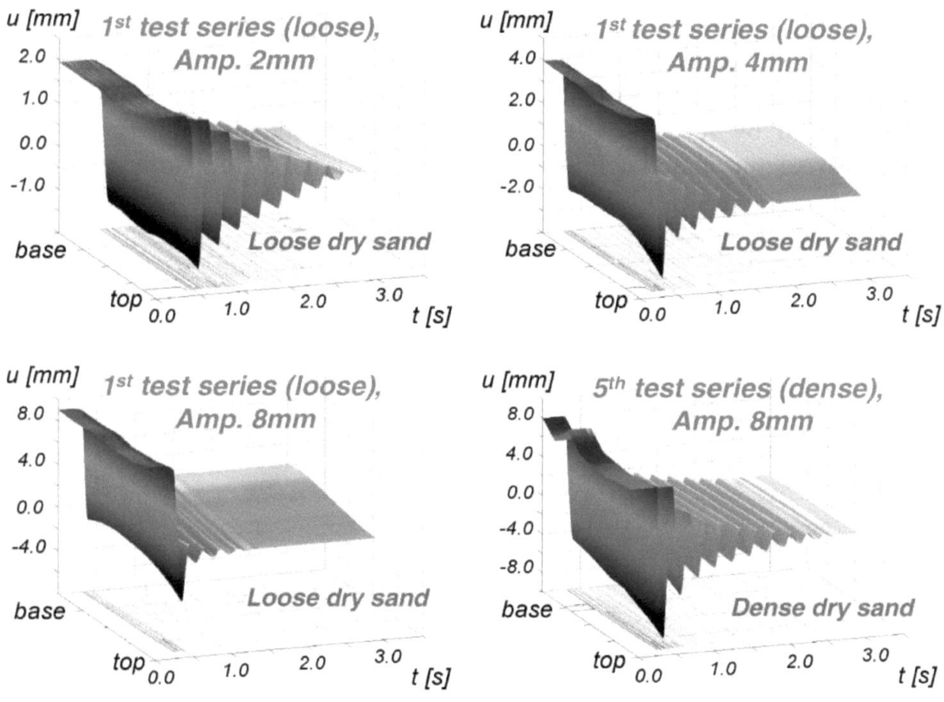

Figure 5: Laboratory results for dry sand

In most the experiments fine quartz sand (d_{50}=0.25 mm, e_{max}=0.98, e_{min}=0.65, U=2.35) was used in both dense and loose initial state. In the first test series, loose dry sand has been investigated for different intensities of excitation of the base by applying initial deflections of the base of 2, 4, and 8mm. These deflections cause peak acceleration values of slight, moderate and strong earthquakes, respectively.

Fig. 5 shows the horizontal displacements of the lamellas over time. As can be seen, damping increases with increasing shaking intensity and decreases with increasing initial density. This demonstrates the hysteretic nature of soil damping, and indicates that the assumption of viscous damping in elastic models is not realistic for moderate and strong earthquakes. During shaking the surface settles due to densification. After about five test series the sample becomes denser and further settlement of the surface became negligibly small.

Subsequent experiments have been carried out with saturated sand (Fig. 6) where the soil showed higher energy dissipation than in a dry state even for small amplitudes. Liquefaction, which is assumed to occur when the measured excess pore water pressure at the base equalled the initial effective stress, took place already for an initial base deflection of 4 mm.

Since the upper half of the soil specimen moved as a rigid body (see Fig. 6, upper right graph) we deduce that liquefaction must have occurred underneath. Whereas for an initial deflection of 8 mm skeleton disaggregation must have extended to the whole specimen since the observed dynamic response of the material resembled that of a viscous suspension (see Fig. 6, bottom left graph). Excess pore pressure dissipates and subsequent settlements of the surface develop about ten times faster than calculated by means of the conventional consolidation theory taking the permeability of the material into account. Observation of the surface during the

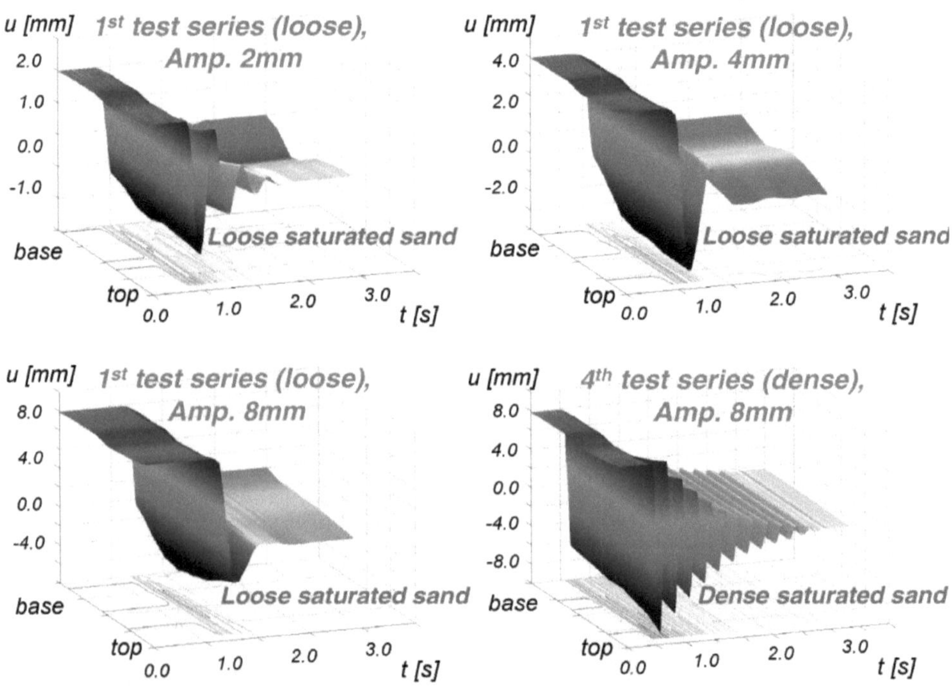

Figure 6: Laboratory results for saturated sand

experiment reveals a system of fine vertical water channels allowing faster drainage and the formation of mini sand boils at the ground surface, as it usually observed at places where liquefaction takes place during strong. After repeating the test, the sample densifies and thus, the liquefaction susceptibility is reduced. The final settlements of the dry and saturated samples are very similar.

5 Modelling of the ground response during earthquakes

Our numerical model for the level ground dynamic response analysis is based on the solution of a one-dimensional boundary value problem for a horizontal soil layers. The unknown variables are the horizontal and vertical material velocities, the nonzero components of the stress tensor (σ_{11}, σ_{22}, σ_{33}, τ_{12}) and the pore pressure. These variables are functions of the depth and time. The governing system of equations consists of the equation of motion, the constitutive equations for the solid skeleton and the pore fluid and the mass balance equation. The constitutive equation for the pore pressure considers the compressibility of the pore fluid, which depends on the degree of saturation.

Both drained and undrained conditions can be considered in the calculations. In the numerical simulations, seepage is taken into account by using the so-called u-p formulation ([9],[10]), which assigns different velocities but same accelerations to solid and fluid phases.

The initial vertical stresses and the initial pore pressures results from the densities of the solid and fluid phases and gravity. The horizontal stress is related to the vertical stress via the earth pressure coefficient at rest. The upper surface is assumed to be free of traction. At the base of the model horizontal and vertical velocities are prescribed as a function of time. In the case of saturated soil, the pore pressure at the water table and above is assumed to be zero and the lower boundary is assumed to be impermeable.

5.1 Ground response at the Port Island site

Major liquefaction-induced damage was reported at the reclaimed Port Island in Kobe during the 1995 Hyogoken-Nanbu earthquake ([11],[12]). At a site in this island, the seismic response was recorded by a four-accelerometer downhole array. The accelerometers were located at the surface and at depths of 16, 32 and 83 m.

Figure 7a shows the soil profile at the site. For the ground response analysis, the soil profile is simplified as shown in Fig. 7b, with the base located at a depth of 83 m. The parameters of the hypoplastic and visco-hypoplastic constitutive law were estimated from granulometric properties of the layers. The void ratios in the cohesionless soil layers were determined from SPT data. Using the acceleration record for 83 m depth from the 1995 Hyagoken-Nanbu earthquake the one-dimensional dynamic response of a soil profile was calculated.

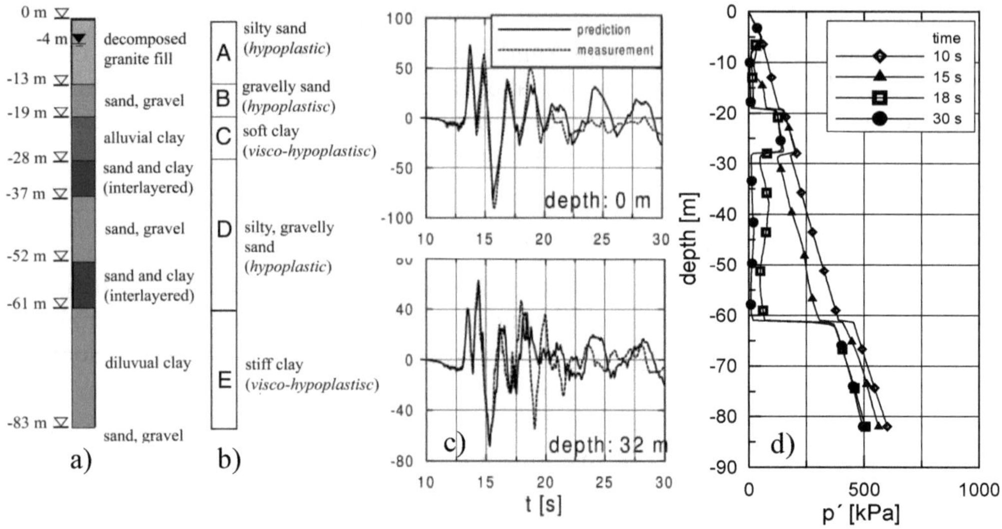

Figure 7: Port Island: real (a) and idealized (b) soil profiles; c) measured and predicted velocities at the surface and at depths of 16 and 32 m in the North-South direction; d) calculated mean effective stresses over depth

The calculated and the experimental velocities versus time at the recording depths for the stronger North-South motion component are compared in Fig. 7c. The numerical results agree well with the measured ground response between 10 and 18 s. The subsequent discrepancies between calculated and experimental velocities may indicate that the motion of the soil at this later stage diverges considerably from a plane-wave motion, for instance through the influence of surface waves or other disturbances.

Figure 7d shows the calculated distribution of the mean effective pressure versus depth at different times between the beginning (t=10 s) and the end of the strong shaking phase (t=30 s). As can be seen, the effective stresses are reduced over the whole profile during the earthquake. In the loose reclaimed layer, liquefaction takes place already 5 to 10 s after the beginning of the earthquake. The liquefaction of this layer must be responsible for widespread sand ejecta, settlements, lateral spreading and major damage of waterfront quay facilities caused by the earthquake.

The onset of liquefaction coincides with the deamplification of waves observed in the velocity records after 18 s. In the dense silty gravely sand layer, longer shaking was necessary to reduce the mean effective pressure to nearly zero. As expected, the effective stress reduction in the fine-grained layers is not as drastic as in the cohesionless layers. This contributes to the amplification of ground motion observed at all recording depths in the first seconds of shaking.

5.2 Treasure Island site

As reported in [13] and [14], geotechnical factors exerted a major influence on the nature and severity of ground shaking during the 1989 Loma Prieta earthquake. An illustration of the influence of local site conditions on ground shaking is provided by the set of strong motion recordings obtained on Yerba Buena Island, and on Treasure Island in San Francisco Bay, at approximately the same distance from the fault rupture. Treasure Island is a man-made island comprised primarily of a loose, dredged hydraulic fill underlain by natural bay sediments. Yerba Buena Island is a large rocky outcrop near the center of the bay. Figure 8a presents a schematic illustration of the soil column underlying the Treasure Island recording station. The hydraulic fill consisting of loose sand and silty sand is underlain by soft to medium stiff normally consolidated silty clay (so-called young Bay Mud). The Bay Mud is underlain by dense, fine sand and silty sand and layers of stiff overconsolidated sandy clays. Beneath, stiff to hard overconsolidated clays extend down to the bedrock. Figure 8b shows the idealized profile used in the response analysis. The hypoplastic parameters of the idealized layers were estimated on the basis of existing geotechnical information on the real soil layers ([13],[14]). The Yerba Buena Island records were used as the bedrock motion (Fig. 8c). The comparison of the measured and predicted E-W velocity component at the surface of Treasure Island is presented in Fig. 8c.

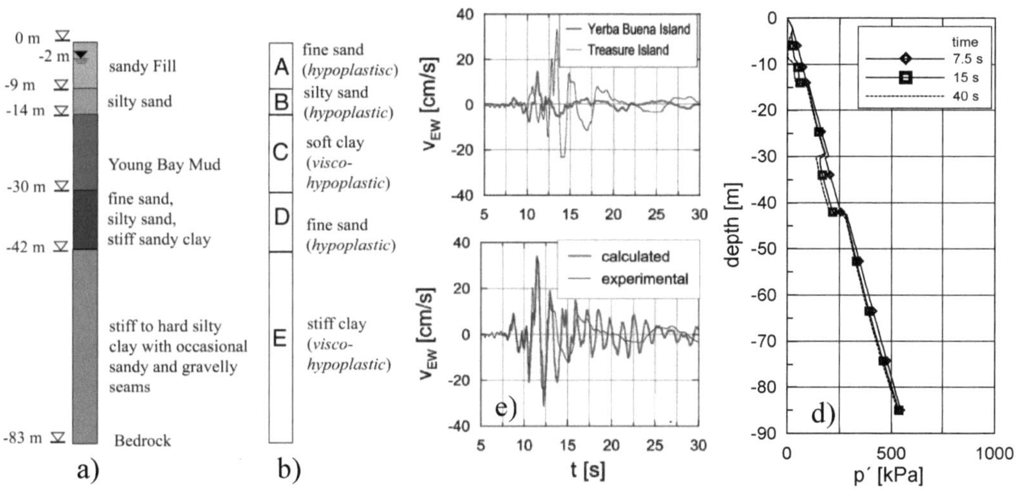

Figure 8: Treasure Island: real (a) and idealized (b) soil profiles; c) measured E-W velocity components at the surface and at the bedrock and measured and predicted velocities at TI; d) calculated mean effective stresses over depth

Figure 8d shows the change in the distribution of the mean effective pressure over depth during the shaking. Other than in the Port Island case, liquefaction concentrates in a small zone comprising both the bottom of the fine sand layer and the upper part of the silty sand layer. Liquefaction of the upper layer must be responsible for sand boils, lateral spreading and damage of coastal facilities observed in Treasure Island. On the other hand, the sand layer (D) and the clayey layers (C) and (E) did not experience any substantial reduction of effective stresses and could mainly have contributed to the amplification of the bedrock motion leading to the extensive structural damage in the marina district of San Francisco.

6 Conclusion

As shown by the numerical analyses presented in this and other contributions (e.g. [3],[7]), the mechanisms of soil liquefaction can be realistically simulated with hypoplastic constitutive models. The block-model of sec. 3 can be used instead of the *Newmark method* [16], which assumes ideal-plastic Coulomb-friction sliding, for more realistic estimations of earthquake-induced displacement of buildings and slopes during earthquakes.

The proposed wave propagation model in sec. 5 can predict the behavior of real soils during strong earthquakes. The application of the model to two seismic events validates its ability to realistically take into account the influence of local site conditions and characteristics of the bedrock motion on seismic response. According to our experience a good soil profile with densities and ground water tables, combined with realistic base shaking, suffices for the free-field ground response analysis.

The results of our block sliding simulations, the ground response analyses and also the experiments with the laminar box confirm the so-called *layer separation* effect, which has been used by Yohida et al. [8] to justify the small damage suffered by some structures founded on liquefiable soils during strong earthquakes: The reduction of the effective pressure (liquefaction) causes a decrease of the shear stiffness which in turn impedes the transmission of further shear waves from the soil to the foundation of the building. Through this mechanism liquefaction could provide a natural and effective isolation for a building as long as base failure, excessive tilting and settlement can be prevented and thus, the stability and serviceability are not endangered ([7],[15]). Clearly, more research is needed in this topic.

7 Acknowledgements

This contribution is mainly based on the research work of the author at the Institute of Soil mechanics and Rock Mechanics, University of Karlsruhe. Prof. Gudehus for his scientific advice, Dr. Osinov for the development of the ground response model and Dr. Bühler and Dr. Weibroer, who carried out the laminar box tests, are gratefully acknowledged.

REFERENCES

[1] Seed, H. B.; Lee, K. E.: Liquefaction of saturated sand during cyclic loading. J. Soil Mech. and Found. Div., ASCE, 92 (SM6), 1966, 105-134.

[2] Osinov, V. A. Wave-induced liquefaction of a saturated sand layer. Continuum Mech. Thermodyn., 12(5), 2000, 325-339.

[3] Cudmani, R. ; Osinov V. A.; Bühler M. M.; Gudehus G.: A model for the evaluation of liquefaction susceptibility in layered soils due to earthquakes. In: Proceedings 12th Panamerican Conference on SMGE, Vol.1, 2003, 969-977.

[4] Seed, H. B.; Idriss, I. M.: Simplified procedure for evaluating soil liquefaction potential; Journal of the Soil Mechanics and Foundations Division, ASCE}, Vol. 107, No. SM9, 1970, 1249-1274.

[5] Niemunis, A.; Herle, I.: Hypoplastic model for cohesionless soils with elastic strain range; Mech. Cohesive-frictional Mater., 2(4); 1997: 279-299.

[6] Niemunis, A.: Extended hypoplastic models for soils; Institut für Grundbau und Bodenmechanik der Ruhr-Universität Bochum, No. 34; 2003.

[7] Gudehus,G; Cudmani, R.O.; Libreros-Bertini, A.B.; Bühler, M.M.:In-plane and anti-plane strong shaking of soil systems and structures; Soil Dynamics and Earthquake Engineering, 24; 2004; 319–342.

[8] Yoshida, N; Tokimatsu, K.; Yasuda, S.; Kokusho, T.; Okimura, T.: Geotechnical aspects of damage in adapazari city during 1999 Kocaeli, Turkey earthquake; Soils and Foundations, Vol. 41, No 4; 2001, 25-45.

[9] Zienkiewicz O. C.; Chang C. T.; Bettess P.: Drained, undrained, consolidating and dynamic behaviour assumptions in soils. Géotechnique, 30(4), 1980, 385-395.

[10] Zienkiewicz O. C.; Chan A. H. C.; Pastor M.; Schrefler B. A.; Shiomi T.: Computational geomechanics with special reference to earthquake engineering. Chichester, Wiley; 1999.

[11] Ishihara, K.: Therzaghi oration: Geotechnical aspects of the 1995 Kobe earthquake; in Proceedings of 14th ICSMFE; 1997, 2047-2073.

[12] Elgamal,,A-W.; Zeghal, M.; Parra, E.: Liquefaction of reclaimed island in Kobe, Japan; J. Geot. Eng. Div., ASCE, 122(1); 1996, 39-49.

[13] Seed, R.B.; Dickenson, S.E.; Idriss, I.M.: Principal geotechnical aspects of the Loma Prieta earthquake; Soil and Foundations, Vol. 31, No 1; 1991. 1-26.

[14] Arulanandan, K.; Muraleetharan K. K.; Yogachandran C.: Seismic Response of Soil Deposits in San Francisco Marina District. J. Geot. and Geoenv. Eng., ASCE, 123(10); 1997, 965-974.

[15] Cudmani, R.; Cudmani, R.O.: Numerical study of the soil-structure interaction during strong earthquakes; in proceedings 13th World Conference on Earthquake Engineering, Vancouver, B.C., Canada; Paper Nr. 2959; 2004.

[16] Newmark, N: Effects of earthquakes on dams and embankments; Geotechnique, Vol.15, No. 2; 1965,139-160.

Seismic Design and Verification of a Nuclear Power Plant Structure for the Storage of Radioactive Waste Components

Davide Kurmann[1], **Zdenek Janda**[2] **and Jan Cervenka**[2]

[1] Nuclear Division / Axpo Power AG
Parkstrasse 23, Baden, Switzerland
davide.kurmann@axpo.com

[2] Cervenka Consulting Ltd.
Na Hrebenkach 55, Prague, Czech Republic
zdenek.janda@cervenka.cz

ABSTRACT:

Seismic design and qualification of safety and/or radiological relevant structures for Swiss NPPs are subjected to rigorous procedures. Structures have to meet high safety standards, be robustly designed and therefore cover a wide range of parameter uncertainties both on seismic action and capacity side. In the framework of a wide Facility Power Retrofit at the Swiss Leibstadt NPP the Project ZENT (acronym of radioactive waste storage building) aims to erect two new structures on the site. Due to operational and radiological reasons, these new structures have to be built very close respectively above to the already mentioned existing structures. One structure has to be founded on piles above the existing NPP main cooling pipelines. Approval of this new seismic foundation system by Swiss Federal Nuclear Safety Inspectorate (ENSI) required extended seismic design. This paper attempts to describe the analyses performed by the Owner for seismic qualification and verification of structural integrity at planning stage.

Keywords: nuclear power plant, pile foundation, pile-structure-interaction

1 Introduction

1.1 Leibstadt Swiss Nuclear Power Plant

The Swiss NPP Leibstadt (abbreviated KKL) is located in the municipality of Leibstadt (canton Aargau) on the Rhine River close to the Aare Delta and in proximity of the German border. After twelve years of construction, Leibstadt NPP was commissioned on May 24, 1984 and is the youngest nuclear power station built in Switzerland. With a boiling water reactor having 1,245 MW of electrical power, Leibstadt NPP is also the most powerful of the five existing nuclear power stations.

S. Klinkel et al. (Eds.), *Seismic Design of Industrial Facilities*,
DOI 10.1007/978-3-658-02810-7_42, © Springer Fachmedien Wiesbaden 2014

1.2 Radioactive Waste Storage Building Project (acronym ZENT)

In the framework of a wide renewal and retrofitting scheme at Leibstadt NPP it was decided to erect two new structures adjacent to the east façade of the turbine building with the primarily scope of storing radioactive waste components (Project ZENT). The new structures have to be built adjacent to many existing one having safety and/or radiological relevance: turbine building, condensate storage tank building, supply channels, auxiliary building complex, radioactive waste building. Moreover, the ZENT north building has to be erected above the existing NPP main cooling pipeline. Nonetheless the irrelevance of the main cooling pipelines for extreme accident management, this core cooling system is of exceptional importance for the electrical power production. Therefore, negative effects due to the new constructions have to be avoided. Based on these considerations, the structural and seismic design for the ZENT north building was carried out by means of extensive analyses. Particular attention was paid to the design and verification of the soil-pile-foundation plate system because due to the fundamental importance of lateral load carrying capacity under earthquake excitations.

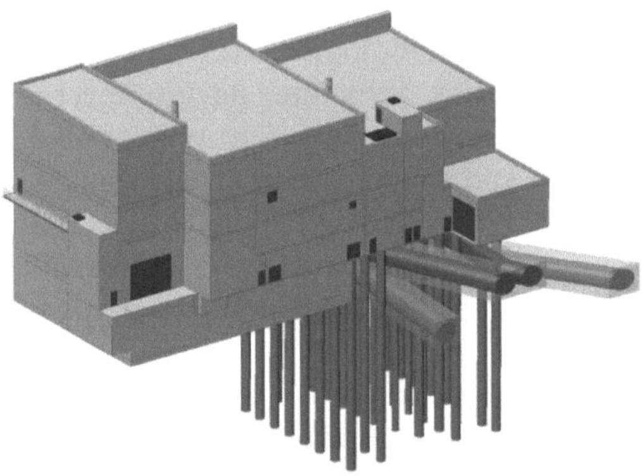

Figure 1: SW perspective view of the projected ZENT south and north buildings above already existing main cooling water pipelines (left) and along east façade of turbine building (right)

2 Probabilistic seismic hazard studies

Starting from later 90's of the past century the Swiss Federal Nuclear Safety Inspectorate ENSI (formerly HSK) identified the need to upgrade the seismic hazard assessments for the Swiss NPPs.

A probabilistic seismic hazard analysis (PSHA) according to the rules first established by the “Senior Seismic Hazard Committee” (SSHAC) on behalf of the US-NRC, Department of Energy and EPRI, was considered to best represent the current state-of-the-art. In response to regulators’ request, Swiss NPP operators (licensees) performed a new hazard study between the years 2000-2004 that satisfied SSHAC Level 4 criteria and become know as PEGASOS Project. As hazard results, ground motion exceedence probabilities including aleatory variability and epistemic uncertainty for the four Swiss NPP Sites were carried out. Such extensive hazard computation with the entire input based on expert elicitations and systematic, quantitative assessment of uncertainties was firstly adopted for NPPs worldwide. PEGASOS results were discussed in professional circles worldwide. Especially the unusual large scatter in the ground motion results was judged with criticism by experts. As a consequence, ENSI and licencees decided in year 2007-08 to start the PEGASOS Refinement Project (PRP, SSHAC Level 4 criteria as well).

2.1 Soil investigations in the framework of PEGASOS Refinement Project

An extensive field investigation campaign and laboratory tests were carried out in years 2008/2010 at the four Swiss NPP Sites in the framework of the PRP Project with the aim to reduce uncertainties in dynamic soil parameters.

Because of the large variability in the grain size distribution and cohesion characteristics of the soft soil at the NPP Site (see Figure 3), numerous applied methods in the field campaign did not work successfully (see Figure 2). A large band width in the S-wave and P-wave velocity profiles and dynamic soil strain dependent properties from laboratory testing were the consequence. Especially the P-wave velocity profile estimation in soft soil was unsatisfactory.

	Vp		Vs	
	Soil	Rock	Soil	Rock
This Study				
Logging (Full Wave Sonic)				
Uphole				
Downhole				
Crosshole				
MASW				
Ambient Noise				
Previous Studies				
Ambient Noise (SED 2001)				
Crosshole (1973)				
Uphole (1973)				

Figure 2: Summary of the methods applied to determine S-wave, P-wave velocities during the 2008-09 field campaign. Light brown (unsuitable), bright green (successful), pale green (unreliable method)

Figure 3: Channel type deposits of gravels and occasional sand lenses

2.2 Dynamic Soil Properties

Based on results of field campaign and after expert elicitations, a set of three S-wave velocity profiles P1 to P3 and a generic P-wave velocity profile has been defined for subsequent site-response analyses. Profile P1 were judged as preferred S-wave velocity profile by PRP experts. In Figure 4 is shown, that a wave velocity variability ranging between 10-25% has been considered in the study. For design purposes, the ensemble of the three S-wave velocity profiles P1-P3 and the generic P-wave velocity profile including their variability were considered in the process.

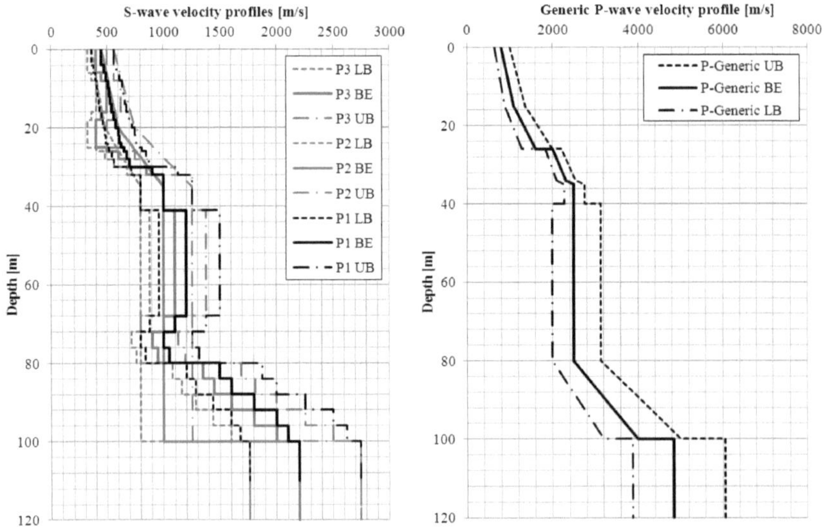

Figure 4: Low strain S-wave (left) and P-wave (right) velocity profiles for considered in the PRP Project at Leibstadt NPP

3 ZENT north building

3.1 Nuclear Building and Earthquake Safety Classification

Structures of Swiss nuclear facilities with safety and/or radiological importance are classified according to the ENSI-Guideline G01 as building class BKI or BKII and earthquake category EK1 or EK2. Classification of structures is governed by either the electrical and/or mechanical systems and components stored in the structure, potential radiological inventory stored or adjacent structures with nuclear classification. Based on the radiological inventory the new ZENT structures fit in the criteria for building class BKII and earthquake class EK 2. This means that the ZENT structures could solely be designed for an Operational-Basis Earthquake (OBE). However, due to the proximity to nuclear structures classified in the highest category BK I/EK 1 requires to design the new structures for a Safe-Shutdown Earthquake (SSE) with a PGA value of 0.28g at the soil surface. In the case of Leibstadt NPP this means a consideration of two-times of the OBE uniform hazard spectra.

3.2 Structural System

The Superstructure consists of a two-storey RC shear wall building with a total height H = 14.80 m above foundation plate at El. +/- 0.00. Planimetric building dimensions of the storage area are L x W = 23.45 m x 21.40 m. An overhanging section with the dimensions L x W = 11.50 m x 5.00 m extends on the east façade and is structurally monolithic connected with the main building (Figure 5).

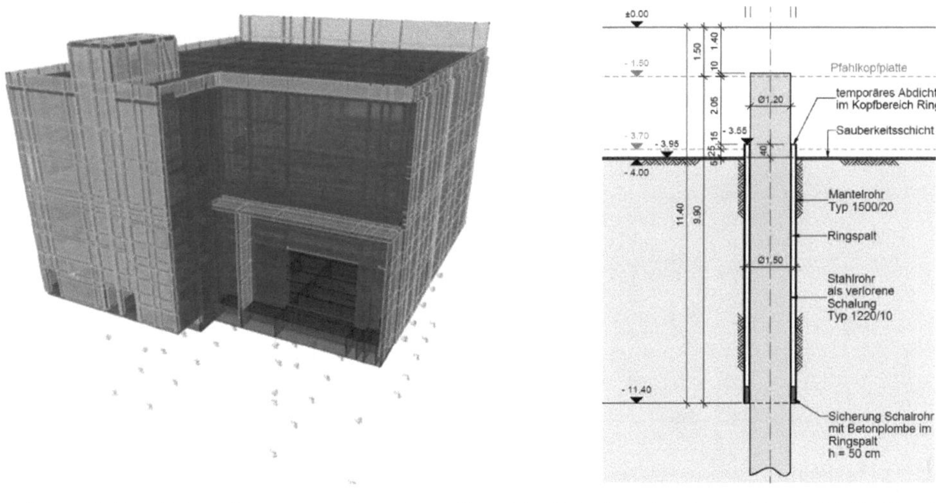

Figure 5: East-Façade perspective view of the 3D FE structural model of ZENT North building (left) and construction detail of a typical steel coated pile from El. -1.50 up to El. -11.40 (right)

The lateral load resisting system mainly consists in four exterior walls that extend along the four façades of the storage area. Primary bearing capacity is also carried by these elements. A massive foundation plate of t = 1.50-2.00 m thickness ties vertical and lateral loads of the superstructure to the pile system consisted of a total of thirty RC piles of D = 1.20 m diameter. In order to protect the already existing main cooling piping system, piles are separated by a 130 mm gap up to an El. -11.40 m from soft soil (Figure 5). RC piles are coated by a steel shaft D/t = 1220/10 mm along this clear height. Considering the flexibility of the foundation piles with respect to the superstructure it can be stated that soil-foundation-structure interaction under seismic actions is governed by the response of the piles. In recognizing the importance of this structural element extensive analyses were carried out with the scope of accurately represent piles' force-deflection behaviour.

4 Single pile analytical studies

4.1 Study on Pile-Soil-Interaction

Extensive analytical and experimental studies on vertical load bearing capacity and flexibility can be found in literature. Nonetheless, extremely reduced information were found on the force-deflection behaviour of the piles of large diameter D > 1.00 m when subjected to lateral loads. As a consequence, extensive analyses using Plaxis software were carried out with an aim to accurately describe soil flexibility under different pile actions. Soil flexibility criteria were modelled starting from results of equivalent-linear soil column response analyses carried out using profiles showed in Figure 4. Soil failure were considered by assuming Mohr-Coulomb-Criteria as well. Results of the analyses shows, that piles behave mostly linear when subjected to lateral loads in the range of V = 1000 kN (point load on top). The reduced horizontal soil deflection due to horizontal pile push-pull confirms the excellent soft soil characteristics at Leibstadt NPP Site.

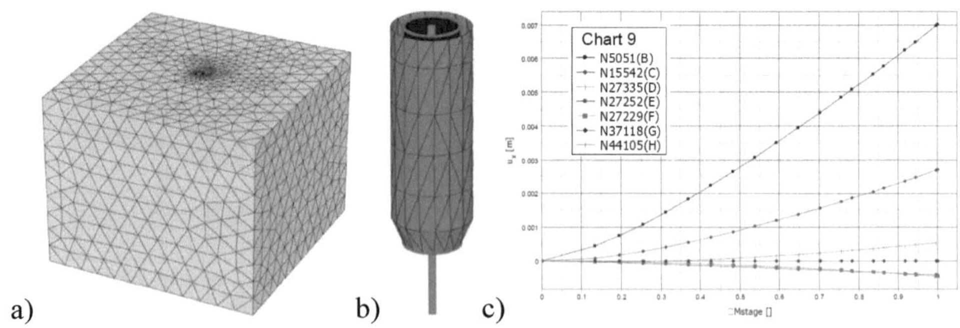

Figure 6: 3D FE-Model of embedded pile in Plaxis: a) Soil Layers and Mesh, b) Embedded RC pile with soil gap of 130 mm and steel tube, c) Incremental horizontal pile deflection at selected nodes (top to bottom, B-F at El. -11.0 up to -15.0 each meter, G-H at El. -18.35 and El. -28.35)

4.2 Study on static nonlinear force-deflection-relation of RC pile

Pushover analyses of a single pile (clamped-clamped restraints) having a clear length L = 10.40 m were performed with the finite element program ATENA. The analyses considered different axial (compression) load conditions for P = 0kN to 3000 kN reflecting the axial load variability acting on the thirty piles. FE-Model took into account concrete cracking, crushing, reinforcement yielding and strain-hardening up to failure. Tensile cracking nonlinear material model was based on fracture mechanics. Concrete compressive crushing and steel yielding were based on plasticity theory. Interface between steel coating and concrete was modeled by Mohr-Coulomb friction criteria. A small friction coefficient equal to $\mu = 0.1$ was used. Results of performed static nonlinear analyses in ATENA can be seen on Figure 7. For the considered axial load range an increasing in compression force leads to an increasing of pre-plateau stiffness and maximal plateau lateral force capacity (10% in maximum). P-Δ-Effects in post-peak load were amplified as well, leading to a sign change in plateau force-deflection-slope. However axial load ratio affects ultimate pile displacement capacity but does not modify yield displacement. Consequently, displacement ductility of RC members is affected by axial load ratio as reported in literature.

Member failure was governed by rupture of longitudinal rebars. An expected positive effect on concrete compressive stress-strain-behavior (concrete confinement) due to steel coating was observed in the analyses. Due to the particular pile construction scheme (Figure 5), pile deflection at El. -3.65 (steel mantels' top elevation) with respect pile-top deflection was estimated as well (~80% of top displacement). Considering a 130 mm gap between RC steel coated pile and steel mantel an approximate maximal displacement of 160 mm (top) is allowed, if dynamic pile-mantel-pounding effects have to be avoided. It means, that potential pile lateral displacement capacity as been reported in Figure 7 can not be achieved in reality due to the presence of a steel mantel around each pile of diameter D = 1500 mm.

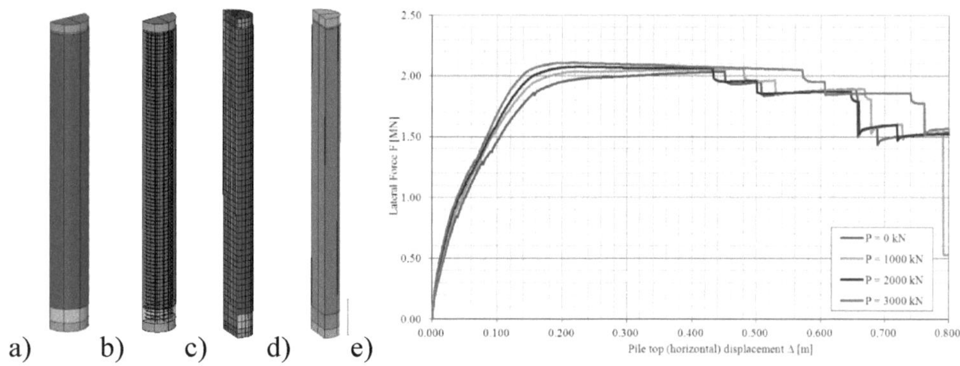

Figure 7: FE-Model in ATENA a) elements, b) concrete, reinforcement, c) FE-Mesh d) steel coat ranging from El. -1.55 to -11.40 in pink and e) F-Δ-curves

4.3 Comparison and discussion of soil pile flexibility

Based on linear theory for a continuous supported beam on elastic foundation, a global horizontal soil stiffness was calculated. For the purpose, best-estimate soil properties and stiffness parameters were used. Similarly, uncracked pile stiffness for a clamped-clamped beam of length L = 10.40 m with circular cross section D = 1.20 m was calculated according to Formula 1.

$$k_{pile} = \frac{12EI}{L^3} = \frac{12E}{L^3} \cdot \frac{\pi D^4}{64} \tag{1}$$

$$\frac{1}{k_{pile-soil}} = \frac{1}{k_{pile}} + \frac{1}{k_{soil}} \tag{2}$$

Even if uncracked pile stiffness were assumed, pile flexibility is considerably higher (~7.6-times) when compared to lateral soil flexibility. Consequently, flexibility of clear pile length governs dynamic behavior of overall ZENT structural system for horizontal seismic actions. Thus, soil stiffness characteristics and their variability barely modified dynamic properties of the structural system. Such effect becomes even greater as seismic demand increases. For vertical flexibility, pile to soil stiffness ratio becomes larger (around ~3.5-times). As a consequence, structural behavior will more be affected by pile-soil-interaction effects in case of vertical seismic actions.

5 Linear and nonlinear Analyses on hole model

5.1 Three Dimensional Finite Elements Model and Dynamic Properties

A realistic 3D FE-Model in SAP2000 was carried out for design verification. Structural elements such as foundation plate, interior and exteriors walls, decks and roofs were modelled by means of shell elements. RC piles were modelled by means of multi-linear plastic link/support elements. Monotonic force-deflection behaviour was obtained from ATENA analyses (Figure 7) while Takeda hysteresis rule was assumed for cyclic behaviour. Soil stiffness was modelled by means of global (translational) linear springs at El. -11.90 (clamped-end of RC pile with soil). Total assembled structural mass for dynamic analyses corresponds to m = 9820 t. Mass portions consists of self-weight, surcharge loads (i.e. façades, roof construction) components' weight and portioned live loads.

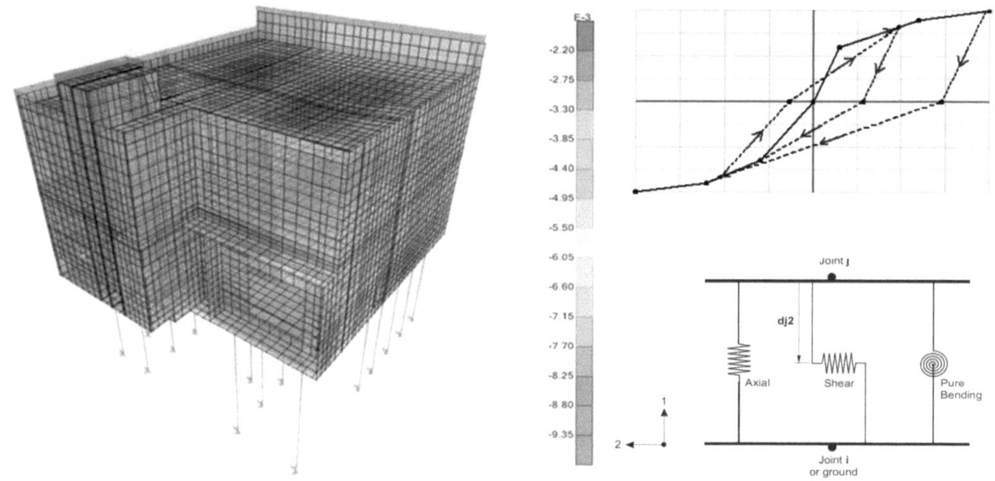

Figure 8: Vertical displacement in [m] of 3D FE model for static load combination (left), link/support element characteristic (right, bottom) and cyclic link behaviour for Takeda-Rule (right, top)

Table 1: Modal results (uncracked pile stiffness assumption). Cartesian X-, Y-dir. in the horizontal plane (NS- resp. EW-dir.), Z-dir. vertical

Mode	Frequency	Damping	$M_{X, mod.}$	$M_{Y, mod.}$	$M_{Z, mod.}$
[-]	[Hz]	[-]	[t]	[t]	[t]
1	1.30	0.07	3347	4916	-
2	1.32	0.07	6376	2770	-
3	1.43	0.07	9	1989	-
4	5.64	0.07	1	143	50
5	6.34	0.07	60	-	3120
6	7.01	0.07	24	1	5563

5.2 Linear Modal Time-History Analysis

Linear modal time-history analysis (LMTHA) was carried out in order to verify results of nonlinear dynamic analyses for SSE earthquake ground motion with a PGA value of 0.21g at El. -11.90. Maximum relative roof displacements in both horizontal directions are smaller than 80 mm and reduces to roughly 90% at foundation plate level (Figure 9). Seismic forces from LMTHA are mostly consistent with design values, obtained response spectral analyses and SRSS modal combination rule assumptions in an equivalent FE-Program.

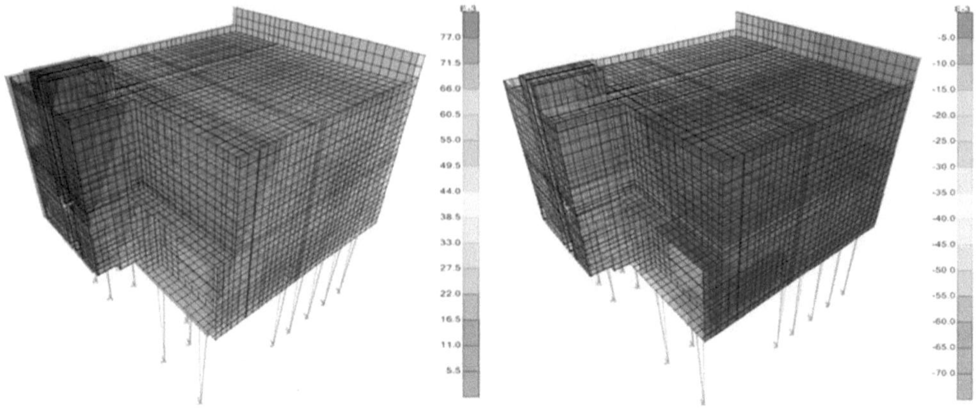

Figure 9: Deformed shape for absolute max. roof displacement in [m] in X-direction (left) and Y-direction (right) from linear modal time history analysis

5.3 Nonlinear Time-History Analysis

Considering the fact that superstructures' stiffness is significantly higher than soil-pile-systems' stiffness, rigid body constraints were applied to all superstructure DOFs. Strongly reduced analysis computation time, small increase in vibration frequencies and therefore reduction of maximal displacements were the results. Nonlinear time-history analysis verification was performed on the rigid-body constrained model. Hilber-Hughes-Taylor time-step integration algorithm ($\alpha = 0.0$) and Rayleigh damping were used. A small elastic damping ratio of $\xi_{el} = 0.04$ was set at frequencies f = 1.00 Hz respectively f = 8.00 Hz in order to avoid overestimation of total modal damping. Results confirm the conservative assumptions made in seismic design and structural robustness of the ZENT north building. Due to particular conservatism, maximum base shear from NLTHA in X- and Y-dir. are significantly lower when compared with design values of response spectra analyses. However, in the vertical Z-dir. almost identical values are obtained. Maximal roof top displacement in the X- and Y-dir. from NLTHA are around 70 mm ($\approx 1.2 \cdot 59$ mm) and therefore at least 30% smaller when compared with design values. As a consequence design values of pile-steel mantel horizontal gap (130 mm) and ZENT building's horizontal gap to adjacent structures (150 mm) are adequate. Especially hysteretic modal damping values of $\xi_{hyst} = 0.065\text{-}0.073$ at peak structural response leaded to total modal damping ratios of $\xi_{tot} = \xi_{el} + \xi_{hyst} = 0.098\text{-}0.108$ for the X- and Y-dir. of loading (see Formula 3, were A_h = loop area, F_m = peak load, Δ_m = peak displacement). Design modal damping equal $\xi_{tot} = 0.07$ for SSE earthquake excitations was conservatively assumed to be for all modes.

$$\xi_{hyst} = \frac{A_h}{2\pi \cdot F_m \cdot \Delta_m} \tag{3}$$

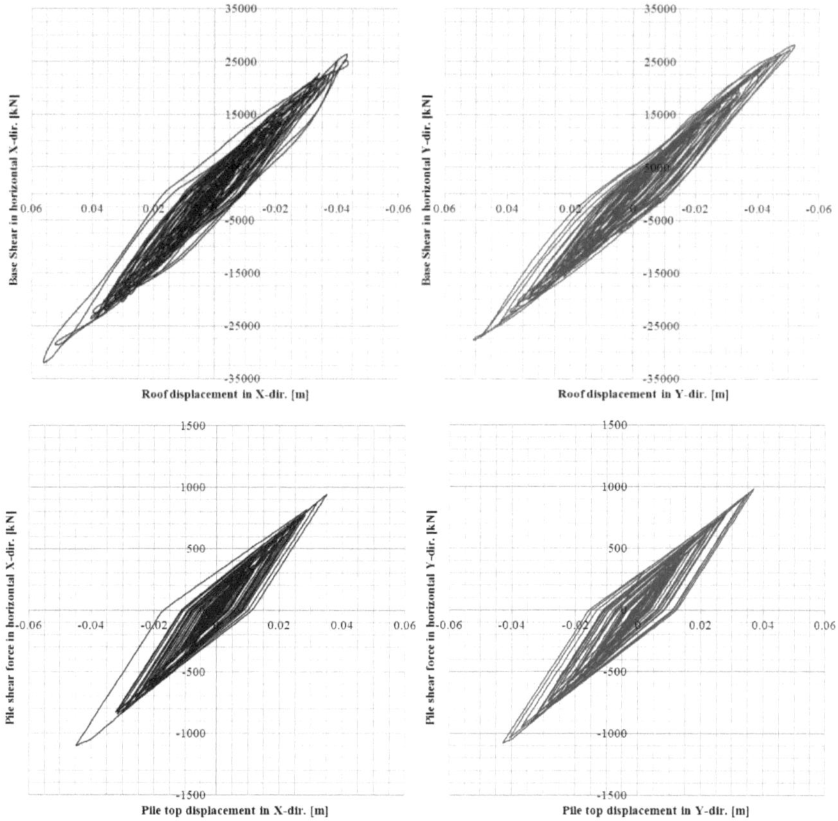

Figure 10: base shear hysteresis loop vs. roof displacement for the X- and Y-dir. from NLTHA (top) and for the controlling corner pile (bottom)

6 Conclusions

Structural design for a new structure with radiological importance on a existing Swiss NPP Site has been presented. Results of soil investigations and experts judgment of SSHAC Level 4 PSHA Studies PEGASOS and PRP has fully been considered in the design process. Due to the particular foundation scheme, extensive analytical investigations on pile-soil-interaction and nonlinear pile force-deflection behaviour were carried out at the design stage. Furthermore, results of single pile analytical studies were implemented into a 3D structural FE-Model and dynamic nonlinear time-history analyses were performed for design verification purposes. Findings of this investigation were compared with design assumptions. Pile reinforcement detailing was finally found to be compliant to the capacity design criteria (ensuring a ductile member behaviour). It could be shown that seismic structural design was well performed and conservative with respect to verification. Consequently, a large seismic safety margin for earthquake demands exceeding safe-shutdown-level can be ensured.

References

[1] Cervenka, J. and Pappanikolau V. (2008). Three dimensional combined fracture-plastic material model for concrete. *Int. Journal of Plasticity*. **Vol. 24:12**, ISSN 0749-6419, pp. 2192-2220

[2] Cervenka, J., Jendele J. and Cervenka V. (2010). ATENA Manual

[3] Chopra, A.K. (2000). Dynamics of Structures – Theory and Applications to Earthquake Engineering. 2nd Ed., Prentice Hall, Upple Saddle River, NJ, USA

[4] Computers and Structuers Inc. (2009). CSI Analysis Reference Manual for SAP2000, ETABS, SAFE and CSiBridge. ISO No. GEN062708M1 Rev.4, Computers and Structures Inc., Berkeley, CA, USA

[5] GLM AG (2012). KKL ZENT, ZL6, Einzelpfahlanalyse. **Rep. 11202A-R-001**

[6] Kramer, S.L. (1996). Geotechnical Earthquake Eng., 1st Ed., Prentice Hall, Upple Saddle River, NJ, USA

[7] Hauptabteilung für die Sicherheit der Kernanlagen HSK (2004). HSK-RT Final Report: Review Approach and Comment and Comment on Probabilistic Seismic Hazard Analysis for Swiss Nuclear Power Plant Sites (PEGASOS Project), Final Report

[8] Idriss, I.M. and Sun J.I. (1992). User's Manual for SHAKE91. Center for Geotechnical Modeling, Dept. of Civil Engineering, University of South California, Davis, California, USA

[9] Interoil (2008). OPAL – Phase 1. Bewertung der bisher in Leibstadt erfassten geologischen und geotechnischen Daten und Konzept zu deren Erweiterung. Report IO08-TA1632, Zürich, CH

[10] Interoil (2009). OPAL–Phase 2. Report IO09-TA1123, Zürich, CH

[11] National Cooperative for Disposal of Radioactive Waste NAGRA (2004). Probabilistic Seismic Hazard Analysis for Swiss Nuclear Power Plant Sites (PEGASOS Project). Wettingen, Aargau, CH

[12] Priestley, M.J.N., Calvi G.M. and Kowalski M.J. (2007). Displacement-Based Seismic Design of Structures. First Edition, ROSE School, IUSS Pavia, I

[13] Resonance SA (2011). Revised Version of Genereation of Synthetic Eartrhquakes for KKL. *Resonsance SA* **Report TB-405-02/ST/MK**

[14] Senior Seismic Hazard Analysis Committee SSHAC (1997). Recommendations for Probabilistic Seismic Hazard Analysis: Guidance on Uncertainty and Use of Experts. **NUREG/CR-6372**

[15] Swissnuclear (2010). PEGASOS Refinement Project: Soil models for the Leibstadt Site (KKL). *Swissnuclear* **Technical Note TP3-TN-1067, Ver. 4.3**

Seismic Analysis of Onshore Wind Turbine Including Soil-Structure Interaction Effects

Francesca Taddei[1], Konstantin Meskouris[1]

[1] Lehrstuhl für Baustatik und Baudynamik, RWTH Aachen University
Mies-van-der-Rohe-Str. 1, D-52074, Aachen, Germany
Francesca.Taddei@lbb.rwth-aachen.de

ABSTRACT:

The structural behavior of modern wind turbines has reached a very high complexity and many factors are involved: slenderness of the structure, excitation environment and operational controls. Moreover, if a project is located at sites with relevant seismic hazard, the wind unit must be designed considering a reasonable likelihood of earthquake occurrence during the operational state or an emergency shutdown. The influence of the subsoil on the seismic response of a wind turbine can be crucial during the seismic design phase and need to be properly included into the computational model. Norms and guidelines need to keep up with technological developments and structural peculiarities. However the dynamic soil-structure interaction is often neglected or roughly mentioned. It is usually suggested to represent the soil through springs. The proposed investigation estimates the seismic response of a soil-turbine system and involves a 1.5-MW, 3-blade wind turbine, grounded on a layered half space. The wind turbine system is modeled by means of Finite Element Method (FEM). The effects of the layering are investigated. The soil is simply idealized as a generalized spring, according to the majority of standard codes. In parallel, the same investigation is performed with a more accurate method, a coupling between finite element and Boundary Element Method (BEM). This allows assessing the applicability and accuracy of the simplified soil representation.

Keywords: Soil Structure Interaction, Wind Turbine, Seismic response, Layered half space, Soil Stiffness

1 Introduction

Annual installations of wind power have increased constantly across Europe over the last 2 decades, expanding the market towards seismically active areas. In Figure 1 a map of the seismic hazard in Europe is represented in combination with statistics of the wind power installed by end of 2011 and the average wind speed at 50 m from the ground. The latter is a key issue for the site suitability assessment for wind power and must be greater than 5-6 m/s.

S. Klinkel et al. (Eds.), *Seismic Design of Industrial Facilities*,
DOI 10.1007/978-3-658-02810-7_43, © Springer Fachmedien Wiesbaden 2014

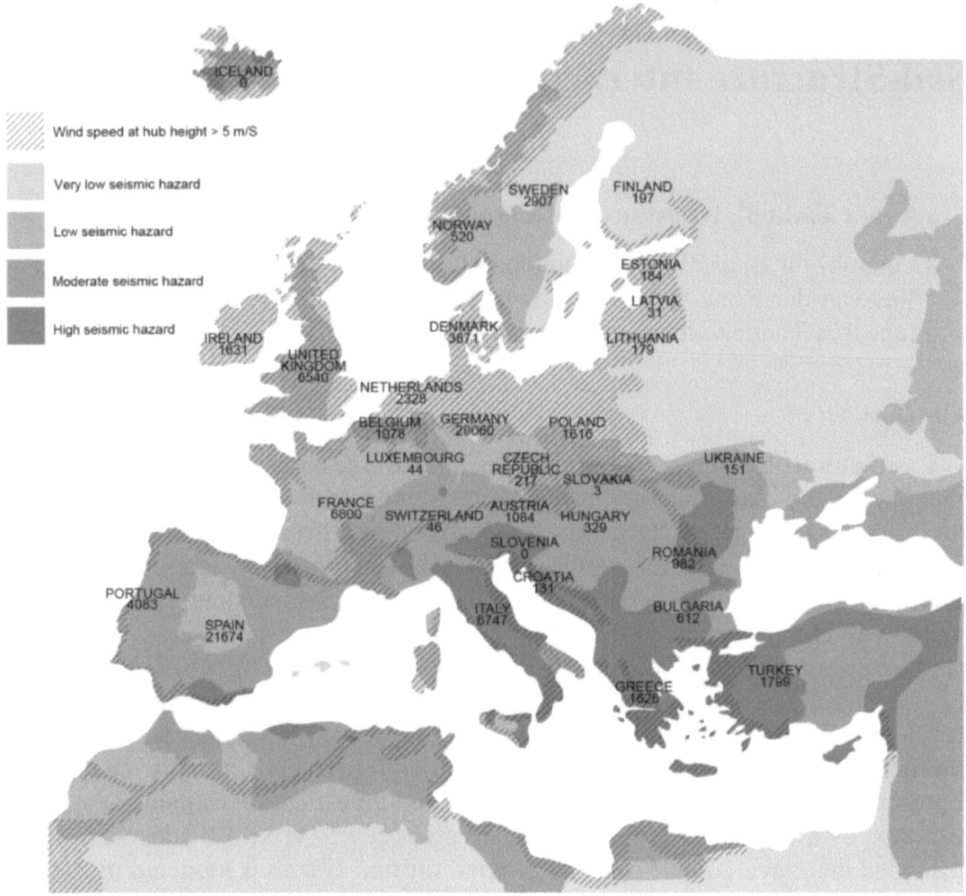

Figure 1: Seismic hazard map of Europe with annotation of suitable site for wind power installation (with average wind speed at 50 m from the ground > 5 m/s) and of wind power installed in each country by end of 2011

As we can see, a large part of the south European coastal areas present high seismic hazard and such wind conditions, which are sufficiently suitable for financial returns possible from modern wind turbines. If a project is located at sites with relevant seismic hazard, the wind unit must be designed considering a reasonable likelihood of earthquake occurrence during the operational state or an emergency shutdown. It is crucial to recognize that, in some cases, seismic plus operational loads may govern tower and foundation design. A look into the current practice for seismic loading determination for wind turbine foundations is discussed by Prowell, I. et al. [1]. They confirmed that, the loads combinations prescribed by the IEC [2] provide a seismic safe design.

Standards codes state that, the supporting soil has a finite stiffness and the structure cannot be assumed to have a fixed support. In fact, the entity of the soil compliance, and consequently of the interaction between soil and structure, plays an important role in the dynamic response of the whole structure. For practical applications, the soil can be represented as a lumped parameter model, which is a package of springs, dashpots and masses. The model coefficients are usually frequency-dependent and this is particularly important for the seismic design of structures. These coefficients can be determined approximating the dynamic stiffness of the soil in the frequency domain through the curve-fitting technique. Alternatively, static values for the coefficients may be assumed. They are the soil stiffnesses for frequencies approaching zero. Assuming static values, the computed response of the foundation-soil system may deviate from the actual one, especially in case of high-frequent seismic excitation. However, depending on the excitation frequency content, the static stiffnesses may be still representative for the foundation-soil subsystem. In order to assess the accuracy of the Spring Model (SM), the same investigation was performed also with a more accurate method, based on a coupling between finite and boundary element methods (FEM/BEM).

2 Soil-structure interaction for wind turbine

First of all, the dynamic natural properties of the structure are modified by the presence of a compliant soil, with particular reference to the frequencies of the tower bending modes. Consequently the minimum frequency separation between the natural frequency of the structure and the operational frequency may be violated and resonance effects may raise. Moreover, the frequency content of the seismic signal may lead to vibration amplification (or attenuation) phenomena, with possible high shear force and overturning moment at the tower base. Last but not least, large motions of the tower may disturb the control processes of the machine leading to an inefficient production or even an emergency shutdown.

A first investigation of wind turbine behavior under the influence of SSI is reported in a work published by Bazeos et al.[3] , where a lumped-parameter model was employed. They confirmed that if the soil compliance is included the natural frequencies of the system decrease with respect to the fixed base system and the most affected frequencies are those related to the second and third bending modes. Ritschel [4] found that, the IEC design loads (see [2]) cover the seismic load combination loads, obtained with both modal and time domain analyses. As far as the tower is concerned, a peak acceleration of 0.3 g may be considered as the limit for this 60m-hub-height wind turbine. Cao et al. [5]considered different aspects of the problem such as P-Δ effect, soil-structure interaction, the vertical seismic action and the rotating rotor. The decrease of the natural frequencies and influence on the time domain seismic response were the main findings of the investigations. Zhao et al. [6] presented a multi-body system model for wind turbine towers to investigate the seismic response properties in time domain. Soil-foundation interaction is represented by a frequency-independent discrete parameter model. They found, the tower bending modes of higher orders than the first one are considerably affected by the SSI effects with relative error 16.5 per cent. A lumped-parameter model for wind turbine footings is proposed by Andersen et al. [7]. A weighted-least-squares fitting process is employed for approximating the dynamic stiffness curves of the foundation-soil interaction. Guidelines for the formulation of such a model are given, with focus on two soil configurations: a soft top layer on stiff clay and a consolidated clay deposit below a top layer of medium sand. The main conclusion is that the second mode (and higher modes) of the tower is damped by geometrical dissipation in the ground. However no such a strong effect is observed for the first frequency which is in general lower than the cutoff frequency. In other words for soil consisting of a stratum on stiff deposit, it if the frequencies of the structure lie below the fundamental frequency of the layer no waves propagation can actually take place.

3 Investigation Model

The case study focuses on a 1.5 MW three-bladed wind turbine, which refers to a design study of the National Renewable Energy Laboratory [8]. The construction site is placed on the westerner cost of Turkey, where suitable wind conditions are observed as well as relevant seismic hazard. According to the Turkish national annex of the EC8 [9] , the chosen area is classified as zone 1. The geometry of the structure is represented in Figure 2 and reported in Table 1.The material properties are collected in Table 2. The rotor blades, the nacelle and the gear box are idealizes as a concentrated mass point at the top of the tower (Table 3). Both the 1st and 2nd bending mode structural damping ratios are set to 3.435 %. As the structural model is symmetrical, the same value is assumed for the side-to-side and fore-aft modes.

Table 1: Geometry of tubular steel tower

Section	d [m]	thick [m]
base	5.6978	0.0174
top	2.8404	0.0087

Table 2: Material properties

Structural part	E $\left[\frac{N}{m^2}\right]$	ρ $\left[\frac{kg}{m^3}\right]$	υ [--]
tower	2.00e11	8240	0.3
foundation	3.00e15	2400	0.2

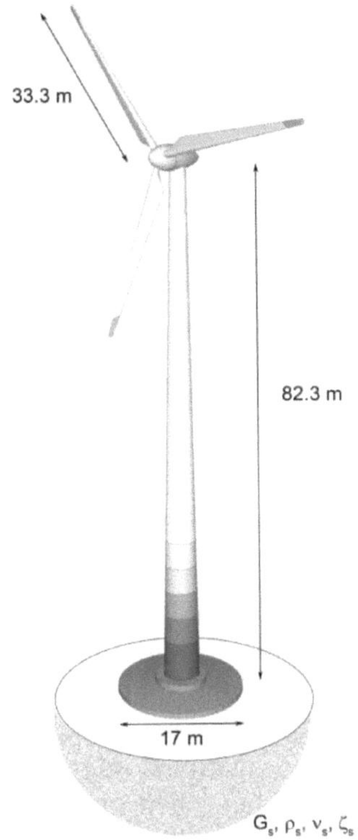

Figure 2: Geometry of the system

Table 3: Rotor mass properties

Overall mass [kg]	$7.806 \cdot 10^4$
Horizontal moment of inertia [m^4]	$9.384 \cdot 10^6$
Torsional moment of inertia [m^4]	$1.857 \cdot 10^7$

The soil is represented by a soft clay layer over a harder clay half space, as shown in Figure 3. The layer thickness D varies from 3R to ¼ R. The Poisson's ratio, υ_S, is assumed equal to 0.3 and the density, ρ_S, equal to 2200 kg/m^3 for both layer and half space. The ratio between the shear wave velocity of layer c_S^L and half space c_S^{HS} is held constant to 0.5. The layer shear wave velocity c_S^{HS} is assumed equal to 200 m/s. These values are chosen so that all four soil configurations fall into the category of the standard ground type D, according to the EC8 site classification. The embedment of the foundation is not taken into consideration.

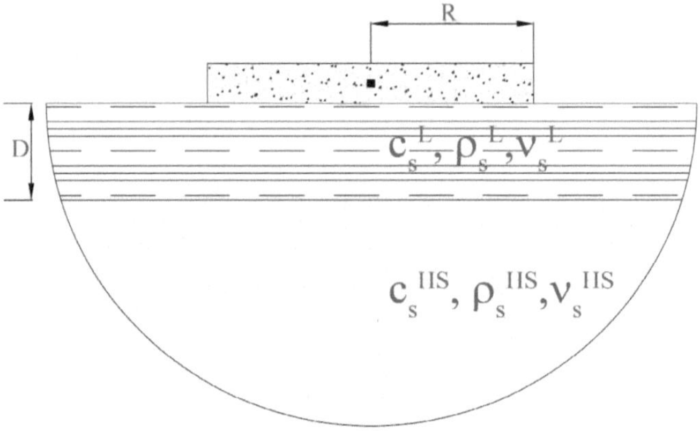

Figure 3: Investigated soil configuration

In the present investigation, the employed lumped parameter model contains only springs, as prescribed by the DNV/Riso Guidelines [10]. The SM is made up of six uncoupled springs, one along each of the six degrees of freedom. This model, also referred to as generalized spring, turns out to be frequency independent and no coupling between translational and rotational degrees of freedom is considered. No dashpots or fictitious masses are added. Therefore, the seismic input can be directly applied to the far end of the springs. The formulas for the soil springs coefficients are given in Table 4 and address the case of stratum over half space. From the formulas, it emerges that, the thinner the layer becomes, the stiffer the generalized spring is.

Table 4: Formulas for the soil springs coefficients [10]

Vertical	Horizontal	Rocking
$K_V = \frac{4G_1R}{1-\upsilon_1}\frac{1+1.28\frac{R}{H}}{1+1.28\frac{R}{H}\frac{G_1}{G_2}}$	$K_H = \frac{8G_1R}{2-\upsilon_1}\frac{1+\frac{R}{2H}}{1+\frac{R}{2H}\frac{G_1}{G_2}}$	$K_R = \frac{8G_1R^3}{3(1-\upsilon_1)}\frac{1+\frac{R}{6H}}{1+\frac{R}{6H}\frac{G_1}{G_2}}$

In parallel, a FEM/BEM model is also used for comparison [11]. The BEM is a very accurate method for wave propagation problems, because it satisfies exactly the energy radiation conditions at infinity. At the heart of the BEM lies the fundamental solution of the wave propagation problem. These solutions, also called Green´s functions, were obtained with the aid of the Thin Layered Method (TLM). The latter is an efficient semi-analytical method, which can also take into account stratification conditions effects like reflection, refraction at the layers boundary, dispersion and geometrical damping [12].

3.1 Modal analysis

As explained before the first step of a dynamic analysis is the determination of the natural period, or natural frequency, of the whole system (Figure 4). Table 5 reports the results of the modal analysis.

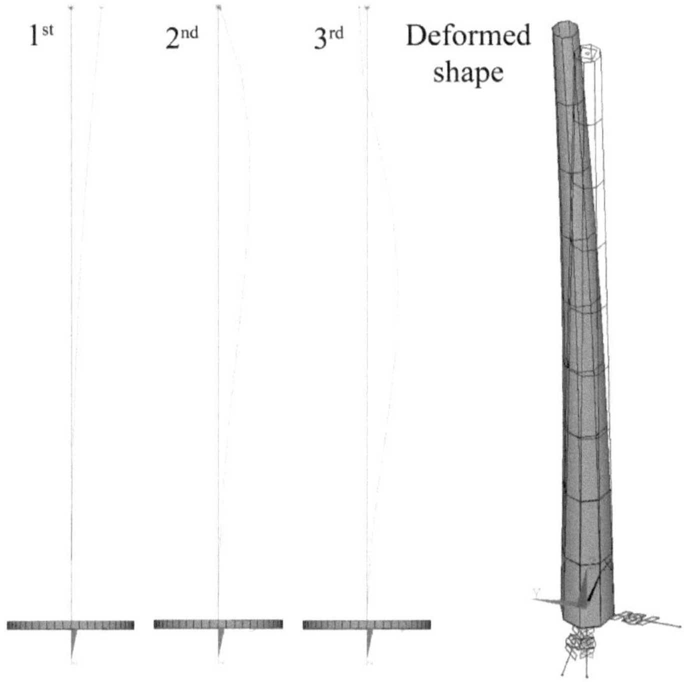

Figure 4: Bending modes and deformed shape of the tower

Table 5: Natural bending frequencies of the system

[Hz]	Deep soft layer 3R	Soft layer R	Thin soft layer ½ R	Very thin soft layer ½ R	Fixed base
1st freq.	0.3959	0.3971	0.3986	0.4008	0.4146
2nd freq.	1.7523	1.7565	1.7615	1.7685	1.8147
3rd freq.	4.0576	4.0791	4.1037	4.1376	4.3809

Table 6 compares the 1st, 2nd and 3rd natural bending frequencies of the structure-soil systems with the fixed base system frequencies. In general, the deviation of the natural frequencies ranges between 2.5% and 7.5%. Moreover, the third natural frequency is more affected than the first two. Increasing the thickness of the soft layer, the natural frequencies decrease.

Table 6: Deviation of the natural frequencies of the whole system with respect to the fixed base system

[%]	Deep soft layer 3R	Soft layer R	Thin soft layer ½ R	Very thin soft layer ¼ R
1st freq.	4.5	4.2	3.9	3.3
2nd freq.	3.4	3.2	2.9	2.5
3nd freq.	7.4	6.9	6.3	5.6

3.2 Seismic response

For the purpose of this study the author used spectrum-compatible synthetic accelererograms. First of all, the design spectra are created according to the Turkish national annex of the EC8 [9]. For the seismic zone 1 the design ground acceleration is $a_g = 0.40$. The local soil class results to be the Z4. Figure 5 shows the resulting spectrum for a local site class Z4, in comparison with other classes.

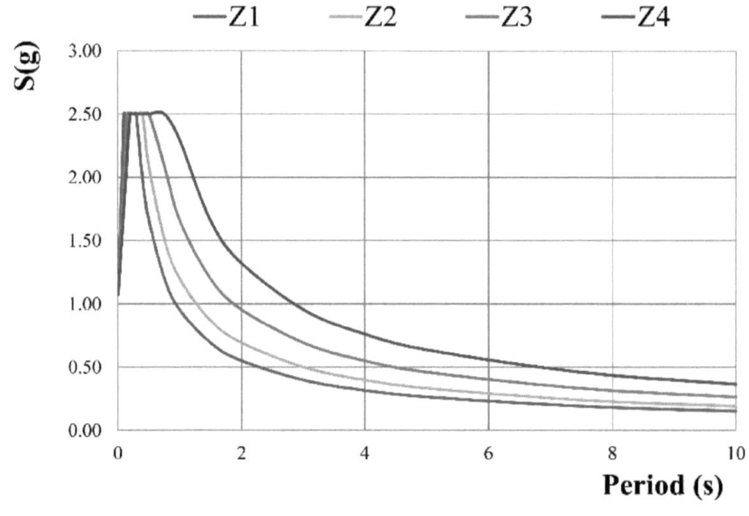

Figure 5: Elastic design acceleration spectra for different type of soil

Before dealing with the time domain analysis, let us firstly evaluate the soil-structure interaction effects on the seismic loads, computed as static equivalent forces. We consider a SDOF system with a lumped mass at the top. If the mass of the oscillator $\mathcal{M}$ is set equal to the head mass plus half of the tower mass and it is positioned at $\mathcal{H} = 82.39$ m above the ground, the following expression gives an approximation of the overturning moment at the base of the tower:

$$M_x = S_{ae}(T_1) \cdot \mathcal{M} \cdot \mathcal{H} \qquad (1)$$

where T_1 is the first period of the structure, as determined previously with the aid of the FEM-SM system. For each of the four soil configurations, a rough estimation of the base bending moment can be read in Table 7.

Table 7: Estimation of the base overturning moment according to IEC [2]

	D=3R	D=R	D=½ R	D=¼ R
T_1 [s]	2.526	2.519	2.509	2.495
M_x [Nm]	49358345	49483989	49634457	49846513

Entering now into the time domain procedure, at least three set of earthquake ground motions shall be selected or generated, satisfying all of the conditions given in 2.9.1 in [9]. The analysis is performed applying the seismic accelerograms to the foundation node. Figure 6 shows the oscillation trend of the seismic accelerations for soil type D. Table 8 compares the maxima of the overturning bending moment and the horizontal top displacements, for the different layer thicknesses and for the two models. Figure 7 and Figure 8 show the SRSS of the most important results for different layer thicknesses. Finally, Figure 9 compares the results for the simplified soil representation and the more accurate FEM/BEM model. The phase shift and the different amplitudes between the two curves are due to the coupling between the translational and rotational foundation motions, which are neglected in the SM model but considered by the FEM/BEM model.

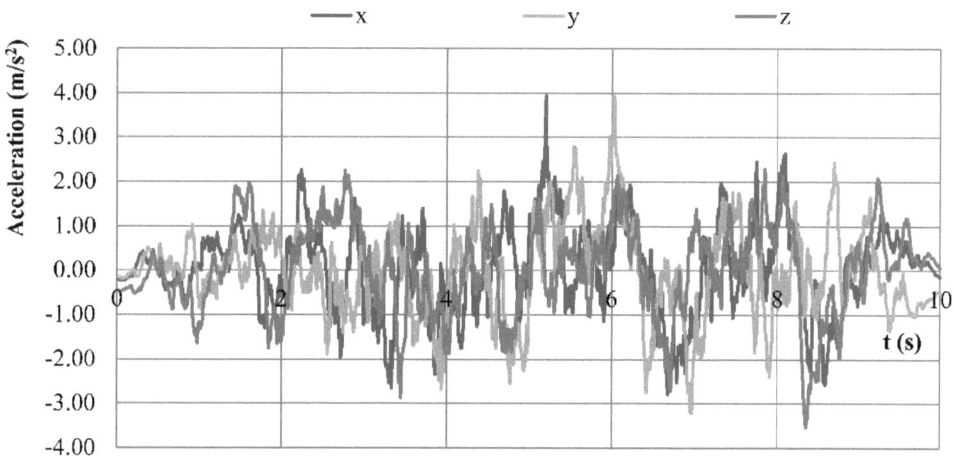

Figure 6: Example of synthetic seismic accelerograms

Table 8: Comparison of the SM and FEM/BEM results

Model	SM		FEM/BEM			
	M_{max}[Nm]	U_{max}[m]	M_{max}[Nm]	U_{max}[m]	M_{max}	U_{max}
D=3R	41714408	0.7674	40222634	0.764602	↓	↑
D=R	43859891	0.7634	42361324	0.760778		
D=½ R	44516856	0.7580	42983858	0.753964		
D=¼ R	45357199	0.7506	42890242	0.745723		

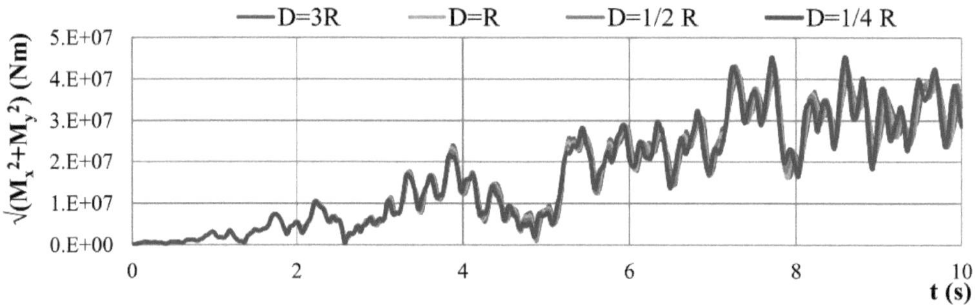

Figure 7: SRSS of the bending moment at the foot of the tower

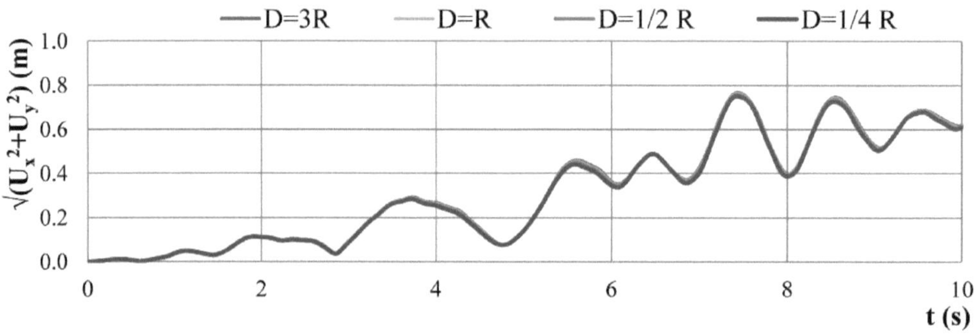

Figure 8: SRSS of the bending moment at the top of the tower

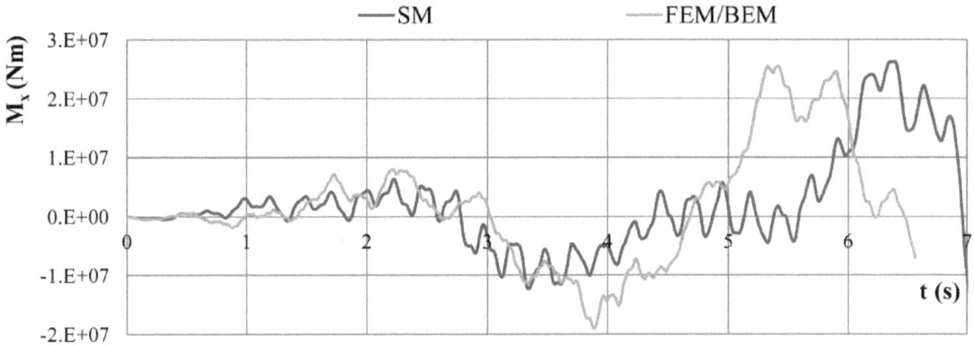

Figure 9: Bending moment at the foot of the tower for D=R. Comparison of SM and FEM/BEM results

4 Conclusion

A simplified model for SSI was compared to an accurate FEM/BEM approach, in order to evaluate its applicability for seismic design of wind turbine. This model is based on a generalized soil spring, as prescribed in several current standard codes. The main assumptions are: 1) frequency independent coefficients and 2) no coupling between translational and rotational degrees of freedom. The investigated soil was a soft stratum over a harder half space.

The main results are:

- Increasing the thickness of the soft layer, the natural frequencies decrease. This suggests that, if the layer is very thick, it becomes a predominant influence on the dynamic response of the soil and the whole system becomes more flexible. Conversely, if the layer is very thin the soil behaves just like the underlying harder half space, which lends more rigidity to the system. This is also confirmed by the transient analysis, where, increasing D, the horizontal displacements increases as well, revealing a more flexible structure.
- However, no such a strong effect is observed for the first frequency. In general, varying the thickness of the layer, a very small variation of the results is noticed. Possible variation due to the layering depend on the ratio between the shear wave velocity of layer c_S^L and half space c_S^{HS}.
- For this specific case, the variation can be considered unimportant. For this specific case the simplified model and the more accurate FEM/BEM model were in perfect agreement. The spring model for SSI was able to account for all the most important dynamic properties of the whole turbine-soil system. This conclusion applies only for the case of softer layer on harder half space. For other configurations additional investigations are necessary.

5 Acknowledgements

This research was supported by the DFG within the framework of a Sino-German joint project about "Complex soil-structure interaction issues" [Web-1].

REFERENCES

[1] Prowell, I. et al.: Assessment of wind turbine seismic risk: Existing literature and simple study of tower moment demand, Sandia National Laboratories, California, 2009

[2] IEC: EN 61400-1:2005+A1:2010 - Design Requirements; 2010

[3] Bazeos, N. et al.: Static, seismic and stability analyses of a prototype wind turbine steel tower, Eng. Struct.,2002, 24(8), 1015–1025

[4] Ritschel, U. et al.: Wind turbines and earthquakes; in: World Wind Energy Conference, Cape Town, 2003

[5] Cao, Q. et al.: The research of the affecting factors on the seismic response of wind turbine tower; in: International Conference on Mechanic Automation and Control Engineering, Nanjing, China, 2010, pp. 5449-5452.

[6] Zhao, X. et al.: Seismic response analysis of wind turbine towers including soil-structure interaction; in: Proceedings of The Institution of Mechanical Engineers Part K-journal of Multi-body Dynamics, 2006, 220(1), 53-61

[7] Anderson, L. et al.: Lumped-parameter models for wind-turbine footings on layered ground, Aalborg University, Department of Civil Engineering DCE Technical Memorandum, 2007, ISSN 1901-7278

[8] Malcolm, D. J. and Hansen, A.C.: WindPACT Turbine Rotor Design Study, National Renewable Energy Laboratory, Subcontract Report, 2002

[9] Teknik Kurul: Turkish Standard TS ENV 1998-1-1; 2005

[10] DNV/Riso: Guidelines for Design of Wind Turbines

[11] Mykoniou, K. et al.: Dynamic Foundation-Soil Interaction: a Comperative Study, Bauingenieur, D-A-CH, 2012

[12] Kausel, E.: Thin-layered-method: formulation in the time domain; International Journal for Numerical Methods in Engineering, 37-6, 927-941, 1994

[Web-1] http://gepris.dfg.de/gepris/OCTOPUS/;jsessionid=C384C24C8E7EDDEAD45DC89EDDF70688?module=gepris&task=showDetail&context=projekt&id=164875965http://www.rwth-aachen.de, publisher: RWTH Aachen University, data gathered on 12. Jan. 2013

Part X

Soil-Structure Interaction: Scientific Approaches

Dynamic Impedance of Foundation on Multi-Layered Half-Space

Gao Lin[1]

[1] Faculty of Infrastructure Engineering, Dalian University of Technology No.2 Linggong Road, Ganjingzi District, Dalian City, China
gaolin@dlut.edu.cn

ABSTRACT:

Dynamic soil-structure interaction (SSI) effects have always been important in the context of assessing the seismic safety and vulnerability of large and complex infrastructures such as bridges, dams, power plants, industrial units etc. Although SSI problem has been under intensive investigations in the past several decades, relatively little is known about the SSI in the case of multi-layered half-space. Majority of previous researches dealing with SSI problems were conducted for rather simple soil profiles: elastic half-space, a soil layer bonded to rigid base, or a soil layer overlying elastic half-space. Very few papers appeared in the literature tackle the SSI problem having two or more soil layers overlying elastic half-space. This is probably due to the substantial computational effort required. Advantages and limitations of widely used current approaches such as finite element method (FEM), boundary element method (BEM) and thin-layer method (TLM) are discussed. Sponsored by the Science Foundation of Sino-German Center researches into SSI on layered soil carried out at Dalian University of Technology are briefly introduced. An advanced approach for dynamic SSI analysis of structures on multi-layered half-space is proposed, which circumvent difficulties encountered by FEM, BEM and TLM with relative ease. The governing wave motion equations are solved in the frequency-wave-number domain analytically. The precise integration method (PIM) is employed to perform integration to obtain numerical results. Very high accuracy can be achieved. Analytical solution of wave motion equation is written in dual vector form, which enables efficient and convenient assembling of two adjacent layers into a new one without losing effective digits. Formulations dealing with dynamic impedance of arbitrary-shaped foundation on isotropic as well as arbitrary anisotropic multi-layered soil are presented. The solution is not difficult to extend to problems dealing with foundation embedment and through-the-soil coupling of two or more foundations. Numerical results validate efficiency and accuracy of the proposed approach.

S. Klinkel et al. (Eds.), *Seismic Design of Industrial Facilities*,
DOI 10.1007/978-3-658-02810-7_44,

Keywords: dynamic soil-structure interaction, multi-layered soil, anisotropic material, precise integration method, dual vector form of the displacement and stress field.

1 Introduction

Dynamic soil-structure interaction (SSI) including structure-soil-structure interaction (SSSI) is a subject of considerable scientific and practical importance. Typical applications concern the seismic safety and vulnerability of infrastructures and the behaviour of industrial facilities subject to vibrations of machine foundations. In the past decades, there has been significant progress in the development of sophisticated numerical methods, but a homogeneous half-space has been tacitly assumed in most of the actually available programs, which may lead to unreliable results for markedly heterogeneous soil. The dynamic impedance constitutes one of the key elements in the formulation of the linear soil-structure interaction problem. Formulation of dynamic impedance for multi-layered soil is still a major challenge. No attempt is made to give a general review of the literature concerning dynamic foundation impedance for multi-layered soil due to the limited space of the paper. Only some important points are addressed.

Closed-form solutions for layered media are difficult to obtain. Green's functions derived by Luco and Apsel [1][2] for layered soil are of considerable complexity and computationally expensive. Currently the finite element method (FEM), boundary element method (BEM) and the thin-layer method (TLM) are in the favour of researchers.

The FEM is simple and suitable for problems of arbitrary material properties and geometric shapes. The major difficulty of FEM arises from proper modelling of the unbounded medium. Wave absorbing boundaries must be imposed on the truncated surface to account for the radiation of energy into the region not included in the model. Truly 3D FEM solutions are expansive and very rarely used in practice.

The BEM is well suited to model infinite medium as the radiation condition is satisfied automatically. It is computationally efficient because only the boundary needs to be discretized. However, fundamental solution is required for BE analysis, and for BEM formulation based on full space fundamental solution, a discretization of the soil-foundation interface and the surrounding free surface as well as the soil layer interface is necessary.

The TLM has become an efficient and versatile technique for the problem of wave propagation in layered soils since its inception in 1970 by Lysmer and Waas [3][4] and the later development by Kausel [5][6] and many other researchers. In the meantime, Tajimi et al. independently published their papers on TLM in Japanese journals [7][8]. TLM is semi-analytical in the sense that it combines a finite element discretization in the direction of layering with an analytical solution in the direction of wave propagation in the frequency-wave-number domain. For layered media, continuity of the displacements and of the tractions at the interfaces between adjacent layers is usually formulated by applying the propagator matrix or transfer matrix proposed by Thomson [9] and Haskell [10]. However, transfer matrices are not symmetric and contain terms of exponential growth that require special treatment. With increasing number of layers, this situation becomes more pronounced. There may be exponential overflow which could make the method unstable [11]. Kausel presented stiffness matrix method (SMM) [12], which has many advantages over the transfer matrix method. Stiffness matrices are symmetric and involve only half as many degrees of freedom as transfer matrices. Stiffness matrices lead naturally to the solution of normal modes (eigenvalue problems) without the need for special manipulations and treatment. TLM is regarded as the discrete versions of stiffness matrices. However, the TLM still has some deficiencies. The solution of TLM is approximate in nature, because the displacements within the layer are assumed having prescribed variations. The assembly of layer stiffness in TLM yields rather large size of global stiffness equation of the layered system, i.e. rather large size of eigenvalue problem has to be solved. There may be some difficulties to deal with large number of layers to be considered. In addition, the presence of an underlying half-space cannot be taken into consideration in a consistent manner. The soil model has to be extended to sufficient depth and a certain radiation condition has to be introduced at the truncated boundary.

Sponsored by the Science Foundation of Sino-German Center under grant No. GZ566 researches carried out at the Dalian University of Technology, China are briefly introduced. An advanced approach for dynamic SSI analysis of structures on multi-layered half-space is proposed, which circumvent difficulties encountered by FEM, BEM and TLM with relative ease. The governing equations of wave motion on multi-layered soil are solved in the frequency-wave-number domain analytically. The integration is performed by precise integration method (PIM). The solution is exact that any desired accuracy can be reached, and the precision of the results is limited only by the computer used. The dual vector form of the resultant wave motion equation enables efficient and convenient assembling of the layers. Formulation for 3D and 2D dynamic impedance of arbitrary-shaped foundation on isotropic as well as arbitrary anisotropic multi-layered soil are presented. Numerical examples validate the efficiency and accuracy of the proposed approach.

2 Governing equation of wave motion in wave-number domain

Only 3D cases with isotropic and anisotropic media are addressed, for 2D isotropic and anisotropic cases readers may refer to [13][14] presented by the author and the co-workers of the author.

2.1 Cases of isotropic media

A multi-layered soil including l layers is considered. The problem is solved in the cylindrical coordinates (Fig. 1). The following stress and displacement vectors are specified as

$$\mathbf{p}=\{\tau_{rz} \quad \tau_{\theta z} \quad \sigma_z\}^T,\ \mathbf{q}=\{u_r \quad u_\theta \quad u_z\}^T \tag{1}$$

with τ , σ and u being the tangential, normal stress, and displacement components in the directions identified by the subscripts in cylindrical coordinates.

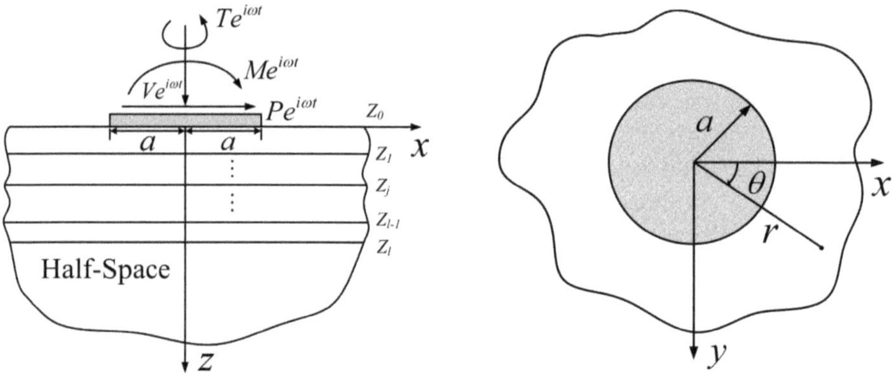

Figure 1: Description of the model

The wave motion equation

$$(\lambda+\mu)\nabla\nabla\mathbf{q}+\mu\nabla\times\nabla\times\mathbf{q}=\rho\ddot{\mathbf{q}} \tag{2}$$

is solved in the frequency-wave-number domain by employing a Fourier transform in time, which changes $\ddot{\mathbf{q}}$ into $-\omega^2\mathbf{q}$, and then applying a Fourier Bessel transform in r and in θ . The formula for the corresponding operations are expressed as follows

$$\tilde{\mathbf{q}}(\kappa,z,\omega)=a_n\int_{r=0}^{\infty}r\mathbf{C}_n(\kappa r)\int_{\theta=0}^{2\pi}\mathbf{D}_n\int_{-\infty}^{+\infty}\mathbf{q}(r,\theta,z,t)e^{-i\omega t}dtd\theta dr \tag{3}$$

which admits the formal inversion

$$\mathbf{q}(r,\theta,z,t)=\frac{1}{2\pi}\int_{0}^{\infty}e^{-i\omega t}\sum_{n=0}^{\infty}\mathbf{D}_n\int_{\kappa=0}^{\infty}\kappa\mathbf{C}_n(\kappa r)\tilde{\mathbf{q}}(\kappa,z,\omega)d\kappa d\omega \tag{4}$$

In the above expressions

$$\mathbf{C}_n(\kappa r)=\frac{1}{kr}\begin{bmatrix} rJ_n(\kappa r)_{,r} & nJ_n(\kappa r) & 0 \\ nJ_n(\kappa r) & rJ_n(\kappa r)_{,r} & 0 \\ 0 & 0 & -J_n(\kappa r) \end{bmatrix} \tag{5}$$

$$\mathbf{D}_n = diag\left[\begin{pmatrix}\cos n\theta \\ \sin n\theta\end{pmatrix}, \begin{pmatrix}-\sin n\theta \\ \cos n\theta\end{pmatrix}, \begin{pmatrix}\cos n\theta \\ \sin n\theta\end{pmatrix}\right] \tag{6}$$

where λ and μ are Lame' constants; a_n is the normalization factor, which is equal to $1/2\pi$ for $n=0$ and $1/\pi$ for $n\neq 0$. The upper set of diagonal matrix $\mathbf{D}_n$ corresponds to the case when the loads and displacements are symmetric with respect to the x-axis and the lower one corresponds to the antisymmetric case.

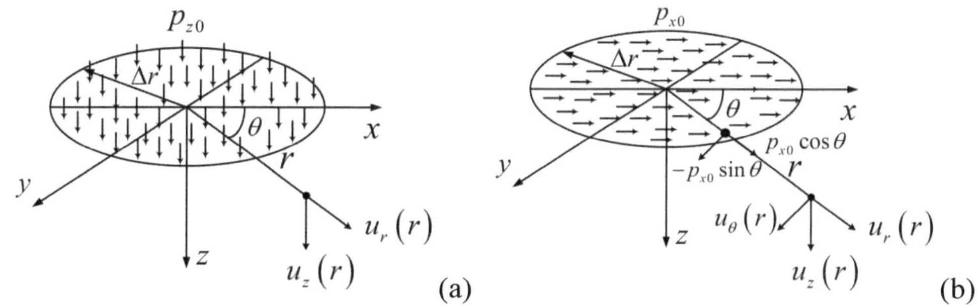

Figure 2: Subdisk-element showing constant load distribution (a) vertical (b) horizontal

The Green's influence functions are evaluated for the subdisk-elements (Fig. 2). Taking advantage of the axisymmetric geometry of it, the displacements are split into components which are either symmetric or antisymmetric about the r-axis at $\theta=0$.

$$\begin{aligned} u_r(r,\theta,z,n) &= \sum_n u_r^s(r,z,n)\cos n\theta + \sum_n u_r^a(r,z,n)\sin n\theta \\ u_\theta(r,\theta,z,n) &= -\sum_n u_\theta^s(r,z,n)\sin n\theta + \sum_n u_\theta^a(r,z,n)\cos n\theta \\ u_z(r,\theta,z,n) &= \sum_n u_z^s(r,z,n)\cos n\theta + \sum_n u_z^a(r,z,n)\sin n\theta \end{aligned} \tag{7}$$

where superscripts s and a denote the symmetric and antisymmetric components respectively. For the evaluation of Green's functions of subdisk-elements, only the cases of $n=0$ and $n=1$ need to be considered.

Applying Fourier-Bessel transform to the wave motion equation leads to a pair of uncoupled equations [12]: a 2-vector equation for the SV-P degrees of freedom and a scalar equation for the SH degree of freedom in the frequency-wave-number domain.

$$\begin{bmatrix} \mu & 0 \\ 0 & \lambda+2\mu \end{bmatrix} \begin{Bmatrix} \tilde{u}_r''(\kappa) \\ \tilde{u}_z''(\kappa) \end{Bmatrix} - i\kappa \begin{bmatrix} 0 & \lambda+\mu \\ \lambda+\mu & 0 \end{bmatrix} \begin{Bmatrix} \tilde{u}_r'(\kappa) \\ \tilde{u}_z'(\kappa) \end{Bmatrix} - \left(\kappa^2 \begin{bmatrix} \lambda+2\mu & 0 \\ 0 & \mu \end{bmatrix} - \rho\omega^2 \mathbf{I} \right) \begin{Bmatrix} \tilde{u}_r(\kappa) \\ \tilde{u}_z(\kappa) \end{Bmatrix} = 0 \tag{8}$$

$$\mu\tilde{u}_\theta''(\kappa) - \left(\kappa^2\mu - \rho\omega^2\right)\tilde{u}_\theta(\kappa) = 0 \tag{9}$$

where the superscript of $\tilde{u}'$ denotes differentiation with respect to z. Note that the SV-P component and the SH component are no longer plane waves.

The two cases of Eqs. (8) and (9) can be unified into a general formula

$$\mathbf{K}_{22}^{m}\left(\tilde{\mathbf{q}}^{m}\right)'' + \left(\mathbf{K}_{21}^{m} - \mathbf{K}_{12}^{m}\right)\left(\tilde{\mathbf{q}}^{m}\right)' - \left(\mathbf{K}_{11}^{m} - \rho\omega^2\mathbf{I}_m\right)\tilde{\mathbf{q}}^{m} = 0 \tag{10}$$

with superscript $m=1$ and $m=2$ corresponding to the in-plane motion and out-of-plane motion respectively; $\mathbf{I}_m$ denotes a unit matrix of the size $(m\times m)$ and the coefficient matrices are defined as

$$\mathbf{K}_{11}^{1} = k^2\begin{bmatrix} \lambda+2\mu & 0 \\ 0 & \mu \end{bmatrix},\ \mathbf{K}_{12}^{1} = \mathbf{K}_{21}^{1\ H} = ik\begin{bmatrix} 0 & \lambda \\ \mu & 0 \end{bmatrix},\ \mathbf{K}_{22}^{1} = \begin{bmatrix} \mu & 0 \\ 0 & \lambda+2\mu \end{bmatrix}$$

$$\mathbf{K}_{11}^{2} = k^2\mu,\ \mathbf{K}_{12}^{1} = \mathbf{K}_{21}^{2\ H} = 0,\ \mathbf{K}_{22}^{2} = \mu \tag{11}$$

For brevity, the superscript m is omitted hereinafter.

2.2 Cases of anisotropic media

The problem may be solved in Cartesian coordinates. The wave motion equation is expressed as

$$\mathbf{D}_{xx}\frac{\partial^2\mathbf{q}}{\partial x^2} + \mathbf{D}_{yy}\frac{\partial^2\mathbf{q}}{\partial y^2} + \mathbf{D}_{zz}\frac{\partial^2\mathbf{q}}{\partial z^2} + \left(\mathbf{D}_{xy} + \mathbf{D}_{yx}\right)\frac{\partial^2\mathbf{q}}{\partial x\partial y} + \left(\mathbf{D}_{yz} + \mathbf{D}_{zy}\right)\frac{\partial^2\mathbf{q}}{\partial y\partial z} + \left(\mathbf{D}_{xz} + \mathbf{D}_{zx}\right)\frac{\partial^2\mathbf{q}}{\partial x\partial z} = \rho\ddot{\mathbf{q}} \tag{12}$$

in which the displacement vector $\mathbf{q}$ and constitutive matrices are defined as follows, note that $\mathbf{D}_{ji} = \mathbf{D}_{ij}\left(i, j = x, y, z\right)$

$$\mathbf{q} = \begin{bmatrix} u_x & u_y & u_z \end{bmatrix}^T \tag{13}$$

$$\mathbf{D}_{xx} = \begin{bmatrix} d_{11} & d_{16} & d_{15} \\ d_{16} & d_{66} & d_{56} \\ d_{15} & d_{56} & d_{55} \end{bmatrix}, \mathbf{D}_{yy} = \begin{bmatrix} d_{66} & d_{26} & d_{46} \\ d_{26} & d_{22} & d_{24} \\ d_{46} & d_{24} & d_{44} \end{bmatrix}, \mathbf{D}_{zz} = \begin{bmatrix} d_{55} & d_{45} & d_{35} \\ d_{45} & d_{44} & d_{34} \\ d_{35} & d_{34} & d_{33} \end{bmatrix}$$

$$\mathbf{D}_{xy} = \begin{bmatrix} d_{16} & d_{12} & d_{14} \\ d_{66} & d_{26} & d_{46} \\ d_{56} & d_{25} & d_{45} \end{bmatrix}, \mathbf{D}_{xz} = \begin{bmatrix} d_{15} & d_{14} & d_{13} \\ d_{56} & d_{46} & d_{36} \\ d_{55} & d_{45} & d_{35} \end{bmatrix}, \mathbf{D}_{yz} = \begin{bmatrix} d_{56} & d_{46} & d_{36} \\ d_{25} & d_{24} & d_{23} \\ d_{45} & d_{44} & d_{34} \end{bmatrix} \tag{14}$$

with the stress and strain vector specified as

$$\begin{aligned} \boldsymbol{\sigma} &= \begin{bmatrix} \sigma_x & \sigma_y & \sigma_z & \tau_{yz} & \tau_{xz} & \tau_{xy} \end{bmatrix}^T \\ \boldsymbol{\varepsilon} &= \begin{bmatrix} \varepsilon_x & \varepsilon_y & \varepsilon_z & \gamma_{yz} & \gamma_{xz} & \gamma_{xy} \end{bmatrix}^T \end{aligned} \tag{15}$$

$$\boldsymbol{\sigma} = \mathbf{D}\boldsymbol{\varepsilon} \quad \text{with} \quad \mathbf{D} = \begin{bmatrix} d_{11} & d_{12} & d_{13} & d_{14} & d_{15} & d_{16} \\ & d_{22} & d_{23} & d_{24} & d_{25} & d_{26} \\ & & d_{33} & d_{34} & d_{35} & d_{36} \\ & & & d_{44} & d_{45} & d_{46} \\ & \text{(sym)} & & & d_{55} & d_{56} \\ & & & & & d_{66} \end{bmatrix}$$

For the transversely anisotropic medium, the elements of constitutive matrix are simplified as follows

$$d_{11} = d_{22} = \lambda + 2G,\ d_{12} = \lambda \text{ and } d_{66} = G \text{ (in the isotropic plane)} \tag{16}$$

$$d_{33} = \lambda_t + 2G_t,\ d_{13} = d_{23} = \lambda_t \text{ and } d_{44} = d_{55} = G_t \text{ (in the transverse direction)} \tag{17}$$

All other elements are zero (except for the symmetric ones, such as d_{12} etc.).

Carrying out double Fourier transformation

$$\tilde{\mathbf{q}}\left(k_x, k_y, z, \omega\right) = \int_{-\infty}^{+\infty}\int_{-\infty}^{+\infty} \mathbf{q}\left(x, y, z, \omega\right) e^{-i\left(k_x x + k_y y\right)} dxdy \tag{18}$$

leads to the wave equation in the frequency-wavenumber domain expressed as

$$\begin{aligned}&\mathbf{D}_{zz}\tilde{\mathbf{q}}'' - i\left[k_x\left(\mathbf{D}_{xz}+\mathbf{D}_{zx}\right)+k_y\left(\mathbf{D}_{yz}+\mathbf{D}_{zy}\right)\right]\tilde{\mathbf{q}}' \\ &-\left[k_x^2\mathbf{D}_{xx}+k_xk_y\left(\mathbf{D}_{xy}+\mathbf{D}_{yx}\right)+k_y^2\mathbf{D}_{yy}\right]\tilde{\mathbf{q}}+\rho\omega^2\tilde{\mathbf{q}}=0\end{aligned} \tag{19}$$

For brevity, Eq. (19) is rewritten as

$$\mathbf{K}_{22}\tilde{\mathbf{q}}'' + \left(\mathbf{K}_{21}-\mathbf{K}_{12}\right)\tilde{\mathbf{q}}' - \left(\mathbf{K}_{11}-\rho\omega^2\mathbf{I}\right)\tilde{\mathbf{q}} = 0 \tag{20}$$

with

$$\mathbf{K}_{22}=\mathbf{D}_{zz}\,,\quad \mathbf{K}_{21}=-\mathbf{K}_{21}{}^{T}=ik_x\mathbf{D}_{xz}+ik_y\mathbf{D}_{yz}$$

$$\mathbf{K}_{11}=k_x^2\mathbf{D}_{xx}+k_y^2\mathbf{D}_{yy}+k_xk_y\left(\mathbf{D}_{xy}+\mathbf{D}_{yx}\right)$$

The same formula as that for the isotropic case Eq. (10) is obtained.

3 Solution produce of the wave motion equation and the precise integration method

For the solution of wave motion equation a dual vector $\mathbf{p}$ of $\mathbf{q}$ is introduced. In the Cartesian coordinates, $\mathbf{p}$ is specified as $\mathbf{p}=\left[\tau_{xz}\quad \tau_{yz}\quad \sigma_z\right]^T$, whereas in cylindrical coordinates $\mathbf{p}=\left[\tau_{xz}\quad \tau_{yz}\quad \sigma_z\right]^T$. It is easy to verify in both cases that

$$\tilde{\mathbf{p}} = -\left(\mathbf{K}_{22}\dot{\tilde{\mathbf{q}}}+\mathbf{K}_{21}\tilde{\mathbf{q}}\right) \tag{21}$$

Then in the frequency-wave-number domain the wave motion Eq. (10) or Eq. (20) is expressed in dual vector form as

$$\tilde{\mathbf{q}}' = \mathbf{A}\tilde{\mathbf{q}}+\mathbf{D}\tilde{\mathbf{p}}\,,\ \tilde{\mathbf{p}}'=\mathbf{B}\tilde{\mathbf{q}}+\mathbf{C}\tilde{\mathbf{p}} \tag{22}$$

with

$$\mathbf{A}=-\mathbf{K}_{22}^{-1}\mathbf{K}_{21}\,,\ \mathbf{B}=-\mathbf{K}_{11}+\mathbf{K}_{12}\mathbf{K}_{22}^{-1}\mathbf{K}_{21}\,,\ \mathbf{C}=\mathbf{K}_{12}\mathbf{K}_{22}^{-1}\,,\ \mathbf{D}=-\mathbf{K}_{22}^{-1} \tag{23}$$

In the state space the dual vector form wave equation (22) is expressed as a first order linear differential equation

$$\mathbf{V}'=\mathbf{H}\mathbf{V} \tag{24}$$

with

$$\mathbf{V}=\begin{Bmatrix}\tilde{\mathbf{q}}\\ \tilde{\mathbf{p}}\end{Bmatrix},\ \mathbf{H}=\begin{bmatrix}\mathbf{A} & \mathbf{D}\\ \mathbf{B} & \mathbf{C}\end{bmatrix} \tag{25}$$

The boundary condition for wave motion equation at the free surface is

$$\tilde{\mathbf{p}}_0 = \tilde{\mathbf{p}}(z=0) = \mathbf{0} \tag{26}$$

At the interface between two adjacent layers, the continuity conditions lead to

$$\tilde{\mathbf{p}}(z_r^+) = \tilde{\mathbf{p}}(z_r^-),\ \tilde{\mathbf{q}}(z_r^+) = \tilde{\mathbf{q}}(z_r^-) \quad (r = 1,2,3...l) \tag{27}$$

In case the multi-layered soil rests on rigid base, we have

$$\tilde{\mathbf{q}}_l = \tilde{\mathbf{q}}(z = z_l) = \mathbf{0} \tag{28}$$

Whereas for the multi-layered soil overlying an elastic half-space, the radiative condition should be considered.

The eigenvalue problem of the half-space (layer $l+1$) is solved

$$\mathbf{H\Phi} = \mathbf{\Phi\Lambda} \tag{29}$$

with the eigenvalues and eigenvectors partitioned as follows

$$\mathbf{\Lambda} = \begin{bmatrix} \boldsymbol{\lambda}_i & \\ & -\boldsymbol{\lambda}_i \end{bmatrix},\ \mathbf{\Phi} = \begin{bmatrix} \mathbf{\Phi}_{11} & \mathbf{\Phi}_{12} \\ \mathbf{\Phi}_{21} & \mathbf{\Phi}_{22} \end{bmatrix} \tag{30}$$

where the real parts of all elements $\boldsymbol{\lambda}_i$ are positive.

Let

$$\mathbf{b} = \mathbf{\Phi}^{-1}\mathbf{V} \tag{31}$$

Then Eq (24) becomes

$$\mathbf{b}' = \mathbf{\Lambda b},\ \mathbf{b}(z) = \begin{bmatrix} \exp(\boldsymbol{\lambda}_i z) & \\ & \exp(-\boldsymbol{\lambda}_i z) \end{bmatrix} \begin{Bmatrix} \mathbf{c}_1 \\ \mathbf{c}_2 \end{Bmatrix} \tag{32}$$

Substituting Eq (32) into Eq (31) yields

$$\mathbf{V} = \begin{bmatrix} \mathbf{\Phi}_{11} & \mathbf{\Phi}_{12} \\ \mathbf{\Phi}_{21} & \mathbf{\Phi}_{22} \end{bmatrix} \begin{bmatrix} \exp(\boldsymbol{\lambda}_i z) & \\ & \exp(-\boldsymbol{\lambda}_i z) \end{bmatrix} \begin{Bmatrix} \mathbf{c}_1 \\ \mathbf{c}_2 \end{Bmatrix} \tag{33}$$

For an unbounded medium, the displacements $\tilde{\mathbf{q}}$ and stresses $\tilde{\mathbf{p}}$ must remain finite. The boundary condition requires that the coefficients $\mathbf{c}_1 = \mathbf{0}$.

Applying Eq (33) yields the boundary condition at the surface of the half-space $(z=0)$

$$\tilde{\mathbf{q}}_l = \mathbf{\Phi}_{12}\mathbf{\Phi}_{22}^{-1}\tilde{\mathbf{p}}_l \tag{34}$$

$$\text{or}\quad \tilde{\mathbf{q}}_l = \mathbf{R}_\infty \tilde{\mathbf{p}}_l \qquad \text{with}\quad \mathbf{R}_\infty = \mathbf{\Phi}_{12}\mathbf{\Phi}_{22}^{-1} \tag{35}$$

The general solution to the differential equation (24) takes the form

$$\mathbf{V} = \exp(\mathbf{H}z)\mathbf{c} \tag{36}$$

where $\mathbf{c}$ is the integration constants.

For a typical layer $(\eta = z_b - z_a)$ of thickness η within the interval of the soil stratum $[z_a, z_b]$, the relationship between the displacements and stresses at the two ends of the layer is found from Eq (36)

$$\mathbf{V}_b = \exp(\mathbf{H}\eta)\mathbf{V}_a \tag{37}$$

Rewrite Eq (37) in the following form

$$\mathbf{V}_b = \mathbf{T}\mathbf{V}_a \tag{38}$$

$$\mathbf{T} = \exp(\mathbf{H}\eta) = \mathbf{I} + \mathbf{H}\eta + \frac{1}{2!}(\mathbf{H}\eta)^2 + \frac{1}{3!}(\mathbf{H}\eta)^3 + \frac{1}{4!}(\mathbf{H}\eta)^4 + \ldots \tag{39}$$

where $\mathbf{I}$ is a unitary matrix.

As the accuracy of evaluating Green's influence function is controlled by the accuracy of calculating matrix $\mathbf{T}$, the precise integration method (PIM) presented by Zhong [15] is applied . The basic idea of PIM is introduced as follows.

For the computation of $\mathbf{T}$, we note the basic characteristics of exponential function

$$\mathbf{T} = \exp(\mathbf{H}\eta) = \left[\exp(\mathbf{H}\eta/b)\right]^b = \left[\exp(\mathbf{H}\tau)\right]^b \tag{40}$$

where $\tau = \eta/b$, b is an arbitrary integer. Let $b = 2^N$ and N may be chosen as $N = 20$. This is equivalent to further divide the layer into 2^N sublayers. Since τ is extremely small, for the computation of $\exp(\mathbf{H}\tau)$, five terms Taylor expansion is sufficient.

$$\exp(\mathbf{H}\tau) \approx \mathbf{I} + \mathbf{H}\tau + \frac{1}{2!}(\mathbf{H}\tau)^2 + \frac{1}{3!}(\mathbf{H}\tau)^3 + \frac{1}{4!}(\mathbf{H}\tau)^4 \tag{41}$$

or

$$\begin{aligned} &\exp(\mathbf{H}\tau) \approx \mathbf{I} + \mathbf{T}_a \\ &\mathbf{T}_a = \mathbf{H}\tau + \frac{1}{2}(\mathbf{H}\tau)\left[(\mathbf{H}\tau) + \frac{1}{3}(\mathbf{H}\tau)^2 + \frac{1}{12}(\mathbf{H}\tau)^3\right] \end{aligned} \tag{42}$$

It can be observed that $\exp(\mathbf{H}\tau)$ deviates from unit matrix $\mathbf{I}$ by only a very small remainder $\mathbf{T}_a$. The computation of $\mathbf{T}$ (Eq (40)) is performed in the following way

$$\mathbf{T} = (\mathbf{I} + \mathbf{T}_a)^{2^N} = (\mathbf{I} + \mathbf{T}_a)^{2^{N-1}} \times (\mathbf{I} + \mathbf{T}_a)^{2^{N-1}} \tag{43}$$

Such factorization should be carried out successively N times. We note the following relationship for matrix multiplication.

$$(\mathbf{I}+\mathbf{T}_b)\times(\mathbf{I}+\mathbf{T}_c)=\mathbf{I}+\mathbf{T}_b+\mathbf{T}_c+\mathbf{T}_b\times\mathbf{T}_c \tag{44}$$

In a case of $\mathbf{T}_b=\mathbf{T}_c=\mathbf{T}_a$, it results in

$$(\mathbf{I}+\mathbf{T}_a)\times(\mathbf{I}+\mathbf{T}_a)=\mathbf{I}+\mathbf{T}_r,\ \mathbf{T}_r=2\mathbf{T}_a+\mathbf{T}_a\times\mathbf{T}_a \tag{45}$$

Since $\mathbf{T}_r$ is quite small, it should be calculated and stored independently to avoid losing effective digits.

From Eqs (43) and (45), we obtain:

$$\mathbf{T}=\mathbf{I}+\mathbf{T}_r^N \tag{46}$$

with

$$\mathbf{T}_r^i=2\mathbf{T}_r^{i-1}+\mathbf{T}_r^{i-1}\times\mathbf{T}_r^{i-1}\quad (i=1,2,...,N)\ \text{and}\ \mathbf{T}_r^0=\mathbf{T}_a$$

where $\mathbf{T}_a$ is calculated by Eq. (42). It is therefore clear that $\mathbf{T}$ is evaluated by applying the recursive formula (46) N times, and the size of matrices are equal to (4×4) and (2×2) for the isotropic medium and (6×6) for the anisotropic medium. That is to say, for $N=20$, the recursive formula (46) is applied 20 times, it is equivalent to that the layer is subdivided into 2^{20}=1048576 mini-layers. So by chosen an appropriate value of N, any desired precision can be achieved. The precision of the numerical results can reach the precision of computer.

4 Assembly of layers

Integration of the wave equation (24) by applying PIM yields the relationship between the displacements and stresses (or loads) at the two ends z_a and z_b of the layer. Writing it in partitioned form leads to

$$\begin{Bmatrix}\tilde{\mathbf{q}}_b\\ \tilde{\mathbf{p}}_b\end{Bmatrix}=\mathbf{T}\begin{Bmatrix}\tilde{\mathbf{q}}_a\\ \tilde{\mathbf{p}}_a\end{Bmatrix},\quad \mathbf{T}=\begin{bmatrix}\mathbf{T}_A & \mathbf{T}_D\\ \mathbf{T}_B & \mathbf{T}_C\end{bmatrix} \tag{47}$$

In order to ease the assembly of layers, rearrange Eq (47) into following dual-vector form:

$$\tilde{\mathbf{q}}_b=\mathbf{M}_F\tilde{\mathbf{q}}_a-\mathbf{M}_G\tilde{\mathbf{p}}_b,\ \tilde{\mathbf{p}}_a=\mathbf{M}_Q\tilde{\mathbf{q}}_a+\mathbf{M}_E\tilde{\mathbf{p}}_b \tag{48}$$

with

$$\mathbf{M}_F=\mathbf{T}_A-\mathbf{T}_D\mathbf{T}_C^{-1}\mathbf{T}_B,\ \mathbf{M}_G=-\mathbf{T}_D\mathbf{T}_C^{-1},\ \mathbf{M}_Q=-\mathbf{T}_C^{-1}\mathbf{T}_B,\ \mathbf{M}_E=-\mathbf{T}_C^{-1} \tag{49}$$

Assembly of layers is performed two by two. Consider the case for combining two adjacent layers $[z_a,z_b]$ and $[z_b,z_c]$ into $[z_a,z_c]$, we denote them by layer 1, layer 2 and layer c respectively. Applying Eq. (48) to layer 1 and layer 2 yields

$$\begin{aligned} \tilde{\mathbf{q}}_b &= \mathbf{M}_F^1\tilde{\mathbf{q}}_a - \mathbf{M}_G^1\tilde{\mathbf{p}}_b, \quad \tilde{\mathbf{p}}_a = \mathbf{M}_Q^1\tilde{\mathbf{q}}_a + \mathbf{M}_E^1\tilde{\mathbf{p}}_b \\ \tilde{\mathbf{q}}_c &= \mathbf{M}_F^2\tilde{\mathbf{q}}_b - \mathbf{M}_G^2\tilde{\mathbf{p}}_c, \quad \tilde{\mathbf{p}}_b = \mathbf{M}_Q^2\tilde{\mathbf{q}}_b + \mathbf{M}_E^2\tilde{\mathbf{p}}_c \end{aligned} \tag{50}$$

Eliminating $\tilde{\mathbf{q}}_b$ and $\tilde{\mathbf{p}}_b$ from Eq. (50) leads to the dual-vector form of the new layer c within the interval $[z_a,z_c]$.

$$\tilde{\mathbf{q}}_c = \mathbf{M}_F^c\tilde{\mathbf{q}}_a - \mathbf{M}_G^c\tilde{\mathbf{p}}_c, \quad \tilde{\mathbf{p}}_a = \mathbf{M}_Q^c\tilde{\mathbf{q}}_a + \mathbf{M}_E^c\tilde{\mathbf{p}}_c \tag{51}$$

with

$$\begin{aligned} \mathbf{M}_F^c &= \mathbf{M}_F^2\left(\mathbf{I}+\mathbf{M}_G^1\mathbf{M}_Q^2\right)^{-1}\mathbf{M}_F^1 \qquad \mathbf{M}_G^c = \mathbf{M}_G^2 + \mathbf{M}_F^2\left[\left(\mathbf{M}_G^1\right)^{-1}+\mathbf{M}_Q^2\right]^{-1}\mathbf{M}_E^2 \\ \mathbf{M}_E^c &= \mathbf{M}_E^1\left(\mathbf{I}+\mathbf{M}_Q^2\mathbf{M}_G^1\right)^{-1}\mathbf{M}_E^2 \qquad \mathbf{M}_Q^c = \mathbf{M}_Q^1 + \mathbf{M}_E^1\left[\left(\mathbf{M}_Q^2\right)^{-1}+\mathbf{M}_G^1\right]^{-1}\mathbf{M}_F^1 \end{aligned} \tag{52}$$

Thus, assembly of layers is proceeded based on matrix algebra with the size of matrices equal to (2×2) and (1×1) for isotropic medium, and (3×3) for anisotropic medium, respectively. The computational effort is reduced to a great extent, while high precision is ensured. And there is no limit for the number of layers to be considered.

Eventually, for layered strata consisting of l layers (Fig. 1), the following relationship holds

$$\tilde{\mathbf{q}}_l = \mathbf{M}_F^s\tilde{\mathbf{q}}_0 - \mathbf{M}_G^s\tilde{\mathbf{p}}_l, \quad \tilde{\mathbf{p}}_0 = \mathbf{M}_Q^s\tilde{\mathbf{q}}_0 + \mathbf{M}_E^s\tilde{\mathbf{p}}_l \tag{53}$$

For layered strata bonded to rigid base, the boundary condition Eq. (28) leads to the relationship between surface displacements and tractions as

$$\tilde{\mathbf{p}}_0 = \left(\mathbf{M}_Q^s + \mathbf{M}_E^s\left(\mathbf{M}_G^s\right)^{-1}\mathbf{M}_F^s\right)\tilde{\mathbf{q}}_0 = \tilde{\mathbf{S}}(\kappa,\omega)\tilde{\mathbf{q}}_0 \tag{54}$$

Whereas for layered strata overlying elastic half-space, substituting Eq. (35) into Eq. (53) yields.

$$\tilde{\mathbf{p}}_0 = \left(\mathbf{M}_Q^s + \mathbf{M}_E^s\mathbf{R}_\infty\left(\mathbf{I}+\mathbf{M}_G^s\mathbf{R}_\infty\right)^{-1}\mathbf{M}_F^s\right)\tilde{\mathbf{q}}_0 = \tilde{\mathbf{S}}(\kappa,\omega)\tilde{\mathbf{q}}_0 \tag{55}$$

Elements of $\tilde{\mathbf{S}}(\kappa,\omega)$ in Eq. (54) and (55) denote the dynamic impedance coefficients condensed at the surface of the strata. The relevant dynamic compliance coefficients are found by inversion of $\tilde{\mathbf{S}}(\kappa,\omega)$.

$$\tilde{\mathbf{q}}_0(\kappa,\omega) = \tilde{\mathbf{F}}(\kappa,\omega)\tilde{\mathbf{p}}_0(\kappa,\omega), \quad \tilde{\mathbf{F}}(\kappa,\omega) = \tilde{\mathbf{S}}(\kappa,\omega)^{-1} \tag{56}$$

For the case of anisotropic medium, $\tilde{\mathbf{q}}_0(\kappa,\omega) = \tilde{\mathbf{q}}_0(\kappa_x,\kappa_y,0,\omega)$, $\tilde{\mathbf{p}}_0(\kappa,\omega) = \tilde{\mathbf{p}}_0(\kappa_x,\kappa_y,0,\omega)$, and the rest can be inferred by analogy.

5 Formulation of Green's influence function in frequency-spatial domain

The Green's influence functions are evaluated for the subdisk-elements of radius r_0 subjected to the uniformly distributed vertical load p_{z0} or horizontal load p_{x0}/p_{y0} as shown in Fig. 2.

5.1 Cases of isotropic medium [16]

The vertical distributed load of intensity $p_{z0}/(\pi r_0^2)$ acting on the subdisk corresponds to the amplitude of the load in the wave number domain.

$$\tilde{p}_z(\kappa,\omega) = \frac{1}{2\pi}\int_0^{r_0} -rJ_0(\kappa r)\int_0^{2\pi}\frac{p_{z0}}{\pi r_0^2}J_1(\kappa r_0)\,d\varphi dr = -\frac{p_{z0}}{\pi\kappa r_0}J_1(\kappa r_0) \tag{57}$$

The corresponding Green's influence function in the frequency-spatial domain is evaluated as

$$\begin{Bmatrix} u_r(r) \\ u_z(r) \end{Bmatrix} = \frac{1}{\pi r_0}\left[\int_{\kappa=0}^{\infty} J_1(\kappa r_0)\begin{Bmatrix} \tilde{F}_{rz}(\kappa)J_1(\kappa r) \\ \tilde{F}_{zz}(\kappa)J_0(\kappa r) \end{Bmatrix} d\kappa\right] p_{z0} \tag{58}$$

For details the readers may refer to [16].

The constant horizontal distributed load of intensity $p_{x0}/(\pi r_0^2)$ acting on the disk is split into radial and circumferential components varying as $p_{x0}\cos\theta$ and $-p_{x0}\sin\theta$, respectively. The corresponding amplitudes of the load in the wave-number domain are evaluated as

$$\begin{aligned} \begin{Bmatrix} \tilde{p}_r(\kappa) \\ \tilde{p}_\theta(\kappa) \end{Bmatrix} &= \frac{1}{\pi}\int_{r=0}^{r_0}\frac{1}{\kappa}\begin{bmatrix} rJ_1(\kappa r)_{,r} & J_1(\kappa r) \\ J_1(\kappa r) & rJ_1(\kappa r)_{,r} \end{bmatrix}\left(\frac{1}{\pi r_0}\int_{\theta=0}^{2\pi}\begin{bmatrix} \cos\theta & \\ & -\sin\theta \end{bmatrix}\begin{Bmatrix} p_{x0}\cos\theta \\ -p_{x0}\sin\theta \end{Bmatrix} d\theta\right) dr \\ &= \frac{p_{x0}r_0}{\kappa}J_1(\kappa r_0)\begin{Bmatrix} 1 \\ 1 \end{Bmatrix} \end{aligned} \tag{59}$$

Analogously, the Green's influence functions in the frequency-spatial domain are evaluated accordingly.

$$\begin{aligned} \begin{Bmatrix} u_r(r,\theta) \\ u_\theta(r,\theta) \\ u_z(r,\theta) \end{Bmatrix} &= \frac{r_0}{2\pi r_0}\begin{bmatrix} \cos\theta & & \\ & -\sin\theta & \\ & & \cos\theta \end{bmatrix} \\ &\left(\int_{\kappa=0}^{\infty} J_1(\kappa r_0)\begin{bmatrix} J_0(\kappa r)-J_2(\kappa r) & J_0(\kappa r)+J_2(\kappa r) & \\ J_0(\kappa r)+J_2(\kappa r) & J_0(\kappa r)-J_2(\kappa r) & \\ & & -2J_1(\kappa r) \end{bmatrix}\begin{Bmatrix} \tilde{F}_{rr}(\kappa) \\ \tilde{F}_{\theta\theta}(\kappa) \\ \tilde{F}_{rz}(\kappa) \end{Bmatrix} d\kappa\right) p_{x0} \end{aligned} \tag{60}$$

In case the distributed horizontal load in the y-direction of intensity $p_{y0}/(\pi r_0^2)$, the same form of Eq. (60) applies, with p_{y0} instead of p_{x0}, and $\cos\theta$ and $-\sin\theta$ replaced by $\sin\theta$ and $\cos\theta$, respectively.

Summarizing all the horizontal and vertical load cases, the frequency domain relationships between the surface tractions and the displacement amplitudes for a subdisk-element are obtained as

$$\begin{Bmatrix} u_r(r,\theta) \\ u_\theta(r,\theta) \\ u_z(r,\theta) \end{Bmatrix} = \begin{bmatrix} F_{xr} & F_{yr} & F_{zr} \\ F_{x\theta} & F_{y\theta} & 0 \\ F_{xz} & F_{yz} & F_{zz} \end{bmatrix} \begin{Bmatrix} p_{x0}(x_0,y_0) \\ p_{y0}(x_0,y_0) \\ p_{z0}(x_0,y_0) \end{Bmatrix} \tag{61}$$

where the load (p_{x0}, p_{y0}, p_{z0}) is applied on the subdisk-element with its center placed at (x_0, y_0).

Carrying out coordinate transformation of the displacements leads to the Green's influence function expressed in the Cartesian coordinates with the origin placed at the centre of the subdisk as follows

$$\begin{Bmatrix} u_x(x,y,0,\omega) \\ u_y(x,y,0,\omega) \\ u_z(x,y,0,\omega) \end{Bmatrix} = \begin{bmatrix} F_{xx} & F_{xy} & F_{xz} \\ F_{yx} & F_{yy} & F_{yz} \\ F_{zx} & F_{zy} & F_{zz} \end{bmatrix} \begin{Bmatrix} p_{x0} \\ p_{y0} \\ p_{z0} \end{Bmatrix} \tag{62}$$

where $\begin{bmatrix} u_x & u_y & u_z \end{bmatrix}^T$ is related with $\begin{bmatrix} u_r & u_\theta & u_z \end{bmatrix}^T$ by the following expression

$$\begin{bmatrix} u_x & u_y & u_z \end{bmatrix}^T = \mathbf{T}\begin{bmatrix} u_r & u_\theta & u_z \end{bmatrix}^T$$

$$\mathbf{T} = \begin{bmatrix} \cos\theta & \sin\theta & 0 \\ -\sin\theta & \cos\theta & 0 \\ 0 & 0 & 1 \end{bmatrix} \tag{63}$$

5.2 Cases of anisotropic medium [17]

Transformation of the compliance coefficients matrices Eq. (56) from wave-number domain into Cartesian space domain involves a double inverse Fourier transformation.

$$\mathbf{q}_0(x,y,0,\omega) = \frac{1}{4\pi^2}\int_{-\infty}^{+\infty}\int_{-\infty}^{+\infty} \tilde{\mathbf{q}}_0(\kappa_x,\kappa_y,0,w)e^{i(\kappa_x x+\kappa_y y)}d\kappa_x d\kappa_y \tag{64}$$

This time-consuming operation can be greatly simplified if the variables κ_x and κ_y of Eq. (64) are expressed in cylindrical coordinate system and the integral is evaluated along the line defined by $\kappa_x = 0$. After some manipulations the expression for double inverse Fourier transformation becomes

$$\mathbf{q}_o(x,y,0,\omega) = \frac{1}{4\pi^2}\int_0^{+\infty}\int_0^{2\pi}\mathbf{R}(\theta)\mathbf{R}(\varphi)\tilde{\mathbf{F}}(0,\kappa,0,\omega)\left[\mathbf{R}(\theta)\mathbf{R}(\varphi)\right]^T\tilde{\mathbf{p}}(0,k_y,0,\omega)e^{i\kappa r\sin\varphi}d\varphi\kappa d\kappa \tag{65}$$

with the transformation matrix $\mathbf{R}$ defined by

$$\mathbf{R}(\varphi) = \begin{bmatrix} \sin\varphi & \cos\varphi & 0 \\ -\cos\varphi & \sin\varphi & 0 \\ 0 & 0 & 1 \end{bmatrix} \quad (\varphi = \varphi, \theta) \tag{66}$$

For vertical distributed load of intensity $p_{z0}/(\pi r_0^2)$ acting on the subdisk

$$\tilde{p}(0,\kappa,0,\omega) = \frac{p_{z0}}{\pi r_0^2}\int_0^{r_0}\int_0^{2\pi}e^{-i\kappa r\sin\varphi}d\varphi r dr = \frac{2p_{z0}}{\kappa r_0^2}J_1(\kappa r_0) \tag{67}$$

Proceed in the similar manner, we also have the same formula for the horizontal distributed load $p_{x0}/(\pi r_0^2)$ and $p_{y0}/(\pi r_0^2)$.

$$\tilde{p}(0,\kappa,0,\omega) = \frac{2p_{j0}}{\kappa r_0^2}J_1(\kappa r_0),\ (j = x, y) \tag{68}$$

Substituting Eqs. (67) and (68) in Eq. (65), the displacement amplitudes in frequency-spatial domain are found as

$$\mathbf{q}_o(x,y,0,\omega) = \begin{bmatrix} u_x & u_y & u_z \end{bmatrix}^T = \frac{1}{\kappa\pi r_0^2}\mathbf{R}(\theta)\int_0^{+\infty}\mathbf{G}\mathbf{R}^T(\theta)J_1(\kappa r_0)\mathbf{p}_0\kappa d\kappa \tag{69}$$

$$\mathbf{G} = \frac{1}{2\pi}\int_0^{2\pi}\mathbf{R}(\varphi)\tilde{\mathbf{F}}(0,\kappa,0,\omega)\mathbf{R}(\varphi)^T e^{i\kappa r\sin\varphi}d\varphi \tag{70}$$

with $\mathbf{p}_0 = \begin{bmatrix} p_{x0} & p_{y0} & p_{z0} \end{bmatrix}^T$

Note that in Eq. (70) due to the fact that the load is applied with rotational symmetry around the z-axis, the vector $\mathbf{p}_0$ may be taken outside the integral over φ. Thus, Eqs. (69) and (70) only involves numerical integration in one dimension. This provides an efficient evaluation of the complex amplitudes of the surface displacements.

The elements of matrix $\mathbf{G}$ in Eq. (70) can be identified as integral representation of Bessel function. For example

$$G_{11} = \frac{1}{2\pi}\int_0^{2\pi}\left(\sin^2\varphi\hat{F}_{11} + \cos^2\varphi\hat{F}_{22}\right)e^{i\kappa r\sin\varphi}d\varphi \tag{71}$$

may be evaluated as

$$\begin{aligned}\frac{1}{2\pi}\int_0^{2\pi}\sin^2\varphi e^{i\kappa r\sin\varphi}d\varphi &= \frac{1}{2}\left[J_0(\kappa r)-J_2(\kappa r)\right]\\ \frac{1}{2\pi}\int_0^{2\pi}\cos^2\varphi e^{i\kappa r\sin\varphi}d\varphi &= \frac{1}{2}\left[J_0(\kappa r)+J_2(\kappa r)\right]\end{aligned} \tag{72}$$

These Bessel functions may be computed in an efficient manner by their series expansions. Coefficients $\hat{F}_{11}$ and $\hat{F}_{22}$ in Eq. (71) are elements of the matrix $\tilde{\mathbf{F}}(0,\kappa,0,\omega)$, which can be taken outside the integral over φ. The rest of the elements in matrix $\mathbf{G}$ is evaluated in a similar manner.

Eventually, the same expression for other elements of Green's influence matrix $\mathbf{G}$ as given in Eq. (72) is obtained.

6 Dynamic impedance of arbitrary-shaped foundation

The interface between the foundation and the soil is discretized into n subdisk-elements of equal radius, such that the total area equals the area of foundation interface. Six cases are studied, i.e. the foundation is subjected to three components of concentrated harmonic forces and three components of harmonic moments with amplitudes equal to P_x, P_y, P_z, M_x, M_y and M_z respectively (see Fig. 3).

Using Eq. (62), the load displacement relationship at the soil-foundation interface may be expressed as

$$\begin{Bmatrix}\mathbf{u}_p^1\\ \mathbf{u}_p^2\\ \vdots\\ \mathbf{u}_p^n\end{Bmatrix} = \begin{bmatrix}\mathbf{F}_u^{11} & \mathbf{F}_u^{12} & \cdots & \mathbf{F}_u^{1n}\\ \mathbf{F}_u^{21} & \mathbf{F}_u^{22} & \cdots & \mathbf{F}_u^{2n}\\ \vdots & \vdots & \ddots & \vdots\\ \mathbf{F}_u^{n1} & \mathbf{F}_u^{n2} & \cdots & \mathbf{F}_u^{nn}\end{bmatrix}\begin{Bmatrix}\mathbf{p}_0^1\\ \mathbf{p}_0^2\\ \vdots\\ \mathbf{p}_0^n\end{Bmatrix} \tag{73}$$

or writing in compact form

$$\mathbf{u}_p = \mathbf{F}_u\mathbf{p}_0 \tag{74}$$

where $\mathbf{u}_p^i\,(i=1,2,...,n)$ denotes surface displacements of subdisk-element i ; $\mathbf{p}_0^j\,(j=1,2,...,n)$ denotes the unknown tractions acting on the subdisk-element j .

$$\begin{aligned}\mathbf{u}_p^i &= \begin{bmatrix}u_x^i & u_y^i & u_z^i\end{bmatrix}^T \quad (i=1,2,...,n)\\ \mathbf{p}_0^j &= \begin{bmatrix}p_{x0}^j & p_{y0}^j & p_{z0}^j\end{bmatrix}^T \quad (j=1,2,...,n)\end{aligned} \tag{75}$$

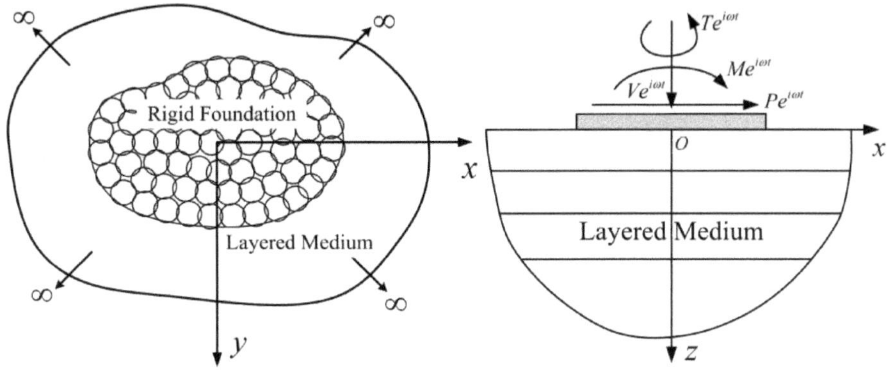

Figure 3: Foundation with subdisk discretization

For bonded condition of the foundation-soil interface, the displacement field must satisfy the following formula

$$\mathbf{u}_p = \mathbf{N}\mathbf{u}_b \tag{76}$$

where $\mathbf{u}_b$ represents the generalized displacement response of rigid foundation including three translational components (u_{xb}, u_{yb}, u_{zb}) and three rotational components $(\theta_{xb}, \theta_{yb}, \theta_{zb})$; the matrix $\mathbf{N}$ is the corresponding shape function. In which

$$\mathbf{u}_b = \begin{bmatrix} u_{xb} & u_{yb} & u_{zb} & \theta_{xb} & \theta_{yb} & \theta_{zb} \end{bmatrix}^T$$

$$\mathbf{N} = \begin{bmatrix} \mathbf{N}_1 & \mathbf{N}_2 & \cdots & \mathbf{N}_n \end{bmatrix}^T$$

$$\mathbf{N}_i = \begin{bmatrix} 1 & 0 & 0 & 0 & 0 & -y_i \\ 0 & 1 & 0 & 0 & 0 & x_i \\ 0 & 0 & 1 & y_i & -x_i & 0 \end{bmatrix} \quad (i = 1, 2, ..., n) \tag{77}$$

Combining Eqs. (74) and (76) leads to

$$\mathbf{F}_u \mathbf{p}_0 = \mathbf{N}\mathbf{u}_b \tag{78}$$

The balance of total loads $(P_x, P_y, P_z, M_x, M_y, M_z)$ acting on the foundation and the surface tractions on the foundation-soil interface yields:

$$P_x = \sum_{i=1}^{n} p_{x0}^i, \quad P_y = \sum_{i=1}^{n} p_{y0}^i, \quad P_z = \sum_{i=1}^{n} p_{z0}^i$$

$$M_x = \sum_{i=1}^{n} p_{z0}^i y_i, \quad M_y = -\sum_{i=1}^{n} p_{z0}^i x_i, \quad M_z = \sum_{i=1}^{n} \left(-p_{x0}^i y_i + p_{y0}^i x_i\right) \tag{79}$$

Rewrite in the matrix form, we have

$$\mathbf{P}_b = \mathbf{N}^T \mathbf{p}_0 \tag{80}$$

where

$$\mathbf{P}_b = \begin{bmatrix} P_x & P_y & P_z & M_x & M_y & M_z \end{bmatrix}^T \tag{81}$$

Substituting Eq. (78) into Eq. (80) results in

$$\mathbf{P}_b = \mathbf{N}^T \mathbf{F}_u^{-1} \mathbf{N} \mathbf{u}_b \tag{82}$$

The dynamic impedance matrix $\mathbf{S}(\omega)$ of the surface foundation is defined as

$$\mathbf{S}(\omega) = \mathbf{N}^T \mathbf{F}_u^{-1} \mathbf{N} \tag{83}$$

7 Numerical Examples

Due to the limited space, a comprehensive numerical example is provided. The dynamic impedance of a rigid circular foundation of radius a on multi-layered half-space is studied. The material properties of the layers and the half-space are listed in Table 1, where μ denotes the shear modulus of elasticity. A damping ratio of $\xi = 0.05$ is specified for all layers and the half-space.

The first part of the example aims at verifying the accuracy of the proposed approach by comparison with the results available in the literature. Isotropic soil medium is considered. This problem has been solved by Kausel [18] on request of Wolf to check the results obtained by approximate cone model. Kausel employed the thin-layer method. The evaluated frequency-dependent dynamic impedance is normalized in the following form

$$S_g(a_0) = (K_{st})_g \left[k_g(a_0) + i a_0 c_g(a_0) \right] \tag{84}$$

where K_{st} denotes the static-stiffness coefficient of a surface disk on a homogeneous half-space with the material properties identical to the first layer; the subscript g denotes either horizontal, vertical, rocking or torsional motion designated by symbols H, V, R and T respectively; k_g and c_g are the real and imaginary parts of the dynamic impedance coefficients. As shown in Fig. 4 very good agreement between the impedance coefficients predicted by the proposed approach and those predicted by thin-layer method is reached. Only a small deviation occurs in the real part of rocking impedance.

As no real example of anisotropic multi-layered soil can be found in the literature, the second part of the example aims at testing the applicability of the proposed approach dealing with the anisotropic medium, the cross-anisotropic soil media are considered. The same soil strata as the aforementioned case is studied, assuming that the coefficients of anisotropy are the same for all the media and equal to $n = 1/3$, 1 and 2 respectively. $n = E_H / E_V$ is defined as the ratio of horizontal to vertical modulus of elasticity. The evaluated dynamic impedance coefficients (Fig. 5) are compared with each other for different value of n to show the effect of anisotropy on the dynamic response of super-structures.

The proposed approach has been applied to the solution of the problems for 3D foundation-soil-foundation interaction on stratified soil [19]. The application to the dynamic impedance of embedded foundation on multi-layered soil is underway. And the time-domain solution procedure to this problem is presented in the conferenceproceedings [20].

Table 1: Material properties of the layers and half-space

layer	E_H	ρ	ν	h
1	2.5μ	ρ	0.25	$1.0a$
2	1.3μ	ρ	0.30	$0.5a$
3	0.533μ	0.89ρ	1/3	Semi-infinite

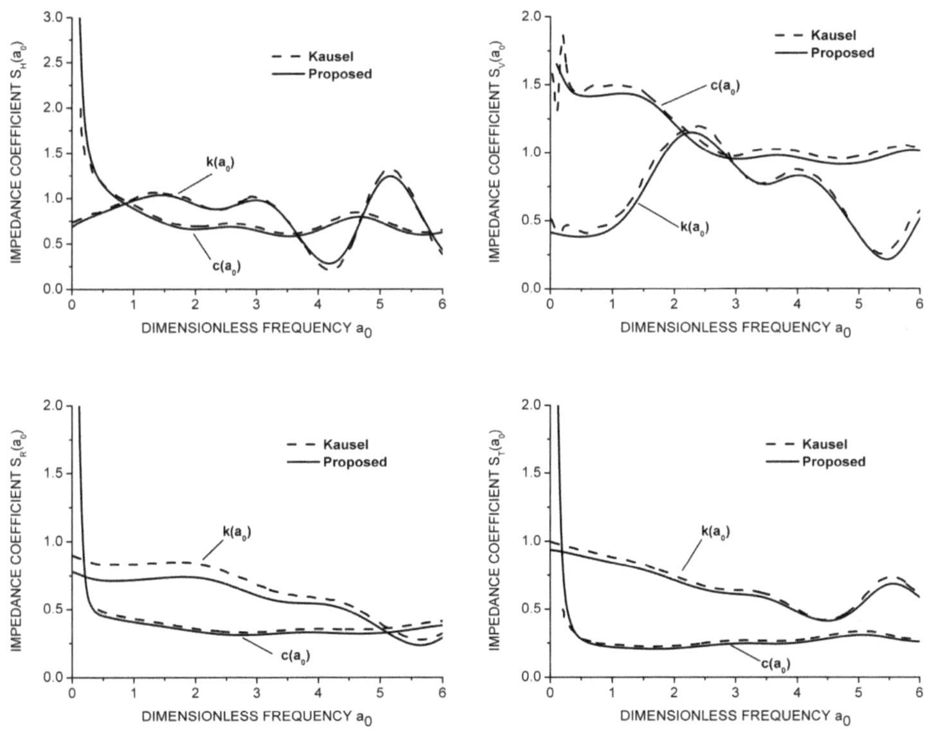

Figure 4: Dynamic impedance coefficients of circular foundation on layered half-space (isotropic media)

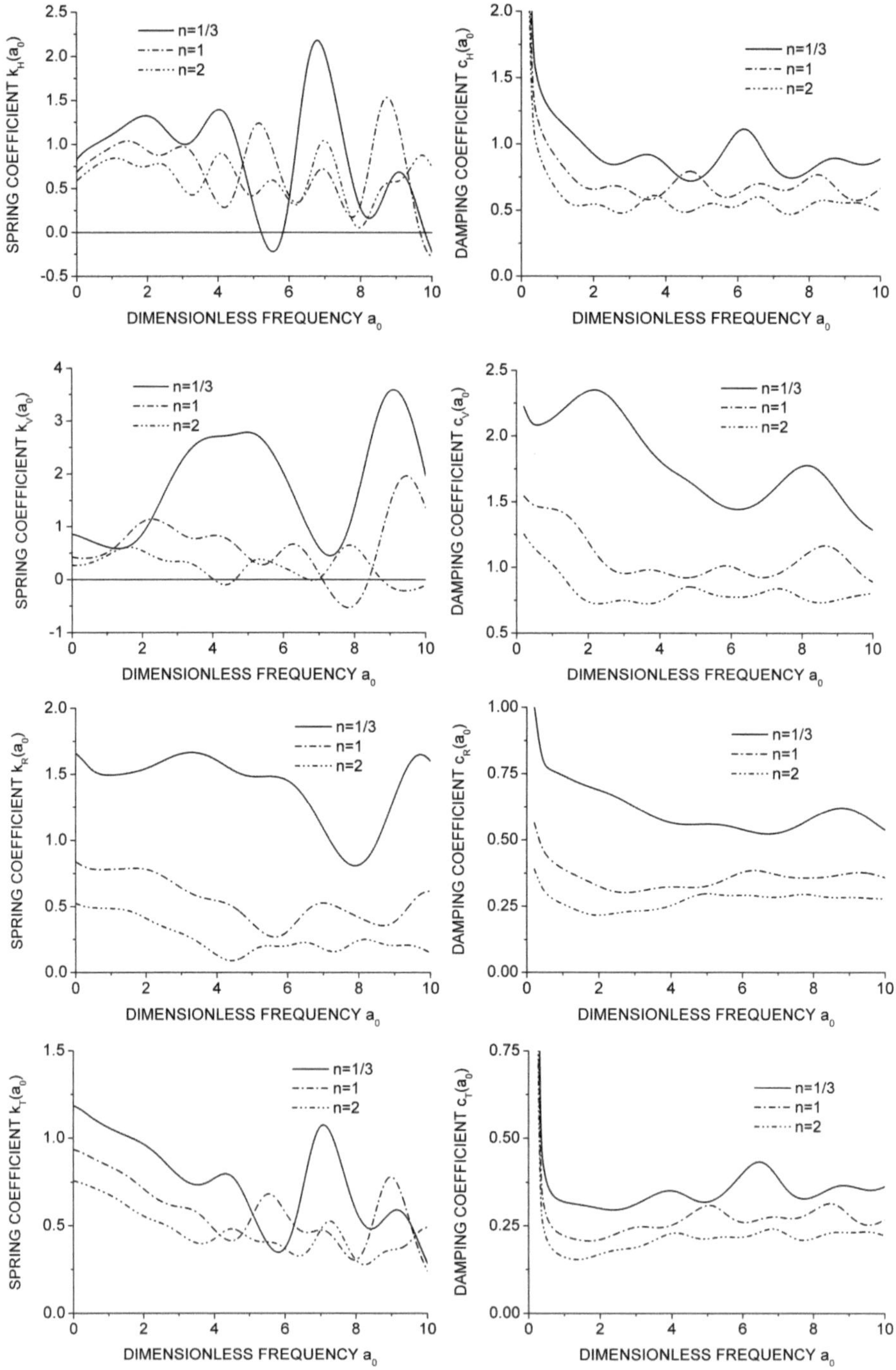

Figure 5: Dynamic impedance coefficients of circular foundation on layered half-space (anisotropic media)

8 Conclusion

The dynamic impedance constitutes one of the key elements in the formulation of SSI problems. An advanced general approach for evaluating dynamic impedance of arbitrary-shaped foundation on multi-layered half-space is developed. The proposed approach finds the general solution of frequency domain Green's influence function for isotropic as well as arbitrary anisotropic layered media, which is exact in the sense that wave propagation in the stratified soil is solved analytically, the resulting displacement field is free from approximations or discretization errors. The precise integration method ensures that the solution is accurate. By applying N times the recursive formula with the size of matrices not greater than (6×6) (for isotropic medium 4×4 and 2×2) it is equivalent to subdividing the layer into 2^N sublayers, any desired accuracy can be achieved. The procedure is efficient, with the aid of dual-vector form of the solution for wave motion equation, the assembly of layers can be carried out quite easily and conveniently, all calculation is based on matrix algebra, with the size of matrices not greater than (3×3) (for isotropic medium 2×2 and 1×1). As a result, the computational effort is reduced to a great extent. The computation is always stable. There is no limit of the number of layers and thickness of the layer to be considered. The efficiency and accuracy of the proposed approach can be validated by the numerical examples.

9 Acknowledgements

The research studies that lead to the material presented in this paper were sponsored by the Sino-German Science Foundation under grant no.GZ566 and the National Natural Science Foundation of China under grant no.51138001. These supports are gratefully acknowledged.

Grateful application is also expressed to the author's current and former students, Doctoral candidate Mr. Zejun Han, Dr. Jianbo Li, Dr. Hong Zhong and Dr. Zhiqiang Hu, whose research provided the basis for most part of the paper presented.

REFERENCES

[1] Luco J.E., Apsel R.J.: On the Green's functions for a layered half-space. Part Ⅰ; Bulletin of the Seismological Society of America; 73(1983):909-929.

[2] Apsel R.J., Luco J.E.: On the Green's functions for a layered half-space. Part Ⅱ; Bulletin of the Seismological Society of America; 73(1983):931-951.

[3] Lysmer J.: Lumped mass method for Rayleigh wave. Bulletin of the Seismological Society of America February 1970; 60(1):89-104.

[4] Waas G.: Linear two-dimensional analysis of soil dynamic problems in semi-infinite layer media, PhD thesis, University of California, Berkeley, 1972.

[5] Kausel E.: and Roesset JM. Semianalytic hyperelement for layered strata. Journal of the Engineering Mechanics Division, 1977; 103(4):569-588.

[6] Kausel E.: Thin-layer method: formulation in the time domain. Int. J. Numer Meth Eng. 1994; 37: 927-941.

[7] Tajimi H. *et al.*: Approximate solution to the vibration of foundation-soil system based on partitioned layers, Lecture Abstracts of Japanese Architecture Society (In Japanese), 1972; 485-486.

[8] Tajimi H. *et al.*: Vibration analysis of soil-structure system based on three-dimensional thin layer-element method, Collected Papers of Japanese Architecture Society, Structural Division (In Japanese), 1976; 243, 41-51.

[9] Thomson W.: Transmission of Elastic Waves through a Layered Medium[J], J. Appl. Phys., 1950, 21:89-93.

[10] Haskell N.A.: The Dispersion of Surface Waves in Multilayered Media[J], Bull. Seis. Soc. America, 1953, 43:17-34.

[11] Chew W.C.: Waves and Fields in Inhomogeneous Media, IEEE Press, New York, 1995.

[12] Kausel E.: Fundamental Solutions in Elastodynamics. A Compendium [M], 2006, Cambridge University Press.

[13] Lin G., Han Z., Li W., Li J.: A precise integration approach for the dynamic-stiffness matrix of strip footings on a layered medium. Chinese Journal of Theoretical and Applied Mechanics, 2012,V(3): 557-567.

[14] Lin G., Han Z., Zhong H. and Li J.: A precise integration approach for dynamic impedance of rigid strip footing on arbitrary anisotropic layered half-space, Soil Dynamics and Earthquake Engineering. 2013; 49:96-108.

[15] Zhong W.X.: On precise integration method, Journal of Computational and Applied Mathematics, 2004; 163(1):59–78.

[16] Gao Lin, Zejun Han, Jianbo Li.: An Efficient Approach for Dynamic Impedance of Surface Footing on Layered Half-space, Soil Dynamics and Earthquake Engineering. Volume 49, June 2013, Pages 39–51.

[17] Lin F. Han Z.: A 3D dynamic impedance of arbitrary-shaped foundation on anisotropic multi-layered half-space, Proc. SeDIF Conference, 2013.

[18] Wolf J.P., Preisig M.: Dynamic stiffness of foundation embedded in layered halfspace based on wave propagation in cones. Earthquake Engineering and Structural Dynamics; 32(2003):1075–1098.

[19] Han Z., Lin G., Li J.: Dynamic 3-D foundation-soil-foundation interaction on stratified soil; Earthquake Engineering and Structural Dynamics (to be published).

[20] Han Z., Lin G., Li J.: A time-domain approach for dynamic response of 3D rigid foundation on multi-layered soil using mixed-variable formulation; Proc. SeDIF Conference, 2013.

SeDIF Conference

International Conference on
Seismic Design of Industrial Facilities
2013, RWTH Aachen University

The Scaled Boundary Finite Element Method for Transient Wave Propagation Problems

Carolin Birk[1], Denghong Chen[2], Chongmin Song[1] and Chengbin Du[3]

[1] School of Civil and Environmental Engineering, University of New South Wales, Sydney NSW 2052, Australia
c.birk@unsw.edu.au; c.song@unsw.edu.au

[2] College of Civil Engineering and Architecture, China Three Gorges University, Yichang 443002 China

[3] Department of Engineering Mechanics, Hohai University, Nanjing 210098 China

ABSTRACT:

A high-order time-domain approach for wave propagation in bounded and unbounded domains is developed based on the scaled boundary finite element method. The dynamic stiffness matrices of bounded and unbounded domains are expressed as continued-fraction expansions. The coefficient matrices of the expansions are determined recursively. This approach leads to accurate results with only about 3 terms per wavelength. A scheme for coupling the proposed high-order time-domain formulation for bounded domains with a high-order transmitting boundary suggested previously is also proposed. In the time-domain, the coupled model corresponds to equations of motion with symmetric, banded and frequency-independent coefficient matrices, which can be solved efficiently using standard time-integration schemes. A numerical example is presented.

Keywords: dynamic soil-structure interaction, wave propagation, scaled boundary finite element method, continued fractions

1 Introduction

The modelling of wave propagation is essential in a dynamic soil-structure interaction analysis. This is associated with two major challenges: the unbounded extent of the soil and fine mesh requirements for high-frequency components. Numerical methods for wave propagation in unbounded domains include absorbing boundaries [1, 2], the boundary element method [3, 4], infinite elements [5], the thin-layer method [6] and perfectly matched layers [7]. For extensive reviews of these methods the reader is referred to References [8, 9]. For wave propagation in bounded domains, spectral elements [10, 11] have been proven to be efficient.

S. Klinkel et al. (Eds.), *Seismic Design of Industrial Facilities*,
DOI 10.1007/978-3-658-02810-7_45, © Springer Fachmedien Wiesbaden 2014

A relatively recent method that combines the advantages of accurately modelling radiation damping and employing spectral element concepts is the scaled boundary finite element method [12]. This semi-analytical technique also excels in modelling singularities and can thus be used to model the propagation of seismic waves in the ground containing faults or discontinuities. The original solution procedure of the scaled boundary finite element method has been developed in the frequency domain [13]. Time-domain solutions have thus been obtained using inverse Fourier transformation and evaluating convolution integrals in early publications.

Recently, efficient direct time-domain formulations of the scaled boundary finite element method have been proposed in References [14, 15] for unbounded and bounded domains, respectively. These are based on continued-fraction solutions of the scaled boundary finite element equation in dynamic stiffness. Although these approaches are conceptually appealing, they have only been applied to problems with a small number of degrees of freedom in References [14, 15]. The extension to large scale problems is challenging, due to potential ill-conditioning of the original continued-fraction algorithms. In Reference [16] an improved, numerically more robust continued-fraction expansion technique has been proposed for unbounded domains by introducing an additional scaling. The improved continued-fraction solution is extended to wave propagation problems in bounded domains in this paper. The coupling of the resulting time-domain model for bounded domains with the transmitting boundary derived in Reference [16] is also addressed. Finally, a robust unified high-order time-domain formulation of the scaled boundary finite element method is established, that can be used for the direct time-domain analysis of complex coupled soil-structure systems containing singularities.

2 Concept of the scaled boundary finite element method

In the scaled boundary finite element method, a so-called scaling centre O is chosen in a zone from which the total boundary, other than the straight surfaces passing through the scaling centre, must be visible (Figures 1(a) and 1(b)). Only the boundary S is discretized. A typical line element to be used in a two-dimensional analysis is shown in Figure 1(c). The scaled boundary transformation (Eq. (1)) relating the Cartesian coordinates $\hat{x}$, $\hat{y}$, $\hat{z}$ to the scaled boundary coordinates ξ, η, ζ is introduced. Here, the symbols $\{x\}$, $\{y\}$, $\{z\}$ and $[N(\eta,\zeta)]$ denote nodal coordinates and shape functions of isoparametric elements, respectively.

$$\begin{aligned}
\{\hat{x}(\xi,\eta,\zeta)\} &= \xi[N(\eta,\zeta)]\{x\} \\
\{\hat{y}(\xi,\eta,\zeta)\} &= \xi[N(\eta,\zeta)]\{y\} \\
\{\hat{z}(\xi,\eta,\zeta)\} &= \xi[N(\eta,\zeta)]\{z\}
\end{aligned} \tag{1}$$

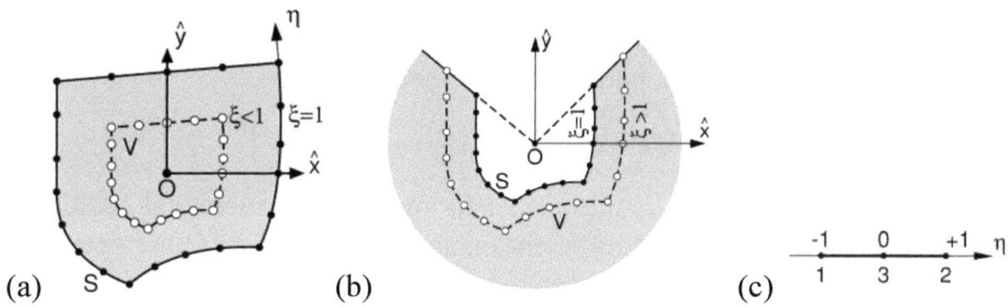

Figure 1: Concept of scaled boundary finite element method: (a) bounded domain, (b) unbounded domain, (c) 3-node line element on boundary

The displacements at a point (ξ, η, ζ) are obtained interpolating nodal displacements $\{u(\xi)\}$ using the same shape functions as for the geometry.

$$\{u(\xi,\eta,\zeta)\} = [N(\eta,\zeta)]\{u(\xi)\} \tag{2}$$

Applying the method of weighted residuals to the governing equations formulated in terms of the scaled boundary coordinates, the scaled boundary finite element equation in displacements $\{u(\xi)\}$ is obtained.

$$[E^0]\xi^2\{u(\xi)\},_{\xi\xi} + \big((s-1)[E^0] - [E^1] + [E^1]^T\big)\xi\{u(\xi)\},_{\xi} + \big((s-2)[E^1]^T - [E^2]\big)\{u(\xi)\} + \omega^2[M^0]\xi^2\{u(\xi)\} = 0 \tag{3}$$

The coefficient matrices $[E^0]$, $[E^1]$, $[E^2]$ and $[M^0]$ are evaluated using standard finite element technologies [13]. The dynamic stiffness $[S(\omega)]$ relates the amplitudes of the nodal forces $\{R(\omega)\}$ to the amplitudes of the nodal displacements $\{u(\omega)\}$ at the boundary.

$$\{R(\omega)\} = [S(\omega)]\{u(\omega)\} \tag{4}$$

Using the relationship between internal nodal forces and nodal displacements, Equation (3) can be transformed into an equivalent differential equation in $[S(\omega)]$, the so-called scaled boundary finite element equation in dynamic stiffness.

$$(\pm[S(\omega)] - [E^1])[E^0]^{-1}(\pm[S(\omega)] - [E^1]^T) \pm (s-2)[S(\omega)] \pm \omega[S(\omega)],_{\omega} - [E^{\{2\}}] + \omega^2[M^0] = 0 \tag{5}$$

Equation (5) is valid for both bounded and unbounded domains, where the upper and lower signs apply in the bounded and unbounded case, respectively.

3 Bounded domains

A high-order time domain formulation for bounded domains can be constructed by expanding the dynamic stiffness $[S^b(\omega)]$ into a series of continued fractions.

3.1 Continued-fraction expansion of dynamic stiffness matrix

The dynamic stiffness at the boundary is expressed as

$$[S^b(\omega)] = [K] - \omega^2[M] - \omega^4[X^{(1)}][S^{(1)}(\omega)]^{-1}[X^{(1)}]^T, \tag{6}$$

where a scaling factor $[X^{(1)}]$ is introduced to improve the numerical condition of the solution. Equations for the coefficient matrices in Equation (6) are obtained by substituting it in Equation (5) and setting individual terms corresponding to powers of ω^2 to zero in ascending order. The constant term yields an equation for the static stiffness matrix $[K]$,

$$([K] - [E^1])[E^0]^{-1}([K] - [E^1]^T) - [E^2] + (s-2)[K] = 0. \tag{7}$$

An equation for the mass matrix $[M]$ is obtained by setting the terms in ω^2 equal to zero.

$$([K] - [E^1])[E^0]^{-1}[M] + [M][E^0]^{-1}([K] - [E^1]^T) + s[M] - [M^0] = 0 \tag{8}$$

The remaining terms yield an equation for the residual $[S^{(i)}(\omega)]$ (with $i = 1$),

$$\begin{aligned}&[S^{(i)}(\omega)][c^{(i)}][S^{(i)}(\omega)] - [S^{(i)}(\omega)]\left[b_0^{(i)}\right]^T - \left[b_0^{(i)}\right][S^{(i)}(\omega)] + \\ &\omega^2\left([S^{(i)}(\omega)]\left[b_1^{(i)}\right]^T + \left[b_1^{(i)}\right][S^{(i)}(\omega)]\right) + \omega[S^{(i)}(\omega)]_{,\omega} + \omega^4[a^{(i)}] = 0,\end{aligned} \tag{9}$$

with the constants

$$\begin{aligned}[a^{(1)}] &= [X^{(1)}]^T[E^0]^{-1}[X^{(1)}], \\ \left[b_0^{(1)}\right] &= [X^{(1)}]^T[E^0]^{-1}([K] - [E^1]^T)[X^{(1)}]^{-T} - (s+2)/2[I], \\ \left[b_1^{(1)}\right] &= [X^{(1)}]^T[E^0]^{-1}[M][X^{(1)}]^{-T}, \\ [c^{(1)}] &= [X^{(1)}]^{-1}[M][E^0]^{-1}[M][X^{(1)}]^{-T}.\end{aligned} \tag{10}$$

The parameter $[X^{(1)}]$ is selected in such a way that $[c^{(1)}]$ is a diagonal matrix with entries +1 or -1.

Similarly, Eq. (9) is solved by postulating

$$[S^{(i)}(\omega)] = \left[S_0^{(i)}\right] - \omega^2\left[S_1^{(i)}\right] - \omega^4[X^{(i+1)}][S^{(i+1)}(\omega)]^{-1}[X^{(i+1)}]^T \tag{11}$$

The solution for $\left[S_0^{(i)}\right]$ is obtained from

$$\left[S_0^{(i)}\right]^{-1}\left[b_0^{(i)}\right] + \left[b_0^{(i)}\right]^T\left[S_0^{(i)}\right]^{-1} = [c^{(i)}]. \tag{12}$$

The solution for $[S_0^{(i)}]$ follows from

$$\left(-[b_0^{(i)}] + [S_0^{(i)}][c^{(i)}]\right)[S_1^{(i)}] + [S_1^{(i)}]\left(-[b_0^{(i)}]^T + [c^{(i)}][S_0^{(i)}]\right) + 2[S_1^{(i)}] = [b_1^{(i)}][S_0^{(i)}] + [S_0^{(i)}][b_1^{(i)}]^T. \tag{13}$$

The equation for $[S^{(i+1)}(\omega)]$ is the same as Eq. (9) with i replacing $i+1$ and the corresponding coefficient matrices

$$\begin{aligned}
[a^{(i+1)}] &= [X^{(i+1)}]^T[c^{(i)}][X^{(i+1)}] \\
[b_0^{(i+1)}] &= [X^{(i+1)}]^T\left(2[I] - [b_0^{(i)}]^T + [c^{(i)}][S_0^{(i)}]\right)[X^{(i+1)}]^{-T} \\
[b_1^{(i+1)}] &= [X^{(i+1)}]^T\left(-[b_1^{(i)}]^T + [c^{(i)}][S_1^{(i)}]\right)[X^{(i+1)}]^{-T} \\
[c^{(i+1)}] &= [X^{(i+1)}]^{-1}\left([a^{(i)}] - [b_1^{(i)}][S_1^{(i)}] - [S_1^{(i)}][b_1^{(i)}]^T \right. \\
&\quad \left. + [S_1^{(i)}][c^{(i)}][S_1^{(i)}]\right)[X^{(i+1)}]^{-T}
\end{aligned} \tag{14}$$

Therefore, Equation (9) can be solved recursively for high-order terms with the coefficient matrices updated by Equation (14). The LDL^T decomposition [17] of the coefficient $[c^{(i)}]$ is used to determine the scaling factor $[X^{(i)}]$. It is chosen as the lower diagonal matrix $[L^{(i)}]$, which can be normalized such that the diagonal entries of $[D^{(i)}] = \pm 1$.

$$[c^{(i)}] = [X^{(i)}]^{-1}[\tilde{c}^{(i)}][X^{(i)}]^{-T}, \quad [\tilde{c}^{(i)}] = [L^{(i)}][D^{(i)}][L^{(i)}]^T \tag{15}$$

3.2 High-order time-domain formulation

Starting from the continued-fraction solutions of the dynamic stiffness matrix, high-order time-domain formulations can be constructed as equations of motion describing bounded domains. Substituting Eq. (6) into Eq. (4), the force-displacement relationship is expressed as

$$\{R(\omega)\} = ([K] - \omega^2[M])\{u(\omega)\} + \omega^2[X^{(1)}]\{u^{(1)}(\omega)\} \tag{16}$$

where the auxiliary variable $\{u^{(1)}(\omega)\}$ is defined as the case $i = 1$ of

$$-\omega^2[X^{(i)}]^T\{u^{(i-1)}(\omega)\} = [S^{(i)}(\omega)]\{u^{(i)}(\omega)\} \tag{17}$$

with $\{u(\omega)\} = \{u^{(0)}(\omega)\}$. Substituting Eq. (11) into Eq. (17) leads to

$$\omega^2\left[X^{(i)}\right]^T\left\{u^{(i-1)}(\omega)\right\} + \left(\left[S_0^{(i)}\right] - \omega^2\left[S_1^{(i)}\right]\right)\left\{u^{(i)}(\omega)\right\} + \omega^2\left[X^{(i+1)}\right]\left\{u^{(i+1)}(\omega)\right\} = 0 \tag{18}$$

Equations (16) and (18) are easily written in the time domain as

$$\begin{aligned} \{R(t)\} &= [K]\{u(t)\} + [M]\{\ddot{u}(t)\} - \left[X^{(1)}\right]\left\{\ddot{u}^{(1)}(t)\right\} \\ 0 &= -\left[X^{(i)}\right]^T\left\{\ddot{u}^{(i-1)}(t)\right\} + \left[S_0^{(i)}\right]\left\{u^{(i)}(t)\right\} + \left[S_1^{(i)}\right]\left\{\ddot{u}^{(i)}(t)\right\} \\ &\quad - \left[X^{(i+1)}\right]\left\{\ddot{u}^{(i+1)}(t)\right\} \end{aligned} \tag{19}$$

An order M_b continued fraction expansion is terminated with the assumption $\left\{u^{(M_b+1)}(t)\right\} = 0$.

4 Unbounded domains

A detailed derivation for the improved continued fraction solution of the dynamic stiffness of an unbounded domain is presented in Reference [16]. It is obtained in the same way as for the bounded domain but at the high frequency limit. The continued fraction solution is postulated as

$$\begin{aligned} \left[S^\infty(\omega)\right] &= i\omega[C_\infty] + [K_\infty] - \left[X_u^{(1)}\right]\left[Y^{(1)}(\omega)\right]^{-1}\left[X_u^{(1)}\right]^T \\ \left[Y^{(i)}(\omega)\right] &= i\omega\left[Y_1^{(i)}\right] + \left[Y_0^{(i)}\right] - \left[X_u^{(i+1)}\right]\left[Y^{(i+1)}(\omega)\right]^{-1}\left[X_u^{(i+1)}\right]^T \end{aligned} \tag{20}$$

Substituting into Eq. (4), the force-displacement relationship is expressed in the time domain as

$$\begin{aligned} \{R(t)\} &= [C_\infty]\{\dot{u}(t)\} + [K_\infty]\{u(t)\} - \left[X_u^{(1)}\right]\left\{v^{(1)}(t)\right\} \\ 0 &= -\left[X_u^{(i)}\right]^T\left\{\dot{v}^{(i-1)}(t)\right\} + \left[Y_1^{(i)}\right]\left\{\dot{v}^{(i)}(t)\right\} + \left[Y_0^{(i)}\right]\left\{v^{(i)}(t)\right\} \\ &\quad - \left[X_u^{(i+1)}\right]\left\{v^{(i+1)}(t)\right\} \end{aligned} \tag{21}$$

where $\{v^{(i)}(t)\}$ are auxiliary variables. An order M_u continued fraction expansion is terminated with the assumption $\left\{u^{(M_u+1)}(t)\right\} = 0$.

5 Coupling of bounded and unbounded domains

The force-displacement relationships (Eqs. (19) and (21)) of the bounded and unbounded domains can be assembled together to formulate the equation of motion of the whole system

$$\{f(t)\} = [M_G]\{\ddot{z}(t)\} + [C_G]\{\dot{z}(t)\} + [K_G]\{z(t)\}. \tag{22}$$

The vector of unknowns $\{z(t)\}$ contains the displacements $\{u(t)\}$ of the coupled soil-structure system, the internal variables $\{u^{(1)}\}$ to $\{u^{(M_b)}\}$ corresponding to the bounded domain and the internal variables $\{v^{(1)}\}$ to $\{v^{(M_u)}\}$ of the unbounded domain. The vector $\{f(t)\}$ contains all external forces acting on the coupled soil-structure system. The high-order mass, damping and stiffness matrices $[M_G]$, $[C_G]$ and $[K_G]$ are banded, symmetric and sparse. Equation (22) can be solved using standard time-integration methods.

6 Numerical example

The coupled soil-structure interaction problem shown in Figure 2 is analysed. It consists of an elastic block of width $2b$ and height h, with $2b/h = 2/3$, resting on a homogeneous soil halfspace with shear modulus G_1, mass density ρ_1 and Poisson's ratio $\nu_1 = 0.25$. The shear modulus, mass density and Poisson's ratio of the elastic block are: $G_2 = 9G_1$, $\rho_2 = \rho_1$ and $\nu_2 = 0.25$. Plain strain is assumed.

A uniformly distributed strip load $P(t)$ is acting on the top surface of the block. It's time-dependence and the corresponding Fourier transform are shown in Figure 3. Here, the dimensionless frequency is defined as $a_0 = \omega b/c_{s,1}$ with $c_{s,1}^2 = G_1/\rho_1$.

In the scaled boundary finite element model, the elastic block and a semi-circular near-field portion of the soil of radius b are modelled as two subdomains and discretized with eight nine-node high-order elements. The scaling centre of the unbounded domain is located at the point O shown in Figure 2.

The bounded domain is modelled using the high-order time-domain formulation proposed in Section 3.1. Considering the requirement of 6 nodes per wavelength, the discretization represents $\lambda = 4/3b$. This wavelength corresponds to a maximum dimensionless frequency $a_0 = 14.1$. In the radial direction, 3 to 4 continued- fraction terms per wavelength are required [15]. The order of continued-fraction expansion is thus chosen as $M_b = 3$. The high-order transmitting boundary

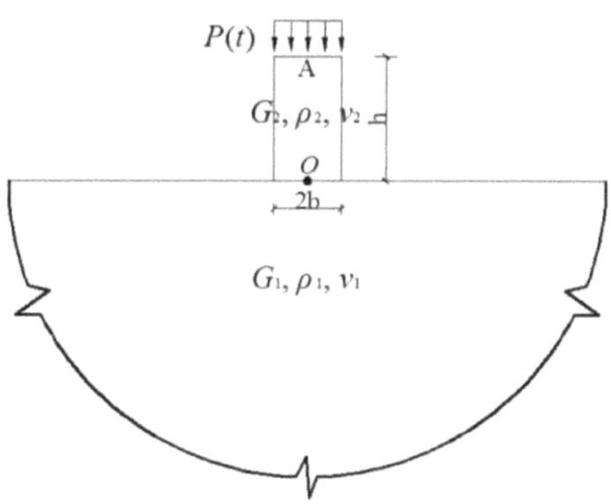

Figure 2: Elastic block resting on homogeneous halfspace

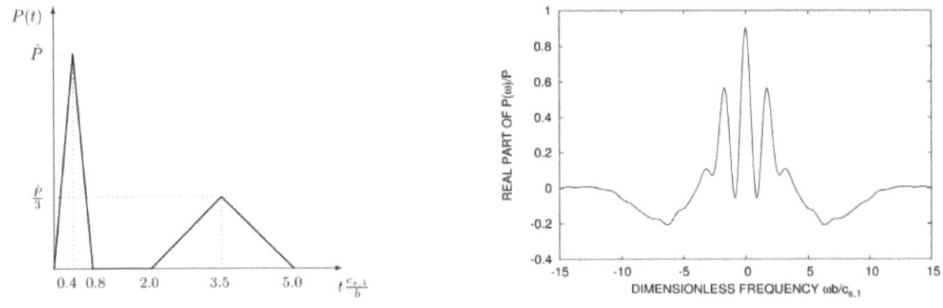

(a) (b)

Figure 3: Uniformly distributed load: (a) time domain, (b) frequency domain

summarized in Section 4 is used to model the far field with $M_u = 9$ and $M_u = 15$. The dimensionless vertical displacements at points A and O (see Figure 2) obtained by solving the coupled Equation (22) with Newmark's method are shown in Figure 4. The time step is $\Delta t = 0.02b/c_{s,1}$.

To verify the proposed method, an extended mesh with a rectangular area of $21b \times 20b$ to the right of the plane of symmetry is analysed using the finite element method (ABAQUS/Standard [18]). Half of the symmetric system is discretized with 6768 eight-node elements of size $0.25b \times 0.25b$, yielding 20657 nodes.

For comparison, a viscous-spring boundary [19] combined with a finite element model of size $8b \times 3b$ is also employed. It consists of parallel connected spring-dashpot systems in the normal and tangential directions, with normal and tangential spring and damping coefficients K_{BN}, C_{BN} and K_{BT}, C_{BT}, respectively.

$$K_{BN} = A\frac{G}{r_b}, \quad C_{BN} = A\rho c_p, \quad K_{BT} = A\frac{G}{2r_b}, \quad C_{BT} = A\rho c_s \tag{23}$$

In Equation (23), the symbols A and r_b denote the total area of all elements around a node at the boundary and the distance from the scattering wave source to the artificial boundary point. Here, r_b is taken as $3b$. The finite element region is discretized with 480 eight-node elements of size $0.25b \times 0.25b$, yielding 1553 nodes.

In Figure 4, the vertical displacements computed using the present coupled method and the viscous-spring boundary agree very well with the extended mesh solution for early times up to $\bar{t} = 5$. After that, the results obtained using the viscous-spring boundary differ considerably from the reference results. On the other hand, the vertical displacements determined using the proposed method agree very well with the reference solution up to $\bar{t} = 10$. The extent of the slight deviations occurring after that depends on the order of continued fraction expansion used in the unbounded domain. The displacements calculated using the present technique

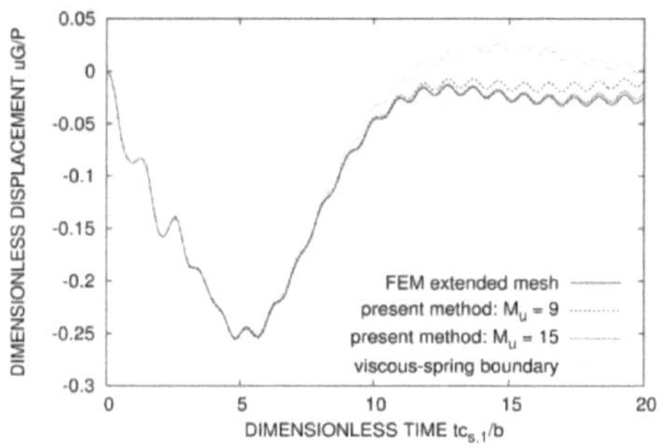

(a)

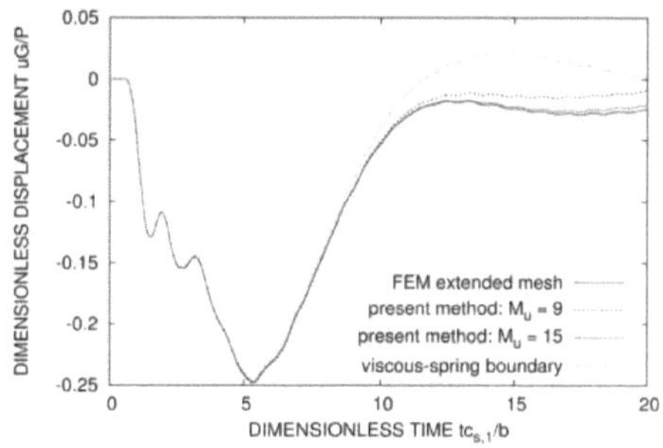

(b)

Figure 4: Dimensionless vertical displacements of elastic block on homogeneous halfspace: (a) point A; (b) point O

converge to the extended mesh results with increasing order of continued fraction M_u. In the given example, excellent agreement is obtained using $M_u = 15$.

7 Conclusion

High-order time-domain formulations for modelling wave propagation in bounded and unbounded domains of arbitrary geometry have been developed. A standard equation of motion of a linear system in the time domain is obtained, which can be solved using established time-stepping schemes, such as Newmark's method. Only the boundaries of the bounded and unbounded domains are discretized, leading to reduced numerical effort. The numerical results demonstrate the accuracy of the proposed coupled method. The approach presented in this paper can easily be extended to three-dimensional problems and applied to investigate influence of faults and other geological discontinuities on structural responses.

References

[1] Givoli, D.: High-order local non-reflecting boundary conditions: a review; Wave Motion; 39 (2004), Pages 319-326

[2] Kausel, E.: Local transmitting boundaries; Journal of Engineering Mechanics; 114 (1988), Pages 1011-1027

[3] Beskos, D.E.: Boundary element methods in dynamic analysis; Applied Mechanics Reviews; 40 (1987), Pages 1-23

[4] Beskos, D.E.: Boundary element methods in dynamic analysis: Part II (1986-1996); Applied Mechanics Reviews; 50 (1997), Pages 149-197

[5] Bettess, P.: Infinite Elements, Penshaw Press, Sunderland, 1992

[6] Kausel, E.: Thin-layer method: formulation in the time domain; International Journal for Numerical Methods in Engineering; 37 (1994), Pages 927-941

[7] Basu U.; Chopra, A.K.: Perfectly matched layers for transient elastodynamics of unbounded domains; International Journal for Numerical Methods in Engineering; 59 (2004), Pages 1039-1074

[8] Tsynkov, S.V.: Numerical solution of problems on unbounded domains. A review; Applied Numerical Mathematics; 27 (1998), Pages 465-352

[9] Lou, M.L.; Wang, H.F.; Chen X.; Zhai, Y.M.: Soil-structure interaction: Literature review; Soil Dynamics and Earthquake Engineering; 31 (2011), Pages 1724-1732

[10] Komatitsch, D.: Spectral-element simulations of global seismic wave propagation – I. Validation; Geophysical Journal International; 149 (2002), Pages 390-412

[11] Mehdizadeh O.; Paraschivoiu, M.: Investigation of two-dimensional spectral element method for Helmholtz's equation; Journal of Computational Physics; 189 (2003), Pages 111-129

[12] Wolf, J.P.: The scaled boundary finite element method, Wiley & Sons, Chichester, 2003

[13] Song, C.; Wolf, J.P.: The scaled boundary finite element method – alias consistent infinitesimal finite-element cell method – for elastodynamics; Computer Methods in Applied Mechanics and Engineering; 147 (1997), Pages 329-355

[14] Bazyar, M.; Song, C.: A continued-fraction based high-order transmitting boundary for wave propagation in unbounded domains of arbitrary geometry; International Journal for Numerical Methods in Engineering; 74 (2008), Pages 209-237

[15] Song, C.: The scaled boundary finite element method in structural dynamics; International Journal for Numerical Methods in Engineering; 77 (2009), Pages 1139-1171

[16] Birk, C.; Prempramote, S.; Song, C.: An improved continued-fraction-based high-order transmitting boundary for time-domain analyses in unbounded domains; International Journal for Numerical Methods in Engineering; 89 (2012), Pages 269-298

[17] Golub, G.H.; Van Loan, C.F.: Matrix Computations. North Oxford Academic, Oxford, 1983

[18] ABAQUS Inc.: ABAQUS Theory Manual, Version 6.10, Providence, 2010

[19] Deeks, A.J.; Randolph, M.F.: Axisymmetric time-domain transmitting boundaries ; Journal of Engineering Mechanics; 120 (1994), Pages 25-42

Attenuation of Ground-borne Vibrations Induced by Underground Dynamic Excitation

Tong Jiang[1]

[1] Research Institute of Structural Engineering and Disaster Reduction, Tongji University
1239, Siping Road, Shanghai, China
jt@tongji.edu.cn

ABSTRACT:

A two-dimensional model based on thin layer method (TLM) has been used to analyze the attenuation of ground-borne vibration induced by the subway in Shanghai. The subway's tunnel was simulated by the finite element method (FEM), and the nodes of FEM are coincident with the layer division of TLM. The frequency response functions on ground surface under action of the acceleration at ballast bed near rail track were calculated. By using the Fourier transform, the vibration acceleration and the vibration level (VL) induced by the subway on ground surface with deferent distance away from subway central line was analyzed. Comparison between the calculated and the measured VL at ground surface in Shanghai showed good agreement. Then the VL on ground surface induced by the subway in Shanghai with deferent distances and with deferent tunnel depths has been calculated. Finally the empirical attenuation equation of ground-borne vibration induced by subway in Shanghai has been proposed.

Keywords: Thin layer method, vibration level, subway tunnel, in-situ measurement, frequency response functions

1 Introduction

Cause of underground dynamic excitations generally can be classified into two categories: geological or environmental activities and human activities. Earthquake is a typical example of the first categories, while excitations caused by machine and traffic belong to the later. Vibrations generated by those activities transmit to nearby buildings in form of waves, causing different kinds of influence on human life and work in the vicinity. In this paper, underground dynamic excitations are confined to vibration induced by subway transit. With the rapid development of the subway transit system in Shanghai, China, the environmental vibration induced by subway becomes a significant problem of great concerns.

S. Klinkel et al. (Eds.), *Seismic Design of Industrial Facilities*,
DOI 10.1007/978-3-658-02810-7_46,

Thin-layer method (TLM) is a semi-analytical and semi-numerical approach for the analysis of elastic wave propagation in stratified media. The TLM has been widely used in the fields of vibration analysis of stratified soils [1], [2], [3]. In the paper, the TLM and its application to calculate the environmental vibration induced by subway transit were studied. A two-dimensional model based on TLM was proposed in the paper to analyze the ground-borne vibration induced by subway. The analytical model is formed by a tunnel embedded in stratified soils as shown in figure 1. The frequency displacement response functions of the stratified soils under action of unit displacement on ballast bed were calculated. The Fourier spectrum of the in-situ measured acceleration time history on ballast bed as the vibration source can be obtained by using the FFT. Based on the Fourier spectrum of the source excitation and the frequency displacement response functions, the environment vibration on the ground surface excited by the subway was analyzed by using the reverse FFT. Comparison between calculated vibration level (VL) and the measured results of the ground surface on the line perpendicular to subway line showed fine agreement [4]. After the adaptation of the TLM was validated, it was used to calculate the ground surface response induced by subway in order to obtain the empirical prediction equation of vibration attenuation. The underground tunnel was assumed to be embedded in a typical horizontal layered subsoil profile in Shanghai urban area. Based on statistical analysis of the results obtained from the numerical model, an empirical prediction equation of the attenuation with distance induced by subway in Shanghai was proposed.

2 Introduction of analysis method

The governing equations of dynamic soil-foundation interaction were derived by flexible volume method and fundamental displacement solutions of TLM. The tunnel-soil system is shown in figure 1.

The displacement response of the $\{u_T\}$ tunnel under action of the exciting forces $\{F_T^F\}$ can be indicated as

$$\left[S(i\omega)\right]\{u_F\}=\{F_F^F\} \tag{1}$$

Where $\left[S(i\omega)\right]$ is the impedance function matrix of the tunnel-soil system and can be indicated as

$$\left[S(i\omega)\right]=\left[K_S\right]-\left[K_S^G\right]+\left[A(i\omega)\right]^{-1}-\omega^2\left(\left[M_S\right]-\left[M_S^G\right]\right) \tag{2}$$

Where $\left[A(i\omega)\right]$ is the dynamic flexible matrix obtained by TLM [1]; $[K_S],\left[K_S^G\right]$ are the stiffness matrices of the tunnel and they excluded soil respectively; $[M_T],\left[M_S^G\right]$ are the mass matrices of the tunnel and they excluded soil respectively.

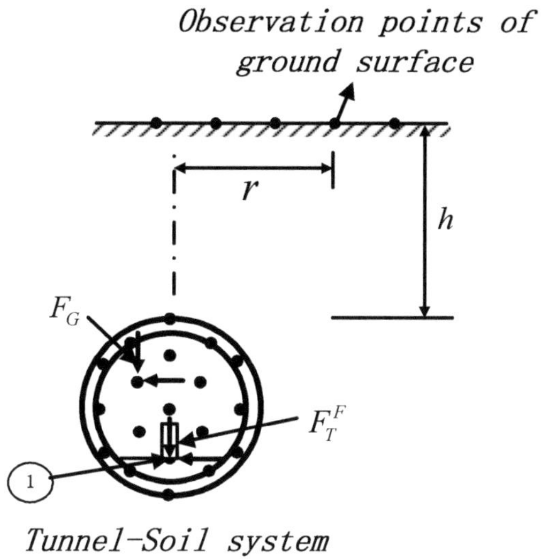

Figure 1: Analytical model

We can obtain the relationship of the force-displacement of the tunnel-soil system as

$$\begin{Bmatrix} \{F_1\} \\ \{0\} \end{Bmatrix} = \begin{bmatrix} [\bar{S}_{11}] & [\bar{S}_{12}] \\ [\bar{S}_{21}] & [\bar{S}_{22}] \end{bmatrix} \begin{Bmatrix} \{u_1\} \\ \{\bar{u}_2\} \end{Bmatrix} \tag{3}$$

Where $\{u_1\}$ is the displacement of the exciting point node 1 on the ballast bed; $\{\bar{u}_2\}$ are the displacements of the rest nodes.

For the problem of environmental vibration induced by subway transit, there is only the train exciting force $\{F_1\}$ on ballast bed, from equation (3) we can obtain:

$$\{u_1\} = \left[\bar{R}(i\omega)\right]^{-1}\{F_1\} \tag{4}$$

$$\{\bar{u}_2\} = -\left[\bar{S}_{22}\right]^{-1}\left[\bar{S}_{21}\right]\{u_1\} \tag{5}$$

Where, $\left[\bar{R}(i\omega)\right] = \left[\bar{S}_{11}\right] - \left[\bar{S}_{12}\right]\left[\bar{S}_{22}\right]^{-1}\left[\bar{S}_{21}\right]$

The dynamic response exciting point node 1 on ballast bed $\{u_1\}$ induced by subway transit can be measured easily, so based on equation (5) we can calculate the response of the rest nodes. Then the interaction forces of all nodes can be obtained as

$$\{F_G\} = [A(i\omega)]^{-1} \begin{Bmatrix} \{u_1\} \\ -\left[\bar{S}_{22}\right]^{-1}\left[\bar{S}_{21}\right]\{u_1\} \end{Bmatrix} \tag{6}$$

Based on the interaction forces $\{F_G\}$, we can calculate the response of any position including the ground surface of soil by using the fundamental solutions of the TLM.

In addition, based on ISO 2631-1[5] and Chinese Standard GB10070-88[6], the vibration acceleration level (VL) is defined by the acceleration root mean square(r.m.s) along z direction whose formula is specified as

$$VL = 20\lg\left(\frac{a_{rms}}{a_0}\right) \text{ (dB)} \tag{7}$$

where: a_{rms} is the frequency-weighted acceleration(r.m.s); a_0 is a basic acceleration $a_0=10^{-6}\text{m/s}^2$.

3 VL attenuation on ground surface induced by subway in Shanghai

In order to investigate the influence of tunnel embedment depth on VL of the ground surface, the tunnel was assumed to be embedded in different depths of the typical subsoil profile in Shanghai. Numerical simulation was carried out here using TLM and the free field responses as well as its attenuation were calculated.

3.1 Description of typical horizontal layered subsoil profile in Shanghai urban area

The typical horizontal layered subsoil profile used in the numerical calculation was specified in Table 1.

Six values of tunnel embedment depth h were considered, i.e. 6.5m, 8.5m, 10.5m, 12.5m, 13.5m and 16.5m. The analytical model depth was extended to 100m(small model) and 230m(large model), respectively, and the computed VL values of the ground surface were nearly the same. So, the depth of the TLM model was taken 100m upon the transfer boundary. The soil behavior was assumed visco-elastic, with the material properties reported in Table 1 and a constant damping ratio 0.05.

Six groups of acceleration records, measured at different sites of subway lines with different fasteners and tracks in Shanghai, were employed as inputs exciting on the ballast bed. Those records were named as 2-S, 9-F, 8-Y, 8-Y2, 8-Q and 8-K for each group, respectively. For a certain group, take 8-K as an example, the VLs of its samples fluctuates lightly and their mean value is used for analysis, as shown in figure 2. However, the VLs among those six groups vary significantly, with a range between 64dB and 85dB, as listed in Table 2.

Table 1: A typical subsoil profile of Shanghai urban area

No.	Thickness of layer (m)	Depth (m)	Soil type	Density (g/cm^3)	Shear wave velocity (m/s)	Poisson ratio
1	1	1	fill	1.89	74	0.35
2	3.3	4.3	silty clay	1.85	89	0.3
3	2.1	6.4	silty clay	1.85	85	0.3
4	9.66	16.06	silty clay	1.79	108	0.3
5	1.6	17.66	silty clay	1.87	111	0.3
6	26.4	44.06	silty clay	1.82	220	0.25
7	2	46.06	silty clay	1.93	189	0.25
8	2	48.06	silty clay	1.94	191	0.25
9	3.4	51.46	clay	2.04	195	0.25
10	8.05	59.51	silty sand	1.92	230	0.25
11	16.96	76.47	sandy silt	1.95	220	0.25
12	2.2	78.67	fine sand	1.92	263	0.25
13	3.78	82.45	gravelly sand	1.96	267	0.25
14	2.99	85.44	fine sand	1.92	272	0.25
15	6.65	92.09	gravel sand	1.94	279	0.25
16	4.36	96.45	fine sand	1.93	287	0.25

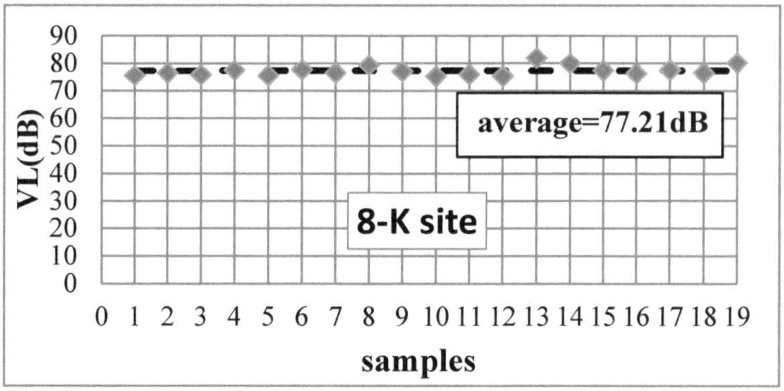

Figure 2: An example of in-situ measured VL on ballast bed

Table 2: In-situ measured VL on ballast bed for six sites (dB)

Site	2-S	8-Y	9-F	8-Q	8-K	8-Y2
Average VL(dB)	79.57	84.64	63.56	75.91	77.21	72.90

3.2 Attenuation results with different embedment depths of the tunnel

Figure 3 shows the computed VL values and VL ratios (the ratios of VL calculated at ground surface to the input VL on ballast bed) attenuating with distances from tunnel axis. It is worth to be noticed from the figure:

- The VL values at ground surface which were calculated using 6 groups of input excitations, vary evidently from each other. Basically, the VL values increase for increasing VL of input on ballast bed, particularly when the distance is short. This trend is diminished as the distance increases.
- It seems that the VL of input on ballast bed has a smaller influence on VL ratios or normalized VL, especially for distance less than 30m.Variability among calculated VL values gets bigger for increasing distance.
- The VL values decrease for increasing values of tunnel embedment depths.

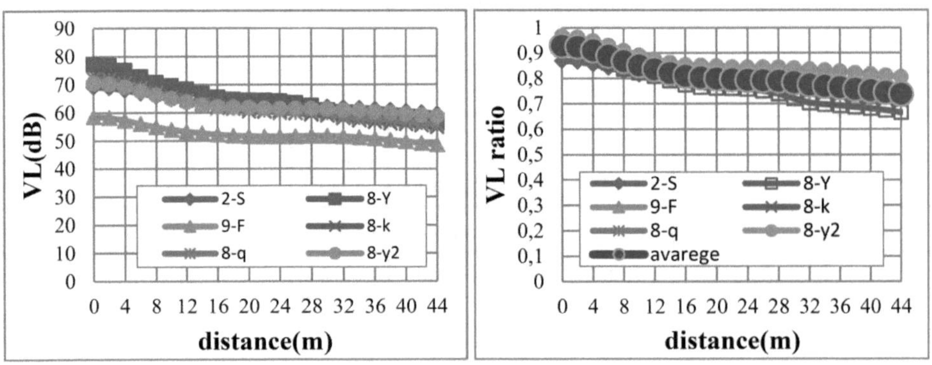

Figure 3: Attenuation of VL and VL ratio for tunnel embedment depth 10.5m

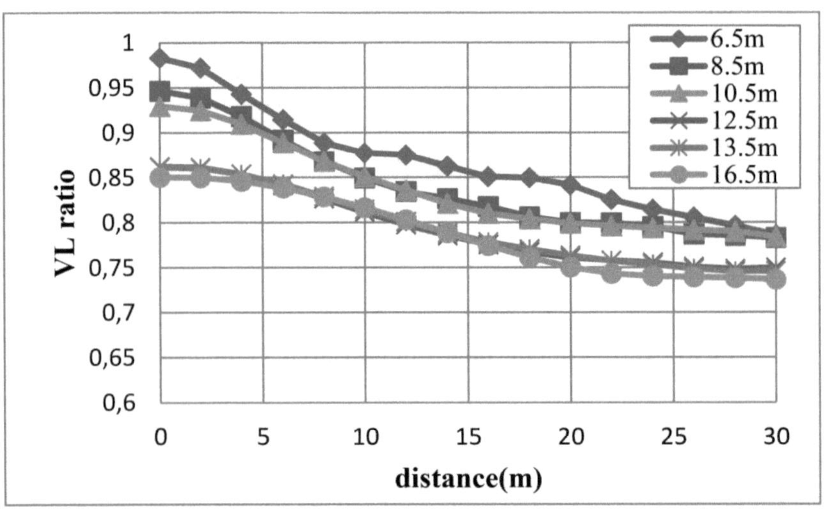

Figure 4: Attenuation of VL ratios for several tunnel embedment depths

Figure 4 plots results provided by numerical simulations, expressed in terms of mean values of normalized VL ratio with different tunnel embedment depth and distances. The attenuation of ground-borne vibration induced by subway is now not only affected by distance, but also by tunnel embedment depth.

3.3 Feasibility analysis on development of empirical prediction equation for vibration attenuation induced by subway

As far as empirical prediction equation for vibration attenuation induced by subway traffic in Shanghai is concerned, the following problems should be taken into consideration.

① It has to be highlighted that the VL input on ballast bed seem to dominate VL at ground surface, as presented in the former section and figure 3. In our case, VLs of ballast bed vary from 65dB to 85dB because the types of track and fasteners are different. It is impossible to develop an empirical prediction equation to calculate the absolute VL induced by subway traffic. However, attempts may be made to develop predictive models when VL of ballast bed is given. Namely, empirical prediction equation can be developed to estimate VL at ground surface for certain type track.

② VL attenuation was calculated with four kinds of subsoil profiles(SL1(as shown in Table 2), TL2, TL3 and TL4). The tunnel was embedded in those profiles at a depth of 13.5m and the acceleration records on the ballast bed of 2-S was selected as input excitations. Figure 5 shows the numerical simulation results using the above parameters and the comparison among those four kinds of subsoil profiles.

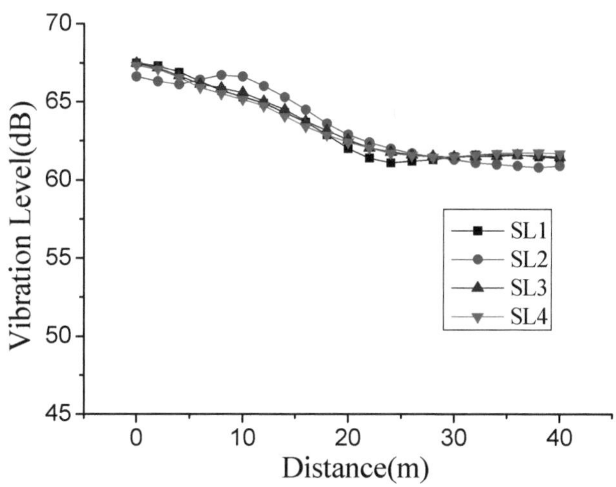

Figure 5: Attenuation of VL for 4 kinds of soil profiles in Shanghai

As can been seen from figure 5, the vibration attenuation calculated from different kinds of subsoil profiles in Shanghai differs by a maximum of 1.5dB. The closeness and similarity of the results indicate that a uniform subsoil profile can be used for numerical simulation of ground-borne vibration attenuation in Shanghai.

4 Empirical prediction method for VL attenuation in Shanghai

According to vibration propagation equation given by Bonitz[7],

$$U_r = U_0 e^{-\alpha(r-r_0)} \left(r / r_0\right)^{-n} \tag{8}$$

where, U_r is effective value of acceleration at distance of r; U_0 is effective value of acceleration at reference point; α is material attenuation coefficient; n is scattering attenuation coefficient. These Coefficients need to be determined according to measured data.

Take the log of both sides of equation (8) and referring to equation (7), we can obtain:

$$VL_r - VL_0 = -20n\lg(r / r_0) - 8.68\alpha(r - r_0) + C \tag{9}$$

Here VL_0 is take as the VL on ballast bed. Normalized with VL_0 we can obtain the VL ratio as,

$$f(r) = VL_r / VL_0 = k_0 + k_1 r^{k_2} + k_3 \lg(r+1) \tag{10}$$

Where, $r+1$ is used instead of r to ensure the expression is meaningful for $r = 0$.

4.1 Regression analysis on the empirical prediction equation

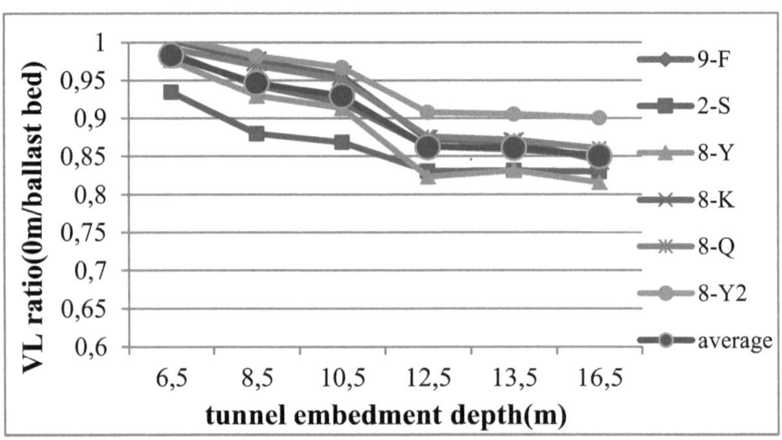

Figure 6: VL ratio (0m/ballast bed) for several tunnel embedment depths

In order to obtain the prediction equation for VL ratio as a function of distance and tunnel embedment depth, the aim of the next step is to determine parameters in the equation (10). If the observation point on top of the tunnel is considered, that means $r = 0$, only one term k_0 is unknown in right side of equation (10). Figure 6 gives plots of calculated VL ratios for those six kinds of ballast bed input excitations, respectively.

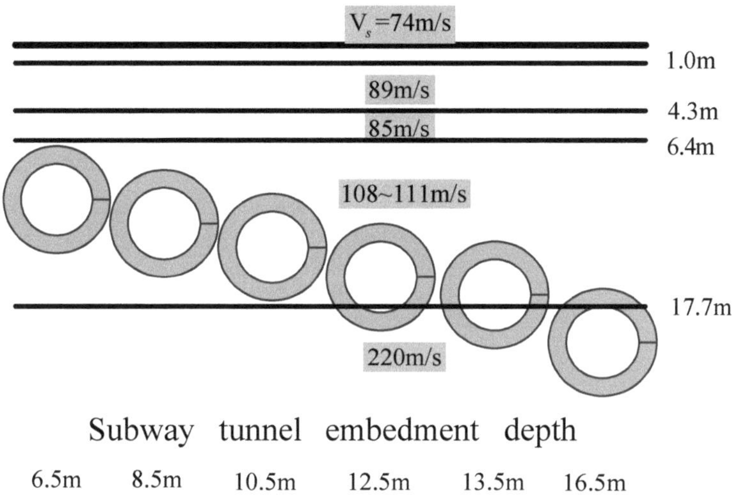

Figure 7: Relation between the tunnel embedment depth and V_s of soil layer

As the tunnel embedment depth is increased, the VL ratio at $r = 0$ drops, as shown in figure 6. However, there is a turning point at depth about 12.5m. Further investigation and discussion is carried out to determine the mechanism background for this characteristics. Figure 7 sketches out the tunnel-in-soil model with deferent tunnel embedment depths from 6.5m to 16.5m. When tunnel embedment depth increases from 10.5m to 12.5m, the bottom of the tunnel embeded into a harder subsoil layer from a softer one, leading to a significant change in tunnel vibration and wave propagation. As a consequence, the VL ratio on the ground surface changed when tunnel embedment depth is about 12.5m.

Linear least-squares fitting procedure was used and k_0 was obtained,

$$\begin{aligned} k_0 &= -0.01901h + 1.111 \qquad && 6.5 \le h \le 12.5 \\ k_0 &= -0.005665h + 0.9442 \qquad && 12.5 \le h \le 16.5 \end{aligned} \tag{11}$$

The mean value of figure 6 and the fitting line are shown in figure 8(a).Numerical results in figure 4 were then used as data for regression analysis. Parameters of k_1, k_2 and k_3 in equation (10) were determined and listed in Table 3.

Table 3: Regression analysis results of k_1, k_2 and k_3

Tunnel depth	k_1	k_2	k_3
6.5m	-0.01069	0.7774	-0.01331
8.5m	-0.01228	0.713555	0.002999
10.5m	-0.01663	0.709174	0.01930
12.5m	-0.01616	0.693446	0.03561
13.5m	-0.01752	0.686267	0.03804
16.5m	-0.02496	0.667148	0.04533

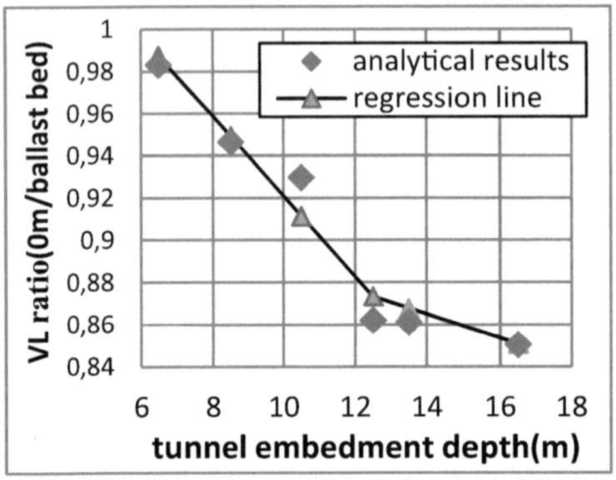

(a)VL ratio at r=0 (k_0)

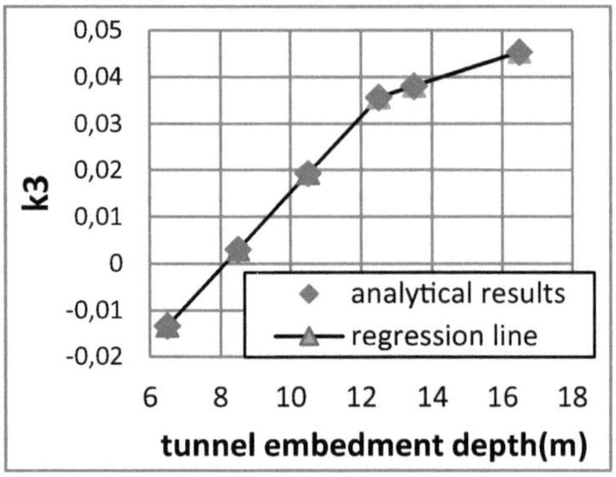

(b) k_3 vs. tunnel depths

Figure 8: Regression of k_0 and k_3

Notice that changes of depths have little effect on value of k_1 and k_2, the mean value in Table 3 was adopted. Substituting $k_1 = -0.01637$ and $k_2 = 0.7078$ into equation (10) yields,

$$f(r) = k_0 - 0.01637 r^{0.7078} + k_3 \lg(r+1) \tag{12}$$

Lastly, k_3 was obtained using linear least-squares fitting procedure, as shown in figure 8(b):

$$\begin{aligned} k_3 &= 8.153\times10^{-3} h - 0.06630 && 6.5 \le h \le 12.5 \\ k_3 &= 2.430\times10^{-3} h + 5.241\times10^{-3} && 12.5 \le h \le 16.5 \end{aligned} \tag{13}$$

The empirical prediction equation was summarized as follows.

$$\begin{aligned} VL_r &= VL_{ballast} \times f(r) \\ f(r) &= k_0 - 0.01637 r^{0.7078} + k_3 \lg(r+1) \end{aligned} \tag{14}$$

Where, k_0 and k_3 are shown in equetions (11) and (13), they have similar features with a broken line with a turning point at depth 12.5m, as shown in figure 8.

4.2 Application and validation of the prediction equation

Finally, figure 9 shows the predicted and measured VL of 2-S and 8-Y2 site. The predicted value of VL, red line in the figure, is similar to the mean values of measured ones.

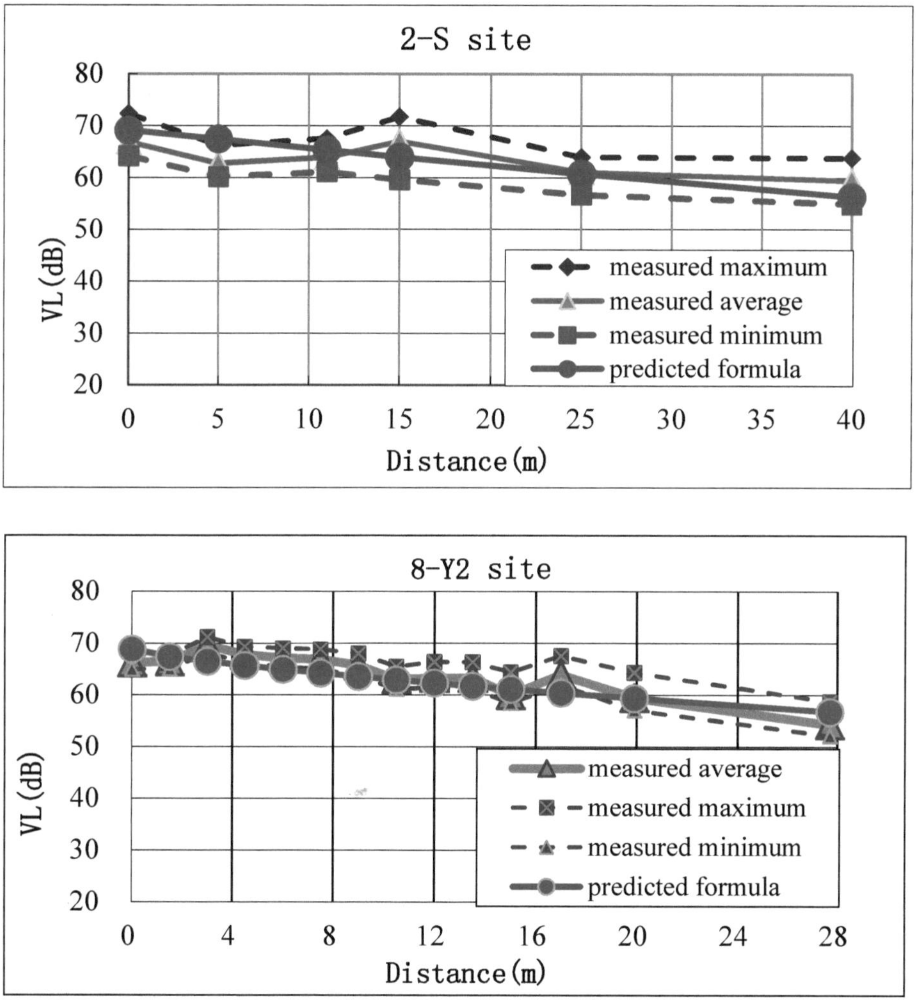

Figure 9: Comparison of predicted and measured VL at 2-S and 8-Y2 sites

5 Conclusion

(1) The TLM can be used to analyze the environmental vibration induced by subway with good agreement of the average of in-situ measured VL and the average of analyzed results.

(2) Six groups of acceleration records, measured at different sites of subway lines with different fasteners and slab tracks in Shanghai, were employed as inputs acting on the ballast bed of the tunnel. Attenuation characteristics of ground vibration were calculated using a typical horizontal layered subsoil profile in Shanghai urban area.

(3) By regression analysis of numerical simulation data obtained by TLM, this paper proposed an empirical prediction equation for environmental vibration attenuation induced by subway in Shanghai.

6 Acknowledgements

This research was supported by Science and Technology Commission of Shanghai Municipality (Grant No.09231201400).

REFERENCES

[1] Jiang, T.; Tajimi, H.: The soil-structure interaction analysis method, Tongji University Press, Shanghai, 2009 (In Chinese).

[2] Jiang, T.; Cheng, C. S.: Environmental vibration induced by elevated railway traffic using thin-layer method, Journal of Vibration Engineering, 20 (2007), Pages 623-627, (In Chinese)

[3] Jiang, T. ; Song, X.X.: Analysis of impedance function of strip foundation embedded in stratified soils by using thin layer method. Chinese Quarterly of Mechanics, 20 (2009), Pages 62-70, (In Chinese)

[4] Jiang, T.; Yue, J.Y.: Analysis of environmental vibration induced by subway in Shanghai and Chengdu by using Thin Layer method; in: Proceedings of 5th International Symposium on Environmental Vibration(ISEV2011), 2011, Pages 113-119.

[5] International Standard Organization: Mechanical Vibration and Shock-Evaluation of Human Exposure to Whole-body Vibration, 2631-1,1997

[6] GB 10070-1988, Standard of vibration in urban area environment, 1988, (In Chinese)

[7] Bornitz, G.: Über die Ausbreitung der von Großkolbenmaschinen erzeugten Bodenschwingungen in die Tiefe, Springer, Berlin,1931

Boundary Effects on Seismic Analysis of Multi-Storey Frames Considering Soil Structure Interaction Phenomenon

Kemal Edip[1], Mihail Garevski[2], Christoph Butenweg[3], Vlatko Sesov[4], Julijana Bojadjieva[5] and Igor Gjorgjiev[6]

[1-6] Institute of Earthquake Engineering and Engineering Seismology
Salvador Aljende 73,Skopje,Macedonia
kemal@pluto.iziis.ukim.edu.mk

[3] Lehrstuhl für Baustatik und Baudynamik,RWTH Aachen
Mies-van-der-Rohe-Str. 1, D-52074 Aachen, Germany
butenweg@lbb.rwth-aachen.de

ABSTRACT:

In conventional modelling of frame structures, the soil medium is usually taken into account as a wide region in order to minimize the reflections of the propagating waves in far field. Fixed conditions at side boundaries lead to enlargement of internal forces of structural seismic response. The sub-soil conditions in this study are represented by 30m soil deposits with four layers which rest on the bedrock. The soil medium is altered as soft, medium and dense soil profile. Side boundaries at the finite element model are composed of fixed, viscous boundaries and infinite elements. Contact between foundation structures and sub-soil is modelled by constraint equations. The results from performed analysis show that the choice of side boundaries plays important role in seismic response of RC frame structures.

Keywords: Soil structure interaction, infinite elements

1 Introduction

The past earthquakes have shown that the seismic response of a structure is considerably influenced by the soil structure interaction. The main difficulty in the soil-structure interaction problems is the correct numerical simulation of the soil media and its interaction with the structure standing on it. In recent years the development of computers has enabled the usage of sophisticated computer programs for numerical simulation of soil media. In this work three types of soil are taken into consideration as hard, medium dense and soft soils as stated in Eurocode 8 part 1. In order to examine the SSI effects on the structural rigidity, RC models of one, three and five storey frames are modelled and time history analysis

S. Klinkel et al. (Eds.), *Seismic Design of Industrial Facilities*,
DOI 10.1007/978-3-658-02810-7_47, © Springer Fachmedien Wiesbaden 2014

is performed. In the analysis soil medium is subjected to acceleration time history of Imperial Valley EQ, El Centro record, 1940-May-18 (El Centro) earthquake. Coupled soil-structure system is subjected to acceleration time history and the results of structural response are compared accordingly. The dynamic analysis is done by using the general finite element program ANSYS where it is possible modelling of both soil and structure and taking into consideration the soil-structure interaction. The variation in structural response for acceleration, displacements and internal forces are tabularly presented and comparisons are made accordingly.

2 Soil modelling

The soil medium is presented as a two dimensional model composed of four layers resting on bedrock. In Table 1 soil layers properties are tabulated in a way that the bottom layers are characterized with better soil characteristics as it is usually seen in nature.

Table 1: Soil properties

Soil medium	Layer number	Thickness (m)	Unit weight (kN/m^3)	Shear velocity (m/s)
Hard	1	3	16	330
	2	7	17	420
	3	6	17.5	510
	4	14	18	690
Medium	1	3	16	160
	2	7	17	210
	3	6	17.5	250
	4	14	18	340
Soft	1	3	16	90
	2	7	17	100
	3	6	17.5	120
	4	14	18	160

The soil medium is assumed to be linear-elastic material and is discretized using four nodded plane strain elements PLANE82. The dynamic analysis is performed by transient analysis using the step by step method. The proportional viscous damping matrix is taken to be proportional to mass and stiffness matrix (Rayleigh damping). The Rayleigh damping factors, alpha and beta are calculated such that the critical damping is 5% for first two modes (α=1.2907, β=0.001405). The bottom boundary of the soil model is fixed while side boundaries are simulated as fixed, viscous and infinite element boundaries. In order to prevent the reflection of the waves viscous and infinite element boundaries are analysed.

2.1 Viscous boundaries

The radiation damping at the side boundaries as given in Cohen [1] is simulated by dashpots in which the radiation coefficient is obtained from the relation:

$$c = A \cdot \rho \cdot V \tag{1}$$

where A is the area between the nodes along the side, ρ is the soil density and V is the shear and/or compression wave velocity depending on the direction of action.

2.2 Infinite element boundaries

The formulation of infinite elements is the same as for the finite elements in addition to the mapping of the domain. Infinite elements are first developed by Zienkiewicz et al. [2] and since then have been developed in both frequency and time domain. In Häggblad et al. [3] infinite elements with absorbing properties have been proposed which can be used in time domain. In this work the development of infinite element has followed the similar techniques as in [3] where the infinite element is obtained from a six noded finite element as shown in Figure 1.

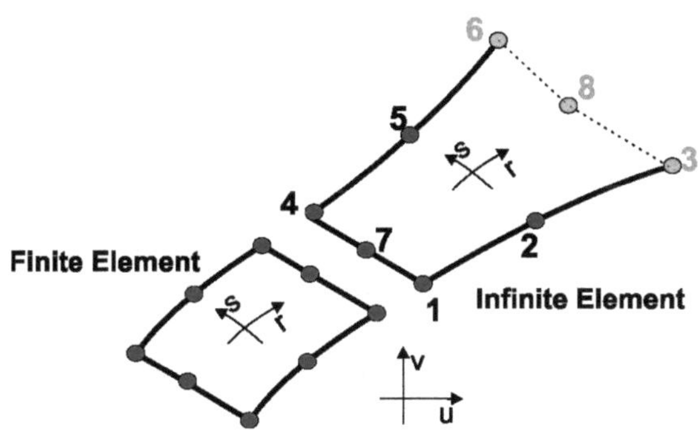

Figure 1: Coupling of finite and infinite elements

Referring to Figure 1 the coupling between finite and infinite element can be presented as follows. The finite element has eight nodes and three nodes on the end side. The infinite element has three nodes on the side which allow for complete coupling with the finite element. The element displacement in **u** and **v** direction is interpolated with the usual shape functions N^1, N^2, N^4, N^5 and N^7:

$$\begin{aligned} u &= [N^1 \quad N^2 \quad 0 \quad N^4 \quad N^5 \quad 0 \quad N^7 \quad 0]\mathbf{u} \\ v &= [N^1 \quad N^2 \quad 0 \quad N^4 \quad N^5 \quad 0 \quad N^7 \quad 0]\mathbf{v} \end{aligned} \tag{2}$$

In expression (2) u and v are vectors with nodal point displacements in global coordinates. The shape functions are given as follows:

$$
\begin{aligned}
N^1 &= (1-r)(-1+s)(s+1+r)/4 \\
N^2 &= (r-1)(1+r)(-1+s)/2 \\
N^4 &= -(r-1)(1+s)(-1-r+s)/4 \\
N^5 &= -\,(r-1)(1+r)(1+s)/2 \\
N^7 &= (-1+s)(1+s)(r-1)/2
\end{aligned} \tag{3}
$$

Based on the isoparametric concept the infinite element in global coordinate is mapped onto an element in local coordinate system using the expression as given in (4).

$$
\begin{aligned}
r &= [M^1 \quad M^2 \quad 0 \quad M^4 \quad M^5 \quad 0 \quad M^7 \quad 0]\mathbf{r} \\
s &= [M^1 \quad M^2 \quad 0 \quad M^4 \quad M^5 \quad 0 \quad M^7 \quad 0]\mathbf{s}
\end{aligned} \tag{4}
$$

The mapping functions are given as follows:

$$
\begin{aligned}
M^1 &= -\frac{(1-s)rs}{1-r} \\
M^2 &= -\frac{(1-s)(1+r)}{2(1-r)} \\
M^4 &= -\frac{(1+s)rs}{1-r} \\
M^5 &= -\frac{(1+s)(1+r)}{2(1-r)} \\
M^7 &= -\frac{2r(1+s)(1-s)}{(1-r)}
\end{aligned} \tag{5}
$$

In expression (5) r and s are vectors of nodal point displacements in local coordinates where it is to be pointed out that on the side of infinity (r=1) no mappings have been assigned to the nodes as it is taken that displacement decays at infinity. The number and location of the nodes connecting finite and infinite elements must coincide to guarantee continuity condition between the elements. The main advantage of the proposed infinite elements is that the number of nodes on the infinite element allows coupling with finite elements with eight nodes which are used for displacement sensitive problems. Construction of element matrices is done by using the usual procedures as described in Bathe[4]. The new coordinate interpolation functions are taken into consideration in the Jacobian matrix as described in Bettess [5]. For the absorbing layer of the infinite element, the Lysmer-Kuhlmeyer approach [6] is used. In all cases, a plane strain two dimensional case is studied. For impact of plane waves on element sides, normal and tangential stresses are derived as follows:

$$
\begin{bmatrix} \sigma^n \\ \tau \end{bmatrix} = \begin{bmatrix} a\rho c^p & 0 \\ 0 & b\rho c^s \end{bmatrix} \begin{bmatrix} \dot{\mathbf{u}}^{\mathbf{n}} \\ \dot{\mathbf{u}}^{\mathbf{t}} \end{bmatrix} \tag{6}
$$

where cp and cs indicate the wave velocities for the P wave (compressional) and S wave (shear) respectively. The term ρ stands for density of soil medium. In order to take into account the directions of the incident waves coefficients a and b are used as multipliers [7]. Transformation from local to global coordinates is done by software ANSYS [8] such that there is no need of defining transformation matrices. Time derivatives are approximated by the Newmark's method. The programming part of the infinite element has been performed using the Programmable Features of the ANSYS software.

3 Coupled soil structure system response

In order to show the influence of the soil boundaries to the structure a comparison of boundary cases has been performed. First the soil side boundary is simulated as a fixed support which is usually done in many application projects. Then the same soil medium is bounded with viscous boundaries which are included into the software ANSYS. Finally the soil is surrounded with the newly programmed infinite elements. The frame structural elements are idealized as two dimensional elastic beam elements BEAM3 having three degrees of freedom at each node, translations in the nodal x and y directions and rotation about the nodal z axis. The behaviour of the frame structure is supposed as elastic and is modelled using two parameters, the modulus of elasticity E=3.15x107 kPa and Poisson's ration n=0.2. The bay length of the frame is taken to be 4.0 m and storey height of 3.0 m. Section of beams is 40 x 50 cm while the column section is 50 x 50cm. A mass of 11 tons is assigned on each node to simulate the real structural behaviour (total 44 tons per floor). There are four different frames that are taken into consideration. For all RC frames the beam and column sections, floor masses and number of bays are kept constant in all cases. The only parameter that is altered is the storey number. The structures are modeled as one, three and five storey RC frames.

Finite element modelling of the coupled soil-structure system is performed by the software ANSYS as shown in Figure 2. The effect of soil-structure interaction is carried out with the acceleration time history of the El Centro earthquake with a

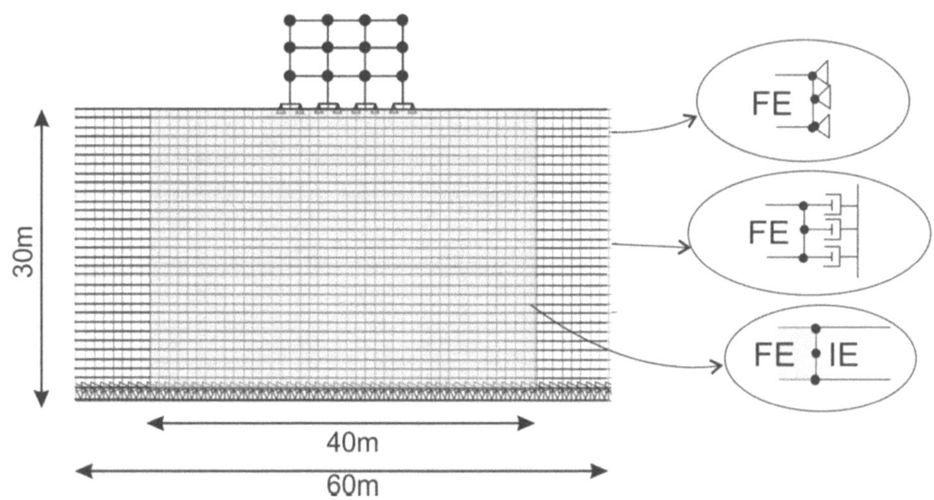

Figure 2: Coupled Soil-structure system of a multi-storey frame

scaled peak ground acceleration of 0.25g. The foundation where the structure is supported is taken to be 8 nodded plane element having two degrees of freedom in each node, translations in the nodal x and y directions. The moment transfer capability between the column and the footing is created by using a constraint equation where the rotation of the beam is transferred as force couples to the plane element.

In Figure 2 the coupled system of soil and structure system is shown. The side boundaries are presented as fixed, viscous and infinite element boundaries. In Table 2 below the difference in the structural response is given.

Table 2: Structural response values

Nr. of Storey	Soil	Boundary	Max. acc. at top of Str. (m/s^2)	Max. displ. at top of Str. (cm)	Max. str. moment at top of Str. (kNm)
1	Hard	Fixed	11.2	0.447	152.1
		Viscous	5.72	0.220	48.7
		Infinite el.	5.57	0.217	48.6
	Medium	Fixed	13.5	0.624	223.2
		Viscous	5.13	0.319	83.5
		Infinite el.	5.01	0.312	82.8
	Soft	Fixed	11.1	1.11	222.2
		Viscous	4.61	0.527	85.3
		Infinite el.	4.29	0.517	81.2
3	Hard	Fixed	8.95	1.87	155.1
		Viscous	8.68	1.93	145.5
		Infinite el.	8.18	1.91	145.1
	Medium	Fixed	10.5	3.45	182.2
		Viscous	7.88	2.96	118.1
		Infinite el.	7.55	2.89	116.9
	Soft	Fixed	10.3	8.22	175.1
		Viscous	7.12	3.65	108.3
		Infinite el.	6.99	3.63	106.3
5	Hard	Fixed	9.74	5.56	153.1
		Viscous	9.15	4.78	145.3
		Infinite el.	8.83	4.72	144.3
	Medium	Fixed	8.51	6.48	158.3
		Viscous	8.04	5.86	149.1
		Infinite el.	7.89	5.68	148.1
	Soft	Fixed	8.80	11.1	131.2
		Viscous	5.85	7.58	81.9
		Infinite el.	5.78	7.53	80.2

According to the acceleration values of the Table 2 the maximum acceleration at the top of structure is considerably decreased when using the viscous boundaries of the commercial software. Moreover, when using infinite elements the values of acceleration is approximated by similar values showing that in case of infinite elements the wave reflection at the boundaries is minimized in a similar manner. The main difference is that by using the infinite elements the size of domain is decreased considerably which can be discretized by smaller number of finite elements. Thus it can be stated that by using infinite elements as a substitution for the viscous boundaries the values of both displacement and structural moments give numerically stable and acceptable results.

4 Conclusion

The usage of side boundaries alters the results greatly. In case of using the fixed boundaries at the far side end increases the amount of computation. Moreover, on the other side the values of internal forces obtained in using fixed boundaries increases the internal forces due to wave reflection of the boundaries. On the other hand, the usage of viscous and infinite elements influences the internal forces of the structural response such that the wave propagation is absorbed at the side boundaries. In this work the infinite elements with absorbing properties are shown to be a promising substitution for the viscous boundaries offered by commercial software in which the number of finite elements decreases while attaining the correctness of the results.

References

[1] M.Cohen (1980) Silent boundary methods for transient wave analysis, PhD Thesis, California Institute of Technology, Pasadena, California

[2] Zienkiewicz, O.C., Emson, C. and Bettess, P. A novel boundary infinite element. Int. J. Num. Meth. Eng. 19 (1983), 393-404MITH L., Limit Analysis and Concrete Plasticity, Prentice Hall, New York, 2003, p. 360.

[3] B. Häggblad and G. Nordgren, Modelling Nonlinear Soil-Structure Interaction Using Interface Elements, Elastic-Plastic Soil elements and Absorbing Infinite Elements , Computers and Structures 26 (1987), 307–324.

[4] K.J.Bathe, Finite Element Procedures in Engineering Analysis, Prentice-Hall, Englewood Cliffs. NJ, 1982.

[5] P. Bettess, Infinite Elements, Penshaw Press. New Castle,England. 1992.

[6] J. Lysmer and R.L. Kuhlmeyer. Finite dynamic model for infinite media. J. Engng Mech. Div., ASCE 95,(1969) 859-877.

[7] W. White, S. Valliappan and I. K. Lee. Unified boundary for finite dynamic models. J. Engng Mech. Div., ASCE 103, (1977) 949-964.

[8] ANSYS. FEM Software. 2006

Response Analysis of a Long-span Arch Bridge under the Seismic Travelling Wave Excitation

Menglin Lou[1], Qiang Li[2] and Shan Gao[3]

[1] State Key Laboratory of Disaster Reduction in Civil Engineering / Tongji University, 1239 Siping Load, Shanghai, China, lml@tongji.edu.cn

[2] Tianjin Municipal Engineering Design & Research Institute, Tianjin, China 403695911@qq.com

[3] Jones Lang LaSalle, Shanghai, China, mountain_gs@yahoo.com.cn

ABSTRACT

Taking a long-span arch bridge as an example, the characteristics of dynamic responses of the arch bridge under the horizontal and vertical travelling seismic wave excitations and the effect of the wave travelling velocity on the responses are discussed based on numerical analysis results. According to the structure symmetry, a simplified computation method is proposed for the travelling seismic response analysis of the bridge in time domain by using the seismic response of two half arch bridges under the uniform excitation. The time history analysis results indicate that the simplified method can achieve a good precision. The comparison of the numerical results describes the phenomena that the seismic response under travelling wave excitation does not linearly change with the seismic wave velocity. Through some further analysis, this article proposes a new concept of the resonant under travelling wave excitation, and the mechanism of the resonant effect in travelling seismic wave excitation is being expatiated.

Keywords: long-span arch bridge, travelling seismic wave excitation, symmetry

1 Introduction

Arch bridge is a common type of the bridge in the world. Evaluating correctly its seismic responses is an important work for the seismic design of the bridge when the arch bridge is constructed in earthquake zone. The uniform seismic input is adopted usually in seismic response analysis of the structures, but the assumption is not suitable for the seismic response analysis of the long-span arch bridge. The difference of the ground motions at two arch bases of the bridge must be considered during the seismic wave propagates in the soil or rock site. Therefore, the dynamic analysis of the arch bridge under the multiple-support seismic excitations should be carried out for the seismic responses computation of the long-

DOI 10.1007/978-3-658-02810-7_48,

span arch bridge, as shown by Kiureghian and Neuenhofer[1], Pan et al[2], Wang and Wang[3], Clough and Pengien[4], Harichadran et al.[5]. In this paper, the response characteristics of a long-span arch bridge under the horizontal and vertical travelling seismic wave excitations respectively will be discussed.

2 Equation of arch bridge under multiple-support seismic excitations

The analysis equation of the seismic response of the arch bridge under the multiple-support excitations can be written as:

$$[M_{ss}]\{\ddot{u}_d(t)\}+[C_{ss}]\{\dot{u}_d(t)\}+[K_{ss}]\{u_d(t)\}=-[M_{ss}][T]\{\ddot{u}_b(t)\} \tag{1}$$

in which $[M_{ss}]$, $[C_{ss}]$ and $[K_{ss}]$ are the mass, damping and stiffness matrices of the bridge with n freedom degrees respectively. $\{\ddot{u}_d(t)\}$, $\{\dot{u}_d(t)\}$ and $\{u_d(t)\}$ are the acceleration, velocity and displacement vectors relative to the bridge's base respectively. $[T]$ is the pseudo-static matrix that is formed by

$$[T]=-[K_{ss}]^{-1}[K_{sb}] \tag{2}$$

where $[K_{sb}]$ is the stiffness matrix corresponding to the structural freedom degrees of the bridge and the freedom degree of the bridge supports where the seismic wave excites the bridge.

In Eq.(1), the seismic input vector can be described as

$$\{\ddot{u}_b(t)\}^T=\{\ddot{u}_{1b}(t),\ddot{u}_{2b}(t),...,\ddot{u}_{ib}(t),...,\ddot{u}_{mb}(t)\} \tag{3}$$

where, $\{\ddot{u}_{ib}(t)\}$ is the ith support input vector with three components. For the travelling seismic input, it can be written as

$$\{\ddot{u}_{ib}(t)\}=\{\ddot{u}_g(t-\frac{\Delta_i}{v_a})\} \tag{4}$$

in which $\{\ddot{u}_g(t)\}$ is the ground motion wave vector with three components. If only one component is considered, other two components are zero. Δ_i is the distance from the ith support to the first support where the seismic wave $\{\ddot{u}_g(t)\}$ arrives at first.

3 Travelling response analysis of an arch bridge

3.1 Brief introduction of the arch bridge

A long-span arch bridge shown in Figure 1 is a steel composite construction with span (L) 420 m and height (h) 84 m. Its finite element model is shown in Figure 2. The mode frequencies of the bridge are listed in third column of Table 1.

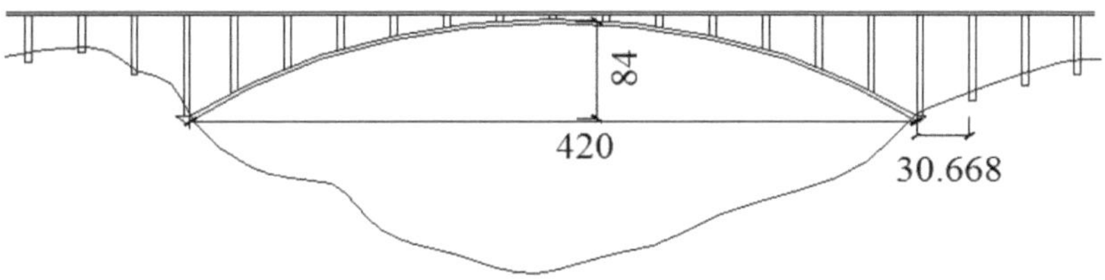

Figure 1: Long-span arch bridge

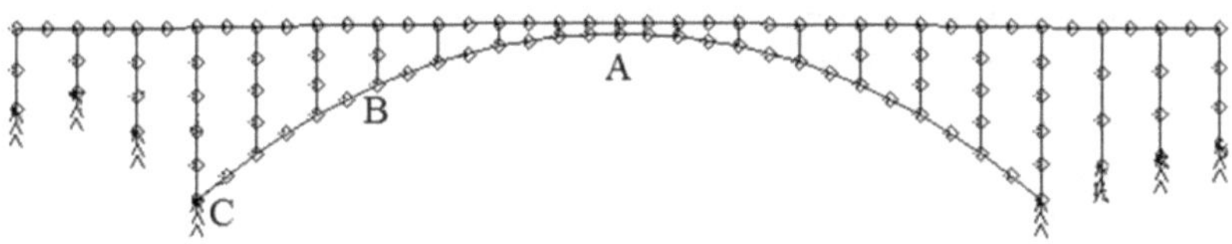

Figure 2: Finite element model

Table 1: Mode frequency f_i (Hz), wave length λ_i (m) and ratio β_i

Mode order	Symmetry	f_i	$c=2000m/s$		$c=1000m/s$		$c=500m/s$	
			λ_i	β_i	λ_i	β_i	λ_i	β_i
1	asymmetrical	0.335	5970	0.07	2985	0.14	1492	0.28
2	symmetrical	0.645	3101	0.14	1550	0.27	775	0.54
3	asymmetrical	1.179	1696	0.25	848	**0.50**	424	0.99
4	symmetrical	1.282	1560	0.27	780	**0.54**	390	1.08
5	asymmetrical	2.442	819	**0.51**	410	1.02	204	2.06
6	symmetrical	2.868	697	**0.60**	349	1.20	174	2.41
7	asymmetrical	3.458	578	0.73	289	1.45	145	2.90
8	symmetrical	3.511	570	0.74	285	1.47	142	2.96

3.2 Resonant under travelling wave excitation

First, two sine waves with exciting frequencies 0.334Hz(f_1) and 0.646Hz(f_2) are used as support excitation. As shown in Table 1, exciting frequencies 0.334Hz and 0.645Hz are equal to the first and second mode frequency of the arch bridge, as well as f_1 and f_2 are the first asymmetrical and symmetrical mode frequency respectively. The quantity β is defined as the ratio of the bridge span L and the seismic travelling wave length λ. The length λ=∞ and β=0 express the case of the uniform excitation obviously. For the travelling wave excitation, the wave travelling velocity is chosen to lead to β=0.25, 0.5 and 0.75 respectively.

The horizontal and vertical relative displacements at middle-span cross section A shown in Figure 2 under different excitation case including uniform excitation and travelling wave excitation with three propagating velocities are shown in Figure 3 and Figure 4.

Due to the symmetry of the structure, the vertical displacement of the cross section A should be zero because the uniform horizontal excitation is asymmetry. As well as the peak value of the seismic response of the cross section A should be same for two excitation cases of β=0.25 and β=0.75. The numerical results shown in Figure 3 and Figure 4 illustrate above conclusions. Because of the resonant, the peak displacement under the uniform excitation with 0.335Hz is larger than the peak displacement under the uniform excitation with 0.646Hz although the input sine wave amplitudes are same.

In particular case, the displacement responses of the cross section A under the travelling wave excitation β=0.5 must be paid more attention. As shown in Figure 3 and Figure 4, the horizontal and vertical displacements of the cross section A display the dynamic response characteristics of the bridge under the symmetrical uniform excitation. The horizontal displacement approaches to zero and the vertical displacement is the largest. It is more important that the peak displacement under exciting frequency 0.646Hz shown in Figure 4 is larger than the peak displacement under the exciting frequency 0.335Hz. When β=0.5, not only the excitations at two arch bases of the bridge are same but also symmetry at case the exciting direction is horizontal. This means that the travelling wave excitation case with β=0.5 transfers to the symmetrical uniform excitation case and the bridge will be in resonant status in symmetrical mode to lead to the structural responses increasing when the excitation frequency is equal to symmetrical mode frequency.

We define this resonant phenomenon as **the resonant under travelling wave excitation** that will influence obviously responses of the symmetrical structures in some particular conditions.

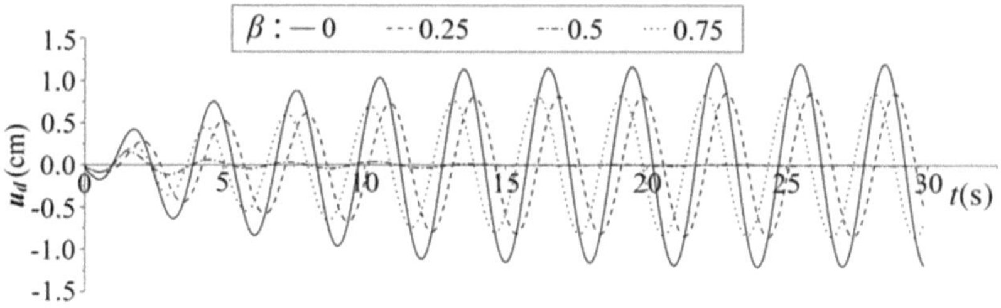

(a) Horizontal displacement

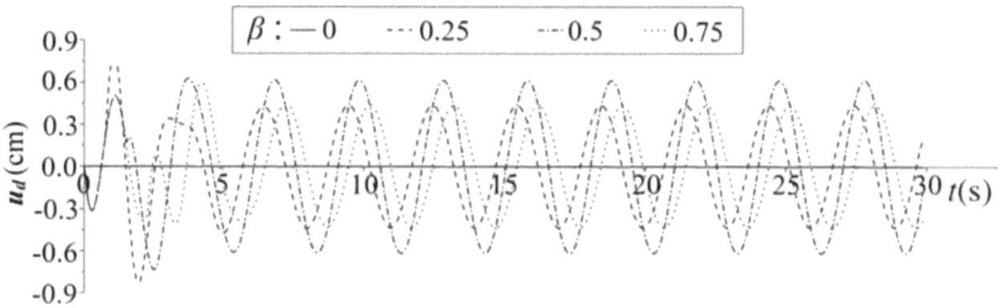

(b) Vertical displacement

Figure 3: Displacement comparison of cross section A (0.335Hz)

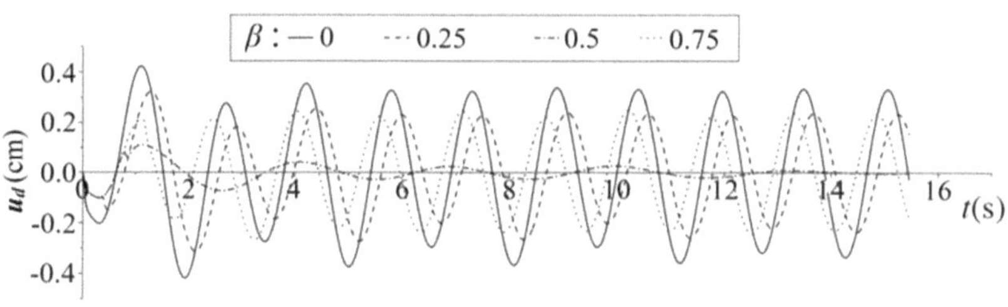

(a) Horizontal displacement

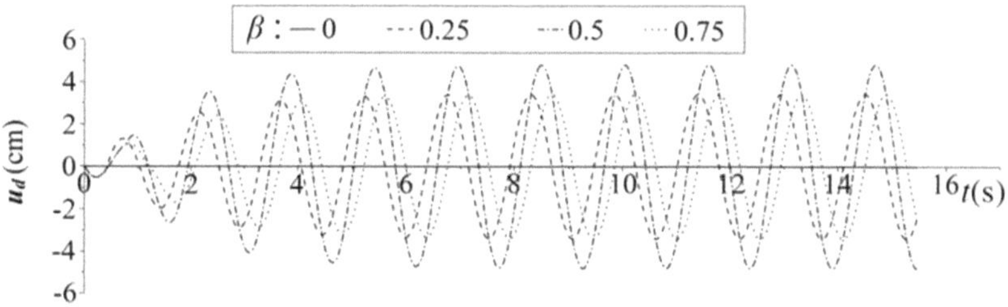

(b) Vertical displacement

Figure 4: Displacement comparison of cross section A (0.645Hz)

3.3 Seismic responses of the bridge under horizontal excitation

The input horizontal seismic waves are shown in Figure 5 and its peak acceleration and displacememnt are $1.139 m/s^2$ and 0.0735m respectively. The propagating direction of the seismic wave is from left bridge base to right bridge base. The propagating velocity c is respectively equal to 500m/s, 1000m/s, 2000m/s and ∞ (expressing uniform excitation), corresponding ratio β values of different mode are listed in Table 1.

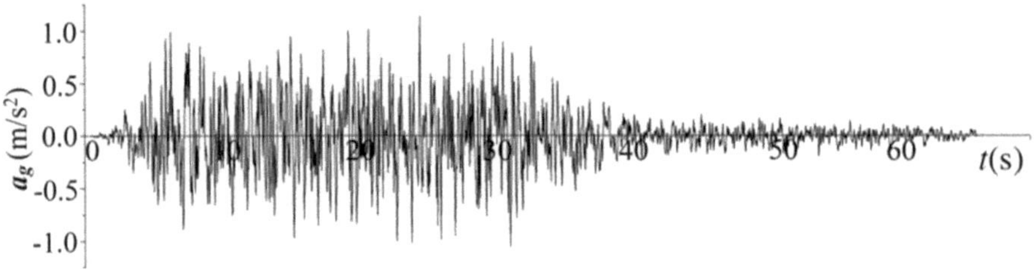

(a) Acceleration wave

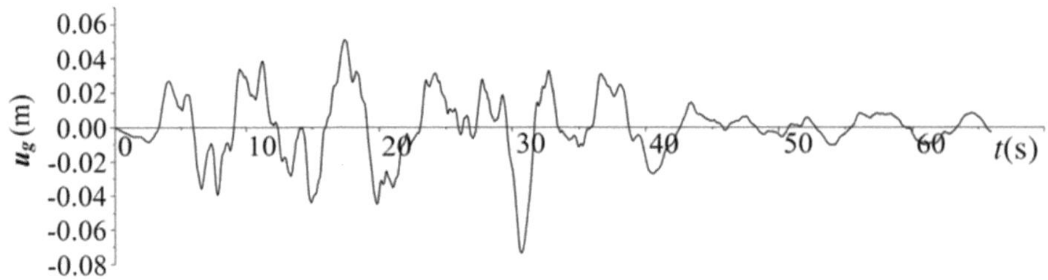

(b) Displacement wave

Figure 5: Input seismic waves

The seismic response peaks of the bridge in above four excitation cases are shown in Table 2 and Table 3.

Table 2: Peak acceleration (m/s^2)

Cross section	Response direction	Seismic wave propagating velocity c (m/s)			
		∞	2000	1000	500
A	horizontal	0.9032	0.6401	0.6674	0.7102
	vertical	0.0000	1.2180	**1.6322**	1.4978
B	horizontal	0.8562	0.6142	0.8331	0.7516
	vertical	0.5993	1.5039	1.3785	1.2543

Table 3: Peaks of axial force (10^6N) **and moment** (10^6N-m)

Force	**Cross section**	**Seismic wave propagating velocity c** (m/s)			
		∞	2000	1000	500
Axial force	A	9.1301	62.606	98.813	108.01
	B	67.737	84.638	94.347	99.830
	C	93.039	120.60	**142.46**	126.26
Moment	A	0.3950	17.145	**23.967**	18.028
	B	4.2615	20.500	**26.233**	20.275
	C	7.6557	24.257	**28.950**	23.780

Data in Tables 2 and 3 show that the seismic responses of the bridge under the travelling wave excitation are larger than the ones under the uniform excitation, so that the travelling wave effect of the seismic input should be considered in the seismic design of the long-span arch bridges.

The bold data in Tables 2 and 3 denote that the influence of the travelling wave excitation on the seismic responses of the bridge does not simply increase with the propagating velocity decreases.

When the propagating velocity is equal to 1000m/s, the vertical displacement of the cross section A, axial force of the cross section C and moment of the cross section A, B and C are large than ones in other excitation cases. There is the effect of the resonant under travelling wave excitation. It can be seen from Table 1 that β value of the forth mode of the bridge (that is the second symmetrical mode) is equal to 0.54 near to 0.50.

3.4 Seismic responses of the bridge under vertical excitation

The acceleration wave shown in Figure 5 is still taken as the vertical seismic input but the acceleration peak is equal to 0.0735m/s^2. The peak values of the seismic responses of the bridge under four vertical excitation cases with c=500m/s, 1000m/s, 2000m/s and ∞ are shown in Figures 6~9.

The figures show that the effect of travelling wave excitation on the seismic responses of the bridge is more important and the resonant under travelling wave excitation is more remarkable in the case of the vertical excitation, when c=1000m/s. It can be found that the β is exactly equal to 0.5 at the third mode of the system that is the second asymmetrical mode. The resonant under travelling excitation changes the general variance trend of the seismic response of the arch bridge under the travelling wave excitation.

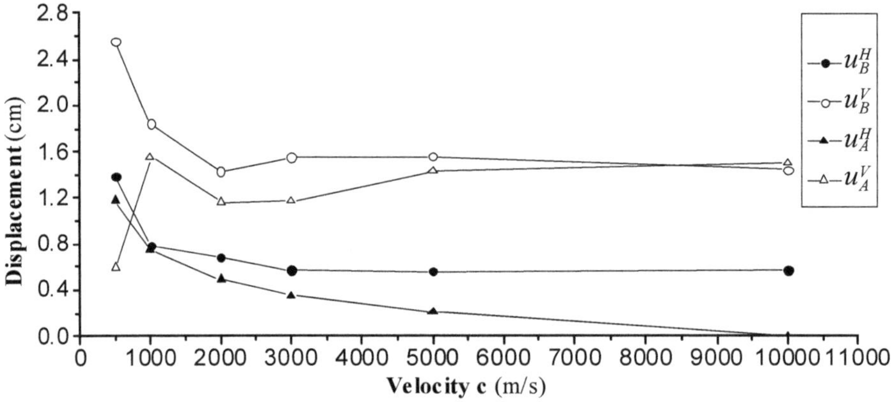

Figure 6: Displacement of the cross sections A and B

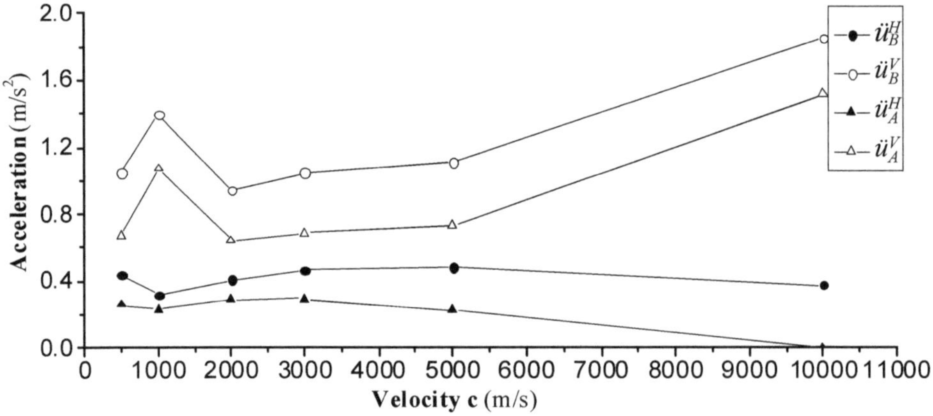

Figure 7: Acceleration of the cross sections A and B

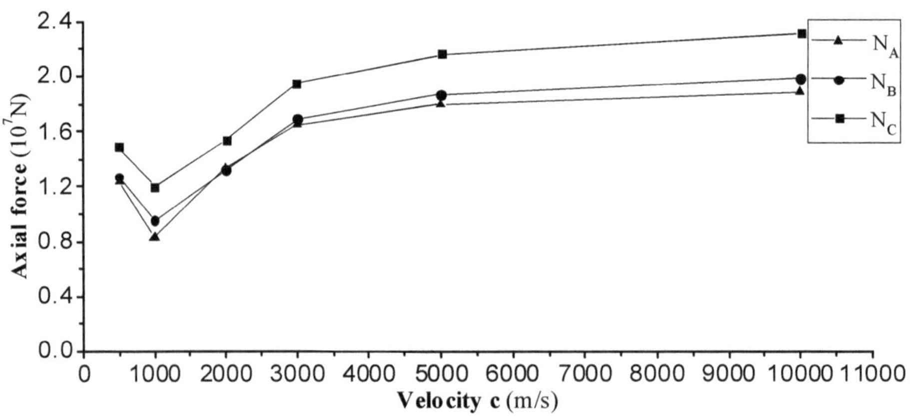

Figure 8: Axial force of the cross sections A and B

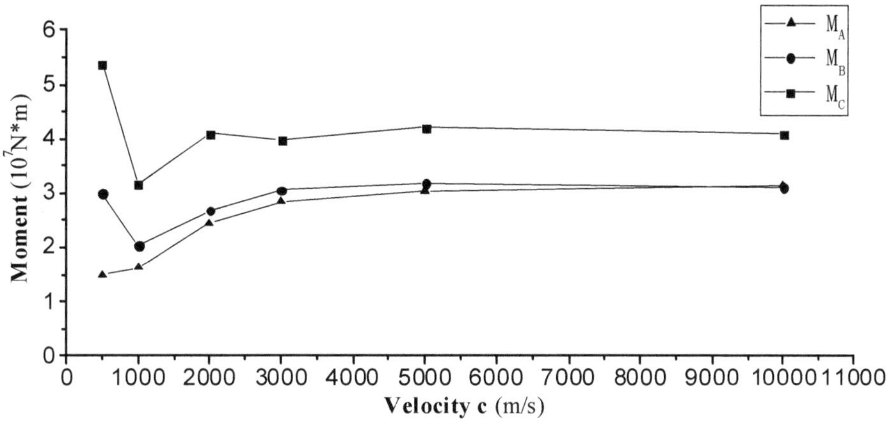

Figure 9: Moment of the cross sections A and B

4 Simplified method for seismic responses analysis of the arch bridge

Although the main arch of the arch bridge has the symmetry, the structural system of the bridge sometime is not completely symmetry because the bridge approaches to two sides of the arch bridge are not absolutely same. Next, a simplified method, in that two half arch bridge models under uniform seismic excitation are instead of the whole arch bridge model under travelling excitation seismic excitation, is discussed.

Neglecting the non-symmetry of the arch bridge, the whole structure system can be decomposed as symmetrical and asymmetrical half bridge as show in Figures 10～12. And then, it is assumed that the seismic inputs at all supports of the half bridge are same with the one at the main arch base.

For the symmetrical half arch bridge and the asymmetrical half arch bridge, the uniform seismic input can be expressed as follow respectively.

For horizontal excitation:

$$u_s(t) = \left(u_l(t)\text{-}u_r(t)\right)/2 \tag{5}$$

$$u_a(t) = \left(u_l(t)+u_r(t)\right)/2 \tag{6}$$

For vertical excitation:

$$u_s(t) = \left(u_l(t)+u_r(t)\right)/2 \tag{7}$$

$$u_a(t) = \left(u_l(t)\text{-}u_r(t)\right)/2 \tag{8}$$

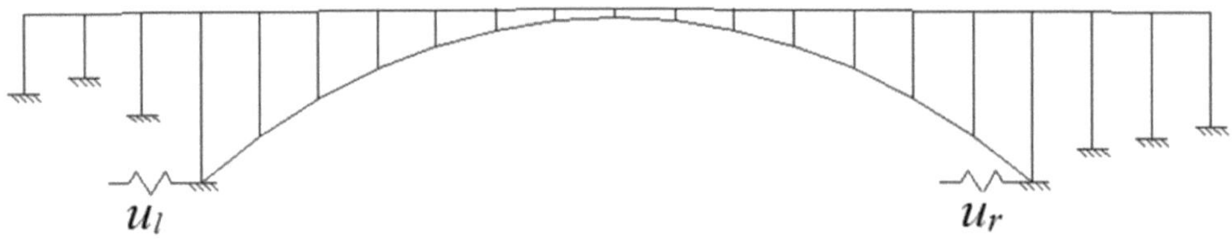

Figure 10: Whole system of the bridge

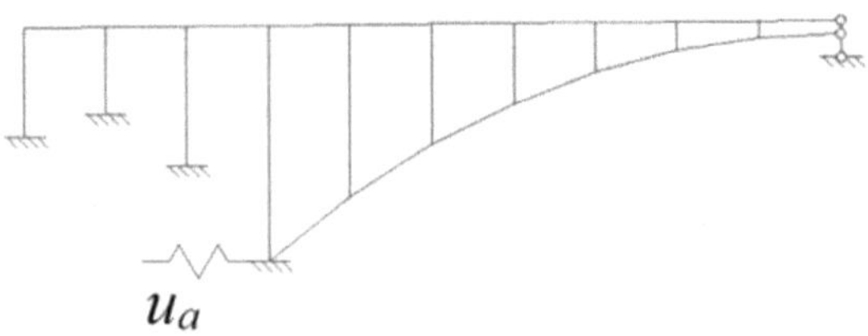

Figure 11: Asymmetrical half bridge

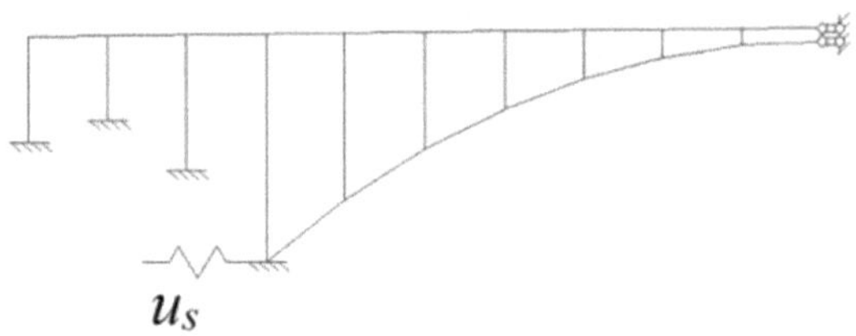

Figure 12: Symmetrical half bridge

The mode frequencies of two half arch bridges are lists in Table 4 those are almost same with the ones of the whole arch bridge listed in Table 1.

Table 4: Mode frequencies of two half arch bridges (Hz)

Half bridge	**Symmetrical half arch bridge**				**Asymmetrical half arch bridge**			
mode order	1	2	3	4	1	2	3	4
frequency	0.334	1.179	2.442	3.460	0.645	1.282	2.869	3.511

The numerical results of the seismic responses of the bridge obtained from whole arch bridge model and half arch bridge model are shown in Tables 5～10 repectively.

Table 5: Comparison of acceleration and displacement (c=500m/s)

Response	Cross section	Horizontal			Vertical		
		Whole bridge	Half bridge	Error (%)	Whole bridge	Half bridge	Error (%)
Acceleration (m/s^2)	A	0.7102	0.7164	0.87	1.4978	1.4996	0.12
	B	0.7516	0.7591	1.00	1.2506	1.2523	0.14
Displacement (m)	A	0.0558	0.0557	-0.18	0.1180	0.1185	0.42
	B	0.0594	0.0595	0.17	0.0763	0.0758	-0.66

Table 6: Comparison of axial force and moment (c=500m/s)

Cross section	Axial force (10^6N)			Moment (10^6N-m)		
	Whole bridge	Half bridge	Error (%)	Whole bridge	Half bridge	Error (%)
A	18.03	18.08	0.27	108.0	109.1	1.05
B	20.28	20.14	-0.65	99.83	100.0	0.17
C	23.78	23.66	-0.50	128.3	126.1	-1.68

Table 7: Comparison of acceleration and displacement (c=1000m/s)

Response	Cross section	Horizontal			Vertical		
		Whole bridge	Half bridge	Error (%)	Whole bridge	Half bridge	Error (%)
Acceleration (m/s^2)	A	0.4095	0.4089	-0.15	0.6476	0.6459	-0.26
	B	0.3809	0.3855	1.21	0.7263	0.7256	-0.10
Displacement (m)	A	0.0107	0.0108	0.93	0.0734	0.0735	0.14
	B	0.0100	0.0099	1.01	0.0771	0.0772	0.13

Table 8: Comparison of axial force and moment (c=1000m/s)

Cross section	Axial force (10^6N)			Moment (10^6N-m)		
	Whole bridge	Half bridge	Error (%)	Whole bridge	Half bridge	Error (%)
A	8.2677	8.2676	-0.31	16.229	17.006	4.79
B	9.5181	9.4897	-0.30	20.369	20.716	1.70
C	12.007	11.983	-0.20	31.540	30.252	-4.08

Table 9: Comparison of acceleration and displacement (c=2000m/s)

Response	Cross section	Horizontal			Vertical		
		Whole bridge	Half bridge	Error (%)	Whole bridge	Half bridge	Error (%)
Acceleration (m/s^2)	A	0.5320	0.5372	0.98	0.6164	0.6150	-0.23
	B	0.5627	0.5700	1.30	0.8768	0.8803	0.40
Displacement (m)	A	0.0070	0.0071	1.43	0.0829	0.0830	0.12
	B	0.0803	0.0805	0.25	0.0073	0.0074	1.27

Table 10: Comparison of axial force and moment (c=2000m/s)

Cross section	Axial force (10^6N)			Moment (10^6N-m)		
	Whole bridge	Half bridge	Error (%)	Whole bridge	Half bridge	Error (%)
A	13.311	13.386	0.56	24.411	24.519	0.44
B	13.219	13.258	0.30	26.744	27.088	1.29
C	15.290	15.317	0.18	40.934	41.518	1.43

The results listed in above tables show that the simplified method can archive good precision and simplify effectively the calculation of the seismic responses of the long-span arch bridge under the multiple-support excitation and the travelling wave excitation.

5 Conclusion

The effect of the seismic travelling wave excitation on the dynamic response of the long-span arch bridge is important. Comparing with the results under the seismic uniform excitation, it will increase the seismic response, especially the seismic forces of the cross section of the arch.

For the long-span symmetrical structures, there is the phenomena of the resonant under travelling wave excitation at the particular cases. Sometime, it will also increase obviously the seismic responses of the symmetrical structures.

The numerical results show that the proposed simplified method is an effective method for calculating the seismic responses of the long-span arch bridge under the travelling wave excitation as well as the multiple-support excitation.

6 Acknowledgements

This research was sponsored by Nation Natural Science Foundation of China through a grant 90915011 and State Key Laboratory Basic Theory Foundation of the Ministry of Science and Technology of China through a grant SLDRCE08-A-07. These supports are gratefully acknowledged.

REFERENCES

[1] Kiureghian, A.D.; Neuenhofer, A.: Response Spectrum Method for Multi-support Seismic Excitations; Earthquake Engineering and Structural Dynamics; 8 (1992), 713-740

[2] Pan, D.; Lou, M.; Fan, L.: Status of Seismic Response Analysis of Long-span Structures under Multiple Support Excitations; Journal of Tongji University; 10 (2001), 1213-1219 (in Chinese)

[3] Wang, J.; Wang Q.: Seismic Response of Long Span Arch Bridge under Spatially Variable Seismic Excitation; Journal of Vibration Engineering; 2 (1995), 119-126 (in Chinese)

[4] Clough, R.W.; Pengien, J.P.: Dynamic of Structures, McGraw-Hill, Inc, New York, 1993

[5] Harichadran, R.S.; Hawwari, A.; Sweidan, B.N.: Responses of Long-span Bridges to Spatially Varying Ground Motion; Journal of Structural Engineering; 5 (1996), 476-484

A 3D Dynamic Impedance of Arbitrary-Shaped Foundation on Anisotropic Multi-Layered Half-Space

Gao Lin[1], **Zejun Han**[1]

[1] Faculty of Infrastructure Engineering, Dalian University of Technology No.2 Linggong Road, Ganjingzi District, Dalian City, China
gaolin@dlut.edu.cn

ABSTRACT:

It is generally recognized that soils and rocks in nature invariably exhibit some degree of anisotropy in their response to static or dynamic loads. An approach based on precise integration method (PIM) and the dual vector formulation (DVF) of wave motion equation is proposed for the evaluation of Green's influence function of anisotropic stratified half-space. Then the problem of an arbitrary-shaped foundation on a multi-layered subsoil is studied. The wave motion equation for a typical horizontally anisotropic layer and the half-space is solved analytically and the integration is performed by PIM. Any desired accuracy can be achieved. The DVF of wave motion equation is suggested for assembling the layers. As a result, the Green's influence function for anisotropic stratified half-space are found based on the standard method in matrix algebra with the size of matrices not greater than (3×3). The computational effort is reduced to a great extent. Special treatment has been taken to preserve the effective digits. The computation is always stable. There is no limit of the thickness and number of soil layers to be considered. To satisfy the mixed boundary condition at the surface of arbitrary-shaped foundation, the interface between the foundation and the multi-layered soils is discretized into a number of uniformly spaced sub-disks as in the usual manner. Numerical examples validate the efficiency and accuracy of the proposed approach.

Keywords: Dynamic impedance; layered anisotropic soil stratum; precise integration method; wave motion; Green's influence function

1 Introduction

Dynamic soil-structure interaction (SSI) effects have always been important in the context of assessing the safety and vulnerability of critical facilities subjected to earthquake excitation. Problems of SSI have been under intensive investigation in the past decades and significant progress has been made. A survey of the published literature has shown that most studies were concentrated on homogeneous and idealized soil profiles, such as the half-space, the uniform stratum on rigid base and

S. Klinkel et al. (Eds.), *Seismic Design of Industrial Facilities*,
DOI 10.1007/978-3-658-02810-7_49, © Springer Fachmedien Wiesbaden 2014

a single layer on top of a half-space. However, the study of SSI problems in practice necessitates tackling SSI problems under more complicated soil situations. Field investigation and laboratory experiments show that soils and rocks in nature invariably exhibit some degree of anisotropy in their response to static or dynamic stresses. To account for soil anisotropy in the study for estimating foundation response subjected to dynamic excitation is a real challenge. Few works can be found in the literature.

Gazetas [1] proposed a semi-analytical approach to study the static and dynamic response of strip foundations supported on a horizontally layered soil deposit. Each layer is modelled as a homogeneous cross-anisotropic medium with a vertical axis of material symmetry. He introduced two potential foundations which uncouple the 2D Navier-type equation of motion. Kausel E. [2]solved the problem of wave propagation in an anisotropic layered media by employing a discrete layer method. The displacement field within each layer is approximated by a linear expansion. As a result, the natural modes of wave propagation in a layered anisotropic stratum can be solved in terms of an algebraic eigenvalue equation involving narrowly bounded matrices.

In a previous paper of the authors [3] the precise integration approach is proposed for calculating dynamic impedance of strip foundations on arbitrary anisotropic layered half-space. In this paper, the technique is extended to solve the problem in three dimensional case. However, much improvement has to be made.

2 Wave motion equations for general anisotropic medium

A multi-layered soil including l layers overlying an elastic half-space is considered. The coordinate system is shown in Fig. 1. The elastic wave motion equation for an anisotropic medium in Cartesian coordinates is given by [4]

$$\mathbf{D}_{xx}\frac{\partial^2\mathbf{q}}{\partial x^2}+\mathbf{D}_{yy}\frac{\partial^2\mathbf{q}}{\partial y^2}+\mathbf{D}_{zz}\frac{\partial^2\mathbf{q}}{\partial z^2}+\left(\mathbf{D}_{xy}+\mathbf{D}_{yx}\right)\frac{\partial^2\mathbf{q}}{\partial x\partial y}+\left(\mathbf{D}_{yz}+\mathbf{D}_{zy}\right)\frac{\partial^2\mathbf{q}}{\partial y\partial z}+\left(\mathbf{D}_{xz}+\mathbf{D}_{zx}\right)\frac{\partial^2\mathbf{q}}{\partial x\partial z}=\rho\ddot{\mathbf{q}} \tag{1}$$

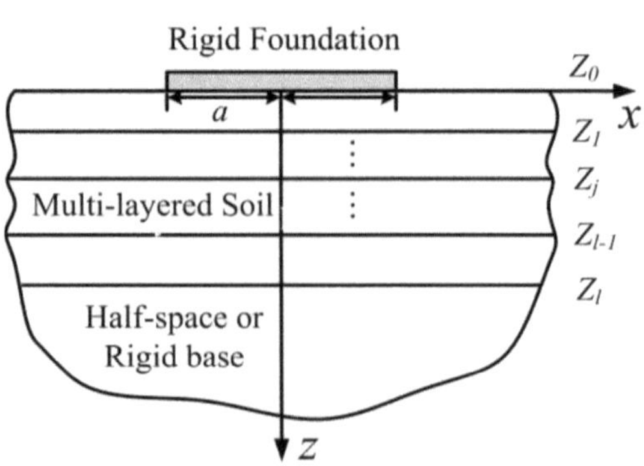

Fig. 1: A multi-layered soil system

in which the displacement vector $\mathbf{q}$ is defined as

$$\mathbf{q} = \begin{bmatrix} u_x & u_y & u_z \end{bmatrix}^T \tag{2}$$

For the transversely anisotropic material, the elements of constitutive matrix are simplified as follows

$$d_{11} = d_{22} = \lambda + 2G\,,\ d_{12} = \lambda \text{ and } d_{66} = G \quad \text{(in the isotropic plane)} \tag{3}$$

$$d_{33} = \lambda_t + 2G_t\,,\ d_{13} = d_{23} = \lambda_t \text{ and } d_{44} = d_{55} = G_t \text{ (in the transverse direction)} \tag{4}$$

Carrying out the Fourier transformation

$$\tilde{\mathbf{q}}(k_x, k_y, z, \omega) = \int_{-\infty}^{+\infty}\int_{-\infty}^{+\infty} \mathbf{q}(x, y, z, \omega) e^{-i(k_x x + k_y y)} dxdy \tag{5}$$

leads to the wave equation in frequency-wavenumber domain expressed as

$$\begin{aligned} &\mathbf{D}_{zz}\ddot{\tilde{\mathbf{q}}} - i\left[k_x(\mathbf{D}_{xz} + \mathbf{D}_{zx}) + k_y(\mathbf{D}_{yz} + \mathbf{D}_{zy})\right]\dot{\tilde{\mathbf{q}}} \\ &-\left[k_x^2\mathbf{D}_{xx} + k_x k_y(\mathbf{D}_{xy} + \mathbf{D}_{yx}) + k_y^2\mathbf{D}_{yy}\right]\tilde{\mathbf{q}} + \rho\omega^2\mathbf{q} = 0 \end{aligned} \tag{6}$$

where the superscript dots of $\ddot{\mathbf{q}}$ denote double differentiation with respect to z.

For brevity, Eq (6) is rewritten as

$$\mathbf{K}_{22}\ddot{\tilde{\mathbf{q}}} + (\mathbf{K}_{21} - \mathbf{K}_{12})\dot{\tilde{\mathbf{q}}} - (\mathbf{K}_{11} - \rho\omega^2\mathbf{I})\mathbf{q} = 0 \tag{7}$$

with

$$\mathbf{K}_{22} = \mathbf{D}_{zz}\,,\quad \mathbf{K}_{21} = -\mathbf{K}_{21}{}^T = ik_x\mathbf{D}_{xz} + ik_y\mathbf{D}_{yz}$$

$$\mathbf{K}_{11} = k_x^2\mathbf{D}_{xx} + k_y^2\mathbf{D}_{yy} + k_x k_y(\mathbf{D}_{xy} + \mathbf{D}_{yx})$$

Finally, a first order linear differential equation in the state space is obtained.

$$\dot{\mathbf{V}} = \mathbf{H}\mathbf{V} \tag{8}$$

where

$$\mathbf{V} = \begin{Bmatrix} \tilde{\mathbf{q}} \\ \tilde{\mathbf{p}} \end{Bmatrix},\ \mathbf{H} = \begin{bmatrix} \mathbf{A} & \mathbf{D} \\ \mathbf{B} & \mathbf{C} \end{bmatrix} \tag{9}$$

and

$$\mathbf{p} = \begin{bmatrix} \tau_{xz} & \tau_{yz} & \sigma_z \end{bmatrix}^T,\ \tilde{\mathbf{p}} = -(\mathbf{K}_{22}\dot{\tilde{\mathbf{q}}} + \mathbf{K}_{21}\tilde{\mathbf{q}}) \tag{10}$$

The boundary condition for wave motion equation at the free surface is

$$\tilde{\mathbf{p}}_0 = \tilde{\mathbf{p}}(z = 0) = \mathbf{0} \tag{11}$$

At the interface between two adjacent layers, the continuity conditions lead to

$$\tilde{\mathbf{p}}(z_r^+)=\tilde{\mathbf{p}}(z_r^-),\quad \tilde{\mathbf{q}}(z_r^+)=\tilde{\mathbf{q}}(z_r^-)\ (r=1,2,3...l) \tag{12}$$

In case the multi-layered soil rests on rigid base, we have

$$\tilde{\mathbf{q}}_l=\tilde{\mathbf{q}}(z=z_l)=\mathbf{0} \tag{13}$$

Whereas for multi-layered soil overlying an elastic half-space, the radiative condition should be considered [6].

$$\tilde{\mathbf{q}}_l=\mathbf{R}_\infty\tilde{\mathbf{p}}_l \text{ with } \mathbf{R}_\infty=\mathbf{\Phi}_{12}\mathbf{\Phi}_{22}^{-1} \tag{14}$$

where $\mathbf{\Phi}$ is the eigenvector of $\mathbf{H}$.

3 The precise integration method

The general solution to the differential equation (8) takes the form

$$\mathbf{V}=\exp(\mathbf{H}z)\mathbf{c} \tag{15}$$

where $\mathbf{c}$ is the integration constants.

For a typical layer $(\eta=z_b-z_a)$ of thickness η within the interval of the soil stratum $[z_a,z_b]$, the relationship between the displacements and stresses at the two ends of the layer is found from Eq (15) as

$$\mathbf{V}_b=\exp(\mathbf{H}\eta)\mathbf{V}_a \tag{16}$$

Rewrite Eq (16) in the following form

$$\mathbf{V}_b=\mathbf{T}\mathbf{V}_a \tag{17}$$

$$\mathbf{T}=\exp(\mathbf{H}\eta)=\mathbf{I}+\mathbf{H}\eta+\frac{1}{2!}(\mathbf{H}\eta)^2+\frac{1}{3!}(\mathbf{H}\eta)^3+\frac{1}{4!}(\mathbf{H}\eta)^4+... \tag{18}$$

where $\mathbf{I}$ is an unitary matrix.

The precise integration method presented by Zhong [5] is applied for the evaluation of $\mathbf{T}$, which takes the form

$$\mathbf{T}=\mathbf{I}+\mathbf{T}_r^N \tag{19}$$

with

$$\mathbf{T}_r^i=2\mathbf{T}_r^{i-1}+\mathbf{T}_r^{i-1}\times\mathbf{T}_r^{i-1}$$

$$\mathbf{T}_r^0=\mathbf{H}\tau+\frac{1}{2}(\mathbf{H}\tau)\left[(\mathbf{H}\tau)+\frac{1}{3}(\mathbf{H}\tau)^2+\frac{1}{12}(\mathbf{H}\tau)^3\right],\quad \tau=\eta/2^N \tag{20}$$

It is therefore clear that $\mathbf{T}$ is evaluated by applying the recursive formula (19) N times ($N=20$), and the size of matrices is (6×6). In this way, any desired precision can be achieved. The numerical result reaches the computer precision.

4 Assembly of layers

Integration of the wave equation (8) by applying PIM yields the relationship between the displacements and stiffness at the two ends z_a and z_b of a layer. Writing it in partitioned form leads to

$$\begin{Bmatrix} \tilde{\mathbf{q}}_b \\ \tilde{\mathbf{p}}_b \end{Bmatrix} = \mathbf{T} \begin{Bmatrix} \tilde{\mathbf{q}}_a \\ \tilde{\mathbf{p}}_a \end{Bmatrix}, \quad \mathbf{T} = \begin{bmatrix} \mathbf{T}_A & \mathbf{T}_D \\ \mathbf{T}_B & \mathbf{T}_C \end{bmatrix} \tag{21}$$

In order to ease the assembly of layers, rearrange Eq (21) into following dual-vector form:

$$\tilde{\mathbf{q}}_b = \mathbf{M}_F \tilde{\mathbf{q}}_a - \mathbf{M}_G \tilde{\mathbf{p}}_b, \; \tilde{\mathbf{p}}_a = \mathbf{M}_Q \tilde{\mathbf{q}}_a + \mathbf{M}_E \tilde{\mathbf{p}}_b \tag{22}$$

with

$$\mathbf{M}_F = \mathbf{T}_A - \mathbf{T}_D \mathbf{T}_C^{-1} \mathbf{T}_B, \; \mathbf{M}_G = -\mathbf{T}_D \mathbf{T}_C^{-1}, \; \mathbf{M}_Q = -\mathbf{T}_C^{-1} \mathbf{T}_B, \; \mathbf{M}_E = -\mathbf{T}_C^{-1} \tag{23}$$

Assembly of layers is performed two by two. For the details the readers may refer to [6]. Assembly of layers is proceeded based on matrix algebra with the size of matrices equal to (3×3). The computational effort is reduced to a great extent, whereas high precision is ensured.

Eventually, for layered strata consisting of l layers, the following relationship holds

$$\tilde{\mathbf{q}}_l = \mathbf{M}_F^s \tilde{\mathbf{q}}_0 - \mathbf{M}_G^s \tilde{\mathbf{p}}_l, \quad \tilde{\mathbf{p}}_0 = \mathbf{M}_Q^s \tilde{\mathbf{q}}_0 + \mathbf{M}_E^s \tilde{\mathbf{p}}_l \tag{24}$$

For layered stratum bonded to rigid base, the boundary condition Eq. (13) leads to the relationship between surface displacements and tractions as

$$\tilde{\mathbf{p}}_0 = \left(\mathbf{M}_Q^s + \mathbf{M}_E^s \left(\mathbf{M}_G^s \right)^{-1} \mathbf{M}_F^s \right) \tilde{\mathbf{q}}_0 = \tilde{\mathbf{S}}\left(k_x, k_y, 0, \omega\right) \tilde{\mathbf{q}}_0 \tag{25}$$

Whereas for layered strata overlying elastic half-space, the boundary condition Eq. (14) results in

$$\tilde{\mathbf{p}}_0 = \left(\mathbf{M}_Q^s + \mathbf{M}_E^s \mathbf{R}_\infty \left(\mathbf{I} + \mathbf{M}_G^s \mathbf{R}_\infty \right)^{-1} \mathbf{M}_F^s \right) \tilde{\mathbf{q}}_0 = \tilde{\mathbf{S}}\left(k_x, k_y, 0, \omega\right) \tilde{\mathbf{q}}_0 \tag{26}$$

Elements of $\tilde{\mathbf{S}}\left(k_x, k_y, 0, \omega\right)$ in Eq. (26) denote the dynamic impedance coefficients of the whole stratum condensed at the surface of the stratum. The relevant dynamic flexibility coefficients are found by inversion of $\tilde{\mathbf{S}}\left(k_x, k_y, 0, \omega\right)$.

$$\begin{aligned} \tilde{\mathbf{q}}_0\left(k_x, k_y, 0, \omega\right) &= \tilde{\mathbf{F}}\left(k_x, k_y, 0, \omega\right) \tilde{\mathbf{p}}_0\left(k_x, k_y, 0, \omega\right) \\ \tilde{\mathbf{F}}\left(k_x, k_y, 0, \omega\right) &= \tilde{\mathbf{S}}\left(k_x, k_y, 0, \omega\right)^{-1} \end{aligned} \tag{27}$$

5 Evaluation of Green's influence function in spatial domain

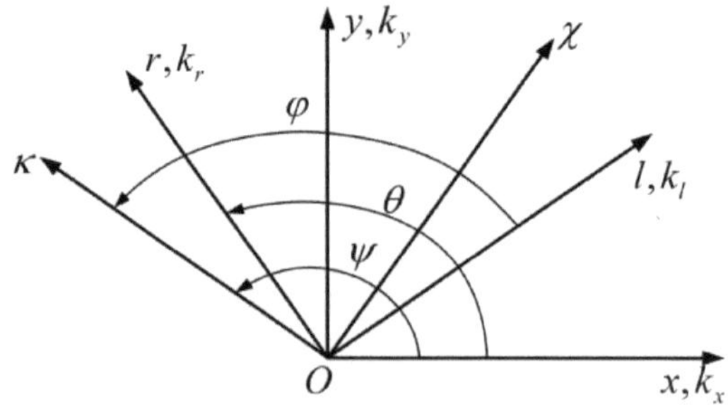

Fig. 2: Definition of coordinate system

The transformation from the matrix of flexibility coefficients in the wave-number domain Eq. (27) into the Cartesian space domain involves a double inverse Fourier transformation, which is a time-consuming operation

$$\mathbf{q}(x,y,0,\omega)=\frac{1}{4\pi^2}\int_{-\infty}^{+\infty}\int_{-\infty}^{+\infty}\tilde{\mathbf{q}}\left(k_x,k_y,0,\omega\right)e^{i\left(k_x x+k_y y\right)}dk_x dk_y \tag{28}$$

The computation time can be greatly reduced, if the variables k_x and k_y of the integral in Eq. (28) are expressed in a cylindrical polar coordinate system

The coordinate transformation is undergone by the following matrix

$$\begin{Bmatrix} k_x \\ k_y \\ z \end{Bmatrix}=\mathbf{R}(\psi)\begin{Bmatrix} \chi \\ \kappa \\ z \end{Bmatrix},\ \mathbf{R}(\psi)=\begin{bmatrix} \sin\psi & \cos\psi & 0 \\ -\cos\psi & \sin\psi & 0 \\ 0 & 0 & 1 \end{bmatrix} \tag{29}$$

Sheng *et. Al* [7], Andersen and Clausen [8] pointed out that the treatment is further simplified if the integral is performed along the line defined by $k_x=0$ such that

$$\tilde{\mathbf{F}}\left(k_x,k_y,0,\omega\right)=\mathbf{R}(\psi)\tilde{\mathbf{F}}(0,\kappa,0,\omega)\mathbf{R}(\psi)^T \tag{30}$$

Similarly to the transformation of the wave-numbers from $\left(k_x,k_y\right)$ into (χ,κ) provided by Eq. (29), the Cartesian coordinate system is rotated around the z axis according to the transformation (see Fig. 2).

$$\begin{Bmatrix} x \\ y \\ z \end{Bmatrix}=\mathbf{R}(\theta)\begin{Bmatrix} l \\ r \\ z \end{Bmatrix},\ \mathbf{R}(\theta)=\begin{bmatrix} \sin\theta & \cos\theta & 0 \\ -\cos\theta & \sin\theta & 0 \\ 0 & 0 & 1 \end{bmatrix} \tag{31}$$

The complex amplitudes of displacements and tractions follow by the transformation

$$\begin{aligned}\mathbf{q}(x,y,0,\omega)&=\mathbf{R}(\theta)\hat{\mathbf{q}}(0,r,0,\omega)\\ \mathbf{p}(x,y,0,\omega)&=\mathbf{R}(\theta)\hat{\mathbf{p}}(0,r,0,\omega)\end{aligned} \tag{32}$$

with

$$x=r\cos\theta,\quad y=r\sin\theta,\ r=\sqrt{x^2+y^2},\ \tan\theta=y/x \tag{33}$$

For carrying out double Fourier transformation in cylindrical polar coordinates, the coordinates transformation Eq. (29) and Eq. (31) are conveniently combined by introducing the angle (the Fig. 2)

$$\varphi=\frac{\pi}{2}+\psi-\theta \tag{34}$$

defining the rotation of the wave number (χ,κ) relatively to the spatial coordinates (l,r). And the transformation matrix takes the form

$$\mathbf{R}(\psi)=\mathbf{R}(\theta)\mathbf{R}(\phi) \tag{35}$$

By applying Eqs. (27), (29), (33) and (35), the double inverse Fourier transform Eq. (28) is undergone by the following expression

$$\begin{aligned}\mathbf{q}_0(x,y,z,\omega)&=\frac{1}{4\pi^2}\int_{-\infty}^{+\infty}\int_{-\infty}^{+\infty}\tilde{\mathbf{F}}(k_x,k_y,0,\omega)\tilde{\mathbf{p}}(k_x,k_y,0,\omega)e^{i(k_xx+k_yy)}dk_xdk_y\\ &=\frac{1}{4\pi^2}\int_0^{+\infty}\int_0^{2\pi}\mathbf{R}(\psi)\tilde{\mathbf{F}}(0,\kappa,0,\omega)\mathbf{R}(\psi)^T\tilde{\mathbf{p}}(0,k_y,0,\omega)e^{i\kappa r\sin\varphi}d\varphi\kappa d\kappa\\ &=\frac{1}{4\pi^2}\int_0^{+\infty}\int_0^{2\pi}\mathbf{R}(\theta)\mathbf{R}(\varphi)\tilde{\mathbf{F}}(0,\kappa,0,\omega)\left[\mathbf{R}(\theta)\mathbf{R}(\varphi)\right]^T\tilde{\mathbf{p}}(0,k_y,0,\omega)e^{i\kappa r\sin\varphi}d\varphi\kappa d\kappa\end{aligned} \tag{36}$$

where $\kappa r\sin\varphi=k_xx+k_yy$ is identified as the scalar product of the two dimensional vectors with lengths κ and r, respectively, and $\pi/2-\varphi$ is the plane angle between these vectors as given by Eq. (34).

Green's influence functions are evaluated for subdisk-elements of radius r_0 subjected to uniformly distribute vertical load p_{z0} and horizontal loads p_{x0} and p_{y0}.

$$\begin{aligned}\mathbf{q}_0(x,y,0,\omega)&=\begin{Bmatrix}u_x\\u_y\\u_z\end{Bmatrix}=\frac{1}{\kappa\pi r_0^2}\mathbf{R}(\theta)\int_0^{+\infty}\mathbf{G}\mathbf{R}(\theta)^TJ_1(\kappa r_0)\mathbf{p}_0\kappa d\kappa\\ \mathbf{G}&=\frac{1}{2\pi}\int_0^{2\pi}\mathbf{R}(\varphi)\tilde{\mathbf{F}}(0,\kappa,0,\omega)\mathbf{R}(\varphi)^Te^{i\kappa r\sin\varphi}d\varphi\end{aligned} \tag{37}$$

with $\mathbf{p}_0=\begin{bmatrix}p_{x0} & p_{y0} & p_{z0}\end{bmatrix}^T$.

Note that in Eq. (37), due to the fact that the load is applied with rotational symmetry around the z-axis, the vector $\mathbf{p}_0$ may be taken outside the integral over ϕ. Thus, Eq. (37) only involves numerical integration in one dimension. This provides an efficient evaluation of the complex amplitudes of the surface displacements.

The elements of matrix $\mathbf{G}$ in Eq. (37) may be identified as integral representations of Bessel functions, which can be computed in an efficient manner by their series expansions.

Summarizing all the horizontal and vertical load cases, we obtain the frequency domain relationship between the surface tractions and the displacement amplitudes in Cartesian coordinates for a subdisk-element as follows

$$\begin{Bmatrix} u_x(x,y,0,\omega) \\ u_y(x,y,0,\omega) \\ u_z(x,y,0,\omega) \end{Bmatrix} = \begin{bmatrix} F_{xx} & F_{xy} & F_{xz} \\ F_{yx} & F_{yy} & F_{yz} \\ F_{zx} & F_{zy} & F_{zz} \end{bmatrix} \begin{Bmatrix} p_{x0} \\ p_{y0} \\ p_{z0} \end{Bmatrix} \tag{38}$$

where $(x,y,0)$ denotes the coordinates of the subdisk centre.

6 Dynamic impedance for arbitrary-shaped foundation

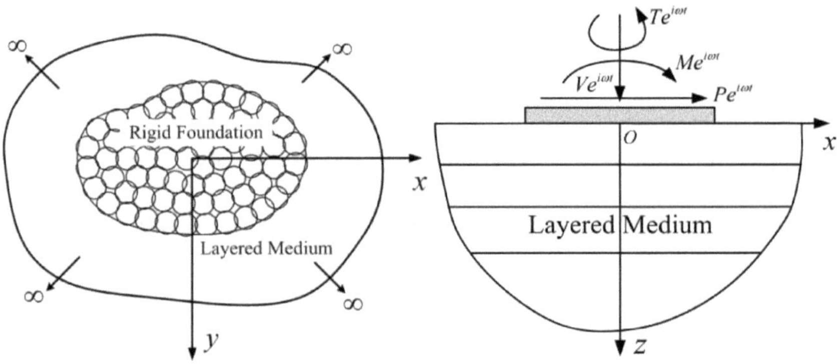

Fig. 3: Foundation with subdisk discretization

The interface between the foundation and the soil is discretized into n subdisk-elements of equal radius, such that the total area equals the area of foundation interface. Six cases are studied, i.e. the foundation is subjected to three components of concentrated harmonic forces and three components of harmonic moments with amplitudes equal to P_x, P_y, P_z, M_x, M_y and M_z respectively (see Fig. 3).

Based on Eq. (38), the dynamic impedance matrix $\mathbf{S}(\omega)$ is found as[6]

$$\mathbf{S}(\omega)=\mathbf{N}^T\mathbf{F}_u^{-1}\mathbf{N} \tag{39}$$

with

$$\mathbf{N}=\begin{bmatrix}\mathbf{N}_1 & \mathbf{N}_2 & \cdots & \mathbf{N}_n\end{bmatrix}^T,\ \mathbf{N}_i=\begin{bmatrix}1 & 0 & 0 & 0 & 0 & -y_i\\ 0 & 1 & 0 & 0 & 0 & x_i\\ 0 & 0 & 1 & y_i & -x_i & 0\end{bmatrix}\ (i=1,2,...,n)$$

$$\mathbf{F}_u=\begin{bmatrix}\mathbf{F}_u^{11} & \mathbf{F}_u^{12} & \cdots & \mathbf{F}_u^{1n}\\ \mathbf{F}_u^{21} & \mathbf{F}_u^{22} & \cdots & \mathbf{F}_u^{2n}\\ \vdots & \vdots & \ddots & \vdots\\ \mathbf{F}_u^{n1} & \mathbf{F}_u^{n2} & \cdots & \mathbf{F}_u^{nn}\end{bmatrix} \tag{40}$$

7 Numerical Examples

Two numerical examples are provided. The first one aims at verifying the accuracy and efficiency of the proposed approach, isotropic soil medium is considered. And the second one is intended to test the applicability of the proposed approach dealing with anisotropic soil stratum and to study the effect of anisotropy on the dynamic response of the foundation, cross-anisotropic material is considered.

7.1 A rigid square foundation on a soil layer overlying an elastic half-space

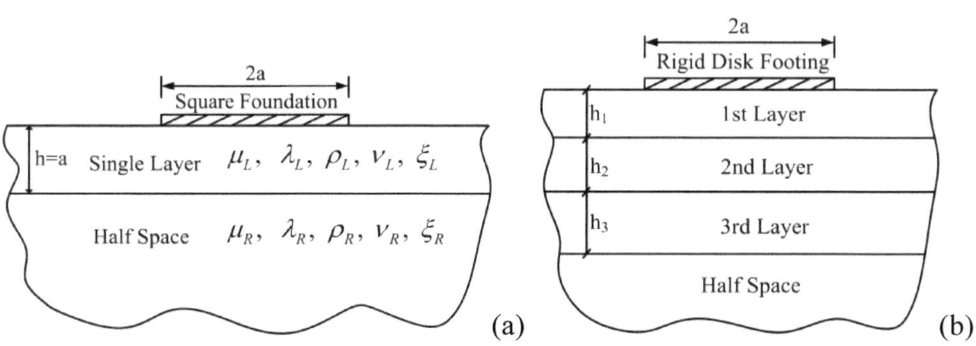

Fig. 4: Arbitrary-shaped foundation on multi-layered soil

A square surface foundation of the dimension 2a×2a on isotropic soil layer and the underlying half-space is considered (Fig. 4a). This case was studied by Wong and Luco [9]. It can be observed (Fig. 5), excellent agreement between the proposed approach and the solutions of Wong and Luco is achieved.

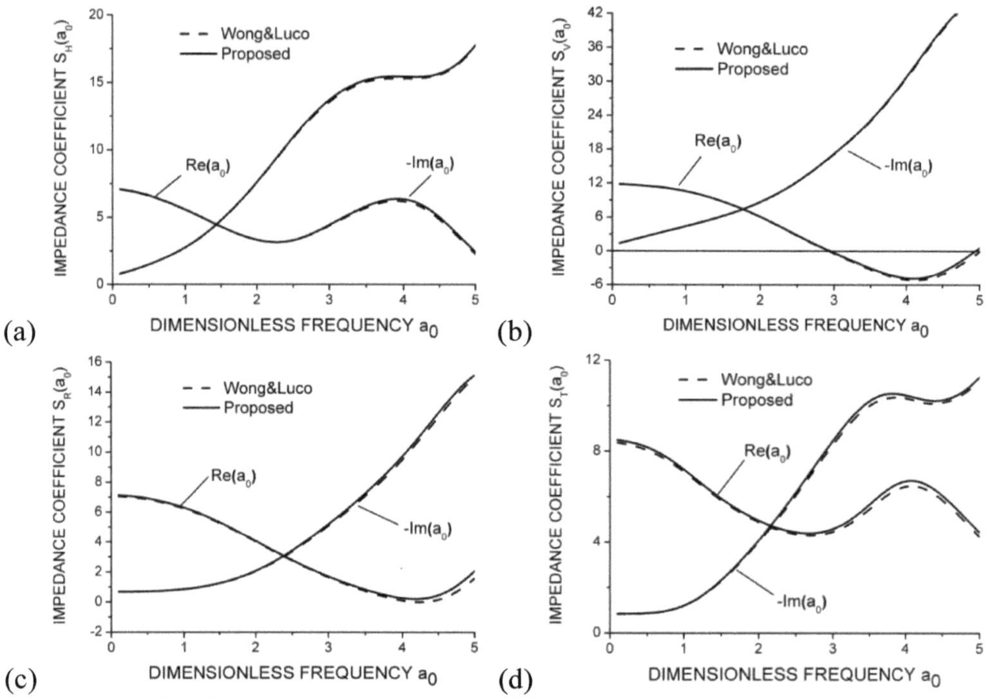

Fig. 5: Dynamic impedance coefficients for square foundations (a) horizontal (b) vertical (c) rocking (d) torsional

7.2 A rigid circular foundation on an anisotropic multi-layered half-space

Table 1 Material properties of the layers and the half-space

layer	E_h	ν_H	ν_r	ρ	ξ	h
1	1	1/3	0.25	1.0	0.05	0.8a
2	1.2	0.30	0.33	1.1	0.05	1.0a
3	1.5	0.25	0.25	1.2	0.05	1.2a
4	2.0	0.301	0.22	1.4	0.05	Semi-infinite

The wave propagation in a real anisotropic three layered stratum and the underlying half-space (Fig. 4b) is studied. The material properties of the layers and the half-space are given in table 1. It is assumed as cross-anisotropic. The evaluated frequency-dependent dynamic compliance coefficients of a rigid circular foundation of radius *a* are shown in Fig. 6. Due to the limited space, only components C_{VV} and C_{RR} are presented. The degree of anisotropy on the dynamic response of the foundation is examined by varying the coefficient of anisotropy n=1/3, 1 and 2 for all layers and the half-space. It is seen, the material anisotropy has great influence on the compliance function and the resulted dynamic response of structures.

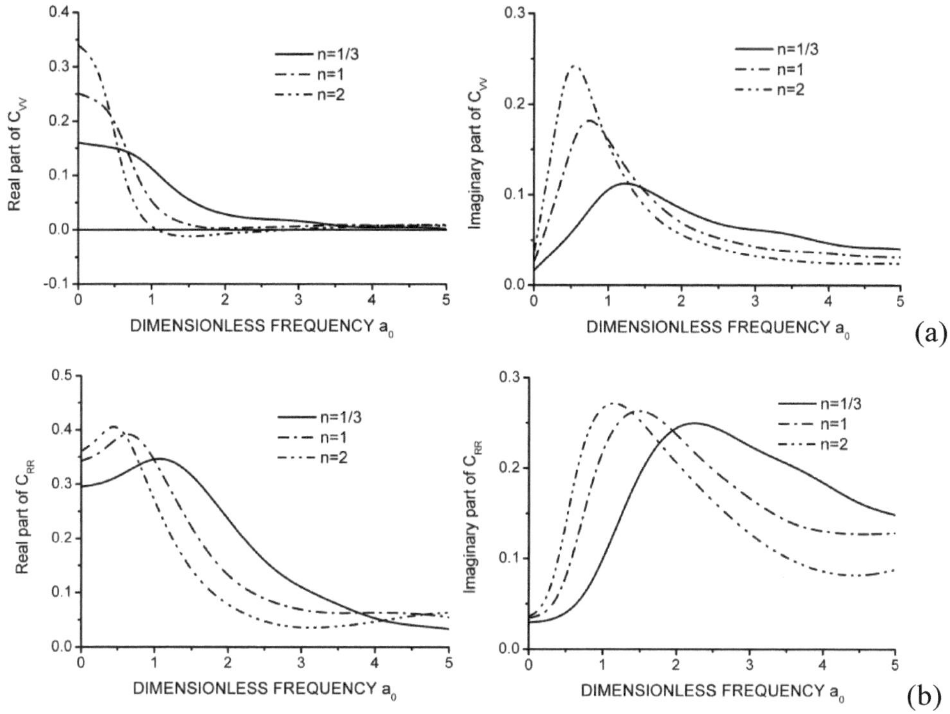

Fig. 6: Dynamic compliance coefficients for rigid circular foundation (a) vertical (b) rocking

8 Conclusion

An approach is proposed for the evaluation of dynamic impedance of arbitrary-shaped foundation on anisotropic multi-layered half-space. The wave-motion equation is solved analytically in the frequency-wavenumber domain, and any anisotropy of the medium can be handled with relative ease. The precise integration method is used to perform the integration of the results, any desired accuracy can be achieved, and the precision is limited only by the precision of the computer used. Dual vector form representation of the wave motion equation makes the assembly of two adjacent layers convenient and efficient. The computation is based on the matrix algebra with the size of matrices equal to (3×3) or (6×6). The computation is always stable. There is no limit of the number of layers and no limit of the thickness of the layer to be considered. No additional effort is needed in case the presence of an underlying half-space. Numerical examples show that the proposed approach is accurate and efficient.

9 Acknowledgements

The authors are grateful for the financial support of the Sino-German Science Foundation under grant no.GZ566 and the National Natural Science Foundation of China under grant no.51138001.

REFERENCES

[1] Gazetas G.: Strip foundations on a cross-anisotropic soil layer subjected to dynamic loading. Geotechnique; 1981, 31: 161-179.

[2] Kausel E.: Wave propagation in anisotropic layered media; Int. J. Numer. Methods Eng.; 1986, 23: 1567-1578.

[3] Lin G.; Han Z.; Zhong H. and Li J.: A precise integration approach for dynamic impedance of rigid strip footing on arbitrary anisotropic layered half-space; Soil Dyn. and Earthquake Eng.; 2013, 49: 96-108.

[4] Kausel D.: Fundamental solutions in elastodynamics, A compendium, Cambridge University Press, New York, 2006, ISBN.

[5] Zhong W. X.: On precise integration method; J. of Computational and Applied Mathematics; 2004, 163: 59-78.

[6] Lin G.: Dynamic Impedance of Foundation on Multi-Layered Half-Space; International Conference on Seismic Design of industrial Facilities; 2013, RWTH Aachen University.

[7] Sheng X, Jones C. J. C., Petyt M.: Ground vibration generated by a harmonic load acting on a railway track. Journal of Sound and Vibration; 1999, 225: 3-28.

[8] Andersen L., Clausen J.: Impedance of surface footings on layered ground; Computers and structures; 2008, 86: 72-87.

[9] Wong H L, Luco J. E.: Tables of impedance functions for square foundations on layered media. Soil Dynamics and Earthquake Engineering; 1985, 4: 64-81.

S e D I F **Conference**

International Conference on
Seismic Design of Industrial Facilities
2013, RWTH Aachen University

Two Parameters to Improve the Accuracy of the Green's Functions Obtained via the Thin Layer Method

Lin Chen

[1] Lehrstuhl für Baustatik und Baudynamik, RWTH Aachen University,
Mies-van-der-Rohe-Str. 1, 52074 Aachen, Germany
lin.chen@lbb.rwth-aachen.de

ABSTRACT:

Two parameters are proposed to improve the accuracy of the Green's functions for a layered half space modelled with the thin layer method (TLM). The parameters, which define the thickness of the thin sub-layer and the buffer layer in the thin layer method, rely on the observation of the Green's functions for a homogeneous half space. Based on them, the convergence of the Green's functions at both high-frequency and low-frequency range can be ensured; and the efficiency of the thin layer method is improved.

Keywords: Green's functions, thin layer method, precise integration method, buffer layer, thin sub-layers

1 Introduction

The Green's functions of a layered half space yields the best description of the dynamic properties of layered medium in many problems in soil-structure interaction, earthquake engineering and seismology. This topic is not new, as evidenced by the well-known works [1-5]. While these solutions have some theoretical appeal in themselves, they are really more important as tools in the solution of the involved boundary value problems arising in seismology and geo-mechanics. Despite of the considerable work that has been done up to this date, the solutions available so far are restricted to medium of relatively simple geometry, such as full space and half space. Close form solutions for a layered medium with arbitrary boundary conditions do not exist. The complexities introduced by layering are formidable because the integral formulations need to be evaluated numerically [6-11]. The formalism, to study the propagation of waves in layered media, was presented by many researchers in different approaches [12-15]. Among them, the thin layer method (TLM) is one of the most powerful tools for the dynamic analysis of laminated media. It is similar to the stiffness matrix method [16], but based on discrete formulation. It consists of a discretization in the

S. Klinkel et al. (Eds.), *Seismic Design of Industrial Facilities*,
DOI 10.1007/978-3-658-02810-7_50, © Springer Fachmedien Wiesbaden 2014

direction of layering (commonly plane layers, but cylindrical or spherical layers can also be modelled) into a number of thin sub-layers [15, 17-21]. In the numerical implementation, the TLM is analogous to the finite element method (FEM); the thin sub-layers together with the layer interfaces and the number of sub-layers can be interpreted as the elements, nodes, and mesh refinement in the FEM. The main advantages of the thin layer method are briefly summarized as: (1) normal modes follow from a quadratic Eigen value problem, not a transcendental one, so standard solution methods can be applied; (2) integral transforms back into space can be carried out analytically—no need for numerical integration. The method is now widely used, for example in soil dynamics and soil-structure interaction. A brief review of the historical development of this method can be found, for example, in References [20, 22, 23]. However, the results of the thin layer method are not stable, like in the FEM, considering different meshing or discretizing approaches [17]. Therefore, it is essential to propose new approach to optimize the thin layer method in order to obtain the stable and accurate solutions.

2 Fundamental concepts

The detailed thin layer method (TLM) has been discussed earlier by Waas [20]. For better illustration, we will briefly introduce the fundamental ideas behind TLM:

- Firstly, the half space under the layered medium is added a paraxial approximation at the base that is preceded by an appropriately thick buffer layer, which has the same properties as the half space [24-26].
- The medium is discretized, i.e. physical layers and 'buffer layer' are divided into thin sub-layers.
- Interpolation functions are used for the variation of displacements in the direction of layering.
- Weighted residual principles are used to manipulate the wave equation and boundary conditions. The resulting discrete equations of motion no longer contain partial derivatives with respect to the direction of discretization, but they are still continuous in the other directions, and also in time.
- The wave motion equations are solved in some fashion for one or more source terms, i.e. for different "right hand sides".

The formulation of the layer stiffness matrix for a single layer in the thin layer method can be obtained as [20, 27]

$$K_n = A_n k^2 + B_n k + G_n - \omega^2 M_n \qquad n = \text{layer index} \tag{1}$$

in which A_n, B_n, G_n and M_n are matrices that depend solely on the material properties: Lame constants λ_i and μ_i, Poisson's ratio v_i, damping ratio ξ_i; and the thickness of the layer h. These matrices can be visualized as resulting from a

Taylor series expansion in the thickness variable, but they are actually obtained via weighted residuals principles using interpolation functions. Explicit expressions for the layer stiffness matrix are given by Kausel [17]. In case of a soil which consists of several layers, the global stiffness matrix $K = \{K_n\}$ is constructed by overlapping the contribution of the layer matrices at each 'node' (interface) of the system.
According to the formulation of the layer stiffness matrix, there are several parameters affecting the accuracy of the thin layer method (TLM). In order to illustrate these parameters, we will present some comparisons with the 'exact' solutions [16] (i.e., formulating the functions with the continuum theory in the wavenumber domain, and integrating numerically). Consider the case of a homogeneous half space, subjected to a unit vertical point load at its free surface. For simplification, here the normalized parameters are used. The material density is $\rho = 1$, the Lame constants are $\lambda = 1.5$ and $\mu = 1$, the Poisson's ratio is $\nu = 0.3$, the damping ratio is $\xi = 0.05$. In the thin layer method, the homogeneous half space is first added an appropriately thick buffer layer with identical properties as the half space; and then the buffer layer is divided into a number of thin sub-layers. Two illustrative cases are calculated for the vertical component of the Green's functions $G_{zz}(r, \theta = 0, z = 0)$, which differ in the thickness of the buffer layer only, with 200 thin sub-layers involved. The excitation frequency is set as $\omega = 0.1 \sim 100$ rad/s. $a_0 = \omega r / V_s$ and $V_s = \sqrt{\mu / \rho}$ denote the dimensionless frequency and the shear wave velocity of the half space.

In the first case, the buffer layer thickness is $D = 20\text{m}$ and the thin sub-layers thickness is $h = 20/200 = 0.1\text{m}$. The results are presented in Fig. 1a. Reasonably good agreement between the results of the thin layer method and the 'exact' solutions can be reported in the low-frequency range, while errors exist in the high-frequency range.

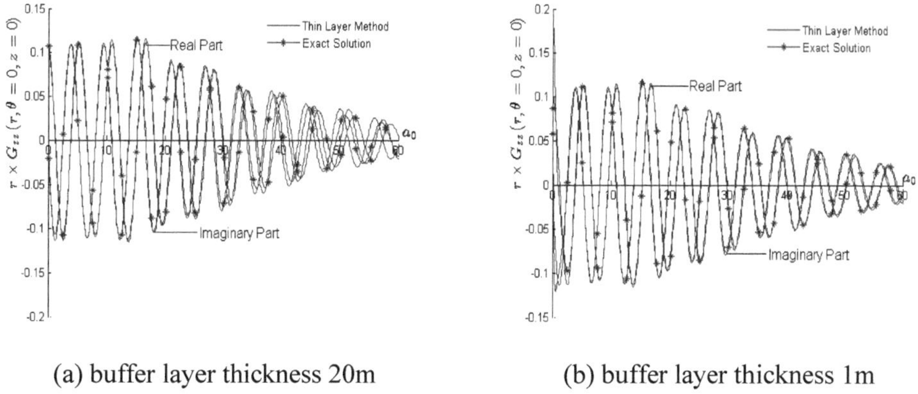

(a) buffer layer thickness 20m (b) buffer layer thickness 1m

Figure 1: Vertical component of Green's functions for the homogenous half space

In the second case, we consider a thinner buffer layer and take the thickness to be $D = 1\text{m}$. The thin sub-layers thickness is $h = 1/200 = 0.005\text{m}$. The results are presented in Fig. 1b. As expected, the agreement of the results is reasonable; however, deviations exist at the low-frequency range.

To sum up, the results of the thin layer method are in good agreement with those obtained by the 'exact' solutions except in some particular range. In the first case, in low-frequency range, the results of the two methods fit well; while differences exist in the high-frequency range. In the second case, opposite comments can be made. Therefore, we can safely infer that the accuracy of the thin layer method is mainly determined by the thickness of the thin sub-layers and the buffer layer.

3 Thickness of the thin sub-layer and the buffer layer

In general, the accuracy in the high-frequency range is determined by the thickness of the thin sub-layer; the accuracy in the low-frequency range is by the buffer layer thickness. In the following, we will present the thickness of the thin sub-layers and the buffer layer to obtain more stable and accurate results in the thin layer method.

3.1 Thickness of the thin sub-layer

In this case, the precise integration method (PIM) [28] will be employed in the comparison with the thin layer method (TLM). For simplification, we consider the case of a homogeneous half space subjected to a unit horizontal point load at its free surface. The density of the soil is $\rho = 1$, the Poisson's ratio is $\nu = 0.3$, and the Lame constants are $\lambda = 1.5$ and $\mu = 1$, the damping ratio is $\xi = 0.05$. For the implementation of the thin layer method (TLM), the thickness of the buffer layer is 75m with 300 thin sub-layers involved. The thickness of the thin sub-layer is $h = 0.25\text{m}$. The excitation frequency is set as $\omega = 0.1 \sim 10$ rad/s. The parameter ζ is defined as the ratio of the shear wavelength δ and the thickness of the thin sub-layer h, $\zeta = \delta/h = 2\pi V_s/\omega h$. The horizontal component of the Green's functions $G_{rr}(r, \theta = 0, z = 0)$ is plotted as a function of the parameter ζ in Fig. 2. The comparison of TLM and PIM shows very good results; only in the lower part of ζ there are some deviations. The deviation between TLM and PIM is redrawn in Fig. 3 for better illustration. Only $\zeta \le 40$ part is shown for easy reference. The deviation is defined as $\text{Diff=abs}((\text{TLM-PIM})/\text{PIM})$. In the engineering field, the deviation within 5% is acceptable. From the figure, when the parameter ζ is greater than 4, the comparison shows very good results.

$$\zeta = \delta/h = 2\pi V_s/\omega h \ge 4 \tag{2}$$

which means there are at least 4 thin sub-layers inside 1 shear wavelength.

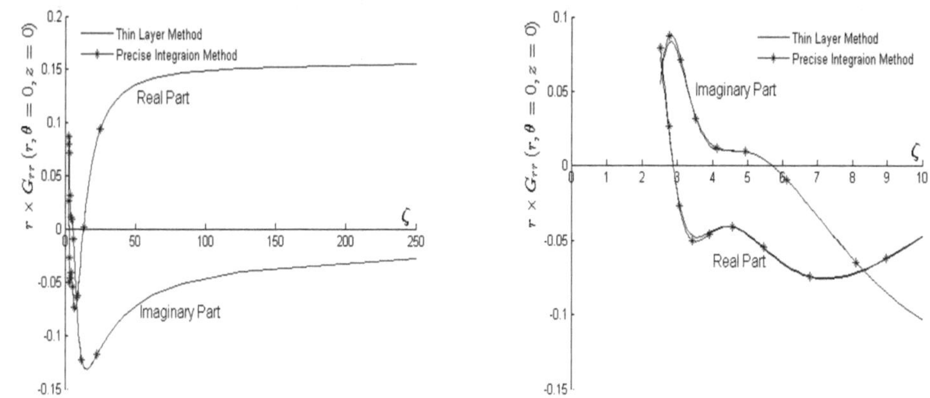

(a) whole range (b) lower part

Figure 2: Horizontal component of the Green's functions

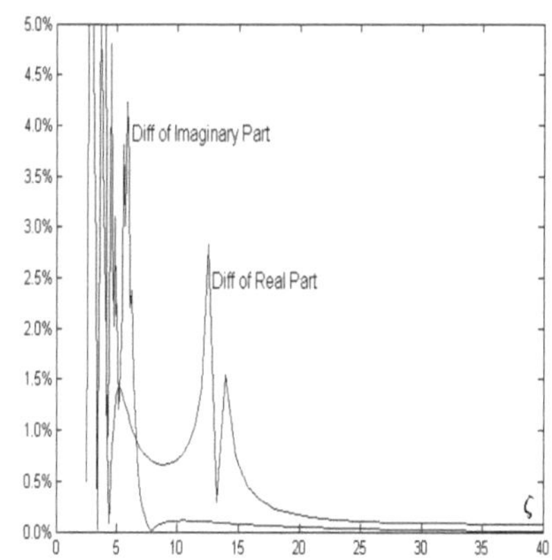

Figure 3: Deviations of the Green's functions between TLM and PIM

3.2 Thickness of the buffer layer

To illustrate this, a purely elastic homogenous half space (Damping ratio $\xi = 0$), subjected to a unit vertical point load at its free surface, is considered with the density $\rho = 1$, the Poisson's ratio $\nu = 0.3$ and the Lame constants $\lambda = 1.5$ and $\mu = 1$. As mentioned above, there is 'exact' solution existing [16], which is formulating in the wavenumber domain and integrating numerically, for the dynamic response of the homogenous half space. However, for the purely elastic solid ($\xi = 0$), the Green's functions in the wavenumber domain obtained by the 'exact' solutions [16] exhibit infinite peaks at certain wavenumber; the integration can not be performed for the singularity. But the thin layer method formulates as algebraic expressions, the integral transforms of the algebraic expressions can readily be evaluated without the numerical integrations. Therefore, we can utilize the thin layer method to deal with this special case. From the above analysis, the accuracy of the Green's functions in the low-frequency range is mainly determined by the thickness of the buffer layer; and the Green's functions tend to be the real value by increasing the buffer layer thickness. In order to obtain the range of the buffer layer thickness, we choose the Green's functions under the buffer layer thickness $D = 100\text{m}$ with 1000 thin sub-layers as the 'exact' value. In reality, the results under such situation are not the real value; however, they are more accurate compared to those of the buffer layer thickness less than 100m. The thickness of the thin sub-layer is 0.1m. The maximum frequency, under which one can obtain good results, is calculated by Eq. (2) as

$$\omega_{\max} = 2\pi V_s / 4h = 2\pi \times 1/(4 \times 0.1) = 5\pi = 15.7 \tag{3}$$

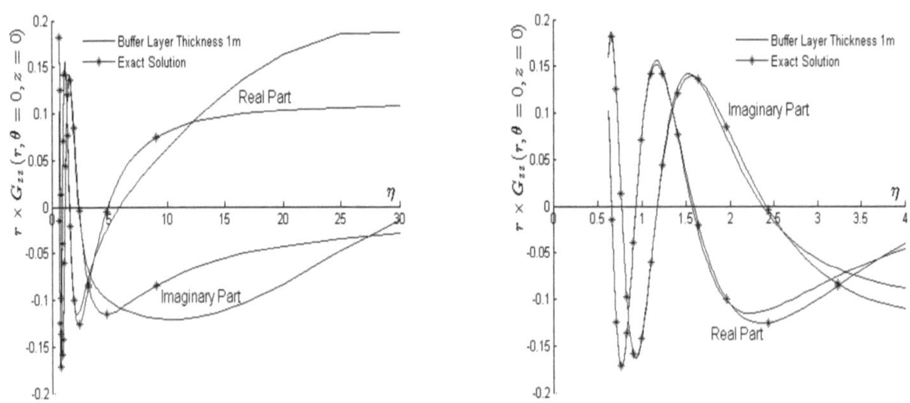

(a) whole range (b) lower part

Figure 4: Vertical component of the Green's functions

Here, we set the excitation frequency as $\omega = 0.1 \sim 10$ rad/s to satisfy Eq.(3). The Green's functions under the buffer layer thickness $D = 1\text{m}$ with the same thin sub-layer thickness $h = 0.1\text{m}$ are presented to compare with the 'exact' value. The parameter η is defined as the ratio of the shear wavelength δ and the thickness of the buffer layer D, $\eta = \delta/D = 2\pi V_s/\omega D$ (in this case, $D = 1\text{m}$). The real parts (solid lines) and the imaginary part (segmented lines) of the vertical component of the Green's functions $G_{zz}(r, \theta = 0, z = 0)$ are plotted as a function of the parameter η in Fig. 4a. The comparison shows very good results in the lower range of η, however, worse in the higher part. For better illustration, we redraw only the results in the lower part in Fig. 4b. In the figure, when the parameter η is smaller than 1, it presents very good comparisons.

$$\eta = \delta / D = 2\pi V_s / \omega D \leq 1 \tag{4}$$

which means the thickness of buffer layer should be at least 1 shear wavelength.

Based on Eq. (2) and (4), the number of the thin sub-layers of the buffer layer should follow the rule

$$n = \frac{D}{h} = \frac{2\pi V_s / \omega h}{2\pi V_s / \omega D} = \frac{\zeta}{\eta} \geq 4 \tag{5}$$

which implies there are at least 4 thin sub-layers to represent the buffer layer in the half space. Normally, we can directly set the number of the thin sub-layers within the buffer layer as 4 to reduce the calculation effort.

In conclusion, for a layered half space, we can set the thickness of the buffer layer in the half space by Eq. (4) for every exciting frequency; and we can determine the thickness of thin sub-layer in the physical layers and buffer layer based on Eq. (2).

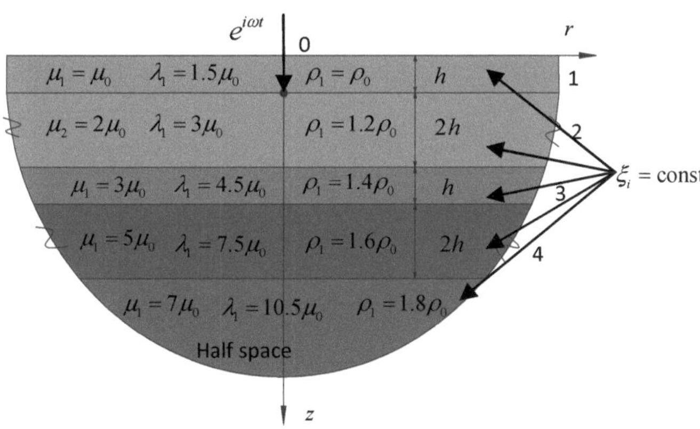

Figure 5: Vertical point source in a four-layer half space

4 Numerical example

The parameters described in the preceding sections have been implemented in a computer program that may be used to evaluate the Green's functions for an arbitrarily layered medium. In order to verify the parameters as well as the program, comparisons were performed for some cases with the results obtained by the precise integration method [28].

Example: A four-layer half space (See Fig. 5), which is subjected to a unit vertical harmonic point load, has been investigated here. Four illustrative examples were considered. We set here $\mu_0 = 1$ and $\rho_0 = 1$, the thickness $h/\lambda_1 = 1/1.5$ and the damping ratio $\xi = 0.05$. The four illustrative examples only differ in the depth of the calculation point: on the plane $z = 0$, $z = h$, $z = 2h$ and $z = 3h$, respectively. The excitation frequency is set as $\omega = 0.1 \sim 10$ rad/s with step 0.1 rad/s. The vertical component of the Green's functions is presented in Fig. 6. They are compared with the results obtained by the precise integration method [28]. In the numerical implementation of the thin layer method, the thickness of the buffer layer for the half space is set based on Eq. (4); and we can calculate the buffer layer thickness for every exciting frequency. However, for simplification in programming, we do not compute the buffer layer thickness for every frequency; we only calculate the maximum thickness in order to satisfy all frequencies.

$$D_{\max} \geq 2\pi V_s / (\omega_{\min} \times 1) = 2\pi \times 1.97 / (0.1 \times 1) = 120\text{m} \tag{6}$$

Therefore, the thickness of the buffer layer is set as $D = 120\text{m}$.

The thickness of the thin sub-layer is calculated by Eq. (2). For simplifying calculation, we choose the minimum thickness to satisfy every frequency. The thickness of the thin sub-layer for every layer is presented as follows.

$$h_{\min} \leq 2\pi V_{s1} / (\omega_{\max} \times 4) = 2\pi \times 1 / (10 \times 4) = 0.157\text{m} \qquad \text{for Layer 1} \tag{7}$$

Set the thickness of thin sub-layer in Layer 1 as $h = 0.1\text{m}$ with 10 thin sub-layers.

$$h_{\min} \leq 2\pi V_{s2} / (\omega_{\max} \times 4) = 2\pi \times 1.29 / (10 \times 4) = 0.2\text{m} \qquad \text{for Layer 2} \tag{8}$$

Choose thickness of thin sub-layer in Layer 2 as $h = 0.2\text{m}$ with 10 thin sub-layers.

$$h_{\min} \leq 2\pi V_{s3} / (\omega_{\max} \times 4) = 2\pi \times 1.464 / (10 \times 4) = 0.23\text{m} \qquad \text{for Layer 3} \tag{9}$$

Select the thickness of thin sub-layer in Layer 3 as $h = 0.2\text{m}$ with 5 thin sub-layers.

$$h_{\min} \leq 2\pi V_{s4} / (\omega_{\max} \times 4) = 2\pi \times 1.768 / (10 \times 4) = 0.28\text{m} \qquad \text{for Layer 4} \tag{10}$$

Set the thickness of thin sub-layer in Layer 4 as $h = 0.2\text{m}$ with 10 thin sub-layers.

The thickness of the thin sub-layer in the buffer layer is

$$h_{\min} \leq 2\pi V_s / (\omega_{\max} \times 4) = 2\pi \times 1.768 / (10 \times 4) = 0.3\text{m} \qquad \text{for Buffer Layer} \tag{11}$$

Choose the thickness of the thin sub-layer in the buffer layer as $h = 0.3\text{m}$ with 400 thin sub-layers. In general, there are 435 thin sub-layers involved in the model of thin layer method.

The comparisons are presented in Fig. 6. $a_0 = \omega r / V_{s1}$ and $V_{s1} = \sqrt{\mu_1 / \rho_1}$ denote the dimensionless frequency and the shear wave velocity of the surface layer. It can be seen from the figures that both methods produce almost identical results.

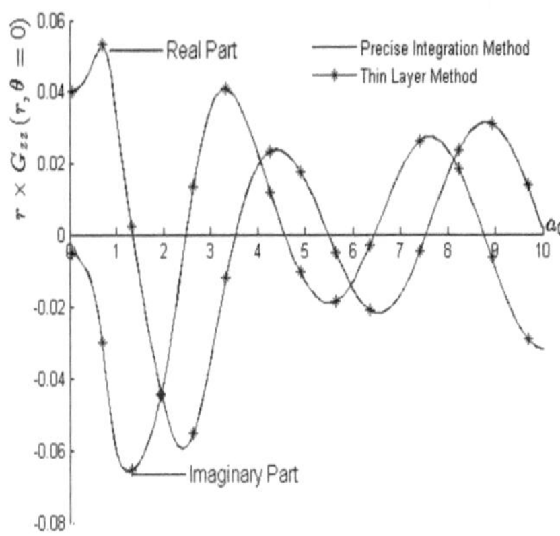

(a) on the free surface

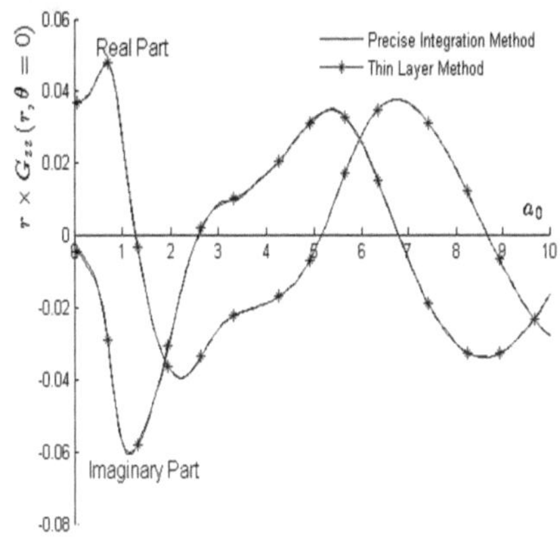

(b) on the plane $z = h$

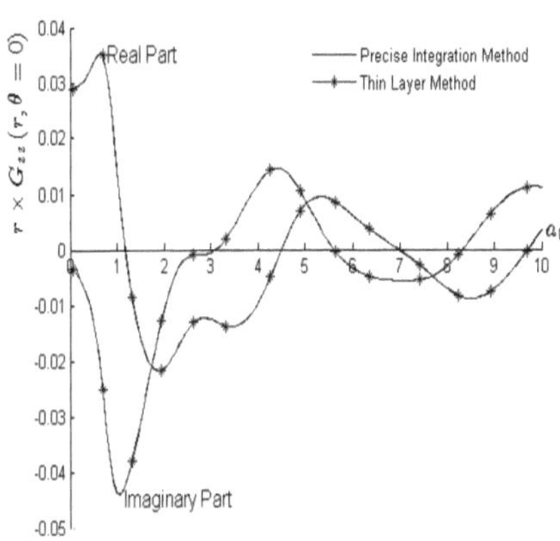

(c) on the plane $z = 2h$

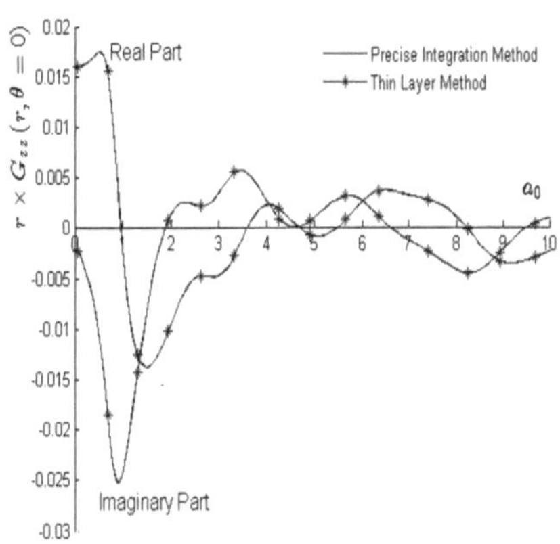

(d) on the plane $z = 3h$

Figure 6: Vertical component of the Green's functions for the layered half space

5 Conclusions

The thin layer method is an efficient approach to analyze the dynamic response of the layered medium in the earthquake engineering, because it formulates in algebraic expressions without numerical integrations. However, the results of the thin layer method are unstable in the applications. In this paper, two parameters to improve the accuracy of the thin layer method are presented. These parameters are proposed to determine the thickness of the buffer layer and the thin sub-layers. They are obtained from the comparison with other method. A numerical example is provided to verify the feasibility of the parameters. Excellent agreement is reached. With these parameters, the accuracy of the calculation of the Green's functions at both high-frequency and low-frequency range is guaranteed. The efficiency of thin layer method is also improved. For the numerical examples described in section 4, which has 435 thin sub-layers in the model of thin layer method, it takes 45 seconds per step based on a 2.8GHz Intel Core2 T9600 laptop with 3.45GB RAM. If we set the thickness of the thin sub-layers and buffer layer according to the exciting frequency, the number of the thin sub-layers will reduce to some extend based on Eq. (5). According to our experience, for the numerical examples in section 4, the number of the thin sub-layers is within 8~39 for $\omega = 0.1 \sim 10$ rad/s. The computation time per step for the Green's functions is 0.5s.

6 Acknoledgements

The work described herein is supported by a grant provided by Siemens AG and DAAD (German Academic Exchange Service). The authors express their sincere appreciation for the support.

References

[1] Kelvin, L., Displacement due to a point load in an indefinitely extended solid. Mathematical and Physical Papers (London), 1848. 1: p. 97.

[2] Boussinesq, J., Citation no. 67 in Love's Treatise on the Mathematical Theory of Elasticity,4th ed. Vol. 16. 1878, New York: Dover Press.

[3] Cerruti, V., Citation no. 68 in Love's Treatise on the Mathematical Theory of Elasticity, 4th ed. Vol. 16. 1882, New York: Dover Press.

[4] Lamb, H., On the propagation of tremors over the surface of an elastic solid. Philosophical Transactions of the Royal Society of London. Series A, Containing Papers of a Mathematical or Physical Character, 1904: p. 1-42.

[5] Mindlin, R.D., Force at a point in the interior of a semi - infinite solid. Physics, 1936. 7(5): p. 195-202.

[6] Harknder, D., Surface waves in multilayer elastic media, Rayleigh and Love waves from a body force. Bulletin of the Seismological Society of America, 1964. 54: p. 53.

[7] Apsel, R.J. and J.E. Luco, On the Green's functions for a layered half-space. Part II. Bulletin of the Seismological Society of America, 1983. 73(4): p. 931-951.

[8] Bouchon, M., A simple method to calculate Green's functions for elastic layered media. Bulletin of the Seismological Society of America, 1981. 71(4): p. 959-971.

[9] Liao, J., C. Wang, and J. Jong, Elastic solutions for a transversely isotropic half-space subjected to a point load. International Journal for Numerical and Analytical Methods in Geomechanics, 1998. 22(6): p. 425-447.

[10] Pan, E., Three-dimensional Green's functions in an anisotropic half-space with general boundary conditions. Journal of Applied Mechanics, 2003. 70(1): p. 101-110.

[11] Park, J. and E. Kausel, Impulse response of elastic half-space in the wave number-time domain. Journal of engineering mechanics, 2004. 130(10): p. 1211-1222.

[12] Fahmy, A. and E. Adler, Propagation of acoustic surface waves in multilayers: A matrix description. Applied Physics Letters, 1973. 22(10): p. 495-497.

[13] Liu, T., C. Zhao, and Y. Duan, Generalized transfer matrix method for propagation of surface waves in layered azimuthally anisotropic half - space. Geophysical Journal International, 2012.

[14] Tan, E.L., Stiffness matrix method with improved efficiency for elastic wave propagation in layered anisotropic media. The Journal of the Acoustical Society of America, 2005. 118: p. 3400.

[15] Kausel, E., Accurate stresses in the thin - layer method. International journal for numerical methods in engineering, 2004. 61(3): p. 360-379.

[16] Kausel, E., Fundamental solutions in elastodynamics: a compendium2006: Cambridge University Press.

[17] Kausel, E., An explicit solution for the Green functions for dynamic loads in layered media. NASA STI/Recon Technical Report N, 1981. 82: p. 29505.

[18] Kausel, E., The thin-layer method in seismology and earthquake engineering. Wave motion in earthquake engineering, 2000: p. 193-213.

[19] Park, J. and E. Kausel, Numerical dispersion in the thin-layer method. Computers & structures, 2004. 82(7): p. 607-625.

[20] Waas, G., Linear two-dimensional analysis of soil dynamics problems in semi-infinite layered media, 1972, University of California, Berkeley.

[21] Park, J., Wave motion in finite and infinite media using the thin-layer method, 2002, Massachusetts Institute of Technology.

[22] Lysmer, J., Lumped mass method for Rayleigh waves. Bulletin of the Seismological Society of America, 1970. 60(1): p. 89-104.

[23] Kausel, E., Early history of soil–structure interaction. Soil Dynamics and Earthquake Engineering, 2010. 30(9): p. 822-832.

[24] Kausel, E., Thin - layer method: Formulation in the time domain. International journal for numerical methods in engineering, 1994. 37(6): p. 927-941.

[25] Kausel, E. and R. Peek, Dynamic loads in the interior of a layered stratum: an explicit solution. Bulletin of the Seismological Society of America, 1982. 72(5): p. 1459-1481.

[26] Seale, S.H., Dynamic loads in layered halfspaces, 1985, Massachusetts Institute of Technology. PhD thesis.

[27] Kausel, E., Dynamic point sources in laminated media via the thin-layer method. International Journal of Solids and Structures, 1999. 36(31): p. 4725-4742.

[28] Zhong, W.X., J. Lin, and Q. Gao, The precise computation for wave propagation in stratified materials. International journal for numerical methods in engineering, 2004. 60(1): p. 11-25.

Time Domain Analysis of Dynamic Response for 3D Rigid Foundation on Multi-layered Soil

Zejun Han[1], **Gao Lin**[1], **Jianbo Li**[1]

[1] Faculty of Infrastructure Engineering, Dalian University of Technology No.2 Linggong Road, Ganjingzi District, Dalian City, P.R.C.
hzjunlq@mail.dlut.edu.cn

ABSTRACT:

Soil-structure interaction is widely recognized as a very important issue that should be considered in dynamic analysis and design of structures subjected to various dynamic disturbances such as earthquake or wind forces. An approach for time-domain response analysis of three-dimensional rigid surface foundations of arbitrary shape bonded to multi-layered soil is presented. The formulation consists of two parts: (a) frequency-spatial domain solution to the dynamic impedance of rigid surface foundation and (b) time-domain analysis by employing interpolating discrete values of dynamic impedance matrices by means of a continued matrix valued rational function. Practical applications compared with the analytical solutions or existing classical results dealing with rigid surface foundations of arbitrary shape demonstrate the accuracy and applicability of the proposed approach.

Keywords: rigid foundation; multi-layered soil; precise integration method, time-domain; mixed-variable formulation

1 Introduction

The dynamic response of rigid foundation is governed by soil-structure interaction (SSI) as well as the dynamic characteristics of the exciting loads, i.e. earthquake, wind, explosion and machinery vibrations. SSI is widely recognized as a very important issue in dynamic analysis and design of structures subjected to various dynamic disturbances. In the field of SSI modelling of the force-displacement relationship has been a major research subjected over the past 40 years. Pioneering efforts, especially for practical engineering purpose, simple physically models have been presented by Wolf and his co-workers [1]. Later, various analytical and numerical methods have been developed to solve the SSI problems. Excellent literature reviews are available in papers [2].

It is well known that most analytical methods are applicable only to foundations of regular shape resting on homogeneous half-space soil in frequency domain [3]. In

S. Klinkel et al. (Eds.), *Seismic Design of Industrial Facilities*,
DOI 10.1007/978-3-658-02810-7_51, © Springer Fachmedien Wiesbaden 2014

recent decades, a number of papers have emerged to deal with time domain analysis of SSI problems. Some of these are based on the application of finite element method (FEM) and boundary element method (BEM) which is well suited to model infinite medium as the radiation condition is satisfied automatically [4][5]. The thin-layer method has evolved into an efficient and versatile technique for the analysis of wave motion in layered soils [6]. It is semi-analytical that it uses exact solution in the horizontal direction and expansion into finite elements in the vertical direction. Apart from the aforementioned theoretical developments various numerical methods have emerged as an important alternative to obtain the dynamic impedance of foundations on multi-layered subsoil in time domain. Some of them consist in calculating the dynamic flexibility or stiffness coefficients in the frequency domain and then transformed into time domain [7].

In this paper, an accurate and efficient approach is presented for the time domain dynamic response analysis of three-dimensional rigid surface foundations of arbitrary shape bonded to multi-layered soil. The technique is based on the discrete dynamic impedance matrix of the rigid surface foundation on multi-layered soil expressed in frequency domain developed by Lin et al. [8]. Then interpolate the discrete dynamic impedance matrix S(ω) by a continuous rational approximation function of denominator degree M. In this way, the coupling between interface degrees of freedom is fully preserved. The mixed-variables which are alternatively forces and displacements are introduced. At last the rational function can be separated into a sequence of linear functions in (iω) and the problem becomes the numerical solution of a first order ordinary differential equations in time domain. The advantages of the proposed method are demonstrated by means of several applications dealing with rigid surface foundations of arbitrary shape. A comparison with the results obtained by other methods validates the accuracy and applicability of the proposed method for the multi-layered soil.

2 Formulation of the problem in frequency-spatial domain

2.1 Statement of the problem

In what follows a study is made for time domain response of a rigid foundation of arbitrary shape resting on the surface of a multi-layered soil. The multi-layered soil consists of l layers overlying an elastic half-space or rigid base. And both the layers and the half-space are assumed to be homogeneous and isotropic, with Lame constants λ_i and μ_i, Poisson's ratio ν_i, density ρ_i, damping ratio ξ_i and thickness $h_i = z_i - z_{i-1}$ ($i = 1,2,...,l$). The foundation is subjected to the action of an impulse or harmonic force or moment. The geometry of the multi-layered soil model and the corresponding cylindrical coordinate system are shown in Fig. 1.

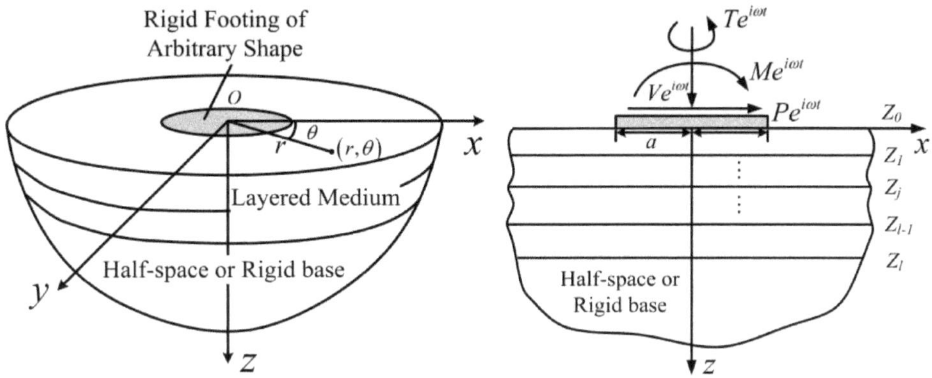

Figure 1: Description of the coordinate system and multi-layered media model

2.2 The Green's influence functions in frequency-wave-number domain

The following stress and displacement vectors are specified as

$$\mathbf{S} = \{\tau_{rz} \quad \tau_{\theta z} \quad \sigma_z\}^T, \; \mathbf{U} = \{u_r \quad u_\theta \quad u_z\}^T \tag{1}$$

with τ , σ and u being the tangential, normal stresses, and displacement components in the direction identified by the subscripts in cylindrical coordinate. It is possible to take advantage of the axisymmetric geometry of the problem, such that, the displacements are split into components which are either symmetric or anti-symmetric about the r-axis at $\theta = 0$. Then the variation of displacements in the circumferential direction is represented by Fourier series as shown below

$$\begin{aligned} u_r(r,\theta,z,n) &= \sum_n u_r^s(r,z,n)\cos n\theta + \sum_n u_r^a(r,z,n)\sin n\theta \\ u_\theta(r,\theta,z,n) &= -\sum_n u_\theta^s(r,z,n)\sin n\theta + \sum_n u_\theta^a(r,z,n)\cos n\theta \\ u_z(r,\theta,z,n) &= \sum_n u_z^s(r,z,n)\cos n\theta + \sum_n u_z^a(r,z,n)\sin n\theta \end{aligned} \tag{2}$$

where the r , θ and z denote radial, circumferential and vertical components, respectively; superscripts s and a denote the symmetric and anti-symmetric components.

The elastic wave motion equation is expressed as

$$(\lambda + 2\mu)\nabla\nabla U - \mu\nabla\times\nabla\times U = -\rho\omega^2 U \tag{3}$$

For computational convenience, the problem is solved in the frequency-wave-number domain. Let the superscript bar of $\overline{\mathbf{U}}$ and $\overline{\mathbf{S}}$ be referred to the values in the frequency-wave-number domain. The displacements and the loadings are assumed to be expanded into Fourier series in the circumferential direction θ and into

Bessel functions involving the wave number k in the radial direction r. As k runs from 0 to infinitely, all types of waves are captured. The corresponding amplitudes of displacements and loadings are related by the following Bessel transformation pairs [9].

$$\begin{aligned} \mathbf{U}(r,\theta) &= \sum_{n=0}^{\infty} \mathbf{D}(n\theta) \int_{k=0}^{\infty} k\, \mathbf{C}_n(kr)\, \overline{\mathbf{U}}(k,n) dk \\ \overline{\mathbf{U}}(k,n) &= a_n \int_{r=0}^{\infty} r\, \mathbf{C}_n(kr) \int_{\theta=0}^{2\pi} \mathbf{D}(n\theta)\, \mathbf{U}(r,\theta) d\theta dr \end{aligned} \tag{4}$$

and

$$\begin{aligned} \mathbf{S}(r,\theta) &= \sum_{n=0}^{\infty} \mathbf{D}(n\theta) \int_{k=0}^{\infty} k\, \mathbf{C}_n(kr)\, \overline{\mathbf{S}}(k,n) dk \\ \overline{\mathbf{S}}(k,n) &= a_n \int_{r=0}^{\infty} r\, \mathbf{C}_n(kr) \int_{\theta=0}^{2\pi} \mathbf{D}(n\theta)\, \mathbf{S}(r,\theta) d\theta dr \end{aligned} \tag{5}$$

with the diagonal matrix $\mathbf{D}(n\theta)$ consists of $\mathbf{cos}\, n\theta$, $-\mathbf{sin}\, n\theta$, and $\mathbf{cos}\, n\theta$ for the symmetric case and $\mathbf{sin}\, n\theta$, $\mathbf{cos}\, n\theta$, and $\mathbf{sin}\, n\theta$ for the anti-symmetric case. The matrix $\mathbf{C}_n(kr)$ contains the Bessel functions. The orthogonalization scalar a_n is the normalization factor, which equals $1/2\pi$ for $n=0$ and $1/\pi$ for $n \neq 0$. The symmetric part of $n=0$ corresponds to an axisymmetric vertical load case, while the symmetric part of $n=1$ can be used to model uniformly distributed horizontal load which are symmetric about the r-axis at $\theta = 0$.

It is important to point out that the three-dimensional waves formulated in cylindrical coordinates can be decoupled into in-plane motion and out-of-plane motion as that arising for plane waves [10]. Making use of Eq. (2) and Eq. (4), after some manipulations, the set of differential equations of wave motion (3) is transformed into the frequency-wave-number domain for in-plane and out-of-plane wave motions.

$$\mathbf{K}_{22}^{m}\, \overline{\mathbf{q}}^{m''} + \left(\mathbf{K}_{21}^{m} - \mathbf{K}_{12}^{m}\right) \overline{\mathbf{q}}^{m'} - \left(\mathbf{K}_{11}^{m} - \rho\omega^2\, \mathbf{I}_m\right) \overline{\mathbf{q}}^{m} = 0 \tag{6}$$

with superscript $m=1$ and $m=2$ corresponding to the in-plane motion and out-of-plane motion respectively; $\overline{\mathbf{q}}^{m}$ is the displacement vector in the frequency-wave-number domain; Hereinafter, the superscript prime of $\overline{\mathbf{X}}'$ denotes differentiation with respect to z, $\overline{\mathbf{X}}' = \partial \overline{\mathbf{X}} / \partial z$; $\mathbf{I}_m$ is a $m \times m$ unit matrix and the coefficient matrices $\mathbf{K}_m$ are defined by the material constants of the soil layers. If internal material damping is considered, the shear modulus μ is replaced by $\mu(1+2i\xi_i)$, where ξ_i represents the damping ratio of layer i.

In order to solve Eq. (6) for layered stratum in a convenient way, it is transformed into a first order ordinary differential equation in the state space and then solved by precise integration method (PIM) [11].

$$\bar{\mathbf{V}}^{m\prime} = \mathbf{H}^m \bar{\mathbf{V}}^m \tag{7}$$

with $\bar{\mathbf{V}}^m = \{\bar{\mathbf{q}}^m \quad \bar{\mathbf{p}}^m\}^T$, and $\mathbf{H}^m$ is related to the coefficient matrices $\mathbf{K}$.

2.3 The precise integration method

The general solution of the state Eq. (7) is an exponential function. Zhong *et al.* [11] presented the PIM for the solution of the state equation which has the advantage that high precision can be achieved. The basic concept for the derivation of PIM is summarized in this section.

A typical interval $[z_a, z_b]$ ($z_a < z_b$) within a layer is addressed. Let $\mathbf{q_a}$, $\mathbf{p_a}$ and $\mathbf{q_b}$, $\mathbf{p_b}$ be the displacement and force vectors at the two ends z_a and z_b, respectively. For linear systems, the following relations stand classically:

$$\mathbf{q_b} = \mathbf{F}\mathbf{q_a} - \mathbf{G}\mathbf{p_b}, \quad \mathbf{p_a} = \mathbf{Q}\mathbf{q_a} + \mathbf{E}\mathbf{p_b} \tag{8}$$

where $\mathbf{F}$, $\mathbf{G}$, $\mathbf{Q}$ and $\mathbf{E}$ are functions of the matrices $\mathbf{K}_{11}$, $\mathbf{K}_{21}$, $\mathbf{K}_{12}$ and $\mathbf{K}_{22}$ determined by the material constants, and they are complex transfer matrices to be evaluated.

In the PIM, in order to obtain the transfer matrices $\mathbf{F}$, $\mathbf{G}$, $\mathbf{Q}$ and $\mathbf{E}$ as exactly as possible, the thickness of every layer $h_r = z_r - z_{r-1}$ ($r = 1,2,\ldots,l$) is firstly divided into 2^{N_1} sublayers of equal thickness. Then each sublayer is further divided into 2^{N_2} mini-layers of equal thickness τ. Since τ is extremely small, the transfer matrices $\mathbf{F}(\tau)$, $\mathbf{G}(\tau)$, $\mathbf{Q}(\tau)$ and $\mathbf{E}(\tau)$ can be found in terms of Taylor series expansion. With increasing terms of Taylor expansion, any desired accuracy of the results can be reached. From experience, four terms of Taylor's series is considered sufficient.

From (8), combination of two adjacent intervals leads to the new transfer matrices as shown below.

$$\begin{aligned} &\mathbf{G}_c = \mathbf{G}_2 + \mathbf{F}_2\left(\mathbf{G}_1^{-1} + \mathbf{Q}_2\right)^{-1}\mathbf{E}_2, \quad \mathbf{F}_c = \mathbf{F}_2\left(\mathbf{I} + \mathbf{G}_1\mathbf{Q}_2\right)^{-1}\mathbf{F}_1 \\ &\mathbf{Q}_c = \mathbf{Q}_1 + \mathbf{E}_1\left(\mathbf{Q}_2^{-1} + \mathbf{G}_1\right)^{-1}\mathbf{F}_1, \quad \mathbf{E}_c = \mathbf{E}_1\left(\mathbf{I} + \mathbf{Q}_2\mathbf{G}_1\right)^{-1}\mathbf{E}_2 \end{aligned} \tag{9}$$

where the subscript 1 and 2 denote the matrices associated with the original two intervals and the subscript c denotes the newly combined matrices. It is therefore important to note that as the combination is proceeded for a mini-layer, $\mathbf{F}(\tau)$ and $\mathbf{E}(\tau)$ are very small because τ is very small. They should be computed and stored independently to avoid losing effective digits. Hence it is necessary to replace $\mathbf{F}$ and $\mathbf{E}$ in Eq. (9) by $\mathbf{I}+\tilde{\mathbf{F}}$ and $\mathbf{I}+\tilde{\mathbf{E}}$ respectively.

In case all intervals having equal thickness and identical material constants, combination of such intervals is performed easily. For each pass of combination, transfer matrices $\mathbf{F}$, $\mathbf{G}$, $\mathbf{Q}$ and $\mathbf{E}$ are merged together to form a new one, and the total number of intervals is reduced by a half. Proceeding in this way, any desired accuracy can be achieved in the sense that its precision is limited only by the precision of the computer acquired.

Finally, for the assembled stratum with l layers (Fig. 1), the following relationship holds

$$\mathbf{q}_l = \mathbf{F}\mathbf{q}_0 - \mathbf{G}\mathbf{p}_l, \quad \mathbf{p}_0 = \mathbf{Q}\mathbf{q}_0 + \mathbf{E}\mathbf{p}_l \tag{10}$$

To evaluate the dynamic stiffness of the layered system, two cases are considered: the layered strata on rigid base and the layered stratum overlying on elastic half-space. In the former case, the following boundary condition stands

$$\bar{\mathbf{q}}_l = \bar{\mathbf{q}}(\mathrm{z}=\mathrm{z}_l) = \mathbf{0} \tag{11}$$

In the latter case, the boundary condition at the surface of elastic half-space is expressed as

$$\bar{\mathbf{p}}_l = \mathbf{R}_\infty \bar{\mathbf{q}}_l \tag{12}$$

where the analytical solution of $\mathbf{R}_\infty$ can be found in [9].

From Eqs (10)-(12), the relationship between the surface tractions and the surface displacements is formulated as

$$\bar{\mathbf{q}}_0^m = \bar{\mathbf{S}}^{-1}(k)^m \bar{\mathbf{p}}_0^m \tag{13}$$

$$\bar{\mathbf{S}}(\mathrm{k}) = \mathbf{Q} + \mathbf{E}\mathbf{G}^{-1}\mathbf{F} \;(\textit{former case}); \quad \bar{\mathbf{S}}(\mathrm{k}) = \mathbf{Q} + \mathbf{E}\mathbf{R}_\infty(\mathbf{I}+\mathbf{G}\mathbf{R}_\infty)^{-1}\mathbf{F} \;\; (\textit{latter case}) \tag{14}$$

The inverse of $\bar{\mathbf{S}}(k)^{-1}$ represents the flexibility matrix of the system $\bar{\mathbf{F}}(k)$ and it is partitioned in the following form for later use.

$$\begin{aligned} &\begin{Bmatrix} \bar{u}_r(k) \\ \bar{u}_z(k) \end{Bmatrix} = \bar{\mathbf{S}}(k)^{-1} \begin{Bmatrix} -\bar{\tau}_{rz}(k) \\ -\bar{\sigma}_z(k) \end{Bmatrix} = \begin{bmatrix} \bar{F}_{rr}(k) & \bar{F}_{rz}(k) \\ \bar{F}_{zr}(k) & \bar{F}_{zz}(k) \end{bmatrix} \begin{Bmatrix} -\bar{\tau}_{rz}(k) \\ -\bar{\sigma}_z(k) \end{Bmatrix}, \quad \textit{for } m=1 \\ &\{\bar{u}_\theta(k)\} = \bar{\mathbf{S}}(k)^{-1}\{-\bar{\tau}_{\theta z}(k)\} = [\bar{F}_{\theta\theta}]\{-\bar{\tau}_{\theta z}(k)\}, \quad \textit{for } m=2 \end{aligned} \tag{15}$$

2.4 The Green's influence functions in frequency-spatial domain

For the evaluation of the Green's influence functions for a subdisk of radius Δr subjected to the uniformly distributed vertical and horizontal loads, the soil-foundation interface is discretized into n subdisk-elements (Fig. 2). Applying Eq. (5) leads to the amplitude of the load in the wave-number-domain

$$\bar{p}_z(k) = -\frac{p_{z0}\Delta r}{k} J_1(k\Delta r) \quad \text{for vertical load} \tag{16}$$

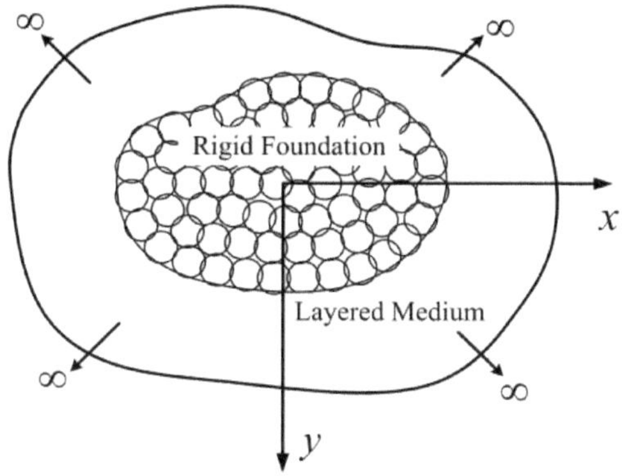

Figure 2: The rigid foundation with subdisk discretization

$$\begin{Bmatrix} \bar{p}_r(k) \\ \bar{p}_\theta(k) \end{Bmatrix} = \frac{p_{x0}\Delta r}{k} J_1(k\Delta r) \begin{Bmatrix} 1 \\ 1 \end{Bmatrix} \quad \text{for horizontal load} \tag{17}$$

Then the displacements in frequency-spatial domain due to the vertical and horizontal (acting in x direction) uniformly distributed loads are calculated from Eq. (4) and expressed as follows.

$$\begin{Bmatrix} u_r(r) \\ u_z(r) \end{Bmatrix} = \Delta r \left[\int_{k=0}^{\infty} J_1(k\Delta r) \begin{Bmatrix} \bar{F}_{rz}(k) J_1(kr) \\ \bar{F}_{zz}(k) J_0(kr) \end{Bmatrix} dk \right] p_{z0} \tag{18}$$

$$\begin{Bmatrix} u_r(r,\theta) \\ u_\theta(r,\theta) \\ u_z(r,\theta) \end{Bmatrix} = \frac{\Delta r}{2} \begin{bmatrix} \cos\theta & & \\ & -\sin\theta & \\ & & \cos\theta \end{bmatrix}$$

$$\left(\int_{k=0}^{\infty} J_1(k\Delta r) \begin{bmatrix} J_0(kr) - J_2(kr) & J_0(kr) + J_2(kr) & \\ J_0(kr) + J_2(kr) & J_0(kr) - J_2(kr) & \\ & & -2J_1(kr) \end{bmatrix} \begin{Bmatrix} \bar{F}_{rr}(k) \\ \bar{F}_{\theta\theta}(k) \\ \bar{F}_{rz}(k) \end{Bmatrix} dk \right) p_{x0} \tag{19}$$

In case the uniformly distributed horizontal load acting on y direction with amplitude p_{y0}, the same form of Eq. (19) applies, but with p_{y0} instead of p_{x0}, $\cos\theta$ and $-\sin\theta$ replaced by $\sin\theta$ and $\cos\theta$, respectively.

Using these Green's influence functions, the dynamic impedance $\mathbf{S}(\omega)$ (6×6) of the rigid foundation in the frequency-spatial domain can be easily determined.

3 Formulation of the problem in time domain

In the following, the process that transmits the dynamic impedance from frequency domain to time domain is summarily presented. Details can be found in Ref. [12]. Assume that N discrete dynamic impedances $\mathbf{S}(i\omega)$ are obtained by the proposed approach described in section 2. Here interface degrees of freedom of the dynamic impedance matrix $\mathbf{S}(i\omega)$ can be fully preserved.

The interpolation of the discrete dynamic impedance $\mathbf{S}(i\omega)$ is carried out by means of a rational approximation in the spectral domain. Then a matrix-valued rational function is split into a series of matrix-valued linear functions in $(i\omega)$

$$\mathbf{S}(i\omega)=\mathbf{L}(i\omega)^{-1}\mathbf{R}(i\omega) \tag{20}$$

where

$$\begin{aligned}\mathbf{L}(i\omega)&=\mathbf{I}+i\omega\mathbf{L}_1+\ldots+(i\omega)^M\mathbf{L}_M\\ \mathbf{R}(i\omega)&=\mathbf{R}_0+i\omega\mathbf{R}_1+\ldots+(i\omega)^{M+1}\mathbf{R}_{M+1}\end{aligned} \tag{21}$$

and $\mathbf{S}(i\omega)$ is the discrete dynamic impedance matrix corresponding to discrete value ω. The coefficient matrices $\mathbf{R}_j\,(j=1,2,\ldots,M+1)$ and $\mathbf{L}_j\;(j=1,2,\ldots,M)$ are determined using a curve fitting technique based on the least squares method. $\mathbf{I}$ is a unit matrix which have the same dimension as $\mathbf{S}(i\omega)$. The rational function of Eq. (20) can also be divided into a linear function in $(i\omega)$ and a strictly proper rational function of numerator degree $(M-1)$. The mixed-variables which are alternatively forces and displacements are introduced. At last the rational function

$$\frac{\mathbf{R}_0+i\omega\mathbf{R}_1+\ldots+(i\omega)^{M+1}\mathbf{R}_{M+1}}{\mathbf{I}+i\omega\mathbf{L}_1+\ldots+(i\omega)^M\mathbf{L}_M}\mathbf{u}_c=\mathbf{f}_c \tag{22}$$

can be separated into a sequence of linear functions in $(i\omega)$. After some manipulations, eventually, the problem becomes the numerical solution of a first order ordinary differential equation in time domain as

$$\mathbf{Az}(t)+\mathbf{B\dot{z}}(t)=\mathbf{f}(t) \tag{23}$$

where $\mathbf{A}$ and $\mathbf{B}$ are related to $\mathbf{R}_j$ and $\mathbf{L}_j$. And the number of degrees of freedom of the matrices $\mathbf{A}$ and $\mathbf{B}$ increase to a total of $6\times(M+1)$. The solution of the Eq. (23) is carried out numerically using a Newmark time-stepping scheme.

4 Numerical Examples

In the published literature, few researchers studied the dynamic response of 3D rigid foundation on multi-layered soil in time-domain. Most of them are focused on simple case of elastic half-space. So comparison of the results is made only with rather simple cases available in the literature. And the third numerical example is provided to validate the applicability of the proposed method for the case of multi-layered soil. Unfortunately, no reference work can be found in the published literature.

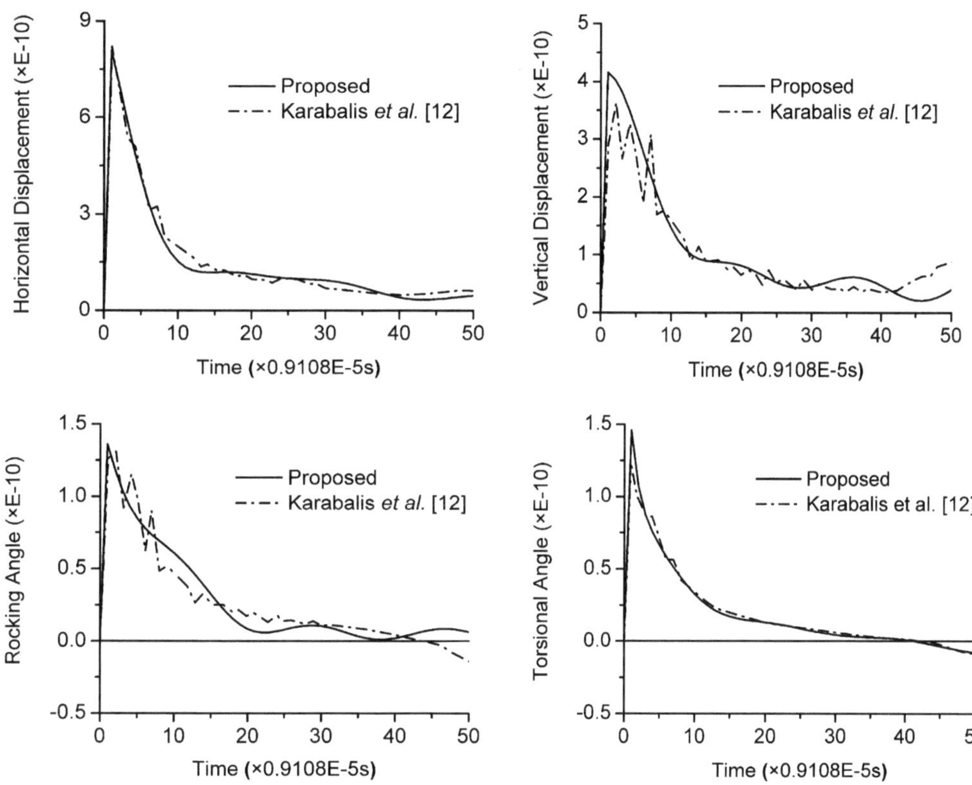

Figure 3: Time history of the impulse response for a massless foundation with internal opening

4.1 Square foundation with internal opening resting on an elastic half-space

A comparison study is presented firstly between the results obtained by the proposed method and results are available in the literature [13]. A square foundation (5×5) with a square concentric opening (3.75×3.75) resting on an elastic half-space is considered. The material properties of the half-space are kept homogeneous, i.e. modulus of elasticity E=2.59E9, Poisson ratio 1/3 and mass density ρ=10.37. The external impulse is defined as:

$$P(t) = \begin{cases} 100, & \textit{during first time step} \\ 0, & \textit{elsewhere} \end{cases} \tag{24}$$

where $P(t)$ represents external forces or moments. The time step Δt is selected as 0.9108E-5 sec. The horizontal, vertical, rocking and torsional impulse responses of the foundation are plotted versus time in Fig. 3 respectively. It can be observed from the figures, the agreement between the results obtained by the proposed method and the reference work is good. There appears some pulsation of the results obtained by Karabalis and Huang [13].

4.2 Square foundation resting on an elastic half-space

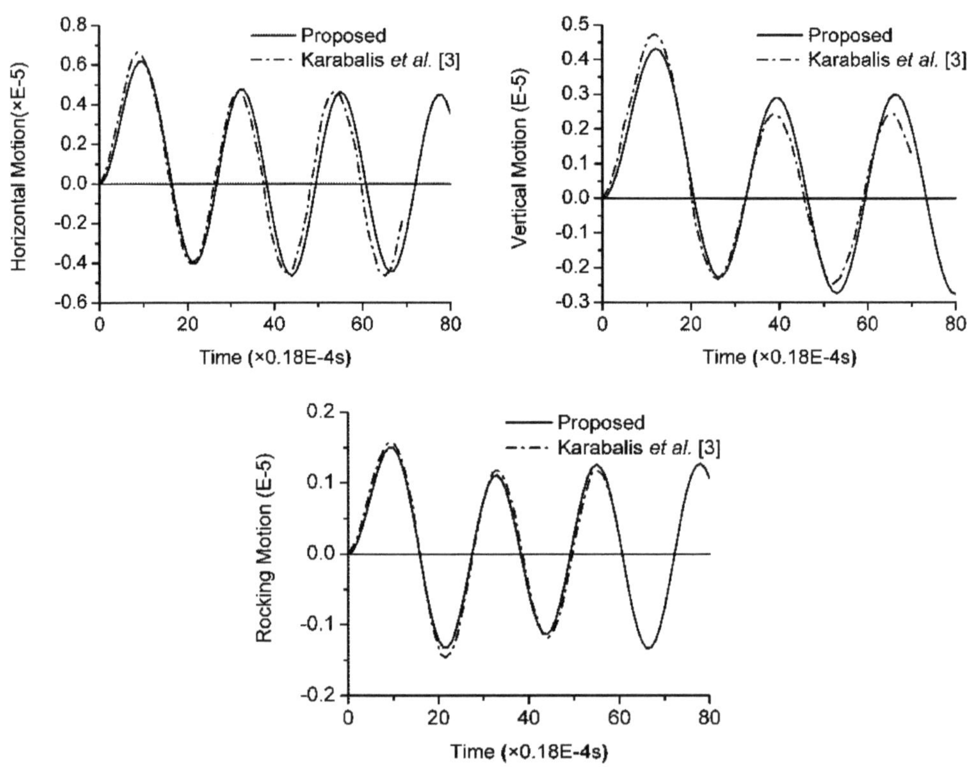

Figure 4: Horizontal, vertical and rocking harmonic force response versus time for a square foundation

A 5×5 square rigid surface massless foundation is chosen to test the proposed method.. The elastic half-space soil is characterized by the same material properties as the above numerical example. The external forces and moment are defined as given in Eq. (25) for specified frequencies.

$$P_x(t)=180\,\mathbf{sin}(15504t),\ P_z(t)=180\,\mathbf{sin}(13000t),\ M_y(t)=180\,\mathbf{sin}(15504t) \quad (25)$$

The time histories showing the harmonic response of the foundation subjected to the external forces and moment are portrayed in Fig. 4. The results obtained by the proposed method are compared with those obtained by Karabalis and Beskos [4]. Perfect agreement is reached, which approves the accuracy of the proposed method.

4.3 Square foundation resting on a multi-layered soil

Table 1: Material constants of the layers and half space

Layer	λ_r	μ_r	ρ_r	v_r	ξ_r	h_r
1	1.00E9	1.00E9	100.0	0.25		2.5
2	7.50E8	5.00E8	100.0	0.30	0.05	1.25
3	4.00E8	2.00E8	89.0	1/3		Infinite

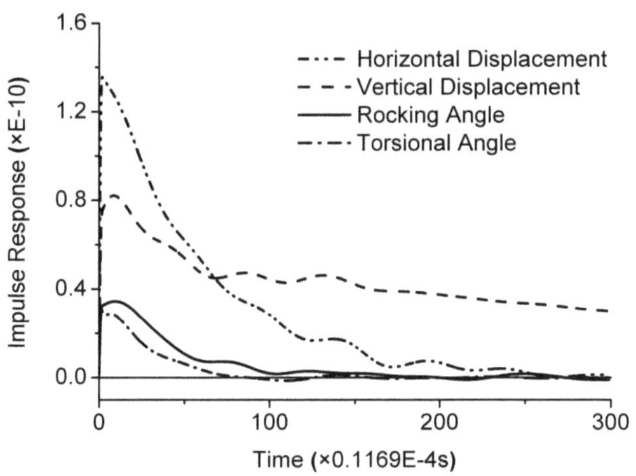

Figure 5: Impulse responses versus time for a square foundation on a multi-layered soil

The case of a circular foundation with radius R=5 resting on a multi-layered soil which consists of two layers and a half-space is considered. The material constants of the layers and half-space are presented in the Table 1. And the external impulse is given in Eq. (24). The horizontal, vertical, rocking and torsional impulse responses of the foundation are plotted versus time in the Fig. 5. As no reference solution is available, this example shows the capability of the proposed method dealing with multi-layered case.

5 Conclusion

An approach for time-domain response analysis of three-dimensional rigid surface foundations of arbitrary shape resting on a multi-layered soil is presented. The foundations are subjected to the action of external forces or moments. Several numerical examples showing dynamic response of rigid surface foundations of arbitrary shape demonstrate the accuracy and applicability of the proposed method to solve SSI problem for the multi-layered soil.

6 Acknowledgements

The authors are grateful for the financial support of the Sino-German Science Foundation under grant No. GZ566 and the National Natural Science Foundation of China under grant No. 51138001.

Reference

[1] Wolf J.P. Foundation Vibration Analysis Using Simple Physical Model. Prentice-Hall: Englewood Cliffs, NJ, 1994.

[2] Kausel, E. Early history of soil-structure interaction, Soil Dynamics and Earthquake Engineering, 2010; 30:822-832.

[3] Luco J.E. Westmann R.A., Dynamic response of circular footing. Journal of the Engineering Mechanics Division, 1971; 97(5): 1381-1395.

[4] Karabalis D.L. Beskos DE. Dynamic response of 3-D rigid surface foundations by time domain boundary element method. Earthquake Engineering and Structural Dynamics, 1984; 12:73-93.

[5] Estorff O.V. and Prabucki M.J. Dynamic response in the time domain by coupled boundary and finite elements, Computational Mechanics, 1990; 6:35-46.

[6] Kausel E. Thin-layer method: formulation in the time domain. International Journal for Numerical Methods in Engineering 1994; 37(6):927–941.

[7] Wolf J.P. and Obernhuber P. Non-linear soil-structure-interaction analysis using dynamic stiffness or flexibility of soil in the time domain. Earthquake Engineering and Structural Dynamics, 1985; 13:195-212.

[8] Lin G, Han Z.J., Li J.B. An Efficient Approach for Dynamic Impedance of Surface Footing on Layered Half-space, Soil Dynamics and Earthquake Engineering. 2013; 49:39–51.

[9] Kausel E. and J. M. Roësset, Stiffness matrices for layered soils, Bulletin of the Seismological Society of America; 1981; 71(6):1743-1761.

[10] Wolf J.P. Dynamic Soil-structure Interaction, Prentice-Hall: Englewood Cliffs, 1985.

[11] Zhong W.X., Lin J.H. and Gao Q., The precise computation for wave propagation in stratified materials. International Journal for Numerical Methods in Engineering, 2004; 60(1):11-25.

[12] Ruge P., Trinks C. and Witte S. Time-domain analysis of unbounded media using mixed-variable formulations, Earthquake Engng Struct. Dyn. 2001; 30:899-925.

[13] Karabalis D.L., Huang C.-F.D. Vibrations of square and circular foundations with concentric openings on elastic half space. Soil Dynamics and Earthquake Engineering, 2005; 25:951-965.

Eta-based Conditional Mean Spectrum, a New Design Spectrum for Industrial Facilities

Alireza Azarbakht[1]

[1] Department of Civil Engineering, Faculty of Engineering,
Arak University, Arak, Iran.
a-azarbakht@araku.ac.ir

ABSTRACT:

The target spectrum which has been used most frequently for the seismic analysis of structures is the Uniform Hazard Response Spectrum (UHRS). The joint occurrence of the spectral values in different periods, in the development of UHRS, is a key assumption which remains questionable. The Conditional Mean Spectrum (CMS) has been recently developed by Baker et al. as an alternative for UHRS. The CMS provides the expected response spectrum conditioned on the occurrence of the target spectral acceleration value in the period of interest which can be accounted as an improvement of the UHRS. In order to enhance the CMS, the correlation between the Peak Ground Velocity (PGV) and the spectral acceleration values has been investigated in the current study, and finally, a newer form of target spectrum has been proposed. It is shown that the emerged new spectrum, named Eta-based Conditional Mean Spectrum (E-CMS), is more efficient than the conventional CMS in order to enhance the UHRS, especially in the case of industrial facilities.

Keywords: Uniform Hazard Response Spectrum, Conditional Mean Spectrum, Epsilon indicator, Eta indicator, record selection

1 Introduction

One of the most important challenges in structural response assessment is the careful Ground Motion Record (GMR) selection before performing dynamic analyses. All of researchers and guidelines emphasize that ground motion records should represent the properties of a given hazard level which can be quantified based on Probabilistic Seismic Hazard Analysis (PSHA) [1]. Most of the design codes use a suitable target spectrum to facilitate ground motion record selection approach and finally use those GMRs as input to dynamic analysis [2]. The Uniform Hazard Response Spectrum (UHRS) is considered to be as a commonly used target in most of design codes and guidelines. However most of recent research results have shown that UHRS is not a good representative of a suitable target [3]. The UHRS is an elastic spectrum at a site with a given hazard level

S. Klinkel et al. (Eds.), *Seismic Design of Industrial Facilities*,
DOI 10.1007/978-3-658-02810-7_52,

which the structure is supposed to be located. The spectral acceleration amplitudes in UHRS would be more than the median predicted spectrum in all periods within a single ground motion. This fact is more highlighted when the UHRS is compared with the spectral shape records in higher hazard levels. Figure 1 shows the UHRS given exceedance of the Spectral acceleration (Sa) values with 2475 years return period. By considering a structure with the first period of one second, only one (non scaled) rare record is found to have Sa value equal to UHRS in the target period. In other words the mentioned record in Figure 1 has an Epsilon value in the target period approximately equal to 1.7 in which Epsilon [3] is defined as the number of standard deviations from the predicted value by an empirical ground motion model. As seen in Figure 1, it is obvious that there is clear observed difference in other periods between the selected record and the UHRS. In other words this fact illustrates why the uniform hazard spectrum is not a good representative of individual ground motion spectrum. As UHRS in lower period range is affected by strong ground motions and weak earthquakes have the most contribution in the UHRS values in lower frequencies, UHRS has not satisfied users to be a suitable target spectrum in ground motion record selection purposes and considered as a conservative target by researchers e.g.[3].

The Conditional Mean Spectrum (CMS) has been introduced by Baker in recent years to decrease the UHRS disadvantages [4]. The Epsilon as a spectral shape indicator is employed in CMS [3,4]. The CMS is a method that accounts for magnitude, distance and Epsilon values likely to cause a given target ground motion intensity at a given site for a specified hazard level. The main assumption in CMS is that the only value which would be exactly equal to the target value (Sa in UHRS) is located at the target period. In fact CMS has a peak value at the target period and decays towards the median spectrum in other periods. The decreasing process is based on a correlation model between the spectral acceleration values for all periods. This correlation is not taken into account in the UHRS concept since UHRS is based on several independent PSHA analyses for each period with no joint occurrences of spectral values.

The spectral acceleration is the only Intensity Measure (IM) which is employed in the Epsilon spectral shape indicator. An alternative indicator, as a more reliable predictor of the non-linear response of structures, is recently proposed by Mousavi et al. which is named Eta [5]. It has been shown that a simple linear combination of different IM Epsilons can result in a robust predictor of non-linear structural response. In addition to the spectral acceleration, the peak ground acceleration, the peak ground velocity and the peak ground displacement are also assumed as IMs in the prediction of the new spectral shape indicator. A new target conditional mean spectrum is presented here which uses the Eta advantages instead of the conventional Epsilon. The Eta-based Conditional Mean Spectrum (E-CMS) provides the mean response spectrum conditioned on occurrence of a target spectral acceleration value in the period of interest by considering of a new correlation model that is based on the new spectral shape indicator.

Replacing Eta indicator instead of the conventional Epsilon in the conditional computation leads to introduction of a new target response spectrum. This issue is discussed in details in the current study.

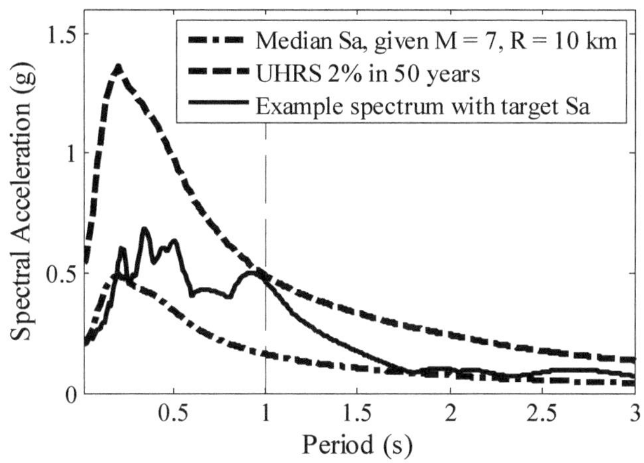

Figure 1: Median predicted spectrum using BA-08 attenuation relationship [6], having M=7 and R=10 km. UHRS for 2 % probability of exceedance in 50 years. The example record spectrum is the Parkfield-Fault Zone 16 recorded from Coalinga event

2 The Eta-based conditional mean spectrum

The potential of the Epsilon indicator encouraged researchers to use it as a suitable predictor of other spectral acceleration values by a given Sa which is representing the target hazard (Sa at the period of T_1 on UHRS obtained based on a specific probability of exceedance). For this purpose an effort has been done to introduce a new elastic spectrum that uses the advantages of the Epsilon spectral shape indicator. The conditional mean spectrum uses the correlation between Epsilon values to predict the Sa values in the whole range of the target spectrum. The aim of the current research is to introduce the Eta-based conditional mean spectrum as a new target spectrum for the record selection purposes. First it is needed to define a target spectral acceleration value at a period of interest. The period of interest can be computed by modal analysis for a particular structure. Usually the target period is chosen equal to the first mode period of vibration. The mean causal magnitude (M), the mean causal distance (R) and the mean causal Epsilon can be obtained by disaggregation analysis based on the probabilistic seismic hazard analysis. The mean predicted spectral acceleration and the corresponding standard deviation of logarithmic spectral acceleration can be computed using the existing ground motion prediction models ([6] in this study). The CMS value in the target period can be calculated easily. The probability calculation shows that the Epsilons in other periods are equal to the original Epsilon value multiply by the correlation coefficient between two Epsilons. The correlation coefficient can be obtained by Baker's prediction equation as a closed-form solution, or using the correlation based on a suitable subset of GMRs (e.g. from NGA database). The GMRs used in this study are given in reference [7].

The target Epsilon (ε*) is needed for the conditional computation as well as the target Eta, but the disaggregation analysis only provides the target Epsilon. In fact the target Eta value (η*) is still unknown. However it is necessary to either perform a new Eta-based disaggregation analysis or normalize the Eta to the target Epsilon in which both can be equal at the target period. For the purpose of simplicity the target Eta value had been normalized to the target Epsilon value in Eq. (3). The target Eta can now be considered to be equal to the target Epsilon which is one of the disaggregation results in addition to the magnitude and distance. The target peak ground velocity Epsilon (ε_{PGV}) can be obtained as written in Eq. (4) by using Eq. (3). Substituting Eq. (1) and (4) into Eq. (3) can produce the conditional mean spectrum based on Eta indicator as written in Eq. (5).

$$\eta = 0.472 + 2.730\varepsilon_{Sa} - 2.247\varepsilon_{PGV} \tag{3}$$

$$\varepsilon^*_{PGV} = \frac{1}{2.247}(1.730\varepsilon^*_{Sa} + 0.472) \tag{4}$$

$$Sa(T) = \exp(\mu_{\ln Sa(T)} + \frac{\eta^* \sigma_{\ln Sa(T)}(\rho_{(\eta(T),\eta(T^*))} + 1.730)}{2.730}) \tag{5}$$

A correlation model can be employed in order to find ρ values in Eq. (5). Baker and Jayaram proposed a model for the correlation coefficients calculation between the two Epsilon values based on the Chiou and Youngs model [8]. This method is consistent enough with other ground motion prediction models with high level of accuracy. In other words the results have shown that the correlation values do not differ appreciably among the different attenuation models. In the current study all parameters including the Epsilon values, the Eta values and the correlation coefficients are computed based on the considered GMR database [7] and BA-08 attenuation model [6] without using any closed-form solution. Figure 4 shows contours of the correlation coefficient, respectively, between each two arbitrary Epsilon and Eta values. The period range is taken from 0.01 to 5 sec in Figure 4. The Epsilon and the Eta values at other periods can be calculated easily by multiplying the target value by the corresponding correlation coefficient value which can be summarized in Eq. (6) and Eq. (7). For comparison of the two correlation coefficients obtained by Eta and Epsilon values, a new correlation parameter is defined in Eq. (8).

$$\varepsilon(T) = \varepsilon^* \times \rho(\varepsilon(T), \varepsilon(T^*)) \tag{6}$$

$$\eta(T) = \eta^* \times \rho(\eta(T), \eta(T^*)) \tag{7}$$

$$\rho'_{(\eta(T),\eta(T^*))} = \frac{\rho_{(\eta(T),\eta(T^*))} + 1.73}{2.730} \tag{8}$$

This parameter named ρ′ expresses the only difference between CMS and E-CMS equations. In fact the parameter ρ′ plays the same role as ρ in CMS computation (Eq. (6)). Therefore Eq. (5) can be rewritten as Eq. (9). Here care should be taken that all correlation coefficient values between two sets of observed Epsilon values are evaluated by using the maximum likelihood estimator that is so-called Pearson product-moment correlation coefficient as written in Eq. (10).

$$Sa(T) = \exp(\mu_{\ln Sa(T)} + \eta * \sigma_{\ln Sa(T)} \rho'_{(\eta(T),\eta(T^*))}) \tag{9}$$

$$\rho(\varepsilon(T), \varepsilon(T^*)) = \frac{\sum_{i=1}^{m} (\varepsilon_i(T) - \mu_{\varepsilon(T)})(\varepsilon_i(T^*) - \mu_{\varepsilon(T^*)})}{\sqrt{\sum_{i=1}^{m} (\varepsilon_i(T) - \mu_{\varepsilon(T)})^2 \sum_{i=1}^{m} (\varepsilon_i(T^*) - \mu_{\varepsilon(T^*)})^2}} \tag{10}$$

where m is the number of observations (GMRs in this study); εi(T) and εi(T*) are the Epsilon values at T and T* respective to the record number i; με(T) and με(T*) represent the sample means. Finally the Epsilon-based conditional mean spectrum can be computed based on [4] and the Eta-based conditional mean spectrum can be obtained by using Eq. (9). It is worth emphasising that the peak ground velocity Epsilon (ε_{PGV}) is a period independent parameter. Therefore ε_{PGV} is a constant value during a period range for a single record. This fact provides an opportunity to obtain a simple predicting equation as expressed in Eq. (4).

3 Comparing CMS and E-CMS spectra by a simple example

In the performance-based approach, the ground motion response spectrum is based on site specific UHRS at the free-field ground surface modified by a design factor to obtain the performance-based site specific response spectrum. The U.S. Geological Survey (USGS) tool is employed to obtain the design spectra [9]. A simple structure located in Riverside with a first-mode period of 0.1 second is assumed, and 1% probability in 100 years is considered as a given hazard level, corresponding to 1E-04 annual probability of exceedance. The median predicted spectral acceleration and the standard deviation values are obtained by BA-08 attenuation model. For the purpose of simplicity, the UHRS is calculated using the predicted median value added by the standard deviation which is multiplied by the target Epsilon. This assumption is accurate for single dominated hazard sites and can be an approximate estimate of UHRS for the sites with multiple seismic hazard sources [10]. CMS and E-CMS can be derived similarity by consideration of the correlation part. The disaggregation results which are considered as the controlling earthquake parameters, are obtained by employing USGS tool updated in 2009 [9]. Figure 2 shows the disaggregation distribution of magnitudes, distances and Epsilons that will cause the occurrence of Sa(0.1sec)=2.0255g at the assumed site. For conditional computations, by using the BA-08 attenuation relationship, the mean magnitude is equal to 7.15, the mean distance is equal to 10.2 and the mean

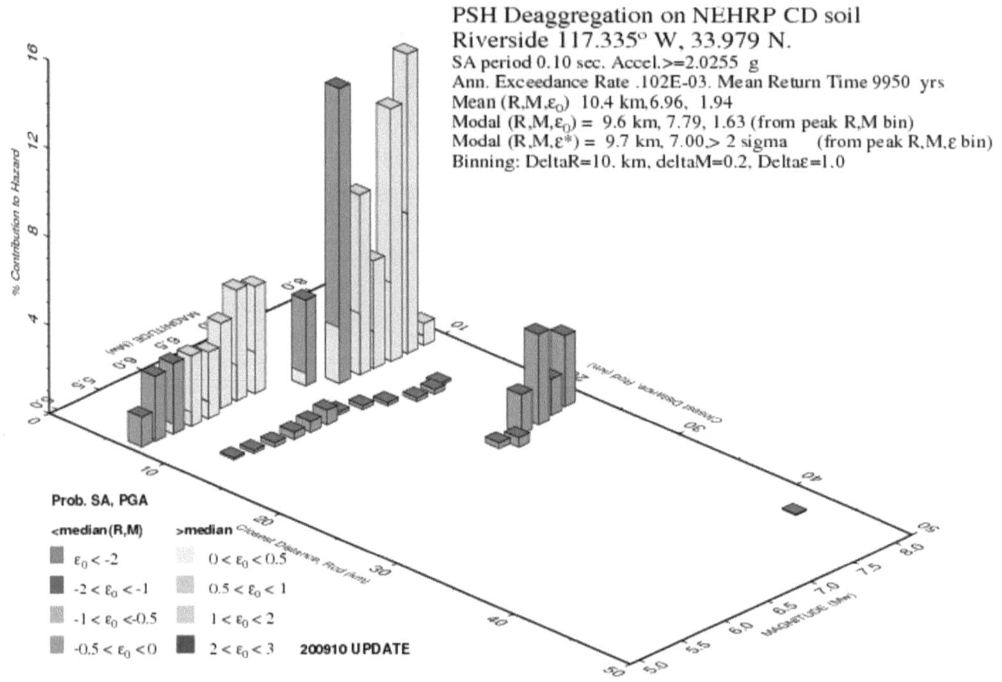

Figure 2: The PSHA disaggregation, obtained by USGS [9]

Epsilon is equal to 2.25. These values are obtained as an earthquake scenario which is most likely to cause Sa(0.1sec)=2.0255g. Note that the shear wave velocity averaged over top 30m is assumed to be 360m/s. The obtained Epsilon from the disaggregation result is assumed to be equal to the target Epsilon and the other Epsilon values at other periods can be calculated as well. The Sa of the conditional mean spectra at the target period is the same as UHRS corresponding to 1% probability of exceedance in 100 years.

Figure 3 compares UHRS with CMS and E-CMS spectra for the given site. As it is expected CMS, E-CMS and UHRS have the same Sa value at period of 0.1 sec. The most interesting finding is that both CMS and E-CMS show a significant reduction in comparison with UHRS. Another arising issue is the significant difference between the CMS and E-CMS. Both CMS and E-CMS have a peak correlation at period of 0.1 second since the correlation coefficient is high near the target period. The correlation coefficients decrease in large and small periods but the reduction process is more significant in CMS from the target period in comparison with the E-CMS. In other words, E-CMS correlation values in other periods are more than the corresponding CMS values. It is clear that using different ground motion prediction models will result in different predicted median spectrum. In fact CMS and E-CMS will be affected by the attenuation model. However the point is that the observed difference will not change because the source of the difference is somewhere else. A comparison between CMS, E-CMS

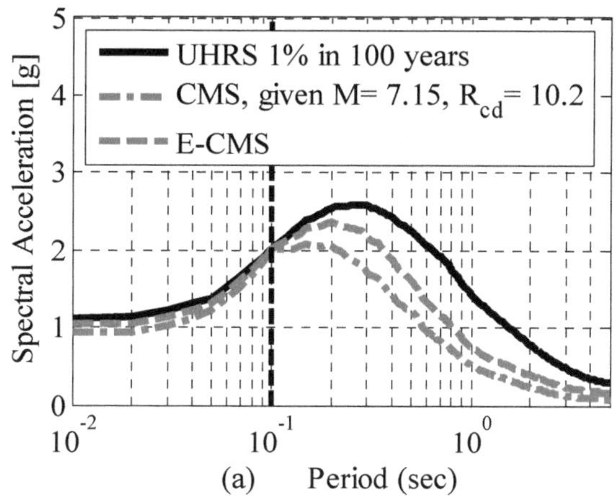

Figure 3: Comparison of the UHRS, CMS, E-CMS for 9950 years return period

and UHRS equations proves that both conditional mean spectra are independent of the spectral acceleration value. In other words the source of the difference is only the correlation part. Although the UHRS uses the correlation coefficient equal to unity for all periods, but both of the conditional mean spectra take the correlation of the spectral values into account. This fact is also shown in Figure 4a where the parameter ρ′ for Eta and ρ for Epsilon are compared versus UHRS. Note that Figure 4a shows the correlation values, and do not reflect the spectral acceleration terms. In other words Figure 4a can justify the differences between CMS, E-CMS and UHRS since CMS is based on ρ and E-CMS is based on ρ′. As a result it is not important what the attenuation model and the design factor are, because the difference is just sourced by the correlation values. Figure 4b shows the correlation values at another target period (T=0.5 sec).

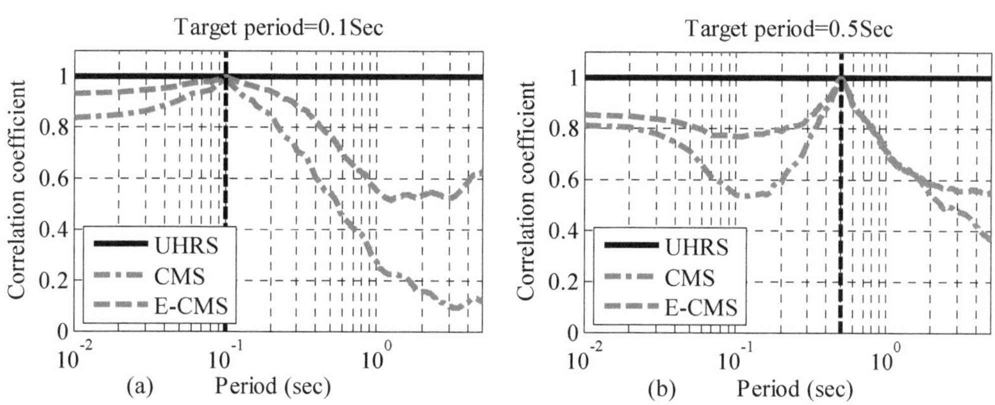

Figure 4: The correlation coefficients over a period range;
(a) Target period=0.1sec; (b) Target period=0.5sec

The higher correlation values between the Eta and the structural response, compared with the corresponding correlation between the Epsilon and the structural response which has been shown briefly in this study (see more details in), is a significant logic that E-CMS is more realistic rather than CMS. However, it is worth to exploring this issue from different viewpoints in a more detailed study.

4 Conclusion

Ground motion selection based on target spectra is currently a timely subject in earthquake engineering society. Therefore considerable efforts have been done to propose a realistic approach to obtain the target spectra. The UHRS, as a result of probabilistic seismic hazard analysis, is the most popular approaches in the design standards since all of the ordinates in UHRS spectrum have a same hazard level. The conditional mean spectrum is one of the recent developments for this purpose which employs the advantages of using the correlation between the spectral values. A new target spectrum, named E-CMS, has been introduced in this paper which uses the Eta indicator advantages and follows the CMS format. The conservation in the estimation of the structural seismic response can be reduced by using the E-CMS since the correlation of Eta and the structural response is greater than the correlation between the conventional Epsilon and the structural response. However the conventional CMS can underestimate the structural response. Therefore the E-CMS is introduced as a realistic target spectrum which can be used in the design procedures of industrial facilities

References

[1] McGuire, R.: Probabilistic seismic hazard analysis and design earthquakes: closing the loop; Bulletin of the Seismological Society of America; 85(5). (1995), p. 1275-1284.

[2] ASCE7-5, Minimum Design Loads for Buildings and Other Structures. 2005, American Society of Civil Engineers/Structural Engineering Institute, Reston ,VA.

[3] Baker, J.W.; Cornell, C.A: Spectral shape, epsilon and record selection; Earthquake Engineering & Structural Dynamics; 35(9) (2006), p. 1077-1095.

[4] Baker, J.W.: Conditional Mean Spectrum: Tool for Ground-Motion Selection; Journal of Structural Engineering; 137(3) (2011), p. 322-331.

[5] Mousavi, M.; Ghafory-Ashtiany, M.; Azarbakht, A.: A new indicator of elastic spectral shape for the reliable selection of ground motion records; Earthquake Engineering & Structural Dynamics; 40(12) (2011), p. 1403–1416.

[6] Boore, D.M.; Atkinson, G.M.: Ground-Motion Prediction Equations for the Average Horizontal Component of PGA, PGV, and 5%-Damped PSA at Spectral Periods between 0.01 s and 10.0 s; Earthquake Spectra; 24(1) (2008), p. 99-138.

[7] Baker, J.W.; Cornell, C.A: A vector-valued ground motion intensity measure consisting of spectral acceleration and epsilon; Earthquake Engineering & Structural Dynamics; 34(10) (2005), p. 1193-1217.

[8] Chiou, B.S.J.; Youngs, R.R: An NGA Model for the Average Horizontal Component of Peak Ground Motion and Response Spectra. Earthquake Spectra; 24(1) (2008), p. 173-215.

[9] USGS. PSHA disaggregation. Available from: https://geohazards.usgs.gov/deaggint/2008/.

[10] Ebrahimian, H.; Azarbakht, A.; Tabandeh, A.; Golafshani, A.A.: The exact and approximate conditional spectra in the multi-seismic-sources regions; Soil Dynamics and Earthquake Engineering; 39. (2012), p. 61-77.

Trade Exhibition SeDIF-Conference

The following associations and companies presented their products and services at the trade exhibition of the SeDIF-Conference 2013:

Deutsche Gesellschaft für Erdbebeningenieurwesen und Baudynamik e.V.

German Society for Earthquake Engineering and Structural Dynamics

c/o Dr.-Ing. Christoph Butenweg (current President)
Chair of Structural Statics and Dynamics
RWTH Aachen University
Mies-van-der-Rohe-Str. 1
52074 Aachen
Germany

Tel.: +49-(0)241 / 80-25088
Fax: +49-(0)241 / 80-22303
Mail: dgeb@lbb.rwth-aachen.de

Dlubal Software GmbH

Am Zellweg 2
93464 Tiefenbach
Germany

Tel.: +49-(0)9673 / 9203-0
Fax: +49-(0)9673 / 9203-51
Mail: info@dlubal.com

DOI 10.1007/978-3-658-02810-7, © Springer Fachmedien Wiesbaden 2014

Hilti Deutschland AG

Hiltistr. 2
86916 Kaufering
Germany

Tel.: +49-(0)800 / 88855-22
Fax: +49-(0)800 / 88855-23
Mail: de.kundenservice@hilti.com
Contact: Dr. Oliver Geibig

KTI-Schwingungstechnik GmbH

Klein Goldberg 23
Postfach 100724
40822 Mettmann
Germany

Tel.: +49-(0)2104 / 8025-75
Fax: +49-(0)2104 / 8025-77
Mail: info@kti-trautmann.com
Contact: Rolf Trautmann

SDA-engineering GmbH

Kaiserstr. 100
52134 Herzogenrath
Germany

Tel.: +49-(0)2407 / 509-4740
Fax: +49-(0)2407 / 509-4077
Mail: info@sda-engineering.de
Contact: Dr. Christoph Butenweg
Dr. Christoph Gellert

sectyelectronics®	**Secty electronics GmbH** Borghagener Str. 145 44581 Castrop-Rauxel Germany Tel.: +49-(0)2305 / 9734-81 Fax: +49-(0)2305 / 9734-82 Mail: Juergen.przybylak@secty-electronics.de Contact: Jürgen Przybylak
Springer Vieweg	**Springer Vieweg** Abraham-Lincoln-Str. 46 65189 Wiesbaden Germany Tel.: +49-(0)611 / 7878-208 Fax: +49-(0)611 / 7878-78208 Mail: ralf.harms@springer.com Contact: Ralf Harms
SYSCOM A member of the BARTEC Group	**SYSCOM Instruments** Rue de l'Industrie 21 1450 Sainte-Croix Switzerland Tel.: +41-(0)24 / 455-4411 Fax: +41-(0)24 / 455-4560 Mail: sales@syscom-instruments.com Contact: Dr. Hans-Jürgen Nitzpon

Wölfel

Max-Planck-Str. 15
97204 Höchberg
Germany

Tel.: +49-(0)931 / 49708-500
Fax: +49-(0)931 / 49708-590
Mail: mueller@woelfel.de
Contact: Dieter Müller

Dlubal
Engineering Software
Structural and Dynamic Analysis
RSTAB 8
The 3D Framework Analysis Program
RFEM 5
The Ultimate FEA Program
Concrete Structures
Steel Structures
Cross-Sections
Finite Elements
Stability
CAD/BIM Integration
Bridge Construction
Eigenmodes
Eigenvibrations
Plant Engineerin
Dynamics
© www.hum-minden.de
Glass Structures
Mechanical Engineerin
© www.ibehlenz.de
Follow us on:
Free Trial at www.dlubal.cor
More Information:
Dlubal Engineering Software
Am Zellweg 2, D-93464 Tiefenbach
Tel.: +49 9673 9203-0
Fax: +49 9673 9203-51
info@dlubal.com
www.dlubal.com
DVD
Multilingual
Enjoy Structural Analysis...

MIX
Papier aus verantwortungsvollen Quellen
Paper from responsible sources
FSC® C105338

If you have any concerns about our products,
you can contact us on
ProductSafety@springernature.com

In case Publisher is established outside the EU,
the EU authorized representative is:
Springer Nature Customer Service Center GmbH
Europaplatz 3, 69115 Heidelberg, Germany

Printed by Libri Plureos GmbH
in Hamburg, Germany